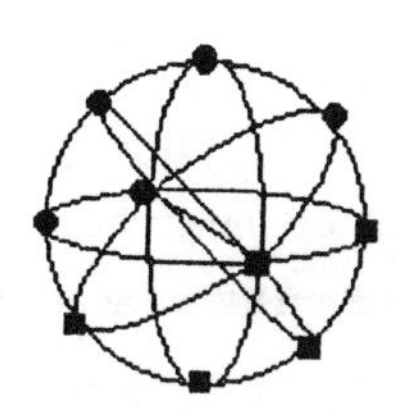

ATOM ONE

Atomic Structure of the Atoms

Glen Clark DeLay

Atom One
Atomic Structure of the Atoms

ISBN 0-9654177-0-0
Library of Congress Card Number: 96-96897

Published by DeLay's Printing
1603 Aviation Blvd., Suite 16
Redondo Beach, California 90278
(310) 374-4782 Fax (310) 376-2787

Preface

The purpose of this text is to present the subject of **Atomic Structure** in as clear, logical and interesting manner as possible. It has been compiled over a thirty six (36) year period by the author.

In order to understand how all of the atoms came into being, I have started when time did not exist and all that existed was pure energy which I believe was magnetism. From the idea that everything came from magnetism and the result of a change in it caused time to exist and from this point on, all mass became created, step by step.

I started this investigation by trying to find out why electrical current flowed from the cathode to the anode within a vacuum tube when there was an opening between them within the vacuum tube. This opening within the vacuum tube should have blocked the flow of electrical current and it did not. With the idea that electrical current is transformer action from atom to atom showed me that magnetism was the thing to investigate. Using this as a line of reasoning, I checked all my physics and chemistry books for leads that would agree with the known facts of how atoms are created. This book is the result of following one line of reasoning after another. The break through came when I realized that the neutron with 1.00867 amu's was given more mass than the proton with 1.00728 amu's and the atom of Hydrogen with 1.00783 amu's, yet the isotopes 1H2 and 1H3 had neutrons within their nucleus that was not of the mass given. The books said that the neutron that was by itself lasted only minutes and was changed into something other than a neutron. With this knowledge, I followed a trail until all atoms and their isotope's construction had been explained. This book is about the Original Neutron, isotopes 1H1, 1H2, 1H3, 2He3, 2He4 and the creation of all other atoms and their isotopes. They became explained with the use of sub-isotopes of an isotope. I hope that the reader understands how to follow this trail as a need to understand some atom or isotope becomes apparent.

In chemistry books, a formula is given of an atom by using different atoms to create it. In this book, I explain how the atoms and their isotopes are created going from left to right using an equal sign. Taking this formula and reading it from right to left will give the atoms used in the chemistry books by a molecule.

All measurements of the electron, proton and neutron comes from measurements found in most chemistry books. The names of these parts have been given the names negatron, positon and first and second neutrons. The name positron for a positive electron is not used because I believe there is no such thing as a positron or anti-matter. The chemistry books say there are two protons found in the nucleus and I call the second proton a positon when the second proton now called a positon has an electron now called a negatron with it. There are two electrons said to be in an orbital. The second electron is said to have an opposite spin to the first electron. This second electron is the negatron and I say they spin in the same direction. The chemistry books say the proton can be released and I say that it is converted to a neutron and the neutron can be released. The electron or negatron can be released by the loss of 0.00100 amu that is emitted as a beta ray and no other atomic mass unit will release either of them. Any other mass lost is said to be an alpha ray if the ray is emitted from a proton or positon and if the ray is emitted from a neutron it is a gamma ray. Their names comes from their atomic mass units.

I have identified the exact position of each electron and their position within an orbital. I also have shown the distance between the electrons and negatron in their orbitals. It is hoped that those reading this book will like these changes as a great deal of effort has been made to find and understand the reason for the existence of all atoms. All math is given five places to the right of the decimal point and never anything less. This converts math to arithmetic. If this number is less than one, no ('s) is used and it will always be used when the atomic mass is one or greater than one.

All events are given in sequential order and each paragraph has or states some line of reasoning and the next part follows this line of reasoning. An effort has been made to prevent referring back through the book to find something that is needed. I have at all times given the name of the parts of an atom and the name of the isotope under consideration to prevent the reader from thinking it is a different part or isotope. This must be done this way because their names are used so often that it would be impossible for the reader to follow the text and know which part or isotope is under investigation.

All of the drawings are of my creation. There are two types. One type is to show the electron and negatron in their orbitals and the result of this gives a rotation of these orbitals like a paddle wheel except for one orbital that goes at right angles to the axis of the orbital given. This orbital is the s subshell. The other type of drawing is one which explains the use of the electron, negatron, proton, positon and any neutrons found in an atom. Their mass is given in atomic mass units (amu) and when this mass is released, it is either mass or a ray. The ray is always given as pure energy which is magnetism. When I first explained how a new isotope is created, a given drawing used words and numbers and when I felt that the words became clutter, I used only numbers. Using only numbers is a form of shorthand.

This book can be ordered from an ordering form found in the back of the book plus if the reader wants older books on this subject they can also be ordered by requesting a catalog from **The Gemmary** found in the back of the book.

I wish to thank my brother, Wilbur Lee DeLay, in Tulare, California for his time spent listening to me talk about this subject. His understanding of science is unending. I also wish to thank all of my customers in DeLay's Printing who gave me encouragement to finish this book. This book will never be really finished but they understand this. This subject is a never ending story.

Glen C. DeLay

Contents

PART 1

PART 2

PART 3

PART 4

PART 5

PART 6

PART 7

PART 1

THE START OF IT ALL

This book will explain how the cosmos, galaxies and our universe came into being, then how our sun was created. After our sun was created, I explain how the elements, atoms and isotopes came into being. I also explain how leftover parts of isotopes form molecules.

Most scientists today believe that a good scientific theory or hypothesis will fit the facts and give connections between apparently unrelated phenomena. When this is done, there is a feeling that the idea is worth believing in. According to David Layzer in his book "Construction of the Universe", he quotes Sir Arthur Stanley Ucenteral as saying "Theories are like balloons floating on the surface of the sea, and facts are like battleships. Occasionally a balloon collides with a battleship and the battleship goes down." I hope the battleship called the "Free Electron" will go down!

THE RULES OF THE COSMOS

In the beginning before time existed, all that existed was in something called space. All of this space was located at a point. It may be hard for the reader to understand but this point in space covers all that exists with no movement at all or a limit to the area covered. We know it as empty space and this empty space is filled with magnetism. Since magnetism can attract or repel other magnetic fields, this is not true here simply because there is only one magnetic field. This space never moves but all the magnetism in it can. For anything in the space to change, there are three rules that must be followed. The term "COSMOS" describes everything that is in this space.

THE FIRST RULE OF THE COSMOS

NOTHING CAN STAY AS IT IS

There is no way to say how long the space filled with magnetism stayed as it is because time did not exist. The change is when the graviton came into existence. When the graviton came into existence, time began.

THE SECOND RULE OF THE COSMOS

TIME CAN NOT START UNTIL A CHANGE HAS BEEN MADE

Before time can exist, the graviton has to come into existence and attract magnetism to it from all directions. When this happened, time came into being. Magnetism is going in one direction only. It is being concentrated by the graviton.

THE THIRD RULE OF THE COSMOS

WHEN ALL THAT EXISTS BECOMES MASS, TIME WILL END

The graviton attracted magnetism to it until mass was formed and this mass exploded and sent mass outward and then attracted it back again. This is repeated until all energy has been converted to mass and then the mass will stay mass and it can not explode because there will be no energy left to cause the explosion and this mass will stay mass and time will end. All pure energy that is magnetism will have been concentrated into mass.

BEFORE TIME EXISTED

THE EQUAL SIGN (=)

Before time could exist, all that is, is pure energy. If all that exists is pure energy and pure energy is magnetism, then everything is the same. The equal sign by itself represents the idea that nothing can be put on the left or right side of it. The graviton and time can not exist because there is no difference between anything. Before time can exist, there must be a difference of some kind. When there is a difference, something can be put on each side of the equal sign.

AFTER TIME CAME INTO EXISTENCE

If all the pure energy lost heat equally, this would not cause all space to decrease. All pure energy within all space would change its temperature.

This changing in temperature would not cause mass to be created because for mass to be created, there must be a point in space that is cooling and there is none. Losing temperature is pure energy cooling. If all pure energy cooled equally until there was no temperature, then there could not be a point in space that would become cooler than any other point. This means that all pure energy could not cool equally for mass to come into being.

From the above, the only way all pure energy could cool is for one point in space to become cooler than the rest. This point in space is where the graviton began to spin or rotate because mass was not formed equally. Since all pure energy is magnetism, the graviton has the ability to pull the magnetic lines of force toward it. This is the way a magnetic field becomes concentrated. As temperature decreases, the graviton gains in its ability to concentrate magnetic lines of force until mass is formed in the form of a Quark.

When pure energy began to lose some of its heat, the point that began to form mass, did so by storing the pure energy as mass. The pure energy that was cooler than the rest of the pure energy was storing the exact amount of pure energy that was lost. This pure energy was never really lost, it was being transformed into a new form of energy that is now called mass.

If all pure energy was moving toward this mass equally, the mass would still have no motion. It would not spin on an axis. Pure energy is moving, not the mass. At the moment pure energy began to move toward the central point in space, the pure energy gained a direction inward only. This gives the pure energy the force of movement even though it has no mass. The force of movement toward a central point created pressure where there was none before. It is because the pressure was equal from all directions that mass could be formed. This line of reasoning will explain why light moves outward equally in all directions. If the movement of all pure energy toward a central point in space is when gravity came into being, then there could be no centrifugal or centripetal forces because all movement is in one direction toward the central point in space.

Once pure energy has been transformed into mass, it can never get all of it back. This means that it is impossible for all the mass that is formed to be returned to pure energy. This is true because it would take more energy than existed in the beginning to return all of it to pure energy.

If all energy cooled equally, then there would not be an axis. If all energy did not cool equally, then an axis would come into being. If an axis came into being, this would be the beginning of the STEADY STATE UNIVERSE. If all energy cooled to one point, this would be the BIG BANG. It is possible for the BIG BANG to become a STEADY STATE UNIVERSE. After the BIG BANG occurred, an axis still remained because the masses moving outward were not equal to each other nor did they move outward with the same speed. Because this axis has a rotation and there is now masses to be attracted to its central point where the graviton exists, the **force with no name** (explained below) came into being and all mass moved outward from the center of the axis and then some of it would go up and some down. That which goes up, goes back to the top of the axis while the rest goes down and back to the bottom of the axis. The mass at the top and bottom would then go to the middle of the axis. When this mass reached the middle of the axis, it would then leave the axis and go outward again. This movement need not have an end until all pure energy has been converted to mass.

THE FORCE WITH NO NAME

There is a force which governs the size of an axis and a sphere and any movement within the sphere and the rotation of the axis itself. It has no name.

The movement inward of mass is toward the graviton and it is called gravity. Gravity is the movement toward the graviton. When mass moves outward from the graviton, the force pushing it outward is called centrifugal force. Centrifugal force pushes mass outward. There is a force that moves mass at an angle that is not straight outward. The force pushing it at an angle is centripetal force. Centripetal force pushes mass around the axis. These forces form a sphere. A sphere is created when the mass can not go outward any more than the graviton will allow. This is also true of the photon which is an electro-magnetic sphere with no mass. The photon has a graviton at its center and magnetism rotating around it.

This new force can turn a sphere into a flat disk which is a sphere with its top and bottom flattened because the graviton does not change while the centrifugal and centripetal forces increase. Its ideal shape is a perfect sphere. What this new force does is to cause that which is in the middle of the axis to leave the axis at its middle and go outward. As it goes outward, it must either go up or down and return to the top or bottom of the axis and then return to the middle point of the axis. If the centrifugal and centripetal forces should be at their maximum, the sphere would have its top and bottom pulled to the middle to form a disk. By losing some of the centrifugal and centripetal force, the disk would begin to return to the shape of a sphere. This means that as the equator of the sphere expands, the top and bottom is being pulled inward toward the middle. Or if the equator begins to shrink, the top and bottom moves away from the middle. The axis therefore gets longer if the centripetal force should decrease. It would get shorter if the centrifugal and centripetal forces should increase and the graviton does not change, therefore, there are forces that give the length of the axis of any spinning body which can change its shape.

THE STEADY STATE UNIVERSE

I will explain the Steady State Universe first because it requires an axis around which all the universe would be rotating. This axis is required to have a top, middle and a bottom. The universe would be rotating around this axis. The universe would be held together by a gravitational field that is rotating. This centripetal force is created by the rotation of the graviton. Because the centripetal force is pushing the gravitational force around the graviton, there is a force pushing away from the graviton. This is a centrifugal force that is the opposite to gravity. The two make for an equal and opposite reaction. In all of the writings that I could find, there was no opposite to gravity. It was there but they did not recognize it. We know it as centrifugal force. Gravity pulls everything inward while the centrifugal force pushes everything outward. Since everything must rotate around an axis, the force pushing it around the axis is called centripetal force.

If there was no "Big Bang", then all the galaxies of the universe must be moving from the central point on an axis and then going up or down to return to the central point. There should be some masses that end up going back where they started from but most should cool to the point where they will remain in the middle area outward from the central point on the axis. There, they will wobble, some times going a little up then a little down. The outer most part should be cool relative to the center. Of that in the central area, it is therefore, the hottest. Centripetal force is the force which forms a spiral arm in space. The area at the top and bottom of the axis would be the coolest of all.

This cooling effect can be seen here on earth. At the equator, it is the hottest point and as we move up or down from it, the climate is cooler. When we reach the poles, it is the coldest it will get. Do not confuse this with the energy received from the sun.

When the pure energy that is magnetism cools enough to form the first mass, it formed a mass called a quark. This quark became a neutrino and it became a meson. This meson became an Original Neutron. From the Original Neutron came all of the other atoms starting with Hydrogen. This Hydrogen becomes the stars or suns. The stars form galaxies. The galaxies leave the axis at the middle point and go outward. As the galaxies go outward, they are spinning and the direction of their spin will govern whether they will go up and back to the axis or if they will go down and back to the axis. As the galaxies go outward, they cool and return to the top and bottom of the axis where they then go to the middle point and outward again. If this happens, then this movement need not end in an explosion. This would then be a steady state universe.

THE BIG BANG THEORY

The Belgian cosmologist Abbe LeMaitre and Dr. George Gamow of George Washington University described the "Big Bang" theory of how it all began. Others described "The Steady State Universe". The two ideas started at the when cosmos went bang, began to create the galaxies as we know them now.

Dr. Gamow believed that the mass of the cosmos went bang about five billion years ago. He believed that the cosmos was in a state of contraction with no light. It ended up with all matter and radiations being pressured together in an inferno of elementary particles of incredible mass and density which he call ylem meaning the primordial, elemental substance of all things. This point in space had a temperature in billions of degrees. There were particles in a state of extreme motion. When the pressure was at its greatest, light and electromagnetic radiations moved outward. He says that when this happened, the temperature fell. When it reached one billion degrees, the particles formed atoms. As the primordial vapor went outward, it cooled, creating turbulences and that gravity shaped it into spiraling galaxies and clusters of galaxies. At first, the point in space was dark but as the material that was forming the galaxies began to radiate, they radiated light.

The above fits if the pure energy that is magnetism becomes a mass that can explode.

THE OLD MIXED WITH THE NEW

It is known that some of the galaxies and their stars (suns) do not have the same mass and that some are older than others. Some of the older galaxies and stars are close to each other. It would seem that the older galaxies should be the farthest out from the axis and they are not. To explain this, one needs to understand that galaxies going outward from the axis without having much energy in their centrifugal and centripetal forces would move slower than those galaxies that have these forces with a greater energy. This would make it possible for a smaller mass to pass a greater mass as the both go outward from the axis. If two masses were both going outward from the axis where one mass goes almost straight outward while the other mass goes outward and around the axis in a greater curved path, the mass going almost straight outward would pass the mass going in a greater curved path.

SPIRALS OF PURE ENERGY

In the beginning when pure energy moved outward from the central part of an axis, it did so unevenly. When this happened, mass and pure energy was sent outward in spirals. This unevenness caused some of the energy to swirl as it moved through space. Each swirl had a central point around which the pure energy moved. It is in these central points in space that galaxies came to be created. Our galaxy is one such swirl. Our sun is in an arm of a swirl.

As the mass and pure energy moved outward from the point where the big bang occurred, it moved in a circular path. This path has a centripetal force going in the direction around the central point of the universe. The force pushing the pure energy away from the point where the big bang occurred is centrifugal force. The force that is slowing down the mass as it moves outward is gravitational force. This slowing down of the outward movement has the effect of pulling the swirling energy inward toward the point where the big bang occurred.

Now put all the above together to form the way we find things today. We know that mass and the pure energy is moving in a swirling motion and that it has an axis. This axis has a top, a middle point and a bottom. The pure energy always moves outward from the middle point into a circular path. Centrifugal forces sends it out and centripetal force sends it in a circular path. When it gets out to a certain point in space, gravity will slow it down until it can go no farther. Some of the pure energy must move upward or downward and toward the top or bottom of the axis and then toward the center of the axis. The part that becomes mass should be fairly stable in the point of space it is created at. It becomes hard for it to move either toward the top or bottom of the axis so it moves up and down around the center of the axis at the point in space it ends up at. When it moves outward from the middle of the axis, some of the swirling mass will go upward toward the top of the axis and some will go to the bottom of the axis. The direction of the swirl will determine which end of the axis the mass will go. While some of the stars are believed to be gases, these gases are made up of atoms which is a form of mass. Any empty space is filled with magnetism which is pure energy.

OUR GALAXY

Our galaxy is called the Milky Way Galaxy. The study of our galaxy shows that there is a central ball of stars with a ring around it. There are stars above and below the ring, but almost all of the stars are in the central part of the ring where the ball of stars is.

Our galaxy has an overall diameter of 100,000 light years and the earth and our sun are 30,000 light years from its center in the part called the Milky Way. It is made up of empty space, stars (suns), dark clouds of gas and dust. The galaxy carries the earth and our sun around a central point. They travel with a velocity of about 600,000 miles an hour.

OUR SUN

In the very beginning in the area where our sun is located, there was a great deal of pure energy spinning through space. It began to spin around at the spot that is the point in space that is now our sun. Because the pure energy stopped passing through space and began to spin around one specific point in space, it began to slow down. As this spinning energy began to slow down, it began to create mass. The pure energy is still going through space but now it has a point in space it is moving around. Both the pure energy and the point in space are moving through space. (The point moves, not the space.) Einstein said that as mass speeds up, it would be converted to pure energy. Therefore, if pure energy slowed down enough, mass would be formed. This happened at the point in space where our sun is located.

GRAVITY MUST INCREASE

Remember the great ball of stars found in our galaxy. Why the ball of stars? Since no new pure energy is being furnished to the universe, the galaxies must be slowing down as they cool. While it may take billions of years to detect any cooling, the fact that no new pure energy is available, the force pushing toward the center of the axis must be decreasing. This decreases the centrifugal and centripetal forces also. When this happens, gravity must increase. If gravity is increasing, the central area must be increasing in mass because the stars can not move outward with the force they once had. This would lead to making one big ball of mass. If we looked at this from the cosmos point of view, then all of the stars would become one single ball of stars which would become one big ball. Gravity would hold this ball together and it would not change. This would be the end of time.

LIGHT

In a Michelson-Morley experiment, light was proved to move at the speed of 186,282 miles per second in all directions. Then Einstein gave the equations and theory that measurements of distance and time vary with the velocity of the observer. When two observers at different places take their measurement of their velocity relative to each other, they will come up with different answers when their relative speeds are close to the speed of light. Thus light is something that moves from one place to another. What is this something? White light has been shown to be composed of many colors. When all of the colors are mixed together, they give us a white light. It has been found the eye can see colors because of this mixing. If we can see white light and colored lights or black which is no light at all, then what is it that we are seeing?

What we are seeing is the mixture of pure energy being sent out from the source of the light. This pure energy is in the form of magnetic pulses being sent from atom to atom until it reaches our eyes. The eye is a device that can separate the colors or it can not see them which is called the black color. This pure energy is being sent out from the atoms at the atom's lowest energy level most of the time and therefore does not have much energy. The atoms pulse at different frequencies according to the way the atoms are made. This pulse is called a photon when it is sent out from the atoms at a frequency that the eye can detect. There are other frequencies being sent out from the atoms that the eye can not detect and these frequencies can become visible when special equipment is used. Later on, I will have some drawings that will show how these pulses are sent out or received. There are three ways an atom can use a pulse.

1. Receive only
2. Send only
3. Receive and send

Because a pulse is emitted or received by an atom as a magnetic field, the term "photon" can be used to show that a magnetic field is sent out or received from an atom at its frequency. There is a door which will let a photon in or out. The photon either goes to or away from the proton or positon. This photon's energy is then used by the proton or positon as magnetic or electrical energy. Some of this energy is transferred to a neutron. This transfer of energy is in the form of mechanical energy. It is also sent to or away from the electrons and negatrons via the transformer action of their magnetic fields. This is in the form of electro-magnetic energy.

SUBATOMIC PARTICLES

Before I go into the masses formed from the pure energy, I would like to point out that the scientific community has labeled many forms of energy or mass.

There are subatomic particles under the heading of Baryons that are strongly interacting fermions that have omega hyperons, cascade hyperons, sigma hyperons, limbda hyperons, and nucleon of protons and neutrons. Sometimes the protons (positons) and neutrons are put under the heading of hadrons.

There are mesons, strongly interacting bosons, eta-mesons, k-mesons and pi-mesons. There is a family of weakly interacting fermions that are muons, electrons (negatrons) and the neutrino and maybe the tau. There is a neutrino-muon and neutrino-electron. The family of bosons have photons and gravitons. The starting of all of these subatomic particles is the quark. Keep in mind that the photons and gravitons do not have any mass.

If I use the term "photon", think of it as pure energy in the form of electro-magnetism going from one place to another at a certain frequency.

THE GRAVITON AND THE QUARK

The graviton is the starting point of everything from pure energy. It concentrates pure energy which is magnetism into mass. The pure energy is magnetism, which had no magnetic lines of force because there was no beginning or end of a line. It gains a beginning and end when mass first becomes formed. Mass is created around the graviton in the form of a Quark. As quarks are created, they combine until they form the neutrino and the neutrino grows until it becomes a meson. The meson gains an internal structure by forming internal parts with the graviton at its center. It is the graviton that holds these internal parts together.

When mass is formed as a quark, the graviton is at its center which gives it a magnetic field which can then attract other quarks. When these quarks become a neutrino, the magnetic field becomes concentrated within them until the meson is created. When the meson reaches 1.00867 amu's, the magnetic field can no longer grow and it becomes the Original Neutron. From this Original Neutron, all other atoms and their isotopes and sub-isotopes are created. The graviton is at the center of all of the internal parts as well as the center of the atom.

SPIN FORMS AN AXIS

As quarks move to and become a larger mass, this movement to the larger mass causes it to spin. This spin brings into being an axis around which all the quarks are rotating. This means that the magnetic field around them is also rotating. The neutrino, the meson and the Original Neutron are all rotating around an axis. The graviton is at the center of every axis that has magnetism.

THE START OF MASS

In the beginning, when pure energy was first turned into mass, the mass thus formed, was built up around a graviton. The graviton is the center of a photon and the photon which is a form of magnetism that has a magnetic field around it. Whatever the first mass is called, it is formed around a graviton until the neutrino is formed and from the neutrino all other masses are created. The magnetic field of the neutrino is described by me as being very powerful until it has gained 1.00867 amu's. When it reaches this mass, it has gained more mass than it needs. Its magnetic field can no longer attract quarks or other masses near it. This description only means that the magnetic field does not interact with other magnetic fields. When the temperature became low enough to form mass, the first quark was formed at the center of this magnetic field. It is the smallest mass known. **It is measured in atomic mass units written as amu or amu's.** It is spinning which creates an axis.

When the quarks reach 1.00757 amu's, they stop growing around the neutrino and begin to grow two masses on each side of the central mass. I call the mass of 1.00757 amu's a positron because it has a positive charge and 0.00029 amu more than a normal proton of 1.00728 amu's. As this mass grows, it is called a meson. The two masses on each side of the positron stops growing when they each reach the mass of 0.00055 amu. I call the mass on the left side of the positron an electron and the mass on its right side a negatron. When this happens, the meson will have 1.00867 amu's. At this point I no longer call this mass a meson but an Original Neutron. This is the mass found in most chemistry books only they just simply call it a neutron. They failed to realize that this neutron was the beginning of all the atoms. They know of the neutron in other atoms and their isotopes. They know that the isotope 1H1 of Hydrogen had less mass than the neutron. They should have paid more attention to this fact. A Hydrogen isotope of 1H1 has 1.00783 amu's compared to the neutron with 1.00867 amu's.

This Original Neutron, when it is found by itself, it changes in twelve and a half minutes into the Hydrogen atom (1H1). It loses 0.00029 amu as pure energy which is magnetism. It is lost as a 29 ray. This 0.00029 amu is the mass which holds the electron and negatron within the Original Neutron. When this mass is lost, either the electron or negatron is released into space and the remaining mass of either an electron or a negatron goes into an orbital around the remaining mass found in the center of the orbital. This mass has 1.00728 amu's and is a normal proton or positon. Their total mass is equal to 1.00783 amu's.

As additional Original Neutrons become atoms of 1H1, it becomes possible for more atoms other than 1H1 to be created.

THE PAULI EXCLUSION PRINCIPLE

The Pauli Exclusion Principle states, "There cannot exist an atom in such a quantum state that two electrons within it have the same set of quantum numbers." The quantum numbers are n, l, ml and ms. The ms is the quantum number which states that an electron has 1/2 . Since there are two electrons used by Pauli, one has a plus sign (+e) and the other has a minus sign (-e). The "e" is the symbol for an electron. This + and - signs mean that one electron is spinning in the clock direction and the other is spinning in the counter-clock direction. Pauli says that to use quantum number for the electron you must put an external magnetic field so strong the couplings among the various electrons are broken and the electrons orient themselves independently with respect to the external magnetic field. The state of each electron is then given by fixing the values of a set of quantum numbers. He gives the value of n to be equal to the total quantum number. The azimuthal quantum number is l and orbital magnetic quantum number is ml. The quantum number ml is the component of the orbital angular momentum in the direction of the external magnetic field. He says the spin quantum number is s and it is either a + or - spin. Pauli states that the most stable orbital is the orbital in the Shell K. Note that he says "orbital". This means any single or two electrons found in the Shell K. An orbital is the path the electron or electrons take as they go around the nucleus of an atom. He is not comparing an orbital with one electron vs an orbital with two electrons in it. He is comparing the orbitals of the different Shells. There are K, L, M, N, O, and P Shells. The Shell K has only one orbital while the Shell L has four. The Shell M has nine orbitals. Shell N has sixteen orbitals. Pauli says that the next most stable orbital is the "s" subshell in the Shell M. What I wish to point out is that these orbitals are filled with one or two electrons and these electrons have a magnetic field around them. Therefore, the orbitals represent a moving magnetic field. Pauli in his exclusion principle talks about the spin of the electron and not the orbitals. He says that when there are two electrons in an orbital, their spins must oppose each other. I ask the reader to look at this last statement very close.

Note that Pauli is comparing two electrons in a strong magnetic field such that they can orient themselves independently. This is not what happens in the real world. In the real world, the electrons are separated by at least one proton and in other cases two protons and one or more neutrons. Note also that his exclusion principle does not say where the North and South Poles are located relative to their spin. The failure to identify the poles of the electrons and the protons plus any neutrons causes a failure in the definition of their spin. Also quantum numbers are independent of the position of the electrons relative to their spin and the spin of the proton and neutron. By this, I mean that Pauli's exclusion principle treats the electrons as if the proton and neutron are not involved in it. Since the proton and neutron both have spins and poles, this affects the way the electron spins. Below is my drawing showing that the electrons have the same spin relative to each other but not in the area between them and the protons and neutrons. The quantum number Pauli found is not the spin of the electrons but the direction of the magnetic field found between the protons and their electrons. The direction of the magnetic field found between the electron and its proton on the left side of the proton goes in the opposite direction to the magnetic field found between the proton and its electron on the right side of the proton. These are the spins that are the + or - sign. Note that the electrons are spinning in the same direction relative to each other, the protons are spinning in the same direction to each other and the two neutrons are spinning in the same direction to each other. Note also the two proton's poles repel each other thus helping to keep them apart and the two neutrons are also helping to keep the two protons apart by physical means. It is the magnetic fields of the two protons that keep the two protons apart plus the fact that the spin of the two protons oppose each other. Note also that the two protons keep the two neutrons apart by physical means. Note that the magnetic field around the entire atom is going in the same direction. This is also true of the Original Neutron.

In the drawings, the electron is called an electron when it is on the left side of the proton and the second electron given by Pauli is called a negatron and it is on the right side of a positon. The electron and negatron are 180 degrees apart. The electron and negatron are going in an orbital from left to right which means that the magnetic field is going in that direction also.

In Pauli's book "The Chemical Bond", he says that there is a chemical bond between two atoms or groups of atoms in the case that the forces acting between them are such as to lead to the formation of an aggregate with sufficient stability to make it convenient for the chemist to consider it as an independent molecular species.

I will now change the word "chemical bond" to "magnetic bond" and add the word "magnetic" to the word "forces". The word "chemical" does not convey a picture of what is meant by that term. The word "forces" does not tell the reader what force is being considered. A magnetic bond is a magnetic force that can hold a given atom to one or more other atoms by magnetic attraction. These magnetic forces are the same forces that hold the parts of an atom to each other. The difference between a molecule and an atom is that the molecule holds atoms together by a magnetic field around all of the atoms and the atom's parts are held together by a magnetic field in the same manner.

Rewritten:

There is a magnetic bond between two atoms or groups of atoms in the case that the magnetic forces acting between them lead to the formation of an aggregate with sufficient stability to make it convenient for the chemist to consider it as an independent molecular species.

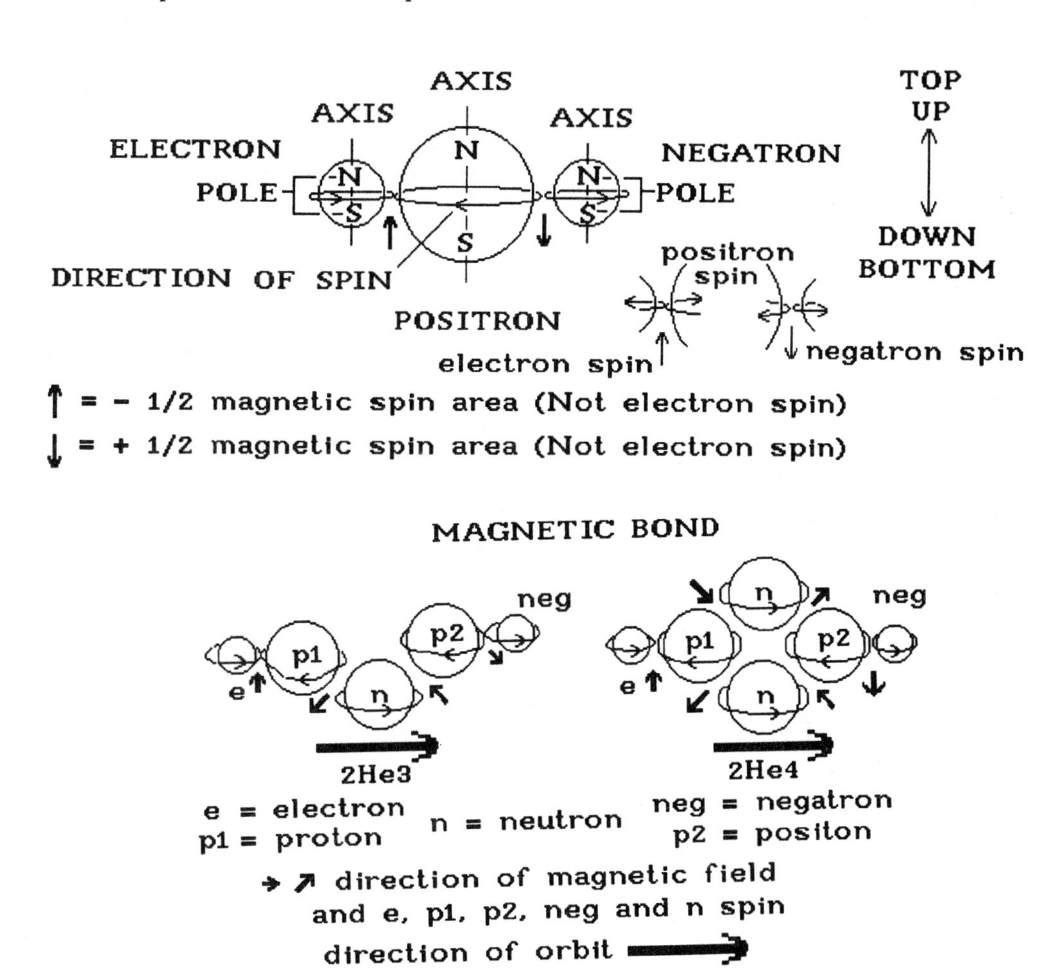

When the proton or positron in the Original Neutron has the same polarity as the electron and negatron, the Original Neutron acts as a very weak magnet because they are part of a neutron with 1.00867 amu's as given in most chemistry books. The two masses on each side of the positron must spin in the directions shown. The chemistry books say that when there are two such masses, they must spin in opposite directions. The trouble with this statement is that they do not identify where this difference really is. Note that the electron and negatron are spinning in the same direction relative to each other but they are spinning in different directions in the area between them and the positron. In the Original Neutron, they are connected to the positron and the positron is spinning in the opposite direction to both of the masses on each side of it. Note also that there is only one mass called the positron for both of the masses on each side of it. There are no orbiting masses. The positron has 0.00029 amu more than a regular proton and it therefore is more positively charged than a regular proton. It does not become a proton until it loses the 0.00029 amu. Only then does a mass orbit the proton. One of the two masses must be released and when it is in space by itself, its pole's direction has no meaning until it comes near another mass's magnetic poles.

ORTHO- AND PARA-HYDROGEN

The Original Neutron has an electron and a negatron growing on each side of the positron. The difference between the electron and negatron is their spin found between them and the positron in the Original Neutron and in an isotope, between them and a proton or positon. If the electron has its magnetic pole with the north pole at the top then the negatron must have its north magnetic pole at the top also. If the electron has its spin going in the counter direction between its proton then the negatron must have its spin going in the counter clock direction between its positon. If both of the electrons and negatron are spinning in the same direction, the proton is spinning in the opposite direction or clockwise. Because the left side of the proton is spinning in the opposite direction of the right side, they are said to be in an anti-parallel spin relative to the electron and negatron when the proton is spinning in the clock direction. If the proton is spinning in the opposite direction or counter clock direction, the protons sides are spinning in the parallel direction. This makes the atom upside down compared to the atom with its protons going in the clock direction. While this may be a good way of describing the difference between an electron, negatron and proton plus any neutrons, there is another description that is found in chemistry books. The chemistry books describe an electron and a negatron as having their spins in an ortho and para direction in a three to one ratio (3:1). Under normal temperatures, there are three (3) ortho-Hydrogen atoms to one (1) para-Hydrogen atom.

Two atoms of Hydrogen form a molecule of Hydrogen, 2(1H1). When they are in the form of a molecule, they neutralize each other and this makes the molecule neutral. While they are in the form of a molecule, they do not have an external magnetic field. A single atom can be found with either an ortho or a para spin. A molecule can never have its electron and negatron both in a parallel or anti-parallel spin. Thus when two Hydrogen atoms combine to form a molecule, one of the Hydrogen atoms must have its proton spinning in the opposite direction to the other. This means that the Original Neutron can release either an atom of Hydrogen going in the clock direction or in the counter clock direction. One out of every three Original Neutrons will be a para-Original Neutron for a 1:3 ratio.

If the above is true, then the proton spinning in the clock direction found with an electron will be an ortho-proton and the positon spinning in the counter clock direction with its negatron will be a para-positon.

Now that a para-Hydrogen atom has been created, at 20 degrees K, the ortho-Hydrogen atom is converted to a para-Hydrogen atom. This means that the proton has turned upside down.

A molecule is never formed with a parallel spin. This means that when an ortho-Hydrogen atom combines with a para-Hydrogen atom, the molecule thus formed has an anti-parallel spin. The single atoms of Hydrogen are not molecules thus they can combine to form a molecule with an anti-parallel spin.

If two atoms of Hydrogen are used to create the isotope 1H2, this does not create a molecule.

When a 100 ray (called a beta ray) is emitted, it comes from either an ortho-proton or a para-positon. This means that the ray will be spinning in the direction of either an ortho or para direction, depending on the emitter. This is also true of the other masses that are lost which do not release an electron or negatron. These are now called alpha or gamma rays. An alpha ray is emitted by a proton only and a gamma ray is emitted by a neutron only. An alpha and a gamma ray are now named according to the amount of mass lost . A 0.00217 amu is lost as a 217 ray and if an ortho-proton lost this amount of mass, then it will be spinning in the ortho direction. This ray will then be an ortho-217 ray. If a para-positon lost the 0.00217 amu, then the ray will be called a para-217 ray.

THE ISOTOPE AND SUB-ISOTOPE

Atoms found naturally in nature will have one or more isotopes. There are two atoms that are not found naturally in nature. They are Technetium (Tc) and Promethium (Pm). They therefore do not have any isotopes. In the books on chemistry, there is an atomic mass that is given to an atom that is a composite of the naturally occurring isotopes of that atom. This mass is not a real atom so d- not look for an isotope belonging to it.

In this book, I give the isotopes and their atomic mass a- chemistry books and use this atomic mass to create other isot- are created from a formula that takes one or more isotopes different steps. I call these isotopes, sub-isotopes. These s same formula but the final isotope created will have differ proton, positon and neutrons. This gives different ways that example of this difference is when a person receives blood from someone with a different formula. The result of the two sub-isotopes that come from the same formula can be found in our blood. The blood looks the same in both cases but a person receiving blood from a person who used a different sub-isotope will reject the new blood.

An isotope used with another isotope to create a new isotope may remain the isotope it is within the isotope they create. If an isotope is found in the new isotope formed without change and the new isotope breaks up, this isotope will be one of the parts found after the break up. If there is a neutron created in the new isotope, it may be one of the parts found after the new isotope breaks up. These parts that are left over are said to have dropped out.

MECHANICAL OR NEUTRAL ENERGY

From the above, there can be found six forces which mold our existences as we know it.

A 1. Magnetic forces
   2. Electrical forces

When two magnetic forces act on each other, the electrical force is created. When the electrical force stops acting as a force, it becomes a single magnetic force.

B 1. Gravitational force (Force pulling toward graviton.)
   2. Centrifugal force (Force pushing away from graviton.)

When the gravitational force equals centrifugal force, the action that pushes or pulls stops.

C 1. Centripetal force (Force going around the graviton.)
   2. No name force (Force governing the axis.)

When the centripetal force becomes greater than the force that pulls the energy to the middle of the axis, the axis will divide into two parts.

From the above six forces, it can be seen that when they are in perfect agreement with each other, no one force will cause an expenditure of energy more than the other and when this happens, they neutralize each other. When they neutralize each other and two such neutralized forces meet, nothing can happen. If the neutralized forces were to have motion, they would act on that which they can act on as mechanical energy or as a mechanical force.

THE NEUTRINO

It is believed that the origin of neutrinos came from subatomic particles that decay and disintegrate. It is believed that the neutrinos carry off much of the energy released by exploding stars known as supernovas. The neutrino does not carry away energy as believed, but it is its formation that uses the energy plus the fact that it has at its center, a graviton, which holds the magnetic field in place. The magnetic field is a form of energy and this energy must be accounted for.

In 1956, two American physicists, Frederick Reins and Clyde L. Cowan Jr. believed they had created an anti-neutrino in a nuclear reactor because they could trace the tracks of a collision between a neutrino and an anti-neutrino. What they really saw was the tracks of the splitting mass that was growing around the neutrino. The neutrino itself, stayed as the center of the two new masses and the two new masses rotated around it.

On February 24th, 1987, an astronomer named Ian Shelton with the University of Toronto, using a telescope at the Cas Campanas Observatory in Chile found a supergiant star in a small galaxy known as the Large Magellanic Cloud, that had just become a supernova. The day before, February 23rd, Carlo Castagnoli of Italy and G.T. Zatsepin of the Soviet Union reported from their neutrino detection facility at Mont Blanc in the Alps, a sudden burst of five (5) neutrinos with a seven second period. It was believed that only a supernova could produce this and that when one was found the next day, the belief that a supernova did cause the burst of five neutrinos was believed to be correct. The term "five pulses" was used and this is in agreement with my way of describing the movement of electrons around the nucleus of an atom. This means that as the neutrino split from the mass that was forming around it, it sent out a magnetic pulse. It took 170,000 years for the magnetic pulse to reach our detecting equipment at Mont Blanc.

If it takes a supernova to cause a neutrino to pulse, then the number of neutrino pulses must be very small because supernovas do not happen very often. It is likely that most of the neutrinos were created when mass was first formed and now only when something like a supernova happens is a neutrino forced to send out a magnetic pulse. Or if the "big bang" is what happened, then they were created at that time. I believe that there is a neutrino for every electron and proton combination plus the fact that a neutron comes from a proton. This means there is a neutrino at the center of every neutron. It is the neutrino that holds the electron and proton together because of the graviton at its center. When the electron is released from the proton, the magnetic field does not control it any more. The proton's magnetic field can only affect another proton when an electron is involved. Because of this, the proton is said to be magnetically neutral and it is now called a neutron. This means the proton is not a proton when it has no electron to control.

If the above is true, then the neutrino is the cooling agent of the stars because of the graviton at its center.

THE MESON BECOMES AN ELECTRON, A NEGATRON AND A POSITRON

MESON

ORIGINAL NEUTRON

1.00867 amu's

1N84

0.00055 amu Electron O(φ)O Negatron 0.00055 amu

Neutrino

Proton has 1.00728 amu's

Lost mass has 0.00029 amu

Positron has 1.00757 amu's

As the neutrinos gained in mass, they become a Meson with 1.00757 **atomic mass units written as amu's.** The neutrinos first grow into a layer of mass with 0.00029 amu. The next layers consist of a series of masses such as 0.00002 amu, 0.00100 amu, 0.00217 amu, 0.00374 amu, 0.00434 amu plus the mass of electrons and negatrons that each have 0.00055 amu. There will be one layer of 0.00110 amu that will become external but still connected to the remaining mass. There are other layers which will be given in the text. Any layer can be combined with another layer or group of layers. An external mass of an electron or negatron is lost as mass while all the rest of the layers are lost as alpha, beta and gamma rays. A alpha ray comes from a proton or positon while a gamma ray comes from a neutron. The beta ray comes from 0.00100 amu only. These internal layers can be transferred from layer to layer internally with no ray emitted. The transferring of 0.00100 amu internally will cause an electron or negatron to be released from within the mass that lost the 0.00100 amu to go into an orbital with an electron or negatron already there. The 0.00100 amu can be transferred internally if it is combined with 0.00055 amu which does not release the electron or negatron. Thus a layer of 0.00155 amu can be transferred internally.

The Meson grows 1.00728 amu's around the 0.00029 amu. This new mass becomes either a proton or positon which is in the center of the Meson. On each side of the 1.00728 amu's, two masses grow to become the electron and negatron. When they reach 0.00055 amu each, the Meson now has 1.00867 amu's and I now call this mass the Original Neutron. When the Original Neutron comes into being, it is by itself and will change within twelve minutes to an isotope of Hydrogen One (1H1) and the 0.00029 amu is lost as a 29 ray. This leaves a proton or positon with an electron and a negatron on each side. Only one electron can stay with a proton or only one negatron can stay with a positon so either an electron or a negatron must be released. The remaining mass is the atom Hydrogen (1H1) with no neutrons.

The two masses that are the electron and negatron are held to the proton or positon with spins of their own. When they reach 0.00055 amu, this spin causes a repulsion between either the proton or positon forcing the 0.00029 amu to be converted to a 29 ray and the repulsion then forces out either an electron or a negatron.

The magnetic fields of the electron and negatron each begin to rotate until the magnetic force breaks to form two magnetic fields of equal strength and spinning in opposite directions which gives them different polarities. When the two masses separated, the internal magnetic field of the Original Neutron became external as the 29 ray is emitted and either the electron or negatron gets released.

When the positron reaches a specific mass, the neutrino can no longer cause the positron to grow and this sets the size of the positron. The positron limit is 1.00757 amu's. As soon as it reaches this limit, it loses 0.00029 amu and becomes a proton. The electron is released to go into an orbital around the new proton with the release of the negatron. This forms an ortho-Hydrogen atom. It is possible for the electron to be released and the negatron to orbit that which is now a positon. This forms a para-Hydrogen atom. There are three (3) ortho-Hydrogens created for every para-Hydrogen atom created.

The positron loses 0.00029 amu and its energy drops below that which is needed to keep the electron and negatron connected to it, the isotope or atom of Hydrogen (1H1) is formed. The isotope 1H1 thus formed can combine with other isotopes to form new atoms or isotopes.

MASS EQUALS FREQUENCY

$$\text{ENERGY} = mc^2 = hf$$

Mass (m) times the square of light (c) is equal to Plank's constant (h) times the frequency (f) of any mass in motion either around a nucleus or an axis. The value of "c^2" never changes, then the value of "m" must change. Since the value of "h" never changes, the value of "f" must change. The value of "hf" has no mass, then it must be pure energy. If the "hf" has no mass, then the "f" must be the frequency of the pure energy going around the point in space which is the graviton. This means that Plank's constant "h" is the graviton. If mass has a frequency, then it must be going around a central point in space. If the quark is the first mass formed from pure energy, then it is at this point in space that pure energy changes from "hf" to "mc^2".

If the isotope 1H1 can be compressed to the point where the electron can touch the proton, then some of the energy that is holding the electron to the proton will be converted to pure energy which is the Hydrogen Bomb. The energy holding the electron and proton together is the graviton. Some of the mass around the graviton is converted to pure energy. This means that when the graviton released the mass around it, this released mass exploded. In the case of the Hydrogen Bomb, the mass that is combined is radiactive and thus unstable. In our everyday life, we encounter mass being converted to pure energy in very small amounts of magnetic energy. This is because the magnetic energy is converted to a photon that can be used over and over again by different atoms or isotopes. A match made of wood and some chemicals will ignite when friction is used to change as stable chemical into a chemical that can emit photons. This causes the chemical that makes up the wood to be above its magnetic energy level at the burning end of the match. As the wood burns, photons are released and this cools that which remains. The friction that ignited the match is the same friction that ignited the atomic bomb.

1H1 AND THE COULOMB

The mass of an electron is 0.00055 amu and in the case of the isotope 1H1 which has only one electron and one proton, the proton has a mass of 1.00728 amu's. This means that the single electron is going around the proton a certain number of times. To find the number of revolutions, the definition of a coulomb is needed.

A coulomb is equal to 6,241,450,000,000,000,000 electrons going past a given point in one second. This statement should be reworded.

A coulomb is equal to one electron going 6,241,450,000,000,000,000 times past a given point in one second.

This is shown in the drawing below.

1H1
Electron
0.00055 amu
orbital
proton
1.00728 amu's

Electron making 6,241,450,000,000,000 revolutions around the proton in its orbital

THE FIRST FIVE ISOTOPES

1H1 1H2 1H3 2He3 2He4

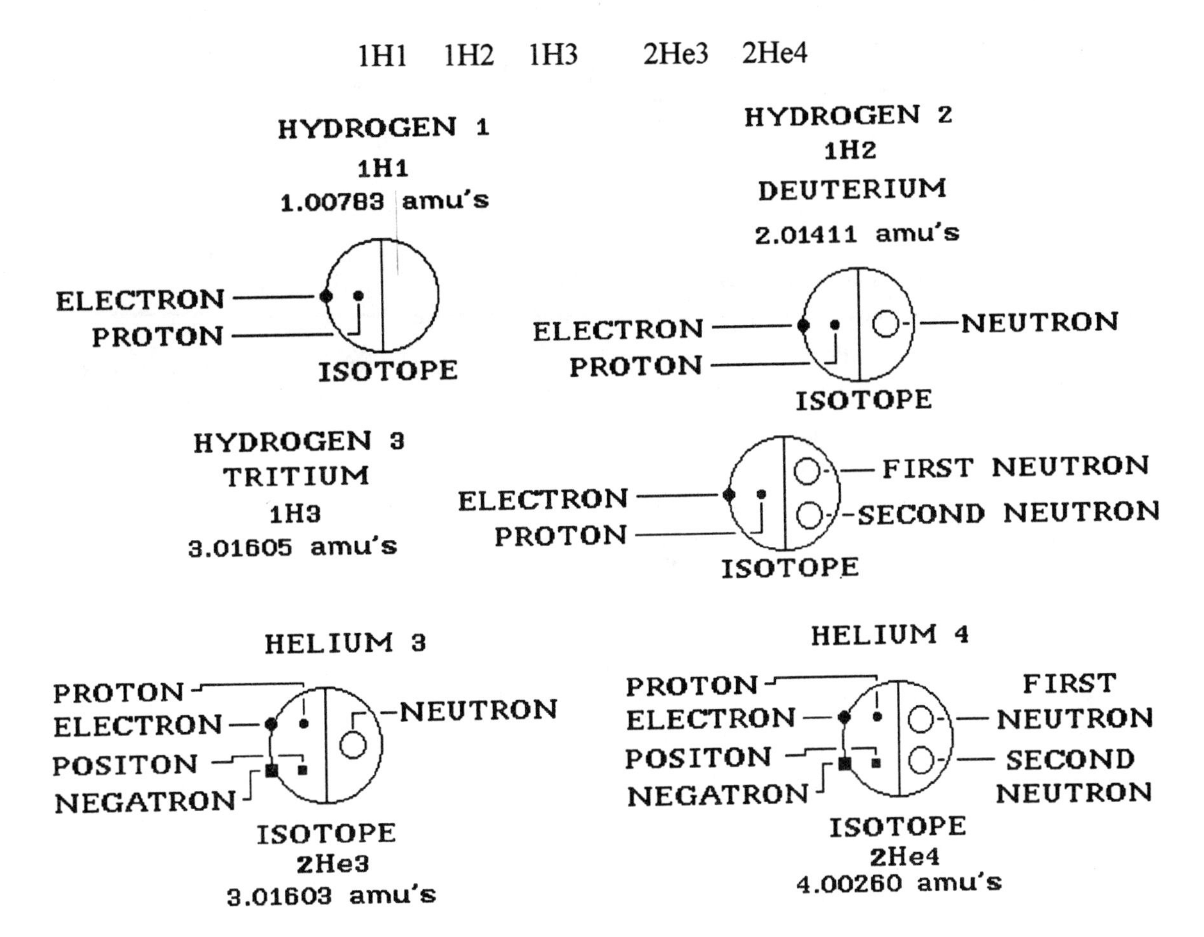

When the isotope 1H1 is formed, the neutrino became the center of the proton. Meanwhile, other neutrinos are growing electrons, negatrons and protons. That is, other neutrinos are growing into more mesons to create a new isotope of 1H1 from the Original Neutron.

Now there are two isotopes of 1H1 and one neutrino at the center of each. The two isotopes of 1H1 begin to rotate around a common central point. When this happens, a graviton comes into being. Now there is a single external magnetic field around the two isotopes of 1H1 holding them together by a graviton that is common to both of them. These two isotopes of 1H1 can not create a new atom or isotope of 1H1 because the isotope 1H1 can only have one electron and one proton when there is no neutron.

One of the isotopes of 1H1 must lose an electron and its proton must become a neutron for the two isotopes of 1H1 to stay together as an isotope of 1H2. One of the isotopes of 1H1's proton must lose 0.00100 amu to become a neutron and release its electron. Both isotopes of 1H1 have their parts spinning or rotating on their axis in the same direction. They can not do this and remain connected by an external magnetic field. Two isotopes of 1H1 will not create the isotopes 1H3, 2He3 or 2He4. They will create the isotope 1H2 by having a proton become a neutron. For this to happen, one of the isotopes has its spin and magnetic field in agreement with the neutrino. This causes that isotope to gain more energy than the other isotope. The result is that the proton that gains more energy than it can have, loses mass. This loss is that amount of mass now needed to maintain a central point between the two isotopes where the new graviton is located. There are now two masses located in the nucleus and they do not have equal mass.

This new graviton must maintain two masses in such a way that they balance each other. When two unequal masses rotate around a common center point, the center point is located in or closer to the largest mass. This would place the center point just inside of the proton because it is the one with the greatest mass. However, both masses are rotating around each other as a single unit and their centrifugal and centripetal forces are trying to keep them apart while the graviton keeps them together.

The reader is to keep in mind that the proton and neutron each have a graviton at their center and there is a third graviton at the central point between them. Every rotating body or unit has a graviton if they have a magnetic field. A graviton is nothing more than a point in space that has the ability to pull mass toward it. (The photon is two magnetic fields rotating at right angles to each other with no mass thus it has a graviton at its center.)

Centripetal force is a force which pushes at right angles to the centrifugal force. When the electron is going around the proton, there is a centrifugal force pushing it away from the graviton. There also is a centrifugal force pushing the proton outward from the proton and graviton. The electron can change its position due to this centrifugal force but the proton can not. There is a centripetal force pushing the electron around the proton and there is a centripetal force pushing the proton around the graviton. In this last case, both can change their speed. The electron can go faster or slower in its orbital and the proton can rotate on its axis faster or slower, depending upon the amount of outside energy it is receiving. This is true of the negatron and positon also.

Now that the neutron is created, the proton begins to move around it in such a way that the neutron can transfer mechanical energy to it. The two masses are rotating around a graviton that is closer to the proton than the neutron. This makes them rotate around each other as a single unit. This means that the neutron is turning with its internal magnetic field acting as a solid against the proton and thus as external energy is supplied to the isotope, mechanical energy is passed to the proton. At the same time this is happening, both the neutron and proton are rotating as a unit because their neutrino is inside each. The remaining electron then orbits them both with a single magnetic field around all of them. Thus is the isotope 1H2 created with one 100 ray emitted when the electron is released and the 0.00100 amu is lost, being the mass that is converted to pure magnetic energy that is lost when the magnetic field became an internal one. Remember a 100 ray is an electromagnetic form of energy that is radiated. In chemistry books it is called a beta ray. A ray is mass converted to pure magnetic energy while a photon is pure energy that is received and emitted by an atom or isotope.

It should be pointed out that the orbital of a single electron around a single proton is different than when the electron takes up an orbital around a proton and a neutron. The radius from the center of the nucleus to the center of the electron has changed.

The isotope 1H1 and 1H2 rotates around their gravitons. Each has its own external magnetic field which combines to form one magnetic field around both isotopes. This prevents them from separating and they become an isotope of 1H3.

There are three isotopes of Hydrogen and I have only explained two. The third isotope is the isotope 1H3 which is radioactive. It has one electron, one proton and two neutrons. I wanted to show the creation of new isotopes that are found naturally in nature and stable. Radioactive isotopes are found naturally in nature but they are not stable. They change into a new isotope by losing a small amount of mass.

It is because the neutrino has a graviton at its center that forces the other subatomic particles to rotate around each other creating another graviton. Thus when two isotopes create a new isotope, the graviton tries to become the center of the new isotope. It is this action that causes the isotope to be created.

Two isotopes of 1H1 create the isotope 1H2. An isotope of 1H2 and one of 1H1 create the isotope 1H3. The isotope 1H3 and 1H1 will create the isotope 2He3 with the release of a neutron or the isotope 1H3 by radioactivity will create the isotope 2He3.

The next isotope to be formed is 2He4. Note that this isotope has gained one more neutron than the isotope 2He3. I explain the isotopes later in a sequence that shows their creation in a way that all elements and their isotopes follow a certain line of reasoning.

THE NUCLEUS

The nucleus has two ways it can be formed. One is to have the proton in the center with the neutrons rotating around it. The second way is for the neutrons to be in the center with the protons rotating around them. Of course they would rotate around each other if only one proton and one neutron are found in the isotope.

If the nucleus has one proton and two neutrons, then the two neutrons must rotate around the single proton. This puts the proton in the center of the nucleus. The two neutrons are supplying mechanical energy to the proton. However if the isotope is combined with other isotopes, this combination may cause the lineup of the proton and its two neutrons to re-orientate themselves to a new position.

When there are two protons and one neutron, then the neutron is in the center and the two protons are rotating around the neutron. This time, the single neutron is supplying mechanical energy to the two protons. When there are two protons and two neutrons in the nucleus, the two protons are at right angles to the two neutrons. The two neutrons supply mechanical energy to both of the protons.

Two protons and three neutrons line the three neutrons up in a row and two neutrons supplies mechanical energy to one of the two protons on one side of the row of neutrons while the other proton does the same thing with the two neutrons on the other side and the other end of the row. As more protons and neutrons are added to the nucleus, they arrange themselves so that mechanical energy can be transferred. Two protons and four neutrons would put the four neutrons where they form a square, each neutron passing mechanical energy to its neighbor while there is a proton above and below the center of the square. Two protons are found above and below a six sided figure when there are two protons and six neutrons. It is the mass of the neutrons that slow down the rotating proton. Note that all the protons are called protons even though some are positons.

THE MAGNETIC CHARGE STARTS WITH THE ELECTRON AND PROTON

The definition of an electrical charge is the quantity of electrical energy stored in a capacitor, battery or insulated object. The term electrical energy as given in this book means magnetic energy caused by either a magnetic field that is larger or smaller than required by the proton or by the electron or negatron going faster or slower than required by the proton. If these conditions exist without the electron or negatron returning to their normal state, the electrical charge is electrical energy being held within the capacitor, battery or insulated object. The difference between magnetic energy and electrical energy is when there is only one magnetic field, there is only magnetic energy and if there are two magnetic fields crossing each other at some angle, then this action creates an electrical charge and thus there is an electrical charge when centrifugal and centripetal forces are not normal.

When the positron reaches 1.00757 amu's, the centrifugal force overcomes the gravitational force holding the electron, negatron and positon together due to the centripetal force pushing the mass faster and faster until the electron and negatron separate from the positron. When they separate, they all gain an external magnetic field. The positron loses 0.00029 amu and becomes a proton. The negatron is released while the electron goes into an orbital around the proton. The electron is controlled by the proton. This makes them the isotope 1H1. The proton now has 1.00728 amu's.

It is at this time that the electron first takes up a point in space where the gravitational force balances out the centrifugal force. The electron's polarity must now become attracted to the proton by the external magnetic field. Because the electron is going around the axis of the proton in the series connection faster than the proton can rotate, a centripetal force is created. The centripetal force is pushing the new electron around the proton at right angles to the axis of the newly created proton. The negatron must have a balancing magnetic charge and it has none and it is therefore ejected.

The location of the electron relative to the proton is found by understanding that the electron can go to whatever point in space that will allow a balance of the magnetic field. A magnetic field does not change directions which is found in direct current while alternating current uses an electrical charge that does change its direction.

TWO MAGNETS

If I take two cigar shaped magnets and hold them in such a way that they can attract each other, they will come together to form one magnetic field around both. Now holding one magnet so that it can not move in any way, rotate the other magnet so that the end that is touching the non-moving magnet will turn. Be sure the magnets are still held together by their magnetic fields. What you are doing is proving that two magnets will still be attracted to each other even though they are turning in different directions. Their magnetic fields will still be one magnetic field around both of them.

If one magnet that is rotating, comes close to another magnet, it can cause the second magnet to rotate without touching the second magnet. An electrical charge will come into being if their rotations are not equal.

As the rotating magnetic fields are making and breaking, they are trying to maintain one nonbroken magnetic field. If the two cigar magnets come together and both rotate in the same direction at the same time, they will act as one magnet. This is not true where one magnet turns or rotates around their common axis opposite to the other magnet even though they are held together by the attraction between them.

What I have here described is two magnets attracted to each other at their magnetic poles. I describe this type of attraction to show their attraction takes place independently of the direction of their rotation.

Now move the two magnets away from each other. Make one magnet the size of an electron and the other the size of the proton. Place them so that each magnet has an axis that is parallel to the other. A magnet will attract a second magnet along its long side as well as their ends. This means that the two magnets will come together.

If one magnet has a much greater magnetic field than the other, the larger magnet becomes the controlling magnet. This means that the smaller magnet will do whatever the magnetic lines of force of the larger magnet forces it to do provided that the second magnet is free to do so.

A single magnet with all of its sides in the shape of a rectangle has two poles at each end. If a second magnet is moved over the top of one end so that it is attracted to the first magnet, it will dip toward this attracting end. Keep the second magnet in hand so that it does not touch the first magnet. Move it slowly over the first magnet. As the second magnet moves over the attracting end, it will meet at some point the repelling end. When this happens, the second magnet will slowly rise. This point where the magnet reaches the repelling end in not half way between the first magnet but at a point closer to the repelling end. Repeat this from the repelling end. The second magnet will be held away from the first magnet. Move the second magnet slowly over the first magnet again. The second magnet will dip toward the attracting end of the first magnet. It will not be at the middle of the first magnet but at some point near the attracting end. These two movements of a second magnet over a first magnet will trace out a teardrop shape with the small end of the teardrop overlapping each other.

The area where the two magnetic poles overlap that is between the two poles of the first magnet is the weakest point in the magnet's magnetic field. In this case, it is the middle of the first magnet that is the weakest attraction or repulsion of the two magnets.

I have magnets that are round, square, spheres and odd shaped. The shape of the magnetic field results from the way the magnet is shaped. Magnetic fields of the other magnets will also cause their magnetic fields to form different shapes.

The microscope shows that very small magnets form a line from one pole to the other. They follow a magnetic line of force. Sometimes there are small magnets that are not completely in the lineup and they can jump from one line to the other if there is a vibration. Each line is independent of any other line and they do not form bridges between them. They can lay across each other as they go from one pole to the other. They can attract any magnetic particle that is free to move. Once that particle becomes part of the line it is now in, it will not be affected by the other magnetic lines near it. Once a magnetic line is completed, it will not attract or repel other completed lines.

HIGH VOLTAGE BULB BY SILVER LABORATORIES INC. OF HONG KONG

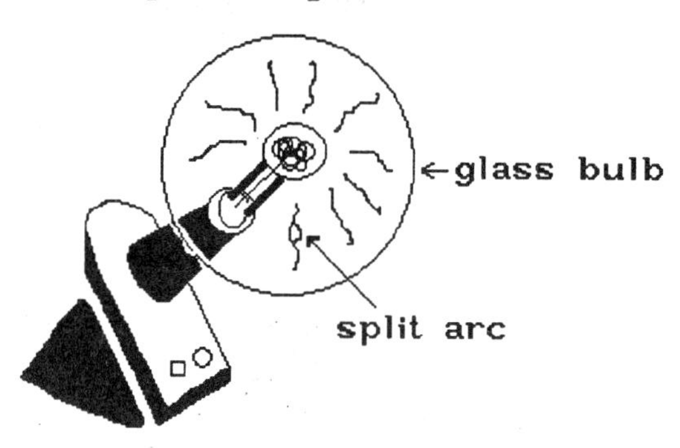

On the atomic level, where there are only a few atoms such as in a vacuum, the magnetic lines of force will travel outward from a source of high voltage in the form of an arc. They will travel along a single magnetic line of force. However, each magnetic line of force will be composed of two magnetic lines of force that are winding around each other as it goes outward as a single line. If they are enclosed in a glass bulb with the source of high voltage in the middle area of the bulb, they form an electrical arc from the high voltage source to the glass if the high voltage is great enough to make an arc that long. When the arc first forms, it is red. Then as they twist around each other going outward, they become bluish white until they reach the glass. There, they become reddish again and by varying the strength of the voltage, the arc can be made to split into two arcs. What has happened is that the point where the high voltage is, the atoms are gathered and sent outward along a line of force. The line of force is made up of two different kinds of charges that are moving around each other. When they first start out, they are cool until they wind around each other. When they begin to wind around each other and are forced outward, they are more energetic and radiate a bluish white light. They do not generate much heat as the glass does not warm up as they reach it. If two lines come near each other, they will vanish and a new line will start out from the high voltage.

If a person puts their finger to the glass, there will be an arc formed from the high voltage source to the finger. If a magnet is used instead of the finger, the color of the arc becomes more intense. At very high voltage, the red area near the magnet will be seen to divide into sheets of red along the glass and move outward along the glass until they vanish. As long as the magnet is held to the glass, the arc will stay there. This is not true of the finger on the glass or a weak voltage. In the case of a weak voltage, the arcs move around until they come near each other and vanish. As long as there is voltage, new arcs will be formed over and over again. Arcs from low voltage will not reach the glass and they can not be forced to separate.

There is one experiment that is very hard to do but when it can be done, it will show that an arc under high voltage can be pulled apart in the middle area. This proves that an arc is composed of two arcs going around each other. One arc is an ortho spinning arc while the other is a para spinning arc. That is to say that one single arc is made up of two arcs that are spinning around each other in opposite directions as they go to the glass. I used a magnet to pull the arc apart.

UNDERSTANDING THE FIRST ATOM OR ISOTOPE OF HYDROGEN (H) OR 1H1

After the meson separates into an electron and a negatron, with a positron as its nucleus, it becomes an Original Neutron. The Original Neutron takes on more mass until it reaches an atomic mass of 1.00867. It has a half-life of twelve minutes. This means that within twelve minutes, one Original Neutron will change from an Original Neutron to the isotope 1H1 with the electron going into an orbital around a proton, the negatron being released and 0.00029 amu is lost. This is a total of 0.00084 amu lost in creating the isotope 1H1. Both the electron and negatron have 0.00055 amu and the proton has 1.00728 amu's. The isotope of 1H1 has 1.00783 amu's.

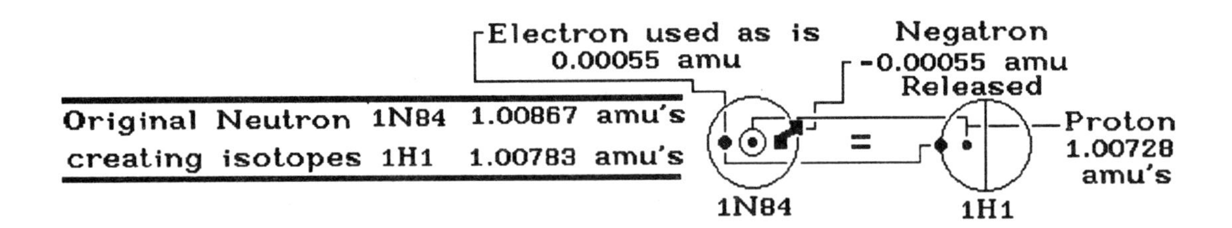

This forms the first atom. It is the isotope of Hydrogen 1H1. The 1 in front of the H is the number of protons found in this isotope. The H is the symbol for Hydrogen and the 1 following the H is the number of protons and neutrons found in the isotope. Therefore, there is one proton and no non-original neutrons found in the isotope 1H1.

When an electron and proton are together with the electron orbiting the proton, it is the isotope 1H1 with an external magnetic field and the neutrino is in the middle of the proton.

THE ISOTOPE 1H2

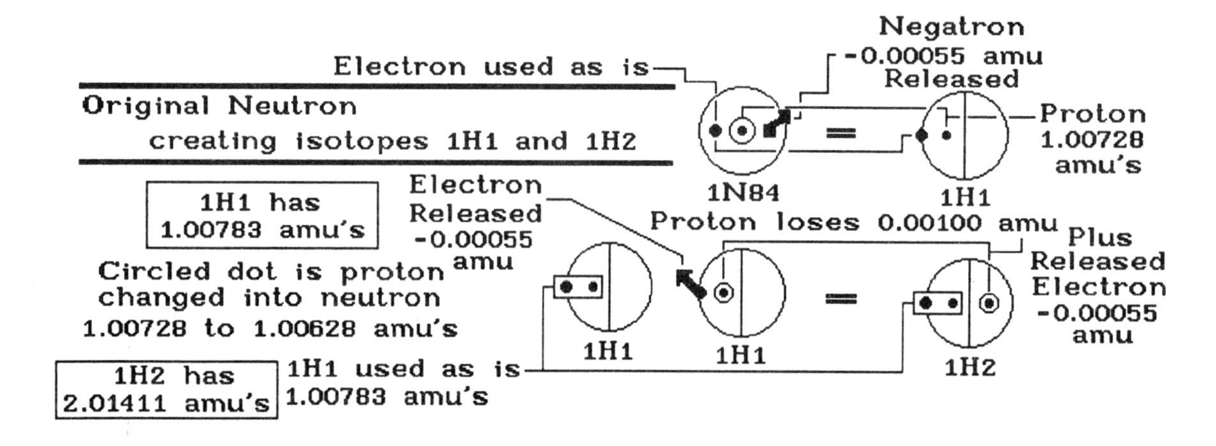

After two isotopes of 1H1 have been created from two Original Neutrons, they come together and one isotope of 1H1 is converted to a neutron. This neutron is different than the Original Neutron because it comes from the isotope 1H1. In order for the isotope 1H1 to become a neutron, it's proton must lose a certain amount of mass that will release the electron that is in an orbital. This mass that is lost will always be 0.00100 amu and no other mass that is lost will release an electron from its orbital. This lost mass is mass converted to pure energy which is magnetism. It is emitted as a 100 ray. (Called a beta ray.) In the neutron, the neutrino is in the center of it. The proton and neutron rotate as a unit around a center point between them where the graviton is located. (The graviton is a point in space around which some pure magnetic energy or mass is rotating. It is more than just a point in space because it is the point in space that has magnetic energy being directed to it.)

Remember that isotopes are numbered by the number of protons found in the nucleus. If the neutron is released as is, then the number of electrons and protons stay the same as the isotope that released it. It keeps its atomic number but the new isotope created from this action has a new isotopic number. However, the nucleus now has one less neutron than before and this will change the way the electron will orbit the proton. This will change the way the proton and neutron will rotate around each other as a unit. This will change the amount of external energy or magnetism the proton can receive and the mechanical energy that can be transferred between them. This means that the overall energy or magnetism of the isotope has changed within the isotope.

The neutron has its magnetic field within itself and it therefore does not by itself change the magnetic field that is found around the proton and neutron. While this is true, its loss causes the magnetic field to change due to the ability of the electron to orbit the nucleus. While there will at a later time be more protons, positons and neutrons in the nucleus, I wish the reader to consider the isotopes 1H1 and 1H2. This will give the reader a feeling for how these changes will affect different isotopes as they are created.

The first thing to get a feeling for is that the Original Neutron created an isotope of 1H1. The reader knows that any two magnets will attract or repel each other. The reader also knows that some objects are unaffected by a magnet. These objects are said to be magnetically neutral. The Original Neutron is unaffected by a magnetic field and it is always by itself in space. The isotope 1H1 is not neutral when it is by itself but it is neutral when two Hydrogen atoms form a molecule. A neutral particle has its magnetic field within itself and does not affect other atoms or molecules near it.

For an isotope of 1H1 to give up its electron, a certain type of energy is needed. It is like a key that can unlock the electron from the proton and thus it can be released. It is this ability to be disconnected from the proton and not the amount of mass that causes energy to be used to disconnect it from the proton. When the electron is released, the isotope of 1H1 is no longer an isotope of 1H1 but a neutron that remains with the item that forced the release of the electron. In the case of two isotopes of 1H1, one isotope gives up its electron to the area near it and the isotope of 1H1 becomes a neutron as part of the remaining isotope of 1H1. Neither isotope of 1H1 remain as such and they become the isotope of 1H2. The magnetic field that was around each isotope of 1H1 is now around the isotope 1H2. This new isotope of 1H2 is not magnetically neutral and it can combine with another isotope of 1H1 to form a molecule of Hydrogen. When the isotope of 1H1 and 1H2 combine to form a molecule, the molecule is neutral. It does not affect other atoms or molecules near it. There are not as many isotopes of 1H2 as there are of the isotope 1H1. Because of this, there are more molecules made of two isotopes of 1H1 than there are of molecules made of 1H1 and 1H2.

The size of the overall magnetic field will change because of the electron that is released from its orbital and the lost mass. Now there is one electron orbiting a proton and a neutron instead of orbiting a single proton. This changes the size of the magnetic field.

The graviton that held the proton in the center of the isotope 1H1 is still there as well as the graviton in the center of the neutron. Since there are now two particles in the nucleus, there is a third graviton that holds them in the center of the new isotope of 1H2. This graviton is located closer to the proton because it has more mass than the neutron.

MAINTAINING AN ATOM OR ISOTOPE

A photon is the source of outside energy that keeps the isotope as it is. This outside energy is received by the proton. Because there is a neutron with the proton in the nucleus, the amount of energy the proton can receive is greater than one proton by itself. The proton can pass some of this energy to the neutron as mechanical energy. What happens is that the mass of the neutron will slow down the rotation of the proton. This will govern the number of revolutions of the electron by the interaction of their magnetic fields. The neutron acts as a governor that keeps the isotope from spinning out of control.

I am only using the photon as an example of how an atom maintains itself. Where there are no photons reaching an atom, the atom is at its lowest energy level and it can stay this way because it is always balanced out by the pure magnetic energy found in all empty space. An atom can maintain itself by other means such as heat, added electrical current and bonding between atoms. The amount of energy an atom can receive is set by the amount of mass found in the nucleus.

Mechanical energy can affect the magnetic field by vibrations. A blow to a non-magnetic bar of iron will cause a magnetic field to come into being. This happens because the blow to the iron bar caused the iron atoms to vibrate and this vibration caused the magnetic field of the proton to move. This movement will, for a short time increase the size of the magnetic field and the iron bar will show a magnetic field. The increased size of the magnetic field will decrease until the vibrations stop.

If a magnet is placed under a paper and iron filings are poured onto the paper, the iron filing will trace out the path of the magnetic field's lines connecting the ends of the magnet. The reader is directed to note that some of the iron filings do not complete a path from one end of the magnet to the other. It is these magnetic lines that are connected to the pure energy that is magnetism. All magnetic fields are connected to the magnetic field found in any empty space.

When the nucleus has an equal number of neutrons and protons, the energy of the proton and positon will be the amount they can give to the electrons and negatrons after the energy has been balanced between the protons and positons with the neutrons or neutron.

ATOMIC MASS UNITS VS MILLION ELECTRON VOLTS
amu vs MeV

Most of my chemistry books give the value of an atom in atomic mass units (amu) or grams (g) and then tell the reader they must convert this to million electron volts (MeV). The readers I have talked to about the use of these two values say it is very confusing to do this. Why don't they stay with amu's and if need for the value in MeV, is found, then learn the conversion at that time. In this book, the amu will be the value used. For those who would like to convert amu to MeV, I will here give these values.

The conversion factor for converting amu's into MeV is 931.48
This value is sometimes rounded off to 931.5. This is the value I will use.

Original Neutron	1.00867 amu's	939.576 MeV
Positron	1.00757 amu's	938.552 MeV
Lost mass	0.00084 amu	0.782 MeV
Lost mass	0.00029 amu	0.270 MeV
Lost mass	0.00100 amu	0.932 MeV
Electron/negatron	0.00055 amu	0.512 MeV
Proton/positon	1.00728 amu's	938.281 MeV
Normal neutron	1.00628 amu's	937.350 MeV
1H1	1.00783 amu's	938.793 MeV
1H2	2.01411 amu's	1876.144 MeV
1H3	3.01605 amu's	2809.451 MeV
2He3	3.01603 amu's	2809.432 MeV
2He4	4.00260 amu's	3728.422 MeV

The value given for the Original Neutron includes the electron, negatron and positron.

Electron in Original Neutron	0.512 MeV
Negatron in Original Neutron	0.512 MeV
Positron in Original Neutron	938.552 MeV
Original Neutron	939.576 MeV
Original Neutron	939.576 MeV
Book Value for Original Neutron	939.570 MeV
Difference	0.006 MeV
Electron	0.00055 amu
Negatron	0.00055 amu
Positron	1.00757 amu's
Original Neutron	1.00867 amu's

When the Original Neutron keeps its electron and changes its positron into a proton, it will release its negatron and lose 0.00029 amu.

Released negatron	0.00055 amu	0.512 MeV
Lost mass	0.00029 amu	0.270 MeV
Negatron and lost mass	0.00084 amu	0.782 MeV

The Original Neutron loses 0.00084 amu which is worth 0.782 MeV. This means that the Original Neutron becomes the isotope 1H1 with 1.00783 amu's which is worth 938.793 MeV.

Electron in 1H1	0.00055 amu	0.512 MeV
Proton in 1H1	1.00728 amu's	938.281 MeV
Isotope 1H1	1.00783 amu's	938.793 MeV

The isotope 1H2 has one electron, one proton and one neutron. Using the above values for the electron and proton with the value of 937.350 MeV for the neutron, the value in amu's and MeV can be found. This value must equal the value for two isotopes of 1H1 because two isotopes of 1H1 are used to create the isotope 1H2 minus the value of the electron that was released and 0.00100 amu that released the electron.

Isotope 1H1	1.00783 amu's	938.793 MeV
Neutron in 1H2	1.00628 amu's	937.350 MeV
Isotope of 1H2	2.01411 amu's	1876.143 MeV

Isotope 1H1	1.00783 amu's	938.7935 MeV
Isotope 1H1	1.00783 amu's	938.7935 MeV
Two isotopes of 1H1	2.01566 amu's	1877.5870 MeV

Electron that is released	0.00055 amu	0.512 MeV	
Lost mass releasing electron	0.00100 amu	0.931 MeV	(931.48)
Total mass lost creating 1H2	0.00155 amu	1.443 MeV	

Two isotopes of 1H1	2.01566 amu's	1877.587 MeV
Total mass lost creating 1H2	0.00155 amu	1.443 MeV
Isotope 1H2	2.01411 amu's	1876.144 MeV

The reader is to know that some of the values given were rounded off and it therefore changed some of the values such that when a number is multiplied by 931.5 which is itself a rounded off number will be off by about 0.002 MeV. By taking the values as given above in their steps, the values are the best values that can be found. Note that the value 931.5 should be 931.48. I therefore made the value of 0.00100 amu into 0.931 MeV. Note also that two isotopes of 1H1 has the value of 1877.587 MeV. This is the correct value and to get this value I added

0.000005 amu to each of the isotopes 1H1 as this will then give me the correct value for two isotopes of 1H1.

The above is given to help those readers who use MeV. These readers can now understand that using MeV can give some answers that need to be adjusted. Voltage is the result of an action between two magnetic fields and the meters used to read it can be off in their calibration when it is measuring MeV.

THE CENTER POINT OF TWO OR MORE ROTATING MASSES

When two or more masses rotate or circle around each other, they have a common central point around which they rotate. Their masses, relative to each other, will determine where this central point is. If two masses are equal, then this central point in space is half way between them.

If the two masses are unequal, then they can not rotate around a point in space that is half way between them. The central point will shift closer to the greater mass. To find this central point, take a board that has two equal masses on each end of it. Place the board with its two masses on a tight wire so that the weights will balance each other. Note the two masses are equal distance from each other on the board. Now make one mass greater than the other and place them again so that they will balance on the wire. Note where the wire is located relative to the two masses. The wire has shifted closer to the larger weight. This point is the central point around which both masses will rotate.

Now take one mass and rotate it around this central point where the two masses are unequal. Note the orbital path it takes. Now take the other mass and rotate it around this same central point and note its orbital path. Now place the greater mass's orbital within the orbital of the lighter one. These two orbital paths have a common central point.

One system has more mass than the other system at this point in time and therefore they rotate around a common point that is the central point for both masses.

After the proton and electron become another Hydrogen atom, the two systems become equal and they rotate around the central point where the graviton is located with one magnetic field around them.

WHY DOES THE LIGHTER MASS ORBIT THE HEAVIER MASS?

If the electron has such small mass compared to the proton, why doesn't the centrifugal force put the proton on the outside and have it orbit the electron?

Four of the six forces are involved in the question. The first two are gravity which pulls mass inward and centrifugal force which pushes mass outward.

When the negatron and electron was first formed by the neutrino, the two masses were equal. The direction of the negatron spin and the neutrinos spin are the same. The neutrino is so powerful that the electron and negatron can not escape. The negatron can not gain more mass than the electron. They balance each other so that the neutrino can maintain itself as the center of the meson.

With an electron and a negatron as part of the meson, the positron could gain more mass than a normal proton. Because it gained more mass, it lost some of its centrifugal force while the electron and negatron did not. Also, the greater mass stayed in one place while the electron and negatron did not. The electron and negatron rotated on their axis and they also rotated around the positron. The electron and negatron, due to their greater centripetal force, could not remain connected to the positron. They could not escape due to the attraction of the magnetic fields between them, the positron and the neutrino. The attraction was divided between an electron, a negatron and a positron whereas the normal attraction is between one electron and one proton. (Keep in mind that all other books on chemistry call the negatron as given here an electron.) When either the electron or negatron went into an orbital, it was still held by the proton and the neutrino's magnetic field.

RAY VS PHOTON

Care should be taken not to misunderstand the difference between a ray and a photon. The ray is a powerful pulse of electro-magnetic energy sent out from the point where something separated. It is like a bullet. When gunpowder explodes, it sends the bullet outward. When the electron and proton separate, it is like an explosion and the electro-magnetic radiation is the bullet. The photon is a form of radiation also except that it is very weak and can be sent out on a continuous basis while the ray is sent out only at the time of the explosion.

TWO RULES FOR MASS AND RAYS

Rule 1.
When mass is converted to pure energy (magnetism), a ray is emitted.

Rule 2.
When mass is not converted to pure energy (magnetism) and it is interchanged between isotopes or within an isotope, no ray is emitted.

TIME SCALE FOR CREATING

It takes 14 billion years to form a Hydrogen atom that has one neutron from the two isotopes of 1H1. Instead of creating a new neutrino from the pure energy found at the center of the sun, the pressure and temperature forces the two isotopes of 1H1 to become one atom. To do this, both atoms recombine to form one new Hydrogen atom, isotope 1H2 with the mass of the proton and neutron at its center. There is one neutrino in the proton and one in the neutron. These two neutrinos rotate around a common central point where the graviton is located.

The next atom that is found naturally in nature and is not radioactive is the isotope 2He3. It takes about one million years for this isotope of Helium to be created.

The time scale for the creation of 1H2 from 1H1 is much greater than that for the creation of Helium. It is a very short time compared to the time it takes a quark to make an Original Neutron. To make an isotope of 1H1 in comparison to the creation of the Original Neutron is even less. The 14 billion years needed to create the isotope 1H2 is by comparison an even shorter time and the creation of the isotope 2He3 is much shorter. In creating the neutrino from quarks, I said that the quark grows by gaining more quarks until a neutrino is formed. I said that the neutrino grows into an Original Neutron. Then the Original Neutron changes into the first Hydrogen atom or isotope, Then two isotopes of 1H1 combine to create the isotope 1H2. The isotope 1H2 combines with an isotope of 1H1 to create the isotope 1H3. Now the isotope 1H3 does not combine every time to create the isotope 2He3. It changes internally by changing a neutron into a proton. Next the isotope 2He3 combines with an isotope of 1H1 to create the isotope 2He4. Growing the Original Neutron takes a greater time than combining different isotopes while changing internally takes even less time than combining. When the Original Neutron is found by itself, it only takes about twelve minutes for it to split. To split is to brake apart while to change internally means to have some internal part grow back into something. Growing takes longer than splitting, splitting takes more time than combining and changing internally takes less time than combining.

There is one more way isotopes and atoms can be created. Sometimes there will be left over parts and these left over parts can combine to create a new isotope. This method takes the same time as the combining of masses already given or it may take less. It depends upon what isotopes or parts are combining. Since there will be more choices as to how to combine isotopes and left over parts, the time to create a lot of them is less because the parts are already there.

THE HUNT FOR THE NEUTRINO

Mr. Raymond Davis, Jr. built a tank filled with 100,000 gallons of cleaning fluid perchlorethylene that was a mile underground in a South Dakota gold mine to capture a neutrino. When the experiment was run long enough to capture a calculated number of neutrinos, the expected amount did not show up and therefore some kind of explanation for this failure was needed. Mr. Friedman says the entire astrophysical theory of the architecture of stellar interiors is at stake. So far, what has been presented has shown how the neutrino is formed and used. Nothing has been said about the neutrino leaving the sun as most scientist now believe. If the neutrino remains part of the creating process found within the sun, they would not always be emitted by the sun. Since there are so few neutrinos found, there must be a reason for this. They are there except they are inside something and therefore not detected. There are no neutrinos detected at the tank, only their pulses are detected there. A neutrino will only pulse when it loses the mass around it.

GAMMA RAY, BETA RAY AND POSITRON EMISSION

In the book "CHEMISTRY, A conceptual Approach" by Charles E. Mortimer of Muhlenberg College, Fourth edition, page 754, the following is given:

Gamma radiation is electromagnetic radiation of very short wavelength; its emission is caused by energy changes within the nucleus. Its emission alone does not cause changes in the mass number or in the atomic number of the nucleus. At times, nuclides are produced in excited states by nuclear reactions (Section 21.5), and such nuclides revert to their ground states by emission of the excess energy in the form of gamma radiation.

The gamma rays emitted by a specific nucleus have a definite energy value or set of energy values because they correspond to transitions between discrete energy levels of the nucleus. Thus, an emission spectrum of gamma radiation is analogous to the line spectrum that results from transitions of electrons between energy levels in an excited atom.

Beta emission is observed for nuclides that have too high a neutron/proton ratio for stability. The beta particle is an electron, which may be considered to result from the transformation of a nuclear neutron into a nuclear proton. Electrons, as such, do not exist in the nucleus. The net effect of beta emission is that the number of neutrons is decreased by 1 and the number of protons is increased by 1. Thus, the neutron/proton ratio is decreased; the mass number does not change.

Positron emission, consists of the ejection of a positive electron, which is called a positron, from the nucleus. A positron has the same mass as an electron but an opposite charge. It arises from the conversion of a nuclear proton into a neutron. Positron emission results in a decrease of one in the number of protons and an increase of one in the number of neutrons; no change in mass number occurs. Hence, positron emission raises the numerical value of the neutron/proton ratio.

From the above, taken directly from the book, the following reasons for making changes are given.

First, beta and gamma radiations are not particles, they are rays. I call the beta ray a 100 ray because it is this ray that releases an electron. There is a belief that the electron is both a particle and a ray. This results from the fact that when a ray is emitted, there is an electron emitted. The 100 ray comes from the conversion of 0.00100 amu into pure energy which is magnetism when the electron is released.

Second, Beta radiations come from a neutron regaining an electron or negatron and the neutron regaining the specific amount of mass it lost. The neutron becomes a proton or a positon. It also is radiated when a proton or positon loses a specific amount of mass and has the electron or negatron that belong to it, released. This mass that is lost or gained is 0.00100 amu. The electron or negatron that is released does not come from within the nucleus. It is released from an orbital. When the electron is regained by a neutron, the total mass before the neutron regained its mass and the electron is still the same within the isotope. This means that the electron and regained mass must come from within the isotope. For this to happen, the electron or negatron must be found in the proton, positon or neutron, and the mass that is added to the neutron must also come from one of them. No mass is lost.

Third, gamma radiations come from the mass that is lost from a neutron and does not release an electron. It must have the ability to be found as one unit of mass per each gamma ray emitted. This single unit of mass may contain the 0.00100 amu when it is combined with other mass that are lost. As such it will not release an electron or negatron. An example of a gamma ray is 0.00217 amu and 0.00434 amu. Each one is emitted as a single ray. If a 0.00217 amu is combined with a 0.00100 amu for a total of 0.00317 amu and emitted as a single ray, it will not release an electron or negatron.

Fourth, the proton or positon distributes its energy between its electron or negatron and any neutrons found in the nucleus. An excited nuclide has more energy than is needed. The electron and negatron have an external magnetic field as well at the proton and positon. Their magnetic fields will interact with each other as they transfer energy from the two nucleons. Both of the protons and positons will cause the neutron to rotate.

Fifth, the positron as given above is a negatron. There is a positron found in the center of the Original Neutron which has more mass than a proton or positon.

Sixth, the lines seen on a spectrograph are caused by the magnetic lines being concentrated as the electrons and negatrons cross their specific point on their orbital at a specific point in time. This is three dimensional space time.

Seventh, an isotope in the excited state has its electrons and negatrons moving in their orbit faster than normal. This means that they would cross their specific point more often than normal if the distance they travel is the same as in an unexcited state.

The magnetic field of a normal isotope has a normal size. When the electrons and negatrons speed up and cross their specific points more often than normal, the energy content of the magnetic field is increased and it becomes concentrated. The spectrograph will show this as dark lines.

If the distance the electron and negatron travels is increased and the magnetic field's energy is not increased, and the electrons and negatrons do not lose their speed and therefore will not cross their specific points as often, then the spectral lines will be lighter than normal. This change in size of the isotope is a mechanical type of change rather than an electrical one.

When the distance the electron and negatron travel is increased and the electrons and negatrons still cross their specific point in the normal time, the spectral lines will either be thicker or they will split into more than one line. The energy content of the magnetic field must be greater than normal plus the magnetic field will be larger in size.

When the electrons and negatrons gain excess energy from their proton or positon, they must either speed up or move out from the nucleus. When the proton or positon loses some of this excess energy by radiating photons, the electrons and negatrons will return to their normal energy level. This energy that is gained or lost comes from photons that enter or leave the atom or isotope when the electrons and negatrons are crossing their specific points. When the electrons or negatrons are at their specific points, they are like a door that is open and photons can either enter or leave the atom. When they are not at the specific points, the doors are closed. If the door is opened at a specific frequency, say one that the eye can detect, we see the object that is emitting or absorbing the photons. This is shown in the drawing below:

When electrons or negatrons are at a specific point a door is open and photons can either enter or leave the isotope.

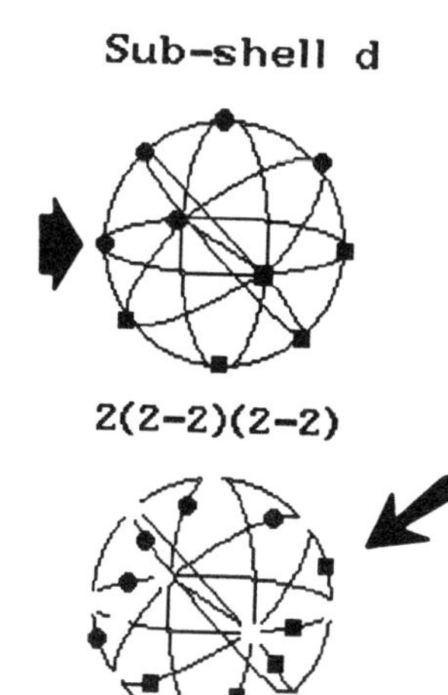

When electrons or negatrons are not at a specific point, the door is closed and no photon can enter or leave the isotope.

RULES FOR THE ELECTRON OR NEGATRON

1. Only 0.00100 amu can release an electron or negatron completely.
2. A released electron or negatron is not converted to pure energy.
3. The difference between an electron and a negatron is the direction of their magnetic field between a proton and a positon.
4. Each electron and negatron has 0.00055 amu.
5. A proton, positon or neutron can have an electron or a negatron within it.
6. When either an electron or a negatron is released, a 100 ray is emitted.
7. An electron in an orbital must have a proton in the nucleus.
8. A negatron in an orbital must have a positon in the nucleus.
9. When either an electron or a negatron is released, its proton or positon must become a neutron.
10. An electron or negatron can be found by itself in space.
11. An electron or a negatron can be found in an orbital by itself.
12. There can not be two electrons or two negatrons in an orbital.
13. An orbital can not have more than one electron and one negatron.
14. An electron is connected to a proton by its magnetic field.
15. A negatron is connected to a positon by its magnetic field.
16. An electron or negatron will generate an electrical current with its magnetic field by transformer action.
17. A free electron will not create an electrical current.
18. An electron or negatron by itself does not have an electrical charge.
19. An electron or negatron by itself has an axis and a north and south pole.
20. The electron or negatron going around the nucleus forms an orbital.
21. All electrons and negatrons form a paddle wheel orbital except the sub-shell s.
22. The paddle wheel action of an orbital with an electron or negatron in it will generate an electrical current between two or more isotopes when the paddle wheel is in motion.
23. All paddle wheels move at right angles to the axis rotation.
24. When an electron and a negatron cross each other's orbital 180 degrees apart, they will create a door which can allow a photon in or out.
25. Only the electrons and negatrons in the p, d and f sub-shells will cross the poles of an axis of an isotope.
26. The electron and negatron in the s sub-shell will not cross the poles of an axis of an isotope.
27. All electrons and negatrons will be found at one point on an orbital relative to other electrons and negatrons at one instant in time that identifies that atom.
28. All electrons and negatrons when not at the one specific point in time will not be able to identify the atom or isotope.
29. All orbitals will rotate in such a manner that they cross the axis of the isotope.
30. When an electron or negatron is at the top position of its orbital, it is in the North Pole position and will connect to other isotopes in the series position.
31. When an electron or negatron is on the left or right side of an orbital, it is in the parallel position.

PERIODIC CLASSIFICATION OF THE ELEMENTS

Dmitri Mendeleev Atomic Chart

IA	IIA	IIIB	IV	V	VI	VII	VIII	VIII	VIII	IB	IIB	IIIA	IV	V	VI	VII	VIIIA
1 H	2 He															1 H	2 He
3 Li	4 Be											5 B	6 C	7 N	8 O	9 F	10 Ne
11 Na	12 Mg											13 Al	14 Si	15 P	16 S	17 Cl	18 Ar
19 K	20 Ca	21 Sc	22 Ti	23 V	24 Cr	25 Mn	26 Fe	27 Co	28 Ni	29 Cu	30 Zn	31 Ga	32 Ge	33 As	34 Se	35 Br	36 Kr
37 Rb	38 Sr	39 Y	40 Zr	41 Nb	42 Mo	43 Tc	44 Ru	45 Rh	46 Pd	47 Ag	48 Cd	49 In	50 Sn	51 Sb	52 Te	53 I	54 Xe
55 Cs	56 Ba	57 La	72 Hf	73 Ta	74 W	75 Re	76 Os	77 Ir	78 Pt	79 Au	80 Hg	81 Tl	82 Pb	83 Bi	84 Po	85 At	86 Rn
87 Fr	88 Ra	89 Ac	104	105	106	107	108	109	110	111	112	113	114	115	116	117	118

Sub-shell s 2

Sub-shell d 2(2-2)(2-2)

Sub-shell p 2(2-2)

58 Ce	59 Pr	60 Nd	61 Pm	62 Sm	63 Eu	64 Gd	65 Tb	66 Dy	67 Ho	68 Er	69 Tm	70 Yb	71 Lu
90 Th	91 Pa	92 U	93 Np	94 Pu	95 Am	96 Cm	97 Bk	98 Cf	99 Es	100 Fm	101 Md	102 No	103 Lr

Sub-shell f 2(2-2)(2-2)(2-2)

I change Sub-shell f

The above Periodic Classification of the Elements is the result of Dmitri Mendeleeve's work which showed that the elements could be listed in an order based on the elements mass. The chart is given in the form that is found in most chemistry books. The Sub-shell and numbers have been added to show that the chart can be shown to be grouped into Sub-shells and these Sub-shells put the electron and negatron in sets of two. Note that element number 71 is given as part of Sub-shell f. My chart shows that the start of Sub-shell f begins with the element 57 and ends with element 70. The second row should begin with element 89 and end with element 102.

TABLE OF ISOTOPES

Isotopes		percentage	electrons or protons	negatrons or positons	neutrons	amu
1.	**H**	**Hydrogen**				**1.00797**
	1H1	99.985%	1	0	0	1.00783
	1H2	00.015%	1	0	1	2.01411
	1H3	00.0	1	0	2	3.00605
2.	**He**	**Helium**				**3.002603**
	2He3	00.00014%	1	1	1	3.01603
	2He4	99.99986%	1	1	2	4.00260
3.	**Li**	**Lithium**				**6.0409**
	3Li6	07.5%	2	1	3	6.015121
	3Li7	92.5%	2	1	4	7.016003
4.	**Be**	**Beryllium**				**9.012182**
	4Be9	100%	2	2	5	9.012182
5.	**B**	**Boron**				**10.81003**
	5B10	19.8%	3	2	5	10.012937
	5B11	80.2%	3	2	6	11.009305
6.	**C**	**Carbon**				**12.0111**
	6C12	98.90%	3	3	6	12.00000
	6C13	01.10%	3	3	7	13.003355
7.	**N**	**Nitrogen**				**14.0067**
	7N14	99.63%	4	3	7	14.003074
	7N15	00.37%	4	3	8	15.000108
8.	**O**	**Oxygen**				**15.9994**
	8O16	99.762%	4	4	8	15.994915
	8O17	00.038%	4	4	9	16.999131
	8O18	00.200%	4	4	10	17.999160
9.	**F**	**Fluorine**				**18.998403**
	9F19	100%	5	4	10	18.998403

10.	**Ne**	**Neon**				**19.992435**
	10**Ne**20	90.51%	5	5	10	19.992435
	10**Ne**21	00.21%	5	5	11	20.993843
	10**Ne**22	09.22%	5	5	12	21.994465
11.	**Na**	**Sodium**				**22.989767**
	11**Na**23	100%	6	5	12	22.989767
12.	**Mg**	**Magnesium**				**22.994124**
	12**Mg**24	78.99%	6	6	12	23.985042
	12**Mg**25	10.00%	6	6	13	24.985837
	12**Mg**26	11.01%	6	6	14	25.982593
13.	**Al**	**Aluminum**				**26.98154**
	13**Al**27	100%	7	6	14	26.98154
14.	**Si**	**Silicon**				**28.0855**
	14**Si**28	92.23%	7	7	14	27.976927
	14**Si**29	04.67%	7	7	15	28.97695
	14**Si**30	03.10%	7	7	16	29.973770
15.	**P**	**Phosphorus**				**30.973762**
	15**P**31	100%	8	7	16	30.973762
16.	**S**	**Sulpher**				**32.066**
	16**S**32	95.02%	8	8	16	31.972070
	16**S**33	00.75%	8	8	17	32.971456
	16**S**34	04.21%	8	8	18	33.967866
17.	**Cl**	**Chlorine**				**45.453**
	17**Cl**35	35.77%	9	8	18	34.968852
	17**Cl**37	24.23%	9	8	19	36.965903
18.	**Ar**	**Argon**				**39.948**
	18**Ar**36	00.337%	9	9	18	35.967545
	18**Ar**38	00.063%	9	9	20	37.962732
	18**Ar**40	99.60%	9	9	22	39.962384
19.	**K**	**Potassium**				**39.0983**
	19**K**39	93.2581%	10	9	20	38.963707
	19**K**40	00.0117%	10	9	21	39.963999
	19**K**41	06.7302%	10	9	22	40.961825

20.	**Ca**	**Calcium**				**40.078**
	20**Ca**40	96.941%	10	10	20	39.962591
	20**Ca**42	00.647%	10	10	22	41.958618
	20**Ca**43	00.135%	10	10	23	42.958766
	20**Ca**44	02.086%	10	10	24	43.955480
	20**Ca**46	00.004%	10	10	26	45.953689
	20**Ca**48	00.187%	10	10	28	47.952533
21.	**Sc**	**Scandium**				**44.055910**
	21**Sc**45	100%	11	10	24	44.955910
22.	**Ti**	**Titanium**				**47.88**
	22**Ti**46	08.0%	11	11	24	45.952629
	22**Ti**47	07.3%	11	11	25	46.951764
	22**Ti**48	73.8%	11	11	26	47.947947
	22**Ti**49	05.5%	11	11	27	48.947871
	22**Ti**50	05.4%	11	11	28	49.944792
23.	**V**	**Vanadium**				**50.9415**
	23**V**50	00.25%	12	11	27	49.947161
	23**V**51	99.75%	12	11	28	50.943962
24.	**Cr**	**Chromium**				**?**
	24**Cr**50	04.35%	12	12	26	49.946046
	24**Cr**52	83.79%	12	12	28	51.940509
	24**Cr**53	09.50%	12	12	29	52.940651
	24**Cr**54	02.36%	12	12	30	53.938882
25.	**Mn**	**Manganese**				**54.938047**
	25**Mn**55	100%	13	12	30	54.938047
26.	**Fe**	**Iron**				**55.847**
	26**Fe**54	05.8%	13	13	28	53.939612
	26**Fe**56	91.72%	13	13	30	55.934939
	26**Fe**57	02.2%	13	13	31	56.935396
	26**Fe**58	00.28%	13	13	32	57.933277
27.	**Co**	**Cobalt**				**58.933198**
	27**Co**59	100%	14	13	32	58.933198

28.	**Ni**	**Nickel**				**58.67**
	28**Ni**58	68.27%	14	14	30	57.935346
	28**Ni**60	26.10%	14	14	32	59.930788
	28**Ni**61	01.13%	14	14	33	60.931058
	28**Ni**62	03.59%	14	14	34	61.928346
	28**Ni**64	00.91%	14	14	36	63.927968
29.	**Cu**	**Copper**				**63.546**
	29**Cu**63	69.17%	15	14	34	62.939598
	29**Cu**65	30.83%	15	14	36	64.927793
30.	**Zn**	**Zinc**				**65.39**
	30**Zn**64	48.6%	15	15	35	63.929145
	30**Zn**66	27.9%	15	15	36	65.926034
	30**Zn**67	04.1%	15	15	37	66.927129
	30**Zn**68	18.8%	15	15	38	67.924846
31.	**Ga**	**Gallium**				**69.723**
	31**Ga**69	60.1%	16	15	38	68.925580
	31**Ga**71	39.9%	16	15	40	70.924700
32.	**Ge**	**Germanium**				**73.941570**
	32**Ge**70	20.5%	16	16	38	69.924250
	32**Ge**72	27.4%	16	16	40	71.922079
	32**Ge**73	07.8%	16	16	41	72.923463
	32**Ge**74	36.5%	16	16	42	73.921177
	32**Ge**76	07.8%	16	16	44	75.921401
33.	**As**	**Arsenic**				**74.921594**
	33**As**75	100%	17	16	42	74.921594
34.	**Se**	**Selenium**				**78.96**
	34**Se**74	00.9%	17	17	40	73.922475
	34**Se**76	09.0%	17	17	42	75.919212
	34**Se**77	07.6%	17	17	43	76.919912
	34**Se**78	23.5%	17	17	44	77.9?
	34**Se**80	49.6%	17	17	46	79.916520
	34**Se**82	09.4%	17	17	48	81.916698
35.	**Br**	**Bromine**				
	35**Br**79	50.69%	18	17	44	78.918336
	35**Br**81	49.31%	18	17	46	80.916289

36.	**Kr**	**Krypton**				**83.80**
	36**Kr**78	00.35%	18	18	42	77.9?
	36**Kr**80	02.25%	18	18	44	79.916380
	36**Kr**82	11.6%	18	18	46	81.913482
	36**Kr**83	11.5%	18	18	47	82.914135
	36**Kr**84	57.0%	18	18	48	83.911507
	36**Kr**86	17.3%	18	18	50	85.910616
37.	**Rb**	**Rubidium**				**85.4678**
	37**Rb**85	72.17%	19	18	48	84.911794
	37**Rb**87	27.83%	19	18	50	86.909187
38.	**Sr**	**Strontium**				**87.62**
	38**Sr**84	00.56%	19	19	46	83.913430
	38**Sr**86	09.86%	19	19	48	85.909267
	38**Sr**87	07.00%	19	19	49	86.908884
	38**Sr**88	82.58%	19	19	50	87.905619
39.	**Y**	**Yttrium**				**88.905849**
	39**Y**89	100%	20	19	50	88.905849
40.	**Zr**	**Zirconium**				**91.224**
	40**Zr**90	51.45%	20	20	50	89.904703
	40**Zr**91	11.27%	20	20	51	90.905644
	40**Zr**92	17.17%	20	20	52	91.905039
	40**Zr**94	17.33%	20	20	54	93.906314
	40**Zr**96	02.78%	20	20	56	95.908275
41.	**Nb**	**Niobium**				**92.906377**
	41**Nb**93	100%	21	20	52	92.906377
42.	**Mo**	**Molybdenum**				**95.94**
	42**Mo**92	14.84%	21	21	50	91.906808
	42**Mo**94	09.25%	21	21	52	93.905085
	42**Mo**95	15.92%	21	21	53	94.905840
	42**Mo**96	16.68%	21	21	54	95.904678
	42**Mo**97	09.55%	21	21	55	96.906020
	42**Mo**98	24.13%	21	21	56	97.905406
	42**Mo**100	09.63%	21	21	58	99.907477
43.	**Tc**	**Technetium**				**none**

44.	**Ru**	**Ruthenium**				**101.07**
	44**Ru**96	05.52%	22	22	52	95.907556
	44**Ru**98	01.88%	22	22	54	97.905287
	44**Ru**99	12.7%	22	22	55	98.905939
	44**Ru**100	12.6%	22	22	56	99.904219
	44**Ru**101	17.0%	22	22	57	100.905582
	44**Ru**102	31.6%	22	22	58	101.904348
	44**Ru**104	18.7%	22	22	60	103.905424
45.	**Rh**	**Rhodium**				**102.90550**
	45**Ru**103	100%	23	22	58	102.90550
46.	**Pd**	**Palladium**				**106.42**
	46**Pd**102	01.02%	23	23	56	101.905634
	46**Pd**104	11.14%	23	23	58	103.904029
	46**Pd**105	22.33%	23	23	59	104.905079
	46**Pd**106	27.33%	23	23	60	105.903478
	46**Pd**108	26.46%	23	23	62	107.903895
	46**Pd**110	11.72%	23	23	64	109.905167
47.	**Ag**	**Silver**				**107.8682**
	47**Ag**107	51.84%	24	23	60	106.905092
	47**Ag**109	48.16%	24	23	62	108.904757
48.	**Cd**	**Cadium**				**112.41**
	48**Cd**106	01.25%	24	24	58	105.906461
	48**Cd**108	00.89%	24	24	60	107.904176
	48**Cd**110	12.49%	24	24	62	109.903005
	48**Cd**111	12.80%	24	24	63	110.904182
	48**Cd**112	24.13%	24	24	64	111.902758
	48**Cd**113	12.22%	24	24	65	112.904400
	48**Cd**114	28.73%	24	24	66	113.903357
	48**Cd**116	07.49%	24	24	68	115.904754
49.	**In**	**Indium**				**114.82**
	49**In**113	04.3%	25	24	64	112.904061
	49**In**115	95.7%	25	24	66	114.903880

50.	**Sn**	**Tin**				**118.710**
	50**Sn**112	01.0%	25	25	62	111.904826
	50**Sn**114	00.7%	25	25	64	113.902784
	50**Sn**115	00.4%	25	25	65	114.903348
	50**Sn**116	14.7%	25	25	66	115.901747
	50**Sn**117	07.7%	25	25	67	116.902956
	50**Sn**118	24.3%	25	25	68	117.901609
	50**Sn**119	08.6%	25	25	69	118.903310
	50**Sn**120	32.4%	25	25	70	119.902200
	50**Sn**122	04.6%	25	25	72	121.903440
	50**Sn**124	05.6%	25	25	74	123.905274
51.	**Sb**	**Antimony**				**108.918143**
	51**Sb**121	57.3%	26	25	70	120.903821
	51**Sb**123	42.7%	26	25	72	122.904216
52.	**Te**	**Tellurium**				**?**
	52**Te**120	00.1%	26	26	68	119.904048
	52**Te**122	02.6%	26	26	70	121.903054
	52**Te**123	00.9%	26	26	71	122.904271
	52**Te**124	04.8%	26	26	72	123.902823
	52**Te**125	07.1%	26	26	73	124.904433
	52**Te**126	19.0%	26	26	74	125.903314
	52**Te**128	31.7%	26	26	76	127.904463
	52**Te**130	33.8%	26	26	78	129.906229
53.	**I**	**Iodine**				**126.904473**
	53**I**127	100%	27	26	74	126.904473
54.	**Xe**	**Xenon**				**131.29**
	54**Xe**124	00.10%	27	27	70	123.905894
	54**Xe**126	00.09%	27	27	72	125.904281
	54**Xe**128	01.91%	27	27	74	127.903531
	54**Xe**129	26.4%	27	27	75	128.904780
	54**Xe**130	04.1%	27	27	76	129.903509
	54**Xe**131	21.2%	27	27	77	130.905072
	54**Xe**132	26.9%	27	27	78	131.904144
	54**Xe**134	10.4%	27	27	80	133.905395
	54**Xe**136	08.9%	27	27	82	135.907214
55.	**Cs**	**Cesium**				**132.905429**
	55**Cs**133	100%	28	27	78	132.905429

56.	**Ba**	**Barium**				**137.33**
	56**Ba**130	00.106%	28	28	74	129.906282
	56**Ba**132	00.101%	28	28	76	131.905042
	56**Ba**134	02.417%	28	28	78	133.904486
	56**Ba**135	06.592%	28	28	79	134.905665
	56**Ba**136	07.854%	28	28	80	135.904553
	56**Ba**137	11.230%	28	28	81	136.905812
	56**Ba**138	71.70%	28	28	82	137.905232
57.	**La**	**Lanthanum**				**138.9055**
	57**La**138	00.09%	29	28	81	137.907105
	57**La**139	99.91%	29	28	82	138.906346
58.	**Ce**	**Cerium**				**140.12**
	58**Ce**136	00.19%	29	29	78	135.907140
	58**Ce**138	00.25%	29	29	80	137.905985
	58**Ce**140	88.48%	29	29	82	139.905433
	58**Ce**142	11.08%	29	29	84	141.909241
59.	**Pr**	**Praseodyium**				**140.907647**
	59**Pr**141	100%	30	29		140.907647
60.	**Nd**	**Neodymium**				**144.24**
	60**Nd**142	27.13%	30	30	82	141.907719
	60**Nd**143	12.18%	30	30	83	142.909810
	60**Nd**144	23.80%	30	30	84	143.910083
	60**Nd**145	08.30%	30	30	85	144.912570
	60**Nd**146	17.19%	30	30	86	145.913113
	60**Nd**148	05.76%	30	30	88	147.916889
	60**Nd**150	05.64%	30	30	90	149.920887
61.	**Pm**	**Promethium**				**none**
	61**Pm**	none	31	30	?	none
62.	**Sm**	**Samarium**				**150.36**
	62**Sm**144	03.1%	31	31	82	143.911998
	62**Sm**147	15.0%	31	31	85	146.914895
	62**Sm**148	11.3%	31	31	86	147.914820
	62**Sm**149	13.8%	31	31	87	148.917181
	62**Sm**150	07.4%	31	31	88	149.917273
	62**Sm**152	26.7%	31	31	90	151.919729
	62**Sm**154	22.7%	31	31	92	153.922206

63.	**Eu**	**Europium**				**?**
	63**Eu**151	47.8%	32	31	88	150.919847
	63**Eu**153	52.2%	32	31	90	152.921225
64.	**Gd**	**Gadolinium**				**157.25**
	64**Gd**152	00.20%	32	32	88	151.919786
	64**Gd**154	02.18%	32	32	90	153.920861
	64**Gd**155	14.80%	32	32	91	154.922618
	64**Gd**156	20.47%	32	32	92	155.922118
	64**Gd**157	15.65%	32	32	93	156.923956
	64**Gd**158	24.84%	32	32	94	157.924099
	64**Gd**160	21.86%	32	32	96	159.927049
65.	**Tb**	**Terbium**				**158.925342**
	65**Tb**159	100%	33	32	94	158.925342
66.	**Dy**	**Dysprosium**				**162.50**
	66**Dy**156	00.06%	33	33	90	155.925277
	66**Dy**158	00.10%	33	33	92	157.924403
	66**Dy**160	02.34%	33	33	94	159.925190
	66**Dy**161	18.9%	33	33	95	160.926930
	66**Dy**162	25.5%	33	33	96	161.926795
	66**Dy**163	24.9%	33	33	97	162.928728
	66**Dy**164	28.2%	33	33	98	163.929171
67.	**Ho**	**Holmium**				**164.930319**
	67**Ho**165	100%	34	33	98	164.930319
68.	**Er.**	**Erbium**				**167.26**
	68**Er**162	00.14%	34	34	94	161.928775
	68**Er**164	01.61%	34	34	96	163.929198
	68**Er**166	33.6%	34	34	98	165.930290
	68**Er**167	22.95%	34	34	99	166.932046
	68**Er**168	26.8%	34	34	100	167.932368
	68**Er**170	14.9%	34	34	102	169.935461
69.	**Tm**	**Thulium**				**168.934212**
	69**Tm**169	100%	35	34	100	168.934212

70.	**Yb**	**Ytterium**				**173.043894**
	70**Yb**168	00.13%	35	35	98	167.933894
	70**Yb**170	03.05%	35	35	100	169.934759
	70**Yb**171	14.3%	35	35	101	170.936323
	70**Yb**172	21.9%	35	35	102	171.936378
	70**Yb**173	16.12%	35	35	103	172.938208
	70**Yb**174	31.8%	35	35	104	173.938859H
	70**Yb**176	12.7%	35	35	106	175.942564
71.	**Lu**	**Lutetium**				**174.967**
	71**Lu**175	97.40%	36	35	104	174.940770
	71**Lu**176	02.6%	36	35	105	175.942679
72.	**Hf**	**Hafnium**				**178.49**
	72**Hf**174	00.16%	36	36	102	173.940044
	72**Hf**176	05.2%	36	36	104	175.941406
	72**Hf**177	18.6%	36	36	105	176.943217
	72**Hf**178	27.1%	36	36	106	177.943696
	72**Hf**179	13.74%	36	36	107	178.945812
	72**Hf**180	35.2%	36	36	108	179.946545
73.	**Ta**	**Tantalum**				**180.9479**
	73**Ta**180	00.012%	37	36	107	179.947462
	73**Ta**181	99.988%	37	36	108	180.747992
74.	**W**	**Tungesten (Wolforam)**				**?**
	74**W**180	00.13%	37	37	106	179.946701
	74**W**182	16.3%	37	37	108	181.948202
	74**W**183	14.3%	37	37	109	182.950220
	74**W**184	30.67%	37	37	110	183.950928
	74**W**186	28.6%	37	37	112	185.954357
75.	**Re**	**Rhenium**				**186.207**
	75**Re**185	37.40%	38	37	110	184.952951
	75**Re**187	62.6%	38	37	112	186.955744
76.	**Os**	**Osmium**				**?**
	76**Os**184	00.02%	38	38	108	183.952488
	76**Os**186	01.58%	38	38	110	185.953830
	76**Os**187	01.6%	38	38	111	186.955741
	76**Os**188	13.3%	38	38	112	187.955860
	76**Os**189	16.1%	38	38	113	188.958137
	76**Os**190	26.4%	38	38	114	189.958436
	76**Os**192	41.0%	38	38	116	191.961467

77.	**Ir**	**Iridium**			**?**
	77**Ir**191	37.3%	39	38	114 190.960584
	77**Ir**193	62.7%	39	38	116 192.962917
78.	**Pt**	**Platinum**			**195.08**
	78**Pt**190	00.01%	39	39	112 189.959917
	78**Pt**192	00.79%	39	39	114 191.961019
	78**Pt**194	32.9%	39	39	116 193.962655
	78**Pt**195	33.8%	39	39	117 194.964766
	78**Pt**196	25.3%	39	39	118 195.964926
	78**Pt**198	07.2%	39	39	120 197.967869
79.	**Au**	**Gold**			**196.966543**
	79**Au**197	100%	40	39	118 196.966543
80.	**Hg**	**Murcury**			**200.59**
	80**Hg**196	00.15%	40	40	116 195.965807
	80**Hg**198	10.1%	40	40	118 197.966743
	80**Hg**199	17.0%	40	40	119 198.968254
	80**Hg**200	23.1%	40	40	120 199.968300
	80**Hg**201	13.2%	40	40	121 200.970277
	80**Hg**202	29.65%	40	40	122 201.970617
	80**Hg**204	06.8%	40	40	124 203.973467
81.	**Tl**	**Thallium**			**204.383**
	81**Tl**203	29.524%	41	40	122 202.972320
	81**Tl**205	70.476%	41	40	124 204.974401
82.	**Pb**	**Lead**			**207.2**
	82**Pb**204	01.4%	41	41	122 203.973020
	82**Pb**206	24.1%	41	41	124 205.974440
	82**Pb**207	22.1%	41	41	125 206.975872
	82**Pb**208	52.4%	41	41	126 207.976627
83.	**Bi**	**Bismuth**			**208.980374**
	83**Bi**209	100%	42	41	126 208.980374
84.	**Po**	**Polonium**			
	84**Po**210				
85.	**At**	**Astatine**			
	85**At**210				
86.	**Rn**	**Radon**			
	86**Rn**222				

87.	**Fr**	**Francium**
88.	**Ra**	**Radium**
89.	**Ac**	**Actinium**
90.	**Th**	**Thorium**
91.	**Pa**	**Protactinum**
92.	**U**	**Uranium**
93.	**Np**	**Neptunium**
94.	**Pu**	**Plutonium**
95.	**Am**	**Americium**
96.	**Cm**	**Curium**
97.	**Bk**	**Berkelium**
98.	**Cf**	**Californium**
99.	**Es**	**Einsteinum**
100.	**Fm**	**Fermium**
101.	**Md**	**Mendelevium**
102.	**No**	**Nobelium**
103.	**Lr**	**Lawrencium**
104.	**Rf**	
105.	**Ha**	

THE NEW PERIODIC CHART OF THE ELEMENTS

CHART ONE

This is not the correct chart - do not use!

	30 degrees				45 degrees				90 degrees			180 degrees		Sub Shell	SHELL	Horizontal Rows	Sub Group	GROUP
14	13	12	11	10	9	8	7	6	5	4	3	2	1	Vertical Rows				
												1	2	s	K	1	1A	A
												3	4	s	L	2	2B	B
												11	12	s	M	3	1C	A
								5	6	7	8	9	10	p	L	4	2C	A
												19	20	s	N	5	1D	B
								13	14	15	16	17	18	p	M	6	2D	B
												37	38	s	N	7	1E	A
								31	32	33	34	35	36	p	N	8	2E	A
				21	22	23	24	25	26	27	28	29	30	d	M	9	3E	A
												55	56	s	O	10	1F	B
								49	50	51	52	53	54	p	N	11	2F	B
				39	40	41	42	43	44	45	46	47	48	d	N	12	3F	B
												87	88	s	P	13	1G	A
								81	82	83	84	85	86	p	O	14	2G	A
				71	72	73	74	75	76	77	78	79	80	d	N	15	3G	A
57	**58**	**59**	**60**	**61**	**62**	**63**	**64**	65	66	67	68	69	70	f	M	16	4G	A
												119	120	s	Q	17	1H	B
								113	114	115	116	117	118	p	P	18	2H	B
				103	104	105	106	107	108	109	110	111	112	d	O	19	3H	B
89	**90**	**91**	**92**	**93**	**94**	**95**	**96**	97	98	99	100	101	102	f	N	20	4H	B

Bold numbers are 30 degrees. There is no 45 degree Sub-shell f.

NEW PERIODIC CHART OF THE ELEMENTS

CHART ONE

This NEW PERIODIC CHART OF THE ELEMENTS was created by me to show that the placement of the atomic elements can be placed so that they look like they form a chart that places the elements is their correct position. However, this is not the case. The elements Hydrogen through Magnesium are placed so that their "s" subshell in Shell K ends up over the element Neon. The element Neon has its subshell as part of a set of four electrons and negatrons in subshell p, Shell L. The elements Boron and Carbon are the elements with their subshells in the "s" subshell, Shell L. These are the elements that are found one hundred and eighty degrees apart while Neon has its electron and negatron found in an orbital that places them ninety degrees apart in subshell p, Shell L.

Also note that the subshells "s" are placed above the subshells p, d and f. This looks like the correct way of listing the elements but it is not. The correct way to list the elements is to invert this way of listing the elements. The subshell f is placed over the subshell d and the d subshell is placed over the p subshell. The subshell s is placed under all of these subshells. By doing it this way all of the element's numbers are given is their correct numerical order. Their electrons and negatrons are found in their correct positions within their orbitals.

THE NEW CORRECT PERODIC CHART OF THE ELEMENTS

CHART TWO

30 degrees				45 degrees				90 degrees				180 degrees		Sub Shell	SHELL	Horizontal Rows	Sub Group	GROUP
14	13	12	11	10	9	8	7	6	5	4	3	2	1	Vertical Rows				
												2	1	s	K	1	1A	A
												4	3	s	L	2	2B	B
								10	9	8	7	6	5	p	L	3	1C	A
												12	11	s	M	4	2C	A
								18	17	16	15	14	13	p	M	5	1D	B
												20	19	s	N	6	2D	B
				30	29	28	27	26	25	24	23	22	21	d	M	7	1E	A
								36	35	34	33	32	31	p	N	8	2E	A
												38	37	s	O	9	3E	A
				48	47	46	45	44	43	42	41	40	39	d	N	10	1F	B
								54	53	52	51	50	49	p	O	11	2F	B
												56	55	s	P	12	3F	B
70	**69**	**68**	**67**	**66**	**65**	**64**	**63**	62	61	60	59	58	57	f	N	13	1G	A
				80	79	78	77	76	75	74	73	72	71	d	O	14	2G	A
								86	85	84	83	82	81	p	P	15	3G	A
												88	87	s	Q	16	4G	A

Bold numbers are 30 degrees. There is no 45 degree in Sub-shell f.

30 degrees				45 degrees				90 degrees				180 degrees		Sub Shell	SHELL	Horizontal Rows	Sub Group	GROUP
102	**101**	**100**	**99**	**98**	**97**	**96**	**95**	94	93	92	91	90	89	f	O	17	1H	B
				112	111	110	109	108	107	106	105	104	103	d	P	18	2H	B
								118	117	116	115	114	113	p	Q	19	3H	B
												120	119	s	R	20	4H	B

THE NEW CORRECT PERIODIC CHART OF THE ELEMENTS

CHART TWO

This NEW PERIODIC CHART OF THE ELEMENTS was created by me to show that there is a periodicity that tells where to put the Hydrogen and Helium atoms. The normal periodic table of the elements puts the Hydrogen atom over the Lithium atom and the Helium atom over the Neon atom. This is done because the Helium atom is considered an inert atom like the atom of Neon. I noticed that there are two sets of the sub-shells. By rearranging the periodic chart of the elements in these two sets of sub-shells, the correct position of all of the atoms will be shown. Notice that the new chart of the elements places all of the atoms on the right side of the page.

There are two sets of the sub-shells numbered one and two as a row of sub-shells is added to a row of sub-shells. This means that the first two rows of sub-shells have atoms 1 and 2 in sub-shell s as the first row of atoms while the second row of atoms are atoms 3 and 4 which is also a sub-shell s. The next row will start out as sub-shell s and add the first sub-shell p found in the normal periodic chart. This is the first set that has a sub-shell p in it. The second set that has sub-shell p in it will be created in this same fashion. The next two sets of sub-shells will add the sub-shell d and the last two sets of sub-shells will add the sub-shell f. The importance of showing the periodic chart in this manner is that by grouping all of the sets of set 1, the atoms found in this set will be related as a different grouping that the atoms found in set 2. The atoms found in set 2 will be related to each other. This breaks the periodic chart of the elements into two sets that should show the properties of each of the two sets that was unknown before.

All of the first two atoms found in sub-shell s will be found under the vertical rows of atoms 1 and 2. These atoms will have their electron and negatron one hundred and eighty degrees apart. The next four atoms to the left of the vertical rows of 1 and 2 will have their electrons and negatrons ninety degrees apart in sub-shell p. The next sub-shell is sub-shell d. It will have four more atoms than the sub-shell p and they are found to the left of the last four atoms in sub-shell p. These atoms will have their electrons and negatrons forty-five degrees apart. The last sub-shell is sub-shell f. It is found having its last four atoms to the left of sub-shell d. All of its electrons and negatrons will be found thirty degrees apart. For this to happen, the atoms representing the forty five degree's will not be found within this sub-shell. These electrons and negatrons are found in orbitals. This means that sub-shells of sub-shell s has only one orbital per atom when there are only two atoms in the sub-shell s. Sub-shell p has two atoms found in an orbital just like sub-shell s plus four more atoms. Therefore, sub-shell p has two electrons and negatrons in the one hundred eighty degree orbit and four electrons and negatrons in the ninety degree orbitals. Sub-shell d has this same way of arranging its electrons and negatron in their orbital plus the last set of four atoms have their electrons and negatrons found in the forty five degree orbitals. The last sub-shell is sub-shell f. It has all of its electrons and negatrons thirty degrees apart. It will be noted that there is a one hundred and eighty degree set of orbitals and a set of ninety degree orbitals but no set of forty-five degree orbitals. This placement of orbitals and the electrons and negatrons explains why the periodic chart is given as shown. The properties found in the isotopes of the atoms come from where the electrons and negatrons orbit their nucleus. However, the properties of the different isotopes vary according to the mass of each isotope's proton, positon and neutron. This happens because the electron and negatron never lose any mass and they stay electrons and negatrons. The proton and positon can each change into a neutron while a neutron can change into a proton or a positon. They do this by either losing or gaining mass. These properties come from the magnetic field found around the electron and negatron. The magnetic field found around the electron and negatron will vary according to the magnetic field found around the proton and positon. The magnetic field of the neutron is found mostly within the neutron and it does not cause changes of the magnetic field of the electron and negatron. The neutron or neutrons found within a nucleus controls the speed of the rotating proton and positon which then controls the speed of the electrons and negatrons. The speed of the electrons and negatrons causes the distance they are from the nucleus to change therefore, the type of orbital the electron and negatron is in will vary its properties.

THE NEW CORRECT PERIODIC CHART OF THE ELEMENTS

CHART THREE

30 degrees					45 degrees				90 degrees			180 degrees		Sub Shell	SHELL	Horizontal Rows	Sub Group	GROUP
14	13	12	11	10	9	8	7	6	5	4	3	2	1	Vertical Rows				
												He	H	s	K	1	1A	A
												Be	Li	s	L	2	2B	A
								Ne	F	O	N	C	B	p	L	3	1C	A
												Mg	Na	s	M	4	2C	A
								Ar	Cl	S	P	Si	Al	p	M	5	1D	B
												Ca	K	s	N	6	2D	B
				Zn	Cu	Ni	Co	Fe	Mn	Cr	V	Ti	Sc	d	M	7	1E	A
								Kr	Br	Se	As	Ge	Ga	p	N	8	2E	A
												Sr	Rb	s	O	9	3E	A
				Cd	Ag	Pd	Rh -	Ru	Tc	Mo	Nb --	Zr	Y	d	N	10	1F	B
								Xe	I	Te	Sb	Sn	In	p	O	11	2F	B
												Ba	Cs	s	P	12	3F	B
Yb	**Tm**	**Er**	**Ho**	**Dy**	**Tb**	**Gd**	**Eu**	Sm	Pm	Nd	Pr	Ce	La	f	N	13	1G	A
				Hg	Au	Pt	Ir	Os	Re	W	Ta	Hf	Lu	d	O	14	2G	A
								Rn	At	Po	Bi	Pb	Tl	p	P	15	3G	A
												Ra	Fr	s	Q	16	4G	A
Bold symbols are 30 degrees. There is no 45 degree in Sub-shell f.																		
No	**Ms**	**Fm**	**Es**	**Cf**	**Bk**	**Cm**	**Am**	Pu	Np	U	Pa	Th	Ac	f	O	17	1H	B
				112	111	110	109	108	107	106	105	104	103	d	P	18	2H	B
								118	117	116	115	114	113	p	Q	19	3H	B
												120	119	s	R	20	4H	B

THE NEW CORRECT PERIODIC CHART OF THE ELEMENTS

CHART THREE

Chart three presents the same information as found in Chart Two except that it gives the symbols of the elements instead of their numbers. This Chart Three lets the reader see the elements in their new correct positions and know which element is above or below an element the reader might like to know about. Chart Two and Chart Three show that there are two sets of the elements as shown in GROUP A and GROUP B. By grouping the elements in this manner, the elements in GROUP A and B will have their properties understood better than if they are just listed in a chart that is the Standard Table of the Elements as given in most chemistry books.

CHART FOUR
and
CHART FIVE

Charts four and five are given to show their number and symbol names as they are found in Group A.

CHART SIX
and
CHART SEVEN

Charts six and seven are given to show their number and symbol names as they are found in Group B.

NEW CORRECT PERIODIC CHART OF THE ELEMENTS

CHART FOUR

30 degrees				45 degrees				90 degrees				180 degrees		Sub Shell	SHELL	Horizontal Rows	Sub Group	GROUP
14	13	12	11	10	9	8	7	6	5	4	3	2	1	Vertical Rows				
												2	1	s	K	1	1A	A
								10	9	8	7	6	5	p	L	3	1C	A
												12	11	s	M	4	2C	A
				30	29	28	27	26	25	24	23	22	21	d	M	7	1E	A
								36	35	34	33	32	31	p	N	8	2E	A
												38	37	s	O	9	3E	A
70	**69**	**68**	**67**	**66**	**65**	**64**	**63**	62	61	60	59	58	57	f	N	13	1G	A
				80	79	78	77	76	75	74	73	72	71	d	O	14	2G	A
								86	85	84	83	82	81	p	P	15	3G	A
												88	87	s	Q	16	4G	A

Bold numbers are 30 degrees. There is no 45 degree in Sub-shell f.

NEW CORRECT PERIODIC CHART OF THE ELEMENTS

CHART FIVE

30 degrees				45 degrees				90 degrees				180 degrees		Sub Shell	SHELL	Horizontal Rows	Sub Group	GROUP
14	13	12	11	10	9	8	7	6	5	4	3	2	1	Vertical Rows				
												4	3	s	L	2	2B	B
								18	17	16	15	14	13	p	M	5	1D	B
												20	19	s	N	6	2D	B
				48	47	46	45	44	43	42	41	40	39	d	N	10	1F	B
								54	53	52	51	50	49	p	O	11	2F	B
												56	55	s	P	12	3F	B
102	**101**	**100**	**99**	**98**	**97**	**96**	**95**	94	93	92	91	90	89	f	O	17	1H	B
				112	111	110	109	108	107	106	105	104	103	d	P	18	2H	B
								118	117	116	115	114	113	p	Q	19	3H	B
												120	119	s	R	20	4H	B

Bold numbers are 30 degrees. There is no 45 degree in Sub-shell f.

NEW CORRECT PERIODIC CHART OF THE ELEMENTS

CHART SIX

30 degrees				45 degrees				90 degrees				180 degrees		Sub Shell	SHELL	Horizontal Rows	Sub Group	GROUP
14	13	12	11	10	9	8	7	6	5	4	3	2	1	Vertical Rows				
												He	H	s	K	1	1A	A
								Ne	F	O	N	C	B	p	L	3	1C	A
												Mg	Na	s	M	4	2C	A
				Zn	Cu	Ni	Co	Fe	Mn	Cr	V	Ti	Sc	d	M	7	1E	A
								Kr	Br	Se	As	Ge	Ga	p	N	8	2E	A
												Sr	Rb	s	O	9	3E	A
Yb	**Tm**	**Er**	**Ho**	**Dy**	**Tb**	**Gd**	**Eu**	Sm	Pm	Nd	Pr	Ce	La	f	N	13	1G	A
				Hg	Au	Pt	Ir	Os	Re	W	Ta	Hf	Lu	d	O	14	2G	A
								Rn	At	Po	Bi	Pb	Tl	p	P	15	3G	A
												Ra	Fr	s	Q	16	4G	A

Bold symbols are 30 degrees. There is no 45 degree in Sub-shell f.

NEW CORRECT PERIODIC CHART OF THE ELEMENTS

CHART SEVEN

	30 degrees				45 degrees				90 degrees			180 degrees		Sub Shell	Horizontal SHELL Rows		Sub Group	GROUP
14	13	12	11	10	9	8	7	6	5	4	3	2	1	Vertical Rows				
												Be	Li	s	L	2	2B	B
								Ar	Cl	S	P	Si	Al	p	M	5	1D	B
												Ca	K	s	N	6	2D	B
				Cd	Ag	Pd	Rh	Ru	Tc	Mo	Nb	Zr	Y	d	N	10	1F	B
								Xe	I	Te	Sb	Sn	In	p	O	11	2F	B
												Ba	Cs	s	P	12	3F	B
No	**Md**	**Fm**	**Es**	**Cf**	**Bk**	**Cm**	**Am**	Pu	Np	U	Pa	Th	Ac	f	O	17	1H	B
				112	111	110	109	108	107	106	105	104	Lr	d	P	18	2H	B
								118	117	116	115	114	113	p	Q	19	3H	B
												119	120	s	R	20	4H	B

Bold symbols are 30 degrees. There is no 45 degree in Sub-shell f.

Notes

ATOMIC DRAWINGS

There are three types of drawings. One type shows the Shell and their sub-shells. This type of drawing shows the proper placement of the electrons and negatrons in their orbitals. It names them and it names their orbitals. They show the atoms in three dimensions. There are three types of frequencies shown and all of these drawings originated with me. These drawings show "space time".

The second type of drawing shows the electrons and negatrons as they create new isotopes. They do not show orbitals. They show protons, positons and neutrons. This type of drawing shows how and why an electron is released. It will give the specific masses of the isotopes and how these masses are used. It will explain how and why mass is lost. It will show left-over parts and what happens to them. It may or may not show a released electron or negatron.

The third type of drawing is the way the Original Neutron is drawn. It shows what it is composed of and what happens to it.

ELECTRON AND NEGATRON PLACEMENT AND DIRECTION IN THEIR ORBITALS

The electron and negatron go in an orbital. These orbitals are found in Shells and sub-shells. The Shells are either called by a letter or a number. I use the letter system. The letters for a Shell are always in capital letters or upper case. The sub-shell letters are always in small letters or lower case.

The Shells are K, L, M, N, O, P and Q.
The sub-shells are s, p, d and f.

Shell K has only sub-shell s.
Shell L has a sub-shell s and p.
Shell M has a sub-shell s, p and d.
Shell N has a sub-shell s, p, d and f.

The remaining Shells may be partially filled. These Shells are filled in a specific manner and some Shells may or may not be full.

There can be one electron in an orbital but no more than two. If there is a second mass to go into an orbital, it is called a negatron. There can only be one electron and one negatron in a given orbital. The difference between them is found in the area between their proton or positon.

The specific filling of a Shell goes from left to right until each orbital has an electron and a negatron in it. The first Shell, Shell K, has only an electron and a negatron in a sub-shell s and no other sub-shells. Next, Shell L receives a sub-shell s and then a sub-shell p. After this is done, there is a check to see if any Shell to the left needs a sub-shell. At this time there is none, all Shells are full. Next, Shell M gets a sub-shell s. When it is in place, again look to the left to see if any sub-shell needs a new sub-shell. They are full. Now put in Shell M its sub-shell p. After this is done, note that Shell K has only sub-shell s then Shell L has one more sub-shell then Shell K. As the Shells are filled going from left to right, the Shell on the right will always have one more sub-shell than the one on its left. Do this checking to see that this is the case when a Shell has its sub-shell s filled. After a Shell to the left has a sub-shell filled that was not filled, ask yourself if the Shell to the right has a sub-shell that needs to be started and filled? This filling of the Shells to the right continues until the sub-shell s is full.

The Shells and their sub-shells are shown below for Shells K, L, M, N, O and P. When these Shells and sub-shells are filled like the example given below, a specific atom will be made. By counting the numbers, their total will tell which atom is created.

ELECTRON AND NEGATRON SHELLS AND SUB-SHELLS

Shell		K	L	M	N	O	P
Sub-shell	s	2	2	2	2	2	
	p		2(2-2)	2(2-2)	2(2-2)		
	d			2(2-2)(2-2)			
	f						

This atom is atom number thirty-eight (38) which is Strontium (Sr).

Note that in Shell L, I filled the first 2 in sub-shell s and I started sub-shell p with its 2 then I put the next set of two 2's in parenthesis. Doing it this way shows that the next sub-shell has one set of two 2's more than the sub-shell above it. This is also true of the sub-shell to its left.

In the example given above, the next electron will go in Shell N, sub-shell d. Sub-shell f is not filled. It will not be started until Shell N and Shell O have their next sub-shells filled. When Shell O has its sub-shell p filled, Shell P sub-shell s will be filled next and after it is full, then Shell N will get its first electron in sub-shell f.

Note that Shell K has a total of one electron and one negatron in it. Shell L has eight of them. Shell M has eighteen and Shell N would have thirty-two.

Now look at the drawing that shows how each sub-shell looks. The arrows show which electron and negatron belong to what orbital. A dot indicates an electron and a square indicates a negatron. The arrow indicates the direction they are going. Because every orbital is going in a specific direction around a central point, each orbital has an axis. However, any given atom or isotope has only one axis and this axis is at right angles to the sub-shell s. All axes of an orbital other than the sub-shell s move like the spokes of a wheel. When the wheel turns, the spokes go in the direction the wheel is turning. The spokes of a wheel are connected to the axis of the wheel at right angles to its rotation. This means that the axis of an orbital goes in the same direction as the rotating sub-shell s axis.

DRAWING OF SUB-SHELLS

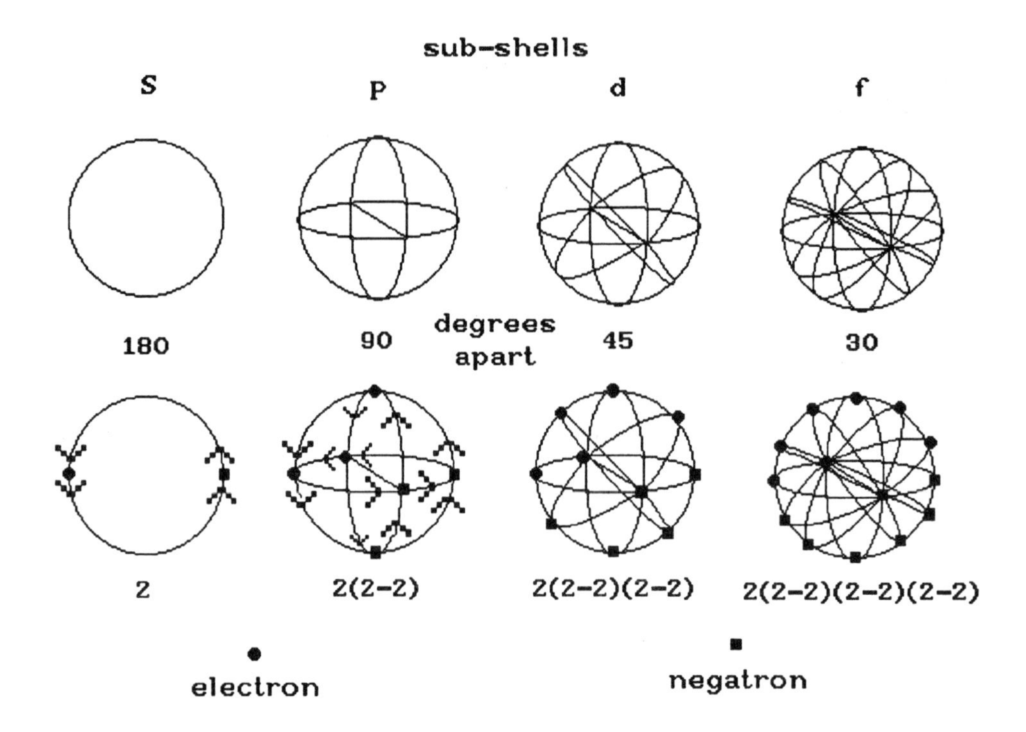

SUB-SHELL ORBITALS

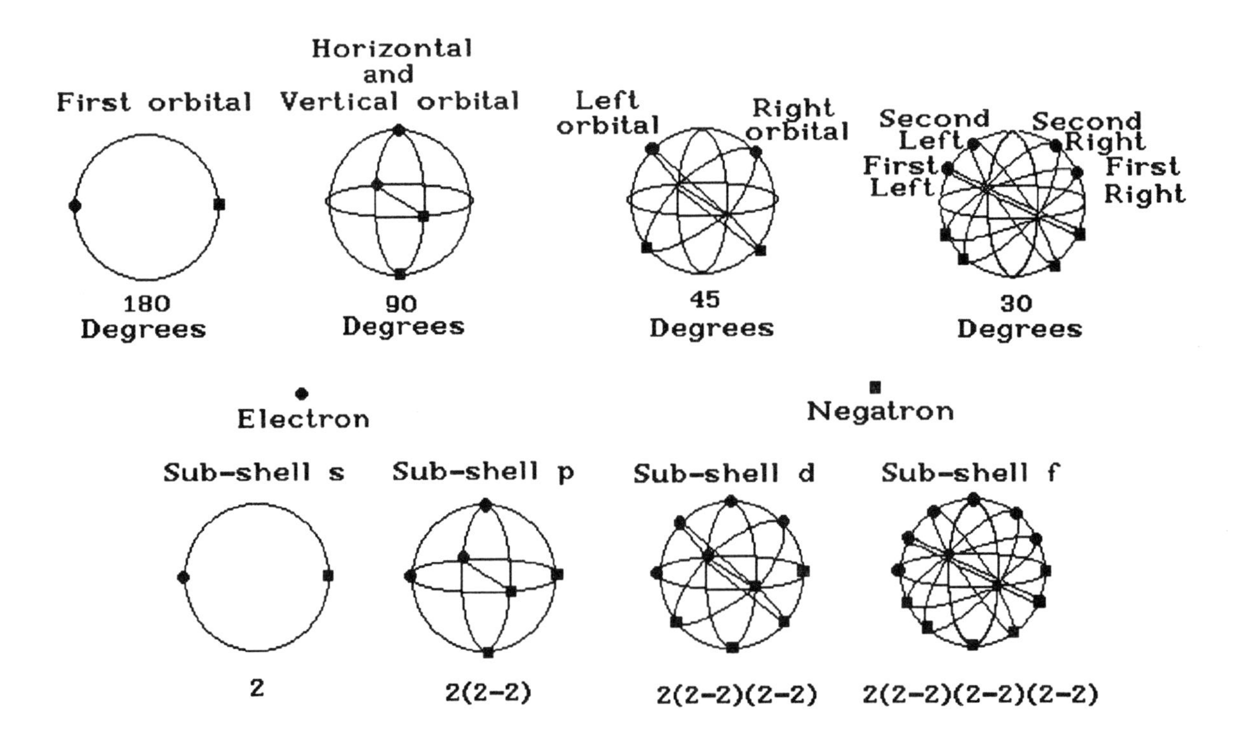

UNDERSTANDING THE ELECTRON, NEGATRON, PROTON, POSITON AND NEUTRON DRAWINGS

The drawings showing how isotopes create a new isotope to give the reader the ability to see how it is done with words and pictures. This is done by drawing a big circle with a vertical line dividing it in half.

The electrons and negatrons are placed on the left side of the big circle. The electrons are drawn as a dot and the negatrons are drawn as a square.

The proton and positon are placed on the left side of the vertical line within the big circle. The proton is drawn as a dot and the positon is drawn as a square.

The neutron or neutrons are drawn on the right side of the vertical line within the big circle. They are drawn as a small circle. Wherever it is possible, the first neutron that has the largest mass is placed on top of the second neutron.

The electron and negatron will always have 0.00055 amu. They never lose any mass. It is possible for an electron or negatron to be released completely from the isotope used to create a new isotope. It is also possible for an electron or negatron to be released from a proton, positon or neutron and go into an orbital.

The proton, positon and neutron will lose or gain mass which will change the isotope being created into sub-isotopes.

It is possible for an isotope to be intact within an isotope used to create a new isotope and when this happens, the intact isotope will be enclosed within an enclosure drawn around it. This enclosure shows that the isotope is used as is and does not lose any mass. It is possible for this enclosed isotope to drop out. When this happens, the enclosed isotope will be drawn on the right side of the equal sign. It is also possible for a neutron to drop out and it will be drawn as a small circle on the right side of the equal sign.

It is possible for parts of an isotope used to create a new isotope to not be used or needed. When this happens they are said to have dropped out.

When mass is lost, it is lost as a ray with a number given to the ray that is the same as the mass lost. An alpha ray is emitted by a proton or positon, a gamma ray is emitted by a neutron. An alpha ray has an external magnetic field while a gamma rays has no external magnetic field. Because a gamma ray has no external magnetic field, it can penetrate obstacles better than an alpha ray.

The lost mass of 0.00100 amu will release an electron or negatron completely from a proton, positon and neutron. Only this lost mass will do this.

Any mass other than 0.00100 amu lost or transferred will not release an electron or negatron.

If the 0.00100 amu is combined with other masses that are lost or transferred, no electron or negatron is released.

The electron or negatron that is released completely from an isotope will have an arrow pointing away from it in the isotope that lost it. This may or may not be shown on the right side of the equal sign.

DRAWINGS SHOWING HOW AN ISOTOPE WILL CREATE A NEW ISOTOPE

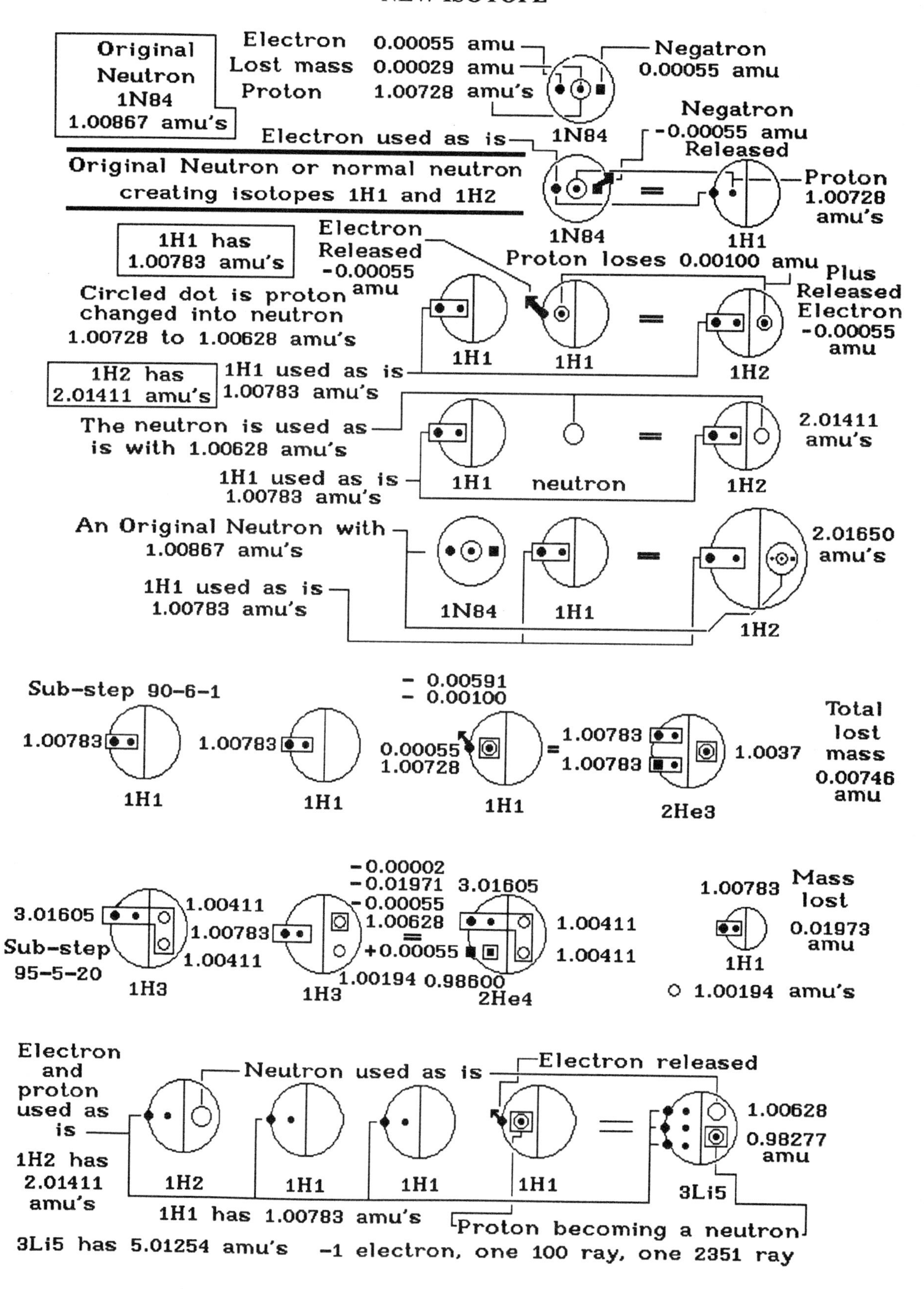

ATOMIC DRAWING FOR THE ORIGINAL NEUTRON

This drawing shows how the Original Neutron is drawn with all of its internal parts. It enables the finding of any errors and to see the correct placing of all of the parts.

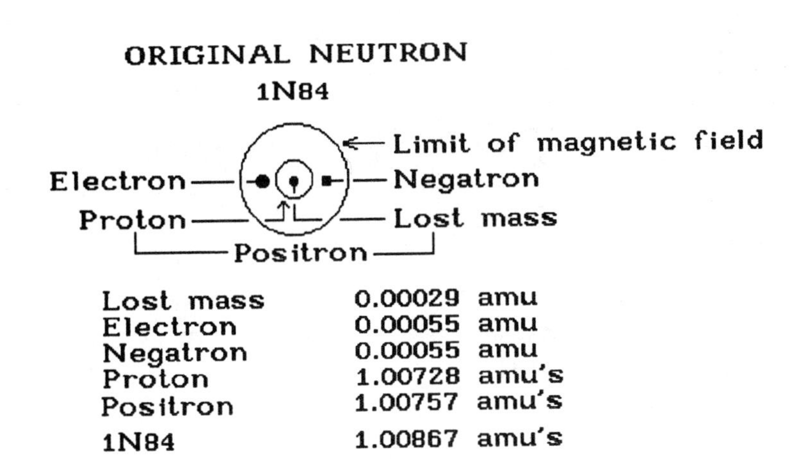

This is the only drawing that looks like this. All other drawings will have a vertical line dividing the circle into two halves.

First draw a circle. This circle will not have any electron or negatron on it. There is an electron drawn as a dot, then a small circle with a dot in its center and to the right of the small circle there is a small square that is the negatron.

The small circle is the positron. This includes the dot in its center which is the lost mass. The small circle will become the proton or positon after the lost mass is lost.

It is known that there is a Hydrogen atom with a proton and an electron that has the spin belonging to these parts. This spin is called an ortho spin. There also is a Hydrogen atom with the spin for the negatron and positon. This Hydrogen atom has a para spin. Therefore, the small circle can either be a proton or a positon after the lost mass is lost.

I have given the symbol name for the Original Neutron as 1N84. The 1 means it is the first neutron and the N means it is a neutron. I can not use the zero or the letter O because in will look like the word ON. The number 84 represents the mass that is lost and the electron or negatron that is released.

The Original Neutron does not have a nucleus. The circle that is drawn shows the limit of the Original Neutron's magnetic field.

The Original Neutron has a neutrino in its center that is not shown. To convert the neutrino to pure energy with the lost mass would require far more energy than is needed to split it into its parts. Therefore, the neutrino must remain with the proton, positon or neutron.

DRAWING OF ATOMS OR ISOTOPES

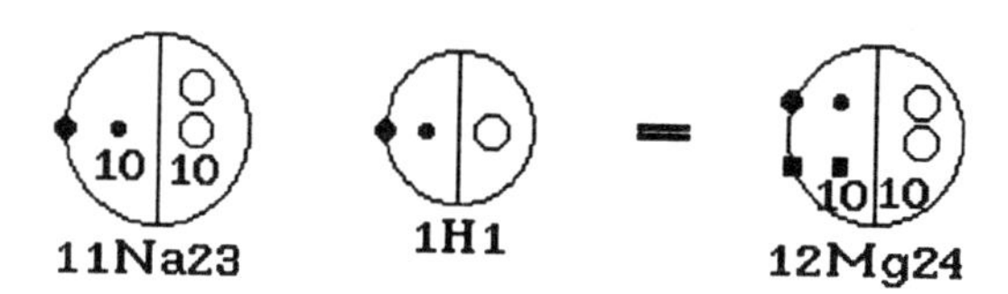

Whenever a drawing is made, the amount of mass on the left side of an equal sign must be shown on the right side of it. The amount of mass shown must be equal on both sides of the equal sign.

To draw any isotope, draw one circle with one vertical line that divides it into two halves. The circle is where the electrons and negatrons will go. Place a dot on the circle's left side to represent the first electron. The second electron called a negatron will be placed under the first dot on the circle in the form of a small square.

When more than two masses are to be put on the circle, look at the isotope being drawn and subtract two from the total number of the isotope. An example is the isotope magnesium (12Mg24). Place an electron and negatron on the circle with a dot and a square. Put the number 10 inside the left part of the circle. The 10 goes to the left of the vertical line. This number will represent the number of electrons or negatrons on the circle that is not shown. It will also represent the number of protons and positons found in the isotope minus the two that are drawn in. These two masses that are drawn in represent a proton and a positon and they are also drawn as a dot for the proton and a square for the positon. The number 10 is a short hand way of drawing dots and squares when the number of them makes drawing them to difficult to place on the circle or inside the left area. If drawing in the dots and squares do not cause any difficulty in understanding the drawing, then put them in. If the short hand is used, then put the same number on the right side of the vertical line that is within the circle. This will represent the number of neutrons not shown. To draw a neutron, you must know whether a proton or a positon created it. If a proton created the neutron, draw it as a small circle without a dot in its center. If the neutron comes from a positon, draw a square without a square in its middle. The square has its middle area not filled in.

When an electron is released from the isotope, draw an arrow pointing away from the dot that represents the electron that is lost. Since this electron is connected magnetically to a proton, this proton has a small circle drawn around it. This looks like a small circle with a dot in its middle. When this proton loses 0.00100 amu, it releases its electron and becomes a neutron. This neutron is drawn as a circle with a dot in its middle. It is placed on the right side of the vertical line. It will look just like the proton that created it. If the negatron is released, an arrow is also drawn pointing away from it. Since it also belongs to a positon, the positon is drawn as a square that represents this positon. Since this positon lost 0.00100 amu also, it becomes a neutron. Draw this neutron on the right side of the vertical line as a small circle with a small square in its middle. It will look just like the positon that created it. Drawn below.

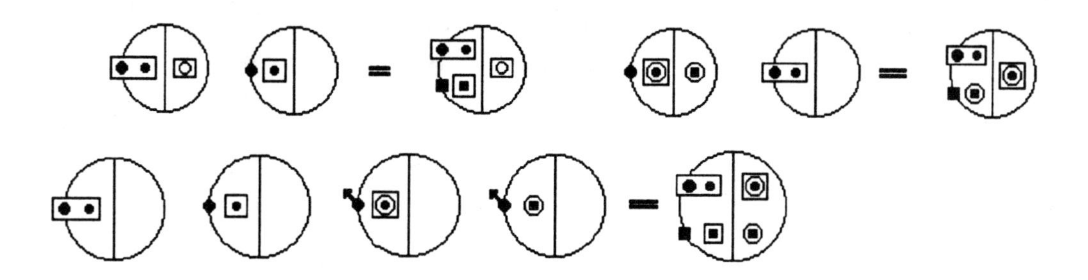

If an isotope that is used in creating a new isotope is used as is within the new isotope, draw a box or an enclosure around the parts that represent the isotope used to create a new isotope and within the isotope being created, draw this same thing. The two isotopes will look alike except that in the new isotope just created there will be the other parts needed to create it and they will not have a box or enclosure around them. The reason for drawing them this way is to show that if the new isotope just created, breaks up into its parts, then the parts that have a box or enclosure around them may not be needed in the new isotope and they drop out of where they came from. They may then be used with some other isotope that can be created. This box or enclosure will always have two or more parts to it. An example is the isotope 1H1. If it is used intact within the newly created isotope, it will have a rectangle drawn around it when it is a single isotope of 1H1 and within the newly created isotope. This isotope of 1H1 is found with a rectangle around it in the new isotope and it can drop out at some later time.

Sometimes a proton or positon will lose mass that is regained by a neutron. If this happens, draw a line from the proton or positon that lost this mass to the neutron that gained it. This may happen within an isotope and this will make a new isotope. Other times it will happen between two different isotopes. The new isotope just created will show this exchange of mass. When either of these types of exchanges of mass take place, no mass is lost. Since a ray represents mass that is lost, no ray can be emitted.

Below are some of the drawings that the reader will encounter which show the many ways to draw the isotopes involved in the creation of a new isotope.

Sometimes there is a transfer between different isotope as well as a transfer between the parts within an isotope. Drawing samples given below.

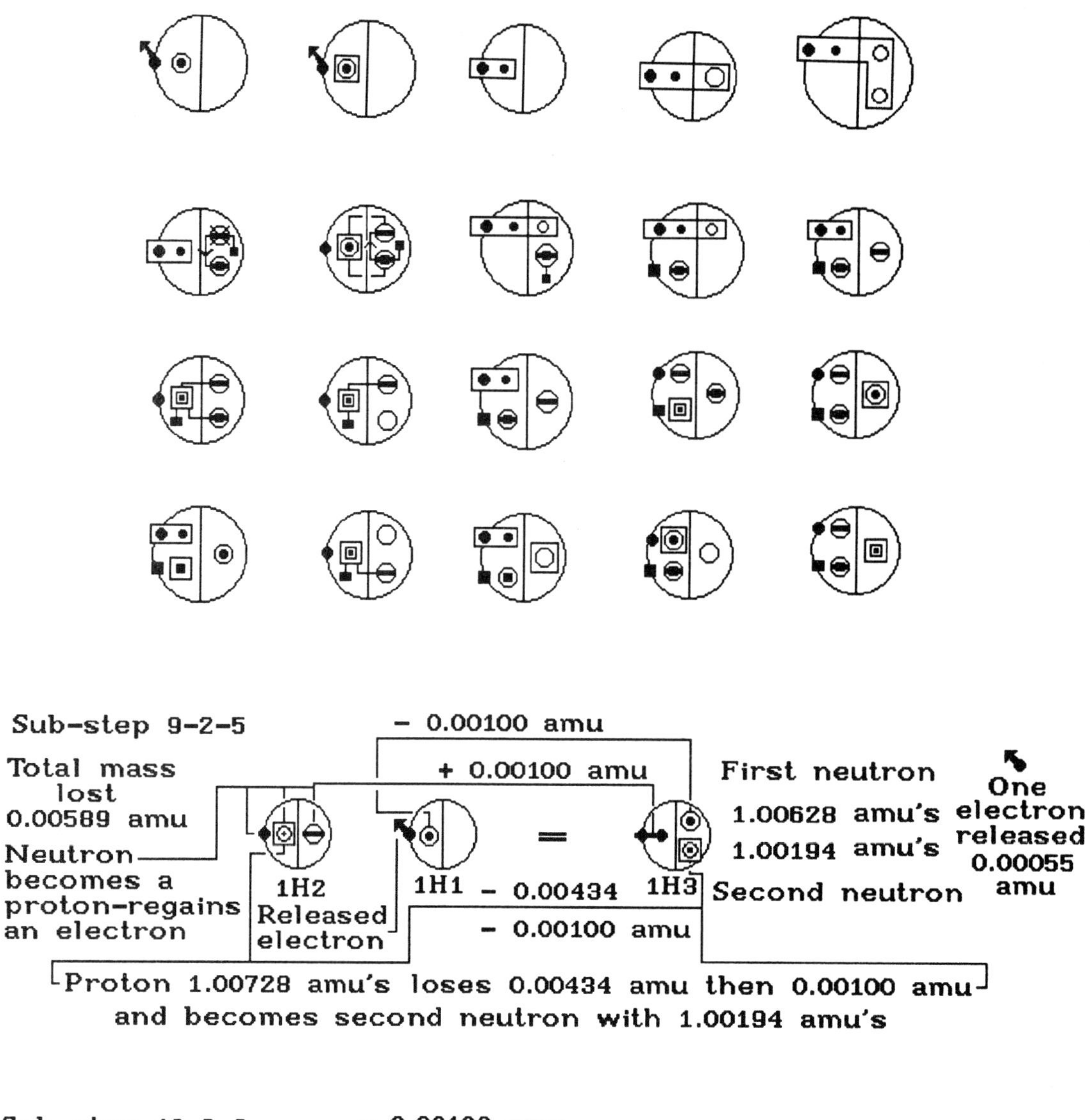

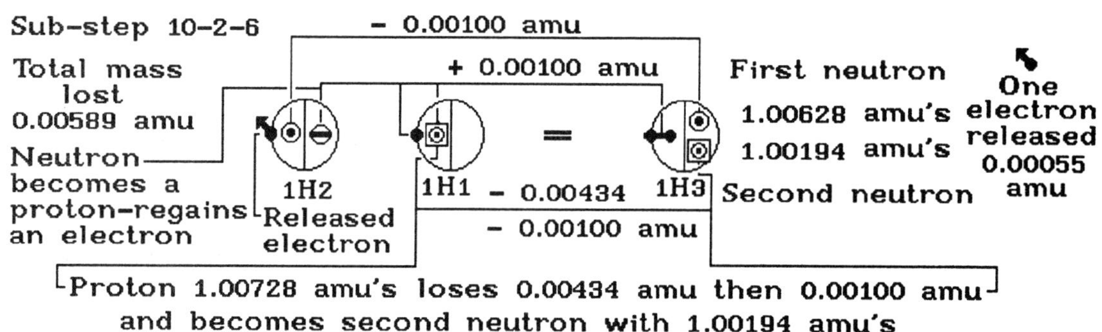

WRITING OUT THE STEPS TO CREATE A NEW ISOTOPE

I found a method whereby the formula for any isotope or sub-isotope can be found.

To find these steps in the creation of a new isotope, always start with one isotope less than the one being created. Sometimes the isotope comes from only one isotope as given in the drawings. Other times it starts with two isotopes. To make a formula, write out the isotopes needed to create a new isotope on the left side of an equal sign. On the right side of the equal sign write the name or symbol for the isotope being created. An example of this is given below.

1H3 = 2He3
3Li5 + 3Li5 = 3Li6

When creating a new isotope from an isotope that can not start with a single isotope, write down the first two isotopes that are alike. These two isotopes combine to create the new isotope shown on the right side of the equal sign. This is written out as a formula to create a new isotope. This is called Step 1. In the case where only one isotope will create a new isotope, Step 2 will be one less isotope than the single isotope. The formula 1H3 = 2He3 will lose 0.00002 amu and emit a #2 ray. The next isotope to create the isotope 2He3 is 1H2 plus what ever other isotope will go with it. The isotope 1H2 needs one isotope of 1H1 to create the isotope 2He3. When this is done, there will be 0.00591 amu lost which is emitted as a 591 ray (A gamma ray). Next will be three isotopes of 1H1. Two of these isotopes of 1H1 are used in a like manner while the third one must lose an electron or negatron. The two isotopes of 1H1 can be written as 2(1H1). When this is done, it will be written as 2(1H1) + 1H1 = 2He3, 591 ray. The next isotope is the single 1H1. It has either a proton or a positon which will lose 0.00591 amu that will not release an electron or negatron. The 0.00591 amu is lost as a 591 ray. After the 0.00591 amu is lost, a 0.00100 amu is lost and emit a 100 ray. This must be shown on the right side of the equal sign. Samples shown below.

1. 1H3 = 2He3, #2 ray
2. 1H2 + 1H1 = 2He3, 591 ray
3. 1H1 + 1H1 + 1H1 = 2He3, 591 ray, 100 ray, 1 electron

For the Lithium isotope 3Li6 to be created, the Step 1 can not be a single isotope of 3Li5 because this will not create the isotope 3Li6. Two isotopes of 3Li5 must be used. Next, there will be one 3Li5 and one 2He4. After this, one 3Li5 and one 2He3. There is no isotope of 3Li4 and there is no 2He2. Because there is no 2He2, the next step will use 3Li5 + 1H3 and then 3Li5 + 1H2 . The last 3Li5 goes with a 1H1 to create the isotope 3Li6. Because the isotope 1H1 is the last possible isotope to be used with 3Li5, the next step will start with 2He4. It will take two isotopes to create the isotope 3Li6. Next is 2He4 + 2He3.

Because there is no isotope of 2He2, the next step will be two isotopes of 1H3. Follow this pattern of adding isotopes to create a new isotope and you will end up with one isotope of 1H2 and four isotopes of 1H1 to create the isotope 3Li6. I do not believe that you can use five isotopes of 1H1 because the magnetic connections only use four such bonds. This is shown below.

1. 3Li5 + 3Li5 = 3Li6, 2He3, 1H1, rays
2. 3Li5 + 2He4 = 3Li6, 2He3, rays
3. 3Li5 + 2He3 = 3Li6, 2(1H1), rays
4. 3Li5 + 1H3 = 3Li6, 1H2, rays
5. 3Li5 + 1H2 = 3Li6, H1, rays
6. 3Li5 + 1H1 = 3Li6, 1 electron, 100 ray
7. 2He4 + 2He4 = 3Li6, 1H2, rays
8. 2He4 + 2He3 = 3Li6, 1H1 rays
9. 2He4 + 1H3 = 3Li6, 1 neutron, rays
10. 2He4 + 1H2 = 3Li6, rays
11. 2He4 + 1H1 + 1H1 = 3Li6, 1 electron, 100 ray
12. 2He3 + 2He3 = 3Li6, 1 electron, 100 ray
13. 2He3 + 1H3 = 3Li6, rays
14. 2He3 + 1H2 + 1H1 = 3Li6, 1 electron, 100 ray
15. 2He3 + 1H1 2(1H1) = 3Li6, 2 electrons, two 100 rays
16. 1H3 +1H3 = 3Li6, neutron gains 100 ray and electron
17. 1H3 +1H2 + 1H1 = 3Li6, rays
18. 1H3 2(1H1) + 1H1 = 3Li6, 1 electron, 100 ray
19. 3(1H2) = 3Li6, rays
20. 2(1H2) + 1H1 + 1H1 = 3Li6, 1 electron, 100 ray

No more steps are possible.

In the steps given above I do not give the values of the rays except where an electron is released. For this to happen, 0.00100 amu must be converted to pure energy and a 100 ray emitted. The value of the rays not given requires a lot of figuring and it is not need here.

Step 16 is the only step that requires an electron or negatron to be regained and the neutron to regain 0.00100 amu to become a proton or positon.

These twenty steps show the progressive steps starting with two isotopes that are alike until no more steps are possible. I say that step 21 is impossible because it would need more than four isotopes of Hydrogen to make the isotope 3Li6.

I chose the isotope 3Li6 to show more detail on how the formulas are created. This book only goes to 2He4 and to understand how formulas are created needs more steps than this isotope will give.

A LIST OF ATOMS AND THEIR ISOTOPES WITH THEIR ATOMIC MASS UNITS (amu's)

1. Hydrogen (H) isotope	1H1	1H2	1H3*				
2. Helium (He)	2He3	2He4	2He5*				
3. Lithium (Li)	3Li5*	3Li6*	3Li7	3Li8*	3Li9*		
4. Beryllium (Be)	4Be6*	4Be7*	4Be8*	4Be9	4Be10*	4Be11*	
5. Boron (B)	5B8*	5B9*	5B10	5B11	5B12*	5B13*	
6. Carbon (C)	6C9*	6C10*	6C11*	6C12	6C13*	6C14*	
	6C15*	6C16*					

* Indicates the isotope is radioactive.
The reader can use this list to check for an isotope that can be created.
All of these isotopes are the result of the Original Neutron.

Note that there is no such isotope of 2H2 or 2He2. Hydrogen can only have one proton therefore 2H2 is impossible. Helium must have at least one neutron and 2He2 has none therefore, it is also impossible. However, when the left over parts are two electrons or two negatrons and two protons or two positons, they will create two isotopes of 1H1.

Isotope	Mass	Isotope	Mass
1H1	1.00783 amu's	5B8	8.02461 amu's
1H2	2.01411 amu's	5B9	9.01333 amu's
1H3	3.01605 amu's	5B10	10.01294 amu's
		5B11	11.00931 amu's
2He3	3.01603 amu's	5B12	12.01435 amu's
2He4	4.00260 amu's	5B13	13.01778 amu's
2He5	5.01222 amu's		
3Li5	5.01254 amu's	6C9	9.03104 amu's
3Li6	6.01512 amu's	6C10	10.01686 amu's
3Li7	7.01600 amu's	6C11	11.01143 amu's
3Li8	8.02248 amu's	6C12	12.00000 amu's
3Li9	9.02678 amu's	6C13	13.00336 amu's
		6C14	14.00324 amu's
4Be6	6.01973 amu's	6C15	15.01060 amu's
4Be7	7.01693 amu's	6C16	16.01470 amu's
4Be8	8.00531 amu's		
4Be9	9.01218 amu's		
4Be10	10.01353 amu's		
4Be11	11.02165 amu's		

PART 2

THE ORIGINAL NEUTRON

1.00867 amu's

THE ORIGINAL NEUTRON
1.00867 amu's

MESON

ORIGINAL NEUTRON

1.00867 amu's
1N84

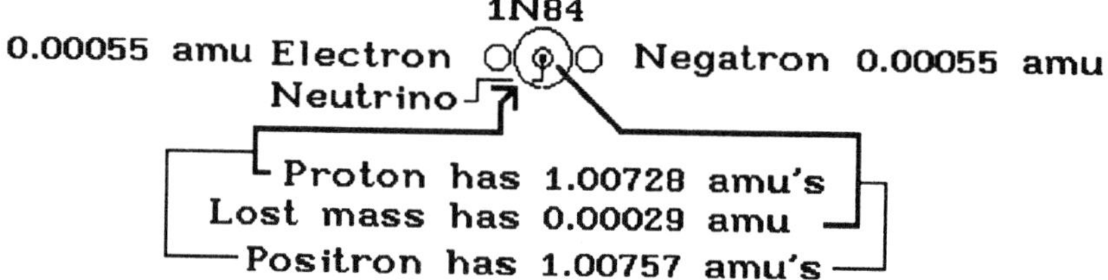

ORIGINAL NEUTRON

1N84

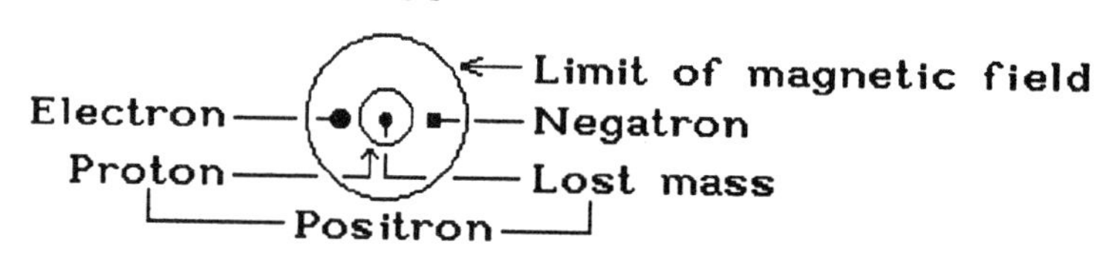

Lost mass	0.00029 amu
Electron	0.00055 amu
Negatron	0.00055 amu
Proton	1.00728 amu's
Positron	1.00757 amu's
1N84	1.00867 amu's

The term "Original Neutron" is of my creation. It is not found in any chemistry or physic books. Neither are the terms positon and negatron. I have also given a different meaning to the term "positron". All of the drawings are of my creation and they will not be found in any chemistry or physic books. The symbol 1N84 is the symbol I have given the Original Neutron. The lost mass can not be found in any chemistry or physic books. The values for the electron, proton, Original Neutron and some of the neutrons found in some of the isotopes are not mine. They are found in chemistry or physic books. The term "sub-isotope" is of my creation because no one as of now knows they exist.

In chemistry and physic books, the term "lost mass" is called "the binding force" or a "bond".

The internal structure of the Original Neutron and a proton, positon and a normal neutron started with the Quark. They are made up of masses that include within them an electron, negatron, 0.00100 amu and many other masses that can be converted to pure energy that is magnetism. As a Quark gains mass, its internal mass forms patterns that can be repeated as more mass is added.

The Original Neutron is composed of one electron which is drawn as a dot on the left side of the small circle. The lost mass is drawn as a dot that is in the center of the small circle. There is a small circle drawn around the lost mass. This small circle is the mass that remains when the lost mass is emitted as a ray and either the electron or negatron is released. On the right side of the small circle is a square that is the negatron.

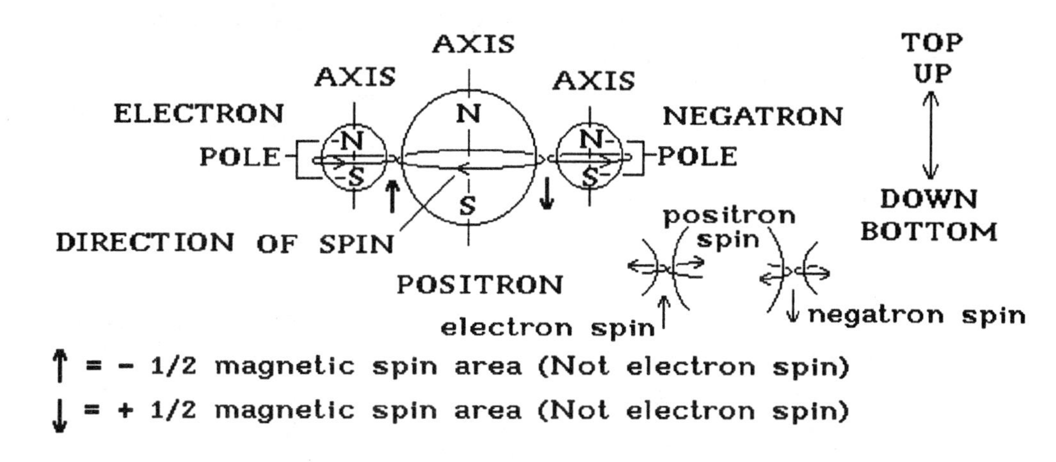

The larger circle that is around all of the parts of the Original Neutron is drawn to show the limits of its magnetic field. The small circle will become the proton in the isotope 1H1. The mass that is the proton or positon plus the lost mass make up the positron. It is called a positron because it has more mass than a normal proton. When the lost mass is lost, the positron becomes a proton or a positon. The lost mass is called lost mass because when an electron or negatron goes into an orbital, this mass is lost. It is converted from mass to pure energy which is magnetism.

There is a graviton and a neutrino in the center of the Original Neutron. They hold all of the parts together and stay with the proton or positon when the isotope 1H1 is created.

The electron and negatron is bonded to the positron because the lost mass has not yet been lost. They differ from each other because of their spin found between the positron and themselves. Both the electron and negatron are spinning in the same direction but the positron is spinning in a different direction. The area between the electron and the positron spins in a counter clock direction while the area between the negatron and the positron spins in the clock direction. If I can get all of the people who read this book to agree that the area between the electron and positron is spinning in the counter clock direction, it will have an ortho spin. The area between the negatron and positron must therefore be spinning in the clock direction and it is a para spin.

When this bond is broken, either the electron or negatron goes into an orbital. The electron or negatron that does not go into an orbital is released and leaves the Original Neutron completely. When they are connected to the positron, they are rotating on their own axis. The positron is also rotating on its axis. They rotate as a single unit. The electron and negatron have the mass of 0.00055 amu each for a total of 0.00110 amu.

The positron has 1.00757 amu's. This mass is composed of a proton or a positon and the mass that is lost. The proton or positon each have a mass of 1.00728 amu's. The mass that is lost has 0.00029 amu. There is also a neutrino but its mass is not given because it is so small.

The electron or negatron that is released has 0.00055 amu and the mass that is lost has 0.00029 amu and these are lost from the Original Neutron for a total lost mass of 0.00084 amu. The 0.00029 amu is lost as a 29 ray.

The Original Neutron is magnetically neutral because it does not have an external magnetic field.

The term "atomic mass unit" or "amu" is defined as one twelfth the mass of the isotope Carbon 12. The Carbon 12 isotope has 12.00000 amu's.

One amu is equal to 931.48 or 931.5 MeV. (MeV is one Million electron Volts.)

Earnest Rutherford in 1920 suggested that the neutron must exist because the mass of the isotopes indicated that there should be something to balance the mass of an isotope. In 1932, James Chadwick did some experiments that established the existence of the neutron which has 1.00867 amu's. It is this value that is found in most chemistry books.

MATHEMATICAL STEPS FOR THE ORIGINAL NEUTRON

In the steps given below, I show how to figure what the Original Neutron is composed of in a sequence of steps.

Mass of normal proton	1.00728 amu's
Mass of normal positon	1.00728 amu's
Mass of electron	0.00055 amu
Mass of negatron	0.00055 amu
Mass of 1H1	1.00783 amu's
Mass of positron	1.00757 amu's
Mass lost	0.00029 amu
Mass lost by Original Neutron	0.00084 amu

Mass lost by Original Neutron	0.00084 amu
Mass lost by electron or negatron	0.00055 amu
Mass lost by positron	0.00029 amu

Mass of proton or positon	1.00728 amu's
Mass lost by positron	0.00029 amu
Mass of positron	1.00757 amu's

Mass of positron	1.00757 amu's
Mass of proton or positon	1.00728 amu's
Mass lost by positron	0.00029 amu

Mass of Original Neutron	1.00867 amu's
Mass lost by Original Neutron	0.00084 amu
Mass of 1H1	1.00783 amu's

Mass of electron or negatron	0.00055 amu
Mass of proton or positon	1.00728 amu's
Mass of 1H1	1.00783 amu's

Mass of electron and negatron	0.00110 amu
Mass of positron	1.00757 amu's
Mass of Original Neutron	1.00867 amu's

Quarks build a neutrino which becomes a meson. When the meson reaches 1.00867 amu's it becomes an Original Neutron. This meson then has a positron with an electron and a negatron on each side. The positron splits into a proton or positon with either an electron or a negatron orbiting one of them. Whichever electron or negatron is not in orbit is released and 0.00029 amu is converted to pure energy which is magnetism. This Original Neutron is in a space by itself and within twelve and a half minutes it becomes a Hydrogen isotope of 1H1.

PART 3

HYDROGEN

1.00798 amu's

There is no isotope with this atomic mass. The atomic mass of 1.00798 amu's is not the mass of a real isotope of Hydrogen. This number comes from the combined masses of the real isotopes found in nature.

There may be one or more isotopes of a given element. Each isotope makes a percentage of all the mass found for that element. To find the combined masses of a given element, multiply the mass of each isotope by its percentage and then add their results. The chemistry books call this an "average" which it is not. It should be called a composite number.

Hydrogen 1H1 and 1H2

1.00768 (99.985) + 2.01411 (0.015) = 1.00798 amu's

1.00768 + 0.00030 = 1.00798 amu's

If I add 1.00768 and 1.00798, I get 2.01566 amu's and this is two isotopes of 1H1. 2(1.00783) = 2.01566 amu's

Note that the isotope 1H3 is not included in this math. This is because there is not enough of a percentage for it.

HYDROGEN

CREATING THE ATOM OF HYDROGEN (H)

ISOTOPES 1H1, 1H2 AND 1H3

HYDROGEN ONE

1H1

1.00783 amu's

CREATING THE ISOTOPE 1H1

1.00783 amu's

HYDROGEN 1
1H1
1.00783 amu's

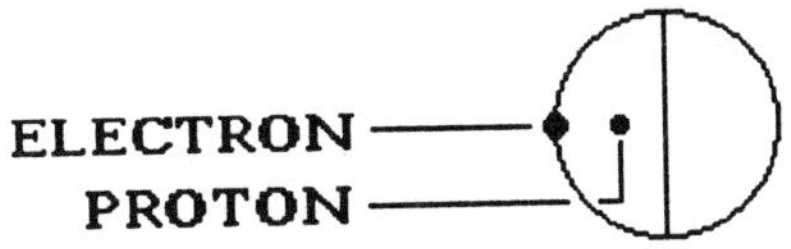

ELECTRON
PROTON

ELECTRON	0.00055 amu
NEGATRON	NONE
PROTON	1.00728 amu's
POSITON	NONE
NEUTRON	NONE

The first isotope of Hydrogen is 1H1. It has either one electron or one negatron in an orbital around the nucleus. The nucleus has one proton or one positon. It does not have a neutron.

The Hydrogen isotope of 1H1 came from the Original Neutron.

An electron or negatron has 0.00055 amu.

A proton or positon has 1.00728 amu's.

The Original Neutron loses 0.00084 amu and the isotope 1H1 comes into being.

Mass of Original Neutron	1.00867 amu's
Mass lost	0.00084 amu
Mass making 1H1	1.00783 amu's

By adding the values of the electron or negatron and proton or positon in the isotope 1H1, the value of the isotope 1H1 is found.

Mass of electron or negatron	0.00055 amu
Mass of proton or positon	1.00728 amu's
Mass of 1H1	1.00783 amu's

This isotope makes up 99.985% of all naturally occurring isotopes of 1H1.

CREATING THE FIRST ISOTOPE OF HYDROGEN FROM THE ORIGINAL NEUTRON

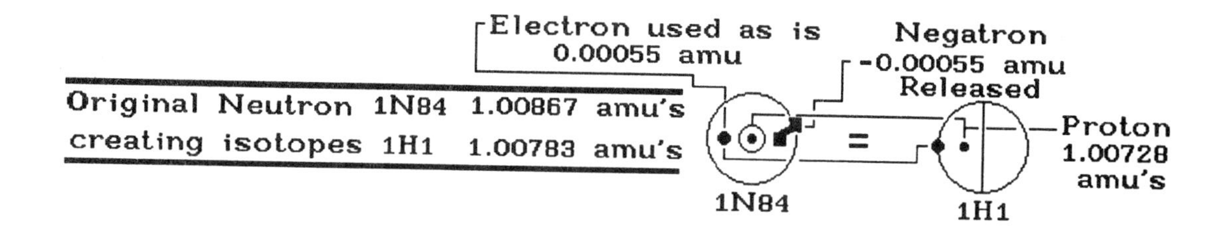

This first isotope of Hydrogen is 1H1. It has one electron and one proton. It does not have a neutron. This isotope is not found within the Original Neutron until the positron loses 0.00084 amu. When this mass is lost, the isotope 1H1 is the mass that is left. When the Original Neutron loses an electron or negatron, (either one has 0.00055 amu) and 0.00029 amu, the positron becomes either a proton or a positon (either one has 1.00728 amu's). Either an electron and a proton or a negatron and a positon will make an isotope of 1H1 with 1.00783 amu's.

The isotope 1H1 that has an electron and a proton with an ortho spin while the isotope that has a negatron and a positon has a para spin. These spins are in opposite directions to each other. The ortho spin is going in the counter clock direction while the para is in the clock direction.

The Original neutron loses 0.00084 amu creating the isotope 1H1.

Mass of Original Neutron	1.00867 amu's
Mass lost	0.00084 amu
Mass remaining (1H1)	1.00783 amu's
Mass of electron or negatron in 1H1	0.00055 amu
Mass of proton or positon in 1H1	1.00728 amu's
Mass of 1H1	1.00783 amu's

This isotope makes up 99.985 % of naturally occurring isotopes of 1H1.

HYDROGEN

1H2

DEUTERIUM

2.01411 amu's

CREATING THE ISOTOPE 1H2

DEUTERIUM

2.01411 amu's

1H1 + 1H1 = 1H2, one 100 ray, electron or negatron

HYDROGEN 2
1H2
DEUTERIUM
2.01411 amu's

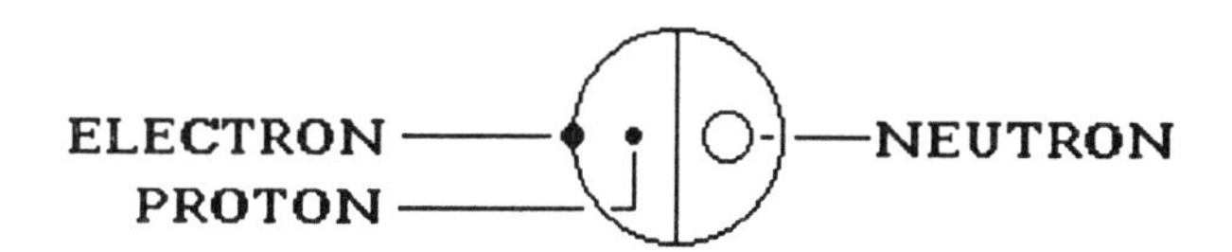

ELECTRON	0.00055 amu
NEGATRON	NONE
PROTON	1.00728 amu's
POSITON	NONE
NEUTRON	1.00628 amu's

This Hydrogen isotope is called Deuterium and it takes two isotopes of 1H1 for its creation. It has one electron or negatron with 0.00055 amu, one proton or positon with 1.00728 amu's with one neutron with 1.00628 amu's. The single neutron can be either an ortho or a para neutron. An ortho spin is in the counter clock direction while the para spins in the clock direction. This means that there is an ortho-1H2 or a para-1H2.

There are other neutrons besides the Original Neutron and the first one is found in the isotope 1H2. This neutron is created when one of the two isotopes of 1H1 has its proton or positon loses 0.00100 amu and loses its electron or negatron. The proton or positon loses its external magnetic field which becomes internal. It can not hold its electron or negatron when its magnetic field becomes internal and either an electron or negatron is released.

The neutron always spins in the same direction as the electron and negatron and in the opposite direction to the proton and positon. If a proton becomes a neutron, then this new neutron changes the direction of its spin so that it will spin in the opposite direction that the proton that is already there. If a positon becomes a neutron, the new neutron will change its spin direction to the spin of the proton already there or if it is a positon that is to go with the new neutron then the new neutron will change its spin direction because the proton and positon always spin in the same direction.

STARTING WITH THE ORIGINAL NEUTRON
CREATING THE ISOTOPES 1H1 AND 1H2

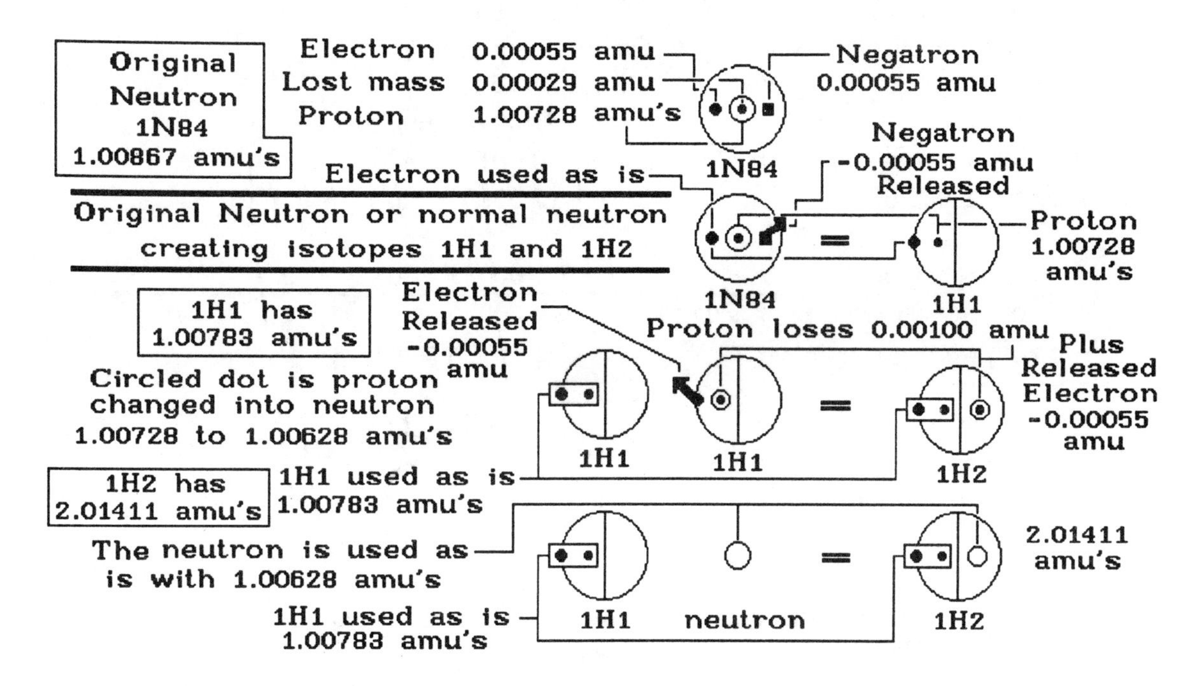

The Original Neutron with 1.00867 amu's lost 0.00084 amu to become an isotope of 1H1. This isotope has either one electron or a negatron orbiting a proton or a positon. The mass of either the electron or negatron is 0.00055 amu while either of the proton or positon has 1.00728 amu's. An electron and a proton make an ortho-Hydrogen isotope of 1H1. A negatron and a positon make a para-Hydrogen isotope of 1H1. They each have 1.00783 amu's.

Original Neutron	1.00867 amu's
Lost mass	0.00084 amu
Mass of 1H1	1.00783 amu's

Mass of electron or negatron	0.00055 amu
Mass of proton or positon	1.00728 amu's
Mass of either ortho or para 1H1	1.00783 amu's

It takes two isotopes of 1H1 to create the isotope 1H2. One of the isotopes of 1H1 is used as is. This means that this isotope of 1H1 does not lose any mass and it is found in the isotope 1H2 as the isotope 1H1. The second isotope of 1H1 loses 0.00100 amu from its proton or positon and then releases its electron or negatron. If a proton loses this mass, then it becomes an ortho-neutron and if a positon loses it, then the neutron is a para-neutron. The isotope 1H2 has only one neutron in its nucleus.

At this point in time, the mass of the electron, negatron, proton and positon is known. The mass of the neutron in the isotope 1H2 is not considered known.

The mass of the isotope 1H2 is given in the chemistry books as 2.01411 amu's. Two isotopes of 1H1 make the isotope 1H2. One is used as is with no loss of mass while the other one is converted to a neutron with the release of an electron or negatron plus the loss of 0.00100 amu as a 100 ray.

Mass of 1H1	1.00783 amu's
Mass lost releasing an electron or negatron	0.00100 amu
Mass remaining of 1H1	1.00683 amu's

Mass remaining of 1H1	1.00683 amu's
Mass of released electron or negatron	0.00055 amu
Mass of new neutron in 1H2	1.00628 amu's

Mass of electron or negatron in 1H2	0.00055 amu
Mass of proton or positon in 1H2	1.00728 amu's
Mass of ortho or para-neutron in 1H2	1.00628 amu's
Mass of ortho or para-1H2	2.01411 amu's

The neutron in the isotope 1H2 here gives proof to the understanding that there is an Original Neutron and other neutrons with different masses.

If the Original Neutron should become part of the isotope 1H2, this new isotope would be greater in mass than the normal isotope of 1H2. It would be very massive.

Mass of proton or positon in 1H2	1.00728 amu's
Mass of electron or negatron in 1H2	0.00055 amu
Mass of Original Neutron	1.00867 amu's
Mass of this 1H2	2.01650 amu's

Mass of this 1H2	2.01650 amu's
Mass of normal 1H2	2.01411 amu's
Mass greater than normal for a 1H2	0.00239 amu

In the isotope 1H2 with 2.01411 amu's, one isotope of 1H1 is used as is while the other isotope of 1H1 creates the neutron. It is therefore possible for the isotope of 1H1 that is used as is to drop out as the isotope 1H1 from the isotope 1H2. This would leave the neutron by itself as the part that is used in the new isotope of 1H2. If the isotope 1H1 should be the mass that stays with the new isotope being created, then the neutron with 1.00628 amu's would drop out. If the isotope 1H1 can drop out at some later time or if it is used as is, it is enclosed in a rectangle to show this. If the neutron drops out, it need not be enclosed because it would be a left over part and it therefore must drop out.

The isotope 1H2 makes up 0.015% of all naturally occurring isotopes of Hydrogen that are not radioactive. There are only two isotopes of Hydrogen that are considered to be found naturally in nature that are not radioactive.

HYDROGEN

1H3

TRITIUM

3.01605 amu's

CREATING THE ISOTOPE 1H3

TRITIUM

3.01605 amu's

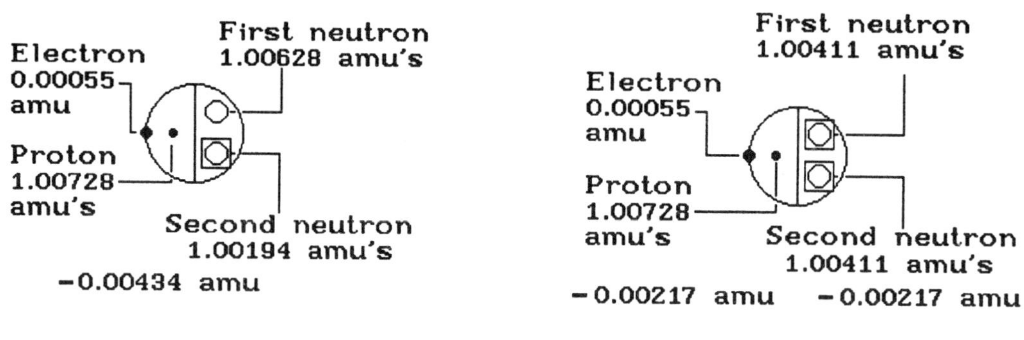

HYDROGEN 3
1H3
TRITIUM

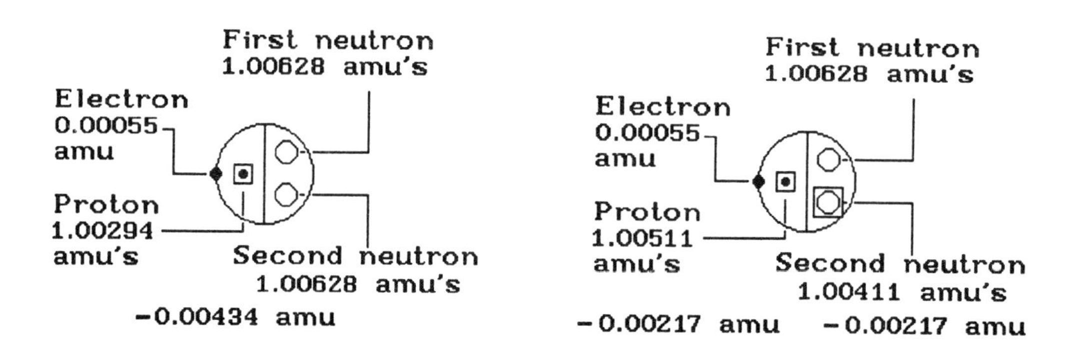

Tritium, with an atomic mass of 3.01605 amu's, is radioactive with a half-life of twelve point twenty six years (12.26 years). It has one electron or negatron in an orbital around the nucleus which has one proton or a positon and two neutrons. Either the proton or positon will have 1.00728 amu's if both neutrons have 1.00411 amu's or the first neutron has 1.00628 amu's and the second one has 1.00194 amu's. If both neutrons have 1.00628 amu's, then the proton or positon will have 1.00294 amu's with either an electron or negatron in an orbital around all of them. If the first neutron has 1.00628 amu's and the second one has 1.00411 amu's, then the proton or positon will have 1.00511 amu's.

The most mass a proton or positon can have is 1.00728 amu's and the most mass a neutron can have is 1.00628 amu's. When either a proton or positon has less mass than this and a proton or positon is needed, it will regain if possible enough mass to equal 1.00728 amu's. If either neutron has less mass than 1.00628 amu's and a neutron is needed, then it will regain if possible enough mass to equal 1.00628 amu's.

If the proton or positon has 1.00294 amu's, it will try to regain 0.00434 amu while if the proton or positon has 1.00511 amu's, it will try to regain 0.00217 amu.

If the neutron has 1.00194 amu's, it will try to regain 0.00434 amu while two neutrons with 1.00411 amu's will have each neutron try to regain 0.00217 amu. An isotope of 1H3 with a proton of 1.00728 amu's can not have two neutrons with 1.00628 amu's and equal 3.001605 amu's. This means that when the two neutrons in the isotope 1H3 become neutrons with 1.00628 amu's, the isotope of 1H3 will not stay an isotope of 1H3 by losing 0.00002 amu to become the isotope 2He3.

The isotopes 1H2, 1H3 and 2He3 are almost non-existent isotopes so if they combine with almost non-existent isotopes there still would not be enough to make many isotopes of 2He4 or other isotopes. There are plenty of isotopes of 1H1 so when they combine with other isotopes of 1H1's, the isotopes thus created will be able to again combine with more isotopes of 1H1.

The isotope 2He3 is one neutron short of having a neutron for each proton to become a balanced isotope of 2He4. While the isotope 2He3 is formed, it is almost non-existent. The isotope 1H3 will combine with other isotopes and lose any excess mass to make the isotope 2He4. When the excess mass is lost, the isotope 2He4 thus formed is not radioactive.

When the isotope 1H3 is subtracted from the mass of the isotope 2He4, the remaining mass is 0.98655 amu. This 0.98655 amu is composed of either a proton with 0.98600 amu and an electron or negatron or it is composed of one neutron with 0.98500 amu and the mass of 0.00157 amu. The 0.00157 amu is composed of 0.00100 amu, 0.00055 amu and 0.00002 amu. This means that the isotope 2He4 can have the isotope 1H3, one neutron of 0.98500 amu, the masses of an electron or negatron and 0.00002 amu. It is a radioactive isotope of 2He4.

If the isotope 1H3 changed into an isotope of 2He3, then the isotope 2He4 would be composed of an isotope of 2He3 plus a neutron with 0.98657 amu. This neutron would be composed of a proton or positon with 0.98600 amu and a mass of 0.00157 amu. This means that there is a proton or positon found within a neutron. It also proves that the mass of 0.00157 can be found within a proton or positon or a neutron as a single unit. This 0.00157 is composed of 0.00100 amu, 0.00002 amu and an electron or negatron. If the 0.00100 amu is transferred to another neutron with 1.00628 amu's and 0.00002 amu is emitted as ray, then the electron would be released to go into an orbital with an electron and negatron already in their orbital. The neutron would make another proton and the electron would go in a new orbital. This would make the isotope Lithium (Li) 3Li5 when there is another neutron gained.

FORMULA CHART ONE
FOR 1H3

The isotope 1H3 is created from three (3) **STEPS.**

1. 1H2 + 1H2 = 1H3
2. 1H2 + 1H1 = 1H3
3. 1H1 + 2(1H1) or 1H1 + 1H1 + 1H1 = 1H3

Step 3 has two ways it can be shown because in one case one isotope of 1H1 is used as is while there are two isotopes of 1H1 used in a like manner but different than the first isotope of 1H1. It could also be used such that all three isotopes of 1H1 are used in a different way. Isotopes used in a like manner can be put in parentheses.

FORMULA CHART TWO
FOR 1H3

The formula chart two uses the three (3) **STEPS** found in FORMULA CHART ONE FOR 1H3 with each **STEP** having additional steps needed to create sub-isotopes of 1H3. These additional steps are called **Sub-steps.**

1. 1H2 + 1H2 = 1H3

1-1-1. 1H2 + 1H2 = 1H3, one 434 ray, 1H1
2-1-2. 1H2 + 1H2 = 1H3, one 434 ray, one 100 ray, -1e, -1n
3-1-3. 1H2 + 1H2 = 1H3, two 217 rays, 1H1
4-1-4. 1H2 + 1H2 = 1H3, one 434 ray, 1H1

2. 1H2 + 1H1 = 1H3

5-2-1. 1H2 + 1H1 = 1H3, one 434 ray, one 100 ray, -1e
6-2-2. 1H2 + 1H1 = 1H3, one 434 ray, one 100 ray, -1e
7-2-3. 1H2 + 1H1 = 1H3, one 434 ray, one 100 ray, -1e
8-2-4. 1H2 + 1H1 = 1H3, one 434 ray, one 100 ray, -1e
9-2-5. 1H2 + 1H1 = 1H3, one 434 ray, one 100 ray, -1e
10-2-6. 1H2 + 1H1 = 1H3, one 434 ray, one 100 ray, -1e
11-2-7. 1H2 + 1H1 = 1H3, two 217 rays, one 100 ray, -1e
12-2-8. 1H2 + 1H1 = 1H3, two 217 rays, one 100 ray, -1e

3. 1H1 + 2(1H1) or 1H1 + 1H1 + 1H1 = 1H3

13-3-1. 1H1 + 1H1 + 1H1 = 1H3, one 434 ray, two100 rays, -2e
14-3-2. 1H1 + 2(1H1) = 1H3, two 217 rays, two 100 rays, -2e
15-3-3. 1H1 + 1H1 + 1H1 = 1H3, two 217 rays, two 100 rays, -2e
16-3-4. 1H1 + 2(1H1) = 1H3, one 434 ray, two 100 rays, -2e

The -1e or -2e means that there is one or two 100 rays emitted with the result that one or two electrons or negatrons are released. If the isotope of 1H1 is an ortho-Hydrogen atom then an electron is released. If the isotope of 1H1 is a para-Hydrogen atom then a negatron is released.

The term alpha ray means a ray has been emitted from a proton.

The term gamma ray means a ray has been emitted from a neutron.

The term beta ray means a 100 ray has been emitted.

When an electron or negatron is lost and one is gained with the result that no electron or negatron is lost, then no 100 ray is emitted.

KNOWN MASSES

The following atomic mass units (amu) have already been given.

The Original Neutron	1.00867 amu's
The positron	1.00757 amu's
The lost mass	0.00029 amu
The electron or negatron	0.00055 amu
The proton or positon	1.00728 amu's
The neutron in 1H2	1.00628 amu's
The isotope 1H1	1.00783 amu's
The isotope 1H2	2.01411 amu's
The isotope 1H3	3.01605 amu's
The mass lost creating 1H2	0.00100 amu

The isotope 1H3 is created in sixteen (16) different ways. The mass lost is as of now unknown and how the electron or negatron is released must be found. There are two neutrons in the isotope 1H3 plus one proton or positon. What happens to them is yet to be explained. Now all of the above steps will be explained with the math proofs given. Now there will be some additional mass lost besides the 0.00100 amu.

STEP 3

1H3

FORMULA

1H1 + 1H1 + 1H1 OR 1H1 + 2(1H1) = 1H3

3. 1H1 + 2(1H1) or 1H1 + 1H1 + 1H1 = 1H3
13-3-1. 1H1 + 1H1 + 1H1 = 1H3, one 434 ray, two 100 rays, -2e
14-3-2. 1H1 + 2(1H1) = 1H3, two 217 rays, two 100 rays, -2e
15-3-3. 1H1 + 1H1 + 1H1 = 1H3, two 217 rays, two 100 rays, -2e
16-3-4. 1H1 + 2(1H1) = 1H3, one 434 ray, two 100 rays, -2e

All rays emitted in this STEP are alpha rays emitted from a proton and all 100 rays are beta rays.

Sub-step 16-3-4

1H3

FORMULA

1H1 + 2(1H1) = 1H3, one 434 ray, two 100 rays, -2e

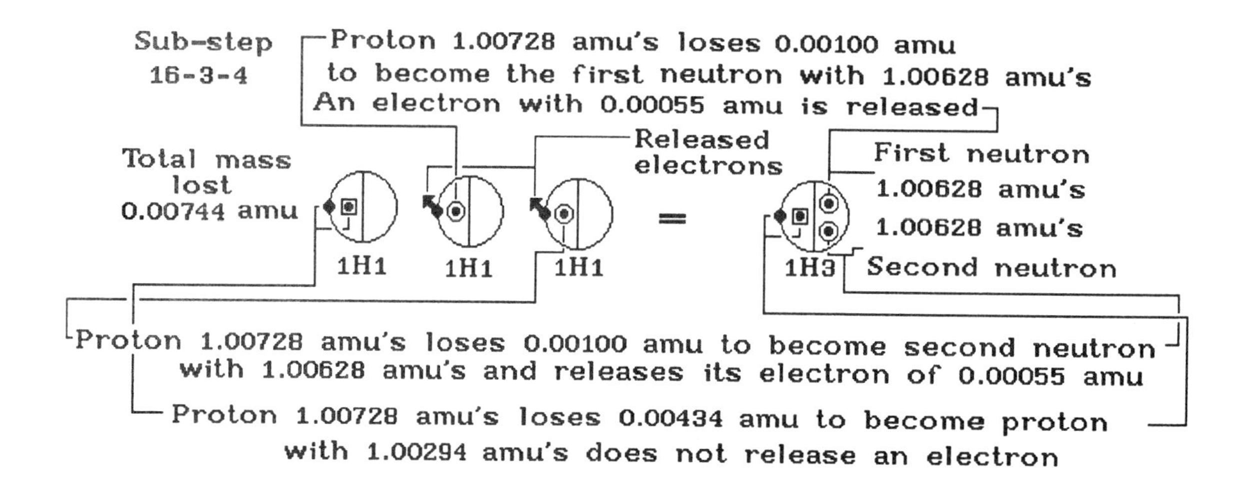

This step has three isotopes of 1H1 used to create the isotope 1H3. The first isotope of 1H1 in the next three steps will be made up of an electron and a proton which will always be the electron and proton found in the isotope 1H3. The other two isotopes of 1H1 will lose 0.00100 amu from each proton or positon and turn each into a neutron with the release of an electron or negatron. Any other mass lost can come from any one or more protons or positons. It will be assumed that all isotopes of 1H1 are composed of an electron and a proton but it is understood that this may not be the case in the real world of isotopes. The drawing will show only electrons and protons.

The drawing shows the first isotope of 1H1's electron as a dot on the right side of the big circle. The proton is drawn as a dot with a square around it on the left side of the vertical line to indicate that it lost some mass. The square means that it lost this mass and did not release its electron. The electron and proton are shown in the isotope created in the same manner- that is they look alike.

The drawing shows the second and third isotope of 1H1's electron as a dot with an arrow pointing away from it. The arrow means that the electron is released. Because the electron is lost, it is not found in the just created isotope of 1H3 but it is given as a -2e in the formula because in this step there are two electrons released. The minus sign in front of the 2 means that two electrons have been released. The proton is drawn to the left of the vertical line as a dot with a small circle around it. The small circle means that this proton lost 0.00100 amu and released its electron to become the first neutron in the just created isotope of 1H3. The drawing shows the proton and the first neutron looking alike with the two neutrons drawn on the right side of the vertical line.

The first isotope of 1H1 has its proton lose 0.00434 amu and it keeps its electron which becomes the electron in the isotope 1H3. This proton with a mass of 1.00728 amu's loses 0.00434 amu's and becomes the proton in the isotope 1H3 with 1.00294 amu's.

The second and third isotope of 1H1 each lose 0.00100 amu from their protons. Each proton then becomes a neutron in the isotope 1H3 with 1.00628 amu's. When the 0.00100 amu is lost, an electron is released. That means that two electrons are released in this step. The 0.00434 amu lost is lost as a 434 ray and the 0.00100 amu lost is lost as a 100 ray. The chemistry books called the 434 ray a gamma ray but there soon will be too many different gamma rays that the reader will not know which gamma ray lost a given mass if it is only called a gamma ray so this term will not be used. In all cases where 0.00100 amu is lost, an electron will be released. The chemistry books say that when this happens, both mass and a ray are emitted. I say that the electron is released and not emitted as a ray while the 0.00100 amu is lost as a 100 ray which is the beta ray. Since it is always 0.00100 amu that is converted to a 100 ray, I will not use the term beta ray. A ray is pure energy which is magnetism which has been decompressed. Any mass is pure energy that is magnetism that has been compressed. Only the proton or positon and a neutron may be decompressed. The electron never loses any mass and it therefore is never decompressed.

The second and third isotope of 1H1 are used in a like manner so they can be put in parenthesis with the number 2 in front.

Someone may wonder why I start with Step 16-3-4 to explain how all the above is done when the FORMULA CHART TWO starts with Step 1-1-1. I made all my charts begin with the largest isotope that can create the new isotope and work my way down to the smallest isotope. To explain how a new isotope is made, I felt that it could be easier to understand if I started explaining by using the smallest isotopes and work my way to the largest.

I start by adding the mass of the three isotopes of 1H1 to know the total amount of mass needed to create the isotope 1H3. I then subtract the mass of the isotope 1H3 from it. The remainder will be the amount of mass lost.

Mass of first 1H1	1.00783 amu's
Mass of second 1H1	1.00783 amu's
Mass of third 1H1	1.00783 amu's
Total mass needed to create 1H3	3.02349 amu's

Total mass needed to create 1H3	3.02349 amu's
Mass of 1H3	3.01605 amu's
Mass lost creating 1H3	0.00744 amu

Each of the last two isotopes of 1H1 lost enough mass from their protons to change each proton into a neutron. Each neutron found in the isotope of 1H3 will therefore have the same mass.

Mass of the second 1H1's proton	1.00728 amu's
Mass lost creating first neutron in 1H3	0.00100 amu
Mass of first neutron in 1H3	1.00628 amu's

Mass of the third 1H1's proton	1.00728 amu's
Mass lost creating second neutron in 1H3	0.00100 amu
Mass of second neutron in 1H3	1.00628 amu's

Each of the second and third isotope of 1H1 lost its electron when it lost 0.00100 amu. Each electron has 0.00055 amu. This means that their mass is lost when creating the isotope 1H3. Their lost mass is added to the mass that is lost releasing them. Each isotope of the second and third isotope of 1H1 then loses the mass of an electron and the mass that released it.

Mass of electron from second 1H1	0.00055 amu
Mass of electron from third 1H1	0.00055 amu
Total mass lost by two electrons	0.00110 amu

Total mass lost by two electrons	0.00110 amu
Total mass lost by two 1H1's protons	0.00200 amu
Total mass lost by second and third 1H1's proton	0.00310 amu

The second and third isotope of 1H1 lost 0.00310 amu. Subtracting this amount from the total mass lost will give the mass lost that does not release an electron.

Total mass lost creating 1H3	0.00744 amu
Total mass lost by second and third 1H1's proton	0.00310 amu
Mass lost that does not release an electron	0.00434 amu

The first isotope of 1H1 lost 0.00434 amu from its proton which did not release its electron. This lost mass is added to the total mass lost by the second and third isotope of 1H1. This is the total mass lost creating the isotope 1H3.

Mass lost by first 1H1's proton	0.00434 amu
Total mass lost by second and third 1H1's proton	0.00310 amu
Total mass lost creating 1H3	0.00744 amu

The mass of the first isotope of 1H1's proton is 1.00728 amu's. By subtracting the 0.00434 amu lost from the proton's mass will give the mass of the proton found in the isotope 1H3. The lost mass of 0.00434 amu is lost as a single unit and emitted as a 434 ray.

Mass of first 1H1's proton	1.00728 amu's
Mass lost by it	0.00434 amu
Mass of proton in 1H3	1.00294 amu's

Now the mass for each part that makes up the isotope 1H3 are known. By adding their masses, the mass of the isotope 1H3 will be known.

Mass of electron in 1H3	0.00055 amu
Mass of proton in 1H3	1.00294 amu's
Mass of first neutron in 1H3	1.00628 amu's
Mass of second neutron in 1H3	1.00628 amu's
Mass of 1H3	3.01605 amu's

Now add the total mass lost to the mass of the isotope 1H3 to find the total mass of all three isotopes of 1H1.

Total mass lost creating 1H3	0.00744 amu
Mass of 1H3	3.01605 amu's
Total mass of three isotopes of 1H1	3.02349 amu's

Sub-step 15-3-3

1H3

FORMULA

1H1 + 1H1 + 1H1 = 1H3, two 217 rays, two 100 rays, -2e

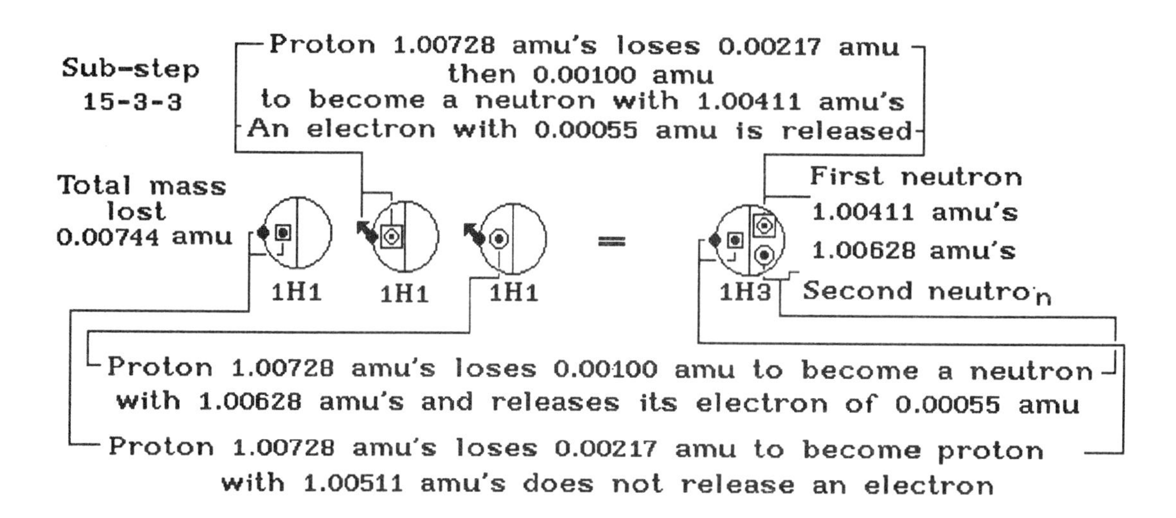

There are three isotopes of 1H1 used to create the isotope 1H3. Each isotope of 1H1 has 1.00783 amu's. The isotope 1H3 has 3.01605 amu's. Each isotope of 1H1 has one electron with 0.00055 amu and one proton with 1.00728 amu's for a total mass of 1.00783 amu's. The isotope 1H3 has one electron with 0.00055 amu, one proton with 1.00511 amu's and two neutrons. The first neutron has 1.00411 amu's and the second one has 1.00628 amu's for a total mass of 3.01605 amu's.

The drawing shows the first isotope of 1H1's electron as a dot on the left side of the big circle. The proton is drawn as a dot with a square around it on the left side of the vertical line. The square around the dot means that this proton has lost some mass. The electron and proton are drawn in the isotope 1H1 and in the isotope 1H3 in a like manner.

The second isotope of 1H1's electron is drawn as a dot with an arrow pointing away from it. The arrow indicates that this electron has been released. Because this electron has been released, it is not shown in the isotope 1H3. The proton is drawn as a dot with a small circle around it and then there is a square around the small circle and the dot. They are drawn to the left of the vertical line. The square means that this proton has lost some mass that does not release its electron. This mass is lost as a single unit. The small circle means that 0.00100 amu has been lost and the electron is released. This proton becomes the first neutron in the isotope 1H3 and it is drawn in the same manner as the proton in the second isotope of 1H1 except it is on the right side of the vertical line and above the second neutron.

The third isotope of 1H1's electron is drawn as a dot on the left side of the big circle with an arrow pointing away from it. The arrow indicates that the electron has been released. Because this electron has been released, it does not show up in the isotope 1H3. The proton is drawn as a dot on the left side of the vertical line with a small circle around it. The small circle means that this proton has lost 0.00100 amu and released its electron. This proton and second neutron are drawn alike in the third isotope of 1H1 and 1H3. The second neutron is drawn on the right side of the vertical line and below the first neutron.

The first isotope of 1H1 loses half of the 0.00434 amu that was lost by the first isotope in the step just given. It loses 0.0217 amu's which does not release its electron. This electron becomes the electron in an orbital in the isotope 1H3. When the proton loses this mass, it becomes the proton with 1.00511 amu's in the isotope 1H3.

The second isotope of 1H1 loses the same mass as the first isotope of 1H1. When it loses 0.00217 amu, it does not release its electron. It then loses 0.00100 amu and this lost mass does release the electron with 0.00055 amu. This isotope of 1H1 has lost 0.00317 amu. It becomes the first neutron in the isotope 1H3 with 1.00411 amu's.

The third isotope of 1H1 has its proton lose only the 0.00100 amu which releases its electron with 0.00055 amu. The proton becomes the second neutron in the isotope 1H3 with 1.00628 amu's.

The first isotope of 1H1 loses 0.00217 amu, the second isotope of 1H1 loses 0.00317 amu and the third isotope of 1H1 loses 0.00100 amu for a total of 0.00744 amu.

The drawings show the isotope 1H3 with its electron on the left side of the big circle and the proton is inside the circle on the left side of the vertical line. The two neutrons are inside the big circle on the right side of the vertical line.

In this step, all of the isotopes of 1H1 are used in a different manner so there is no way any two isotopes of 1H1 can be put in parenthesis.

By adding the masses of the three isotopes of 1H1, the total mass needed to create the isotope 1H3 will be found.

Mass of first 1H1	1.00783 amu's
Mass of second 1H1	1.00783 amu's
Mass of third 1H1	1.00783 amu's
Total mass creating 1H3	3.02349 amu's

By subtracting the mass of the isotope 1H3 from the total mass needed to create the isotope 1H3, the mass that is lost can be found.

Total mass creating 1H3	3.02349 amu's
Mass of 1H3	3.01605 amu's
Total mass lost creating 1H3	0.00744 amu

This total mass lost creating the isotope 1H3 is not lost as a single unit. It is composed of smaller sub-units. To find these sub-units of mass, the different masses lost by each isotope of 1H1 must be found.

In this step, I have changed the first isotope of 1H1 lost mass from 0.00434 amu to half of this amount. The other half is lost by the second isotope of 1H1. I have kept the second and third isotope of 1H1 losing their electrons and losing 0.00100 amu from each. At this point in my write-up, the only thing known is that two electrons are lost and with them 0.00100 amu is lost from each proton from the second and third isotope of 1H1. I therefore know that there are 0.00200 amu lost plus the 0.00110 amu lost by the two electrons. This is equal to 0.00310 amu lost from the total mass that is lost creating the isotope 1H3.

Lost mass releasing electron from second 1H1	0.00100 amu
Lost mass releasing electron from third 1H1	0.00100 amu
Electron mass lost by both 1H1's	0.00110 amu
Total mass lost by both 1H1's	0.00310 amu

Total mass lost creating 1H3	0.00744 amu
Total mass lost by both 1H1's	0.00310 amu
Mass lost not releasing an electron	0.00434 amu

This mass that does not release an electron is lost by the first and second isotope of 1H1's protons. The first isotope lost half of the total mass lost that does not release an electron while the second isotope of 1H1 lost the other half.

$0.5(0.00434) = 0.00217$ amu lost by each proton in the first and second isotopes of 1H1.

The two lost masses of 0.00217 amu are emitted as two 217 rays and the two 0.00100 amu are emitted as two 100 rays.

Now add all of the lost masses of the three isotopes of 1H1 and they should agree with the mass lost as rays creating the isotope of 1H3.

Mass lost by first 1H1	0.00217 amu
Mass lost by second 1H1	0.00317 amu
Mass lost by third 1H1	0.00100 amu
Total mass lost creating 1H3	0.00634 amu

Since the first isotope of 1H1's proton lost 0.00217 amu, subtract this amount from the proton to find the mass of the proton in the isotope 1H3.

Mass of proton in first 1H1	1.00728 amu's
Mass lost by it	0.00217 amu
Mass of proton in 1H3	1.00511 amu's

The second isotope of 1H1's proton lost 0.00217 amu which did not release its electron and it then lost 0.00100 amu which did. Add these lost masses and subtract their total from the mass of the proton to find the mass of the first neutron in the isotope 1H3. Because there is some mass lost which did not release an electron, this mass must be shown as a single unit and because there is also some mass that did release an electron, it must also be shown as a single unit of mass. Do not combine these masses as a single unit because as a single unit, no electron will be released.

Mass lost not releasing an electron by second 1H1	0.00217 amu
Mass lost releasing an electron by second 1H1	0.00100 amu
Total mass lost by second 1H1	0.00317 amu

Mass of proton in second 1H1	1.00728 amu's
Mass lost by it	0.00317 amu
Mass of first neutron in 1H3	1.00411 amu's

The third isotope of 1H1's proton lost only the mass needed to release its electron. This proton becomes the second neutron in the isotope 1H3.

Mass of proton in third 1H1	1.00728 amu's
Mass lost by it	0.00100 amu
Mass of second neutron in 1H3	1.00628 amu's

All of the masses for the parts that make up the isotope 1H3 are now known. By adding these masses, the mass of the isotope 1H3 can be found.

Mass of electron in 1H3	0.00055 amu
Mass of proton in 1H3	1.00511 amu's
Mass of first neutron in 1H3	1.00411 amu's
Mass of second neutron in 1H3	1.00628 amu's
Mass of 1H3	3.01605 amu's

Now add the total lost mass to the mass of the isotope 1H3 and it will be the same as the total mass of the three isotopes of 1H1.

Total mass lost creating 1H3	0.00744 amu
Mass of 1H3	3.01605 amu's
Total mass of three 1H1's	3.02349 amu's

Sub-step 14-3-2

1H3

FORMULA

1H1 + 2(1H1) = 1H3, two 217 rays, two 100 rays, -2e

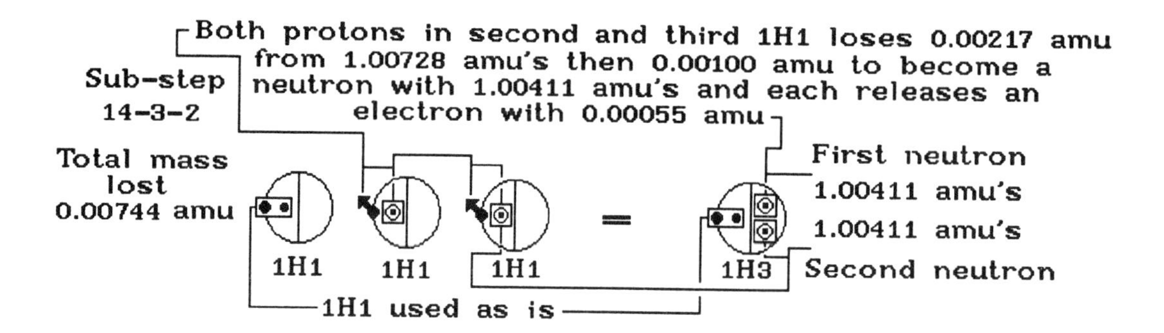

The first isotope of 1H1 has its electron and proton drawn as dots with a rectangle drawn around them. The electron dot is on the left side of the big circle and the proton is drawn to the left of the vertical line. The rectangle means that the electron and proton is used as the isotope 1H1 in the isotope 1H3 and that it can be dropped out at some later time if the isotope 1H3 is split up. The rectangle is drawn in the isotope 1H1 and 1H3 to indicate that they are used in a like manner.

The second and third isotope of 1H1's electron is drawn as a dot on the left side of the big circle. They each have an arrow pointing away from them to indicate that they are released. Each proton is drawn on the left side of the vertical line with a small circle around it, then a square around both of them. The square indicates that some mass is lost that does not release its electron. The small circle indicates that 0.00100 amu has been lost and the electron has been released. The protons in the second and third isotope of 1H1 and the first and second neutron in the isotope 1H3 are drawn in the isotope 1H3 in a like manner to indicate that the protons became the neutrons. These two neutrons are drawn to the right of the vertical line with the first neutron above the second one.

Because the second and third isotopes of 1H1 are used in the same manner, they can be put in parenthesis with a number two in front.

The first thing to know is the amount of mass needed to create the isotope 1H3. To find this mass, add the masses of the three isotopes of 1H1.

Mass of first 1H1	1.00783 amu's
Mass of second 1H1	1.00783 amu's
Mass of third 1H1	1.00783 amu's
Mass needed to create 1H3	3.02349 amu's

To find the amount of mass lost creating the isotope 1H3, subtract the mass of the isotope 1H3 from the mass needed to create the isotope 1H3.

Mass needed to create 1H3	3.02349 amu's
Mass of 1H3	3.01605 amu's
Total mass lost creating 1H3	0.00744 amu

The first isotope of 1H1 is used as is, which means it's proton is not changed into a neutron. To find the mass of the two neutrons found in the isotope 1H3, subtract the mass of the first isotope of 1H1.

Mass of 1H3	3.01605 amu's
Mass of first 1H1	1.00783 amu's
Mass of two neutrons in 1H3	2.00822 amu's

Because the second and third isotope of 1H1's proton each lost the same amount of mass and each became a neutron in the isotope 1H3, the neutrons have the same mass. Because the neutrons have the same mass, the 2.00822 amu's can be divided in half and this will give the mass of each neutron found in the isotope 1H3.

0.5(2.00822) = 1.00411 amu's for each neutron in the isotope 1H3.

Now add the masses of the two neutrons and the isotope of 1H1 to show that they equal the mass of the isotope 1H3

Mass of 1H1 in 1H3	1.00783 amu's
Mass of first neutron in 1H3	1.00411 amu's
Mass of second neutron in 1H3	1.00411 amu's
Mass of 1H3	3.01605 amu's

Since the electron and proton in the first isotope of 1H1 lost no mass, all the lost mass must come from the other two isotopes of 1H1. They both lose the same amount of mass. Therefore, divide the total amount of mass lost in half to find the amount of mass each isotope of 1H1 lost.

0.5(0.00744) = 0.00372 amu lost by each 1H1.

Each proton in the second and third isotopes of 1H1 lost 0.00100 amu to release their electrons. The 0.00100 amu can be subtracted from the mass lost by each proton and the remaining mass will be the mass that must be lost before the electron can be released. This remaining lost mass will include the mass of the electron that is released. Remember that the total mass lost included the released electrons because their mass is also lost.

Mass lost by proton in second and third 1H1	0.00372 amu
Mass lost that releases an electron	0.00100 amu
Remaining mass lost by each 1H1	0.00272 amu

Remaining mass lost by each 1H1	0.00272 amu
Mass of released electron from each 1H1	0.00055 amu
Mass lost that will not release an electron	0.00217 amu

The 0.00217 amu is lost as a 217 ray first and then the 0.00100 amu is lost as a 100 ray which releases the electron. This is done twice so there are two 217 rays and two 100 rays emitted.

By subtracting the lost mass from the second or third isotope of 1H1's proton, the remaining mass will be the first or second neutron in the isotope 1H3.

Mass of second or third 1H1	1.00783 amu's
Mass lost by second or third 1H1's proton	0.00372 amu
Mass of first or second neutron in 1H3	1.00411 amu's

Each of the second and third isotopes of 1H1 lost 0.00317 amu from their protons and the two added together give the mass lost by the protons. This will be the real mass lost by each of the second and third isotopes of 1H1. It is converted to pure energy which is magnetism. The electrons are released as electrons and they are therefore not converted to pure energy.

Mass lost by proton in second 1H1	0.00317 amu
Mass lost by proton in third 1H1	0.00317 amu
Total mass lost by protons in both 1H1's	0.00634 amu

Before an electron can be released, the magnetic field around the second and third isotopes of 1H1 must be decreased. This is why the 0.00217 amu is lost first. After this mass has been converted to pure energy, the 0.00100 amu is then lost by converting it to pure energy. This is done by deconcentrating the mass that is to be lost. The remaining mass of 1.00411 amu's stays concentrated.

Now add the different masses of the parts that make the isotope 1H3, including the real lost mass plus the mass of the electrons that have been released. This will give the total mass of the three isotopes of 1H1 that were used to create the isotope 1H3.

Mass of 1H3's single electron	0.00055 amu
Mass of 1H3's single proton	1.00728 amu's
Mass of 1H3's first neutron	1.00411 amu's
Mass of 1H3's second neutron	1.00411 amu's
Mass lost creating 1H3	0.00634 amu
Mass lost by the two electrons released	0.00110 amu
Total mass used in creating 1H3	3.02349 amu's

Now add the single isotope of 1H1 that was used as is to the mass of the two isotopes of 1H1 that lost the same amount of mass. Next subtract the total mass lost from the total mass needed to create the isotope of 1H3 to find the mass of the isotope 1H3. Seeing the figures in this manner will help the reader understand how the isotope 1H3 was created.

Mass of 1H1 used as is	1.00783 amu's
Mass of two 1H1's that lost mass, became neutrons	2.01566 amu's
Total mass needed to create 1H3	3.02349 amu's

Total mass needed to create 1H3	3.02349 amu's
Total mass lost creating 1H3	0.00744 amu
Mass of 1H3	3.01605 amu's

Sub-step 13-3-1

1H3

FORMULA

1H1 + 1H1 + 1H1 = 1H3, one 434 ray, two 100 rays, -2e

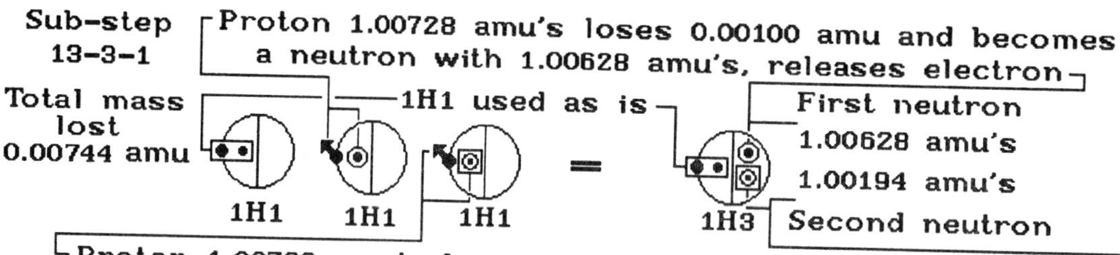

There are three isotopes of 1H1 used to create the isotope 1H3. Each isotope has 1.00783 amu's. The isotope 1H3 has 3.01605 amu's. Each isotope has one electron with 0.00055 amu and one proton with 1.00728 amu's. The isotope 1H3 has one electron and one proton plus there are two neutrons. The first neutron has 1.00628 amu's and the second one has 1.00194 amu's.

The drawing shows the first isotope of 1H1 used as is with the electron drawn on the left side of a big circle and the proton drawn to the left of a vertical line. The proton is inside the big circle.

The electron and proton are drawn in a rectangle to show that they form the isotope 1H1 within the isotope 1H3. This rectangle also indicates that the isotope 1H1 can drop out of the isotope 1H3 at some later time.

The second isotope of 1H1's electron is drawn as a dot on the big circle with an arrow pointing away from it. The arrow indicates the electron has been released. The proton is drawn as a dot with a small circle around it. It is drawn to the left of the vertical line inside of the big circle. It is drawn the same way in the isotope 1H3 except now it is on the right side of the vertical line as the first neutron.

The third isotope of 1H1's electron is drawn on the left side of the big circle with an arrow pointing away from it. The arrow indicates that the electron has been released. The proton is drawn as a dot with a small circle around it plus a square is around both. It is drawn to the left of the vertical line inside the big circle. It is also drawn the same way in the isotope 1H3 except it is now drawn on the right side of the vertical line within the big circle. This means that this proton in the third isotope 1H1 has changed from a proton to the second neutron in the isotope 1H3.

The square represents the loss of some mass that does not release an electron and the small circle represents the loss of 0.00100 amu which will release an electron.

The first isotope of 1H1 does not lose any mass and it is found in the isotope 1H3 as is. It has the mass of 1.00783 amu's.

The second isotope of 1H1 has its proton lose 0.00100 amu and releases its electron. It becomes the first neutron in the isotope 1H3 with 1.00628 amu's.

The third isotope of 1H1 has its proton lose 0.00434 amu which does not release an electron. This proton then loses 0.00100 amu and this time it does release its electron. The proton then becomes the second neutron in the isotope 1H3 with 1.00194 amu's.

All three isotopes of 1H1 are used differently which means that no two isotopes of 1H1 can be put in parenthesis.

The first thing is to find the mass needed to create the isotope 1H3. There are three isotopes of 1H1. Add these masses to find this needed mass.

Mass of first of 1H1	1.00783 amu's
Mass of second 1H1	1.00783 amu's
Mass of third 1H1	1.00783 amu's
Mass needed to create 1H3	3.02349 amu's

Now subtract the isotope 1H3 from the mass needed to create the isotope 1H3 to find the total mass lost.

Mass needed to create 1H3	3.02349 amu's
Mass of 1H3	3.01605 amu's
Mass lost creating 1H3	0.00744 amu

This mass lost creating the isotope 1H3 is not lost as a single unit but it is lost as smaller sub-units. To find these smaller sub-units, the mass lost by each isotope of 1H1 must be found.

The first isotope of 1H1 lost no mass so it does not have any sub-units.

The second isotope of 1H1 lost 0.00100 amu and it also lost the mass of the electron it released. The third isotope of 1H1 lost some of its mass that did not release its electron and as of now this mass is unknown. This isotope also lost 0.00100 amu and released its electron. This mass is known. Now add all the known masses and subtract their total from the total mass lost creating the isotope 1H3.

Mass lost by second 1H1's proton	0.00100 amu
Mass lost by third 1H1's proton	0.00100 amu
Mass lost by second and third 1H1's electrons	0.00110 amu
Total mass lost by second and third 1H1's	0.00310 amu

Total mass lost creating 1H3	0.00744 amu's
Total known mass lost by second and third 1H1's	0.00310 amu
Remaining mass to be lost creating 1H3	0.00434 amu

The first isotope of 1H1 is used as is so its mass is 1.00783 amu in the isotope 1H3.

The second isotope of 1H1 lost 0.00100 amu by its proton and this proton then becomes the first neutron in the isotope 1H3.

Mass of second 1H1's proton	1.00728 amu's
Mass lost by second 1H1's proton	0.00100 amu
Mass of first neutron in 1H3	1.00628 amu's

The third isotope of 1H1 lost some mass that did not release its electron. This mass is now known to be 0.00434 amu. It is lost first to weaken the magnetic field around the isotope 1H1 so that when this isotope of 1H1 loses 0.00100 amu, it will release its electron. This isotope of 1H1's proton has now lost 0.00534 amu to become the second neutron in the isotope 1H3.

Mass lost not releasing electron in third 1H1's proton	0.00434 amu
Mass lost releasing electron in third 1H1's proton	0.00100 amu
Mass lost by third 1H1's proton creating 1H3	0.00534 amu

Now subtract the mass lost by the third isotope of 1H1's proton to find the mass of the second neutron in the isotope 1H3.

Mass of third 1H1's proton	1.00728 amu's
Mass lost by third 1H1's proton	0.00534 amu
Mass of second neutron in 1H3	1.00194 amu's

All of the parts that make the isotope 1H3 are now known. Add all of these parts to find the mass of the isotope 1H3.

Mass of 1H1 in 1H3	1.00783 amu's
Mass of first neutron in 1H3	1.00628 amu's
Mass of second neutron in 1H3	1.00194 amu's
Mass of 1H3	3.01605 amu's

By adding the total lost mass to the mass of the isotope 1H3, the total mass of the three isotopes of 1H1 will be found.

Mass of 1H3	3.01605 amu's
Total mass lost creating 1H3	0.00744 amu
Total mass of three 1H1's needed to create 1H3	3.02349 amu's

Drawings showing **STEP 3** using three isotopes of 1H1 to create the isotope 1H3.

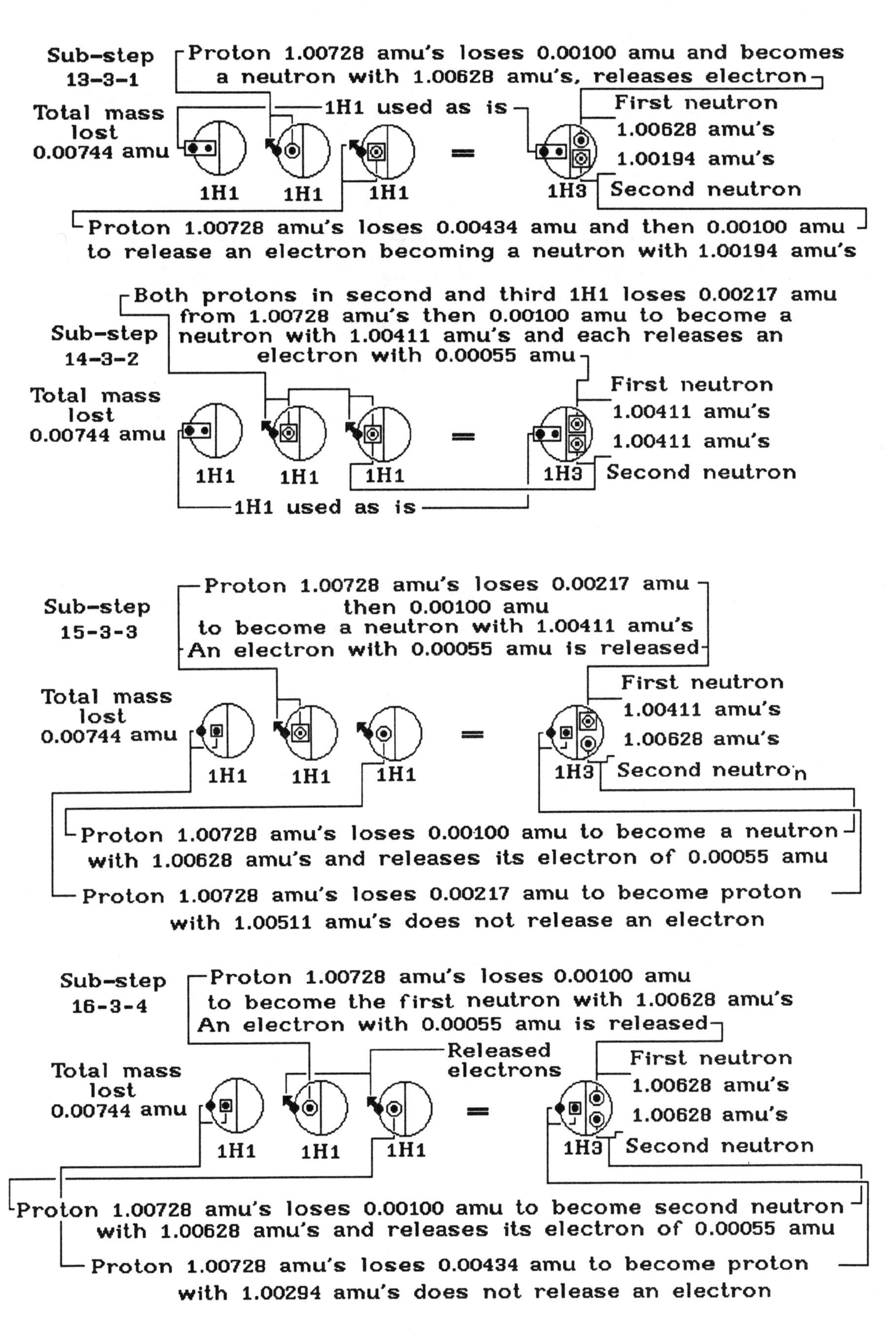

STEP 2

1H3

FORMULA

1H2 + 1H1 = 1H3

2. 1H2 + 1H1 = 1H3

5-2-1. 1H2 + 1H1 = 1H3, one 434 ray, one 100 ray, -1e
6-2-2. 1H2 + 1H1 = 1H3, one 434 ray, one 100 ray, -1e
7-2-3. 1H2 + 1H1 = 1H3, one 434 ray, one 100 ray, -1e
8-2-4. 1H2 + 1H1 = 1H3, one 434 ray, one 100 ray, -1e
9-2-5. 1H2 + 1H1 = 1H3, one 434 ray, one 100 ray, -1e
10-2-6. 1H2 + 1H1 = 1H3, one 434 ray, one 100 ray, -1e
11-2-7. 1H2 + 1H1 = 1H3, two 217 rays, one 100 ray, -1e
12-2-8. 1H2 + 1H1 = 1H3, two 217 rays, one 100 ray, -1e

Sub-step 12-2-8

1H3

FORMULA

1H2 + 1H1 = 1H3, gamma 217 ray, alpha 217 ray, one 100 ray, -1e

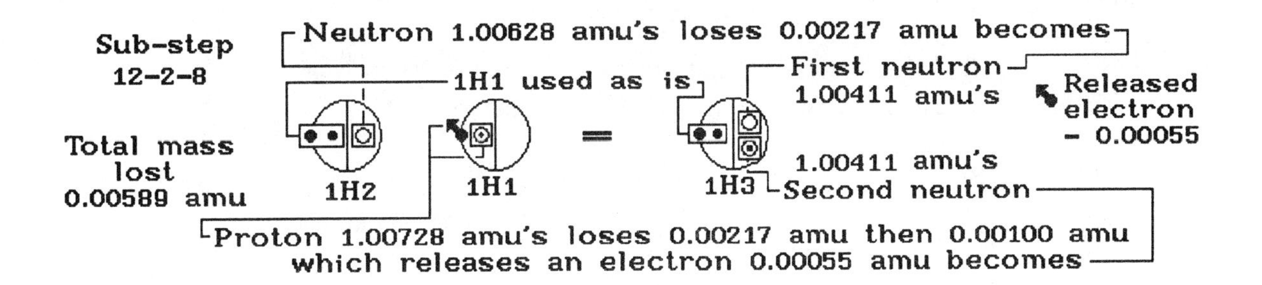

The isotope 1H2 has its electron and proton used as is while its neutron with 1.00628 amu's loses 0.00217 amu to become the first neutron in the isotope 1H3 with 1.00411 amu's. In the isotope 1H1, the proton with 1.00728 amu's loses 0.00217 amu which does not release its electron but it then loses 0.00100 amu and now it does lose its electron. This proton now is the second neutron in the isotope 1H3 with 1.00194 amu's. This makes a loss of 0.00434 amu and 0.00100 amu that is really lost plus the lost mass of the electron which is 0.00055 amu for a total of 0.00589 amu.

In the case of the neutron in the isotope 1H2 losing mass, it remains a neutron with a new mass of 1.00411 amu's in the isotope 1H3. This is not true for the isotope 1H1. Here an isotope of 1H1 has its proton becoming a neutron by losing enough mass to create the second neutron with a different mass of 1.00194 amu's.

Because the neutron in the isotope 1H2 lost 0.00217 amu, the 217 ray that is emitted has no external magnetic field. When the proton in the isotope 1H1 emitted a 217 ray, it has an external magnetic field.

The drawing shows the electron and proton in the isotope 1H2 with a rectangle drawn around them and it is drawn the same way in the isotope 1H3 to show that they are used in a like manner and that they can drop out of the isotope 1H3 at some later time.

The single neutron in the isotope 1H2 loses some of its mass to become the first neutron in the isotope 1H3. The neutron is drawn as a small circle on the right side of the vertical line in the isotope 1H2. To show that it lost some mass, a square is drawn around it and it is drawn the same way in the isotope 1H3.

The isotope 1H1 has its electron drawn on the big circle as a dot with an arrow pointing away from it. The arrow indicates that the electron has been released. The proton found in this isotope is drawn as a dot on the left side of the vertical line. The vertical line is inside the big circle. The proton has a small circle drawn around the dot and a square drawn around both of them. The square shows that this proton has lost some of its mass that does not release its electron. The small circle shows that 0.00100 amu has been lost which releases its electron. This drawing for the proton in the isotope 1H1 is drawn in the same manner in the isotope 1H3 except it now is on the right side of the vertical line to show that this proton is now a neutron.

To find the amount of mass needed to create the isotope 1H3 from one isotope of 1H2 and one isotope of 1H1 add their masses.

Mass of 1H2	2.01411 amu's
Mass of 1H1	1.00783 amu's
Mass needed to create 1H3	3.02194 amu's

Now subtract the mass of the isotope 1H3 from the mass needed to create the isotope 1H3. This will give the total mass lost doing this.

Mass needed to create 1H3	3.02194 amu's
Mass of 1H3	3.01605 amu's
Total mass lost creating 1H3	0.00589 amu

This lost mass of 0.00589 amu is not a single unit of mass. It is made up of sub-units. To find these sub-units, the mass that is lost by each of the isotopes of 1H2 and 1H1 must be found. I do not know the mass of the two neutrons in the isotope 1H3 at this time. In this step, I specify that these two neutrons have the same mass. This is done this way because the later steps will not have the two neutrons with the same mass and in all the steps to be given, these parts will have different masses. Therefore, in this step, I state that the two neutrons in the isotope 1H3 have the same mass.

Since the mass of the electron and proton are known and the mass of the two neutrons are equal to each other, I can subtract the mass of the isotope 1H1 that is found intact within the isotope 1H3 from the mass of the isotope 1H3. This will give me the mass of the two neutrons.

Mass of 1H3	3.01605 amu's
Mass of 1H1 found within 1H3	1.00783 amu's
Mass of the two neutrons in 1H3	2.00822 amu's

The mass of the two neutrons in the isotope 1H3 are equal to each other. I can therefore divide their mass in half to find the mass of each neutron.

0.5(2.00822) = 1.00411 amu's for each neutron in the isotope 1H3.

The mass of the single neutron in the isotope 1H2 is 1.00628 amu's. This neutron lost some of this mass and remained a neutron in the isotope 1H3. The proton in this isotope did not release its electron therefore it did not lose 0.00100 amu. It could lose some mass but in this step it does not.

The mass of the first neutron in the isotope 1H3 is 1.00411 amu's. I can therefore subtract its mass from the mass of the neutron found in the isotope 1H2 to find the mass this neutron lost.

Mass of single neutron in 1H2	1.00628 amu's
Mass of first neutron in 1H3	1.00411 amu's
Mass lost by single neutron in 1H2	0.00217 amu

The proton in the isotope 1H1 lost some of its mass and became the second neutron in the isotope 1H3. This proton lost some of its mass and released its electron. This lost mass is known to be 0.00100 amu. I can now add the mass lost by the isotope 1H2 and this proton and then subtract this total lost mass from the total mass lost creating the isotope 1H3.

Mass lost by neutron in 1H2	0.00217 amu
Mass lost by proton in 1H1 releasing electron	0.00100 amu
Total known mass lost by 1H2 and 1H1	0.00317 amu

Total mass lost creating 1H3	0.00589 amu
Total known mass lost by 1H2 and 1H1	0.00317 amu
Remaining lost mass by 1H1	0.00272 amu

This remaining lost mass by the isotope 1H1 includes the electron that has been released. Now subtract the mass of this electron.

Remaining lost mass by proton in 1H1	0.00272 amu
Mass of released electron from 1H1	0.00055 amu
Mass lost from 1H1 that does not release an electron	0.00217 amu

I could also find this lost mass by subtracting the 0.00100 amu from the proton in the isotope 1H1 and then subtract this new mass of the proton from the known mass of the second neutron in the isotope 1H3.

Mass of proton in 1H1	1.00728 amu's
Mass lost by proton releasing electron	0.00100 amu
Remaining mass of proton in 1H1	1.00628 amu's

Remaining mass of proton in 1H1	1.00628 amu's
Mass of second neutron in 1H3	1.00411 amu's

Last remaining mass lost by proton in 1H1	0.00217 amu

Note doing it this way causes the proton in the isotope 1H1 to become a neutron with 1.00628 amu's before it loses the last lost mass. This way does not allow the magnetic field to be decreased so that the electron can be released. Here the electron is released first and then the remaining mass of the proton is lost.

The last way to find the mass lost by the proton in the isotope 1H1 is to subtract the known mass of the second neutron in the isotope 1H3 from the mass of the proton in the isotope 1H1.

Mass of proton in 1H1	1.00728 amu's
Mass of second neutron in 1H3	1.00411 amu's
Mass lost by proton in 1H1	0.00317 amu

This mass lost by the proton in the isotope 1H1 includes the lost mass that released the electron. Now subtract this lost mass from the remaining lost mass of the isotope 1H1.

Mass lost by proton in 1H1	0.00317 amu
Mass lost releasing electron in 1H1	0.00100 amu
Last remaining mass lost by proton in 1H1	0.00217 amu

Now all of the parts of the isotope 1H3 are known as well as all of the sub-units of lost mass. Now add all of these masses to find the total mass of the two isotopes of 1H2 and 1H1.

Mass of 1H1 in 1H3	1.00783 amu's
Mass of first neutron in 1H3	1.00411 amu's
Mass of second neutron in 1H3	1.00411 amu's
Mass lost releasing electron in 1H1	0.00100 amu
Mass lost by neutron in 1H2	0.00217 amu
Mass lost by proton in 1H1 not releasing neutron	0.00217 amu
Mass of electron that is released	0.00055 amu
Total mass of 1H2 and 1H3	3.02194 amu's

Sub-step 11-2-7

1H3

FORMULA

1H2 + 1H1 = 1H3, alpha 217 ray, gamma 217 ray, one 100 ray, -1e

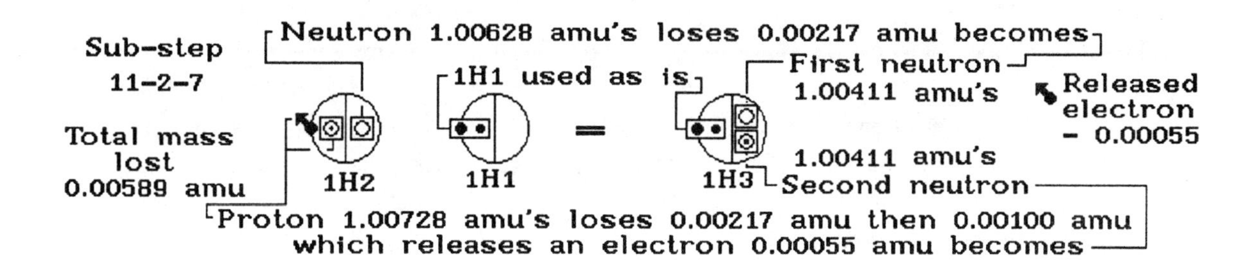

There is one isotope of 1H2 with 2.01411 amu's and one isotope of 1H1 with 1.00783 amu's creating an isotope of 1H3 with 3.01605 amu's. The isotope 1H2 has one electron with 0.00055 amu, one proton with 1.00728 amu's and one neutron with 1.00628 amu's. The isotope 1H1 has one electron with 0.00055 amu and one proton with 1.00728 amu's. The isotope 1H3 has one electron with 0.00055 amu, one proton with 1.00728 amu's and two neutrons. Both neutrons have 1.00411 amu's.

The electron in the isotope 1H2 is drawn as a dot on the left side big circle with an arrow pointing away from it. The arrow indicates that the electron has been released. The proton in this isotope is drawn on the left side of the vertical line inside the big circle as a dot with a small circle around it then a square around both. The square indicates that this proton has lost some mass and did not release its electron. The small circle indicates that 0.00100 amu has been lost and when this happens, the electron is released. When this proton lost the 0.00100 amu, it became the second neutron in the isotope 1H3. This proton is drawn the same way in the isotope 1H3 to show that this proton becomes the second neutron in the isotope 1H3. The single neutron found in this isotope of 1H2 is drawn on the right side of the vertical line inside the big circle. It is drawn with a square around it to indicate that it has lost some mass. It is also drawn this way in the isotope 1H3 to show that the neutron in the isotope 1H2 is the first neutron in the isotope 1H3.

Because the proton in the isotope 1H2 lost 0.00217 amu, the 217 ray that is emitted has an external magnetic field. When the neutron in the isotope 1H2 emits a 217 ray, it has no external magnetic field.

The isotope 1H1 has its electron drawn on the left side of the big circle and the proton is drawn on the left side of the vertical line inside the big circle. Both of them have a rectangle drawn around them in the isotope 1H1 and 1H3. The rectangle shows that the electron and proton are both used in a like manner and the isotope 1H1 is intact within the isotope 1H3 and that it can drop out at some later time.

To find the amount of mass needed to create the isotope 1H3, add the mass of the isotope 1H2 and 1H1.

Mass of 1H2	2.01411 amu's
Mass of 1H1	1.00783 amu's
Total mass needed to create 1H3	3.02194 amu's

Now subtract the mass of the isotope 1H3 from the total mass needed to create 1H3. The remainder will be the total mass lost creating the isotope 1H3.

Total mass needed to create 1H3	3.02194 amu's
Mass of 1H3	3.01605 amu's
Mass lost creating 1H3	0.00589 amu

As given in the last step, I specify that the two neutrons in the isotope have the same mass of 1.00411 amu's. In the last step, the isotope 1H1 release its electron and the proton became the first neutron in the isotope 1H3 while in this step, it is the proton in the isotope 1H2 which becomes the second neutron.

This lost mass of 0.00589 amu is not a single unit of mass. It is made up of sub-units. To find these sub-units, the mass that is lost by the isotopes of 1H2 must be found. I do not know the mass of the two neutrons in the isotope 1H3 at this time. In this step, I specify that these two neutrons have the same mass. This is done this way because the later steps will not have the two neutrons with the same mass and in all the steps to be given, these parts will have different masses. Therefore in this step, I state that the two neutrons in the isotope 1H3 have the same mass.

Since the mass of the electron and proton are known and the mass of the two neutrons are equal to each other, I can subtract the mass of the isotope 1H1 that is found intact within the isotope 1H3 from the mass of the isotope 1H3. This will give the mass of the two neutrons.

Mass of 1H3	3.01605 amu's
Mass of 1H1 found within 1H3	1.00783 amu's
Mass of the two neutrons in 1H3	2.00822 amu's

The mass of the two neutrons in the isotope 1H3 are equal to each other. I can therefore divide their mass in half to find the mass of each neutron.

$0.5(2.00822) = 1.00411$ amu's for each neutron in the isotope 1H3.

The mass of the single neutron in the isotope 1H2 is known. It is 1.00628 amu's and it lost some of this mass and remained a neutron in the isotope 1H3. I also know that the proton in this isotope did release its electron therefore it lost 0.00100 amu. I now know the mass of the first neutron in the isotope 1H3 to be 1.00411 amu's. Therefore, subtract its mass from the mass of the neutron found in the isotope 1H2 to find the mass this neutron lost.

Mass of single neutron in 1H2	1.00628 amu's
Mass of first neutron in 1H3	1.00411 amu's
Mass lost by single neutron in 1H1	0.00217 amu

The proton in the isotope 1H2 lost some of its mass and became the second neutron in the isotope 1H3. This proton lost some of its mass and released its electron. This lost mass is known to be 0.00100 amu. Now add the mass lost by the isotope 1H2's neutron and this proton, then subtract their total lost mass from the total mass lost creating the isotope 1H3.

Mass lost by neutron in 1H2	0.00217 amu
Mass lost by proton in 1H2 releasing electron	0.00100 amu
Total known mass lost by 1H2	0.00317 amu

Total mass lost creating 1H3	0.00589 amu
Total known mass lost by 1H2	0.00317 amu
Remaining lost mass by 1H2	0.00272 amu

This remaining lost mass in the isotope 1H2 includes the electron that has been released. Now subtract the mass of this electron.

Remaining lost mass by 1H2	0.00272 amu
Mass of released electron from 1H2	0.00055 amu
Mass lost from 1H2 that does not release an electron	0.00217 amu

I can also find this lost mass by subtracting the 0.00100 amu from the proton in the isotope 1H2 and then subtract this new mass of the proton from the known mass of the second neutron in the isotope 1H3.

Mass of proton in 1H2	1.00728 amu's
Mass lost by proton releasing electron	0.00100 amu
Remaining mass of proton in 1H2	1.00628 amu's

Remaining mass of proton in 1H2	1.00628 amu's
Mass of first neutron in 1H3	1.00411 amu's
Last remaining mass lost by proton in 1H2	0.00217 amu

Note doing it this way causes the proton in the isotope 1H2 to become a neutron with 1.00628 amu's before it loses the last lost mass. This way does not allow the magnetic field to be decreased so that the electron can be released. Here the electron is released first and then the remaining mass of the proton is lost.

The last way to find the mass lost by the proton in the isotope 1H2 is to subtract the known mass of the second neutron in the isotope 1H3 from the mass of the proton in the isotope 1H2.

Mass of proton in 1H2	1.00728 amu's
Mass of first neutron in 1H3	1.00411 amu's
Mass lost by proton in 1H2	0.00317 amu

This mass lost by the proton in the isotope 1H2 includes the lost mass that released the electron. Now subtract this lost mass from the remaining lost mass of the isotope 1H2.

Mass lost by proton in 1H2	0.00317 amu
Mass lost releasing electron in 1H2	0.00100 amu
Last remaining mass lost by proton in 1H2	0.00217 amu

Now all of the parts of the isotope 1H3 are known as well as all of the sub-units of lost mass. Now add all of these masses to find the total mass of the two isotopes of 1H2 and 1H1.

Mass of 1H1 in 1H3	1.00783 amu's
Mass of first neutron in 1H3	1.00411 amu's
Mass of second neutron in 1H3	1.00411 amu's
Mass lost releasing electron in 1H2	0.00100 amu
Mass lost by neutron in 1H2	0.00217 amu
Mass lost by proton in 1H2 not releasing neutron	0.00217 amu
Mass of electron that is released	0.00055 amu
Total mass of 1H2 and 1H3	3.02194 amu's

Sub-step 10-2-6

1H3

FORMULA

1H2 + 1H1 = 1H3, one alpha 434 ray, one 100 ray, -1e

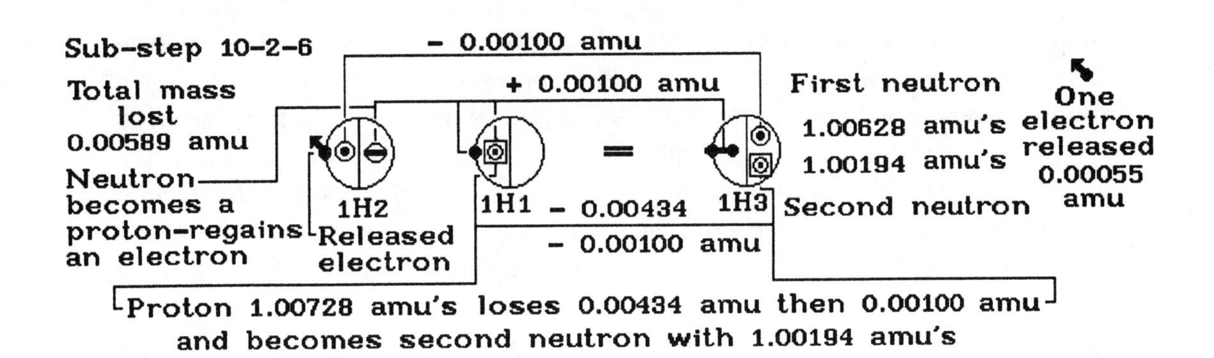

There is one isotope of 1H2 with 2.01411 amu's and one isotope of 1H1 with 1.00783 amu's creating an isotope of 1H3 with 3.01605 amu's. The isotope 1H2 has one electron with 0.00055 amu, one proton with 1.00728 amu's and one neutron with 1.00628 amu's. The isotope 1H1 has one electron with 0.00055 amu and one proton with 1.00728 amu's. The isotope 1H3 has one electron with 0.00055 amu, one proton with 1.00728 amu's and two neutrons. The first neutron has 1.00628 amu's and the second neutron has 1.00194 amu's.

It is possible for all protons to be changed into neutrons with the release of their electrons. The neutron in the isotope 1H2 will regain its lost mass to become a proton in the isotope 1H3 and the proton to then regain its electron. This means that the electron that is regained must belong to the proton in the isotope 1H1 because the proton in the isotope 1H2 will lose the mass that will release the electron that is not found in the isotope 1H3. This means that the electron lost by the isotope 1H1 is the electron found in the isotope 1H3. This would mean that the proton in the isotope 1H2 becomes the first neutron in the isotope 1H3. The neutron in the isotope 1H2 regains the lost mass from the proton in the isotope 1H1 to become the proton found in the isotope 1H3. This regained mass is the mass that released the electron from the proton in the isotope 1H1. Because the neutron in the isotope 1H2 regained this mass lost by the proton in the isotope 1H1, this mass is not lost. However the proton in the isotope 1H1 also loses some mass that will not release its electron which changes this proton into the second neutron found in the isotope 1H3. When the neutron in the isotope 1H2 changes into a proton, it regains the electron released by the proton in the isotope 1H1.

When this happens, the proton and neutron in the isotope 1H2 switched places in the isotope 1H3.

The electron in the isotope 1H2 is drawn on the left side of the big circle with an arrow pointing away from it. The arrow indicates that the electron has been released. The proton in this isotope is drawn as a dot on the left side of the vertical line within the big circle. The dot has a small circle drawn around it to show that this proton lost 0.00100 amu and released its electron. It leaves the proton when this proton becomes the first neutron in the isotope 1H3.

The neutron in the isotope 1H2 is drawn as a small circle with a line drawn through it on the right side of the vertical line within the big circle. The electron and proton in the isotope 1H1 have a line connecting them. The line drawn through the neutron in the isotope 1H2 represents the line connecting the electron and proton in the isotope 1H3 to show that the mass lost by the proton in the isotope 1H1 is regained by the neutron in the isotope 1H2. It show that the proton in the isotope 1H1 lost 0.00100 amu that is regained by the neutron in the isotope 1H2 which has 1.00628 amu's and this changes the neutron into a proton with the mass of 1.00728 amu's in the isotope 1H3. The electron is connected to the proton in the isotope 1H3 by the line to show that it came from the isotope 1H1 because the electron in the isotope 1H2 has been released and is not part of the isotope 1H3.

The proton with 1.00728 amu's in the isotope 1H1 also lost 0.00434 amu which does not release its electron. This lost mass is lost as an alpha 434 ray which is pure energy that is magnetism. The alpha 434 ray has an external magnetic field. The square drawn around the small circle of the dot that is the proton shows that 0.00434 amu has been lost and the small circle drawn around the dot represents the 0.00100 amu that is transferred to the neutron found in the isotope 1H2. The proton in the isotope 1H1 is drawn the same way in the isotope 1H3 as a neutron found in the isotope 1H3 to the right of the vertical line within the isotope 1H3. When the proton in the isotope 1H1 loses 0.00100 amu and 0.00434 amu, it changes from a proton with 1.00728 amu's to a neutron with 1.00194 amu's in the isotope 1H3. The electron in the isotopes 1H1 and 1H3 has 0.00055 amu and the electron released from the isotope 1H2 causes a loss of 0.00055 amu.

For a ray to be emitted, mass must be lost and none has been lost when the 0.00100 amu was transferred to the neutron in the isotope 1H2 from the isotope 1H1 so no ray can be emitted. The proton in the isotope 1H2 loses 0.00100 amu which causes a 100 ray to be emitted. The isotope 1H1 loses 0.00434 amu that causes an alpha 434 ray to be emitted.

To find the amount of mass needed to create the isotope 1H2, add the mass of the isotope 1H2 and 1H1.

Mass of 1H2	2.01411 amu's
Mass of 1H1	1.00783 amu's
Total mass needed to create 1H3	3.02194 amu's

Now subtract the mass of the isotope 1H3 from the total mass needed to create 1H3. The remainder will be the total mass lost creating the isotope 1H3.

Total mass needed to create 1H3	3.02194 amu's
Mass of 1H3	3.01605 amu's
Mass lost creating 1H3	0.00589 amu

I have stated that the proton in the isotope 1H1 will lose 0.00434 amu keep its electron. This decreases the magnetic bond between the electron and it proton but not enough to release the electron from the proton. The proton then transfers 0.00100 amu to the neutron in the isotope 1H2 which does break the magnetic bond between the electron and its proton. This electron is then regained by the neutron in the isotope 1H2 when it becomes a proton in the isotope 1H3. When the 0.00100 amu is transferred, the proton in the isotope 1H1 has its magnetic field become internal, making the proton in the isotope 1H1 into a neutron in the isotope 1H3.

Mass of proton in 1H2	1.00728 amu's
Mass releasing electron in 1H2	0.00100 amu
Mass of first neutron in 1H3	1.00628 amu's

The total mass lost is known to be 0.00589 amu and the mass lost by the proton in the isotope 1H2 can now be subtracted from it as well as the electron that has been released. The remainder will be the mass lost by the proton in the isotope 1H1.

Mass releasing electron from 1H2	0.00100 amu
Mass of released electron	0.00055 amu
Mass lost by 1H2	0.00155 amu

Total mass lost creating 1H3	0.00589 amu
Mass lost by 1H2	0.00155 amu
Mass lost by 1H1	0.00434 amu

This lost mass by the proton in the isotope 1H1 does not release the proton's electron. To release this electron, 0.00100 amu must be lost. When it is lost, this proton becomes the second neutron in the isotope 1H3. By adding the lost masses of 0.00434 amu and 0.00100 amu and then subtracting their total from the proton will give the mass of the second neutron in the isotope 1H3.

Mass lost by proton in 1H1 not releasing electron	0.00434 amu
Mass lost by proton in 1H1 releasing electron	0.00100 amu
Mass lost by proton becoming neutron in 1H3	0.00534 amu

Mass of proton in 1H1	1.00728 amu's
Mass lost by proton becoming neutron in 1H3	0.00534 amu
Mass of second neutron in 1H3	1.00194 amu's

The mass of the two neutrons in the isotope 1H3 is now known. By subtracting their mass from the mass of the isotope 1H3, the mass of the electron and proton can be found.

Mass of first neutron in 1H3	1.00628 amu's
Mass of second neutron in 1H3	1.00194 amu's
Total mass of two neutrons in 1H3	2.00822 amu's

Mass of 1H3	3.01605 amu's
Total mass of two neutrons in 1H3	2.00822 amu's
Mass of electron and proton in 1H3	1.00783 amu's

Since two electrons have been released and only one is needed, and both protons have lost mass, neither proton can be the proton found in the isotope 1H3. The only place this proton can come from is from the single neutron in the isotope 1H2. This neutron is known to have the mass of 1.00628 amu's. The mass of the electron in the isotope 1H3 is known to be 0.00055 amu which means that the proton must have 1.00728 amu's.

Mass of electron and proton in 1H3	1.00783 amu's
Mass of electron in 1H3	0.00055 amu
Mass of proton in 1H3	1.00728 amu's

Because the neutron in the isotope 1H2 has 1.00628 amu's and it is to become the proton in the isotope 1H3, it must regain 0.00100 amu and when it does this, it becomes a proton and regains an electron.

Mass of neutron in 1H2	1.00628 amu's
Mass regained from 1H1's proton	0.00100 amu
Mass of proton in 1H3	1.00728 amu's

This regained 0.00100 amu must come from the proton in the isotope 1H1 and the electron regained by the proton in the isotope 1H3 must also come from the isotope 1H1.

Now the mass of all parts of the isotope of 1H3 are known and how they got where they are. Add these known mass to find the mass of the isotope 1H3.

Mass of electron in 1H3	0.00055 amu
Mass of proton in 1H3	1.00728 amu's
Mass of first neutron in 1H3	1.00628 amu's
Mass of second neutron in 1H3	1.00194 amu's
Mass of 1H3	3.00605 amu's

It should be noted that the regained mass did not come from outside the two isotopes of 1H2 and 1H1. This step shows there is an internal exchange of mass and therefore when the 0.00434 amu is lost, it is the only 434 ray emitted and when only one 0.00100 amu is lost, only one 100 ray is emitted. There is no 0.00100 amu lost when it is regained and the total mass lost of 0.00589 amu lost indicates this. This is also the proof that only one electron has been released. The 434 ray is a gamma ray and the 100 ray is a beta ray. The electron that is released is the mass that is lost as mass. The two rays represent the wave function and the mass lost as mass represents the mass function.

Sub-step 9-2-5

1H3

FORMULA

1H2 + 1H1 = 1H3, one alpha 434 ray, one 100 ray, -1e

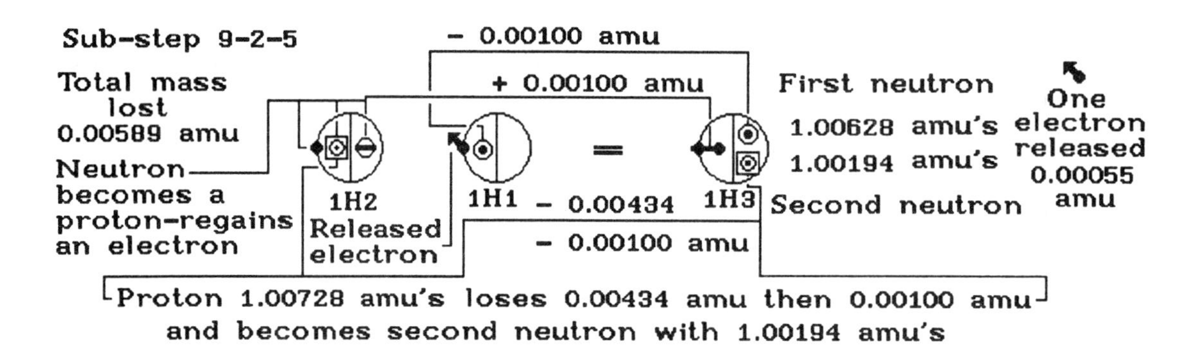

There is one isotope of 1H2 with 2.01411 amu's and one isotope of 1H1 with 1.00783 amu's creating an isotope of 1H3 with 3.01605 amu's. The isotope 1H2 has one electron with 0.00055 amu, one proton with 1.00728 amu's and one neutron with 1.00628 amu's. The isotope 1H1 has one electron with 0.00055 amu and one proton with 1.00728 amu's. The isotope 1H3 has one electron with 0.00055 amu, one proton with 1.00728 amu's and two neutrons. The first neutron has 1.00628 amu's and the second neutron has 1.00194 amu's.

It is possible for all protons to be changed into neutrons with the release of their electrons and the neutron in the isotope 1H2 will regain its lost mass to become a proton in the isotope 1H3 and the proton to then regain its electron. This means that only one electron is released. It must belong to the proton in the isotope 1H1 because this proton must lose 0.00100 amu to become the first neutron in the isotope 1H3.

This would mean that the proton and neutron in the isotope 1H2 switched places in the isotope 1H3. If the proton in the isotope 1H2 loses its electron and it is then regained by the neutron when this neutron becomes a proton in the isotope 1H3, then the electron in the isotope 1H1 is the only electron lost.

The electron in the isotope 1H2 is drawn as a dot on the left side of the big circle. The proton in this isotope is drawn as a dot on the left side of the vertical line within the big circle. The dot has a small circle drawn around it to show that this proton lost 0.00100 amu and released its electron. There is a square drawn around both of them to show that 0.00434 amu has been lost that does not release an electron. The 0.00434 amu is lost as an alpha 434 ray that has an external magnetic field. The single neutron has a small line drawn through the small circle that represents the neutron. The small circle is drawn on the right side of the vertical line within the big circle. The small line shows that this neutron has regained all of its lost mass to become a proton in the isotope 1H3. The proton and its electron have a small line drawn through them also to indicate that they came from the single neutron in the isotope 1H2. The electron in the isotope 1H3 must come from the isotope 1H2. It is not released because it is regained when the isotope 1H3 is created.

The isotope 1H1 has its electron with 0.00055 amu drawn on the left side of the big circle with an arrow pointing away from it. The arrow indicates that this electron has been released from its proton. The proton in this isotope has 1.00728 amu's and it is drawn as a dot on the left side of the big circle. It has a small circle drawn around it to indicate that 0.00100 amu has been lost. It is lost as a 100 ray.

The proton with 1.00728 amu's in the isotope 1H2 has a small circle drawn around it plus a square around both and this same drawing is found in the isotope 1H3 except it is now on the right side of the big circle. This shows that the proton in the isotope 1H2 has become the second neutron in the isotope 1H3. The proton in the isotope 1H1 that has a small circle around the proton dot is drawn the same way in the isotope 1H3. This shows that the proton in the isotope 1H1 has become the first neutron in the isotope 1H3 except it is drawn on the right side of the vertical line inside the big circle.

To find the amount of mass needed to create the isotope 1H3, add the mass of the isotope 1H2 and 1H1.

Mass of 1H2	2.01411 amu's
Mass of 1H1	1.00783 amu's
Total mass needed to create 1H3	3.02194 amu's

Now subtract the mass of the isotope 1H3 from the total mass needed to create 1H3. The remainder will be the total mass lost creating the isotope 1H3.

Total mass needed to create 1H3	3.02194 amu's
Mass of 1H3	3.01605 amu's
Mass lost creating 1H3	0.00589 amu

I have stated that the proton with its electron will make a neutron from the isotope 1H1. This means that the proton will lose some of its mass and break its magnetic field so that it can no longer hold its electron. This electron is then released. The proton now has its magnetic field as an internal one and it becomes the first neutron in the isotope 1H3. Since I know the mass of the proton and that it loses 0.00100 amu, I can subtract the lost mass that releases the electron from the proton.

Mass of proton in 1H1	1.00728 amu's
Mass releasing electron in 1H1	0.00100 amu
Mass of first neutron in 1H3	1.00628 amu's

The total mass lost is known to be 0.00589 amu and the mass lost by the proton in the isotope 1H1 can now be subtracted from it as well as the electron that has been released. The remainder will be the mass lost by the proton in the isotope 1H2.

Mass releasing electron from 1H1	0.00100 amu
Mass of released electron	0.00055 amu
Mass lost by 1H1	0.00155 amu

Total mass lost creating 1H3	0.00589 amu
Mass lost by 1H1	0.00155 amu
Mass lost by 1H2	0.00434 amu

This lost mass by the proton in the isotope 1H2 does not release the proton's electron. To release this electron, 0.00100 amu must be lost. When this 0.00100 amu is lost, this proton becomes the second neutron in the isotope 1H3. By adding both of these lost masses and then subtracting their total from the proton will give the mass of the second neutron in the isotope 1H3.

Mass lost by proton in 1H2 not releasing electron	0.00434 amu
Mass lost by proton in 1H2 releasing electron	0.00100 amu
Mass lost by proton becoming neutron in 1H3	0.00534 amu

Mass of proton in 1H2	1.00728 amu's
Mass lost by proton becoming neutron in 1H3	0.00534 amu
Mass of second neutron in 1H3	1.00194 amu's

The masses of the two neutrons in the isotope 1H3 are now known. By subtracting their mass from the mass of the isotope 1H3, the mass of the electron and proton can be found.

Mass of first neutron in 1H3	1.00628 amu's
Mass of second neutron in 1H3	1.00194 amu's
Total mass of two neutrons in 1H3	2.00822 amu's

Mass of 1H3	3.01605 amu's
Total mass of two neutrons in 1H3	2.00822 amu's
Mass of electron and proton in 1H3	1.00783 amu's

Since two electrons have been released and only one is needed, and both protons have lost mass, neither proton can be the proton found in the isotope 1H3. The only place this proton can come from is from the single neutron in the isotope 1H2. This neutron is known to have the mass of 1.00628 amu's. The mass of the electron in the isotope 1H3 is known to be 0.00055 amu which means that the proton must have 1.00728 amu's.

Mass of electron and proton in 1H3	1.00783 amu's
Mass of electron in 1H3	0.00055 amu
Mass of proton in 1H3	1.00728 amu's

Because the neutron in the isotope 1H2 has 1.00628 amu's and it is to become the proton in the isotope 1H3, it must regain 0.00100 amu and when it does this, it becomes a proton and regains its electron.

Mass of neutron in 1H2	1.00628 amu's
Mass regained from 1H2's proton	0.00100 amu
Mass of proton in 1H3	1.00728 amu's

This regained 0.00100 amu must come from the proton in the isotope 1H2 and the electron regained by the proton in the isotope 1H3 must also come from the isotope 1H2.

Now the mass of all parts of the isotope of 1H3 are known and how they got where they are. It should be noted that the regained mass did not come from outside the two isotopes of 1H2 and 1H1. This step is an internal exchange of mass and therefore when the 0.00434 amu is lost, it is the only 434 ray emitted and when only one 0.00100 amu is lost, only one 100 ray is emitted. There is no 0.00100 amu lost when it is regained and the total mass lost of 0.00589 amu lost indicates this. This is also the proof that only one electron has been released.

The difference between this step and the one given before it is how the 0.00434 amu is lost. In the step before this one, the isotope 1H1 lost this mass plus 0.00100 amu. In this step, it is the proton in the isotope 1H2 that loses it. This means that there is no exchange of mass between the isotope 1H1 and 1H2 in this step. Since this is true, the isotope 1H1's proton becomes a neutron by losing only its mass. The isotope 1H2 therefore must lose the mass from its proton and regained by its neutron causing this neutron to become a proton in the isotope 1H3. The proton and neutron trade places.

Now add all of the known parts of the isotope 1H3 to find the mass of the isotope 1H3.

Mass of electron in 1H3	0.00055 amu
Mass of proton in 1H3	1.00728 amu's
Mass of first neutron in 1H3	1.00628 amu's
Mass of second neutron in 1H3	1.00194 amu's
Mass of 1H3	3.00605 amu's

Sub-step 8-2-4

1H3

FORMULA

1H2 + 1H1 = 1H3, one alpha 434 ray, one 100 ray, -1e

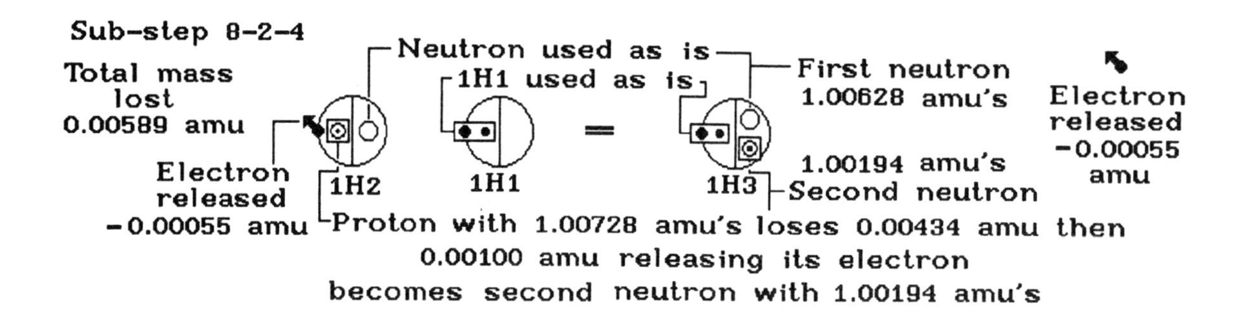

There is one isotope of 1H2 with 2.01411 amu's and one isotope of 1H1 with 1.00783 amu's creating an isotope of 1H3 with 3.01605 amu's. The isotope 1H2 has one electron with 0.00055 amu, one proton with 1.00728 amu's and one neutron with 1.00628 amu's. The isotope 1H1 has one electron with 0.00055 amu and one proton with 1.00728 amu's. The isotope 1H3 has one electron with 0.00055 amu, one proton with 1.00728 amu's and two neutrons. The first neutron has 1.00628 amu's and the second neutron has 1.00194 amu's. All of the lost mass is lost by the isotope 1H2.

The drawing shows the isotope 1H2 having its electron drawn as a dot on the left side of the big circle with an arrow pointing away from it. The arrow indicates that the electron has been released when the proton loses 0.00100 amu. The proton is drawn as a dot with a small circle around it and a square around both. They are located on the left side of the vertical line inside the big circle. The square shows that this proton has lost 0.00434 amu which does not release its electron. The alpha 434 ray has an external magnetic field. After it loses the mass that does not release its electron, it then loses this mass which does release its electron then it becomes the second neutron in the isotope 1H3. The small circle shows that 0.00100 amu has been lost and then the electron is released. This proton with its square and small circle is drawn the same way in the isotope 1H3 as the second neutron.

The isotope 1H2 has a single neutron on the right side of the vertical line inside the big circle. It is drawn as a small circle in the isotope 1H2 which is also drawn as a small circle on the right side of the vertical line inside the isotope 1H3. This neutron is used as is in the isotopes 1H2 and 1H3.

The isotope 1H1 has its electron drawn as a dot on the left side of the big circle and the proton is drawn as a dot on the left side of the vertical line inside the big circle. There is a rectangle drawn around both of them. It is also drawn the same way in the isotope 1H3. This indicates that the electron and proton are used as the isotope 1H1 inside the isotope 1H3. It also indicates that the isotope 1H1 can drop out of the isotope 1H3 at some later time.

To find the amount of mass needed to create the isotope 1H3, add the mass of the isotope 1H2 and 1H1.

Mass of 1H2	2.01411 amu's
Mass of 1H1	1.00783 amu's
Total mass needed to create 1H3	3.02194 amu's

Now subtract the mass of the isotope 1H3 from the total mass needed to create 1H3. The remainder will be the total mass lost creating the isotope 1H3.

Total mass needed to create 1H3	3.02194 amu's
Mass of 1H3	3.01605 amu's
Mass lost creating 1H3	0.00589 amu

All of this lost mass is lost by the isotope 1H2. This isotope loses 0.00100 amu and releases its electron. Add the mass of the electron and the lost mass to get the mass that is known to be lost by the isotope 1H2. Then subtract this total from the mass lost creating 1H3 to find the mass that is lost that does not release an electron.

Mass of released electron from 1H2	0.00055 amu
Mass releasing electron from 1H2	0.00100 amu
Known mass lost by 1H2	0.00155 amu

Mass lost creating 1H3	0.00589 amu
Known mass lost by 1H2	0.00155 amu
Mass lost that does not release electron in 1H2	0.00434 amu

The proton in the isotope 1H2 now is known to lose 0.00434 amu and 0.00100 amu. Add these lost masses and subtract their total from the mass of the proton to find the mass of the second neutron in the isotope 1H3.

Mass lost by 1H2 releasing its electron	0.00100 amu
Mass lost that does not release electron from 1H2	0.00434 amu
Total mass lost by 1H2's proton	0.00534 amu

Mass of proton in 1H2	1.00728 amu's
Total mass lost by 1H2's proton	0.00534 amu
Mass of second neutron in 1H3	1.00194 amu's

The mass of the second neutron can also be found by adding the masses in the isotope 1H3 that did not lose any mass and subtracting their total from the mass of the isotope of 1H3.

Mass of 1H1 found intact in 1H3	1.00783 amu's
Mass of first neutron in 1H3	1.00628 amu's
Known mass of 1H3	2.01411 amu's

Mass of 1H3	3.01605 amu's
Known mass of 1H3	2.01411 amu's
Mass of second neutron in 1H3	1.00194 amu's

Note that in the isotope 1H3, the known mass of the isotope 1H1 and the single neutron in the isotope 1H3 equals the known mass of the isotope 1H2. While this has same mass of the isotope 1H2, the parts that equal this mass in the isotope 1H3 do not create the isotope 1H2 within the isotope 1H3. The isotope 1H1 is inside the isotope 1H3 in such a manner that it can drop out at a later time. This means that the isotope 1H2 is not intact within the isotope 1H3 and it therefore can not drop out at some later time.

The electron is lost as an electron while the 0.00434 amu is lost as a 434 ray and the 0.00100 amu is lost as a 100 ray. A ray is pure energy which is magnetism.

Now all of the parts of the isotope 1H3 are known. Add all of these parts to find the mass of the isotope 1H3.

Mass of 1H1 in 1H3	1.00783 amu's
Mass of first neutron in 1H3	1.00628 amu's
Mass of second neutron in 1H3	1.00194 amu's
Mass of 1H3	3.01605 amu's

Sub-step 7-2-3

1H3

FORMULA

1H2 + 1H1 = 1H3, one alpha 434 ray, 100 ray, -1e

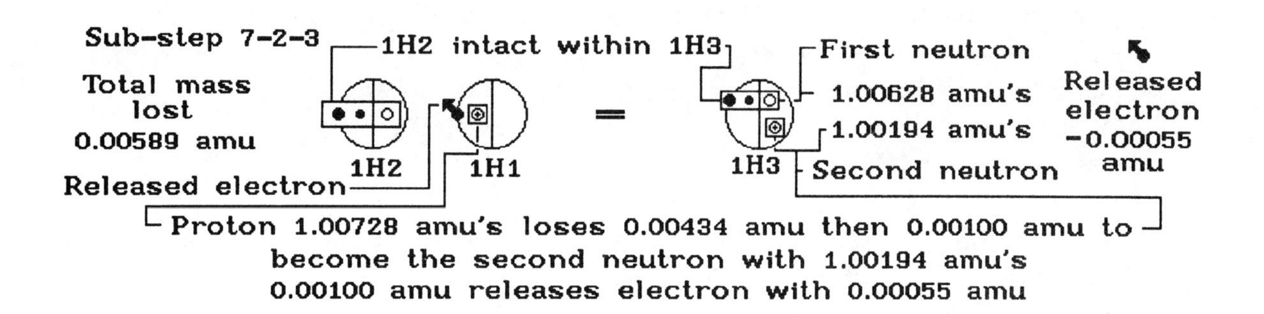

There is one isotope of 1H2 with 2.01411 amu's and one isotope of 1H1 with 1.00783 amu's creating an isotope of 1H3 with 3.01605 amu's. The isotope 1H2 has one electron with 0.00055 amu, one proton with 1.00728 amu's and one neutron with 1.00628 amu's. The isotope 1H1 has one electron with 0.00055 amu and one proton with 1.00728 amu's. The isotope 1H3 has one electron with 0.00055 amu, one proton with 1.00728 amu's and two neutrons. The first neutron has 1.00628 amu's and the second neutron has 1.00194 amu's.

The electron in the isotope 1H2 is drawn as a dot on the left side of the big circle. The proton in this isotope is drawn as a dot on the left side of a vertical line within the big circle. The single neutron is drawn on the right side of the vertical line inside the big circle. All three of these parts of the isotope 1H2 are enclosed within a rectangle and they are drawn this way in the isotope 1H3. This rectangle shows that the isotope 1H2 is intact within the isotope 1H3 and that it can drop out at some later time. The neutron has the mass of 1.00628 amu's.

The isotope 1H1 has its electron drawn as a dot on the left side of the big circle. The proton in this isotope is drawn as a dot with a small circle around it. There is a square drawn around both of them. The square indicates that the proton has lost 0.00434 amu that does not release its electron. The 0.00434 amu is emitted as an alpha 434 ray which has an external magnetic field. The small circle indicates that 0.00100 amu has been lost which does release the electron. When the 0.00100 amu is lost, this proton becomes a neutron in the isotope 1H3. This proton in the isotope 1H1 is drawn the same way in the isotope 1H3 except it is on the right side of the vertical line inside the big circle to show it is now a neutron.

To find the amount of mass needed to create the isotope 1H3, add the mass of the isotope 1H2 and 1H1.

Mass of 1H2	2.01411 amu's
Mass of 1H1	1.00783 amu's
Total mass needed to create 1H3	3.02194 amu's

Now subtract the mass of the isotope 1H3 from the total mass needed to create 1H3. The remainder will be the total mass lost creating the isotope 1H3.

Total mass needed to create 1H3	3.02194 amu's
Mass of 1H3	3.01605 amu's
Mass lost creating 1H3	0.00589 amu

The isotope 1H1 loses 0.00100 amu and releases its electron. Both of these masses are part of the mass lost creating 1H3. Add them and then subtract their total from the mass lost creating 1H3.

Mass of electron released from 1H1	0.00055 amu
Mass lost releasing electron from 1H1	0.00100 amu
Total known mass lost by 1H1	0.00155 amu

Mass lost creating 1H3	0.00589 amu
Total known mass lost by 1H1	0.00155 amu
Mass lost that does not release an electron	0.00434 amu

The mass lost that does not release an electron from the isotope 1H1 and the mass that does release the electron is lost from the isotope 1H1. Add their mass and subtract it from the proton to find the mass of the second neutron in the isotope 1H3.

Mass lost releasing electron from 1H1	0.00100 amu
Mass lost that does not release electron from 1H1	0.00434 amu
Total mass lost by proton in 1H1	0.00534 amu

Mass of proton in 1H1	1.00728 amu's
Total mass lost by proton in 1H1.	0.00534 amu
Mass of second neutron in 1H3	1.00194 amu's

The mass of the second neutron can also be found by adding the mass of the known parts of the isotope 1H3 and subtracting their mass from the isotope 1H3.

Mass of 1H3	3.01605 amu's
Mass of 1H2 inside 1H3	2.01411 amu's
Mass of second neutron in 1H3	1.00194 amu's

Now add all of the known masses to find the mass of the isotope 1H3.

Mass of 1H2 inside 1H3	2.01411 amu's
Mass of second neutron in 1H3	1.00194 amu's
Mass of 1H3	3.01605 amu's

The lost mass of 0.00434 amu is lost as a 434 ray and the 0.00100 amu is lost as a 100 ray. Both of these rays are lost by the isotope 1H1. The electron with 0.00055 amu is released as an electron therefore 0.00055 amu is lost from the isotope 1H1 and not found in the isotope 1H3.

Sub-step 6-2-2

1H3

FORMULA

1H2 + 1H1 = 1H3, one alpha 434 ray, 100 ray, -1e

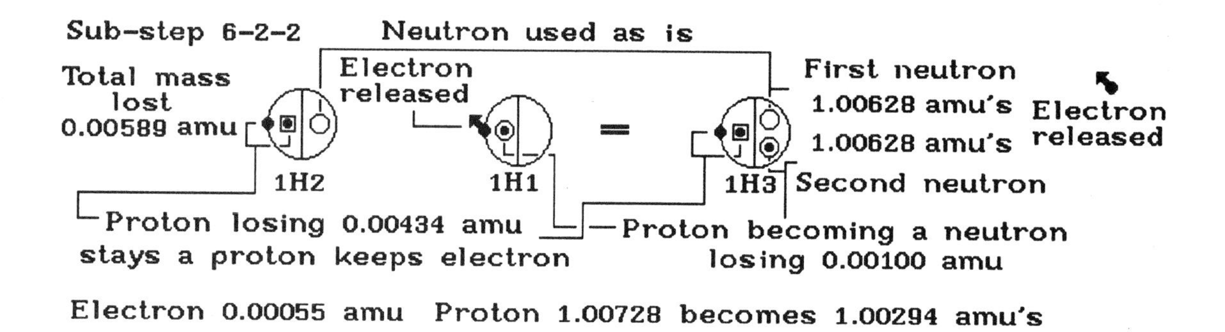

There is one isotope of 1H2 with 2.01411 amu's and one isotope of 1H1 with 1.00783 amu's creating an isotope of 1H3 with 3.01605 amu's. The isotope 1H2 has one electron with 0.00055 amu, one proton with 1.00728 amu's and one neutron with 1.00628 amu's. The isotope 1H1 has one electron with 0.00055 amu and one proton with 1.00728 amu's. The isotope 1H3 has one electron with 0.00055 amu, one proton with 1.00294 amu's and two neutrons. Both neutrons have the same mass of 1.00628 amu's.

The electron in the isotope 1H2 is drawn as a dot on the left side of the big circle. The proton in this isotope is drawn as a dot on the left side of a vertical line within the big circle. This proton has a square drawn around it. Because it does not have a small circle around it, the electron is not released. The square is drawn to indicate that 0.00434 amu has been lost. The 0.00434 amu us lost as an alpha ray that has an external magnetic field. The electron and proton are drawn the same way in the isotope 1H3. The single neutron in the isotope 1H2 is drawn as a small circle and it is drawn this same way in the isotope 1H3 as the first neutron.

The isotope 1H1 has its electron drawn on the left side of the big circle as a dot with an arrow pointing away from it. The arrow indicates that the electron has been released. The proton in this isotope is drawn as a dot with a small circle around it on the left side of the vertical line within the big circle. The small circle indicates that 0.00100 amu has been lost and the electron has been released.

The electron in the isotopes 1H2 and 1H3 each have 0.00055 amu. The proton in the isotope 1H2 has 1.00728 amu's and loses 0.00434 amu which becomes the proton in the isotope 1H3 with 1.00294 amu's. The single neutron with 1.00628 amu's in the isotope 1H2 stays a neutron with the same mass in the isotope 1H3.

The electron in the isotope 1H1 is released as an electron with 0.00055 amu. The proton with 1.00728 amu's loses only 0.00100 amu to become the second neutron in the isotope 1H3 with 1.00628 amu's. This means that both neutrons in the isotope 1H3 have the same mass of 1.00628 amu's.

To find the amount of mass needed to create the isotope 1H3, add the mass of the isotope 1H2 and 1H1.

Mass of 1H2	2.01411 amu's
Mass of 1H1	1.00783 amu's
Total mass needed to create 1H3	3.02194 amu's

Now subtract the mass of the isotope 1H3 from the total mass needed to create 1H3. The remainder will be the total mass lost creating the isotope 1H3.

Total mass needed to create 1H3	3.02194 amu's
Mass of 1H3	3.01605 amu's
Mass lost creating 1H3	0.00589 amu

The isotope 1H1's proton loses 0.00100 amu and releases its electron. Both of these masses are part of the mass lost creating 1H3 but not all of it. Add the lost mass of 0.00100 amu to the mass of the electron that is released and then subtract their total from the total mass lost creating 1H3 to find the remaining mass that is lost. This remaining lost mass is lost from the proton in the isotope 1H2 and does not release the electron.

Mass of electron released from 1H1	0.00055 amu
Mass lost releasing electron from 1H1	0.00100 amu
Total known mass lost by 1H1	0.00155 amu

Total mass lost creating 1H3	0.00589 amu
Known mass lost by 1H1	0.00155 amu
1H2's lost mass from proton that does not release an electron	0.00434 amu

To find the mass of the proton in the isotope 1H3, subtract the lost mass by the proton in the isotope 1H2 from the mass of this proton.

Mass of proton in 1H2	1.00728 amu's
Mass lost by proton in 1H2.	0.00434 amu
Mass of proton in 1H3	1.00294 amu's

To find the mass of the second neutron in the isotope 1H3, subtract the mass lost by the proton in the isotope 1H1 which released its electron.

Mass of proton in 1H1	1.00728 amu's
Mass lost releasing electron in 1H1	0.00100 amu
Mass of second neutron in 1H3	1.00628 amu's

Add the mass lost that does not release an electron from the isotope 1H2's proton and the mass that does release the electron that is lost from the isotope 1H1's proton to find the total mass lost by both protons. The 0.00100 amu is emitted as a 100 ray and the 0.00434 amu is emitted as a 434 ray. Now add the mass of the electron released from the isotope 1H1 to find the total mass lost by the two isotopes to create the isotope 1H3.

Mass lost releasing electron from 1H1	0.00100 amu
Mass lost that does not release electron from 1H2	0.00434 amu
Total mass lost by protons in 1H1 and 1H2	0.00534 amu

Total mass lost by proton's in 1H1 and 1H2	0.00534 amu
Mass of released electron from 1H1	0.00055 amu
Total mass lost creating 1H3	0.00589 amu

Remember that the single neutron in the isotope 1H2 is used as is and it has 1.00628 amu's. It is now shown why both neutrons have the same mass in the isotope 1H3. It also is shown that the proton in the isotope 1H3 does not have 1.00728 amu's which would be the normal mass of a proton but that it has the mass of 1.00294 amu's.

Now all mass of the parts of the isotope are known.

Mass of electron in 1H3	0.00055 amu
Mass of proton in 1H3	1.00294 amu's
Mass of first neutron in 1H3	1.00628 amu's
Mass of second neutron in 1H3	1.00628 amu's
Mass of 1H3	3.01605 amu's

Sub-step 5-2-1

1H3

FORMULA

1H2 + 1H1 = 1H3, one alpha 434 ray, one 100 ray, -1e

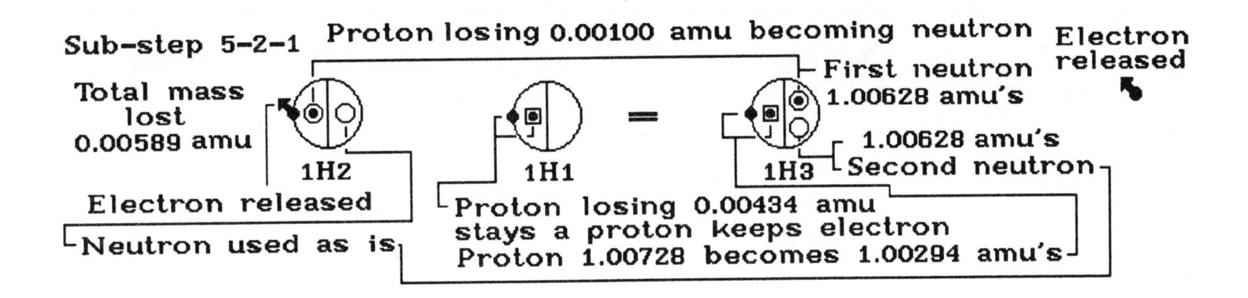

There is one isotope of 1H2 with 2.01411 amu's and one isotope of 1H1 with 1.00783 amu's creating an isotope of 1H3 with 3.01605 amu's. The isotope 1H2 has one electron with 0.00055 amu, one proton with 1.00728 amu's and one neutron with 1.00628 amu's. The isotope 1H1 has one electron with 0.00055 amu and one proton with 1.00728 amu's. The isotope 1H3 has one electron with 0.00055 amu, one proton with 1.00294 amu's and two neutrons. Both neutrons have the same mass of 1.00628 amu's.

The electron in the isotope 1H2 is drawn as a dot on the left side of the big circle with an arrow pointing away from it. The arrow indicates that the electron has been released. The proton in this isotope is drawn as a dot on the left side of a vertical line within the big circle with a small circle around it. Because it has a small circle around it, the electron is released and 0.00100 amu is lost as a 100 ray. The proton is drawn the same way in the isotope 1H3 except the proton becomes the first neutron and it is drawn to the right of the vertical line. The single neutron in the isotope 1H2 is drawn as a small circle to the right of the vertical line within the big circle and it is drawn this same way in the isotope 1H3 as the second neutron. It is used as is and does not lose any mass.

The isotope 1H1 has its electron drawn on the left side of the big circle as a dot and it is drawn this way in the isotope 1H3. The proton in this isotope is drawn as a dot with a square drawn around it on the left side of the vertical line within the big circle. The square indicates that 0.00434 amu has been lost as an alpha 434 ray which has an external magnetic field and the electron has not been released.

The electron in the isotope 1H2 and 1H3 has 0.00055 amu. The proton in the isotope 1H2 has 1.00728 amu's and loses 0.00100 amu which becomes the first neutron in the isotope 1H3 with 1.00628 amu's. The single neutron with 1.00628 amu's in the isotope 1H2 stays a neutron with the same mass in the isotope 1H3 as the second neutron.

The electron in the isotope 1H2 is released as an electron with 0.00055 amu. The proton with 1.00728 amu's loses only 0.00100 amu to become the first neutron in the isotope 1H3 with 1.00628 amu's. This means that both neutrons in the isotope 1H3 have the same mass of 1.00628 amu's.

To find the amount of mass needed to create the isotope 1H3, add the mass of the isotope 1H2 and 1H1.

Mass of 1H2	2.01411 amu's
Mass of 1H1	1.00783 amu's
Total mass needed to create 1H3	3.02194 amu's

Now subtract the mass of the isotope 1H3 from the total mass needed to create 1H3. The remainder will be the total mass lost creating the isotope 1H3.

Total mass needed to create 1H3	3.02194 amu's
Mass of 1H3	3.01605 amu's
Mass lost creating 1H3	0.00589 amu

The isotope 1H2's proton loses 0.00100 amu and releases its electron. Both of these masses are part of the mass lost creating 1H3. Add them and then subtract their total from the mass lost creating 1H3.

Mass of electron released from 1H2	0.00055 amu
Mass lost releasing electron from 1H2	0.00100 amu
Total known mass lost by 1H2	0.00155 amu

Mass lost creating 1H3	0.00589 amu
Total known mass lost by 1H2	0.00155 amu
Mass lost that does not release an electron	0.00434 amu

The mass lost that does not release an electron from the isotope 1H1's proton and the mass that does release the electron that is lost from the isotope 1H2's proton. Add their mass to find the total mass lost by both protons. The 0.00100 amu is emitted as a 100 ray and the 0.00434 amu is emitted as a 434 ray.

Mass lost releasing electron from 1H2	0.00100 amu
Mass lost that does not release electron from 1H1	0.00434 amu
Total mass lost by protons in 1H2 and 1H1	0.00534 amu

Mass of proton in 1H1	1.00728 amu's
Total mass lost by proton in 1H1	0.00434 amu
Mass of proton in 1H3	1.00294 amu's

To find the mass of the first neutron in the isotope 1H3, subtract the mass lost by the proton in the isotope 1H2 which released its electron.

Mass of proton in 1H2	1.00728 amu's
Mass lost releasing electron in 1H2	0.00100 amu
Mass of first neutron in 1H3	1.00628 amu's

Remember that the single neutron in the isotope 1H2 is used as is and it has 1.00628 amu's. It is now shown why both neutrons have the same mass in the isotope 1H3. It also is shown that the proton in the isotope 1H3 does not have 1.00728 amu's which would be the normal mass of a proton but that it has the mass of 1.00294 amu's.

Now all mass of the parts of the isotope 1H3 are known. Add all of these known parts of the isotope 1H3 to find its mass.

Mass of electron in 1H3	0.00055 amu
Mass of proton in 1H3	1.00294 amu's
Mass of first neutron in 1H3	1.00628 amu's
Mass of second neutron in 1H3	1.00628 amu's
Mass of 1H3	3.01605 amu's

All of the drawings in **STEP 2** using one isotope of 1H2 and 1H1 creating the isotope 1H3.

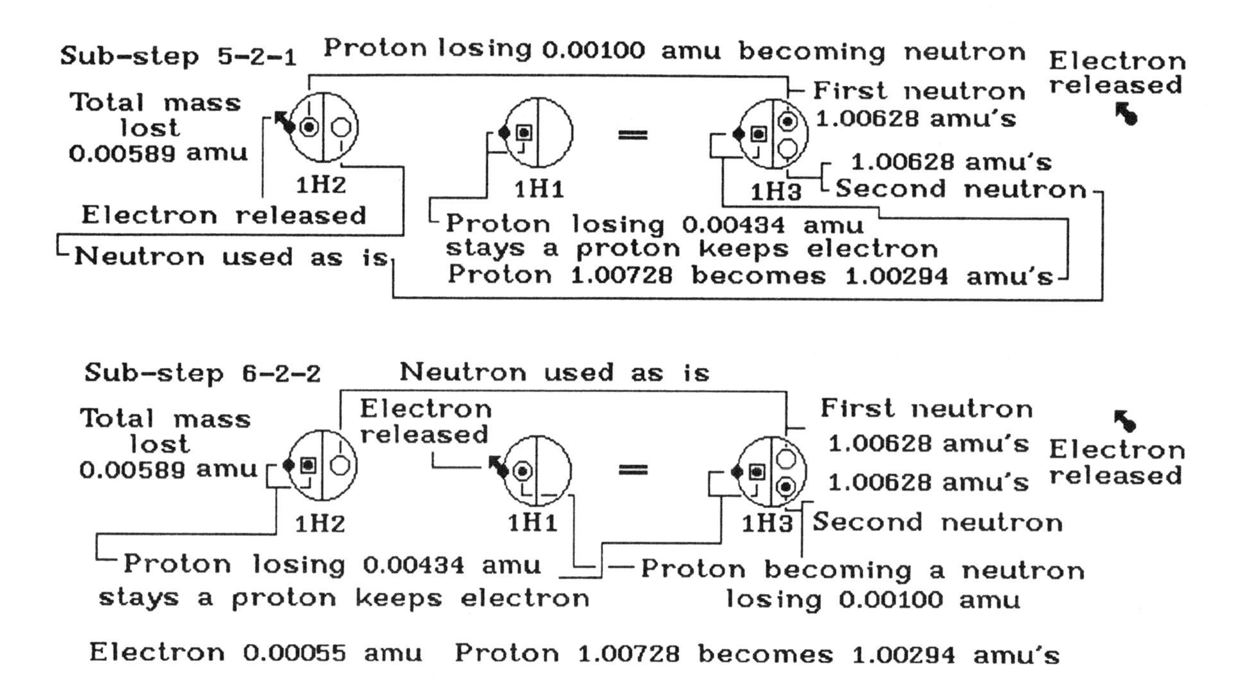

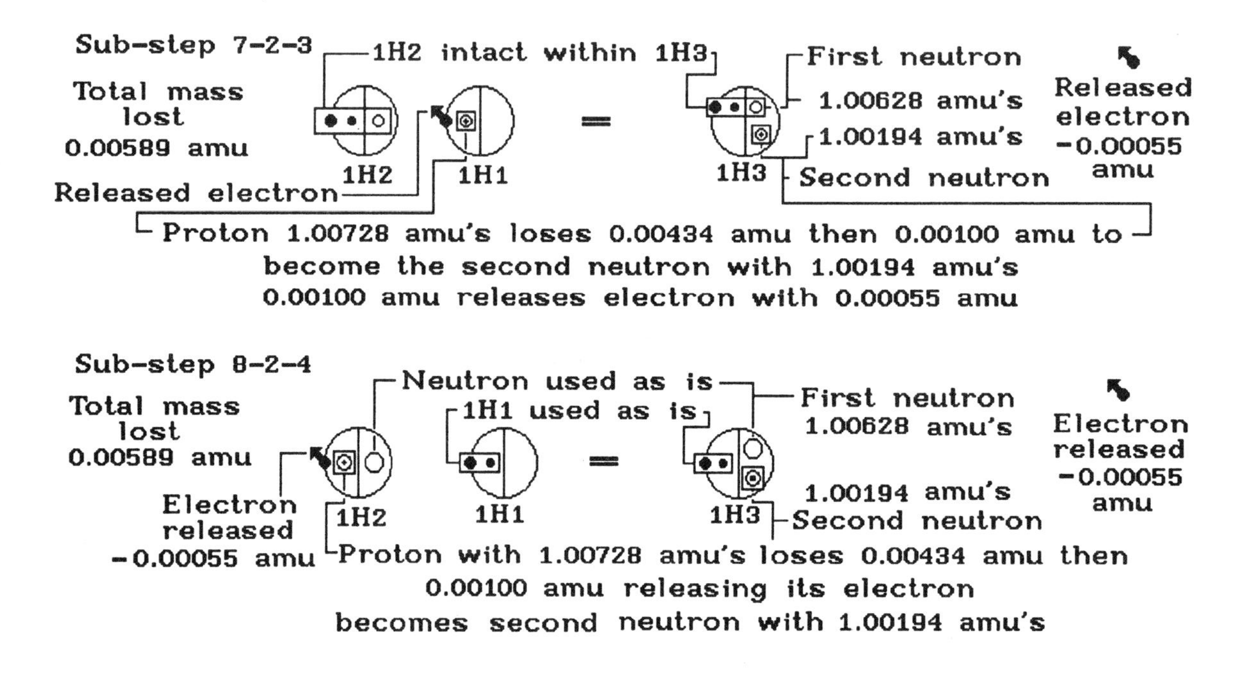

All of the drawings in **STEP 2** using one isotope of 1H2 and 1H1 creating the isotope 1H3.

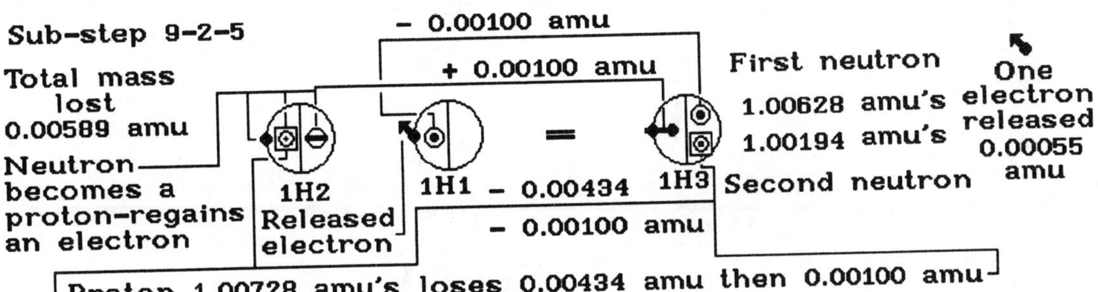

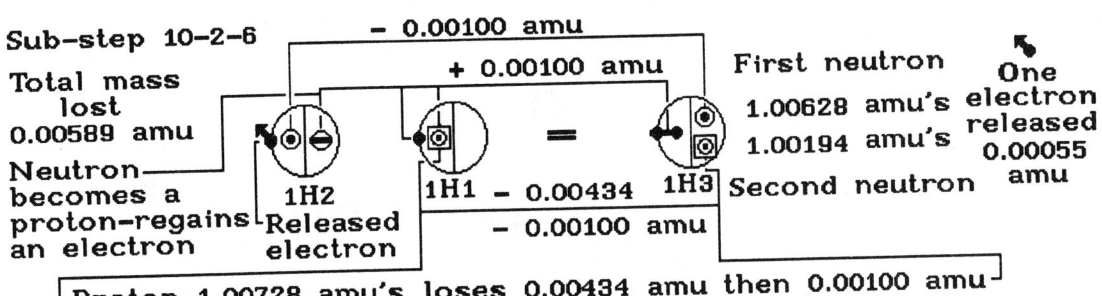

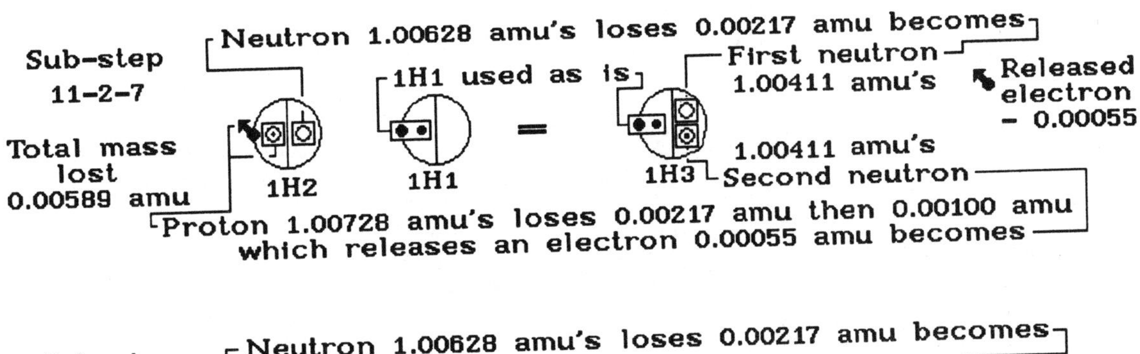

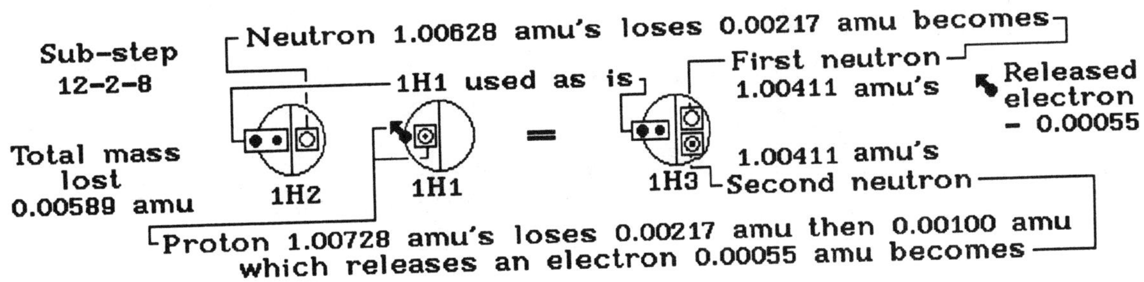

STEP 1

1H3

FORMULA

1H2 + 1H2 = 1H3

1. 1H2 + 1H2 = 1H3

1-1-1. 1H2 + 1H2 = 1H3, one 434 ray, 1H1
2-1-2. 1H2 + 1H2 = 1H3, one 434 ray, one 100 ray, -1e, -1n
3-1-3. 1H2 + 1H2 = 1H3, two 217 rays, 1H1
4-1-4. 1H2 + 1H2 = 1H3, one 434 ray, 1H1

Sub-step 4-1-4

1H3

FORMULA

1H2 + 1H2 = 1H3, one alpha 434 ray, one 1H1

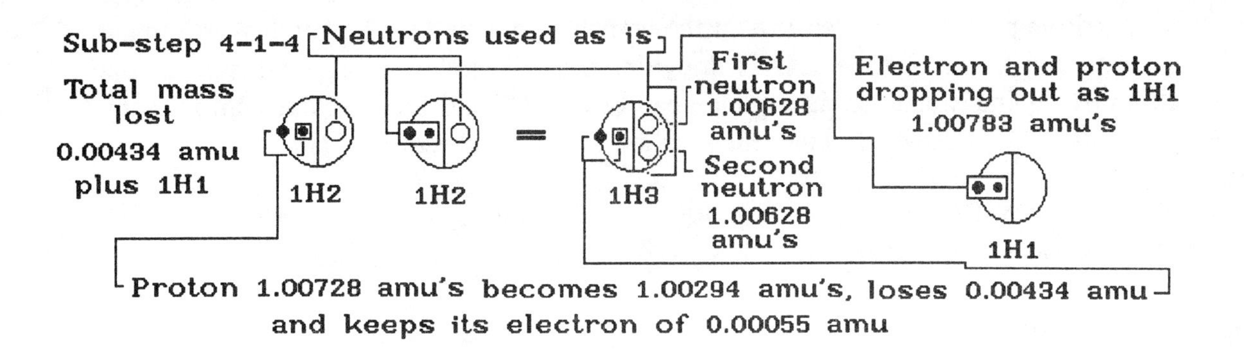

There are two isotopes of 1H2 creating the isotope 1H3. Both isotopes have 2.01411 amu's. Each isotope has one electron with 0.00055 amu, one proton with 1.00728 amu's and one neutron with 1.00628 amu's. The isotope 1H3 has one electron with 1.00055 amu, one proton with 1.00294 amu's and two neutrons with 1.00628 amu's each.

The drawing shows that each isotope of 1H2 has its electron on the left side of the big circle as a dot. They each have their proton on the left side of the vertical line within the big circle. The single neutron is drawn on the right side of the vertical line within the big circle as a small circle. This will be true of all drawings using this formula.

I start out this series of steps with the first isotope of 1H2's proton losing 0.00434 amu and it then becomes the proton in the isotope 1H3 with 1.00294 amu's. The 0.00434 is lost as an alpha 434 ray that has an external magnetic field. This lost mass does not release its electron which means that the electron belonging to this proton is found in the isotope 1H3 with 0.00055 amu. The single neutron is used as is and it is found in the isotope 1H2 and 1H3 with 1.00628 amu's.

The second isotope of 1H2 has its electron with 0.00055 amu and proton with 1.00728 amu's used as the isotope 1H1 with 1.00783 amu's within the isotope 1H2 and it drops out as the isotope 1H1 which means the isotope 1H1 is not part of the isotope 1H3. This leaves the single neutron with 1.00628 amu's and it is used as is in the isotope 1H3 and it does not lose any mass.

The first isotope of 1H2 has its electron drawn on left side of the vertical line as a dot while the proton is drawn as a dot with a square around it. The electron and proton are both drawn the same way in the isotope 1H3. The square means that this proton has lost some mass which will not release its electron. The single neutron is drawn on the right side of the vertical line as a small circle. It does not lose any mass. It is drawn the same way in the isotope 1H3 as the first neutron.

The second isotope of 1H2 has its electron and proton used as the isotope 1H1 and they are drawn with a rectangle around both of them in the isotope 1H2 and the isotope 1H1 is shown on the right side of the equal sign as the isotope 1H1. It is drawn as a big circle with its electron on the left side of the big circle and the proton on the left side of the vertical line within the big circle. The single neutron in the isotope 1H2 is drawn as a small circle on the right side of the vertical line within the big circle in the isotope 1H2 and 1H3 and does not lose any mass. This neutron becomes the second neutron in the isotope 1H3.

To find the total mass lost creating the isotope 1H3, add the mass of each isotope of 1H2 and subtract the mass of the isotope 1H3 from this total.

Mass of first 1H2	2.01411 amu's
Mass of second 1H2	2.01411 amu's
Total mass needed to create 1H3	4.02822 amu's

Total mass needed to create 1H3	4.02822 amu's
Mass of 1H3	3.01605 amu's
Total mass lost creating 1H3	1.01217 amu's

This total mass lost includes the mass of the isotope 1H1 that dropped out of the second isotope of 1H2. This isotope of 1H1 must therefore be subtracted from the total mass lost creating the isotope 1H3 to find the real mass that is lost creating the isotope 1H3.

Total mass lost creating 1H3	1.01217 amu's
Mass of 1H1 dropping out of second 1H1	1.00783 amu's
Real mass lost creating 1H3	0.00434 amu

Since the second isotope of 1H2 has its electron and proton used as the isotope 1H1 which dropped out and the single neutron is used as is in the isotope 1H3, none of these parts can lose the real mass that is lost creating the isotope 1H3. Also it can not come from the electron in the first isotope 1H2 because an electron never loses any mass. It can not be lost by the single neutron in this isotope because it is used as is in the isotope 1H3. This leaves the proton found in the first isotope of 1H2 as the only mass that can lose the 0.00434 amu. Now subtract the 0.00434 amu that is lost from this proton to find the mass of the proton found in the isotope 1H3.

Mass of proton in first 1H2	1.00728 amu's
Mass lost by it	0.00434 amu
Mass of proton in 1H3	1.00294 amu's

Because both isotopes of 1H2 have their neutrons used as is, they both have 1.00628 amu's. By adding their mass to the mass of the electron and then subtracting their total mass from the mass of the isotope of 1H3, the mass of the proton in this isotope can be found.

Mass of electron in 1H3	0.00055 amu
Mass of first neutron in 1H3	1.00628 amu's
Mass of second neutron in 1H3	1.00628 amu's
Total of electron and two neutrons in 1H3	2.01311 amu's

Mass of 1H3	3.01605 amu's
Total of electron and two neutrons in 1H3	2.01311 amu's
Mass of proton in 1H3	1.00294 amu's

All of the parts masses that make up the isotope 1H3 are now known. Add all of these parts to find the mass of the isotope 1H3.

Mass of electron from first 1H2 in 1H3	0.00055 amu
Mass of proton from first 1H2 in 1H3	1.00294 amu's
Mass of first neutron in 1H3	1.00628 amu's
Mass of second neutron in 1H3	1.00628 amu's
Mass of 1H3	3.01605 amu's

From the above, I have proven that the two isotopes of 1H2 will create the isotope 1H3 with the loss of 0.00434 amu from the proton in the first isotope of 1H2 making the proton in the isotope 1H3 with 1.00294 amu's. The lost mass of 0.00434 amu is lost as a 434 ray. To do this, the isotope 1H1 must drop out of the second isotope of 1H2. The electron in the isotope 1H3 must come from the first isotope of 1H1 and the two neutrons must come from each of the two isotopes of 1H2 with no loss of mass.

Sub-step 3-1-3

1H3

FORMULA

1H2 + 1H2 = 1H3, two alpha 217 rays, one 1H1

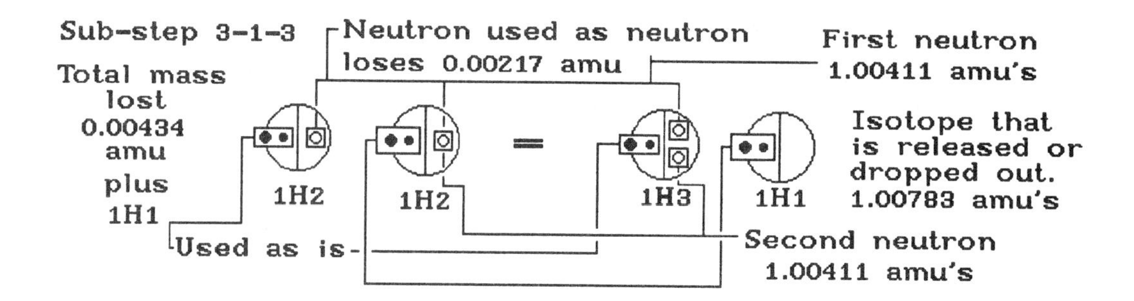

The two isotopes of 1H2 with 2.01411 amu's are used to create the isotope 1H3 with 3.01605 amu's.

Both isotopes of 1H2 have the isotope 1H1 used as the isotope 1H1 within each isotope 1H2. The first isotope of 1H2 has its electron and proton used as the isotope 1H1 within the isotope 1H3. The second isotope of 1H2 has its electron and proton used as the isotope 1H1 which drops out as the isotope 1H1 with 1.00783 amu's.

Each of the isotopes of 1H2 has their single neutron with 1.00628 amu's lose 0.00217 amu to become the first and second neutrons in the isotope 1H3 with each having 1.00411 amu's. This is a total lost mass of 0.00434 amu. Each 0.00217 amu is lost as an alpha 217 ray that has an external magnetic field.

The drawing of the first isotope of 1H2 shows the electron and proton drawn with a rectangle around both. The rectangle indicates that they are used as a single unit and that they can drop out at some later time. It also indicates that they are used as is with no loss of mass. The single neutron in this isotope is drawn as a small circle that has a square drawn around it to indicate that it has lost some mass and it is drawn this same way in the isotope 1H3 as the first neutron.

The second isotope of 1H2 has its electron and proton also drawn with a rectangle around both of them as the isotope 1H1. This time the isotope 1H1 drops out as the isotope 1H1. This leaves the single neutron which is drawn as a small circle with a square around it. The square indicates that this neutron has also lost some mass and it is drawn this way in the isotope 1H3 as the second neutron.

To find the amount of mass needed to create the isotope 1H3, add the mass of the two isotopes of 1H2. Then subtract the isotope 1H3 from this total to find the total mass lost creating the isotope 1H3.

Mass of first 1H2	2.01411 amu's
Mass of second 1H2	2.01411 amu's
Total mass needed to create 1H3	4.02822 amu's

Total mass needed to create 1H3	4.02822 amu's
Mass of 1H3	3.01605 amu's
Total mass lost creating 1H3	1.01217 amu's

This total mass lost includes the mass of the isotope 1H1 that dropped out of the second isotope of 1H2. This isotope of 1H1 must therefore be subtracted from the total mass lost creating the isotope 1H3 to find the real mass that is lost creating the isotope 1H3.

Total mass lost creating 1H3	1.01217 amu's
Mass of 1H1 dropping out of second 1H2	1.00783 amu's
Real mass lost creating 1H3	0.00434 amu

To refresh ones memory as to how the neutron found in the isotope 1H2 got its mass, subtract the mass of the electron and proton from the mass of this isotope. The mass that is left will be the mass of the single neutron.

Mass of 1H2	2.01411 amu's
Mass of 1H1 in 1H2	1.00783 amu's
Mass of single neutron in 1H2	1.00628 amu's

This real mass lost creating the isotope 1H3 is lost equally by each neutron found in both of the neutrons in the two isotopes of 1H2. Since each neutron has its mass known, divide the lost mass in half and then subtract each half from each neutron to find the mass of the two neutrons found in the isotope 1H3.

$0.5(0.00434) = 0.00217$ amu lost by each neutron in 1H2.

Mass of neutron in first 1H2	1.00628 amu's
Mass lost by it	0.00217 amu
Mass of first neutron in 1H3	1.00411 amu's

Mass of neutron in second 1H2	1.00628 amu's
Mass lost by it	0.00217 amu
Mass of second neutron in 1H3	1.00411 amu's

The mass of the electron and proton in the isotope 1H3 is known because they came intact from the first isotope of 1H2. The electron has 0.00055 amu and the proton has 1.00728 amu's. Their mass could be found by adding the mass of the two neutrons in the isotope 1H3 and subtracting this total from the isotope 1H3.

Mass of first neutron in 1H3	1.00411 amu's
Mass of second neutron in 1H3	1.00411 amu's
Total mass of two neutrons in 1H3	2.00822 amu's

Mass of 1H3	3.01605 amu's
Total mass of two neutrons in 1H3	2.00822 amu's
Mass of 1H1 inside 1H3	1.00783 amu's

The masses that makes up the isotope 1H3 are now known. Add these masses to find the mass of the isotope 1H3.

Mass of 1H1 from first 1H2 in 1H3	1.00783 amu's
Mass of first neutron in 1H3	1.00411 amu's
Mass of second neutron in 1H3	1.00411 amu's
Mass of 1H3	3.01605 amu's

The above is proof that the two neutrons found in the isotope 1H3 do not always have 1.00628 amu's. It also shows that the lost mass of 0.00434 amu is composed of two sub-units of 0.00217 amu and that these two sub-units are lost as two 217 rays. It also shows that the total lost mass of 1.01217 amu's includes the isotope 1H1 which must be lost as the isotope 1H1. If the isotope 1H1 is lost as the isotope 1H1, then it must be intact within the second isotope of 1H2. If the isotope 1H1 can be found intact within the second isotope 1H2, then it is possible for it to be intact within the first isotope of 1H2. If it is intact within the first isotope 1H2 and it does not drop out, then it could be found within the isotope 1H3 intact. If the isotope 1H1 does drop out at some later time from the isotope 1H3, then the two neutrons will also be free to combine with some other molecule.

Sub-step 2-1-2

1H3

FORMULA

1H2 + 1H2 = 1H3, one alpha 434 ray, one neutron, one 100 ray, -1e

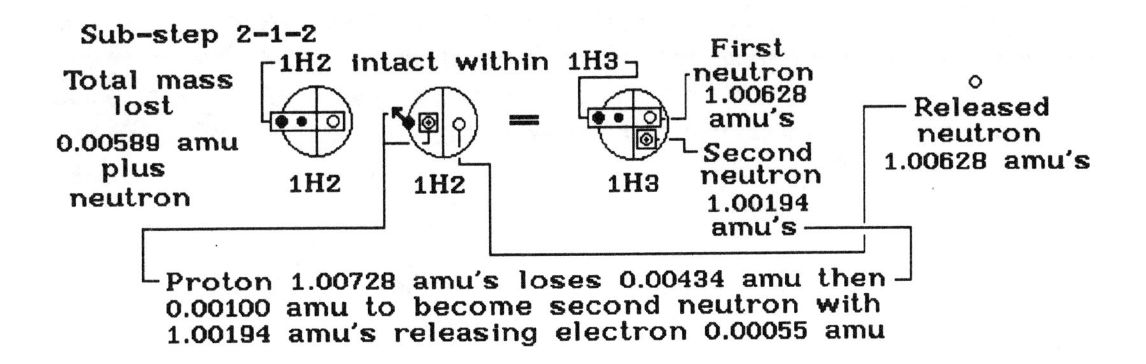

There are two isotopes of 1H2 with 2.01411 amu's each. Both of them are used to create the isotope 1H3 with 3.01605 amu's. Each isotope of 1H2 has one electron with 0.00055 amu, one proton with 1.00728 amu's and one neutron with 1.00628 amu's. The isotope of 1H3 has one electron with 0.00055 amu, one proton with 1.00728 amu's and two neutrons. The first neutron has 1.00628 amu's and the second neutron has 1.00194 amu's.

The first isotope of 1H2 is used as is which means that its electron, proton and neutron are used in the isotope 1H3 with no loss of mass. To show this, there is a rectangle around all three of these parts in the isotope 1H2 and 1H3. The rectangle indicates that the isotope of 1H2 can drop out at some later time. It also indicates that all the parts are used as is with no loss of mass.

The second isotope of 1H2 must lose all the mass that is lost. In this step, I state that it is the proton that does this. Note that the isotope of 1H3 can only have one electron and its electron comes from the first isotope of 1H2. This means that the electron in the second isotope of 1H2 is not needed. For it to be released from its proton, the proton must lose 0.00100 amu. When it does this, it becomes the second neutron with 1.00194 amu's in the isotope 1H3. Since the proton in the isotope 1H2 has 1.00728 amu's and becomes the second neutron in the isotope 1H3 with 1.00194 amu's, this proton must lose more mass than 0.00100 amu. Also since the proton in the second isotope 1H2 becomes a neutron in the isotope 1H3 and the electron is released, the single neutron with 1.00628 amu's must drop out because it is not used.

The only mass needed in the isotope 1H3 that comes from the second isotope of 1H2's proton is the mass that remains after all of the lost mass it must lose is lost. The proton loses 0.00434 amu as a 434 ray which has an external magnetic field. This proton in the second isotope of 1H2 is drawn as a dot on the left side of the vertical line with a small circle around it and then a square around both of them. It is drawn this same way in the isotope 1H3 except it is now a neutron on the right side of the vertical line.

The arrow pointing away from the electron in the second isotope of 1H2 indicates that the electron has been released completely from the isotope of 1H2.

The square that is drawn around the small circle indicates that 0.00434 amu has been lost that does not release the electron.

The neutron that has been released from the second isotope of 1H2 is drawn as a small circle on the right side of the equal sign.

It is ok if the released electron is not shown on the right side of the equal sign because the arrow drawn pointing away from the electron in the second isotope of 1H2 shows that it has been released. When this is done, the mass of the released electron must be part of the total mass lost.

To find the amount of mass needed to create the isotope 1H3, add the mass of the two isotopes of 1H2. Then subtract the isotope 1H3 from this total to find the total mass lost creating the isotope 1H3.

Mass of first 1H2	2.01411 amu's
Mass of second 1H2	2.01411 amu's
Total mass needed to create 1H3	4.02822 amu's

Total mass needed to create 1H3	4.02822 amu's
Mass of 1H3	3.01605 amu's
Total mass lost creating 1H3	1.01217 amu's

This total mass lost includes the mass of the electron that has been released and the neutron that dropped out of the second isotope of 1H2. The mass of the electron and neutron must therefore be subtracted from the total mass lost creating the isotope 1H3 to find the real mass that is lost creating the isotope 1H3.

Mass of electron that is released by second 1H2	0.00055 amu
Mass of neutron released from second 1H2	1.00628 amu's
Total mass lost by electron and neutron in 1H2	1.00683 amu's

Total mass lost creating 1H3	1.01217 amu's
Mass lost by electron and neutron in 1H2	1.00683 amu's
Real mass lost creating 1H3	0.00534 amu

This 0.00534 amu includes the mass that is lost releasing the electron from the second isotope of 1H2 and it must be subtracted to find the rest of the mass the proton in the second isotope of 1H2 lost.

Real mass lost creating 1H3	0.00534 amu
Mass lost by second 1H2 releasing electron	0.00100 amu
Remaining mass lost by proton in 1H2	0.00434 amu

Now subtract the real mass lost creating the isotope 1H3 from the proton found in the second isotope of 1H2 to find the second neutron in the isotope 1H3.

Mass of proton in second 1H2	1.00728 amu's
Real mass lost creating the isotope 1H3	0.00534 amu
Mass of second neutron in 1H3	1.00194 amu's

All of the parts of the isotope 1H3 are now known. Adding the mass of the first isotope 1H2 that is intact within the isotope 1H3 and the mass of the second neutron will give the mass of the isotope 1H3.

Mass of 1H2 in 1H3	2.01411 amu's
Mass of second neutron in 1H3	1.00194 amu's
Mass of 1H3	3.01605 amu's

All of the above is proof that two isotopes of 1H2 will create the isotope 1H3 with the release of an electron and that a neutron drops out from the second isotope of 1H2. It shows that the loss of 0.00534 amu is composed of two sub-units of 0.00434 amu and 0.00100 amu. These two sub-units are emitted as a 434 ray and a 100 ray. It also shows that the isotope 1H2 is intact within the isotope of 1H3.

Sub-step 1-1-1

1H3

FORMULA

1H2 + 1H2 = 1H3, one gamma 434 ray, one 1H1

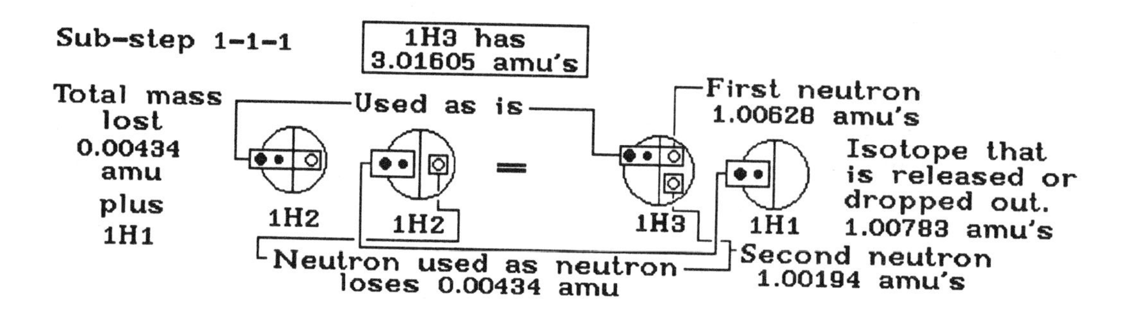

There are two isotopes of 1H2 with 2.01411 amu's each. Both of them are used to create the isotope 1H3 with 3.01605 amu's. Each isotope of 1H2 has one electron with 0.00055 amu, one proton with 1.00728 amu's and one neutron with 1.00628 amu's. The isotope of 1H3 has one electron with 0.00055 amu, one proton with 1.00728 amu's and two neutrons. The first neutron has 1.00628 amu's and the second neutron has 1.00194 amu's.

The first isotope of 1H2 is used as is which means that its electron, proton and neutron are used in the isotope 1H3 with no loss of mass. To show this, there is a rectangle around all three of these parts in the isotope 1H2 and 1H3. The rectangle indicates that the isotope of 1H2 can drop out at some later time. It also indicates that all the parts are used as is with no loss of mass.

The second isotope of 1H2 must lose all the mass that is lost. In this step, I state that it is the neutron that does this. The electron and proton are used as the isotope 1H1 with 1.00783 amu's in the second isotope of 1H2 which drops out because only the neutron is needed from this isotope when the isotope 1H3 is created. The electron and proton have a rectangle drawn around them to show that they are used as the isotope 1H1 and they are drawn on the right side of the equal sign to show that the isotope 1H1 has dropped out. The single neutron in the second isotope of 1H2 has a square drawn around the small circle that is the neutron and it is drawn this way in the isotope 1H3. The square indicates that this neutron has lost 0.00434 amu as an gamma 434 ray which does not have an external magnetic field .

When a single unit of mass has a square drawn around it, it must be the mass that loses some of its mass. When two or more masses have a rectangle drawn around them, then they are used as is and they do not lose any mass which means they can drop out of the isotope they are in.

In the last step, the proton in the second isotope of 1H2 lost all the mass that was lost and it lost this mass as two sub-units. In this step the single neutron in the second isotope of 1H2 loses all the mass that is lost and it stays a neutron in the isotope 1H3. Because no electron has been released, all of the lost mass is lost as a single unit. The single neutron in the second isotope of 1H2 has 1.00628 amu's and becomes the second neutron in the isotope 1H3 with 1.00194 amu's.

To find the amount of mass needed to create the isotope 1H3, add the mass of the two isotopes of 1H2. Then subtract the isotope 1H3 from this total to find the total mass lost creating the isotope 1H3.

Mass of first 1H2	2.01411 amu's
Mass of second 1H2	2.01411 amu's
Total mass needed to create 1H3	4.02822 amu's

Total mass needed to create 1H3	4.02822 amu's
Mass of 1H3	3.01605 amu's
Total mass lost creating 1H3	1.01217 amu's

This total mass lost includes the mass of the isotope 1H1 in the second isotope of 1H2 that dropped out. The mass of the isotope 1H1 must therefore be subtracted from the total mass lost creating the isotope 1H3 to find the real mass that is lost creating the isotope 1H3.

Total mass lost creating 1H3	1.01217 amu's
Mass of 1H1 that dropped out of second 1H2	1.00783 amu's
Real mass lost creating 1H3	0.00434 amu

Now subtract the real mass lost creating the isotope 1H3 from the mass of the neutron found in the second isotope of 1H2 to find the mass of second neutron in the isotope 1H3.

Mass of neutron in second 1H2	1.00628 amu's
Real mass lost creating the isotope 1H3	0.00434 amu
Mass of second neutron in 1H3	1.00194 amu's

All of the parts of the isotope 1H3 are now known. Adding the mass of the first isotope 1H2 that is intact within the isotope 1H3 and the mass of the second neutron will give the mass of the isotope 1H3.

Mass of 1H2 in 1H3	2.01411 amu's
Mass of second neutron in 1H3	1.00194 amu's
Mass of 1H3	3.01605 amu's

All of the above is proof that two isotopes of 1H2 will create the isotope 1H3 with the release of the isotope 1H1 that drops out from the second isotope of 1H2. It shows that the loss of 0.00434 amu is composed of a single unit of 0.00434 amu. This single unit is emitted as a 434 ray. It also shows that the first isotope 1H2 is intact within the isotope of 1H3.

Now that all of the STEPS and Sub-steps have been given creating the isotope 1H1, 1H2 and 1H3, I must point out that there is something that will show up later in the creation of the isotopes 2He3 and 2He4.

When there is a sub-isotope that has its proton with less than its maximum value of 1.00728 amu's, this isotope will not be considered to be stable even if it is not radioactive. This proton will regain the mass to make it have 1.00728 amu's. If there are more than one proton with less than the protons maximum, all of these protons will regain the needed mass to become stable. This regained mass will come from a proton or one or more neutrons used in the creation of the new isotope.

Sub-step 9-2-5 and 10-2-6 shows how the proton in the isotope 1H3 regains the mass to make the proton stable.

This completes all the steps creating the isotope 1H3. Next is all of the drawings used in the creation of the isotope 1H3 in STEP 1. Following these drawings will be **all** of the steps creating the isotopes 1H1, 1H2 and 1H3 from the Original Neutron. All of the drawings that show how this is done are now drawn to show their sequence.

All the drawings in **STEP 1** using two isotopes of 1H2 creating the isotope 1H3.

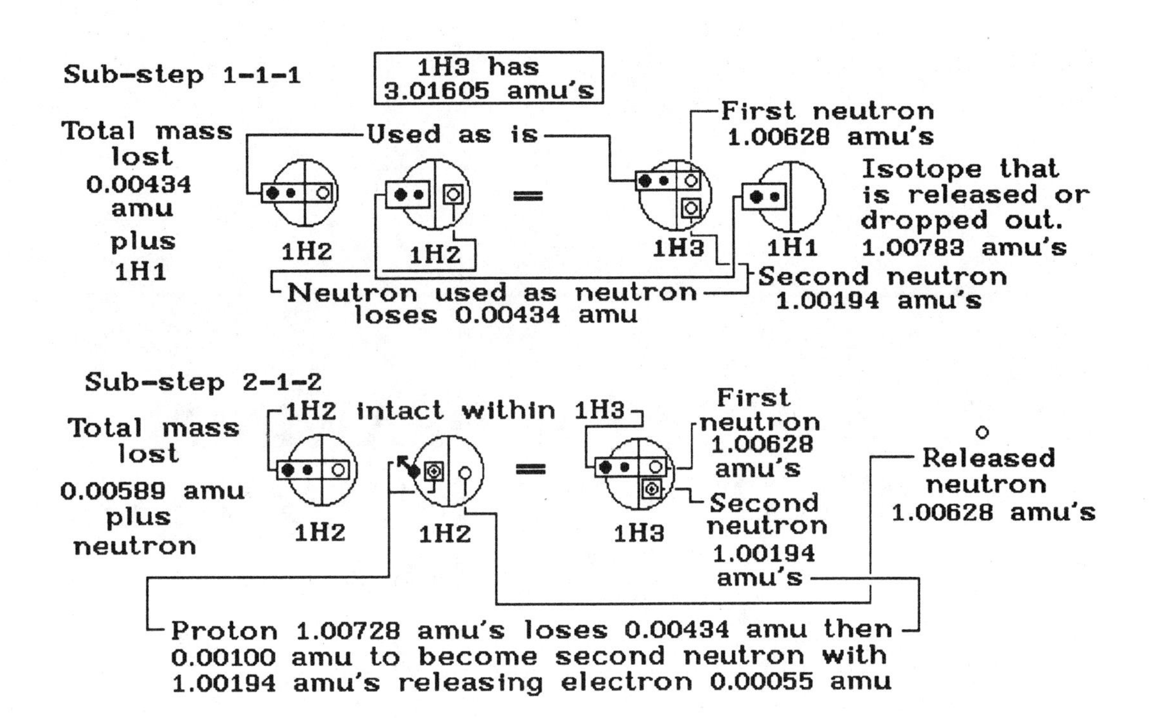

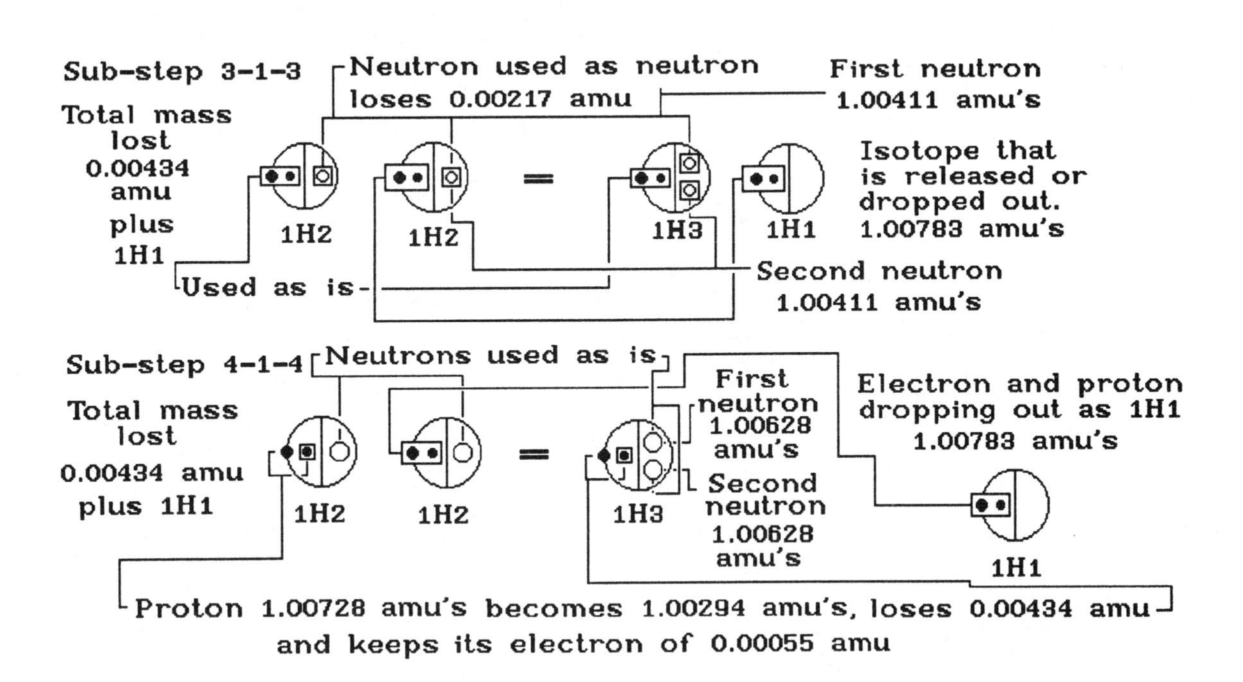

ALL OF THE DRAWINGS CREATING THE ISOTOPES OF HYDROGEN

ALL OF THE DRAWINGS CREATING THE ORIGINAL NEUTRON, 1H1, 1H2 AND 1H3

MESON

ORIGINAL NEUTRON

1.00867 amu's

1N84

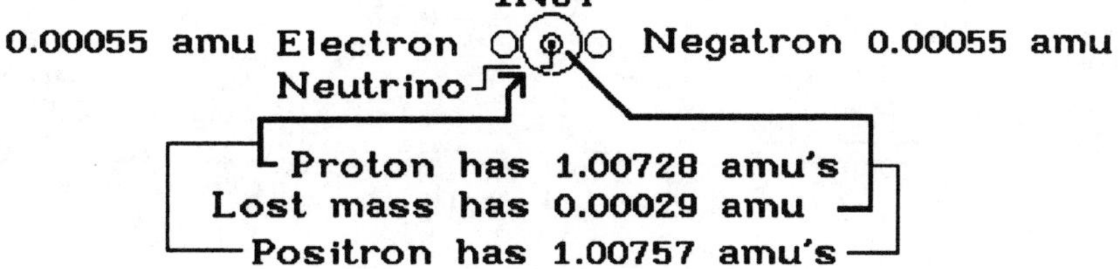

ORIGINAL NEUTRON

1N84

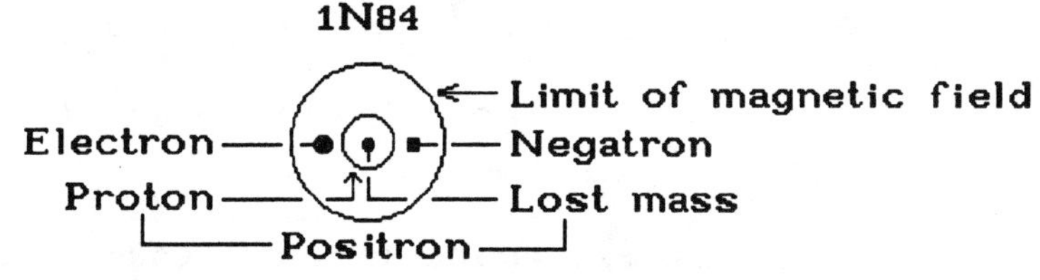

Lost mass	0.00029 amu
Electron	0.00055 amu
Negatron	0.00055 amu
Proton	1.00728 amu's
Positron	1.00757 amu's
1N84	1.00867 amu's

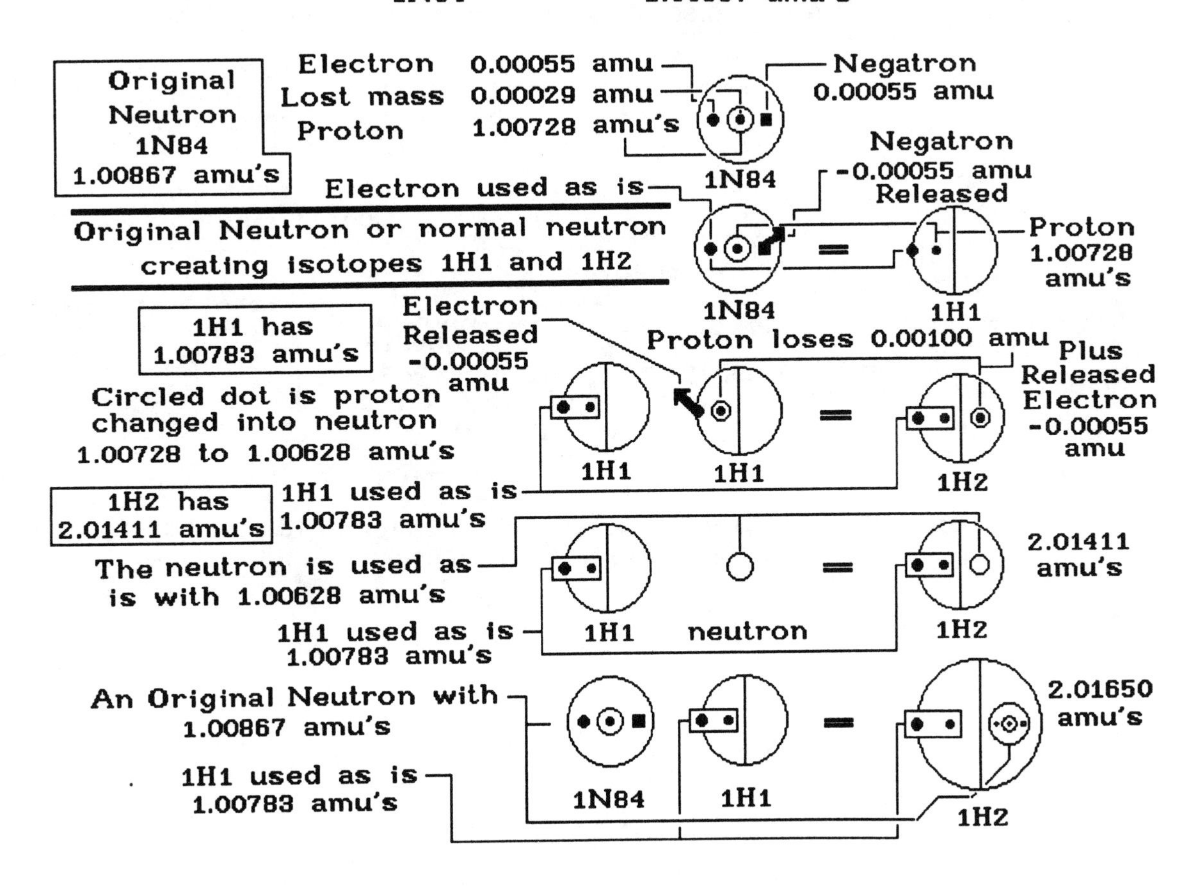

ALL OF THE DRAWINGS CREATING THE ORIGINAL NEUTRON, 1H1, 1H2 AND 1H3

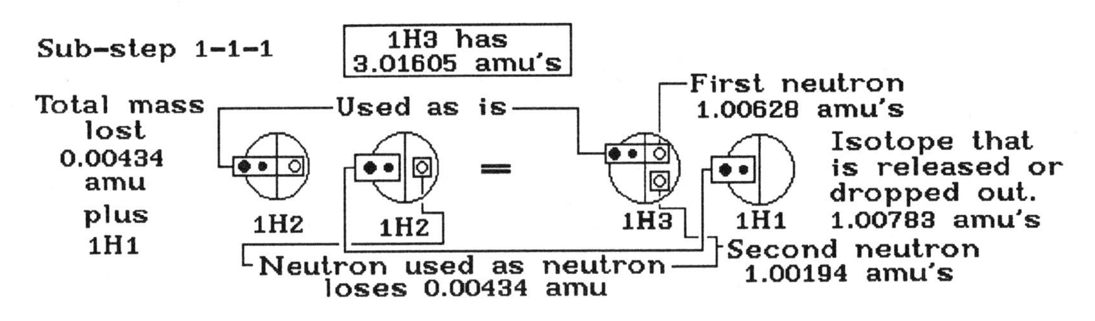

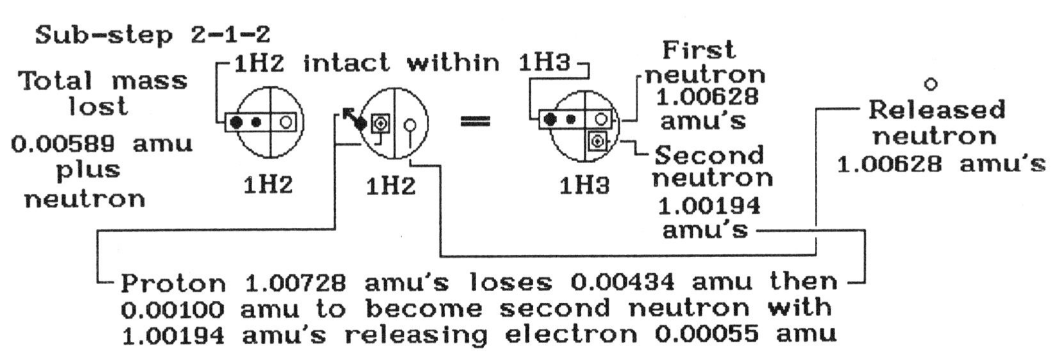

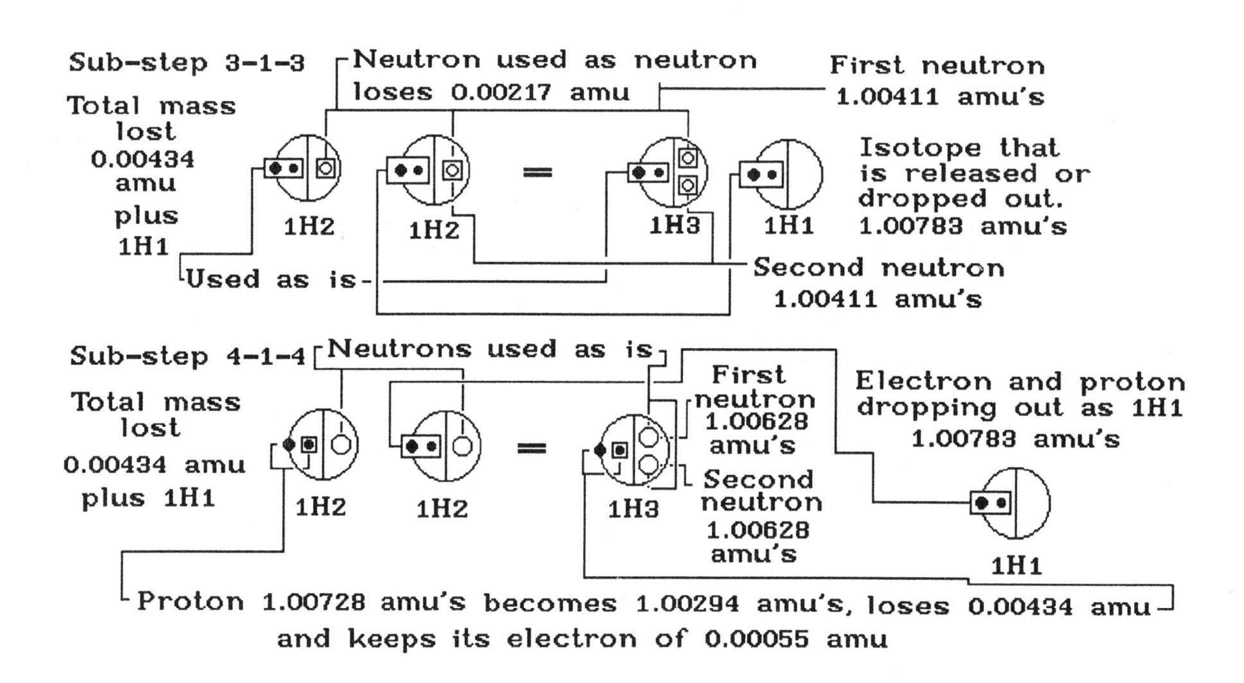

ALL OF THE DRAWINGS CREATING THE ORIGINAL NEUTRON, 1H1, 1H2 AND 1H3

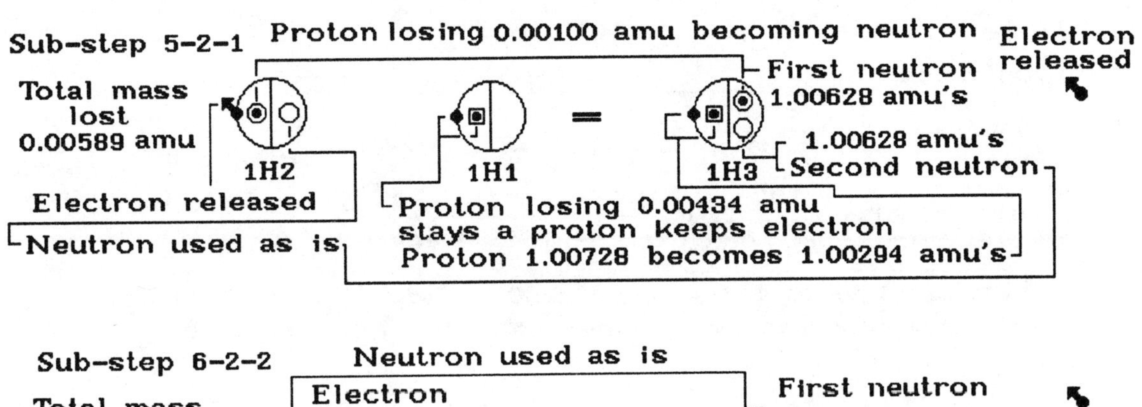

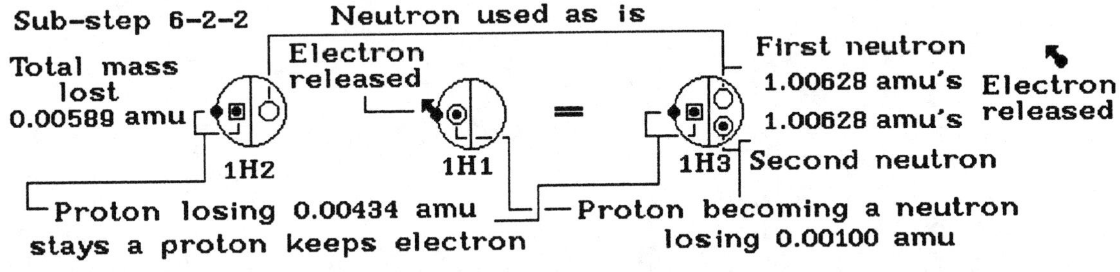

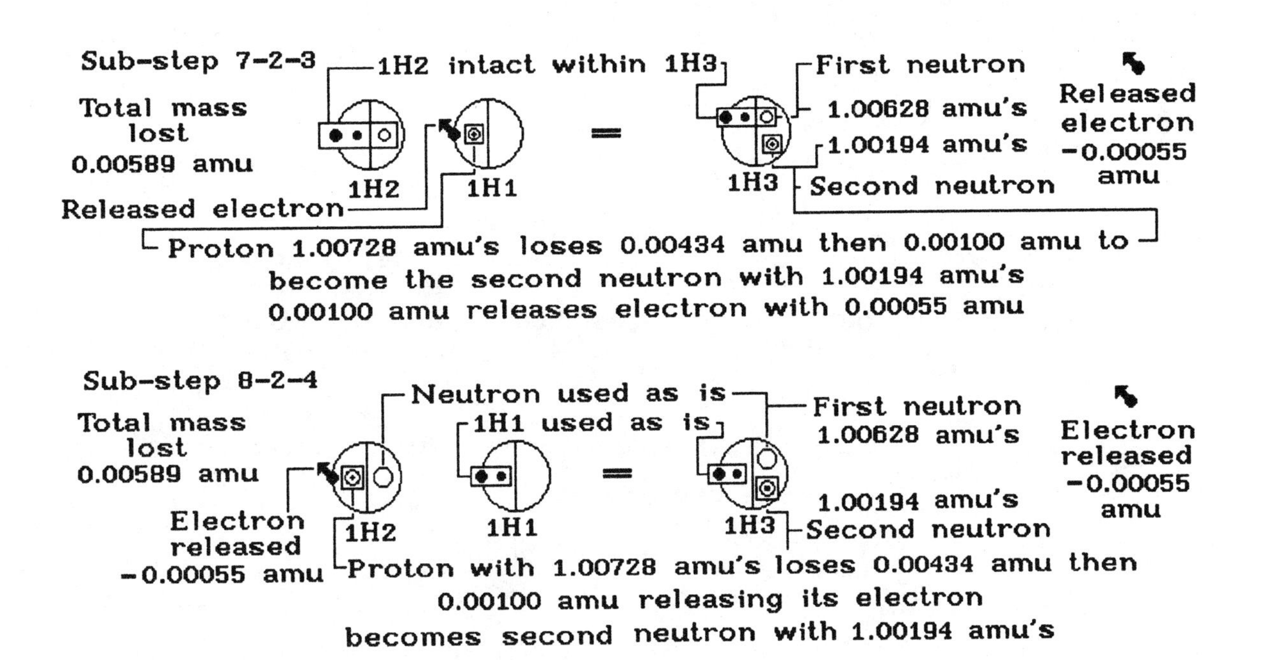

ALL OF THE DRAWINGS CREATING THE ORIGINAL NEUTRON, 1H1, 1H2 AND 1H3

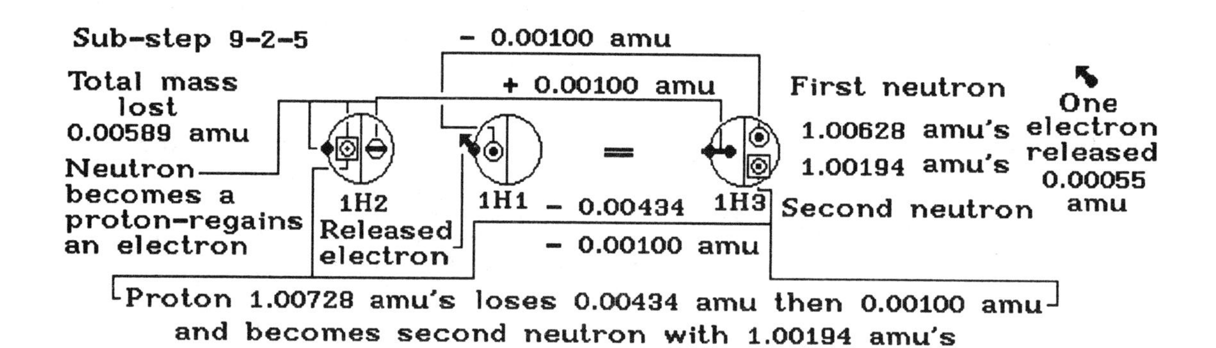

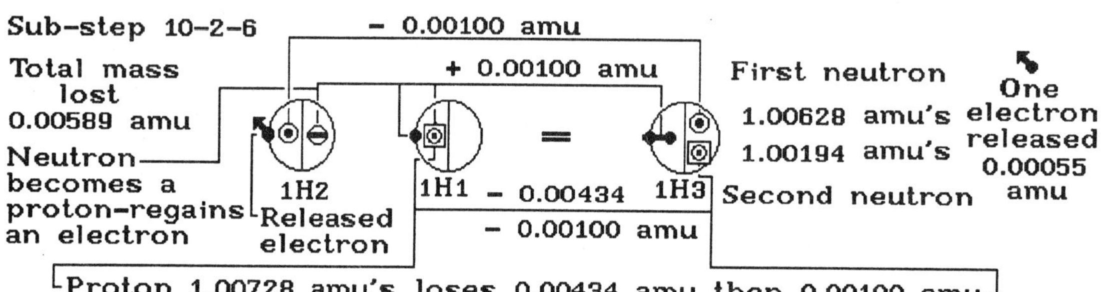

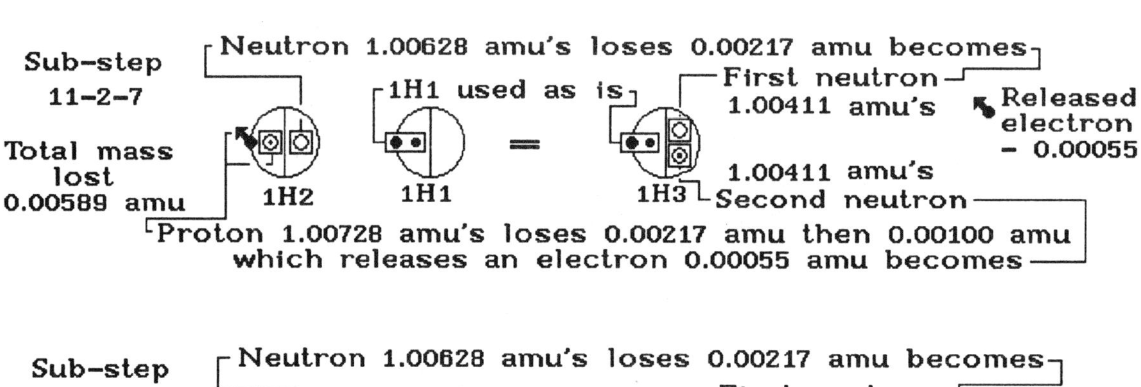

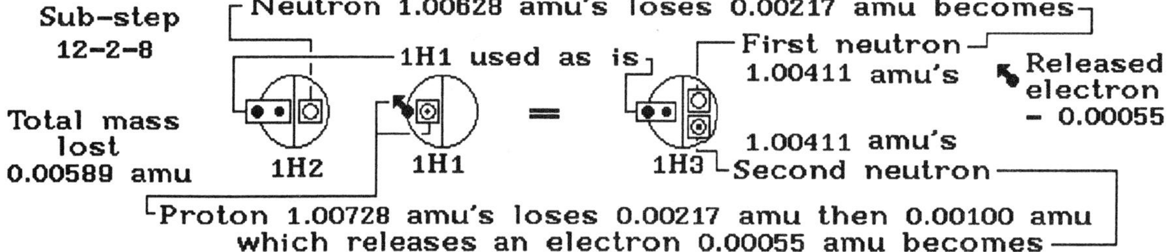

ALL OF THE DRAWINGS CREATING THE ORIGINAL NEUTRON, 1H1, 1H2 AND 1H3

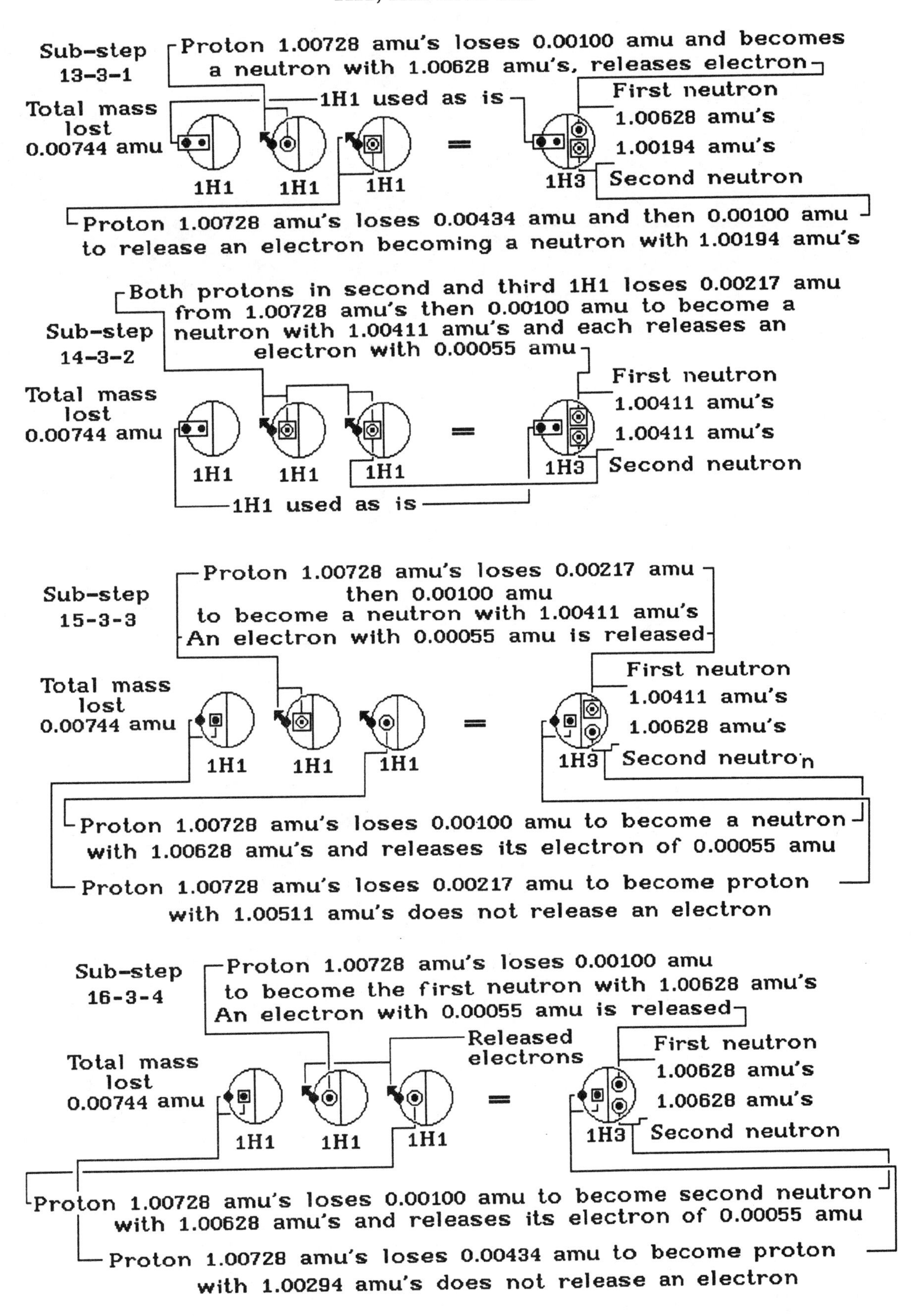

PART 4

HELIUM

ISOTOPES

2He3 and 2He4

HELIUM

4.00260 amu's
(4.0025986 amu's)

The value of 4.0025986 amu's is rounded off to five places to the right of the decimal point. When this is done, it gives the real value of the isotope 2He4 of 4.00260 amu's. This means that the values of 2He3 and 2He4 multiplied by their percentages gives the value of the isotope 2He4 only. The percentage of the isotope 2He3 is 0.0000014 which is too small to change the value for the isotope 2He4. The percentage for 2He4 is 99.99986. When this value is multiplied by the mass of the isotope 2He4 which is 4.00260 amu's, it gives 4.0025944. The atomic mass of the isotope 2He3 is 3.01603 amu's and multiplying this value by 0.0000014 gives 0.0000042. Adding these two values gives the atomic mass of the isotope 2He4 when rounded off to five decimal places of 4.00260 amu's. The chemistry books call the 4.00260 an "average" which it is not. I call this number a composite number.

2He3 2He4 Average value of Helium

3.01603(0.00014%) + 4.00260(99.99986%) = 4.0025986 amu's

3.01603(0.0000014) + 4.00260(99.9999986) = 4.0025986 amu's

0.0000042 + 4.0025944 = 4.0025986 amu's

0.0000042 + 4.00260 = 4.00260 amu's

Always write the number 4.00260 five places to the right of the decimal.
Never write it as 4.0026!!!!

CREATING THE ISOTOPE OF HELIUM (He) 2He3 AND 2H4

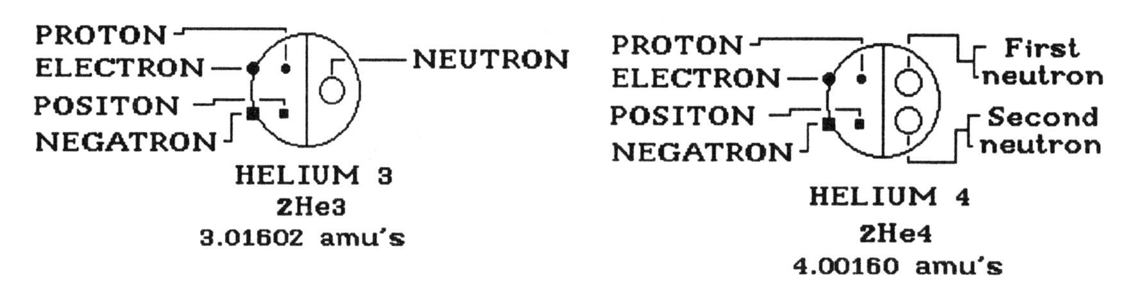

Helium three (2He3) has an atomic mass of 3.01603 amu's. It has an electron, negatron, proton, positon and a neutron.

Helium three (2He3) makes up 0.00014% of all Helium found naturally in nature. It is not radioactive.

Helium four (2He4) has an atomic mass of 4.00260 amu's. It has an electron, negatron, proton, positon and two neutrons.

Helium four (2He4) makes up 99.99986% of all Helium found naturally in nature. It is not radioactive.

Helium five (2He5) has such a small percentage that it is not given and it is radioactive.

HELIUM

2He3

3.01603 amu's

FORMULAS

STEPS

1. 1H3 = 2He3
2. 1H3 + 1H2 = 2He3
3. 1H3 + 1H1 = 2He3
4. 1H2 + 1H2 = 2He3
5. 1H2 + 1H1 = 2He3
6. 1H1 + 1H1 + 1H1 = 2He3

   or

   2(1H1) + 1H1 = 2He3

CHART CREATING THE ISOTOPE 2He3

1. 1H3 = 2He3

1-1-1	1H3 = 2He3, #2 ray, transfer 0.00100 amu, neg in orbit
2-1-2	1H3 = 2He3, #2 ray, transfer 0.00100 amu, 0.00374 amu, neg in orbit
3-1-3	1H3 = 2He3, #2 ray, transfer 0.00100 amu, 0.00217 amu, neg in orbit
4-1-4	1H3 = 2He3, #2 ray, transfer 0.00100 amu, neg in orbit
5-1-5a	1H3 = 2He3, #2 ray, transfer 0.00100 amu, 0.00434 amu, neg in orbit
5-1-5b	2He3 = 2He3, transfer 0.00434 amu
6-1-6	1H3 = 2He3, #2 ray, transfer 0.00100 amu, 0.00434amu, neg in orbit
7-1-7	1H3 = 2He3, #2 ray, transfer 2(0.00100 amu), neg in orbit
8-1-8	1H3 = 2He3, #2 ray, transfer 0.00100 amu, 0.00317, neg in orbit
9-1-9	1H3 = 2He3, #2 ray, 2(0.00100 amu), 0.00434, neg in orbit
10-1-10	1H3 = 2He3, #2 ray, transfer 0.00100 amu, 0.00534 amu, neg in orbit
11-1-11	1H3 = 2He3, #2 ray, neg in orbit
12-1-12	1H3 = 2He3, #2 ray, transfer 0.00100 amu, neg in orbit
13-1-13	1H3 = 2He3, #2 ray, transfer 0.00100 amu, 0.00217 amu, neg in orbit
14-1-14	1H3 = 2He3, #2 ray, transfer 0.00217/0.00317/0.00100, neg in orbit
15-1-15	1H3 = 2He3, #2 ray, transfer 0.00100 amu, neg in orbit
16-1-16	1H3 = 2He3, #2 ray, transfer 0.00100 amu, 0.00217 amu, neg in orbit

2. 1H3 + 1H2 = 2He3

17-2-1	1H3 + 1H2 = 2He3, 591 ray, 1.00628/1.00194 neutrons
18-2-2	1H3 + 1H2 = 2He3, 591 ray, 1.00628/1.00194 neutrons
19-2-3	1H3 + 1H2 = 2He3, 591 ray, 1.00628/1.00194 neutrons
20-2-4	1H3 + 1H2 = 2He3, 591 ray, 1.00628/1.00194 neutrons
21-2-5	1H3 + 1H2 = 2He3, 591 ray, 1.00628/1.00194 neutrons
22-2-6	1H3 + 1H2 = 2He3, 591 ray, 1.00628/1.00194 neutrons
23-2-7	1H3 + 1H2 = 2He3, 434 ray, 157 ray, 1.00628/1.00194 neutrons
24-2-8	1H3 + 1H2 = 2He3, 374 ray, 217 ray, 1.00628/1.00194 neutrons
25-2-9	1H3 + 1H2 = 2He3, 434 ray, 157 ray, 1.00628/1.00194 neutrons
26-2-10	1H3 + 1H2 = 2He3, 374 ray, 217 ray, 1.00628/1.00194 neutrons
27-2-11	1H3 + 1H2 = 2He3, 157 ray, 434 ray, 1.00628/1.00194 neutrons
28-2-12	1H3 + 1H2 = 2He3, 217 ray, 374 ray, 1.00628/1.00194 neutrons
29-2-13	1H3 + 1H2 = 2He3, 157 ray, 434 ray, 1.00628/1.00194 neutrons
30-2-14	1H3 + 1H2 = 2He3, 217 ray, 374 ray, 1.00628/1.00194 neutrons
31-2-15	1H3 + 1H2 = 2He3, 434 ray, 157 ray, 1.00628/1.00194 neutrons
32-2-16	1H3 + 1H2 = 2He3, 157 ray, 434 ray, 1.00628/1.00194 neutrons
33-2-17	1H3 + 1H2 = 2He3, 157 ray, 434 ray, 1.00628/1.00194 neutrons
34-2-18	1H3 + 1H2 = 2He3, 374 ray, 217 ray, 1.00628/1.00194 neutrons
35-2-19	1H3 + 1H2 = 2He3, 217 ray, 374 ray, 1.00628/1.00194 neutrons
36-2-20	1H3 + 1H2 = 2He3, 434 ray, 157 ray, 1.00628/1.00194 neutrons
37-2-21	1H3 + 1H2 = 2He3, 591 ray, 1.00628/1.00194 neutrons
38-2-22	1H3 + 1H2 = 2He3, 591 ray, 1.00628/1.00194 neutrons
39-2-23	1H3 + 1H2 = 2He3, 157 ray, 434 ray, 1.00628/1.00194 neutrons

40-2-24 1H3 + 1H2 = 2He3, 374 ray, 1.00411/1.00628 neutrons
41-2-25 1H3 + 1H2 = 2He3, 157 ray, 217 ray, 1.00411/1.00628 neutrons
42-2-26 1H3 + 1H2 = 2He3, 217 ray, 157 ray, 1.00411/1.00628 neutrons
43-2-27 1H3 + 1H2 = 2He3, 591 ray, 1.00411/1.00411 neutrons
44-2-28 1H3 + 1H2 = 2He3, 157 ray, 434 ray, 1.00411/1.00411 neutrons
45-2-29 1H3 + 1H2 = 2He3, 217 ray, 374 ray, 1.00411/1.00411 neutrons
46-2-30 1H3 + 1H2 = 2He3, 374 ray, 217 ray, 1.00411/1.00411 neutrons
47-2-31 1H3 + 1H2 = 2He3, 434 ray, 157 ray, 1.00411/1.00411 neutrons
48-2-32 1H3 + 1H1 = 2He3, 157 ray, 434 ray, 1.00411/1.00411 neutrons
49-2-33 1H3 + 1H1 = 2He3, 217 ray, 374 ray, 1.00411/1.00411 neutrons
50-2-34 1H3 + 1H1 = 2He3, 374 ray, 217 ray, 1.00411/1.00411 neutrons
51-2-35 1H3 + 1H2 = 2He3, 434 ray, 157 ray, 1.00411/1.00411 neutrons

3. 1H3 + 1H1 = 2He3

52-3-1 1H3 + 1H1 = 2He3, 591 ray, 1.00194 neutron
53-3-2 1H3 + 1H1 = 2He3, 591 ray, 1.00194 neutron
54-3-3 1H3 + 1H1 = 2He3, 591 ray, 1.00194 neutron
55-3-4 1H3 + 1H1 = 2He3, 434 ray, 157 ray, 1.00194 neutron
56-3-5 1H3 + 1H1 = 2He3, 157 ray, 434 ray, 1.00194 neutron
57-3-6 1H3 + 1H1 = 2He3, 434 ray, 157 ray, 1.00194 neutron
58-3-7 1H3 + 1H1 = 2He3, 217 ray, 374 ray, 1.00194 neutron
59-3-8 1H3 + 1H1 = 2He3, 374 ray, 217 ray, 1.00194 neutron
60-3-9 1H3 + 1H1 = 2He3, 374 ray, 217 ray, 1.00194 neutron
61-3-10 1H3 + 1H1 = 2He3, 374 ray, 217 ray, 1.00194 neutron
62-3-11 1H3 + 1H1 = 2He3, 374 ray, 1.00411 neutron
63-3-12 1H3 + 1H1 = 2He3, 374 ray, 1.00411 neutron
64-3-13 1H3 + 1H1 = 2He3, 157 ray, 1.00628 neutron
65-3-14 1H3 + 1H1 = 2He3, 217 ray, 157 ray, 217 ray, 1.00194 neutron

4. 1H2 + 1H2 = 2He3

66-4-1a 1H2 + 1H2 = 2He3, 436 ray, 1H1 isotope
66-4-1b 1H2 + 1H2 = 2He3, 436 ray, 1H1 isotope
66-4-1c 1H2 + 1H2 = 2He3, 436 ray, 1H1 isotope
67-4-2 1H2 + 1H2 = 2He3, 591 ray, 1.00628 neutron
68-4-3 1H2 + 1H2 = 2He3, 591 ray, 1.00628 neutron
69-4-4 1H2 + 1H2 = 2He3, 434 ray, 157 ray, 1.00628 neutron
70-4-5 1H2 + 1H2 = 2He3, 434 ray, 157 ray, 1.00628 neutron
71-4-6 1H2 + 1H2 = 2He3, 217 ray, 374 ray, 1.00628 neutron
72-4-7 1H2 + 1H2 = 2He3, 217 ray, 374 ray, 1.00628 neutron
73-4-8 1H2 + 1H2 = 2He3, 591 ray, 1.00628 neutron
74-4-9 1H2 + 1H2 = 2He3, 217 ray, 157 ray, 217 ray, 1.00628 neutron

5. 1H2 + 1H1 = 2He3

75-5-1 1H2 + 1H1 = 2He3, 591 ray
76-5-2 1H2 + 1H1 = 2He3, 591 ray
77-5-3 1H2 + 1H1 = 2He3, 591 ray
78-5-4 1H2 + 1H1 = 2He3, 591 ray
79-5-5 1H2 + 1H1 = 2He3, 591 ray
80-5-6 1H2 + 1H1 = 2He3, 217 ray, 374 ray
81-5-7 1H2 + 1H1 = 2He3, 374 ray, 217 ray
82-5-8 1H2 + 1H1 = 2He3, 217 ray, 217 ray, 157 ray
83-5-9 1H2 + 1H1 = 2He3, 157 ray, 217 ray, 217 ray
84-5-10 1H2 + 1H1 = 2He3, 217 ray, 157 ray, 217 ray
85-5-11 1H2 + 1H1 = 2He3, 217 ray, 374 ray
86-5-12 1H2 + 1H1 = 2He3, 374 ray, 217 ray
87-5-13 1H2 + 1H1 = 2He3, 157 ray, 434 ray
88-5-14 1H2 + 1H1 = 2He3, 157 ray, 434 ray
89-5-15 1H2 + 1H1 = 2He3, 434 ray, 157 ray

6. 2(1H1) + 1H1 or 1H1 + 1H1 + 1H1 = 2He3

90-6-1 2(1H1) + 1H1 = 2He3, 591 ray, 100 ray, -1e
91-6-2 1H1 + 1H1 + 1H1 = 2He3, 591 ray, 100 ray, -1e
92-6-3 1H1 + 1H1 + 1H1 = 2He3, 434 ray, 157 ray, 100 ray, -1e
93-6-4 1H1 + 1H1 + 1H1 = 2He3, 434 ray, 157 ray, 100 ray, -1e
94-6-5 1H1 + 1H1 + 1H1 = 2He3, 374 ray, 217 ray, 100 ray, -1e
95-6-6 1H1 + 1H1 + 1H1 = 2He3, 374 ray, 217 ray, 100 ray, -1e
96-6-7 1H1 + 1H1 + 1H1 = 2He3, 217 ray, 217 ray, 157 ray, 100 ray, -1e

The above chart creating the isotope 1H3 has some steps looking alike but they differ because of the way their parts are used. The -1e means that an electron has been released from one of the isotopes used to create the isotope 1H3. The isotopes found within parentheses are used as is and in a like manner.

KNOWN FACTS FOR 2He3

1.	Atomic mass 2He3	3.01603 amu's
2.	Atomic mass 1H3	3.01605 amu's
3.	Atomic mass 1H2	2.01411 amu's
4.	Atomic mass 1H1	1.00783 amu's
5.	Atomic mass proton or positon	1.00728 amu's
6.	Atomic mass neutron	1.00628 amu's
7.	Atomic mass neutron	1.00411 amu's
8.	Atomic mass neutron	1.00194 amu's
10.	Atomic mass electron or negatron	0.00055 amu
11.	Mass lost releasing electron or negatron	0.00100 amu
12.	Mass lost not releasing electron or negatron	0.00591 amu
13.	Mass lost not releasing electron or negatron	0.00434 amu
14.	Mass lost not releasing electron or negatron	0.00217 amu

All of the atomic masses listed above have already been given or found.

STEP 6

2He3

FORMULA

2(1H1) + 1H1 = 2He3

OR

1H1 + 1H1 + 1H1 = 2He3

STEPS

6. 2(1H1) + 1H1 or 1H1 + 1H1 + 1H1 = 2He3

90-6-1 2(1H1) + 1H1 = 2He3, 591 ray, 100 ray, -1e
91-6-2 1H1 + 1H1 + 1H1 = 2He3, 591 ray, 100 ray, -1e
92-6-3 1H1 + 1H1 + 1H1 = 2He3, 434 ray, 157 ray, 100 ray, -1e
93-6-4 1H1 + 1H1 + 1H1 = 2He3, 434 ray, 157 ray, 100 ray, -1e
94-6-5 1H1 + 1H1 + 1H1 = 2He3, 374 ray, 217 ray, 100 ray, -1e
95-6-6 1H1 + 1H1 + 1H1 = 2He3, 374 ray, 217 ray, 100 ray, -1e
96-6-7 1H1 + 1H1 + 1H1 = 2He3, 217 ray, 217 ray, 157 ray, 100 ray, -1e

All rays in this **STEP** are alpha rays.

Sub-step 96-6-7

2He3

1H1 + 1H1 + 1H1 = 2He3, two alpha 217 rays, alpha 157 ray, 100 ray, -1e

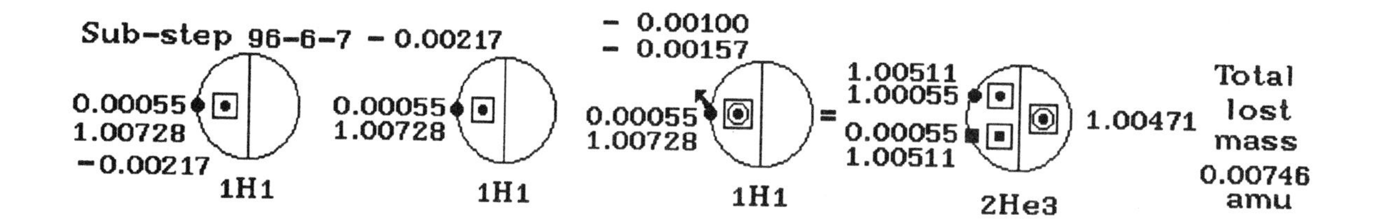

Three isotopes of 1H1 are needed to create the isotope 2He3. Each isotope of 1H1 has 1.00783 amu's. Each one has one electron or negatron with 0.00055 amu and one proton or positon with 1.00728 amu's.

The isotope 2He3 has 3.01603 amu's. It has one electron with 0.00055 amu, one negatron with 0.00055 amu, one proton with 1.00511 amu's, one positon with 1.00511 amu's and one neutron with 1.00471 amu's.

There is a total of 0.00746 amu lost creating the isotope 2He3 which includes one released electron. There are two 0.00217 amu with each emitted as a 217 ray, one 0.00157 amu emitted as a 157 ray and a 0.00100 amu emitted as a 100 ray. All alpha rays have an external magnetic field.

The first isotope of 1H1 has its electron used as the electron in the isotope 2He3 with 0.00055 amu. It is drawn on the big circle as a dot in the isotope 1H1 and 2He3.

The proton with 1.00728 amu's is drawn as a dot on the left side of the vertical line within the big circle. It has a square drawn around it to show that it loses 0.00217 amu to become the proton in the isotope 2He3 with 1.00511 amu's. The loss of 0.00217 amu will not release its electron.

The second isotope of 1H1 has its electron with 0.00055 amu drawn as a dot on the big circle and it is drawn as a square on the big circle in the isotope 2He3 to show that the electron in the isotope 1H1 has become the negatron in the isotope 2He3.

The proton in the second isotope of 1H1 with 1.00728 amu's is drawn as a dot on the left side of the vertical line within the big circle. This proton in the second isotope of 1H1 becomes the positon in the isotope 2He3 and changes from a dot to a small square to show that it is now a positon. It is drawn with a square around the dot in the second isotope of 1H1 and square is also drawn around the small square in the isotope 2He3 to show that it has lost 0.00217 amu to become the positon in the isotope 2He3 with 1.00511 amu's. The loss of 0.00217 amu will not release its electron.

The third isotope of 1H1 has it electron with 0.00055 amu drawn as a dot on the big circle with an arrow pointing away from it which means that this electron has been released. The proton has a small circle drawn around the dot that is the proton. There is a square drawn around both of them. The square shows that the proton has lost 0.00157 of its mass that did not release its electron and then the small circle shows that there is 0.00100 amu that does release the electron. When the electron is released and the 0.00100 amu is lost, the proton with 1.00728 amu's becomes a neutron with 1.00471 amu's in the isotope 2He3 and it is drawn the same way as the proton found in the third isotope of 1H1.

Adding all the masses of the three isotopes of 1H1 will give the total mass needed to create the isotope 2He3

Mass of first 1H1	1.00783 amu's
Mass of second 1H1	1.00783 amu's
Mass of third 1H1	1.00783 amu's
Total mass needed to create 2He3	3.02349 amu's

Now subtract the mass of the isotope 2He3 from this total to find the total mass lost creating the isotope 2He3.

Total mass needed to create 2He3	3.02349 amu's
Mass of 2He3	3.01603 amu's
Total mass lost creating 2He3	0.00746 amu

In this step, I have stated that the single neutron in the isotope 2He3 has 1.00471 amu's. Since this neutron came from the third isotope of 1H1's proton, subtract the mass of this neutron from the mass of this proton plus the 0.00100 amu that is lost creating the neutron.

Mass of proton in third 1H1	1.00728 amu's
Mass of neutron in 2He3	1.00471 amu's
Mass lost creating neutron in 2He3	0.00257 amu

Mass lost creating neutron in 2He3	0.00257 amu
Mass lost releasing electron in third 1H1	0.00100 amu
Mass lost not releasing electron in third 1H1	0.00157 amu

The proton in the third isotope of 1H1 lost first the 0.00157 amu as a 157 ray and then lost 0.00100 amu as a 100 ray to become the neutron in the isotope 2He3. By subtracting the 0.00257 amu from this proton, the mass of the neutron can be found.

Mass of proton in third 1H1	1.00728 amu's
Mass lost creating neutron in 2He3	0.00257 amu
Mass of neutron in 2He3	1.00471 amu's

Now that the total mass lost by the third isotope of 1H1 is known, this lost mass can be subtracted from the total mass needed to create the isotope 2He3 to find the mass lost by the first and second isotope of 1H1 and the released electron of the third 1H1.

Total mass lost creating 2He3	0.00746 amu
Mass lost creating neutron in 2He3	0.00257 amu
Lost mass by first and second 1H1 plus electron of third 1H1	0.00489 amu

This remaining lost mass includes the mass of the electron that has been released. Subtract the mass of the electron to find the mass lost by the protons in the first and second isotope of 1H1.

Lost mass by second 1H1 plus electron of third 1H1	0.00489 amu
Mass of electron released by third 1H1	0.00055 amu
Mass lost by first and second 1H1's proton	0.00434 amu

The lost mass of 0.00434 is composed of two sub-units of 0.00217 amu. Neither the 0.00434 amu or the 0.00217 amu will release the electron from the proton. Note that when I ended the sentence with amu I did not put a 's with it as proper grammar would indicate. I have made it a rule never to put the 's with any mass that is less one atomic mass unit.

Since the 0.00434 amu is composed of two 0.00217 amu, then the protons in the first and second isotopes of 1H1's protons will each lose this mass. When the first isotope's proton loses this mass, it stays a proton in the isotope 2He3. Subtract the 0.00217 from this proton.

Mass of proton in first 1H1	1.00728 amu's
Mass lost by it	0.00217 amu
Mass of proton in 2He3	1.00511 amu's

When the second isotope's proton loses 0.00217 amu, it becomes the positon in the isotope 2He3. Subtract the 0.00217 amu from it to find the mass of the positon in the isotope 2He3.

Mass of proton in second 1H1	1.00728 amu's
Mass lost by it	0.00217 amu
Mass of positon in 2He3	1.00511 amu's

Now that the masses of the electron, negatron, proton, positon and neutron are known, the mass of the isotope 2He3 can be found by adding all of these parts.

Mass of electron and negatron in 2He3	0.00110 amu
Mass of proton in 2He3	1.00511 amu's
Mass of positon in 2He3	1.00511 amu's
Mass of neutron in 2He3	1.00471 amu
Mass of 2He3	3.01603 amu's

Sub-step 95-6-6

2He3

FORMULA

1H1 + 1H1 + 1H1 = 2He3, one alpha 374 ray, alpha 217 ray, 100 ray, -1e

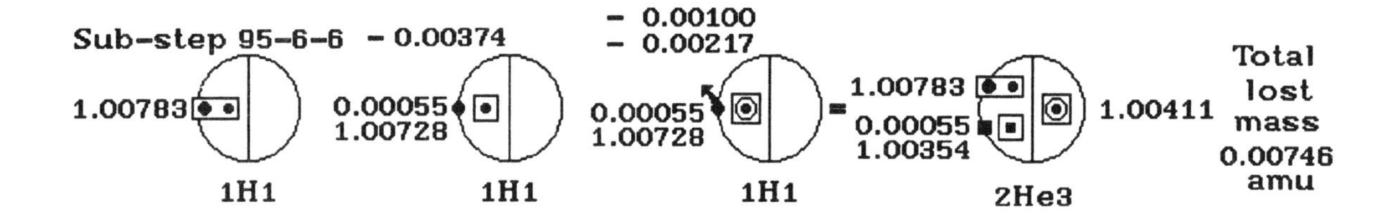

Three isotopes of 1H1 are needed to create the isotope 2He3. Each isotope of 1H1 has 1.00783 amu's. Each one has one electron or negatron with 0.00055 amu and one proton or positon with 1.00728 amu's. The isotope 2He3 has 3.01603 amu's. It has one electron with 0.00055 amu, one negatron with 0.00055 amu, one proton with 1.00728 amu's, one positon with 1.00354 amu's and one neutron with 1.00411 amu's. There is a total of 0.00746 amu lost creating the isotope 2He3 which includes one released electron. All alpha rays have an external magnetic field.

It is understood that the isotope 1H1 can have either an electron or a negatron and a proton or a positon. I will only use the electron and proton in this write-up. It will be understood that the electron and negatron are found on the big circle, the proton and positon are found within the big circle to the left of the vertical line while the neutrons are found to the right of the vertical line. Any mass left over will be found on the right side of the equal sign. The released electron or negatron may or may not be shown on the right side of the equal sign. A square around any part means that this part has lost mass and does not release an electron. A rectangle means that there is more than one part used as is and they are found drawn in the same way in the isotope being created. A small circle around a proton or negatron means that an electron or negatron has been released and that 0.00100 amu has been emitted as a 100 ray. The dot of the proton can also be found drawn in small circle of the neutron to show that this neutron came from the proton or positon that also has a small circle. The neutron may have a square around it to indicate that it has lost some of its mass and it is found as a neutron in the isotope being created.

Anytime a ray is emitted when mass is lost, the mass that is converted to a ray is lost as a unit, not a sub-unit.

If there are sub-units, then these sub-units are part of the total mass that has been converted to a ray.

The "Total mass lost" is composed of all the mass that is lost creating a new isotope. It includes the electron or negatron that has been released and any mass that has been converted to a ray. Are they lost as one combined single unit or does each sub-unit get emitted as a ray individually? I have checked all of my books that should give me this information and they do not. In fact, one book stated that the isotope 1H3 changed into 2He3 with the release of an electron. This is impossible because when this happens, only 0.00002 amu is lost. There is not enough mass lost to include the electron. However, in this step, 0.00374 amu is lost. This can be broken down into one sub-unit of 0.00217 amu, one sub-unit of 0.00100 amu, one sub-unit of 0.00055 amu and a sub-unit of 0.00002 amu. None of these sub-units can be broken down further. When the 0.00374 amu is emitted as a ray, the 0.00100 amu that is part of it does not release an electron or negatron.

The big question is, did the sub-unit of 0.00055 amu get released as an electron when the 0.00100 amu was emitted as a 100 ray? I could not find an answer to this question. The loss of 0.00591 amu, 0.00374 amu and 0.00157 amu all have this question in them. If the answer is they are each lost as a single unit, then the electron is converted to pure energy and I am not sure this can happen. If they each break down into their sub-units, with the electron released, then there would be a 434 ray, a 217 ray, a 100 ray and a #2 ray emitted and the rays should have been detected.

When a neutron or an isotope has dropped out, they are not part of the "Total mass lost" but it is mass that is not part of the new isotope just created. This is mass that has been lost in the sense that it is part of the lost mass from the isotopes that are used to create the new isotope.

In this step, the first isotope of 1H1 has a rectangle drawn around its electron and proton to show that it is used as is and that it is drawn the same way in the isotope 2He3. This rectangle means that the isotope 1H1 is found intact within the isotope 2He3 and can drop out as such at some later time.

The second isotope of 1H1 has its electron with 0.00055 amu drawn as a dot on the big circle and it is drawn as a square on the big circle in the isotope 2He3 to show that the electron in the isotope 1H1 has become the negatron in the isotope 2He3.

The proton in the second isotope of 1H1 with 1.00728 amu's is drawn as a dot on the left side of the vertical line within the big circle. This proton in the second isotope of 1H1 becomes the positon in the isotope 2He3 and changes from a dot to a small square to show that it is now a positon. It is drawn with a square around the dot in the second isotope of 1H1 and square is also drawn around the small square in the isotope 2He3 to show that it has lost 0.00374 amu to become the positon in the isotope 2He3 with 1.00354 amu's. The loss of 0.00374 amu will not release its electron.

The third isotope of 1H1 has it electron drawn with an arrow pointing away from it which means that this electron has been released. The proton has a small circle drawn around the dot that is the proton. There is a square drawn around both of them. The square shows that the proton lost 0.00217 amu that did not release its electron and then the small circle shows that there is 0.00100 amu that does release the electron. When the electron is released and the 0.00100 amu is lost, the proton becomes a neutron with 1.00411 amu's in the isotope 2He3 and neutron is drawn the same way as the proton found in the third isotope of 1H1.

Adding all the masses of the three isotopes of 1H1 will give the total mass needed to create the isotope 2He3

Mass of first 1H1	1.00783 amu's
Mass of second 1H1	1.00783 amu's
Mass of third 1H1	1.00783 amu's
Total mass needed to create 2He3	3.02349 amu's

Now subtract the mass of the isotope 2He3 from this total to find the total mass lost creating the isotope 2He3.

Total mass needed to create 2He3	3.02349 amu's
Mass of 2He3	3.01603 amu's
Total mass lost creating 2He3	0.00746 amu

In this step, I have stated that the single neutron in the isotope 2He3 has 1.00411 amu's. Since this neutron came from the third isotope of 1H1's proton, subtract the mass of this neutron from the mass of this proton plus the 0.00100 amu that is lost creating the neutron.

Mass of proton in third 1H1	1.00728 amu's
Mass of neutron in 2He3	1.00411 amu's
Mass lost creating neutron in 2He3	0.00317 amu

Mass lost creating neutron in 2He3	0.00317 amu
Mass lost releasing electron in third 1H1	0.00100 amu
Mass lost not releasing electron in third 1H1	0.00217 amu

The proton in the third isotope of 1H1 lost first the 0.00217 amu as a 217 ray and then lost 0.00100 amu as a 100 ray to become the neutron in the isotope 2He3. By subtracting the 0.00317 amu from this proton, the mass of the neutron can be found.

The lost mass of 0.00217 amu and 0.00100 amu are units. They are both emitted as rays. They are not sub-units of the 0.00317 amu that has been lost. There is no ray emitted for the 0.00317 amu that is lost.

Mass of proton in third 1H1	1.00728 amu's
Mass lost creating neutron in 2He3	0.00317 amu
Mass of neutron in 2He3	1.00411 amu's

Now that the total mass lost by the third isotope of 1H1 is known, this lost mass can be subtracted from the total mass needed to create the isotope 2He3 to find the mass lost by the second isotope of 1H1.

Total mass lost creating 2He3	0.00746 amu
Mass lost creating neutron in 2He3	0.00317 amu
Remaining lost mass	0.00429 amu

This remaining lost mass includes the mass of the electron that has been released by the third isotope of 1H1. Subtract the mass of the electron to find the mass lost by the second isotope of 1H1.

Remaining lost mass	0.00429 amu
Mass of electron released by third 1H1	0.00055 amu
Mass lost by second 1H1's proton	0.00374 amu

The lost mass of the second isotope of 1H1's proton is now known. This lost mass is subtracted from this proton to find the mass of the positon found in the isotope 2He3.

Mass of second 1H1's proton	1.00728 amu's
Mass lost by second 1H1's proton	0.00374 amu
Mass of positon in 2He3	1.00354 amu's

All of the parts of the isotope 2He3 are now known. Adding these parts will give the mass of the isotope 2He3. The first isotope of 1H1 was used as is so its electron and proton are found in the isotope 2He3 intact with 1.00783 amu's.

First isotope of 1H1 in 2He3	1.00783 amu's
Mass of negatron in 2He3	0.00055 amu
Mass of positon in 2He3	1.00354 amu's
Mass of neutron in 2He3	1.00411 amu's
Mass of 2He3	3.01603 amu's

The value for the neutron in the isotope 2He3 was chosen because there is already a neutron known to have this mass. Thus making this neutron have this mass enables one to find the mass of the positon from the knowledge that 0.00100 amu will release an electron from the third isotope of 1H1.

The 0.00374 amu is emitted as a 374 ray. The 0.00217 amu is emitted as a 217 ray and the 0.00100 amu is emitted as a 100 ray. The electron is released as an electron.

The value 0.00374 amu is composed of one sub-unit of 0.00217 amu and a sub-unit of 0.00157 amu. The value of 0.00157 amu is composed of three more sub-units of 0.00100 amu, 0.0055 amu and the third sub-unit of 0.00002 amu.

Sub-step 94-6-5

2He3

FORMULA

1H1 + 1H1 + 1H1 = 2He3, one alpha 374 ray, alpha 217 ray, one 100 ray, -1e

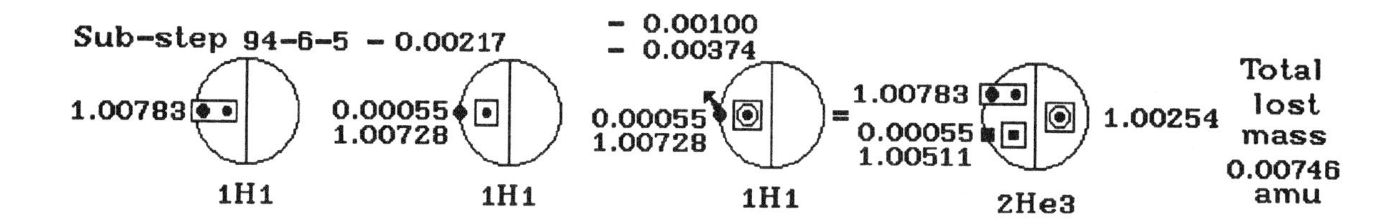

Three isotopes of 1H1 are needed to create the isotope 2He3. Each isotope of 1H1 has 1.00783 amu's. Each isotope has one electron or negatron with 0.00055 amu and one proton or positon with 1.00728 amu's. The isotope 2He3 has 3.01603 amu's. It has one electron with 0.00055 amu, one negatron with 0.00055 amu, one proton with 1.00728 amu's, one positon with 1.00511 amu's and one neutron with 1.00254 amu's. There is a total of 0.00746 amu lost creating the isotope 2He3 which includes one released electron. All alpha rays have an external magnetic field.

The first isotope of 1H1 has a rectangle drawn around its electron and proton to show that it is used as is and that it is drawn the same way in the isotope 2He3. This rectangle means that the isotope 1H1 is found intact within the isotope 2He3 and can drop out as such at some later time.

The second isotope of 1H1 has its electron with 0.00055 amu drawn as a dot on the big circle and it is drawn as a square on the big circle in the isotope 2He3 to show that the electron in the isotope 1H1 has become the negatron in the isotope 2He3.

The proton in the second isotope of 1H1 with 1.00728 amu's is drawn as a dot on the left side of the vertical line within the big circle. This proton in the second isotope of 1H1 becomes the positon in the isotope 2He3 and changes from a dot to a small square to show that it is now a positon. It is drawn with a square around the dot in the second isotope of 1H1. A square is also drawn around the small square in the isotope 2He3 to show that it has lost 0.00217 amu to become the positon in the isotope 2He3 with 1.00511 amu's. The loss of 0.00217 amu will not release its electron.

The third isotope of 1H1 has it electron drawn with an arrow pointing away from it which means that this electron has been released. The proton has a small circle drawn around the dot that is the proton. There is a square drawn around both of them. This square shows that the proton lost some of its mass that did not release its electron and then the small circle shows that there is 0.00100 amu that does release the electron. The square around the small circle shows that 0.00374 amu has been lost. When the electron is released and the 0.00100 amu is lost, the proton becomes a neutron with 1.00254 amu's in the isotope 2He3 and neutron is drawn the same way as the proton found in the third isotope of 1H1.

Adding all the masses of the three isotopes of 1H1 will give the total mass needed to create the isotope 2He3.

Mass of first 1H1	1.00783 amu's
Mass of second 1H1	1.00783 amu's
Mass of third 1H1	1.00783 amu's
Total mass needed to create 2He3	3.02349 amu's

Now subtract the mass of the isotope 2He3 from this total to find the total mass lost creating the isotope 2He3.

Total mass needed to create 2He3	3.02349 amu's
Mass of 2He3	3.01603 amu's
Total mass lost creating 2He3	0.00746 amu

In this step, I have stated that the single neutron in the isotope 2He3 has 1.00254 amu's. Since this neutron came from the third isotope of 1H1's proton, subtract its mass from the mass of this proton creating the neutron will have a remaining mass that is lost.

Mass of proton in third 1H1	1.00728 amu's
Mass of neutron in 2He3	1.00254 amu's
Mass lost creating neutron in 2He3	0.00474 amu

This remaining mass of the proton in the third isotope of 1H1 includes the mass that is lost which released the electron. Subtract this electron releasing mass to find the mass which did not release the electron.

Mass lost creating neutron in 2He3	0.00474 amu
Mass lost releasing electron in third 1H1	0.00100 amu
Mass lost not releasing electron in third 1H1	0.00374 amu

The proton in the third isotope of 1H1 lost first the 0.00374 amu as a 374 ray and then lost 0.00100 amu as a 100 ray to become the neutron in the isotope 2He3. By subtracting the 0.00474 amu from this proton, the mass of the neutron can be found.

Mass of proton in third 1H1	1.00728 amu's
Mass lost creating neutron in 2He3	0.00474 amu
Mass of neutron in 2He3	1.00254 amu's

Now that the total mass lost by the third isotope of 1H1 is known, this lost mass can be subtracted from the total mass needed to create the isotope 2He3 to find the mass lost by the second isotope of 1H1 and electron of third 1H1.

Total mass lost creating 2He3	0.00746 amu
Mass lost creating neutron in 2He3	0.00474 amu
Lost mass by second 1H1 plus electron of third 1H1	0.00272 amu

This remaining lost mass includes the mass of the electron that has been released. Subtract the mass of the electron to find the mass lost by the second isotope of 1H1.

Remaining lost mass	0.00272 amu
Mass of electron released by third 1H1	0.00055 amu
Mass lost by second 1H1's proton	0.00217 amu

By subtracting the mass lost by the second 1H1's proton, the mass of the positon can be found.

Mass of proton in second 1H1	1.00728 amu's
Mass lost by it	0.00217 amu
Mass of positon in 2He3	1.00511 amu's

Now that the mass of the positon is known and the first isotope of 1H1 is intact within the isotope 2He3, the mass of the single neutron can be found. When the proton in the second isotope of 1H1 lost 0.00217 amu, it did not release its electron and this electron is the negatron found in the isotope 2He3. Subtract the total mass of the positon with its negatron and the first isotope of 1H1 from the mass of the isotope 2He3 to find the mass of this neutron.

Mass of positon in 2He3	1.00511 amu's
Mass of negatron in 2He3	0.00055 amu
Mass of first 1H1 intact in 2He3	1.00783 amu's
Total known mass in 2He3	2.01349 amu's

Mass of 2He3	3.01603 amu's
Total known mass in 2He3	2.01349 amu's
Mass of neutron in 2He3	1.00254 amu's

All of the parts of the isotope 2He3 are now known. Adding these parts will give the mass of the isotope 2He3. The first isotope of 1H1 was used as is so its electron and proton are found in the isotope 2He3 intact with 1.00783 amu's.

First isotope of 1H1 in 2He3	1.00783 amu's
Mass of negatron in 2He3	0.00055 amu
Mass of positon in 2He3	1.00511 amu's
Mass of neutron in 2He3	1.00254 amu's
Mass of 2He3	3.01603 amu's

Sub-step 93-6-4

2He3

1H1 + 1H1 + 1H1 = 2He3, one alpha 434 ray, alpha 157 ray, one 100 ray, -1e

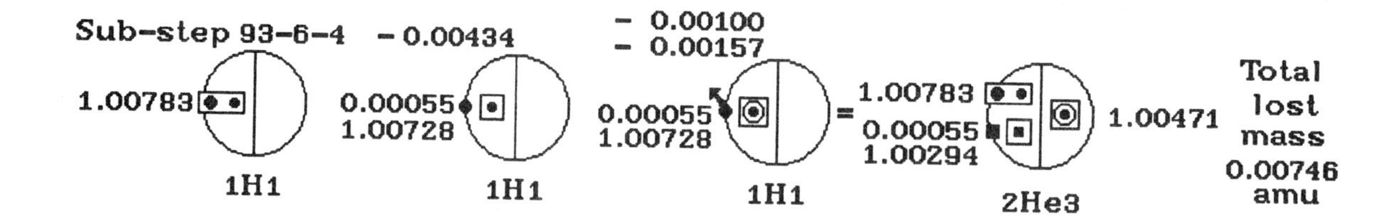

Three isotopes of 1H1 are needed to create the isotope 2He3. Each isotope of 1H1 has 1.00783 amu's. Each isotope has one electron or negatron with 0.00055 amu and one proton or positon with 1.00728 amu's. The isotope 2He3 has 3.01603 amu's. It has one electron with 0.00055 amu, one negatron with 0.00055 amu, one proton with 1.00728 amu's, one positon with 1.00294 amu's and one neutron with 1.00471 amu's. There is a total of 0.00746 amu lost creating the isotope 2He3 which includes one released electron. All alpha rays have an external magnetic field.

The first isotope of 1H1 has a rectangle drawn around its electron and proton to show that it is used as is and that it is drawn the same way in the isotope 2He3. This rectangle means that the isotope 1H1 is found intact within the isotope 2He3 and can drop out as such at some later time.

The second isotope of 1H1 has its electron with 0.00055 amu drawn as a dot on the big circle and it is drawn as a square on the big circle in the isotope 2He3 to show that the electron in the isotope 1H1 has become the negatron in the isotope 2He3.

The proton in the second isotope of 1H1 with 1.00728 amu's is drawn as a dot on the left side of the vertical line within the big circle. This proton in the second isotope of 1H1 becomes the positon in the isotope 2He3 and changes from a dot to a small square to show that it is now a positon. It is drawn with a square around the dot in the second isotope of 1H1. A square is also drawn around the small square in the isotope 2He3 to show that it has lost 0.00434 amu to become the positon in the isotope 2He3 with 1.00294 amu's. The loss of 0.00434 amu will not release its electron.

The third isotope of 1H1 has it electron drawn with an arrow pointing away from it which means that this electron has been released. The proton has a small circle drawn around the dot that is the proton. There is a square drawn around both of them. This square shows that the proton lost some of its mass that did not release its electron and then the small circle shows that there is 0.00100 amu lost that does release the electron. When the electron is released and the

0.00100 amu is lost, the proton becomes a neutron with 1.00471 amu's in the isotope 2He3 and it is drawn the same way as found in the third isotope of 1H1.

Adding all the masses of the three isotopes of 1H1 will give the total mass needed to create the isotope 2He3

Mass of first and second 1H1	2.01566 amu's
Mass of third 1H1	1.00783 amu's
Total mass needed to create 2He3	3.02349 amu's

Now subtract the mass of the isotope 2He3 from this total to find the total mass lost creating the isotope 2He3.

Total mass needed to create 2He3	3.02349 amu's
Mass of 2He3	3.01603 amu's
Total mass lost creating 2He3	0.00746 amu

In this step, I have stated that the single neutron in the isotope 2He3 has 1.00471 amu's. Since this neutron came from the third isotope of 1H1's proton, subtract its mass from the mass of this proton creating the neutron which will give a remaining mass that is lost.

Mass of proton in third 1H1	1.00728 amu's
Mass of neutron in 2He3	1.00471 amu's
Mass lost creating neutron in 2He3	0.00257 amu

This remaining mass of the proton in the third isotope of 1H1 includes the mass that is lost which released the electron. Subtract this electron releasing mass to find the mass which did not release the electron.

Mass lost creating neutron in 2He3	0.00257 amu
Mass lost releasing electron in third 1H1	0.00100 amu
Mass lost not releasing electron in third 1H1	0.00157 amu

The proton in the third isotope of 1H1 lost first the 0.00257 amu as a 157 ray and then lost 0.00100 amu as a 100 ray to become the neutron in the isotope 2He3. By subtracting the 0.00257 amu from this proton, the mass of the neutron can be found.

Mass of proton in third 1H1	1.00728 amu's
Mass lost creating neutron in 2He3	0.00257 amu
Mass of neutron in 2He3	1.00471 amu's

Now that the total mass lost by the third isotope of 1H1 is known, this lost mass can be subtracted from the total mass needed to create the isotope 2He3 to find the mass lost by the second isotope of 1H1 and electron of third 1H1.

Total mass lost creating 2He3	0.00746 amu
Mass lost creating neutron in 2He3	0.00257 amu
Lost mass by second 1H1 plus electron of third 1H1	0.00489 amu

This remaining lost mass includes the mass of the electron that has been released. Subtract the mass of the electron to find the mass lost by the second isotope of 1H1.

Lost mass by second 1H1 plus electron of third 1H1	0.00489 amu
Mass of electron released by third 1H1	0.00055 amu
Mass lost by second 1H1's proton	0.00434 amu

By subtracting the mass lost by the second 1H1's proton, the mass of the positon can be found.

Mass of proton in second 1H1	1.00728 amu's
Mass lost by it	0.00434 amu
Mass of positon in 2He3	1.00294 amu's

Now that the mass of the positon is known and the first isotope 1H1 is intact within the isotope 2He3, the mass of the single neutron can be found. When the proton in the second isotope of 1H1 lost 0.00434 amu, it did not release its electron and this electron is the negatron found in the isotope 2He3. Subtract the total mass of the positon with its negatron and the first isotope of 1H1 from the mass of the isotope 2He3 to find the mass of this neutron.

Mass of positon in 2He3	1.00294 amu's
Mass of negatron in 2He3	0.00055 amu
Mass of first 1H1 intact in 2He3	1.00783 amu's
Total known mass in 2He3	2.01132 amu's

Mass of 2He3	3.01603 amu's
Total known mass in 2He3	2.01132 amu's
Mass of neutron in 2He3	1.00471 amu's

The 0.00434 amu is emitted as a 434 ray. The 0.00157 amu is emitted as a 157 ray and the 0.00100 amu is emitted as a 100 ray. The electron is released as an electron.

All of the parts of the isotope 2He3 are now known. Adding these parts will give the mass of the isotope 2He3. The first isotope of 1H1 was used as is so its electron and proton are found in the isotope 2He3 intact with 1.00783 amu's.

First isotope of 1H1 in 2He3	1.00783 amu's
Mass of negatron in 2He3	0.00055 amu
Mass of positon in 2He3	1.00294 amu's
Mass of neutron in 2He3	1.00471 amu's
Mass of 2He3	3.01603 amu's

Sub-step 92-6-3

2He3

FORMULA

1H1 + 1H1 + 1H1 = 2He3, one alpha 434 ray, alpha 157 ray, one 100 ray, -1e

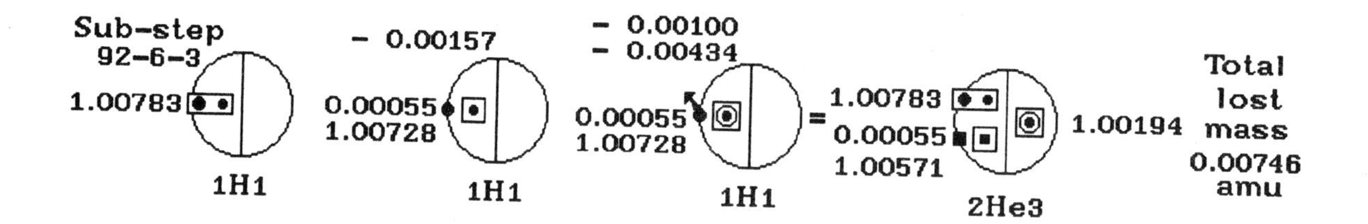

Three isotopes of 1H1 are needed to create the isotope 2He3. Each isotope of 1H1 has 1.00783 amu's. Each isotope has one electron or negatron with 0.00055 amu and one proton or positon with 1.00728 amu's. The isotope 2He3 has 3.01603 amu's. It has one electron with 0.00055 amu, one negatron with 0.00055 amu, one proton with 1.00728 amu's, one positon with 1.00571 amu's and one neutron with 1.00194 amu's. There is a total of 0.00746 amu lost creating the isotope 2He3 which includes one released electron. All alpha rays have an external magnetic field.

The first isotope of 1H1 has a rectangle drawn around its electron and proton to show that it is used as is and that it is drawn the same way in the isotope 2He3. This rectangle means that the isotope 1H1 is found intact within the isotope 2He3 and can drop out as such at some later time.

The second isotope of 1H1 has its electron used as a negatron in the isotope 2He3. The proton is drawn with a square around it and it is also drawn this way in the isotope 2He3 to show that it has lost some mass which does not release its electron.

The third isotope of 1H1 has it electron drawn with an arrow pointing away from it which means that this electron has been released. The proton has a small circle drawn around the dot that is the proton. There is a square drawn around both of them. This square shows that the proton lost some of its mass that did not release its electron and then the small circle shows that there is 0.00100 amu that does release the electron. When the electron is released and the 0.00100 amu is lost, the proton becomes a neutron in the isotope 2He3 and it is drawn the same way as found in the third isotope of 1H1.

Adding all the masses of the three isotopes of 1H1 will give the total mass needed to create the isotope 2He3

Mass of first 1H1	1.00783 amu's
Mass of second 1H1	1.00783 amu's
Mass of third 1H1	1.00783 amu's
Total mass needed to create 2He3	3.02349 amu's

Now subtract the mass of the isotope 2He3 from this total to find the total mass lost creating the isotope 2He3.

Total mass needed to create 2He3	3.02349 amu's
Mass of 2He3	3.01603 amu's
Total mass lost creating 2He3	0.00746 amu

In this step, I have stated that the single neutron in the isotope 2He3 has 1.00194 amu's. Since this neutron came from the third isotope of 1H1's proton, subtract its mass from the mass of this proton creating the neutron will have a remaining mass that is lost.

Mass of proton in third 1H1	1.00728 amu's
Mass of neutron in 2He3	1.00194 amu's
Mass lost creating neutron in 2He3	0.00534 amu

This remaining mass of the proton in the third isotope of 1H1 includes the mass that is lost which released the electron. Subtract this electron releasing mass to find the mass which did not release the electron.

Mass lost creating neutron in 2He3	0.00534 amu
Mass lost releasing electron in third 1H1	0.00100 amu
Mass lost not releasing electron in third 1H1	0.00434 amu

The proton in the third isotope of 1H1 lost first the 0.00434 amu as a 434 ray and then lost 0.00100 amu as a 100 ray to become the neutron in the isotope 2He3. By subtracting the 0.00534 amu from this proton, the mass of the neutron can be found.

Mass of proton in third 1H1	1.00728 amu's
Mass lost creating neutron in 2He3	0.00534 amu
Mass of neutron in 2He3	1.00194 amu's

Now that the total mass lost by the third isotope of 1H1 is known, this lost mass can be subtracted from the total mass needed to create the isotope 2He3 to find the mass lost by the second isotope of 1H1.

Total mass lost creating 2He3	0.00746 amu
Mass lost creating neutron in 2He3	0.00534 amu
Remaining lost mass	0.00212 amu

This remaining lost mass includes the mass of the electron that has been released. Subtract the mass of the electron to find the mass lost by the second isotope of 1H1.

Remaining lost mass	0.00212 amu
Mass of electron released by third 1H1	0.00055 amu
Mass lost by second 1H1's proton	0.00157 amu

By subtracting the mass lost by the second 1H1's proton, the mass of the positon can be found.

Mass of proton in second 1H1	1.00728 amu's
Mass lost by it	0.00157 amu
Mass of positon in 2He3	1.00571 amu's

Now that the mass of the positon is known and the isotope 1H1 is intact within the isotope 2He3, the mass of the single neutron can be found. When the proton in the second isotope of 1H1 lost 0.00157 amu, it did not release its electron and this electron is the negatron found in the isotope 2He3. Subtract the total mass of the positon with its negatron and the isotope 1H1 from the mass of the isotope 2He3 to find the mass of this neutron.

Mass of positon in 2He3	1.00571 amu's
Mass of negatron in 2He3	0.00055 amu
Mass of 1H1 intact in 2He3	1.00783 amu's
Total known mass in 2He3	2.01409 amu's

Mass of 2He3	3.01603 amu's
Total known mass in 2He3	2.01409 amu's
Mass of neutron in 2He3	1.00194 amu's

All of the parts of the isotope 2He3 are now known. Adding these parts will give the mass of the isotope 2He3. The first isotope of 1H1 was used as is so its electron and proton are found in the isotope 2He3 intact with 1.00783 amu's.

First isotope of 1H1 in 2He3	1.00783 amu's
Mass of negatron in 2He3	0.00055 amu
Mass of positon in 2He3	1.00571 amu's
Mass of neutron in 2He3	1.00194 amu's
Mass of 2He3	3.01603 amu's

The value for the lost mass of the proton in the second isotope of 1H1 was chosen because there is already a proton known to have lost the mass it lost. Thus making the positon in the isotope 2He3's mass known which enables one to find the mass of the neutron from the knowledge that 0.00100 amu will release an electron in from the third isotope of 1H1.

The 0.00434 amu is emitted as a 434 ray. The 0.00157 amu is emitted as a 157 ray and the 0.00100 amu is emitted as a 100 ray. The electron is released as an electron.

In the first and second steps creating the isotope 2He3, I stated that the lost mass of 0.00374 amu was composed of two sub-units, one of 0.00217 amu and one of 0.00157 amu. In these last two steps, the lost mass of 0.00157 amu is lost as a unit and not as a sub-unit of the lost mass of 0.00374 amu.

Note that the value of 0.00157 amu includes the value of 0.00055 amu which is the mass of an electron or negatron. Since the 0.00157 amu is emitted as a 157 ray, the electron or negatron is converted to pure energy along with the sub-units of 0.00100 amu and 0.00002 amu. This also means that there is an electron or negatron within the proton found in either the second or third isotope of 1H1.

Because there is no electron emitted from the second isotope of 1H1 in the step that lost 0.00157 amu, this is proof that when the electron is part of a larger unit, it can be converted to pure energy. It is not released as an electron or negatron.

The reader is to be advised that either the proton, positon or neutron has within each, an electron and a negatron. If both are found within the proton, positon or neutron and either an electron or a negatron is released or emitted, then the one that is not emitted as part of the lost mass stays behind because the proton or neutron must have it in the nucleus to match the needs of the isotope that is created when this isotope becomes part of a molecule. This line of reasoning puts a restriction on what can be emitted as mass and as a ray.

Sub-step 91-6-2

2He3

FORMULA

1H1 + 1H1 + 1H1 = 2He3, one alpha 591 ray, one 100 ray, -1e

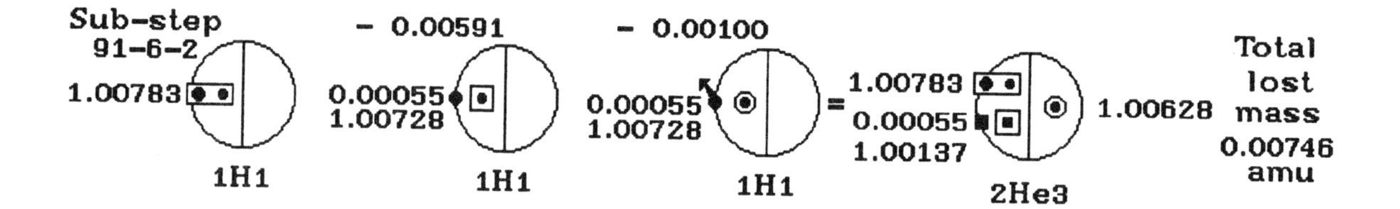

Three isotopes of 1H1 are needed to create the isotope 2He3. Each isotope of 1H1 has 1.00783 amu's. Each isotope has one electron or negatron with 0.00055 amu and one proton or positon with 1.00728 amu's. The isotope 2He3 has 3.01603 amu's. It has one electron with 0.00055 amu, one negatron with 0.00055 amu, one proton with 1.00728 amu's, one positon with 1.00137 amu's and one neutron with 1.00628 amu's. There is a total of 0.00746 amu lost creating the isotope 2He3 which includes one released electron. All alpha rays have an external magnetic field.

The first isotope of 1H1 has a rectangle drawn around its electron and proton to show that it is used as is and that it is drawn the same way in the isotope 2He3. This rectangle means that the isotope 1H1 is found intact within the isotope 2He3 and can drop out as such at some later time.

The second isotope of 1H1 has its electron used as a negatron in the isotope 2He3. The proton is drawn with a square around it and it is also drawn this way in the isotope 2He3 to show that it has lost some mass which does not release its electron.

The third isotope of 1H1 has it electron drawn with an arrow pointing away from it which means that this electron has been released. The proton has a small circle drawn around the dot that is the proton. When the electron is released and the 0.00100 amu is lost, the proton becomes a neutron in the isotope 2He3 and it is drawn the same way as found in the third isotope of 1H1.

Adding all the masses of the three isotopes of 1H1 will give the total mass needed to create the isotope 2He3.

Mass of first 1H1	1.00783 amu's
Mass of second 1H1	1.00783 amu's
Mass of third 1H1	1.00783 amu's

Now subtract the mass of the isotope 2He3 from this total to find the total mass lost creating the isotope 2He3.

Total mass needed to create 2He3	3.02349 amu's
Mass of 2He3	3.01603 amu's
Total mass lost creating 2He3	0.00746 amu

The proton in the third isotope of 1H1 loses 0.00100 amu as a 100 ray to become the neutron in the isotope 2He3. By subtracting the 0.00100 amu from this proton, the mass of the neutron can be found.

Mass of proton in third 1H1	1.00728 amu's
Mass lost creating neutron in 2He3	0.00100 amu
Mass of neutron in 2He3	1.00628 amu's

Now that the total mass lost by the third isotope of 1H1 is known, this lost mass can be subtracted from the total mass needed to create the isotope 2He3 to find the mass lost by the second isotope of 1H1 plus the electron released by the third isotope of 1H1.

Total mass lost creating 2He3	0.00746 amu
Mass lost creating neutron in 2He3	0.00100 amu
Lost mass of second 1H1 plus electron of third 1H1	0.00646 amu

This remaining lost mass includes the mass of the electron that has been released by the third isotope of 1H1. Subtract the mass of the electron to find the mass lost by the second isotope of 1H1.

Lost mass of second 1H1 plus electron of third 1H1	0.00646 amu
Mass of electron released by third 1H1	0.00055 amu
Mass lost by second 1H1's proton	0.00591 amu

By subtracting the lost mass lost by the second 1H1's proton, the mass of the positon can be found.

Mass of proton in second 1H1	1.00728 amu's
Mass lost by it	0.00591 amu
Mass of positon in 2He3	1.00137 amu's

Now that the mass of the positon is known and the first isotope of 1H1 that is intact in the isotope 2He3 is known, the mass of the single neutron can be found. When the proton in the second isotope of 1H1 lost 0.00591 amu, it did not release its electron and this electron is the negatron found in the isotope 2He3. Subtract the total mass of the positon with its negatron and the first isotope 1H1 from the mass of the isotope 2He3 to find the mass of this neutron.

Mass of positon in 2He3	1.00137 amu's
Mass of negatron in 2He3	0.00055 amu
Mass of first 1H1 intact in 2He3	1.00783 amu's
Total known mass in 2He3	2.00975 amu's

Mass of 2He3	3.01603 amu's
Total known mass in 2He3	2.00975 amu's
Mass of neutron in 2He3	1.00628 amu's

All of the parts of the isotope 2He3 are now known. Adding these parts will give the mass of the isotope 2He3. The first isotope of 1H1 was used as is so its electron and proton are found in the isotope 2He3 intact with 1.00783 amu's.

First isotope of 1H1 in 2He3	1.00783 amu's
Mass of negatron in 2He3	0.00055 amu
Mass of positon in 2He3	1.00137 amu's
Mass of neutron in 2He3	1.00628 amu's
Mass of 2He3	3.01603 amu's

The value for the lost mass of the proton in the second isotope of 1H1 was chosen because there is already a proton known to have lost the mass it lost. Thus making the positon in the isotope 2He3's mass known which enables one to find the mass of the neutron from the knowledge that 0.00100 amu will release an electron from the third isotope of 1H1.

The 0.00591 amu is emitted as a 591 ray. The 0.00100 amu is emitted as a 100 ray and the electron is released as an electron.

In the previous steps creating the isotope 2He3, the total mass lost was made up of an electron that was released plus the mass that was converted to pure energy. This included the mass that was lost as a unit and the sub-units that made up that unit. Now there is the 0.00100 amu that is lost which released the electron and a unit of mass which includes all of the other units as sub-units. This unit is 0.00591 amu. The difference here is that this unit has two ways it can be lost. It can be lost as two units where one unit is 0.00374 amu and one unit is 0.00217 amu. It can also be lost where one unit is 0.00434 amu and the second unit is 0.00157 amu. Because of this, knowing the value of the rays emitted will tell how the isotopes of 1H1 were used. It does not tell which proton lost a given unit, only that certain two units are the units lost. This means that the unit 0.00374 amu can not be combined with a unit of 0.00434 amu. Their combined value would be 0.00808 amu and not 0.00591 amu.

Sub-step 90-6-1

2He3

FORMULA

2(1H1) + 1H1 = 2He3, one alpha 591 ray, one 100 ray, -1e

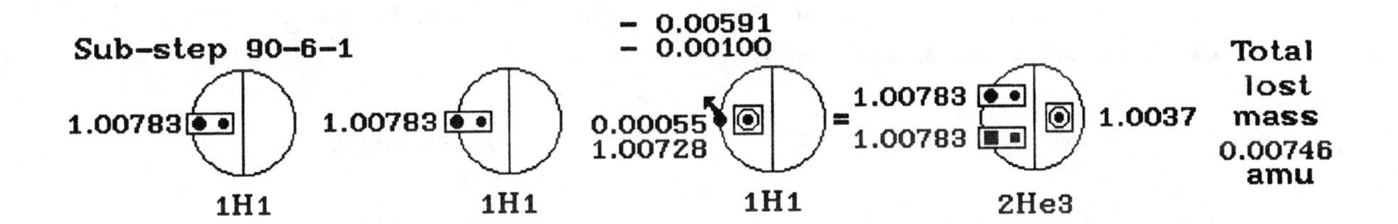

Three isotopes of 1H1 are needed to create the isotope 2He3. Each isotope of 1H1 has 1.00783 amu's. Each isotope has an electron or negatron with 0.00055 amu and one proton or positon with 1.00728 amu's. The isotope 2He3 has 3.01603 amu's. It has one electron with 0.00055 amu, one negatron with 0.00055 amu, one proton with 1.00728 amu's, one positon with 1.00728 amu's and one neutron with 1.00037 amu's. There is a total of 0.00746 amu lost creating the isotope 2He3 which includes one released electron. In this step the first and second isotopes of 1H1 are used in a like manner so they are put in parenthesis in the formula. The third isotope of 1H1 loses all the mass that is lost. All alpha rays have an external magnetic field.

The first and second isotope of 1H1 have a rectangle drawn around their electron and proton to show that they are used as is and that they are drawn the same way in the isotope 2He3. This rectangle means that each isotope 1H1 is found intact within the isotope 2He3 and can drop out as such at some later time.

The third isotope of 1H1 loses all the mass that is lost. The electron in this isotope has an arrow pointing away from it to show that the electron has been released. This isotope has its proton drawn as a dot with a small circle around it. There is a square around both of them. This square indicates that some mass has been lost that does not release its electron. The small circle indicates that the electron is released when 0.00100 amu has been lost. When the electron is released and the 0.00100 amu is lost, this proton becomes the single neutron found in the isotope 2He3. The neutron in the isotope 2He3 is drawn the same way as the proton in the third isotope of 1H1. The third isotope of 1H1 emits two rays. The first one is a 591 ray and the second one is a 100 ray.

Adding all the masses of the three isotopes of 1H1 will give the total mass needed to create the isotope 2He3

Mass of first 1H1	1.00783 amu's
Mass of second 1H1	1.00783 amu's
Mass of third 1H1	1.00783 amu's
Total mass needed to create 2He3	3.02349 amu's

Now subtract the mass of the isotope 2He3 from this total to find the total mass lost creating the isotope 2He3.

Total mass needed to create 2He3	3.02349 amu's
Mass of 2He3	3.01603 amu's
Total mass lost creating 2He3	0.00746 amu

The mass of the first and second isotope of 1H1 is known so add their mass and subtract their mass from the mass of the isotope 2He3 to find the mass of the isotope 2He3's single neutron.

Mass of first 1H1	1.00783 amu's
Mass of second 1H1	1.00783 amu's
Mass of first and second 1H1	2.01566 amu's

Mass of 2He3	3.01603 amu's
Total mass of first and second 1H1's	2.01566 amu's
Mass of neutron in 2He3	1.00037 amu's

Since the third isotope of 1H1 lost all the mass that is lost, with the release of its electron, its proton must lose 0.00100 amu and the mass that does not release it electron. By adding the mass of the electron that is released and 0.00100 amu and then subtracting this from the total mass that is lost, the mass that does not release the electron will be found.

Mass of electron released from third 1H1	0.00055 amu
Mass releasing electron from third 1H1	0.00100 amu
Total lost mass that is known from third 1H1	0.00155 amu

Total mass lost creating 2He3	0.00746 amu
Total lost mass that is known from third 1H1	0.00155 amu
Lost mass not releasing electron from third 1H1	0.00591 amu

This 0.00591 amu that is lost from the proton in the third isotope of 1H1 is lost before the 0.00100 amu is lost that releases the electron. Therefore, subtract this mass first from the mass of the proton. The remainder will then have the mass

that is lost releasing the electron. Then and only then can the 0.00100 amu be subtracted from the mass that remains of the proton.

Mass of proton in third 1H1	1.00728 amu's
Mass lost that does not release electron from it	0.00591 amu
Remaining mass of proton in third 1H1	1.00137 amu's

Remaining mass of proton in third 1H1	1.00137 amu's
Mass lost releasing electron from third 1H1	0.00100 amu
Mass of neutron in 2He3	1.00037 amu's

If I subtracted the lost mass of 0.00100 amu first, this would leave the remaining mass with the mass of a neutron with 1.00628 amu's and the mass of the neutron in the isotope 2He3 must be 1.00037 amu's.

All of the mass is known that makes up the isotope 2He3. By adding these masses, the mass of the isotope 2He3 will be found.

Mass of first 1H1 in 2He3	1.00783 amu's
Mass of second 1H1 in 2He3	1.00783 amu's
Mass of neutron in 2He3	1.00037 amu's
Mass of 2He3	3.01603 amu's

Because two isotopes of 1H1 are found intact within the isotope 2He3, they could both drop out and if this should happen, then the single neutron would be the only mass left. This neutron would then combine with some atom found in a molecule.

Since this is the last step using three isotopes of 1H1, I would like to point out that the lost mass of 0.00591 amu is lost as a 591 ray. This lost mass of 0.00591 amu is composed of two sub-units. One sub-unit is 0.00434 amu and the other sub-unit is 0.00157 amu which were used in some of the steps just given. The 0.00434 amu is composed of two sub-units of 0.00217 amu and the 0.00157 amu is composed of three sub-units. The first one is 0.00100 amu. The second one is 0.00002 amu while the third one is 0.00055 amu that is the mass of an electron. Thus when the mass of 0.00591 amu is lost as a single mass, there is an electron that is converted to pure energy which is magnetism. This means that the proton found in the third isotope of 1H1 has an electron within it which can be converted to pure energy.

All drawings in **STEP 6** using three isotopes of 1H1 to create the isotope 2He3

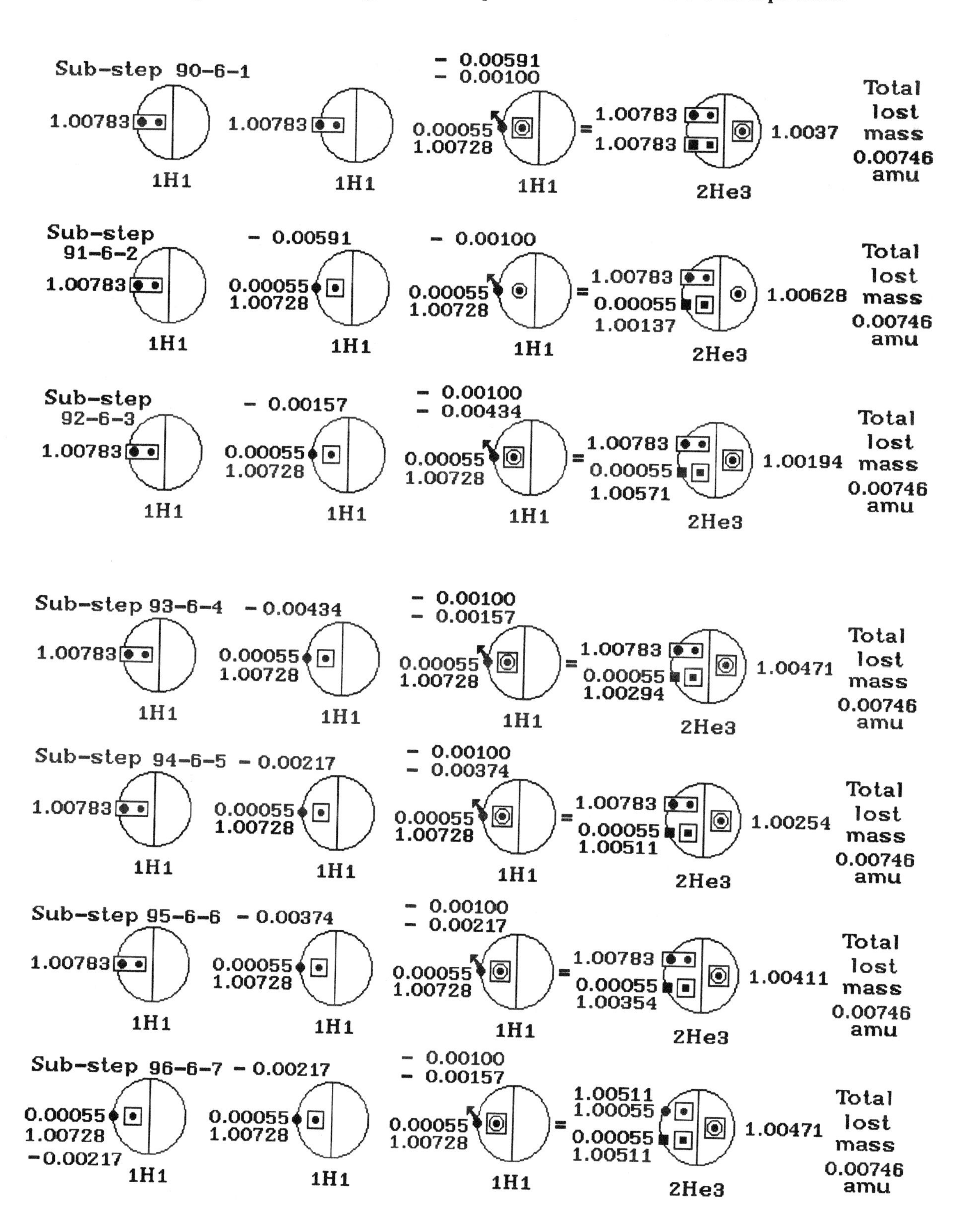

STEP 5

2He3

FORMULA

1H2 + 1H1 = 2He3

STEPS

5. 1H2 + 1H1 = 2He3

75-5-1	1H2 + 1H1 = 2He3, 591 ray
76-5-2	1H2 + 1H1 = 2He3, 591 ray
77-5-3	1H2 + 1H1 = 2He3, 591 ray
78-5-4	1H2 + 1H1 = 2He3, 591 ray
79-5-5	1H2 + 1H1 = 2He3, 591 ray
80-5-6	1H2 + 1H1 = 2He3, 217 ray, 374 ray
81-5-7	1H2 + 1H1 = 2He3, 374 ray, 217 ray
82-5-8	1H2 + 1H1 = 2He3, 217 ray, 217 ray, 157 ray
83-5-9	1H2 + 1H1 = 2He3, 157 ray, 217 ray, 217 ray
84-5-10	1H1 + 1H1 = 2He3, 217 ray, 157 ray, 217 ray
85-5-11	1H2 + 1H1 = 2He3, 217 ray, 374 ray
86-5-12	1H2 + 1H1 = 2He3, 374 ray, 217 ray
87-5-13	1H2 + 1H1 = 2He3, 157 ray, 434 ray
88-5-14	1H2 + 1H1 = 2He3, 157 ray, 434 ray
89-5-15	1H2 + 1H1 = 2He3, 434 ray, 157 ray

USING THE WRITE-UPS AND DRAWINGS

All of the previous steps creating the isotope 2He3 used three isotopes of 1H1. This set of steps will use one isotope of 1H2 and one isotope of 1H1 to create the isotope 2He3. The isotope 1H2 is created from two isotopes of 1H1. The isotope 1H2 has 2.01411 amu's and the isotope 1H1 has 1.00783 amu's. The isotope 2He3 has 3.01603 amu's.

It is understood that the dot in the drawings representing an electron could also represent a negatron and the dot representing a proton could also represent a positon. The proton or positon and the neutron will have more than one possible atomic mass. The electron or negatron will always have 0.00055 amu.

The electron is drawn as a dot, the negatron as a square. The proton is drawn as a dot and the positon is drawn as a square. The neutron is drawn as a small circle.

A small square drawn around a proton, positon or neutron indicates that this mass has lost some of its mass but did not release its electron or negatron.

A small circle drawn around a proton or positon means that this mass has lost 0.00100 amu and released its electron.

If the proton or positon loses its electron or negatron, it will become a neutron. This neutron will be drawn as a dot with a small circle around it. It is possible for a neutron to lose mass and stay a neutron in the isotope being created. When this is the case, the neutron that stays a neutron in the isotope being created will have a square drawn around it. This square means that the neutron has lost some mass and did not release an electron or negatron.

A rectangle drawn around any two or more masses will show that these masses are used as a single unit which is an isotope in the isotope that is used to create a new isotope and in the new isotope being created. The rectangle also means that the isotope that is enclosed in the rectangle can drop out of the isotope being created.

There will be an arrow drawn pointing away from the electron or negatron to show that this electron or negatron has been released. This electron or negatron with its arrow may or may not be drawn on the right side of the equal sign. If it is not shown on the right side of the equal sign, then it must be found in the total mass lost creating the new isotope.

There will be some isotopes that will release its electron or negatron and then regain it. When this happens, the electron or negatron does not lose its mass and the 0.00100 amu must also be regained which means that no ray will be emitted in creating a new isotope. No mass is lost thus no ray can be emitted.

When some mass has been lost, this mass is lost as a single unit. The 0.00100 amu is always lost as a single unit when an electron or negatron has been released and a proton or positon becomes a neutron.

When a mass is lost that does not release an electron or negatron, this lost mass may or may not be composed of smaller masses. These smaller masses are called "sub-units" of the larger unit that is lost. The larger unit may have a sub-unit of 0.00100 amu and 0.00055 amu. When this happens, the larger unit is lost as a ray and because all of the larger unit is emitted as a ray, an electron or negatron is lost as part of the ray. This means that there is an electron or negatron within a proton or positon which can be lost as an emitted ray.

Any mass not used in creating a new isotope will be dropped out of the isotopes that are used in creating the new isotope. It is possible for a neutron to go unused and it can also drop out as a neutron.

If there are two electrons or negatrons and two protons or positons left over after the new isotope has been created, they will form two isotopes of 1H1.

The terms beta ray and gamma ray will not be used.

Hereafter, use the drawings to tell if an electron is an electron or negatron and the proton is a proton or a positon. The drawings will show how a neutron is used and what mass is lost plus the total mass lost and any mass that drops out.

The total lost mass will be lost either as a single unit or it will be lost as a total of several units. If and electron or negatron is released, the total mass lost will include this electron or negatron.

The reader is to understand that an electron in an orbit belongs to a proton in the nucleus and the negatron in the same orbital belongs to a positon found in the same nucleus.

The positioning of the neutron in a drawing is used to show which proton belongs to the neutron. By noting where the neutron is placed within the left side of the big circle, it can be determined whether the proton is a proton or positon in the isotope being created. This is important because the proton in the isotope 1H2 and 1H1 may be drawn alike so the position of the neutron tells which proton stays a proton or which changes into a positon.

In the STEPS just given, all of the rays were alpha ray which have an external magnetic field because only isotopes of 1H1 were used to create the isotope 2He3. The isotopes of 1H1 only have protons and alpha rays only come from protons. Now there will be isotopes with neutrons and when a neutron loses mass that does not release an electron or negatron, the ray emitted will be a gamma ray. An alpha ray with an external magnetic field can not penetrate mass like a gamma ray. An alpha ray can be detected on photographic plates but a gamma ray can not. The energy of a gamma ray may be the same as an alpha ray but because of the lack of a magnetic field, the gamma ray will have more penetrating power than an alpha ray.

Sub-step 89-5-15

2He3

FORMULA

1H2 + 1H1 = 2He3, alpha 434 ray, gamma 157 ray

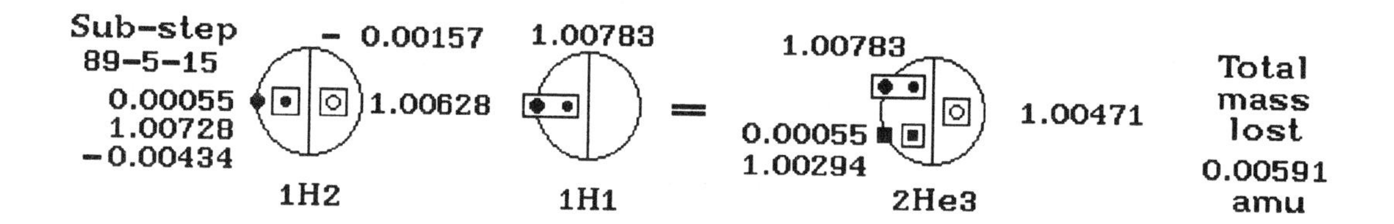

The isotope 1H2 has one electron with 0.00055 amu, one proton with 1.00728 amu's and one neutron with 1.00628 amu's for a total of 2.01411 amu's.

The isotope 1H1 has one electron with 0.00055 amu and one proton with 1.00728 amu's for a total of 1.00783 amu's.

The isotope of 2He3 with 3.01603 amu's has one electron and one negatron with each having 0.00055 amu. Its proton has 1.00728 amu's and the positon has 1.00294 amu's. The single neutron has 1.00471 amu's.

The isotope 1H2 with 2.01411 amu's has its proton with 1.00728 amu's lose 0.00434 amu as a 434 ray which has an external magnetic field. The proton has a square around a dot that is the proton and this proton becomes a positon that is a square in the isotope 2He3. Both have a square drawn around them to indicate that 0.00434 amu has been lost. The electron with 0.00055 amu in the isotope 1H2 is drawn as a dot on the big circle and it is drawn as a square on the big circle that is the isotope 2He3. The change from a dot to a square shows that the electron has become a negatron.

The single neutron in the isotope 1H2 has 1.00628 amu's and it loses 0.00157 amu as a gamma 157 ray to become the single neutron in the isotope 2He3 with 1.00471 amu's. The 157 gamma ray does not have an external magnetic field. It is drawn as a small circle with a square around it in the isotope 1H2 and also in the isotope 2He3. When it loses 0.00157 amu, it stays a neutron in the isotope 2He3.

The isotope 1H1 with 1.00783 amu's is used as the isotope 1H1 in the isotope 2He3 with no loss of mass. Its electron with 0.00055 amu and proton with 1.00728 amu's have a rectangle drawn around them to show that the isotope 1H1 is used as the isotope 1H1 in the isotope 2He3.

The total mass lost is 0.00591 amu and this is composed of two units of 0.00434 amu and 0.00157 amu. While the 0.00434 amu is lost as a single unit of

0.00434 amu, it is composed of two sub-units of 0.00217 amu. The 0.00157 amu is lost as a single unit and it is composed of a sub-unit of 0.00100 amu, a sub-unit of 0.00002 amu and a sub-unit of 0.00055 amu. The 0.00055 amu is the same mass of an electron or negatron but because it is part of the single unit of 0.00157 amu, it is emitted as part of the 157 ray.

To find the amount of mass needed to create the isotope 2He3, add the masses of the isotope 1H2 and 1H1 and then subtract the mass of the isotope 2He3 from their total.

Mass of 1H2	2.01411 amu's
Mass of 1H1	1.00783 amu's
Total mass needed to create 2He3	3.02194 amu's

Total mass needed to create 2He3	3.02194 amu's
Mass of 2He3	3.01603 amu's
Total mass lost creating 2He3	0.00591 amu

The drawings show that the mass of the isotope 1H1 is intact within the isotope 2He3 and there is a negatron in this isotope also. These masses are known. The mass of the positon and single neutron are not known. However, I have already given steps that have a proton with 1.00294 amu's and a neutron with 1.00471 amu's. I can therefore choose either of these masses and thereby know the mass of the other by adding all of the known masses and subtracting their total from the isotope 2He3.

Mass of 1H1 in 2He3	1.00783 amu's
Mass of negatron in 2He3	0.00055 amu
Mass of neutron in 2He3	1.00471 amu's
Total known mass in 2He3	2.01309 amu's

Mass of 2He3	3.01603 amu's
Total known mass in 2He3	2.01309 amu's
Mass of positon in 2He3	1.00294 amu's

Mass of 1H1 in 2He3	1.00783 amu's
Mass of negatron in 2He3	0.00055 amu
Mass of positon in 2He3	1.00294 amu's
Total known mass in 2He3	2.01132 amu's

Mass of 2He3	3.01603 amu's
Total known mass in 2He3	2.01132 amu's
Mass of single neutron in 2He3	1.00471 amu's

With the mass of the positon in the isotope 2He3 known, its mass can be subtracted from the mass of the proton in the isotope 1H2 to find the mass it lost.

Mass of proton in 1H2	1.00728 amu's
Mass of positon in 2He3	1.00294 amu's
Mass lost by proton in 1H2 becoming positon in 2He3	0.00434 amu

With the mass of the single neutron in the isotope 2He3 known, its mass can be subtracted from the mass of the single neutron in the isotope 1H2.

Mass of neutron in 1H2	1.00628 amu's
Mass of neutron in 2He3	1.00471 amu's
Mass lost by neutron in 1H2 becoming neutron in 2He3	0.00157 amu

Adding the mass lost by the proton and neutron in the isotope 1H2 will give the total mass lost creating the isotope 2He3.

Mass lost by proton in 1H2	0.00434 amu
Mass lost by neutron in 1H2	0.00157 amu
Total mass lost creating 2He3	0.00591 amu

The reason I know that the values used above can be substituted is that three isotopes of 1H1 use the same values of the isotope 1H2 and 1H1 because the isotope 1H2 comes from two isotopes of 1H1. Subtracting the total mass lost creating 2He3 from the total mass lost by three isotopes of 1H1 will give the mass lost creating the isotope 1H2.

Total mass lost by three isotopes of 1H1	0.00746 amu
Total mass lost by 1H2 and 1H1	0.00591 amu
Mass lost creating 1H2 from two 1H1's	0.00155 amu

The lost mass of 0.00157 amu is composed of three sub-units. When the unit 0.00157 amu is lost as a 157 ray, it is composed of a sub-unit of 0.00055 amu and a 0.00100 amu plus a sub-unit of 0.00002 amu.

Later on, the isotope 1H3 will create the isotope 2He3 and the only mass lost will be 0.00002 amu which is lost as a #2 ray. I therefore know that there is such a lost mass as the 0.00002 amu. The mass of the isotope 1H3 is 3.01605 amu's and the mass of the isotope 2He3 is 3.01603 amu's for a difference in mass of 0.00002 amu.

The isotope 1H2 also lost 0.00434 amu by its proton which is composed of two sub-units of 0.00217 amu. Because the 0.00434 amu is lost as a single unit, a 434 ray is emitted.

Sub-step 88-5-14

2He3

FORMULA

1H2 + 1H1 = 2He3, gamma 157 ray, alpha 434 ray

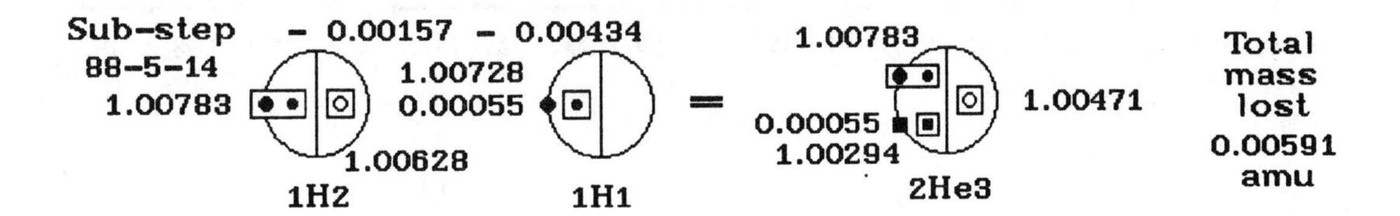

The isotope 1H2 has one electron with 0.00055 amu, one proton with 1.00728 amu's and one neutron with 1.00628 amu's for a total of 2.01411 amu's.

The isotope 1H1 has one electron with 0.00055 amu and one proton with 1.00728 amu's for a total of 1.00783 amu's.

The isotope of 2He3 with 3.01603 amu's has one electron and one negatron with each having 0.00055 amu. It has a proton with 1.00728 amu's and a positon with 1.00294 amu's. The neutron has 1.00471 amu's.

The isotope 1H2 with 2.01411 amu's has its electron with 0.00055 amu and proton with 1.00728 amu's used as the isotope 1H1 with 1.00783 amu's in the isotope 2He3. The electron and proton is drawn with a rectangle around them in the isotope 1H2 and 2He3. They are found as the isotope 1H1 in the isotope 1H2 and 2He3.

The single neutron with 1.00628 amu's in the isotope 1H2 loses 0.00157 amu and stays a neutron with 1.00471 amu's in the isotope 2He3. The 0.00157 amu is emitted as a gamma ray which has no external magnetic field. The neutron in the isotope 1H2 and 2He3 both have a square drawn around them to show that the neutron in the isotope 1H2 lost 0.00157 amu to become the neutron in the isotope 2He3.

The isotope 1H1 with 1.00783 amu's has its electron with 0.00055 amu drawn as a dot on the big circle and it is drawn as a square on the big circle that is the isotope 2He3. The electron in the isotope 1H1 becomes a negatron in the isotope 2He3. The proton in the isotope 1H1 is drawn as a dot that has a square around it and it is drawn as a square with another square drawn around it in the isotope 2He3. The square around the proton in the isotope 1H1 indicates this proton with 1.00728 amu's has lost 0.00434 amu to become the positon with 1.00294 amu's in the isotope 2He3. The 0.00434 amu is emitted as a 434 alpha ray which has an external magnetic field.

The total mass lost is 0.00591 amu and it is composed of a single unit of 0.00434 amu and one unit of 0.00157 amu. The 0.00157 amu is lost as a single unit and it is composed of a sub-unit of 0.00100 amu, a sub-unit of 0.00002 amu and a sub-unit of 0.00055 amu. The 0.00055 amu is the same mass of an electron or negatron but because it is part of the single unit of 0.00157 amu, it is emitted as part of the 157 ray. The 0.00434 amu is lost as a single unit and emitted as a 434 ray. The 0.00434 amu that is lost as a 434 ray is composed of two sub-units of 0.00217 amu.

To find the amount of mass needed to create the isotope 2He3, add the masses of the isotopes 1H2 and 1H1 and then subtract the mass of the isotope 2He3 from their total.

Mass of 1H2	2.01411 amu's
Mass of 1H1	1.00783 amu's
Total mass needed to create 2He3	3.02194 amu's

Total mass needed to create 2He3	3.02194 amu's
Mass of 2He3	3.01603 amu's
Total mass lost creating 2He3	0.00591 amu

The electron and proton in the isotope 1H2 has a rectangle drawn around them to show that they are used as the isotope 1H1 in the isotopes 1H2 and 2He3. This means that the 1.00783 amu's is known in the isotope 2He3.

The single neutron in the isotope 1H2 is known to be 1.00628 amu's and it has a square drawn around it in the isotopes 1H2 and 2He3. The single neutron in the isotope 2He3 is known to be 1.00471 amu's. With this known mass, subtract the mass of the single neutron in the isotope 2He3 from the known mass of the neutron in the isotope 1H2 to find the mass lost by the neutron in the isotope 1H2 when it became the single neutron in the isotope 2He3.

Mass of neutron in 1H2	1.00628 amu's
Mass of neutron in 2He3	1.00471 amu's
Mass lost by neutron in 1H2	0.00157 amu

Now subtract this lost mass from the total mass lost creating the isotope 2He3 to find the mass lost by the proton in the isotope 1H1 that is by itself.

Total lost mass creating 2He3	0.00591 amu
Mass lost by neutron in 1H2	0.00157 amu
Mass lost by proton in 1H1 that is by itself	0.00434 amu

By subtracting the mass lost by the proton in the isotope 1H1 that is by itself from this proton, the mass of the positon in the isotope 2He3 will be found.

Mass of proton in 1H1 that is by itself	1.00728 amu's
Mass lost by it	0.00434 amu
Mass of positon in 2He3	1.00294 amu's

Because there is an isotope of 1H1 used as is in the isotope 2He3 and the mass of the single neutron plus the negatron are all known, by adding their masses and subtracting their total mass from the mass of the isotope of 2He3, the mass of the positon can be found.

Mass of 1H1 in 2He3	1.00783 amu's
Mass of neutron in 2He3	1.00471 amu's
Mass of negatron in 2He3	0.00055 amu
Total known mass in 2He3	2.01309 amu's

Mass of 2He3	3.01603 amu's
Total known mass in 2He3	2.01309 amu's
Mass of positon in 2He3	1.00294 amu's

I could have chosen the mass of the positon as the known mass instead of the neutron found in the isotope 2He3 because this value is also known to exist as shown in some of the steps already given. The values of these parts of the isotopes 1H2, 1H1 and 2He3 are being shown to be interchangeable and the lost mass is also interchangeable. By showing that these parts and their lost masses are interchangeable means that the isotope that is created can have the same mass for its parts as shown in each step but they can come from different isotopes.

When an isotope of 2He3 is found in a molecule, this isotope of 2He3 will be shown in the formula as just an isotope of 2He3 but the mass of its parts will make it different than just an isotope of 2He3. This is why some molecules have the same parts as another isotope of 2He3 but the molecules show different properties. These properties may be temperature, size or the way they connect to other molecules.

Now add all of the masses to find the mass of the isotope 2He3

Mass of 1H1 within 2He3	1.00783 amu's
Mass of negatron in 2He3	0.00055 amu
Mass of positon in 2He3	1.00294 amu's
Mass of neutron in 2He3	1.00471 amu's
Mass of 2He3	3.01603 amu's

Sub-step 87-5-13

2He3

FORMULA

1H2 + 1H1 = 2He3, alpha 157 ray, gamma 434 ray

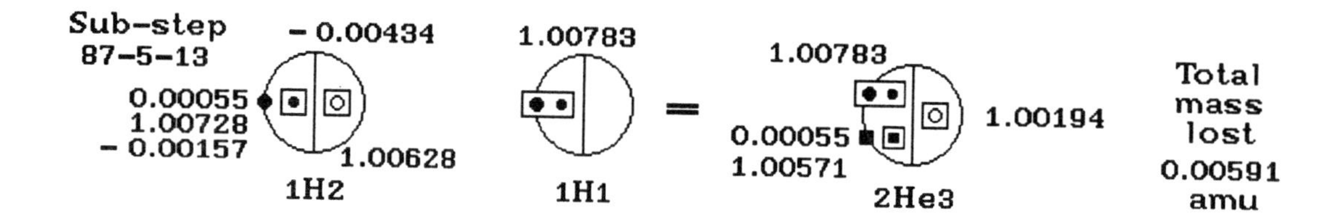

The isotope 1H2 has one electron with 0.00055 amu, one proton with 1.00728 amu's and one neutron with 1.00628 amu's for a total of 2.01411 amu's.

The isotope 1H1 has one electron with 0.00055 amu and one proton with 1.00728 amu's for a total of 1.00783 amu's.

The isotope of 2He3 with 3.01603 amu's has one electron and one negatron with each having 0.00055 amu. It has one proton with 1.00728 amu's and one positon with 1.00571 amu's. The single neutron will have 1.00194 amu's.

The isotope 1H2 with 2.01411 amu's has its proton with 1.00728 amu's lose 0.00157 amu and become the positon with 1.00571 amu in the isotope 2He3. The 0.00157 amu is emitted as a 157 alpha ray which has an external magnetic field. The proton is drawn as a dot with a square around it to show that it has lost some mass and the positon is drawn as a square with another square drawn around it to show that it is the proton that has lost 0.00157 amu. The electron with 0.00055 amu is drawn as a dot in the isotope 1H2 and as a square in the isotope 2He3 to show that the electron is now a negatron.

The single neutron with 1.00628 amu's in the isotope 1H2 loses 0.00434 amu and stays a neutron with 1.00194 amu's in the isotope 2He3. They both have a square drawn around them to show that the neutron in the isotope 1H2 lost 0.00434 amu to become the neutron in the isotope 2He3.

The isotope 1H1 with 1.00783 amu's has its electron with 0.00055 amu and proton with 1.00728 amu's used as the isotope 1H1 the isotope 2He3. There is a rectangle drawn around the electron and proton to show that they are used as the isotope 1H1 in the 2He3 and that the isotope 1H1 is found intact within the isotope 2He3.

The total mass lost is 0.00591 amu and this is composed of a single unit of 0.00434 amu and one unit of 0.00157 amu. The 0.00157 amu is lost as a single unit and it is composed of a sub-unit of 0.00100 amu, a sub-unit of 0.00002 amu and a sub-unit of 0.00055 amu. The 0.00055 amu is the same mass of an electron or negatron but because it is part of the single unit of 0.00157 amu, it is emitted as part of the 157 alpha ray. The 0.00434 amu is lost as a single unit and emitted as a 434 gamma ray. The 0.00434 amu is composed of two sub-units of 0.00217 amu.

To find the amount of mass needed to create the isotope 2He3, add the masses of the isotopes 1H2 and 1H1 and then subtract the mass of the isotope 2He3 from their total.

Mass of 1H2	2.01411 amu's
Mass of 1H1	1.00783 amu's
Total mass needed to create 2He3	3.02194 amu's

Total mass needed to create 2He3	3.02194 amu's
Mass of 2He3	3.01603 amu's
Total mass lost creating 2He3	0.00591 amu

The electron and proton in the isotope 1H1 has a rectangle drawn around them to show that they are used as the isotope 1H1 in the isotope 2He3. This means that the 1.00783 amu's is known in the isotope 2He3.

The electron in the isotope 1H2 is drawn as a dot and it becomes the negatron in the isotope 2He3 and drawn as a square. It has the mass of 0.00055 amu because an electron never loses any mass. The proton is drawn as a dot with a square around it to show that it has lost 0.00157 amu to become the positon in the isotope 2He3. The positon is drawn as a square with another square drawn around it to show that it now has the mass of 1.00571 amu's.

The single neutron in the isotope 1H2 is known to be 1.00628 amu's and it has a square drawn around it in the isotopes 1H2 and 2He3. The single neutron in the isotope 2He3 is known to be 1.00194 amu's. With this known mass, subtract the mass of the single neutron in the isotope 2He3 from the known mass of the neutron in the isotope 1H2 to find the mass lost by the neutron in the isotope 1H2 when it became the single neutron in the isotope 2He3.

Mass of neutron in 1H2	1.00628 amu's
Mass of neutron in 2He3	1.00194 amu's
Mass lost by neutron in 1H2	0.00434 amu

Now subtract this lost mass from the total mass lost creating the isotope 2He3 to find the mass lost by the proton in the isotope 1H2.

Total lost mass creating 2He3	0.00591 amu
Mass lost by neutron in 1H2	0.00434 amu
Mass lost by proton in 1H2	0.00157 amu

By subtracting the mass lost by the proton in the isotope 1H2 from this proton's mass, the mass of the positon in the isotope 2He3 will be found.

Mass of proton in 1H2	1.00728 amu's
Mass lost by it	0.00157 amu
Mass of positon in 2He3	1.00571 amu's

Because there is an isotope of 1H1 used as is in the isotope 2He3 and the mass of the single neutron plus the negatron are all known, by adding their masses and subtracting their total mass from the mass of the isotope of 2He3, the mass of the positon can be found.

Mass of 1H1 in 2He3	1.00783 amu's
Mass of neutron in 2He3	1.00194 amu's
Mass of negatron in 2He3	0.00055 amu
Total known mass in 2He3	2.01032 amu's

Mass of 2He3	3.01603 amu's
Total known mass in 2He3	2.01032 amu's
Mass of positon in 2He3	1.00571 amu's

I could have chosen the mass of the positon as the known mass instead of the neutron found in the isotope 2He3 because this value is also known to exist as shown in some of the steps already given. The values of these parts of the isotopes 1H2, 1H1 and 2He3 are being shown to be interchangeable and the lost mass is also interchangeable. By showing that these parts and their lost mass can lose this lost mass means that the isotope that is created can have the same mass for its parts as shown in each step but they can come from different isotopes.

Now add all of the known masses to find the isotope 2He3.

Mass of 1H1 within 2He3	1.00783 amu's
Mass of negatron in 2He3	0.00055 amu
Mass of positon in 2He3	1.00571 amu's
Mass of neutron in 2He3	1.00194 amu's
Mass of 2He3	3.01603 amu's

Sub-step 86-5-12

2He3

FORMULA

1H2 + 1H1 = 2He3, gamma 374 ray, alpha 217 ray

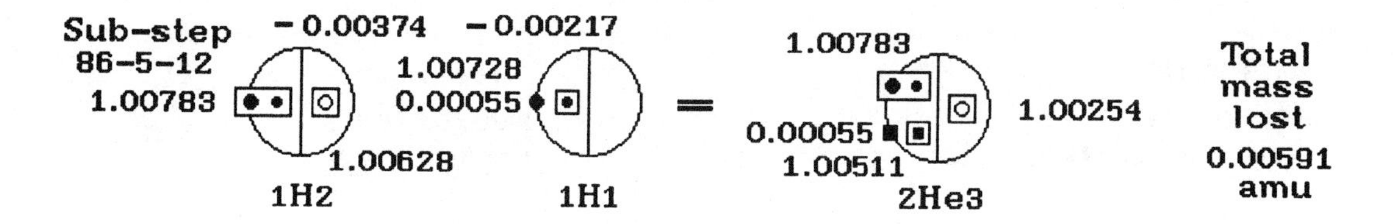

The isotope 1H2 has one electron with 0.00055 amu, one proton with 1.00728 amu's and one neutron with 1.00628 amu's for a total of 2.01411 amu's.

The isotope 1H1 has one electron with 0.00055 amu and one proton with 1.00728 amu's for a total of 1.00783 amu's.

The isotope of 2He3 with 3.01603 amu's has one electron and one negatron with each having 0.00055 amu. There is a proton with 1.00728 amu's and a positon with 1.00511 amu's. The single neutron will have 1.00254 amu's.

The isotope 1H2 with 2.01411 amu's has its electron with 0.00055 amu and proton with 1.00728 amu's used as the isotope 1H1 with 1.00783 amu's in the isotope 2He3 which means they do not lose any mass and that they are drawn with a rectangle around them in the isotope 1H2 and 2He3. They are found as the isotope 1H1 in the isotope 1H2 and 2He3. The single neutron with 1.00628 amu's in the isotope 1H2 loses 0.00374 amu and stays a neutron in the isotope 2He3 with 1.00254 amu's. The 0.00374 amu is emitted as a gamma ray which has no external magnetic field. The neutron in the isotope 1H2 and 2He3 both have a square drawn around them to show that the neutron in the isotope 1H2 lost 0.00374 amu and it is the neutron found in the isotope 2He3.

The isotope 1H1 with 1.00783 amu's has its electron with 0.00055 amu drawn as a dot on the big circle and it is drawn as a square on the big circle that is the isotope 2He3. The electron in the isotope 1H1 becomes a negatron in the isotope 2He3. The proton with 1.00728 amu's in the isotope 1H1 is drawn as a dot that has a square around it and it is drawn as a square with another square drawn around it in the isotope 2He3. The square around the proton in the isotope 1H1 indicates this proton with 1.00728 amu's has lost 0.00217 amu to become the positon with 1.00511 amu's in the isotope 2He3.

The total mass lost is 0.00591 amu and this is composed of a single unit of 0.00374 amu and one unit of 0.00217 amu. The single unit of 0.00374 amu is composed of two sub-units of 0.00217 amu and 0.00157 amu and it is emitted as a 374 gamma ray.

The 0.00217 amu is emitted as a 217 alpha ray. Note that the 0.00217 amu found in the 0.00374 amu is not emitted as a single ray because it is part of the 0.00374 amu that is lost as a single unit.

To find the amount of mass needed to create the isotope 2He3, add the masses of the isotopes 1H2 and 1H1 and then subtract the mass of the isotope 2He3 from their total.

Mass of 1H2	2.01411 amu's
Mass of 1H1	1.00783 amu's
Total mass needed to create 2He3	3.02194 amu's

Total mass needed to create 2He3	3.02194 amu's
Mass of 2He3	3.01603 amu's
Total mass lost creating 2He3	0.00591 amu

The electron and proton in the isotope 1H2 has a rectangle drawn around them to show that they are used as the isotope 1H1 in the isotopes 1H2 and 2He3. This means that the 1.00783 amu's is known in the isotope 2He3.

The single neutron in the isotope 1H2 is known to be 1.00628 amu's and it has a square drawn around it in the isotopes 1H2 and 2He3. The single neutron in the isotope 2He3 is known to be 1.00254 amu's. With this known mass, subtract the mass of the single neutron in the isotope 2He3 from the known mass of the neutron in the isotope 1H2 to find the mass lost by the neutron in the isotope 1H2 when it became the single neutron in the isotope 2He3.

Mass of neutron in 1H2	1.00628 amu's
Mass of neutron in 2He3	1.00254 amu's
Mass lost by neutron in 1H2	0.00374 amu

Now subtract this lost mass from the total mass lost creating the isotope 2He3 to find the mass lost by the proton in the isotope 1H1 that is by itself.

Total lost mass creating 2He3	0.00591 amu
Mass lost by neutron in 1H2	0.00374 amu
Mass lost by proton in 1H1 that is by itself	0.00217 amu

By subtracting the mass lost by the proton in the isotope 1H1 that is by itself, the mass of the positon in the isotope 2He3 will be found.

Mass of proton in 1H1 that is by itself	1.00728 amu's
Mass lost by it	0.00217 amu
Mass of positon in 2He3	1.00511 amu's

Because there is an isotope of 1H1 used as is in the isotope 2He3 and the mass of the single neutron plus the negatron are all known, by adding their masses and subtracting their total mass from the mass of the isotope of 2He3, the mass of the positon can be found.

Mass of 1H1 in 2He3	1.00783 amu's
Mass of neutron in 2He3	1.00254 amu's
Mass of negatron in 2He3	0.00055 amu
Total known mass in 2He3	2.01092 amu's

Mass of 2He3	3.01603 amu's
Total known mass in 2He3	2.01092 amu's
Mass of positon in 2He3	1.00511 amu's

I could have chosen the mass of the positon as the known mass instead of the neutron found in the isotope 2He3 because this value is also known to exist as shown in some of the steps already given. The values of these parts of the isotopes 1H2, 1H1 and 2He3 are being shown to be interchangeable and the lost mass is also interchangeable. By showing that these parts and their lost mass can lose this lost mass means that the isotope that is created can have the same mass for its parts as shown in each step but they can come from different isotopes.

Note that in the other steps given, the total mass lost is 0.00591 amu which is composed of either one unit of 0.00434 amu or two units of 0.00217 amu plus a single unit of 0.00157 amu. In this step the total lost mass is composed of a single lost unit of 0.00374 amu and a single unit of 0.00217 amu. This shows that the lost masses are also interchangeable.

Now add all of the known masses to find the mass of the isotope 2He3.

Mass of 1H1 within 2He3	1.00783 amu's
Mass of negatron in 2He3	0.00055 amu
Mass of positon in 2He3	1.00511 amu's
Mass of neutron in 2He3	1.00254 amu's
Mass of 2He3	3.01603 amu's

Sub-step 85-5-11

2He3

FORMULA

1H2 + 1H1 = 2He3, alpha 217 ray, gamma 374 ray

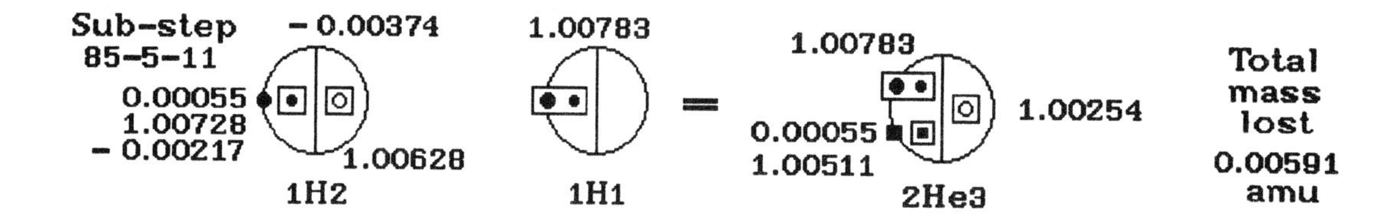

The isotope 1H2 has one electron with 0.00055 amu, one proton with 1.00728 amu's and one neutron with 1.00628 amu's for a total of 2.01411 amu's.

The isotope 1H1 has one electron with 0.00055 amu and one proton with 1.00728 amu's for a total of 1.00783 amu's.

The isotope of 2He3 with 3.01603 amu's has one electron and one negatron with each having 0.00055 amu. Its proton will have 1.00728 amu's, the positon will have 1.00511 amu's and the single neutron will have 1.00254 amu's.

The isotope 1H2 with 2.01411 amu's has its proton with 1.00728 amu's lose 0.00217 amu and become the positon in the isotope 2He3 with 1.00511 amu's. The 0.00217 amu is emitted as a 217 alpha ray which has an external magnetic field. The proton is drawn with a square around it to show that it has lost 0.00217 amu and the positon is drawn as a square with another square drawn around it to show that it is the positon. The electron with 0.00055 amu is drawn as a dot in the isotope 1H2 and as a square in the isotope 2He3 to show that the electron is now a negatron.

The single neutron with 1.00628 amu's in the isotope 1H2 loses 0.00374 amu and stays a neutron with 1.00254 amu's in the isotope 2He3. The 0.00374 amu is emitted as a 374 gamma ray which does not have an external magnetic field. The neutron in the isotopes 1H2 and 2He3 both have a square drawn around them to show that the neutron in the isotope 1H2 lost 0.00374 amu to become the neutron in the isotope 2He3.

The isotope 1H1 with 1.00783 amu's has its electron with 0.00055 amu and proton with 1.00728 amu's used as is in the isotope 2He3. There is a rectangle drawn around both of them in the isotope 1H1 and 2He3 to show that the isotope 1H1 is found intact within the isotope 2He3.

The total mass lost is 0.00591 amu and this is composed of a single unit of 0.00217 amu and one unit of 0.00374 amu. The 0.00374 amu is lost as a single unit and it is composed of a sub-unit of 0.00217 amu and 0.00157 amu. The sub-unit of 0.00157 amu is composed of a sub-unit of 0.00100 amu, a sub-unit of 0.00002 amu and a sub-unit of 0.00055 amu. The 0.00055 amu is the same mass of an electron or negatron but because it is part of the sub-unit of 0.00157 amu, that is part of the unit 0.00374 amu, it is emitted as part of the 374 gamma ray. Note that there are two 0.00217 amu used in this step but only one is emitted as a 217 ray. The 0.00217 amu that is lost as a single unit from the proton in the isotope 1H2 is emitted as a 217 alpha ray while the other 0.00217 is part of the unit 0.00374 and it is not emitted as a unit but it is emitted as a sub-unit.

To find the amount of mass needed to create the isotope 2He3, add the masses of the isotopes 1H2 and 1H1 and then subtract the mass of the isotope 2He3 from their total.

Mass of 1H2	2.01411 amu's
Mass of 1H1	1.00783 amu's
Total mass needed to create 2He3	3.02194 amu's

Total mass needed to create 2He3	3.02194 amu's
Mass of 2He3	3.01603 amu's
Total mass lost creating 2He3	0.00591 amu

The electron and proton in the isotope 1H1 has a rectangle drawn around them to show that they are used as the isotope 1H1 in the isotope 2He3. This means that the 1.00783 amu's is known in the isotope 2He3.

The single neutron in the isotope 1H2 is known to be 1.00628 amu's and it has a square drawn around it in the isotopes 1H2 and 2He3. The single neutron in the isotope 2He3 is known to be 1.00254 amu's. With this known mass, subtract the mass of the single neutron in the isotope 2He3 from the known mass of the neutron in the isotope 1H2 to find the mass lost by the neutron in the isotope 1H2 when it became the single neutron in the isotope 2He3.

Mass of neutron in 1H2	1.00628 amu's
Mass of neutron in 2He3	1.00254 amu's
Mass lost by neutron in 1H2	0.00374 amu

Now subtract this lost mass from the total mass lost creating the isotope 2He3 to find the mass lost by the proton in the isotope 1H2.

Total lost mass creating 2He3	0.00591 amu
Mass lost by neutron in 1H2	0.00374 amu
Mass lost by proton in 1H2	0.00217 amu

By subtracting the mass lost by the proton in the isotope 1H2 from this proton's mass, the mass of the positon in the isotope 2He3 will be found.

Mass of proton in 1H2	1.00728 amu's
Mass lost by it	0.00217 amu
Mass of positon in 2He3	1.00511 amu's

Because there is an isotope of 1H1 used as is in the isotope 2He3 and the mass of the single neutron plus the negatron are all known, by adding their mass and subtracting their total mass from the mass of the isotope of 2He3, the mass of the positon can be found.

Mass of 1H1 in 2He3	1.00783 amu's
Mass of neutron in 2He3	1.00254 amu's
Mass of negatron in 2He3	0.00055 amu
Total known mass in 2He3	2.01092 amu's

Mass of 2He3	3.01603 amu's
Total known mass in 2He3	2.01092 amu's
Mass of positon in 2He3	1.00511 amu's

I could have chosen the mass of the positon as the known mass instead of the neutron found in the isotope 2He3 because this value is also known to exist as shown in some of the steps already given. The values of these parts of the isotopes 1H2, 1H1 and 2He3 are being shown to be interchangeable and the lost mass is also interchangeable. By showing that these parts and their lost mass can lose this lost mass means that the isotope that is created can have the same mass for its parts as shown in each step but they can come from different isotopes.

Now add all of the known masses to find the mass of the isotope 2He3.

Mass of 1H1 within 2He3	1.00783 amu's
Mass of negatron in 2He3	0.00055 amu
Mass of positon in 2He3	1.00511 amu's
Mass of neutron in 2He3	1.00254 amu's
Mass of 2He3	3.01603 amu's

Sub-step 84 -5-10

2He3

FORMULA

1H2 + 1H1 = 2He3, alpha 217 ray, gamma 157 ray, alpha 217 ray

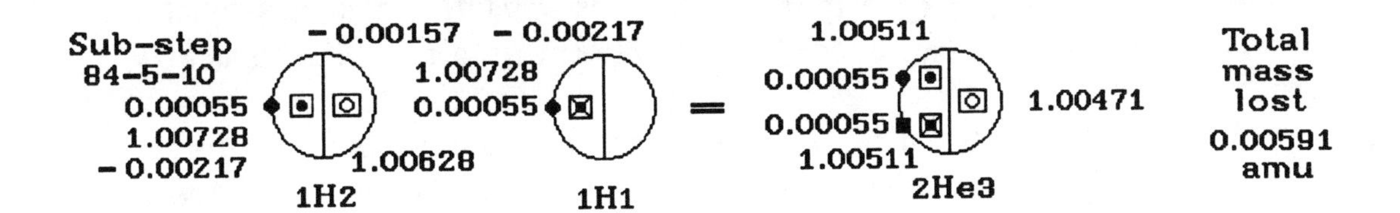

The isotope 1H2 has one electron with 0.00055 amu, one proton with 1.00728 amu's and one neutron with 1.00628 amu's for a total of 2.01411 amu's.

The isotope 1H1 has one electron with 0.00055 amu and one proton with 1.00728 amu's for a total of 1.00783 amu's.

The isotope of 2He3 with 3.01603 amu's has one electron and one negatron with each having 0.00055 amu. It has its proton and positon each having 1.00511 amu's. The single neutron has 1.00471 amu's.

The isotope 1H2 with 2.01411 amu's will have its proton lose 0.00217 amu as a 217 alpha ray which has an external magnetic field. The proton has a square around a dot that is the proton and this proton becomes a proton that is drawn the same way in the isotope 2He3. Both have a square drawn around them to indicate that 0.00217 amu has been lost. The electron is drawn as a dot on the big circle in the isotope 1H2 and it is drawn as a dot on the big circle that is the isotope 2He3. The electron in the isotope 1H2 stays an electron in the isotope 2He3.

The single neutron in the isotope 1H2 is drawn as a small circle with a square around it to show that it has lost 0.00157 amu to become the single neutron in the isotope 2He3 with 1.00471 amu's and it also has a square around it. The 0.00157 amu is emitted as a 157 gamma ray which has no external magnetic field.

The isotope 1H1 with 1.00783 amu's has its electron with 0.00055 amu drawn as a dot on the big circle and it is drawn as a square on the big circle that is the isotope 2He3. The electron in the isotope 1H1 becomes a negatron in the isotope 2He3.

The proton with 1.00728 amu's in the isotope 1H1 is drawn as a dot that has a square around it and it is drawn as a square with another square drawn around it in the isotope 2He3. The square around the proton in the isotope 1H1 indicates this proton with 1.00728 amu's has lost 0.00217 amu to become the positon with 1.00511 amu's in the isotope 2He3. The 0.00217 amu is emitted as a 217 alpha ray which has an external magnetic field. By drawing it this way, the proton in the isotope 1H2 will look just like the proton in the isotope 1H1 and since the proton and positon have the same mass in the isotope 2He3, the drawings must indicate which proton makes the positon in the isotope 2He3. To do this, two lines are drawn within the square of the proton so that they cross each other in 1H1's proton and the positon it becomes in the isotope 2He3.

The total mass lost is 0.00591 amu and this is composed of two units of 0.00217 amu and one unit of 0.00157 amu. The 0.00157 amu is lost as a single unit and it is composed of a sub-unit of 0.00100 amu, a sub-unit of 0.00002 amu and a sub-unit of 0.00055 amu. The 0.00055 amu is the same mass of an electron or negatron but because it is part of the single unit of 0.00157 amu, it is emitted as part of the 157 gamma ray. Each of the 0.00217 amu is lost as a 217 alpha ray.

To find the amount of mass needed to create the isotope 2He3, add the masses of the isotopes 1H2 and 1H1 and then subtract the mass of the isotope 2He3 from their total.

Mass of 1H2	2.01411 amu's
Mass of 1H1	1.00783 amu's
Total mass needed to create 2He3	3.02194 amu's

Total mass needed to create 2He3	3.02194 amu's
Mass of 2He3	3.01603 amu's
Total mass lost creating 2He3	0.00591 amu

The single neutron in the isotope 2He3 has 1.00471 amu's. This is the same as in the step just given. If this is known, then the amount of mass lost creating it from the single neutron in the isotope 1H2 can be found. This neutron has 1.00628 amu's. By subtracting the mass of the neutron in the isotope 2He3 from the neutron in the isotope 1H2, the lost mass will be found.

Mass of neutron in 1H2	1.00628 amu's
Mass of neutron in 2He3	1.00471 amu's
Mass lost by neutron in 1H2	0.00157 amu

Subtract the lost mass from the single neutron in the isotope 2He3 from the total mass lost creating the isotope 2He3 to find the remaining lost mass. This remaining lost mass is lost by each of the protons in the isotopes 1H2 and 1H1.

Total mass lost creating 2He3	0.00591 amu
Mass lost by neutron in 1H2	0.00157 amu
Mass remaining from total mass lost	0.00434 amu

This remaining mass that is lost is composed of two sub-units of 0.00217 amu. Each of the protons in the isotopes 1H2 and 1H1 lose some of their mass. The lost mass of 0.00217 amu will not have any sub-units so this must be the mass each of the protons lost. Because they are the lost masses and they do not have any sub-units, they are each called a unit. By subtracting this unit from each of the protons in the isotopes 1H2 and 1H1, the remaining mass will be the proton and positon in the isotope 2He3.

Mass of each proton in 1H2 and 1H1	1.00728 amu's
Mass lost by each proton in 1H2 and 1H1	0.00217 amu
Mass of proton or positon in 2He3	1.00511 amu's

The electron in the isotope 1H2 becomes the electron in the isotope 2He3 while the electron in the isotope 1H1 becomes the negatron. The proton in the isotope 1H2 becomes the proton in the isotope 2He3 while the proton in the isotope 1H1 becomes the positon. The electron and proton are drawn as dots while the negatron and positon are drawn as squares. They both have a square drawn around them to show that they lost 0.00217 amu and the proton in the isotope 1H1 and the positon in the isotope 2He3 have two lines crossing within the square. The neutrons each have a square drawn around them to show that 0.00157 amu have been lost.

Each of the lost units of 0.00217 amu are emitted as a 217 ray and the single unit of 0.00157 amu is emitted as a 157 ray.

Now add all of the known masses to find the mass of the isotope 2He3.

Mass of electron and negatron in 2He3	0.00110 amu
Mass of proton in 2He3	1.00511 amu's
Mass of positon in 2He3	1.00511 amu's
Mass of neutron in 2He3	1.00471 amu's
Mass of 2He3	3.01603 amu's

Sub-step 83-5-9

2He3

FORMULA

1H2 + 1H1 = 2He3, alpha 157 ray, gamma 217 ray, alpha 217 ray

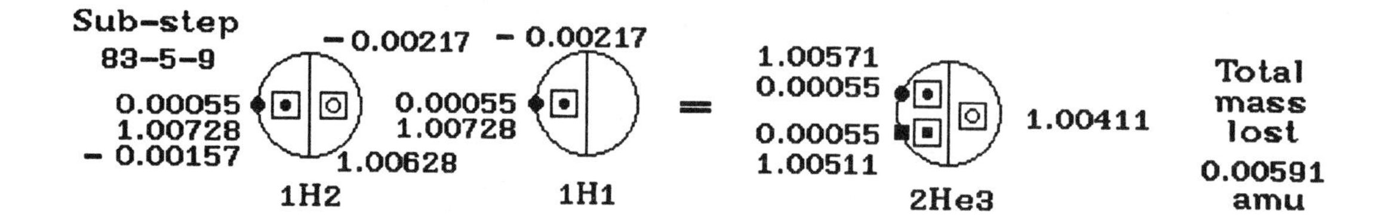

The isotope 1H2 has one electron with 0.00055 amu, one proton with 1.00728 amu's and one neutron with 1.00628 amu's for a total of 2.01411 amu's.

The isotope 1H1 has one electron with 0.00055 amu and one proton with 1.00728 amu's for a total of 1.00783 amu's.

The isotope of 2He3 with 3.01603 amu's has one electron and one negatron with each having 0.00055 amu. Its proton will have 1.00571 amu's, the positon will have 1.00511 amu's and the single neutron will have 1.00411 amu's.

The isotope 1H2's proton with 1.00728 amu's will lose 0.00157 amu as a 157 alpha ray to become the proton in the isotope 2He3 with 1.00571 amu's. The 157 alpha ray has an external magnetic field. The neutron with 1.00628 amu's will lose 0.00217 amu as a 217 gamma ray to become the neutron in the isotope 2He3 with 1.00411 amu's. The 217 gamma ray does not have an external magnetic field. The proton is drawn as a dot in the isotope 1H2 and it has a square drawn around the dot and this proton becomes a proton that is drawn the same way in the isotope 2He3. Both have a square drawn around them to indicate that 0.00157 amu has been lost. The neutron in the isotope 1H2 will be drawn as a small circle with a square drawn around it to show that it has lost 0.00217 amu. It will be drawn this same way in the isotope 2He3 to show that the neutron in the isotope 2He3 came from the isotope 1H2.

The electron with 0.00055 amu in the isotope 1H2 is drawn as a dot on the big circle and it is drawn as a dot on the big circle that is the isotope 2He3. The electron in the isotope 1H2 stays an electron in the isotope 2He3.

The isotope 1H1 has its electron drawn as a dot on the big circle and it is drawn as a square on the big circle that is the isotope 2He3. The electron with 0.00055 amu in the isotope 1H1 becomes a negatron in the isotope 2He3.

The proton in the isotope 1H1 is drawn as a dot that has a square around it and it is drawn as a small square to show that it is a positon in the isotope 2He3. There is another square drawn around the small square in the isotope 2He3. The square around the proton in the isotope 1H1 indicates this proton with 1.00728 amu's has lost 0.00217 amu to become the positon with 1.00511 amu's in the isotope 2He3. The 0.00217 amu is emitted as a 217 alpha ray which has an external magnetic field.

The total mass lost is 0.00591 amu and this is composed of two units of 0.00217 amu and one unit of 0.00157 amu. The 0.00157 amu is lost as a single unit and it is composed of a sub-unit of 0.00100 amu, a sub-unit of 0.00002 amu and a sub-unit of 0.00055 amu. The 0.00055 amu is the same mass of an electron or negatron but because it is part of the single unit of 0.00157 amu, it is emitted as part of the 157 gamma ray. Each unit of 0.00217 amu is emitted as a 217 alpha ray.

To find the amount of mass needed to create the isotope 2He3, add the masses of the isotopes 1H2 and 1H1 and then subtract the mass of the isotope 2He3 from their total.

Mass of 1H2	2.01411 amu's
Mass of 1H1	1.00783 amu's
Total mass needed to create 2He3	3.02194 amu's

Total mass needed to create 2He3	3.02194 amu's
Mass of 2He3	3.01603 amu's
Total mass lost creating 2He3	0.00591 amu

The single neutron in the isotope 2He3 has 1.00411 amu's. If this is known, then the amount of mass lost creating it from the single neutron in the isotope 1H2 can be found. This neutron has 1.00628 amu's. By subtracting the mass of the neutron in the isotope 2He3 from the neutron in the isotope 1H2, the lost mass will be found.

Mass of neutron in 1H2	1.00628 amu's
Mass of neutron in 2He3	1.00411 amu's
Mass lost by neutron in 1H2	0.00217 amu

Subtract the lost mass from the single neutron in the isotope 2He3 from the total mass lost creating the isotope 2He3 to find the remaining lost mass. This remaining lost mass is lost by each of the protons in the isotopes 1H2 and 1H1.

Total mass lost creating 2He3	0.00591 amu
Mass lost by neutron in 1H2	0.00217 amu
Mass remaining from total mass lost	0.00374 amu

This remaining mass that is lost is composed of two units of 0.00217 amu and 0.00157 amu. In this step, I have chosen the proton in the isotope 1H2 to lose the 0.00157 amu. This means that the isotope 1H1's proton will lose the 0.00217 amu. The mass of the proton in the isotope 1H2 can be found by subtracting the lost mass of 0.00157 amu from it.

Mass of proton in 1H2	1.00728 amu's
Mass lost by proton in 1H2	0.00157 amu
Mass of proton in 2He3	1.00571 amu's

The mass of the positon in the isotope 2He3 can be found by subtracting the lost mass of 0.00217 that is lost from the proton in the isotope 1H1.

Mass of proton in 1H1	1.00728 amu's
Mass lost by it	0.00217 amu
Mass of positon in 2He3	1.00511 amu's

Now the mass of all of the masses of the isotope 2He3 are known.

Mass of electron in 2He3	0.00055 amu
Mass of negatron in 2He3	0.00055 amu
Mass of proton in 2He3	1.00571 amu's
Mass of positon in 2He3	1.00511 amu's
Mass of neutron in 2He3	1.00411 amu's
Mass of 2He3	3.01603 amu's

The electron in the isotope 1H2 becomes the electron in the isotope 2He3 while the electron in the isotope 1H1 becomes the negatron. The proton in the isotope 1H2 becomes the proton in the isotope 2He3 while the proton in the isotope 1H1 becomes the positon. The electron and proton are drawn as dots while the negatron and positon are drawn as squares with a square drawn around each to show they have lost mass. The neutron in the isotope 1H2 has a square drawn around it to show that 0.00217 amu has been lost and around the neutron in the isotope 2He3 to show that the neutron in the isotope 2He3 was the neutron in the isotope 1H2.

Each of the lost units of 0.00217 amu are emitted as a 217 ray but one is an alpha ray while the other is a gamma ray and the single unit of 0.00157 amu is emitted as a 157 alpha ray.

Note the position of the neutron in the isotopes 1H2 and 2He3. The proton in the isotope 1H2 looks just like the drawing for the proton in the isotope 1H1. By noting the position of the neutron in the isotope 2He3, the proton next to it is the proton that comes from the isotope 1H2. Both protons are drawn as a dot with a square around each in the isotopes 1H2 and 1H1. This means that the proton in the isotope 1H2 is the proton in the isotope 2He3 and it can not be a positon.

Sub-step 82-5-8

2He3

FORMULA

1H2 + 1H1 = 2He3, alpha 217 ray, gamma 217 ray, alpha 157 rays

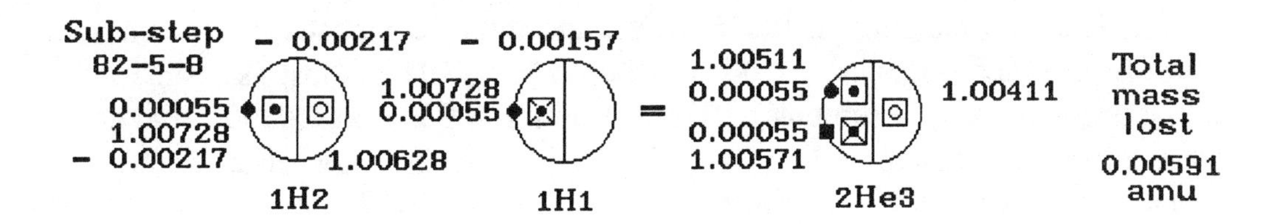

The isotope 1H2 has one electron with 0.00055 amu, one proton with 1.00728 amu's and one neutron with 1.00628 amu's for a total of 2.01411.

The isotope 1H1 has one electron with 0.00055 amu and one proton with 1.00728 amu's for a total of 1.00783 amu's.

The isotope of 2He3 with 3.01603 amu's has one electron and one negatron with each having 0.00055 amu. Its proton will have 1.00511 amu, the positon will have 1.00571 amu's and the single neutron will have 1.00411 amu's.

The electron with 0.00055 amu in the isotope 1H2 is drawn as a dot and it is also a dot in the isotope 2He3. Its proton with 1.00728 amu's will lose 0.00217 amu as a 217 ray to become the proton in the isotope 2He3. The 217 alpha ray has an external magnetic field. It will be drawn as a dot with a square around it in the isotopes 1H2 and 2He3 to indicate that this proton has lost 0.00217 amu to become the proton in the isotope 2He3 with 1.00511 amu's.

The neutron with 1.00628 amu's will also lose 0.00217 amu as a 217 ray. The 0.00217 amu is emitted as a 217 gamma ray which does not have an external magnetic field. The proton and neutron have a square drawn around them to indicate that 0.00217 amu has been lost. The neutron with 1.00628 amu's becomes the neutron in the isotope 2He3 with 1.00411 amu's.

The proton in the isotope 1H1 has 1.00728 amu's and when it loses 0.00157 amu it becomes the positon with 1.00571 amu's in the isotope 2He3. The 0.00157 amu is emitted as a 157 alpha ray which has an external magnetic field. The proton in the isotope 1H1 is drawn as a dot with a square around it the same way as the proton in the isotope 1H2 and to show that this is the proton that becomes the positon in the isotope 2He3, lines are added within the square of the proton in the isotope 1H1 and the positon in the isotope 2He3. The positon is drawn as a small square with another square around it plus the lines within the square to indicate that it came from the isotope 1H1.

The proton is drawn as a dot in the isotope 1H2 and as a dot in the isotope 2He3 to show that the proton in the isotope 1H2 stays a positon in the isotope 2He3. They both have a square drawn around each to show that the proton in the isotope 1H2 lost 0.00217 amu and the proton in the isotope 2He3 was the proton in the isotope 1H2. Note also that the electron in the isotope 1H2 is drawn as a dot on the big circle. It is also drawn as a dot on the big circle in the isotope 2He3. The electron in the isotope 1H2 stays an electron in the isotope 2He3. The electron in the isotope 1H1 is drawn as a dot and in the isotope 2He3 it is drawn as a square to show that it is now a negatron.

The total mass lost is 0.00591 amu and this is composed of two units of 0.00217 amu and one unit of 0.00157 amu. The 0.00157 is lost as a single unit and it is composed of a sub-unit of 0.00100 amu, a sub-unit of 0.00002 amu and a sub-unit of 0.00055 amu. The 0.00055 amu is the same mass of an electron or negatron but because it is part of the single unit of 0.00157 amu, it is emitted as part of the 157 alpha ray. Each of the 0.00217 amu is lost as a 217 ray but one is a 217 alpha ray and the other is a 217 gamma ray.

To find the amount of mass needed to create the isotope 2He3, add the masses of the isotopes 1H2 and 1H1 and then subtract the mass of the isotope 2He3 from their total.

Mass of 1H2	2.01411 amu's
Mass of 1H1	1.00783 amu's
Total mass needed to create 2He3	3.02194 amu's

Total mass needed to create 2He3	3.02194 amu's
Mass of 2He3	3.01603 amu's
Total mass lost creating 2He3	0.00591 amu

The single neutron in the isotope 2He3 has 1.00411 amu's. If this is known, then the amount of mass lost creating it from the single neutron in the isotope 1H2 can be found. This neutron has 1.00628 amu's. By subtracting the mass of the neutron in the isotope 2He3 from the neutron in the isotope 1H2, the lost mass will be found.

Mass of neutron in 1H2	1.00628 amu's
Mass of neutron in 2He3	1.00411 amu's
Mass lost by neutron in 1H2	0.00217 amu

Subtract the lost mass from the single neutron in the isotope 2He3 from the total mass lost creating the isotope 2He3 to find the remaining lost mass. This remaining lost mass is lost by each of the protons in the isotope 1H2 and 1H1.

Total mass lost creating 2He3	0.00591 amu
Mass lost by neutron in 1H2	0.00217 amu
Mass remaining from total mass lost	0.00374 amu

This remaining mass that is lost is composed of a unit of 0.00217 amu and a unit of 0.00157 amu. In this step, the proton in the isotope 1H2 loses the 0.00217 amu. This means that the isotope 1H1's proton will lose the 0.00157 amu. The mass of the proton in the isotope 1H2 can be found by subtracting the lost mass of 0.00217 amu from it.

Mass of proton in 1H2	1.00728 amu's
Mass lost by proton in 1H2	0.00217 amu
Mass of proton in 2He3	1.00511 amu's

The mass of the positon in the isotope 2He3 can be found by subtracting the lost mass of 0.00157 amu that is lost from the proton in the isotope 1H1.

Mass of proton in 1H1	1.00728 amu's
Mass lost by it	0.00157 amu
Mass of positon in 2He3	1.00571 amu's

The electron in the isotope 1H2 becomes the electron in the isotope 2He3 while the electron in the isotope 1H1 becomes the negatron. Each of the lost units of 0.00217 amu are emitted as a 217 ray and the single unit of 0.00157 amu is emitted as a 157 ray.

Note that in the step before this step, the mass lost by the protons in the isotopes 1H2 and 1H1 have been switched while the neutron in the isotope 1H2 has lost the same mass. These two steps show that the different lost masses of 0.00217 amu and 0.00157 amu can be lost by either isotope of 1H2 or 1H1.

Note the position of the neutron in the isotopes 1H2 and 2He3. The proton in the isotope 1H2 looks just like the drawing for the proton in the isotope 1H1. By noting the position of the neutron in the isotope 2He3, the proton next to it is the proton that comes from the isotope 1H2. Both protons are drawn as a dot with a square around each in the isotopes 1H2 and 1H1. This means that the proton in the isotope 1H2 is the proton in the isotope 2He3 and it can not be a positon.

Now add all of the known masses to find the mass of the isotope 2He3.

Mass of electron and negatron in 2He3	0.00110 amu
Mass of proton in 2He3	1.00511 amu's
Mass of positon in 2He3	1.00571 amu's
Mass of neutron in 2He3	1.00411 amu's
Mass of 2He3	3.01603 amu's

Sub-step 81-5-7

2He3

FORMULA

1H2 + 1H1 = 2He3, alpha 374 ray, gamma 217 ray

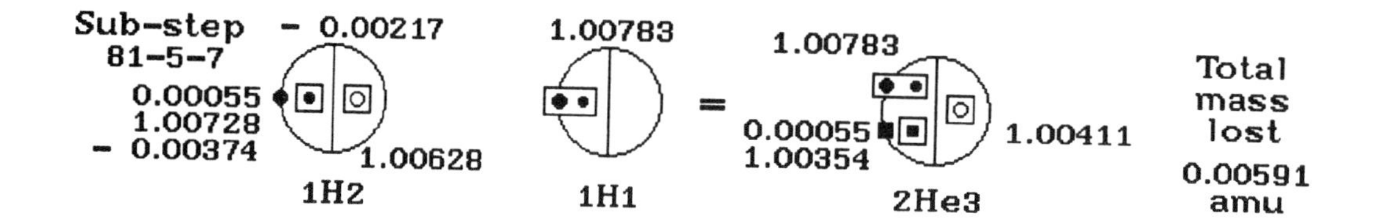

The isotope 1H2 has one electron with 0.00055 amu, one proton with 1.00728 amu's and one neutron with 1.00628 amu's for a total of 2.01411 amu's.

The isotope 1H1 has one electron with 0.00055 amu and one proton with 1.00728 amu's for a total of 1.00783 amu's.

The isotope of 2He3 with 3.01603 amu's has one electron and one negatron with each having 0.00055 amu. It has a proton with 1.00728 amu's and a positon with 1.00354 amu's. The single neutron has 1.00411 amu's.

The isotope 1H2's proton will lose 0.00374 amu as a 374 alpha ray and the neutron will lose 0.00217 amu as a 217 gamma ray. The 374 alpha ray has an external magnetic field while the 217 gamma ray does not. The proton with 1.00728 amu's is drawn as a dot and this proton becomes a positon with 1.00354 amu's that is drawn as a square in the isotope 2He3. The proton in the isotope 1H2 has a square drawn around it and in the isotope 2He3 to indicate that 0.00374 amu has been lost. The electron with 0.00055 amu in the isotope 1H2 is drawn as a dot on the big circle and it is drawn as a square on the big circle that is the isotope 2He3. The change from a dot to a square shows that the electron has become a negatron. The single neutron in the isotope 1H2 is drawn as a small circle with a square drawn around it and it is drawn this same way in the isotope 2He3. The square shows that the neutron in the isotope 1H2 lost 0.00217 amu to become the neutron in the isotope 2He3 with 1.00411 amu's.

The isotope 1H1 with 1.00783 amu's is used as the isotope 1H1 in the isotope 2He3 with no loss of mass. Its electron and proton are each drawn as a dot in the isotopes 1H2 and 2He3. They are enclosed in a rectangle to show that the isotope 1H1 is intact within the isotope 2He3.

The total mass lost is 0.00591 amu and this is composed of two units. One unit of 0.00374 amu and the other is 0.00217 amu. While the 0.00374 amu is lost as a single unit, it is composed of two sub-units also. One sub-unit of 0.00217 amu and another sub-unit of 0.00157 amu. The 0.00157 amu is lost as a single unit and it is composed of a sub-unit of 0.00100 amu, a sub-unit of 0.00002 amu and a sub-unit of 0.00055 amu. The 0.00055 amu is the same mass of an electron or negatron but because it is part of the single unit of 0.00374 amu, it is emitted as part of the 374 alpha ray. The unit 0.00217 is emitted as a 217 gamma ray.

To find the amount of mass needed to create the isotope 2He3, add the masses of the isotopes 1H2 and 1H1 and then subtract the mass of the isotope 2He3 from their total.

Mass of 1H2	2.01411 amu's
Mass of 1H1	1.00783 amu's
Total mass needed to create 2He3	3.02194 amu's

Total mass needed to create 2He3	3.02194 amu's
Mass of 2He3	3.01603 amu's
Total mass lost creating 2He3	0.00591 amu

The drawings show that the mass of the isotope 1H1 is intact within the isotope 2He3 and there is a negatron in this isotope also. These masses are known. The mass of the positon and single neutron are not known. However, I have already given a step with a neutron having 1.00411 amu's. I can therefore use this as the mass of the neutron found in the isotope 2He3. By adding all of the known masses and subtracting their total from the isotope 2He3, the mass of the positon can be found.

Mass of 1H1 in 2He3	1.00783 amu's
Mass of negatron in 2He3	0.00055 amu
Mass of neutron in 2He3	1.00411 amu's
Total known mass in 2He3	2.01249 amu's

Mass of 2He3	3.01603 amu's
Total known mass in 2He3	2.01249 amu's
Mass of positon in 2He3	1.00354 amu's

Mass of 1H1 in 2He3	1.00783 amu's
Mass of negatron in 2He3	0.00055 amu
Mass of positon in 2He3	1.00354 amu's
Total known mass in 2He3	2.01192 amu's

Mass of 2He3	3.01603 amu's
Total known mass in 2He3	2.01192amu's
Mass of single neutron in 2He3	1.00411 amu's

With the mass of the positon in the isotope 2He3 known, its mass can be subtracted from the mass of the proton in the isotope 1H2 to find the mass it lost to find the mass it lost.

Mass of proton in 1H2	1.00728 amu's
Mass of positon in 2He3	1.00354 amu's
Mass lost by proton in 1H2 becoming positon in 2He3	0.00374 amu

With the mass of the single neutron in the isotope 2He3 known, its mass can be subtracted from the mass of the single neutron in the isotope 1H2.

Mass of neutron in 1H2	1.00628 amu's
Mass of neutron in 2He3	1.00411 amu's
Mass lost by neutron in 1H2 becoming neutron in 2He3	0.00217 amu

Adding the mass lost by the proton and neutron in the isotope 1H2 will give the total mass lost creating the isotope 2He3.

Mass lost by proton in 1H2	0.00374 amu
Mass lost by neutron in 1H2	0.00217 amu
Total mass lost creating 2He3	0.00591 amu

Now add all of the known masses to find the mass of the isotope 2He3.

Mass of 1H1 within 2He3	1.00783 amu's
Mass of negatron in 2He3	0.00055 amu
Mass of positon in 2He3	1.00354 amu's
Mass of neutron in 2He3	1.00411 amu's
Mass of 2He3	3.01603 amu's

The next step will have the electron and proton in the isotope 1H2 used as is which means that the lost mass of 0.00374 amu will now be lost by the proton in the isotope 1H1.

Sub-step 80-5-6

2He3

FORMULA

1H2 + 1H1 = 2He3, gamma 217 ray, alpha 374 ray

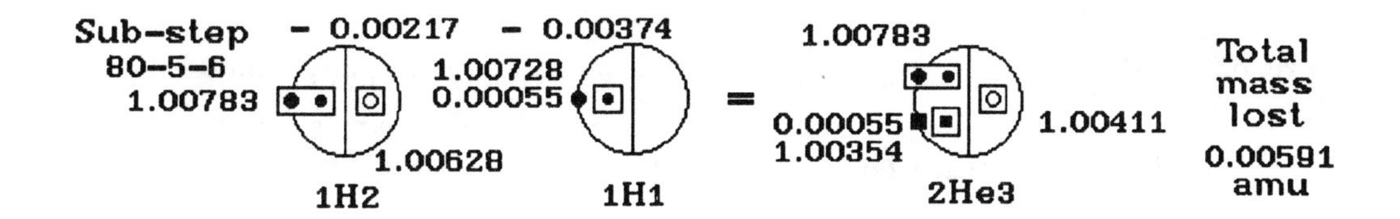

The isotope 1H2 has one electron with 0.00055 amu, one proton with 1.00728 amu's and one neutron with 1.00628 amu's for a total of 2.01411 amu's.

The isotope 1H1 has one electron with 0.00055 amu and one proton with 1.00728 amu's for a total of 1.00783 amu's.

The isotope of 2He3 with 3.01603 amu's has one electron and one negatron with each having 0.00055 amu. The proton will have 1.00728 amu's and the positon will have 1.00354 amu's. The single neutron will have 1.00411 amu's. This means that the value of the proton, positon and neutron will not always be the same in each of the steps.

The isotope 1H2 has its electron and proton drawn with a rectangle around them to show that they are used as is with no loss of mass. They are used as the isotope 1H1 in the isotopes 1H2 and 2He3 with 1.00783 amu's. The electron and proton are each drawn as a dot in the isotopes 1H2 and 2He3.

The neutron in the isotope 1H2 will lose 0.00217 amu as a 217 gamma ray which does not have an external magnetic field. The neutron in the isotopes 1H2 and 2He3 are drawn as a small circle with a square around each to show that the neutron in the isotope 1H2 lost 0.00217 amu to become the neutron in the isotope 2He3 with 1.00411 amu's.

The isotope 1H1's electron will become the negatron in the isotope 2He3 and this will be shown by drawing the electron as a dot and the negatron as a square. The proton will lose 0.00374 amu as a 374 ray. The proton in the isotope 1H1 will be drawn as a dot with a square around it and this proton will become the positon as a square with another square drawn around it. The square drawn around the proton and the positon shows that the proton lost 0.00374 amu to become the positon with 1.00354 amu's.

The total mass lost is 0.00591 amu and this is composed of two units. One of 0.00374 amu and the other is 0.00217 amu. While the 0.00374 amu is lost as a single unit, it is composed of two sub-units. One sub-unit is 0.00217 amu and the second one is 0.00157 amu. The sub-unit of 0.00157 amu is lost as a single unit that is composed of a sub-unit of 0.00100 amu, a sub-unit of 0.00002 amu and a sub-unit of 0.00055 amu. The 0.00055 amu is the same mass of an electron or negatron but because it is part of the single unit of 0.00374 amu, it is emitted as part of the 374 alpha ray. The unit of 0.00217 amu is emitted as a 217 gamma ray.

To find the amount of mass needed to create the isotope 2He3, add the masses of the isotopes 1H2 and 1H1 and then subtract the mass of the isotope 2He3 from their total.

Mass of 1H2	2.01411 amu's
Mass of 1H1	1.00783 amu's
Total mass needed to create 2He3	3.02194 amu's

Total mass needed to create 2He3	3.02194 amu's
Mass of 2He3	3.01603 amu's
Total mass lost creating 2He3	0.00591 amu

The drawings show that the mass of the isotope 1H1 is intact within the isotope 1H2 and the isotope 2He3 The mass of the positon and single neutron are not known. However, I have already given a step with a neutron having 1.00411 amu's. I can therefore use this as the mass of the neutron found in the isotope 2He3. By adding all of the known masses and subtracting their total from thc isotope 2He3, the mass of the positon can be found.

Mass of 1H1 in 1H2 and 2He3	1.00783 amu's
Mass of negatron in 2He3	0.00055 amu
Mass of neutron in 2He3	1.00411 amu's
Total known mass in 2He3	2.01249 amu's

Mass of 2He3	3.01603 amu's
Total known mass in 2He3	2.01249 amu's
Mass of positon in 2He3	1.00354 amu's

Mass of 1H1 in 1H2 and 2He3	1.00783 amu's
Mass of negatron in 2He3	0.00055 amu
Mass of positon in 2He3	1.00354 amu's
Total known mass in 2He3	2.01192 amu's

Mass of 2He3	3.01603 amu's
Total known mass in 2He3	2.01192 amu's
Mass of single neutron in 2He3	1.00411 amu's

With the mass of the positon in the isotope 2He3 known, its mass can be subtracted from the mass of the proton in the isotope 1H1 to find the mass it lost.

Mass of proton in 1H1	1.00728 amu's
Mass of positon in 2He3	1.00354 amu's
Mass lost by proton in 1H1 becoming positon in 2He3	0.00374 amu

With the mass of the single neutron in the isotope 2He3 known, its mass can be subtracted from the mass of the single neutron in the isotope 1H2.

Mass of neutron in 1H2	1.00628 amu's
Mass of neutron in 2He3	1.00411 amu's
Mass lost by neutron in 1H2 becoming neutron in 2He3	0.00217 amu

Adding the mass lost by the proton and neutron in the isotope 1H2 will give the total mass lost creating the isotope 2He3.

Mass lost by proton in 1H2	0.00374 amu
Mass lost by neutron in 1H2	0.00217 amu
Total mass lost creating 2He3	0.00591 amu

The proton's lost mass of 0.00374 amu has traded places with the isotopes 1H1 and 1H2 given in the step before this one. The lost mass of the neutron in the isotope 1H2 is the same in both steps.

Now add all of the known masses to find the mass of the isotope 2He3.

Mass of 1H1 within 2He3	1.00783 amu's
Mass of negatron in 2He3	0.00055 amu
Mass of positon in 2He3	1.00354 amu's
Mass of neutron in 2He3	1.00411 amu's
Mass of 2He3	3.01603 amu's

Sub-step 79-5-5

2He3

FORMULA

1H2 + 1H1 = 2He3, alpha 591 ray

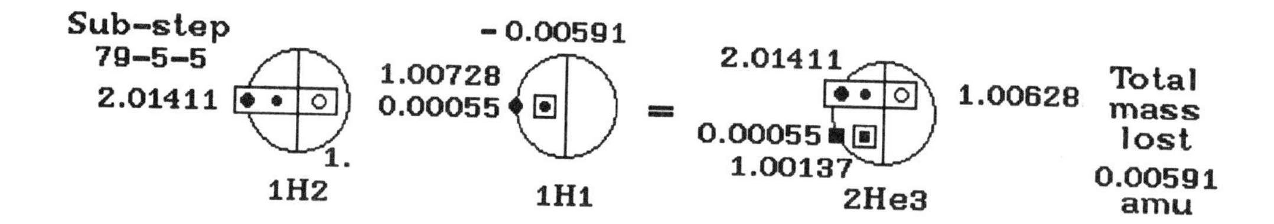

The isotope 1H2 has one electron with 0.00055 amu, one proton with 1.00728 amu's and one neutron with 1.00628 amu's for a total of 2.01411 amu's.

The isotope 1H1 has one electron with 0.00055 amu and one proton with 1.00728 amu's for a total of 1.00783 amu's.

The isotope of 2He3 with 3.01603 amu's has one electron and one negatron with each having 0.00055 amu. There is one proton with 1.00728 amu's and one positon with 1.00137 amu's. The single neutron will have 1.00628 amu's.

The isotope 1H2 has its electron, proton and neutron drawn with a rectangle around them to show that they are used as is with no loss of mass. The electron and proton are each drawn as a dot and the single neutron is drawn as a small circle. They are used as the isotope 1H2 in the 2He3 with 2.01411 amu's. In other steps already given, it is the isotope 1H1 which is enclosed in a rectangle which can drop out at some later time or it does "drop out". In this step, it is the isotope 1H2 that is enclosed by a rectangle which can drop out at some later time.

The isotope 1H1's electron will become the negatron in the isotope 2He3 and this will be shown with the electron as a dot and the negatron as a square. The proton will lose 0.00591 amu as a 591 alpha ray which has an external magnetic field. The proton in the isotope 1H1 will be drawn as a dot with a square around it and this proton will become the positon as a square with another square drawn around it in the isotope 2He3. The square drawn around the proton and the positon shows that the proton lost 0.00591 amu to become the positon with 1.00137 amu's.

The total mass lost is 0.00591 amu and which is lost as one unit. This is composed of two sub-units. One has 0.00374 amu and the other is 0.00217 amu. While the 0.00374 amu is a single sub-unit, it is composed of two sub-units. One has 0.00217 amu and the other is 0.00157 amu. The sub-unit of 0.00157 amu is composed of a sub-unit of 0.00100 amu, a sub-unit of 0.00002 amu and a sub-unit but because it is part of the sub-unit of 0.00374 amu and 0.00157 amu and they are a sub-unit of 0.00591 amu, it is part of the 591 alpha ray that is emitted.

While the above is correct, this lost mass could be composed of either one sub-unit of 0.00434 amu (or two sub-units of 0.00217 amu) and a sub-unit of 0.00157 amu.

To find the amount of mass needed to create the isotope 2He3, add the masses of the isotopes 1H2 and 1H1 and then subtract the mass of the isotope 2He3 from their total.

Mass of 1H2	2.01411 amu's
Mass of 1H1	1.00783 amu's
Total mass needed to create 2He3	3.02194 amu's

Total mass needed to create 2He3	3.02194 amu's
Mass of 2He3	3.01603 amu's
Total mass lost creating 2He3	0.00591 amu

The drawings show that the mass of the isotope 1H2 with 2.01411 amu's is intact within the isotope 2He3. The mass of the negatron is known to be 0.00055 amu. The mass of the positon is not known. To find the mass of the positon, add all of the known masses of the isotope 2He3 and subtract this total from the mass of the isotope 2He3.

Mass of 1H2 in 2He3	2.01411 amu's
Mass of negatron in 2He3	0.00055 amu
Total known mass in 2He3	2.01466 amu's

Mass of 2He3	3.01603 amu's
Total known mass in 2He3	2.01466 amu's
Mass of positon in 2He3	1.00137 amu's

To find the mass that proton in the isotope 1H1 lost, subtract the mass of the positon in the isotope 2He3 from the proton in the isotope 1H1.

Mass of proton in 1H1	1.00728 amu's
Mass of positon in 2He3	1.00137 amu's
Mass lost by proton in 1H1	0.00591 amu

Now add all of the known masses to find the mass of the isotope 2He3.

Mass of 1H2 within 2He3	2.01411 amu's
Mass of negatron in 2He3	0.00055 amu
Mass of positon in 2He3	1.00137 amu's
Mass of 2He3	3.01603 amu's

Sub-step 78-5-4

2He3

FORMULA

1H2 + 1H1 = 2He3, alpha 591 ray

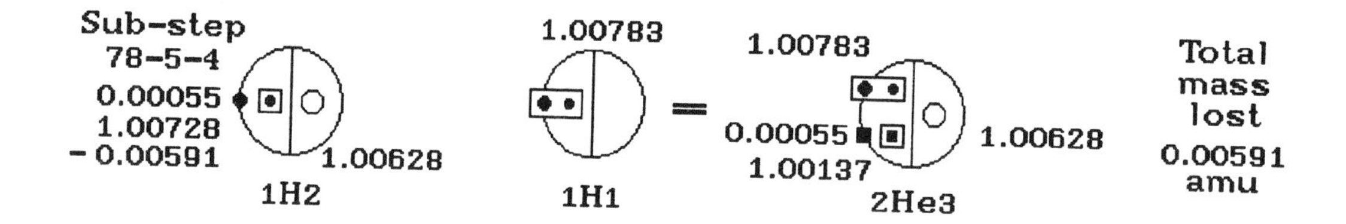

The isotope 1H2 has one electron with 0.00055 amu, one proton with 1.00728 amu's and one neutron with 1.00628 amu's for a total of 2.01411 amu's.

The isotope 1H1 has one electron with 0.00055 amu and one proton with 1.00728 amu's for a total of 1.00783 amu's.

The isotope of 2He3 has one electron and one negatron with each having 0.00055 amu. There is one proton with 1.00728 amu's, one positon with 1.00137 amu's and one neutron with 1.00628 amu's.

The isotope 1H2 has its electron drawn as a dot and this electron becomes a negatron that is drawn as a square in the isotope 2He3. The proton with 1.00728 amu's loses 0.00591 amu as a 591 alpha ray which has an external magnetic field. The proton in the isotope 1H2 is also drawn as a dot and it becomes the positon in the isotope 2He3 that is drawn as a square. This proton with 1.00728 amu's loses 0.00591 amu and this is shown by drawing a square around the dot in the isotope 1H2 and a second square around the first square of the positon with 1.00137 amu's in the isotope 2He3. The neutron is used as is in the isotope 2He3 and it is drawn as a small circle. It has 1.00628 amu's.

The isotope 1H1's electron and proton are drawn with a rectangle around them to show that they are used as is and as the isotope 1H1 within the isotope 2He3. This isotope has 1.00783 amu's.

The total mass lost is 0.00591 amu and this is lost as one unit. This is composed of two sub-units. One unit is 0.00374 amu and the other is 0.00217 amu. While the 0.00374 amu is a single sub-unit, it is composed of two sub-units. One is 0.00217 amu and the other is 0.00157 amu. The sub-unit of 0.00157 amu is composed of a sub-unit of 0.00100 amu, a sub-unit of 0.00002 amu and a sub-unit of 0.00055 amu. The 0.00055 amu is the same mass of an electron or negatron but because it is part of the sub-units of 0.00374 amu and 0.00157 amu, it is emitted as part of the 591 alpha ray.

While the above is correct, this lost mass could be composed of one sub-unit of 0.00434 amu (or two sub-units of 0.00217 amu) with a sub-unit of 0.00157 amu.

To find the amount of mass needed to create the isotope 2He3, add the masses of the isotopes 1H2 and 1H1 and then subtract the mass of the isotope 2He3 from their total.

Mass of 1H2	2.01411 amu's
Mass of 1H1	1.00783 amu's
Total mass needed to create 2He3	3.02194 amu's

Total mass needed to create 2He3	3.02194 amu's
Mass of 2He3	3.01603 amu's
Total mass lost creating 2He3	0.00591 amu

The drawings show that the mass of the isotope 1H1 with 1.00783 amu's is intact within the isotope 2He3. The mass of the negatron is known to be 0.00055 amu. The mass of the positon is not known. To find the mass of the positon, add all of the known masses of the isotope 2He3 and subtract this total from the mass of the isotope 2He3.

Mass of 1H1 in 2He3	1.00783 amu's
Mass of negatron in 2He3	0.00055 amu
Mass of neutron in 2He3	1.00628 amu's
Total known mass in 2He3	2.01466 amu's

Mass of 2He3	3.01603 amu's
Total known mass in 2He3	2.01466 amu's
Mass of positon in 2He3	1.00137 amu's

To find the mass that the proton in the isotope 1H2 lost, subtract the mass of the positon in the isotope 2He3 from the proton in the isotope 1H2.

Mass of proton in 1H2	1.00728 amu's
Mass of positon in 2He3	1.00137 amu's
Mass lost by proton in 1H2	0.00591 amu

Now add all of the known masses to find the mass of the isotope 2He3.

Mass of 1H1 within 2He3	1.00783 amu's
Mass of negatron in 2He3	0.00055 amu
Mass of positon in 2He3	1.00137 amu's
Mass of neutron in 2He3	1.00628 amu's
Mass of 2He3	3.01603 amu's

Sub-step 77-5-3

2He3

FORMULA

1H2 + 1H1 = 2He3, gamma 591 ray

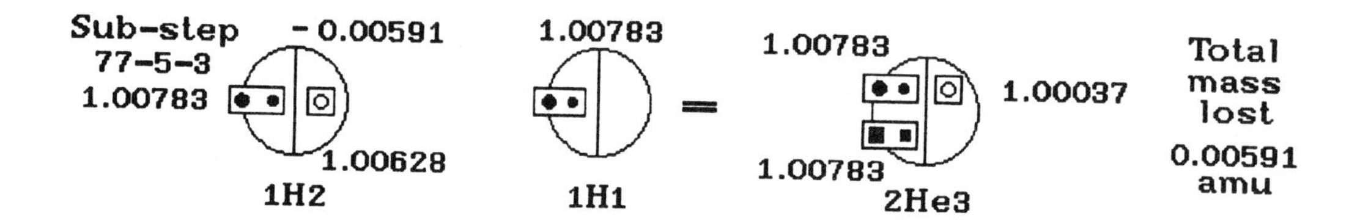

The isotope 1H2 has one electron with 0.00055 amu, one proton with 1.00728 amu's and one neutron with 1.00628 amu's for a total of 2.01411 amu's.

The isotope 1H1 has one electron with 0.00055 amu and one proton with 1.00728 amu's for a total of 1.00783 amu's.

The isotope of 2He3 with 3.01603 amu's has one electron and one negatron with each having 0.00055 amu. There is one proton and one positon with each having 1.00728 amu's. The single neutron will have 1.00037 amu's.

The isotope 1H2 has its electron and proton drawn with a rectangle around both to show that they are used as the isotope 1H1 within the isotope 1H2 and 2He3 with 1.00783 amu's. The electron and proton are each drawn as a dot in the isotopes 1H2 and 2He3. The single neutron with 1.00628 amu's is drawn as a small circle with a square around it in the isotopes 1H2 and 2He3 to show that it has lost all of the 0.00591 amu. The 0.00591 amu is emitted as a 591 gamma ray which does not have an external magnetic field. It becomes the single neutron in the isotope 2He3 with 1.00037 amu's.

The isotope 1H1's electron and proton are drawn with a rectangle around them to show that they are used as is and as the isotope 1H1 within the isotope 2He3. This isotope has 1.00783 amu's

The total mass lost is 0.00591 amu and which is lost as one unit. This is composed of two sub-units. One is 0.00374 amu and the other is 0.00217 amu. While the 0.00374 amu a single sub-unit, it is composed of two sub-units. One is 0.00217 amu and the other is 0.00157 amu. The sub-unit of 0.00157 amu is composed of a sub-unit of 0.00100 amu, a sub-unit of 0.00002 amu and a sub-unit of 0.00055 amu. The 0.00055 amu has the same mass of an electron or negatron but because it is part of the sub-unit of 0.00374 amu and 0.00157 amu, it is emitted as part of the 591 gamma ray.

While the above is correct, this lost mass could be composed of one sub-unit of 0.00434 amu (or two sub-units of 0.00217 amu) with a sub-unit of 0.00157 amu.

To find the amount of mass needed to create the isotope 2He3, add the masses of the isotope 1H2 and 1H1 and then subtract the mass of the isotope 2He3 from their total.

Mass of 1H2	2.01411 amu's
Mass of 1H1	1.00783 amu's
Total mass needed to create 2He3	3.02194 amu's

Total mass needed to create 2He3	3.02194 amu's
Mass of 2He3	3.01603 amu's
Total mass lost creating 2He3	0.00591 amu

The drawings show that the mass of the isotope 1H1 with 1.00783 amu's is intact within the isotope 2He3. The isotope 1H1 that is by itself is also found within the isotope 2He3 intact. The mass of the single neutron in the isotope 2He3 is unknown. To find the mass of this neutron, add all of the known masses of the isotope 2He3 and subtract this total from the mass of the isotope 2He3.

Mass of 1H1 in 1H2 and 2He3	1.00783 amu's
Mass of 1H1 in 2He3	1.00783 amu's
Total known mass in 2He3	2.01566 amu's

Mass of 2He3	3.01603 amu's
Total known mass in 2He3	2.01566 amu's
Mass of neutron in 2He3	1.00037 amu's

To find that the mass the neutron in the isotope 1H2 lost, subtract the mass of the neutron in the isotope 2He3 from the neutron in the isotope 1H2.

Mass of neutron in 1H2	1.00628 amu's
Mass of neutron in 2He3	1.00037 amu's
Mass lost by neutron in 1H2	0.00591 amu

Now add all of the known masses to find the mass of the isotope 2He3.

Mass of 1H1 within 2He3 from 1H2	1.00783 amu's
Mass of 1H1 within 2He3	1.00783 amu's
Mass of neutron in 2He3	1.00037 amu's
Mass of 2He3	3.01603 amu's

Sub-step 76-5-2

2He3

FORMULA

1H2 + 1H1 = 2He3, alpha 591 ray

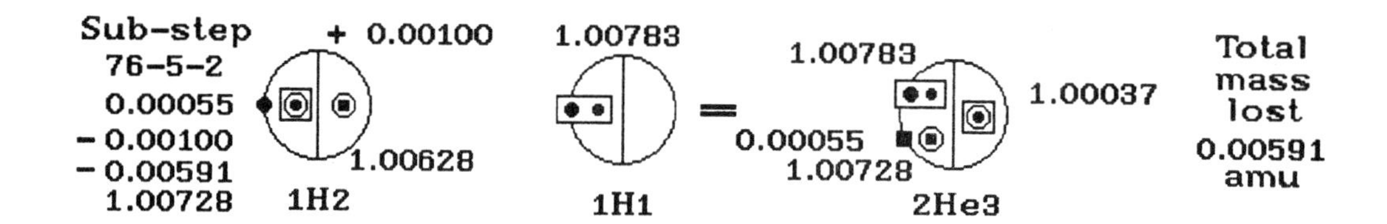

The isotope 1H2 has one electron with 0.00055 amu, one proton with 1.00728 amu's and one neutron with 1.00628 amu's for a total of 2.01411 amu's.

The isotope 1H1 has one electron with 0.00055 amu and one proton with 1.00728 amu's for a total of 1.00783 amu's.

The isotope of 2He3 with 3.01603 amu's has one electron and one negatron with each having 0.00055 amu. There is one proton and one positon with each having 1.00728 amu's. The single neutron has 1.00037 amu's.

The proton in the isotope 1H2 is drawn as a dot with a small circle around it and then a square around both of them. The proton has 1.00728 amu's which loses 0.00591 amu first and then it loses 0.00100 amu to release its electron. The square shows the loss of 0.00591 amu and the small circle shows the loss of 0.00100 amu. The 0.00591 amu is emitted as a 591 alpha ray which has an external magnetic field.

The electron is drawn as a dot on the big circle in the isotope 1H2 and it is drawn in the isotope 2He3 as a square to show that it has been regained by the isotope 2He3.

The neutron in the isotope 1H2 is drawn as a small circle with a square in its middle to show that it has regained 0.00100 amu to become the positon in the isotope 2He3. The neutron and positon are drawn the same in both isotopes.

Note that the neutron is normally drawn as a small circle and when it loses 0.00100 amu, it has a dot in its middle. It is also drawn this way when a proton or positon loses 0.00100 amu to become a neutron. This is drawn differently when the neutron regains mass. When it regains 0.00100 amu, the middle is drawn as a square when the neutron becomes a positon.

The isotope 1H1 has its electron and proton drawn with a rectangle around both to show that they are used as the isotope 1H1 within the isotope 2He3 with 1.00783 amu's.

The total mass lost is 0.00591 amu and which is lost as one unit. This is composed of two sub-units. One is 0.00374 amu and 0.00217 amu. While the 0.00374 amu is a single sub-unit, it is composed of two sub-units. One is 0.00217 amu and the other is 0.00157 amu. The sub-unit of 0.00157 amu is composed of a sub-unit of 0.00100 amu, a sub-unit of 0.00002 amu and a sub-unit of 0.00055 amu. The 0.00055 amu is the same mass of an electron or negatron but because it is part of the sub-unit of 0.00374 amu and 0.00157 amu, it is emitted as part of the 591 alpha ray.

While the above is correct, this lost mass could be composed of one sub-unit of 0.00434 amu (or two sub-units of 0.00217 amu) with a sub-unit of 0.00157 amu.

To find the amount of mass needed to create the isotope 2He3, add the masses of the isotopes 1H2 and 1H1 and then subtract the mass of the isotope 2He3 from their total.

Mass of 1H2	2.01411 amu's
Mass of 1H1	1.00783 amu's
Total mass needed to create 2He3	3.02194 amu's

Total mass needed to create 2He3	3.02194 amu's
Mass of 2He3	3.01603 amu's
Total mass lost creating 2He3	0.00591 amu

The drawings show that the proton in the isotope 1H2 loses 0.00591 amu. It also shows that this proton lost its electron plus 0.00100 amu. When this mass is lost, the proton becomes the single neutron in the isotope 2He3. By adding all of the lost mass of the proton and subtracting this total from the mass of this proton, the mass of the neutron in the isotope 2He3 will be found.

Mass lost first by proton in 1H2	0.00591 amu
Mass lost second by proton in 1H2	0.00100 amu
Total mass lost by proton in 1H2	0.00691 amu

Mass of proton in 1H2	1.00728 amu's
Total mass lost by proton in 1H2	0.00691 amu
Mass of neutron in 2He3	1.00037 amu's

The single neutron in the isotope 1H2 becomes the positon in the isotope 2He3. To do this, it must regain enough mass to become the positon. The isotope 1H1 is used as is and does not lose any mass. It therefore can not lose the mass that is regained by the neutron in the isotope 1H2. This means that the mass that is regained by the neutron must come from the proton in the isotope 1H2. The mass of the positon at this time is unknown. To find its mass, add all of the known masses in the isotope 2He3 and then subtract this total from the isotope 2He3.

Mass of 1H1 in 2He3	1.00783 amu's
Mass of negatron in 2He3	0.00055 amu
Mass of neutron in 2He3	1.00037 amu's
Total known mass in 2He3	2.00875 amu's

Mass of 2He3	3.01603 amu's
Total known mass in 2He3	2.00875 amu's
Mass of positon in 2He3	1.00728 amu's

The mass of the neutron in the isotope 1H2 is known to be 1.00628 amu's. Subtract this mass from the mass of the positon in the isotope 2He3 to find the mass needed to make the neutron in the isotope 1H2, the positon in the isotope 2He3.

Mass of positon in 2He3	1.00728 amu's
Mass of neutron in 1H2	1.00628 amu's
Mass needed by neutron in 1H2	0.00100 amu

This 0.00100 amu is then added to the neutron to make it the positon in the isotope 2He3. This mass comes from the proton in the isotope 1H2.

Mass of neutron in 1H2	1.00628 amu's
Mass added to neutron in 1H2 making it a positon	0.00100 amu
Mass of positon in 2He3	1.00728 amu's

The electron that was released by the isotope 1H2 is the negatron in the isotope 2He3. Since no electron was lost, the mass of the electron is not part of the total mass lost creating the isotope 2He3. The only mass lost then is the 0.00591 amu and it is lost as a 591 ray.

Now add all of the known masses to find the mass of the isotope 2He3.

Mass of 1H1 within 2He3	1.00783 amu's
Mass of negatron in 2He3	0.00055 amu
Mass of positon in 2He3	1.00728 amu's
Mass of neutron in 2He3	1.00037 amu's
Mass of 2He3	3.01603 amu's

Sub-step 75-5-1

2He3

FORMULA

1H2 + 1H1 = 2He3, alpha 591 ray

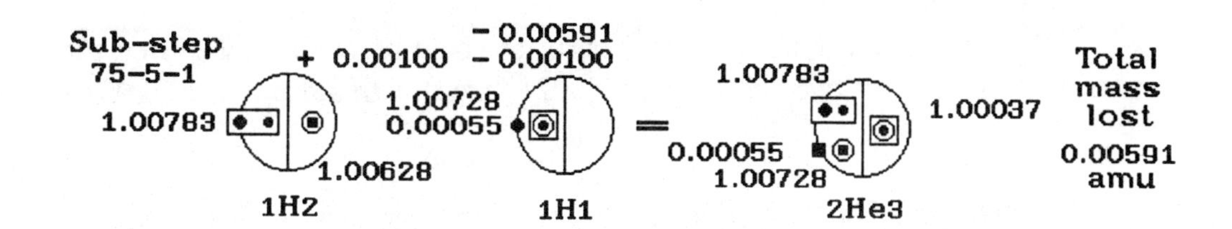

The isotope 1H2 has one electron with 0.00055 amu, one proton with 1.00728 amu's and one neutron with 1.00628 amu's for a total of 2.01411 amu's.

The isotope 1H1 has one electron with 0.00055 amu and one proton with 1.00728 amu's for a total of 1.00783 amu's.

The isotope of 2He3 with 3.01603 amu's has one electron and one negatron with each having 0.00055 amu. There is one proton and one positon with each having 1.00728 amu's. The single neutron has 1.00037 amu's.

The isotope 1H2 has its electron and proton drawn with a rectangle around both to show that it is used as the isotope 1H1 in the isotopes 1H2 and 2He3 with 1.00783 amu's.

The single neutron with 1.00628 amu's in the isotope 1H2 regains 0.00100 amu to become the positon in the isotope 2He3 with 1.00728 amu's. The neutron in the isotope 1H2 is drawn as a small square with a small circle drawn around it to show that the neutron has lost 0.00100 amu and became the positon in the isotope 2He3. The neutron is drawn on the right side of the vertical line in the isotope 1H2 and it is drawn the same way as the positon on the left side of the vertical line in the isotope 2He3.

The proton in the isotope 1H1 is drawn as a dot with a circle around it and then a square is drawn around both. The square shows that 0.00591 amu have been lost first and the circle around the proton means that 0.00100 amu has been transferred to the neutron in the isotope 1H2. When the 0.00100 amu has been lost, the proton releases the electron with 0.00055 amu to be regained by the neutron in the isotope 1H2 when this neutron becomes the positon in the isotope 2He3. The 0.00591 amu is emitted as a 591 alpha ray which has an external magnetic field.

Because the 0.00100 amu is transferred from the proton in the isotope 1H1 to the neutron in the isotope 1H1, the 0.00100 amu is not lost and no ray is emitted. Because the electron in the isotope 1H1 is regained when the isotope 2He3 is created, it mass is not lost.

The total mass lost is 0.00591 amu and which is lost as one unit. This is composed of two sub-units. One is 0.00374 amu and the other is 0.00217 amu. While the 0.00374 amu as a single sub-unit, it is composed of two sub-units. One is 0.00217 amu and there is 0.00157 amu. The sub-unit of 0.00157 amu is composed of a sub-unit of 0.00100 amu, a sub-unit of 0.00002 amu and a sub-unit of 0.00055 amu. The 0.00055 amu has the same mass of an electron or negatron but because it is part of the sub-unit of 0.00374 amu and 0.00157 amu, it is emitted as part of the 591 alpha ray.

While the above is correct, this lost mass could be composed of one sub-unit of 0.00434 amu (or two sub-units of 0.00217 amu) with a sub-unit of 0.00157 amu.

To find the amount of mass needed to create the isotope 2He3, add the masses of the isotopes 1H2 and 1H1 and then subtract the mass of the isotope 2He3 from their total.

Mass of 1H2	2.01411 amu's
Mass of 1H1	1.00783 amu's
Total mass needed to create 2He3	3.02194 amu's

Total mass needed to create 2He3	3.02194 amu's
Mass of 2He3	3.01603 amu's
Total mass lost creating 2He3	0.00591 amu

The drawing shows that the mass of the isotope 1H1 loses 0.00591 amu by its proton. It also shows that this proton lost its electron plus 0.00100 amu. When this mass is lost, the proton becomes the single neutron in the isotope 2He3. By adding all of the lost masses of the proton and subtracting this total from the mass of this proton, the mass of the neutron in the isotope 2He3 will be found.

Mass lost first by proton in 1H1	0.00591 amu
Mass lost second by proton in 1H1	0.00100 amu
Total mass lost by proton in 1H1	0.00691 amu

Mass of proton in 1H1	1.00728 amu's
Total mass lost by proton in 1H1	0.00691 amu
Mass of neutron in 2He3	1.00037 amu's

The single neutron in the isotope 1H2 becomes the positon in the isotope 2He3. To do this, it must regain enough mass to become the positon. The isotope 1H2 has its proton used as is and does not lose any mass. It therefore can not lose the mass that is regained by the neutron in the isotope 1H2. This means that the mass that is regained by the neutron must come from the proton in the isotope

1H1. The mass of the positon at this time is unknown. To find its mass, add all of the known masses in the isotope 2He3, subtract this total from the isotope 2He3.

Mass of 1H1 in 1H2 and 2He3	1.00783 amu's
Mass of negatron in 2He3	0.00055 amu
Mass of neutron in 2He3	1.00037 amu's
Total known mass in 2He3	2.00875 amu's

Mass of 2He3	3.01603 amu's
Total known mass in 2He3	2.00875 amu's
Mass of positon in 2He3	1.00728 amu's

The mass of the neutron in the isotope 1H2 is known to be 1.00628 amu's. Subtract this mass from the mass of the positon in the isotope 2He3 to find the mass needed to make the neutron in the isotope 1H2 the positon in the isotope 2He3.

Mass of positon in 2He3	1.00728 amu's
Mass of neutron in 1H2	1.00628 amu's
Mass needed by neutron in 1H2	0.00100 amu

This 0.00100 amu is then added to the neutron to make it the positon in the isotope 2He3. This mass comes from the proton in the isotope 1H1.

Mass of neutron in 1H2	1.00628 amu's
Mass added to neutron in 1H2 making it a positon	0.00100 amu
Mass of positon in 2He3	1.00728 amu's

The electron that was released by the isotope 1H1 is the negatron in the isotope 2He3. Since no electron was lost, the mass of the electron is not part of the total mass lost creating the isotope 2He3. The only mass lost then is the 0.00591 amu and it is lost as a 591 ray.

Now add all of the known masses to find the mass of the isotope 2He3.

Mass of 1H1 within 2He3	1.00783 amu's
Mass of negatron in 2He3	0.00055 amu
Mass of positon in 2He3	1.00728 amu's
Mass of neutron in 2He3	1.00037 amu's
Mass of 2He3	3.01603 amu's

It should be noted that in this step, the proton and positon each have their maximum mass of 1.00728 amu's. When they have their maximum mass, this isotope of 2He3 is to be considered stable.

All drawings in **STEP 5** using isotopes of 1H2 and 1H1 creating the isotope 2He3

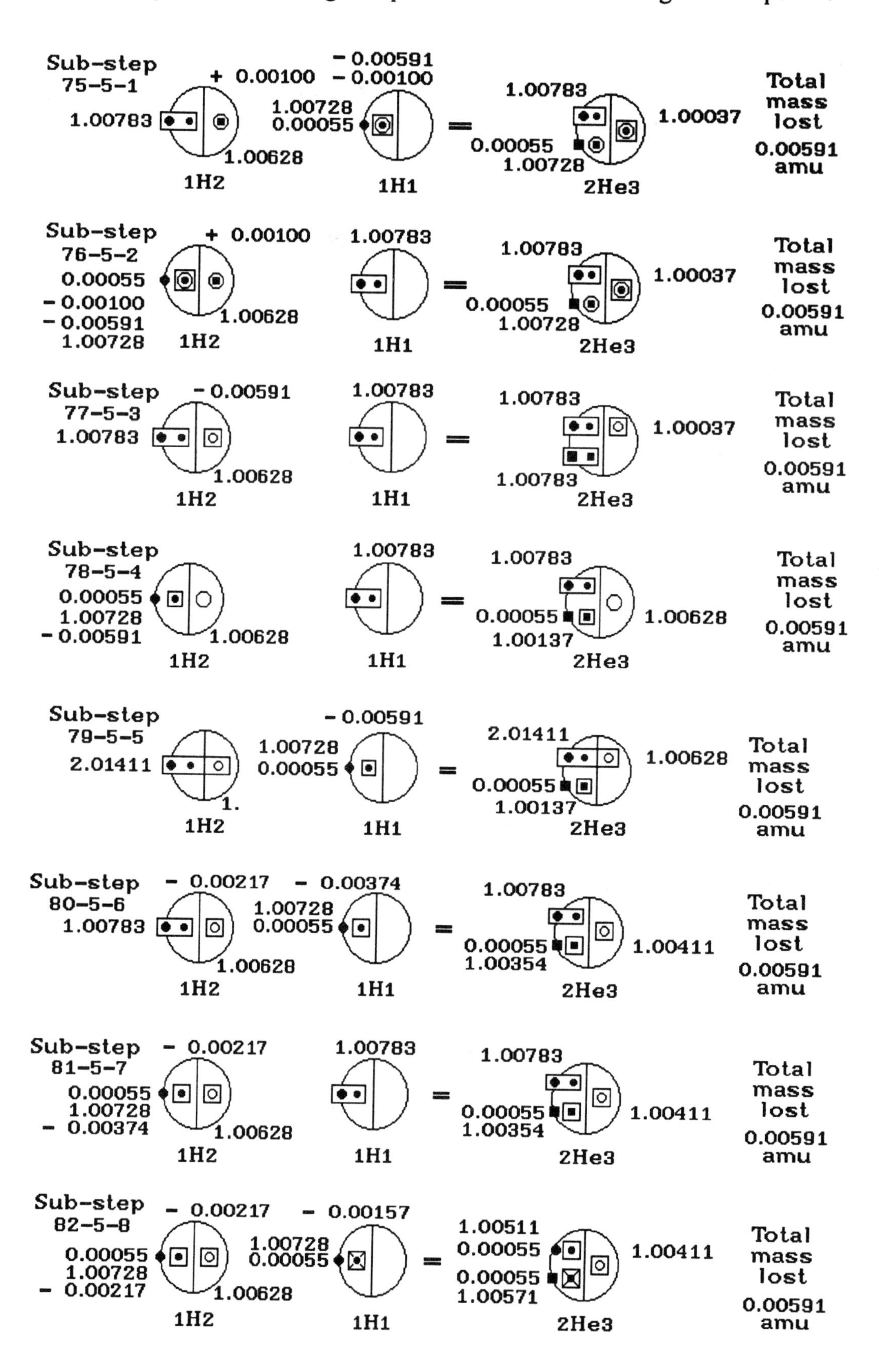

All drawings in **STEP 5** using isotopes of 1H2 and 1H1 creating the isotope 2He3

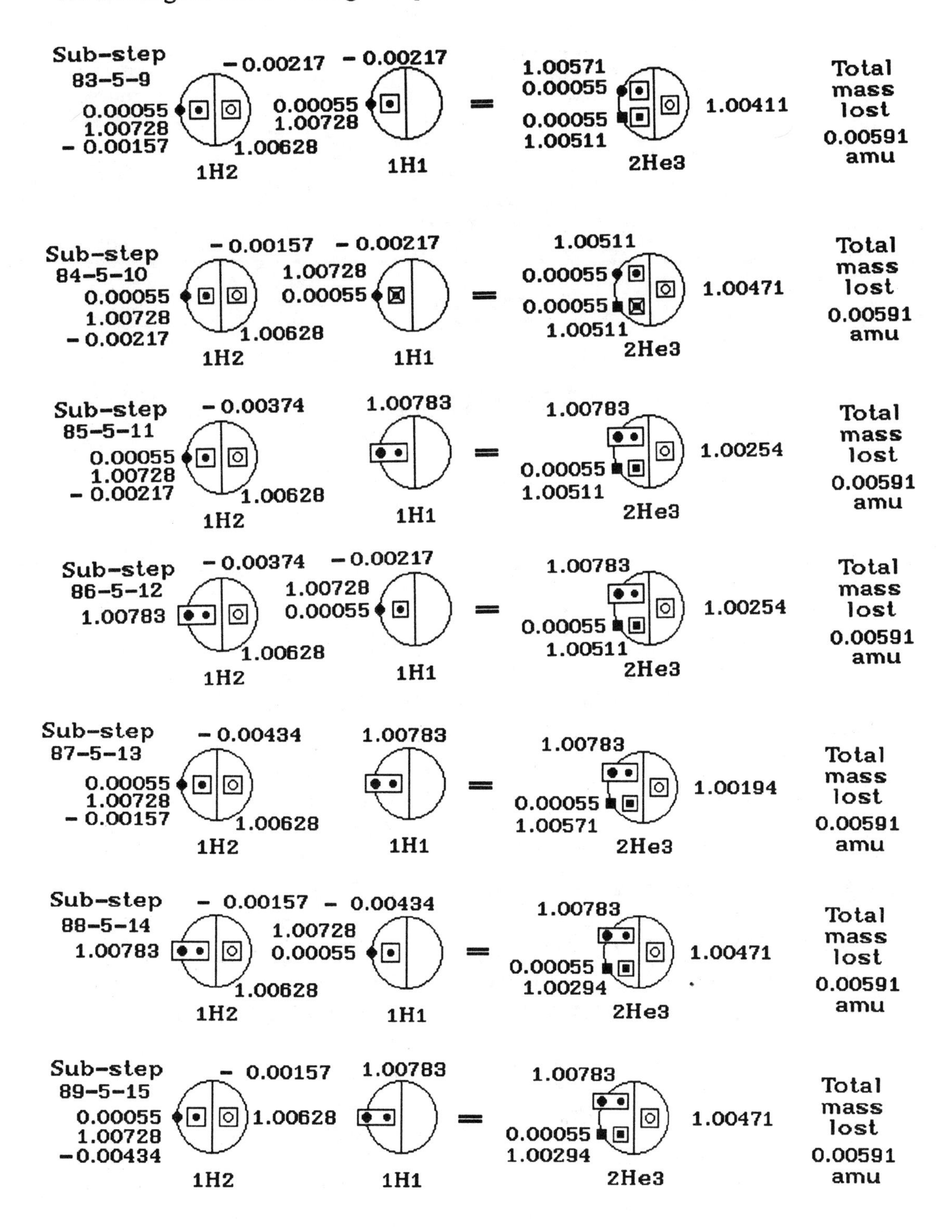

STEP 4

2He3

FORMULA

1H2 + 1H2 = 2He3

STEPS

4. 1H2 + 1H2 = 2He3

66-4-1a	1H2 + 1H2	= 2He3, 436 ray, 1.00783 1H1 isotope
66-4-1b	1H2 + 1H2	= 2He3, 436 ray, 1.00783 1H1 isotope
66-4-1c	1H2 + 1H2	= 2He3, 436 ray, 1.00783 1H1 isotope
67-4-2	1H2 + 1H2	= 2He3, 591 ray, 1.00628 neutron
68-4-3	1H2 + 1H2	= 2He3, 591 ray, 1.00628 neutron
69-4-4	1H2 + 1H2	= 2He3, 434 ray, 157 ray, 1.00628 neutron
70-4-5	1H2 + 1H2	= 2He3, 434 ray, 157 ray, 1.00628 neutron
71-4-6	1H2 + 1H2	= 2He3, 217 ray, 374 ray, 1.00628 neutron
72-4-7	1H2 + 1H2	= 2He3, 217 ray, 374 ray, 1.00628 neutron
73-4-8	1H2 + 1H2	= 2He3, 591 ray, 1.00628 neutron
74-4-9	1H2 + 1H2	= 2He3, 217 ray, 157 ray, 217 ray, 1.00628 neutron

Sub-step 74-4-9

2He3

FORMULA

1H2 + 1H2 = 2He3, alpha 217 ray, gamma 157 ray, alpha 217 ray, 1.00628 neutron

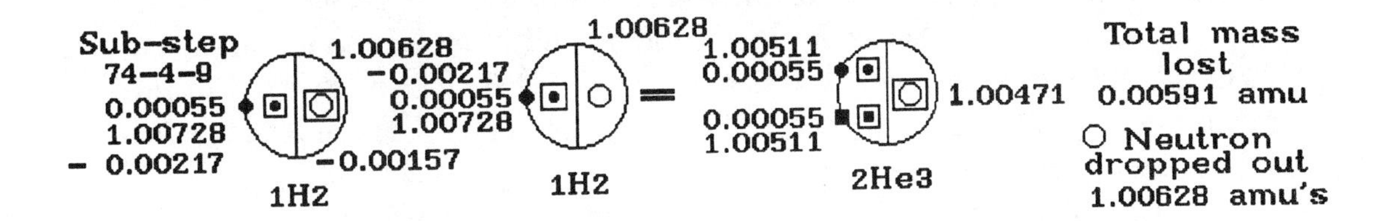

There are two isotopes of 1H2 with each having 2.01411 amu's creating the isotope 2He3 with 3.01603 amu's. There is one neutron that dropped out with 1.00628 amu's and 0.00591 amu is emitted as two 217 rays, one 157 ray.

Each isotope of 1H2 with 2.01411 amu's has one electron with 0.00055 amu, one proton with 1.00728 amu's and one neutron with 1.00628 amu's.

The isotope of 2He3 with 3.01603 amu's has one electron and one negatron with each having 0.00055 amu. There is one proton with 1.00511 amu's and one positon with 1.00511 amu's. The single neutron has 1.00471 amu's.

The first isotope of 1H2 with 2.01411 amu's has its electron drawn as a dot with 0.00055 amu and it is the negatron drawn as a square in the isotope 2He3.

The proton with 1.00728 amu's lose 0.00217 amu to become the positon in the isotope 2He3 with 1.00511 amu's. The proton is drawn as a dot with a square around it in the isotope 1H2 and as a square with another square drawn around it in the isotope 2He3. The square around the dot and the second square around the square of the positon shows that 0.00217 amu has been lost.

The single neutron in the first isotope 1H2 with 1.00628 amu's is drawn as a small circle with a square around it to show that it has lost 0.00157 amu to become the single neutron in the isotope 2He3 with 1.00471 amu's.

When the first isotope of 1H2 loses 0.00217 amu from its proton, the lost mass of 0.00217 amu is converted to a 217 alpha ray which has an external magnetic field. When the single neutron in this isotope loses 0.00157 amu, the 0.00157 amu is converted to a 157 gamma ray which has no external magnetic field. This first isotope of 1H2 loses a total lost mass of 0.00374 amu as two different rays. All lost mass that is emitted as a ray is converted to pure energy which is magnetism.

The second isotope of 1H2 with 2.01411 amu's has its electron with 0.00055 amu drawn as a dot on the big circle and it is also drawn as a dot on the big circle in the isotope 2He3 to show that the electron in the isotope 1H2 remains an electron in the isotope 2He3.

The proton in the second isotope of 1H2 with 1.00728 amu's drawn with a dot to the left of the vertical line within the big circle. It is drawn this same way in the isotope 2He3. It has a square drawn around it in the isotopes 1H2 and 2He3 to show that it has lost 0.00217 amu. The 0.00217 amu is emitted as a 217 alpha ray which has an external magnetic field.

The single neutron with 1.00628 amu's in the second isotope 1H2 is drawn as a small circle to the right of the vertical line within the big circle and it is not needed in the isotope 2He3 which means it drops out of the second isotope with 1.00628 amu's. It is a left over part of the second isotope of 1H2. Its mass is included in the total mass lost creating the isotope 2He3. The mass of this neutron remains mass while the mass of 0.00217 amu is emitted as a 217 alpha ray. A ray is pure energy that is magnetism.

The total mass converted to pure energy is 0.00591 amu as two 0.00217 amu and one 0.00157 amu. Note that I do not spell the amu with an 's. The 's is only put at the end of amu when the mass indicated is one (1.00000) or more such a 1.00728 amu's.

To find the amount of mass needed to create the isotope 2He3, add the masses of the two isotopes of 1H2 and then subtract the mass of the isotope 2He3 from this total.

Mass of first 1H2	2.01411 amu's
Mass of second 1H2	2.01411 amu's
Total mass needed to create 2He3	4.02822 amu's

Total mass needed to create 2He3	4.02822 amu's
Mass 2He3	3.01603 amu's
Total mass lost creating 2He3	1.01219 amu's

This mass lost creating the isotope 2He3 includes the neutron that dropped out of the second isotope of 1H2. By subtracting the mass of this neutron from the total mass lost creating the isotope 2He3, the real mass lost can be found.

Total mass lost creating 2He3	1.01219 amu's
Mass of neutron dropping out of second 1H2	1.00628 amu's
Real mass lost creating 2He3	0.00591 amu

The single unit of lost mass is 0.00591. This lost mass is composed of a sub-unit of 0.00434 amu and another sub-unit of 0.00157 amu. In this step the neutron in the first isotope of 1H2 loses 0.00157 amu and each proton in the two isotopes of 1H2 lose some of their mass. The 0.00434 amu is composed of two sub-units of 0.00217 amu and these two sub-units can not be broken down into smaller sub-units. Therefore the two proton each lose one unit of 0.00217 amu.

I now know that each proton loses 0.00217 amu and this lost mass can be subtracted from each proton to find the mass of the proton and positon found in the isotope 2He3.

Mass of proton in first isotope of 1H2	1.00728 amu's
Mass lost by it	0.00217 amu
Mass of positon in 2He3	1.00511 amu's
Mass of proton in second 1H2	1.00728 amu's
Mass lost by it	0.00217 amu
Mass of proton in 2He3	0.00511 amu's

The single neutron in the first isotope of 1H2 with 1.00628 amu's loses 0.00157 amu to become the single neutron in the isotope 2He3 with 1.00471 amu's. By subtracting mass lost of 0.00157 amu from its mass, the mass of the neutron in the isotope 2He3 will be known.

Mass of neutron in first 1H2	1.00628 amu's
Mass lost by neutron in 1H2	0.00157 amu
Mass of neutron in 2He3	1.00471 amu's

To find the mass the proton lost in the first isotope of 1H2, subtract the mass lost by the neutron in the first isotope of 1H2 from the real mass lost creating the isotope 2He3.

Real mass lost creating 2He3	0.00591 amu
Mass lost by neutron in first 1H2	0.00374 amu
Mass lost by proton in first 1H2	0.00217 amu

The total mass lost is 0.00591 amu and which is lost as three units. They are two units of 0.00217 amu and a single unit of 0.00157 amu. Each of the two units of 0.00217 amu can not be broken down any further. While the 0.00157 amu is lost as a single unit, it is composed of three sub-units of 0.00100 amu, 0.00055 amu and 0.00002 amu. Because the 0.00100 amu is part of the single unit of 0.00374 amu, it can not cause the release of an electron. The 0.00055 amu is the same mass of an electron or negatron but because it is part of the single unit of 0.00157 amu, it is emitted as part of the 157 gamma ray.

When the isotope 1H3 is converted to the isotope 2He3, there is 0.00002 amu lost. In the 0.00591 amu that is lost, the unit of 0.00157 amu includes this sub-unit of 0.00002 amu.

Mass of 1H3	3.01605 amu's
Mass of 2He3	3.01603 amu's
Mass lost creating 2He3 from 1H3	0.00002 amu

The mass of the isotope 2H3 is equal to the mass of all of its parts.

Mass of positon 2He3	1.00511 amu's
Mass of negatron in 2He3	0.00055 amu
Mass of proton in 2He3	1.00511 amu's
Mass of electron in 2He3	0.00055 amu
Mass of neutron in 2He3	1.00471 amu's
Mass of 2He3	3.01603 amu's

The mass of the isotope 2He3, the neutron that dropped out and the real mass lost equals the mass of the two isotopes of 1H2.

Mass of 2He3	3.01603 amu's
Mass of neutron that dropped out	1.00628 amu's
Real mass lost	0.00591 amu
Mass of two isotopes of 1H2	4.02822 amu's

Sub-step 73-4-8

2He3

FORMULA

1H2 + 1H2 = 2He3, one alpha 591 ray, one neutron

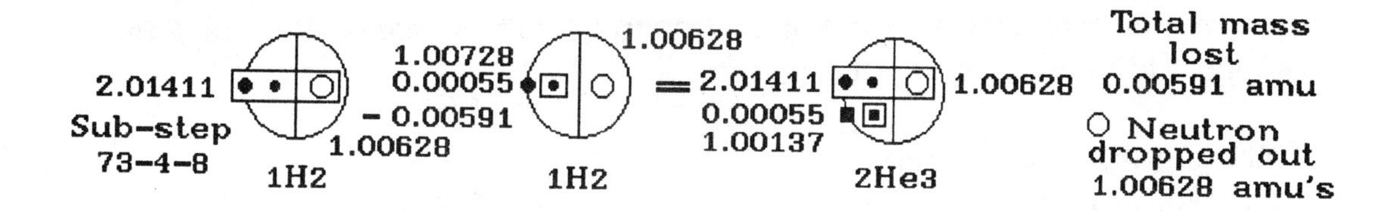

There are two isotopes of 1H2 with each having 2.0411 amu's creating the isotope 2He3 with 3.01603 amu's. There is one neutron that dropped out with 1.00628 amu's and 0.00591 amu is emitted as a 591 alpha ray which has an external magnetic field.

Each isotope of 1H2 with 2.01411 amu's has one electron with 0.00055 amu, one proton with 1.00728 amu's and one neutron with 1.00628 amu's.

The isotope of 2He3 with 3.01603 amu's has one electron and one negatron with each having 0.00055 amu. There is one proton with 1.00728 amu's and one positon with 1.00137 amu's. The single neutron has 1.00628 amu's.

The first isotope of 1H2 with 2.01411 amu's is used as is with its electron drawn as a dot with 0.00055 amu, it proton with 1.00728 amu's and the neutron has 1.00628 amu's. None of these parts lose any mass. They are all drawn with a rectangle around them in the isotopes 1H2 and 2He3 to show that they are used as is and that the isotope 1H2 can drop out at some later time.

The second isotope of 1H2 with 2.01411 amu's has its electron drawn on the big circle as a dot and it is drawn as a square on the big circle in the isotope 2He3. The proton with 1.00728 amu's loses 0.00591 amu to become the positon in the isotope 2He3 with 1.00137 amu's. The proton in the isotope 1H2 is drawn as a dot on the left side of the vertical line within the big circle and it is drawn as a square in the isotope 2He3 to the left of the vertical line within the big circle.

The neutron in the second isotope of 1H2 is not needed so it drops out with the mass of 1.00628 amu's. It is drawn as a small circle to the right of the vertical line within the big circle and it is drawn as a small circle to the right of the equal sign to show that it has dropped out.

To find the amount of mass needed to create the isotope 2He3, add the masses of the two isotopes of 1H2 and then subtract the mass of the isotope 2He3 from this total.

Mass of first 1H2	2.01411 amu's
Mass of second 1H2	2.01411 amu's
Total mass needed to create 2He3	4.02822 amu's

Total mass needed to create 2He3	4.02822 amu's
Mass 2He3	3.01603 amu's
Total mass lost creating 2He3	1.01219 amu's

This mass lost creating the isotope 2He3 includes the neutron that dropped out. By subtracting the mass of this neutron from the total mass lost creating the isotope 2He3, the real mass lost can be found.

Total mass lost creating 2He3	1.01219 amu's
Mass of neutron dropping out of second 1H2	1.00628 amu's
Real mass lost creating 2He3	0.00591 amu

The proton with 1.00728 amu's in the second isotope of 1H2 loses all of the real mass lost creating the isotope 2He3 so this mass can be subtracted from this proton to find the positon it becomes in the isotope 2He3.

Mass of proton in second 1H2	1.00728 amu's
Mass lost by it	0.00591 amu
Mass of positon in 2He3	1.00137 amu's

The first isotope of 1H2 is used as is and the electron in the second isotope of 1H2 is used as the negatron in the isotope 2He3 so with the mass of the positon now known, all of the parts of the isotope 2He3 are known.

Mass of first 1H2 within 2He3	2.01411 amu's
Mass of negatron in 2He3	0.00055 amu
Mass of positon in 2He3	1.00137 amu's
Mass of 2He3	3.01603 amu's

The total mass lost is 0.00591 amu and it is lost as one unit. This unit is composed of two sub-units. One is 0.00434 amu and the other is 0.00157 amu. While the sub-unit of 0.00434 amu is a sub-unit, it is composed of two sub-units of 0.00217 amu. The sub-unit of 0.00157 amu is composed of a sub-unit of 0.00100 amu, a sub-unit of 0.00002 amu and a sub-unit of 0.00055 amu. Because the sub-unit of 0.00100 amu is part of the sub-unit of 0.00157 amu, it can not cause the release of an electron. The 0.00055 amu is the same mass of an electron or negatron but because it is part of the sub-unit of 0.00157 amu, it is not emitted.

While the above is correct, this lost mass could be composed of one sub-unit of 0.00374 amu with a sub-unit of 0.00217 amu. The sub-unit of 0.00374 amu is composed of a sub-unit of 0.00217 amu and a sub-unit of 0.00157 amu.

When the isotope 1H3 is converted to the isotope 2He3, there is 0.00002 amu lost. In the 0.00591 amu that is lost, the unit of 0.00157 amu includes this sub-unit of 0.00002 amu.

Mass of 1H3	3.01605 amu's
Mass of 2He3	3.01603 amu's
Mass lost creating 2He3 from 1H3	0.00002 amu

The mass of the isotope 2He3, the neutron that dropped out and the real mass lost equals the mass of the two isotopes of 1H2.

Mass of 2He3	3.01603 amu's
Mass of neutron that dropped out	1.00628 amu's
Real mass lost	0.00591 amu
Mass of two isotopes of 1H2	4.02822 amu's

Sub-step 72-4-7

2He3

FORMULA

1H2 + 1H2 = 2He3, one 374 alpha ray, one gamma 217 ray, one neutron

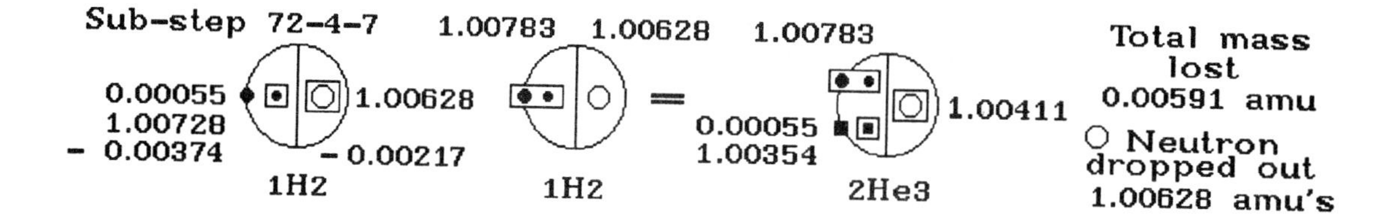

There are two isotopes of 1H2 with each having 2.01411 amu's creating the isotope 2He3 with 3.01603 amu's. There is one neutron that dropped out with 1.00628 amu's and 0.00591 amu is emitted as a 374 alpha ray which has an external magnetic field and a 217 gamma ray which does not have an external magnetic field.

Each isotope of 1H2 with 2.01411 amu's has one electron with 0.00055 amu, one proton with 1.00728 amu's and one neutron with 1.00628 amu's.

The isotope of 2He3 with 3.01603 amu's has one electron and one negatron with each having 0.00055 amu. There is one proton with 1.00728 amu's and one positon with 1.00354 amu's. The single neutron has 1.00411 amu's.

The first isotope of 1H2 with 2.01411 amu's has its electron drawn as a dot with 0.00055 amu and it is the negatron drawn as a square in the isotope 2He3.

The proton with 1.00728 amu's lose 0.00374 amu to become the positon in the isotope 2He3 with 1.00354 amu's. The proton is drawn as a dot with a square around it in the isotope 1H2 and as a square with another square drawn around it in the isotope 2He3. The square around the dot and the second square around the square of the positon shows that 0.00374 amu has been lost.

The single neutron in the isotope 1H2 with 1.00628 amu's is drawn as a small circle with a square around it to show that it has lost 0.00217 amu to become the single neutron in the isotope 2He3 with 1.00411 amu's.

The second isotope of 1H2 with 2.01411 amu's has its electron and proton drawn with a rectangle around them to show that they are used as the isotope 1H1 with 1.00783 amu's in the isotope 1H2 and 2He3.

The single neutron drawn as a small circle drops out with 1.00628 amu's because it is not needed to create the isotope 2He3. It is a leftover part of the second isotope of 1H2.

When the first isotope of 1H2 loses 0.00374 amu from its proton, the lost mass of 0.00374 is converted to a 374 ray. When the single neutron in this isotope loses 0.00217 amu, the 0.00217 amu is converted to a 217 ray. All lost mass that is emitted as a ray is converted to pure energy which is magnetism.

To find the amount of mass needed to create the isotope 2He3, add the masses of the two isotopes of 1H2 and then subtract the mass of the isotope 2He3 from this total.

Mass of first 1H2	2.01411 amu's
Mass of second 1H2	2.01411 amu's
Total mass needed to create 2He3	4.02822 amu's

Total mass needed to create 2He3	4.02822 amu's
Mass 2He3	3.01603 amu's
Total mass lost creating 2He3	1.01219 amu's

This mass lost creating the isotope 2He3 includes the neutron that dropped out. By subtracting the mass of this neutron from the total mass lost creating the isotope 2He3, the real mass lost can be found.

Total mass lost creating 2He3	1.01219 amu's
Mass of neutron dropping out of second 1H2	1.00628 amu's
Real mass lost creating 2He3	0.00591 amu

I have stated that the single neutron in the first isotope of 1H2 lost 0.00217 amu and the proton lost 0.00374 amu. This is equal to 0.00591 amu lost by this isotope. In the next step, the proton will lose the 0.00217 amu and the neutron will lose 0.00374 amu. I do this to show that if I do it one way and get a result then doing it again only switching the lost masses, then the isotopes created will still be an isotope of 2He3 but the internal parts will not have the same masses. Also this is why I can say that the neutron in the first isotope of 1H2 will lose 0.00217 amu. The reader may wonder how I could say this and not know that in the next step I will have the lost masses of the neutron and proton trade places.

The mass of the neutron in the isotope 2He3 is unknown. The single neutron in the first isotope of 1H2 is the neutron that loses some of its mass to become the single neutron in the isotope 2He3. By subtracting the mass this neutron lost from the single neutron in the first isotope of 1H2, the mass of the neutron in the isotope of 2He3 will be known.

Mass of neutron in first 1H2	1.00628 amu's
Mass lost by neutron in 1H2	0.00217 amu
Mass of neutron in 2He3	1.00411 amu's

To find the mass the proton lost in the first isotope of 1H2, subtract the mass lost by the neutron in the first isotope of 1H2 from the real mass lost creating the isotope 2He3.

Real mass lost creating 2He3	0.00591 amu
Mass lost by neutron in first 1H2	0.00217 amu
Mass lost by proton in first 1H2	0.00374 amu

After the proton in the first isotope 1H2 loses 0.00374 amu, it becomes the positon in the isotope 2He3. To find the mass of the positon in the isotope 2He3, subtract the lost mass of the proton in the isotope 1H2 from this proton.

Mass of proton in first 1H2	1.00728 amu's
Mass lost by it	0.00374 amu
Mass of positon in 2He3	1.00354 amu's

The second isotope of 1H2 has its electron and proton used as the isotope 1H1 in the second isotope of 1H2 and in the isotope 2He3. They are enclosed in a rectangle to show that they are used as the isotope 1H1 and that they do not lose any mass. This rectangle also means that the isotope 1H1 can drop out of the isotope 2He3 at some later time.

The single neutron in the second isotope 1H2 is not used and it therefore is not needed in the isotope 2He3 which means it drops out of the second isotope with 1.00628 amu's. Its mass is included in the total mass lost creating the isotope 2He3. The mass of this neutron remains mass while the mass of 0.00374 amu is emitted as a 374 ray and the 0.00217 amu is emitted as a 217 ray. A ray is pure energy that is magnetism. The total mass converted to pure energy is 0.00591 amu.

The total mass lost is 0.00591 amu and which is lost as two units. This is composed of a unit of 0.00374 amu and 0.00217 amu. While the 0.00374 amu is lost as a single unit, it is composed of two sub-units. One is 0.00217 amu and the other is 0.00157 amu. The sub-unit of 0.00157 amu is composed of a sub-unit of 0.00100 amu, a sub-unit of 0.00002 amu and a sub-unit of 0.00055 amu. Because the 0.00100 amu and 0.00055 amu are parts of the sub-unit of 0.00157 amu, it can not cause the release of an electron. When the isotope 1H3 is converted to the isotope 2He3, there is 0.00002 amu lost. In the 0.00591 amu that is lost, the sub-unit of 0.00157 amu includes this sub-unit of 0.00002 amu.

Mass of 1H3	3.01605 amu's
Mass of 2He3	3.01603 amu's
Mass lost creating 2He3 from 1H3	0.00002 amu

The mass of the isotope 2H3 is equal to the mass of all of its parts.

Mass of 1H1 intact within 2He3	1.00783 amu's
Mass of negatron in 2He3	0.00055 amu
Mass of positon in 2He3	1.00354 amu's
Mass of neutron in 2He3	1.00411 amu's
Mass of 2He3	3.01603 amu's

The mass of the isotope 2He3, the neutron that dropped out and the real mass lost equals the mass of the two isotopes of 1H2.

Mass of 2He3	3.01603 amu's
Mass of neutron that dropped out	1.00628 amu's
Real mass lost	0.00591 amu
Mass of two isotopes of 1H2	4.02822 amu's

Sub-step 71-4-6

2He3

FORMULA

1H2 + 1H2 = 2He3, one alpha 217 ray, one gamma 374 ray, one neutron

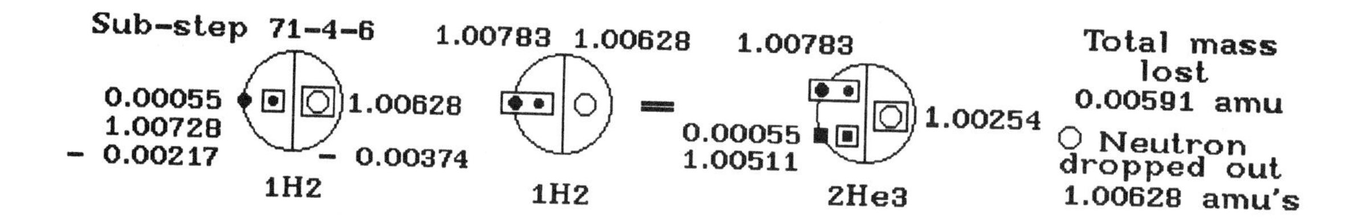

There are two isotopes of 1H2 with each having 2.01411 amu's creating the isotope 2He3 with 3.01603 amu's. There is one neutron that dropped out with 1.00628 amu's and the 0.00591 amu is emitted as a 217 alpha ray which has an external magnet and a 374 gamma ray which does not have an external magnetic field ray.

Each isotope of 1H2 with 2.01411 amu's has one electron with 0.00055 amu, one proton with 1.00728 amu's and one neutron with 1.00628 amu's.

The isotope of 2He3 with 3.01603 amu's has one electron and one negatron with each having 0.00055 amu. There is one proton with 1.00728 amu's and one positon with 1.00511 amu's. The single neutron has 1.00254 amu's.

The first isotope of 1H2 with 2.01411 amu's has its electron drawn as a dot with 0.00055 amu and it is the negatron drawn as a square in the isotope 2He3.

The proton with 1.00728 amu's lose 0.00217 amu to become the positon in the isotope 2He3 with 1.00511 amu's. The proton is drawn as a dot with a square around it in the isotope 1H2 and as a square with another square drawn around it in the isotope 2He3. The square around the dot and the second square around the square of the positon shows that 0.00217 amu have been lost.

The single neutron in the first isotope 1H2 with 1.00628 amu's is drawn as a small circle with a square around it to show that it has lost 0.00374 amu to become the single neutron in the isotope 2He3 with 1.00254 amu's. It is drawn the same way in the isotope 2He3.

When the first isotope of 1H2 loses 0.00217 amu from its proton, the lost mass of 0.00217 amu is converted to a 217 ray. When the single neutron in this isotope loses 0.00374 amu, the 0.00374 amu is converted to a 374 ray. All lost mass that is emitted as a ray is converted to pure energy which is magnetism.

The second isotope of 1H2 with 2.01411 amu's has its electron and proton drawn with a rectangle around them to show that they are used as the isotope 1H1 with 1.00783 amu's in the isotopes 1H2 and 2He3. They are enclosed in a rectangle to show that they are used as the isotope 1H1 and that they do not lose any mass. This rectangle also means that the isotope 1H1 can drop out of the isotope 2He3 at some later time.

The single neutron in the second isotope 1H2 is not used and it therefore is not needed in the isotope 2He3 which means it drops out of the second isotope with 1.00628 amu's. It is a leftover part of the second isotope of 1H2. Its mass is included in the total mass lost creating the isotope 2He3. The mass of this neutron remains mass.

To find the amount of mass needed to create the isotope 2He3, add the masses of the two isotopes of 1H2 and then subtract the mass of the isotope 2He3 from this total.

Mass of first 1H2	2.01411 amu's
Mass of second 1H2	2.01411 amu's
Total mass needed to create 2He3	4.02822 amu's

Total mass needed to create 2He3	4.02822 amu's
Mass 2He3	3.01603 amu's
Total mass lost creating 2He3	1.01219 amu's

This mass lost creating the isotope 2He3 includes the neutron that dropped out. By subtracting the mass of this neutron from the total mass lost creating the isotope 2He3, the real mass lost can be found.

Total mass lost creating 2He3	1.01219 amu's
Mass of neutron dropping out of second 1H2	1.00628 amu's
Real mass lost creating 2He3	0.00591 amu

In the step before this one, the neutron lost 0.00217 amu and the proton lost 0.00374 amu. In this step, these lost masses are switched.

The mass of the neutron in the isotope 2He3 is unknown. The single neutron in the first isotope of 1H2 with 1.00628 amu's loses 0.00374 amu to become the single neutron in the isotope 2He3 with 1.00254 amu's. By subtracting the lost mass of 0.00373 amu from the neutron in the isotope 1H2, the mass of the neutron in the isotope 2He3 will be known.

Mass of neutron in first 1H2	1.00628 amu's
Mass lost by it	0.00374 amu
Mass of neutron in 2He3	1.00254 amu's

To find the mass the proton lost in the first isotope of 1H2, subtract the mass lost by the neutron in the first isotope of 1H2 from the real mass lost creating the isotope 2He3.

Real mass lost creating 2He3	0.00591 amu
Mass lost by neutron in first 1H2	0.00374 amu
Mass lost by proton in first 1H2	0.00217 amu

After the proton in the first isotope 1H2 loses 0.00217 amu, it becomes the positon in the isotope 2He3. To find the mass of the positon in the isotope 2He3, subtract the lost mass of the proton in the isotope 1H2 from this proton.

Mass of proton in first 1H2	1.00728 amu's
Mass lost by it	0.00217 amu
Mass of positon in 2He3	1.00511 amu's

The total mass lost is 0.00591 amu and which is lost as two units. One is 0.00374 amu and the other is 0.00217 amu. While the 0.00374 amu is lost as a single unit, it is composed of two sub-units. One is 0.00217 amu and the other is 0.00157 amu. The sub-unit of 0.00157 amu is composed of a sub-unit of 0.00100 amu, a sub-unit of 0.00002 amu and a sub-unit of 0.00055 amu. Because the 0.00100 amu is part of the single unit of 0.00374 amu, it can not cause the release of an electron. The 0.00055 amu is the same mass of an electron or negatron but because it is part of the single unit of 0.00374 amu, it is emitted as part of the 374 ray. When the isotope 1H3 is converted to the isotope 2He3, there is 0.00002 amu lost. In the 0.00591 amu that is lost, the unit of 0.00374 amu includes this sub-unit of 0.00002 amu.

Mass of 1H3	3.01605 amu's
Mass of 2He3	3.01603 amu's
Mass lost creating 2He3 from 1H3	0.00002 amu

The mass of the isotope 2H3 is equal to the mass of all of its parts.

Mass of 1H1 intact within 2He3	1.00783 amu's
Mass of negatron in 2He3	0.00055 amu
Mass of positon in 2He3	1.00511 amu's
Mass of neutron in 2He3	1.00254 amu's
Mass of 2He3	3.01603 amu's

The mass of the isotope 2He3, the neutron that dropped out and the real mass lost equals the mass of the two isotopes of 1H2.

Mass of 2He3	3.01603 amu's
Mass of neutron that dropped out	1.00628 amu's
Real mass lost	0.00591 amu
Mass of two isotopes of 1H2	4.02822 amu's

Sub-step 70-4-5

2He3

FORMULA

1H2 + 1H2 = 2He3, one alpha 434 ray, one gamma 157 ray, one neutron

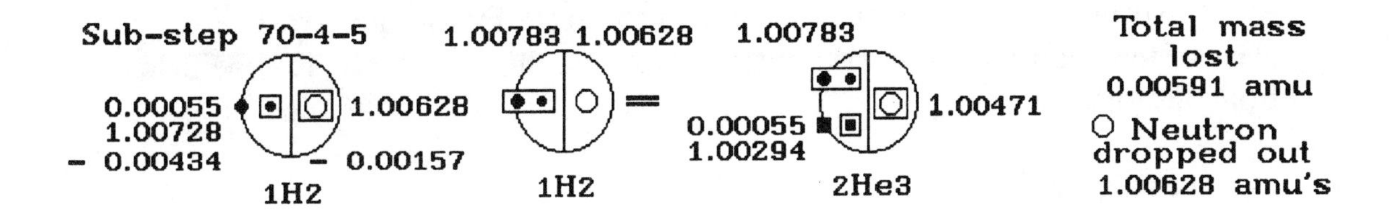

There are two isotopes of 1H2 with each having 2.01411 amu's creating the isotope 2He3 with 3.01603 amu's. There is one neutron that dropped out with 1.00628 amu's and 0.00591 amu lost. The lost mass of 0.00591 amu is emitted as a 434 alpha ray which has an external magnetic field and a 157 gamma ray which does not have an external magnetic field.

Each isotope of 1H2 with 2.01411 amu's has one electron with 0.00055 amu, one proton with 1.00728 amu's and one neutron with 1.00628 amu's.

The isotope of 2He3 with 3.01603 amu's has one electron and one negatron with each having 0.00055 amu. There is one proton with 1.00728 amu's and one positon with 1.00294 amu's. The single neutron has 1.00471 amu's.

The first isotope of 1H2 with 2.01411 amu's has its electron drawn as a dot with 0.00055 amu and it is the negatron drawn as a square in the isotope 2He3.

The proton with 1.00728 amu's lose 0.00434 amu to become the positon in the isotope 2He3 with 1.00471 amu's. The proton is drawn as a dot with a square around it in the isotope 1H2 and as a square with another square drawn around it in the isotope 2He3. The square around the dot is the isotope 1H2 and the second square around the square of the positon in the isotope 2He3 shows that 0.00434 amu has been lost from the isotope 1H2 and emitted as a 434 alpha ray.

The single neutron in the first isotope 1H2 with 1.00628 amu's is drawn as a small circle with a square around it to show that it has lost 0.00157 amu to become the single neutron in the isotope 2He3 with 1.00471 amu's.

When the first isotope of 1H2 loses 0.00434 amu from its proton, the lost mass of 0.00434 amu is converted to a 434 alpha ray. When the single neutron in this isotope loses 0.00157 amu, the 0.00157 amu is converted to a 157 gamma ray. All lost mass that is emitted as a ray is converted to pure energy which is magnetism.

The second isotope of 1H2 with 2.01411 amu's has its electron and proton drawn with a rectangle around them to show that they are used as the isotope 1H1 with 1.00783 amu's in the isotopes 1H2 and 2He3. They are enclosed in a rectangle to show that they are used as the isotope 1H1 and that they do not lose any mass. This rectangle also means that the isotope 1H1 can drop out of the isotope 2He3 at some later time.

The single neutron in the second isotope 1H2 is not used and it therefore is not needed in the isotope 2He3 which means it drops out of the second isotope with 1.00628 amu's. It is a leftover part of the second isotope of 1H2. Its mass is not included in the total mass lost creating the isotope 2He3 because it is shown separately. The mass of this neutron remains mass.

To find the amount of mass needed to create the isotope 2He3, add the mass of the two isotopes of 1H2 and then subtract the mass of the isotope 2He3 from this total.

Mass of first 1H2	2.01411 amu's
Mass of second 1H2	2.01411 amu's
Total mass needed to create 2He3	4.02822 amu's

Total mass needed to create 2He3	4.02822 amu's
Mass 2He3	3.01603 amu's
Total mass lost creating 2He3	1.01219 amu's

This mass lost creating the isotope 2He3 includes the neutron that dropped out. By subtracting the mass of this neutron from the total mass lost creating the isotope 2He3, the real mass lost can be found. These are the total masses that are emitted as a 434 ray and a 157 ray.

Total mass lost creating 2He3	1.01219 amu's
Mass of neutron dropping out of second 1H2	1.00628 amu's
Total mass lost creating 2He3	0.00591 amu

The mass of the positon and neutron in the isotope 2He3 are unknown. The proton with 1.00728 amu and single neutron with 1.00628 amu in the first isotope of 1H2 each loses some of their mass to become the positon and single neutron in the isotope 2He3. Their total mass lost is now known to be 0.00591 amu. This lost mass is composed of 0.00434 amu and 0.00157 amu.

By subtracting the lost mass of 0.00157 amu from the neutron in the isotope 1H2, the mass of the neutron in the isotope 2He3 will be known.

Mass of neutron in first 1H2	1.00628 amu's
Mass lost by it	0.00157 amu
Mass of neutron in 2He3	1.00471 amu's

To find the mass the proton lost in the first isotope of 1H2, subtract the mass lost by the neutron in the first isotope of 1H2 from the total mass lost creating the isotope 2He3.

Total mass lost creating 2He3	0.00591 amu
Mass lost by neutron in first 1H2	0.00157 amu
Mass lost by proton in first 1H2	0.00434 amu

After the proton in the first isotope 1H2 loses 0.00434 amu, it becomes the positon in the isotope 2He3. To find the mass of the positon in the isotope 2He3, subtract the lost mass of the proton in the isotope 1H2 from this proton.

Mass of proton in first 1H2	1.00728 amu's
Mass lost by it	0.00434 amu
Mass of positon in 2He3	1.00294 amu's

The total mass lost is 0.00591 amu and which is lost as two units. This is composed of a unit of 0.00434 amu and a unit of 0.00157 amu. While the 0.00434 amu is lost as a single unit, it is composed of two sub-units of 0.00217 amu. The unit of 0.00157 amu is composed of a sub-unit of 0.00100 amu, a sub-unit of 0.00002 amu and a sub-unit of 0.00055 amu. Because the sub-unit 0.00100 amu is part of the single unit of 0.00157 amu, it can not cause the release of an electron. The 0.00055 amu is the same mass of an electron or negatron but because it is part of the unit of 0.00157 amu, it is emitted as part of the 157 gamma ray. When the isotope 1H3 is converted to the isotope 2He3, there is 0.00002 amu lost. In the 0.00591 amu that is lost, the unit of 0.00157 amu includes this sub-unit of 0.00002 amu.

Mass of 1H3	3.01605 amu's
Mass of 2He3	3.01603 amu's
Mass lost creating 2He3 from 1H3	0.00002 amu

The mass of the isotope 2H3 is equal to the total mass of all of its parts.

Mass of 1H1 intact within 2He3	1.00783 amu's
Mass of negatron in 2He3	0.00055 amu
Mass of positon in 2He3	1.00294 amu's
Mass of neutron in 2He3	1.00471 amu's
Mass of 2He3	3.01603 amu's

The mass of the isotope 2He3, the neutron that dropped out plus the real mass lost equals the mass of the two isotopes of 1H2.

Mass of 2He3	3.01603 amu's
Mass of neutron that dropped out	1.00628 amu's
Real mass lost	0.00591 amu
Mass of two isotopes of 1H2	4.02822 amu's

Sub-step 69-4-4

2He3

FORMULA

1H2 + 1H2 = 2He3, one alpha 157 ray, one gamma 434 ray, one neutron

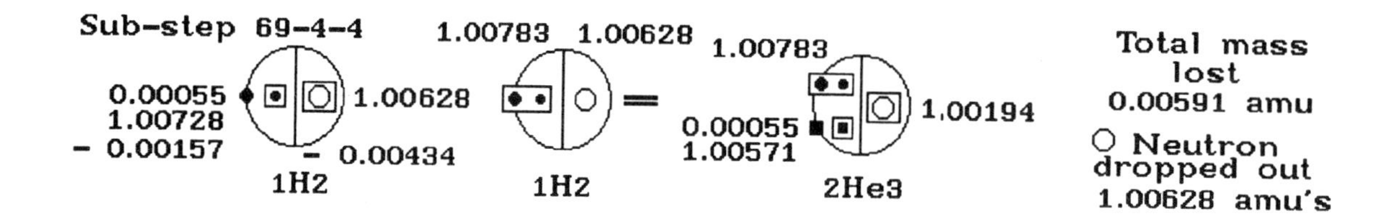

There are two isotopes of 1H2 with each having 2.01411 amu's creating the isotope 2He3 with 3.01603 amu's. There is one neutron that dropped out with 1.00628 amu's and 0.00591 amu lost as a 157 alpha ray which has an external magnetic field and a 434 gamma ray which does not have an external magnetic field..

Each isotope of 1H2 with 2.01411 amu's has one electron with 0.00055 amu, one proton with 1.00728 amu's and one neutron with 1.00628 amu's.

The isotope of 2He3 with 3.01603 amu's has one electron and one negatron with each having 0.00055 amu. There is one proton with 1.00728 amu's and one positon with 1.00571 amu's. The single neutron has 1.00194 amu's.

The first isotope of 1H2 with 2.01411 amu's has its electron drawn as a dot with 0.00055 amu and it is the negatron drawn as a square in the isotope 2He3.

The proton with 1.00728 amu's loses 0.00157 amu to become the positon in the isotope 2He3 with 1.00571 amu's. The proton is drawn as a dot with a square around it in the isotope 1H2 and as a square with another square drawn around it in the isotope 2He3. The square around the dot of the proton in the first isotope 1H2 and the second square around the square of the positon in the isotope 2He3 show that 0.00157 amu has been lost.

The single neutron in the first isotope 1H2 with 1.00628 amu's is drawn as a small circle with a square around it to show that it has lost 0.00434 amu to become the single neutron in the isotope 2He3 with 1.00194 amu's.

When the first isotope of 1H2 loses 0.00157 amu from its proton, the lost mass of 0.00157 amu is converted to a 157 alpha ray. When the single neutron in this isotope loses 0.00434 amu, the 0.00434 amu is converted to a 434 gamma ray. All lost mass that is emitted as a ray is converted to pure energy which is magnetism.

The second isotope of 1H2 with 2.01411 amu's has its electron and proton drawn with a rectangle around them to show that they are used as the isotope 1H1 with 1.00783 amu's in the isotopes 1H2 and 2He3. They are enclosed in a rectangle to show that they are used as the isotope 1H1 and that they do not lose any mass. This rectangle also means that the isotope 1H1 can drop out of the isotope 2He3 at some later time.

The single neutron in the second isotope 1H2 is not used and it therefore is not needed in the isotope 2He3 which means it drops out of the second isotope with 1.00628 amu's. It is a leftover part of the second isotope of 1H2. Its mass is included in the total mass lost creating the isotope 2He3. The mass of this neutron remains mass.

To find the amount of mass needed to create the isotope 2He3, add the masses of the two isotopes of 1H2 and then subtract the mass of the isotope 2He3 from this total.

Mass of first 1H2	2.01411 amu's
Mass of second 1H2	2.01411 amu's
Total mass needed to create 2He3	4.02822 amu's

Total mass needed to create 2He3	4.02822 amu's
Mass 2He3	3.01603 amu's
Total mass lost creating 2He3	1.01219 amu's

This mass lost creating the isotope 2He3 includes the neutron that dropped out. By subtracting the mass of this neutron from the total mass lost creating the isotope 2He3, the real mass lost can be found.

Total mass lost creating 2He3	1.01219 amu's
Mass of neutron dropping out of second 1H2	1.00628 amu's
Real mass lost creating 2He3	0.00591 amu

The mass of the neutron in the isotope 2He3 is unknown. The single neutron in the first isotope of 1H2 with 1.00628 amu's loses 0.00434 amu to become the single neutron in the isotope 2He3 with 1.00194 amu's. By subtracting the lost mass from it, the mass of the neutron in the isotope 2He3 will be known.

Mass of neutron in first 1H2	1.00628 amu's
Mass lost by it	1.00434 amu
Mass of neutron in 2He3	1.00194 amu's

To find the mass the proton lost in the first isotope of 1H2, subtract the mass lost by the neutron in the first isotope of 1H2 from the real mass lost creating the isotope 2He3.

Real mass lost creating 2He3	0.00591 amu
Mass lost by neutron in first 1H2	0.00434 amu
Mass lost by proton in first 1H2	0.00157 amu

After the proton in the first isotope 1H2 loses 0.00157 amu, it becomes the positon in the isotope 2He3. To find the mass of the positon in the isotope 2He3, subtract the lost mass of the proton in the isotope 1H2 from this proton.

Mass of proton in first 1H2	1.00728 amu's
Mass lost by it	0.00157 amu
Mass of positon in 2He3	1.00571 amu's

The total mass lost is 0.00591 amu and which is lost as two units. It is composed of a unit of 0.00434 amu and 0.00157 amu. While the 0.00434 amu is lost as a single unit that is composed of two sub-units of 0.00217 amu. The unit of 0.00157 amu is composed of a sub-unit of 0.00100 amu, a sub-unit of 0.00002 amu and a sub-unit of 0.00055 amu. Because the sub-unit 0.00100 amu is part of the single unit of 0.00157 amu, it can not cause the release of an electron. The 0.00055 amu is the same mass of an electron or negatron but because it is part of the unit of 0.00157 amu, it is emitted as part of the 157 alpha ray.

When the isotope 1H3 is converted to the isotope 2He3, there is 0.00002 amu lost. In the 0.00591 amu that is lost, the unit of 0.00157 amu includes this sub-unit of 0.00002 amu.

Mass of 1H3	3.01605 amu's
Mass of 2He3	3.01603 amu's
Mass lost creating 2He3 from 1H3	0.00002 amu

The mass of the isotope 2H3 is equal to the total mass of all of its parts.

Mass of 1H1 intact within 2He3	1.00783 amu's
Mass of negatron in 2He3	0.00055 amu
Mass of positon in 2He3	1.00571 amu's
Mass of neutron in 2He3	1.00194 amu's
Mass of 2He3	3.01603 amu's

Adding the mass of the isotope 2He3, the neutron that dropped out plus the real mass lost equals the mass of the two isotopes of 1H2.

Mass of 2He3	3.01603 amu's
Mass of neutron that dropped out	1.00628 amu's
Real mass lost	0.00591 amu
Mass of two isotopes of 1H2	4.02822 amu's

In the next step, the proton in the first isotope of 1H2 lose all of the lost mass and its single neutron will be used as is in the isotope 2He3.

Sub-step 68-4-3

2He3

FORMULA

1H2 + 1H2 = 2He3, one alpha 591 ray, one neutron

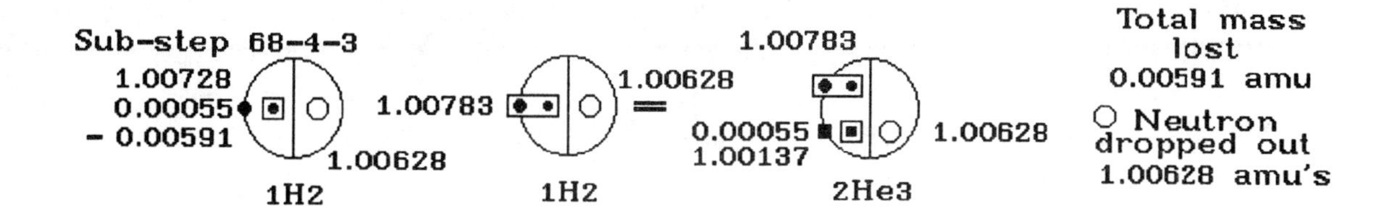

There are two isotopes of 1H2 with each having 2.01411 amu's creating the isotope 2He3 with 3.01603 amu's. There is one neutron that dropped out with 1.00628 amu's and 0.00591 amu lost by the proton in the first isotope of 1H2.

Each isotope of 1H2 with 2.01411 amu's has one electron with 0.00055 amu, one proton with 1.00728 amu's and one neutron with 1.00628 amu's.

The isotope of 2He3 with 3.01603 amu's has one electron and one negatron with each having 0.00055 amu. There is one proton with 1.00728 amu's and one positon with 1.00137 amu's. The single neutron has 1.00628 amu's. The total mass lost is 0.00591 amu and it is lost as a single unit. The 0.00591 amu is emitted as a 591 alpha ray which has an external magnetic field.

The first isotope of 1H2 with 2.01411 amu's has its electron drawn as a dot with 0.00055 amu and it is the negatron drawn as a square in the isotope 2He3.

The proton with 1.00728 amu's loses 0.00591 amu to become the positon in the isotope 2He3 with 1.00137 amu's. The proton is drawn as a dot with a square around it in the isotope 1H2 and as a square with another square drawn around it in the isotope 2He3. The square around the dot and the second square around the square of the positon show that 0.00591 amu has been lost.

The single neutron in the first isotope 1H2 with 1.00628 amu's is drawn as a small circle to become the single neutron in the isotope 2He3 with 1.00628 amu's.

When the first isotope of 1H2 loses 0.00591 amu from its proton, the lost mass of 0.00591 amu is converted to a 591 alpha ray. All lost mass that is emitted as a ray is converted to pure energy which is magnetism.

The second isotope of 1H2 with 2.01411 amu's has its electron and proton drawn with a rectangle around them to show that they are used as the isotope 1H1 with 1.00783 amu's in the isotopes 1H2 and 2He3. They are enclosed in a rectangle to show that they are used as the isotope 1H1 and that they do not lose any mass. This rectangle also means that the isotope 1H1 can drop out of the isotope 2He3 at some later time.

The single neutron in the second isotope 1H2 is not used and it therefore is not needed in the isotope 2He3 which means it drops out of the second isotope with 1.00628 amu's. It is a leftover part of the second isotope of 1H2. Its mass is included in the total mass lost creating the isotope 2He3. The mass of this neutron remains mass while the mass of 0.00591 amu is emitted as a 591 alpha ray which is the total mass lost creating the isotope 2He3.

To find the amount of mass needed to create the isotope 2He3, add the masses of the two isotopes of 1H2 and then subtract the mass of the isotope 2He3 from this total.

Mass of first 1H2	2.01411 amu's
Mass of second 1H2	2.01411 amu's
Total mass needed to create 2He3	4.02822 amu's

Total mass needed to create 2He3	4.02822 amu's
Mass 2He3	3.01603 amu's
Total mass lost creating 2He3	1.01219 amu's

This mass lost creating the isotope 2He3 includes the neutron that dropped out. By subtracting the mass of this neutron from the total mass lost creating the isotope 2He3, the real mass lost can be found.

Total mass lost creating 2He3	1.01219 amu's
Mass of neutron dropping out of second 1H2	1.00628 amu's
Real mass lost creating 2He3	0.00591 amu

In this step, there is enough parts with known masses to find the unknown mass. There is an isotope of 1H1 found within the isotope 2He3, a negatron and the single neutron that does not lose any mass. By adding the masses of these parts and subtracting their total mass from the isotope of 2He3, the remaining mass will be the positon.

Mass of 1H1 within 2He3	1.00783 amu's
Mass of negatron in 2He3	0.00055 amu
Mass of neutron in 2He3	1.00628 amu's
Total known mass in 2He3	2.01466 amu's

Mass of 2He3	3.01603 amu's
Total known mass in 2He3	2.01466 amu's
Mass of positon in 2He3	1.00137 amu's

To find the mass the proton lost in the first isotope of 1H2, subtract the mass of the positon in the isotope 2He3 from the proton in the isotope 1H2.

Mass of proton in 1H2	1.00728 amu's
Mass of positon in 2He3	1.00137 amu's
Mass lost by proton in 1H2	0.00591 amu

The total mass lost is 0.00591 amu which is lost as a single unit. This is composed of two sub-units of 0.00434 amu and 0.00157 amu. While the 0.00434 amu is a single sub-unit, it is composed of two sub-units of 0.00217 amu. The sub-unit of 0.00157 amu is composed of a sub-unit of 0.00100 amu, a sub-unit of 0.00002 amu and a sub-unit of 0.00055 amu. Because the sub-unit of 0.00100 amu is part of the sub-unit of 0.00157 amu, it can not cause the release of an electron or negatron. The 0.00055 amu is the same mass of an electron or negatron but because it is part of the sub-unit of 0.00157 amu, it is not emitted.

While the above is correct, this lost mass could be composed of one sub-unit of 0.00374 amu with a sub-unit of 0.00217 amu. The sub-unit of 0.00374 amu is composed of a sub-unit of 0.00217 amu and a sub-unit of 0.00157 amu.

When the isotope 1H3 is converted to the isotope 2He3, there is 0.00002 amu lost. In the unit 0.00591 amu that is lost, the sub-unit of 0.00157 amu includes this sub-unit of 0.00002 amu.

Mass of 1H3	3.01605 amu's
Mass of 2He3	3.01603 amu's
Mass lost creating 2He3 from 1H3	0.00002 amu

The mass of the isotope 2H3 is equal to the mass of all of its parts.

Mass of 1H1 intact within 2He3	1.00783 amu's
Mass of negatron in 2He3	0.00055 amu
Mass of positon in 2He3	1.00137 amu's
Mass of neutron in 2He3	1.00628 amu's
Mass of 2He3	3.01603 amu's

The mass of the isotope 2He3, the neutron that dropped out and the real mass lost equals the mass of the two isotopes of 1H2.

Mass of 2He3	3.01603 amu's
Mass of neutron that dropped out	1.00628 amu's
Real mass lost	0.00591 amu
Mass of two isotopes of 1H2	4.02822 amu's

The next step will have the neutron in the first isotope of 1H2 lose the 0.00591 amu instead of the proton.

Sub-step 67-4-2

2He3

FORMULA

1H2 + 1H2 = 2He3, one gamma 591 ray, one neutron

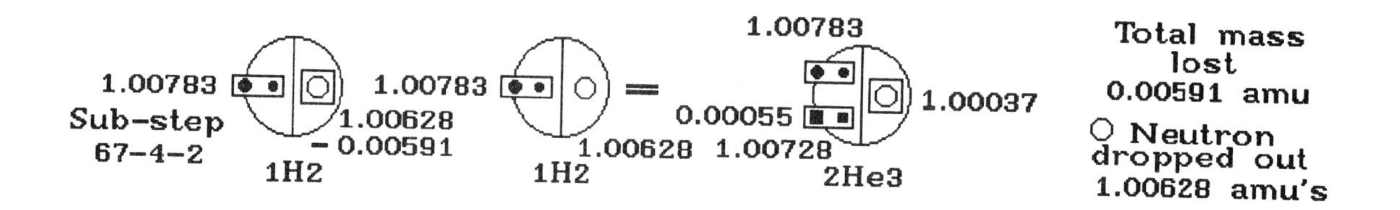

There are two isotopes of 1H2 with each having 2.01411 amu's creating the isotope 2He3 with 3.01603 amu's. There is one neutron that dropped out with 1.00628 amu's and 0.00591 amu lost by the neutron in the first isotope of 1H2.

Each isotope of 1H2 with 2.01411 amu's has one electron with 0.00055 amu, one proton with 1.00728 amu's and one neutron with 1.00628 amu's.

The isotope of 2He3 with 3.01603 amu's has one electron and one negatron with each having 0.00055 amu. There is one proton with 1.00728 amu's and one positon with 1.00728 amu's. The single neutron has 1.00037 amu's.

The first isotope of 1H2 with 2.01411 amu's has its electron with 0.00055 amu and its proton with 1.00728 amu's drawn as a dot in the isotopes 1H2 and 2He3. The electron and proton have a rectangle drawn around them to show that they are used as the isotope 1H1 in the isotopes 1H2 and 2He3. The isotope 1H1 can drop out of the isotope 2He3 at some future time.

The single neutron in the first isotope 1H2 with 1.00628 amu's is drawn as a small circle with a square around it to show that it has lost 0.00591 amu to become the single neutron in the isotope 2He3 with 1.00037 amu's. When the single neutron in this isotope loses 0.00591 amu, the 0.00591 amu is converted to a 591 gamma ray which does not have an external magnetic field. All lost mass that is emitted as a ray is converted to pure energy which is magnetism.

The second isotope of 1H2 with 2.01411 amu's has its electron and proton drawn with a rectangle around them to show that they are used as the isotope 1H1 with 1.00783 amu's in the isotopes 1H2 and 2He3. They are enclosed in a rectangle to show that they are used as the isotope 1H1 and that they do not lose any mass. This rectangle also means that the isotope 1H1 can drop out of the isotope 2He3 at some later time.

The single neutron in the second isotope 1H2 is not used and it therefore is not needed in the isotope 2He3 which means it drops out of the second isotope with 1.00628 amu's. It is a leftover part of the second isotope of 1H2. Its mass is included in the total mass lost creating the isotope 2He3.

To find the amount of mass needed to create the isotope 2He3, add the masses of the two isotopes of 1H2 and then subtract the mass of the isotope 2He3 from this total.

Mass of first 1H2	2.01411 amu's
Mass of second 1H2	2.01411 amu's
Total mass needed to create 2He3	4.02822 amu's

Total mass needed to create 2He3	4.02822 amu's
Mass 2He3	3.01603 amu's
Total mass lost creating 2He3	1.01219 amu's

This mass lost creating the isotope 2He3 includes the neutron that dropped out. By subtracting the mass of this neutron from the total mass lost creating the isotope 2He3, the real mass lost can be found.

Total mass lost creating 2He3	1.01219 amu's
Mass of neutron dropping out of second 1H2	1.00628 amu's
Real mass lost creating 2He3	0.00591 amu

The lost mass of 0.00591 amu is lost completely from the single neutron in the first isotope of 1H2. By subtracting this lost mass from this neutron, the mass of the single neutron in the isotope 2He3 will be found.

Mass of neutron in first 1H2	1.00628 amu's
Mass lost by it	0.00591 amu
Mass of the neutron in 1He3	1.00037 amu's

The total mass lost is 0.00591 amu and which is lost as one unit. This is composed of a sub-unit of 0.00434 amu and sub-unit of 0.00157 amu. While the sub-unit of 0.00434 amu is lost as a sub-unit, it is composed of two sub-units of 0.00217 amu. The sub-unit of 0.00157 amu is composed of a sub-unit of 0.00100 amu, a sub-unit of 0.00002 amu and a sub-unit of 0.00055 amu. Because the sub-unit of 0.00100 amu is part of the sub-unit of 0.00157 amu, it can not cause the release of an electron. The 0.00055 amu is the same mass of an electron or negatron but because it is part of the sub-unit of 0.00157 amu, it is not emitted.

When the isotope 1H3 is converted to the isotope 2He3, there is 0.00002 amu lost. In the 0.00591 amu that is lost, the unit of 0.00157 amu includes this sub-unit of 0.00002 amu.

Mass of 1H3	3.01605 amu's
Mass of 2He3	3.01603 amu's
Mass lost creating 2He3 from 1H3	0.00002 amu

The lost mass of 0.00591 amu could also be composed of a sub-unit of 0.00374 amu and a sub-unit of 0.00217 amu. The sub-unit of 0.00374 amu is composed of a sub-unit of 0.00217 amu and the sub-unit of 0.00157 amu with the sub-unit of 0.00157 amu composed of a sub-unit of 0.00100 amu, a sub-unit 0.00002 amu and a sub-unit of 0.00055 amu.

The mass of the isotope 2H3 is equal to the total mass of all of its parts.

Mass of two 1H1's intact within 2He3	2.01566 amu's
Mass of neutron in 2He3	1.00037 amu's
Mass of 2He3	3.01603 amu's

Adding the mass of the isotope 2He3, the neutron that dropped out plus the real mass lost equals the mass of the two isotopes of 1H2.

Mass of 2He3	3.01603 amu's
Mass of neutron that dropped out	1.00628 amu's
Real mass lost	0.00591 amu
Mass of two isotopes of 1H2	4.02822 amu's

The next step will have the neutron in the first isotope of 1H2 lose 0.00436 amu and 0.00100 amu with the 0.00100 amu transferred to the neutron in the second isotope of 1H2. When this is done, the neutron in the first isotope of 1H2 releases a negatron to go into an orbital with an electron already there. There will be an isotope of 1H1 dropping out instead of a neutron.

Sub-step 66-4-1a

2He3

FORMULA

1H2 + 1H2 = 2He3, one gamma 436 ray, one 1H1

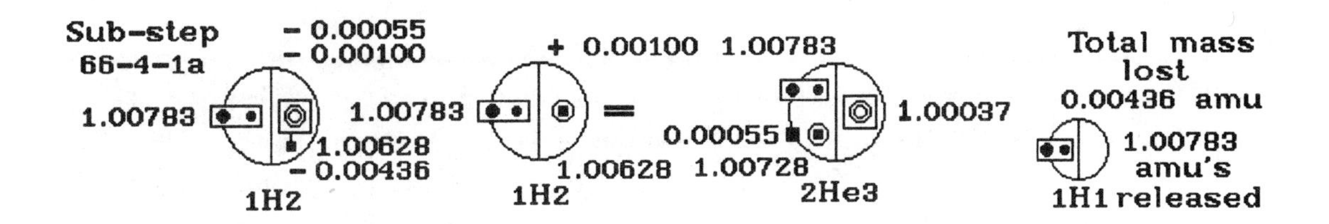

There are two isotopes of 1H2 with each having 2.01411 amu's creating the isotope 2He3 with 3.01603 amu's. There is one isotope of 1H1 that dropped out with 1.00783 amu's and 0.00436 amu lost plus 0.00100 amu that is transferred from the neutron in the first isotope of 1H2 to the neutron in the second isotope of 1H2. When the 0.00100 amu is transferred, the neutron in the first isotope of 1H2 releases a negatron that goes into an orbital with an electron already there.

Each isotope of 1H2 with 2.01411 amu's has one electron with 0.00055 amu, one proton with 1.00728 amu's and one neutron with 1.00628 amu's.

The isotope of 2He3 with 3.01603 amu's has one electron and one negatron with each having 0.00055 amu. There is one proton with 1.00728 amu's and one positon with 1.00728 amu's. The single neutron has 1.00037 amu's. The total mass lost is 0.00436 amu. The 0.00436 amu is emitted as a 436 gamma ray which does not have an external magnetic field.

The first isotope of 1H2 with 2.01411 amu's has its electron and proton each drawn as a dot with 0.00055 amu and its proton with 1.00728 amu's. They have a rectangle drawn around them to show that they are used as the isotope 1H1 within the isotopes 1H2 and 2He3. The isotope of 1H1 can drop out of the isotope 2He3 at some later time.

The single neutron in the first isotope 1H2 with 1.00628 amu's is drawn as a small circle with another circle drawn around it. A square is drawn around both of them. This time the square around the small circle shows that the lost mass is composed of a unit of 0.00436 amu that is composed of a sub-unit of 0.00434 amu and a sub-unit of 0.00002 amu which does not release an electron or negatron from the neutron. The circle drawn around the small circle means that there is 0.00100 amu transferred to the neutron in the second isotope of 1H2 that does release a negatron from within the neutron in the first isotope of 1H2. This means that there is a negatron with 0.00055 amu found inside the neutron and that it can be released. Because this transferred mass of 0.00100 amu is also regained, no transferred mass of 0.00100 amu is really lost creating the isotope 2He3.

The neutron in the first isotope of 1H2 loses 0.00434 amu and 0.00002 amu as a single unit of 0.00436 amu first. Then it loses a unit 0.00100 amu and releases a negatron with 0.00055 amu for a total lost mass of 0.00591 amu to become the single neutron in the isotope 2He3 with 1.00037 amu's. This means that the neutron in the first isotope of 1H1 emits a 436 gamma ray only because the 0.00100 amu is regained by the neutron in the second isotope of 1H2 and the negatron it releases becomes the negatron in the isotope 2He3. Since the two lost masses of 0.00055 amu and 0.00100 amu are not lost, the total mass lost is not 0.00591 amu but 0.00436 amu.

The second isotope of 1H2 with 2.01411 amu's has its electron and proton drawn with a rectangle around them to show that they are used as the isotope 1H1 with 1.00783 amu's within the second isotope of 1H2. When the isotope 1H1 is found within the isotope 1H2 intact it could drop out at some later time. Now is the time when the isotope 1H1 drops out of the isotope of 1H2 and the mass of 1.00783 amu's is part of the total mass lost.

The neutron in the second isotope of 1H2 has 1.00628 amu's and when it regains 0.00100 amu, it becomes the positon with 1.00728 amu's in the isotope 2He3. To show this, the positon in the isotope 2He3 is drawn as a square to the left of the vertical line within the big circle. There is small circle drawn around the square that is the positon. This small circle is the neutron in the second isotope of 1H2 and it also represents the 0.00100 amu that is regained from the neutron in the first isotope 1H2. This means that the 0.00100 amu is not lost creating the isotope 2He3. Because the neutron in the second isotope of 1H2 becomes the positon in the isotope 2He3, it must have a negatron. This negatron is the negatron lost by the neutron in the first isotope of 1H2. Therefore, a neutron can have a negatron within it.

To find the amount of mass needed to create the isotope 2He3, add the mass of the two isotopes of 1H2 and then subtract the mass of the isotope 2He3 from this total.

Mass of first 1H2	2.01411 amu's
Mass of second 1H2	2.01411 amu's
Total mass needed to create 2He3	4.02822 amu's

Total mass needed to create 2He3	4.02822 amu's
Mass 2He3	3.01603 amu's
Total mass lost creating 2He3	1.01219 amu's

This mass lost creating the isotope 2He3 includes the isotope 1H1 that dropped out. Subtract the mass of isotope 1H1 that dropped out from the total mass lost creating the isotope 2He3.

Total mass lost creating 2He3	1.01219 amu's
Mass of 1H1 that dropped out of second 1H2	1.00783 amu's
Mass lost creating 2He3	0.00436 amu

The mass of the electron and proton in the first isotope of 1H2 is used as is in the isotope 2He3 and the isotope 2He3 is known to have an electron and a negatron, therefore, their mass is known. At this point, the isotope 1H1 is known to drop out of the second isotope of 1H2 taking with it the electron that would have been the negatron in the isotope 2He3. None the less, the isotope 2He3 must have a negatron. All of the known masses of the isotope 2He3 can be added to find the mass of the positon and neutron which have their mass unknown. By subtracting the total known mass of the isotope 2He3 from the mass of the isotope 2He3, the remainder will be the mass of the positon and neutron.

Mass of 1H1 in 2He3	1.00783 amu's
Mass of negatron in 2He3	0.00055 amu
Known mass in 2He3	1.00838 amu's

Mass of 2He3	3.01603 amu's
Known mass in 2He3	1.00838 amu's
Mass of positon and neutron in 2He3	2.00765 amu's

Since none of the protons in the two isotopes of 1H2 can become the positon in the isotope 2He3, and the proton in the first isotope of 1H2 is known to have lost 0.00436 amu, it can not become the positon in the isotope 2He3. The only part left that can become the positon is the neutron in the second isotope 1H2. Its known mass is 1.00628 amu's. It therefore must regain 0.00100 amu to become the positon in the isotope 2He3. The only place this 0.00100 amu can come from is the neutron in the first isotope of 1H2. This neutron therefore loses 0.00436 amu and 0.00100 amu at this point in time.

When the neutron in the second isotope of 1H2 regains the 0.00100 amu to become the positon in the isotope 2He3, the positon must also have a negatron. This negatron must also come from the neutron in the first isotope of 1H2. Now the neutron in the first isotope of 1H2 has lost 0.00055 amu and 0.00100 amu plus the 0.00436 amu for a total lost mass of 0.00591 amu. Note that the 0.00436 amu did not release the electron.

Mass lost releasing negatron from neutron in first 1H2	0.00100 amu
Mass of negatron released by first neutron in 1H2	0.00055 amu
Mass lost not releasing electron by neutron in first 1H2	0.00436 amu
Total mass lost by neutron in first 1H2	0.00591 amu

This total mass lost by the first neutron in the isotope 1H2 is subtracted from this neutron to find the neutron in the isotope 2He3. This means that the neutron in the first isotope of 1H2 becomes the neutron found in the isotope 2He3.

Mass of first 1H2's neutron	1.00628 amu's
Mass lost by it	0.00591 amu
Mass of neutron in 2He3	1.00037 amu's

The negatron that is released and the 0.00100 amu that released it is not lost when the isotope 2He3 is created. Therefore, their total mass can be subtracted from the total mass lost by the neutron in the first isotope of 1H2 to find the real mass lost creating the isotope 2He3.

Total mass lost creating 2He3	0.00591 amu
Mass lost by neutron in first 1H2	0.00155 amu
Real mass lost creating 2He3	0.00436 amu

This remaining mass lost by the first isotope of 1H2's neutron is lost as a 436 ray. It is composed of a sub-unit of 0.00434 amu and a sub-unit of 0.00002 amu.

The neutron in the first isotope of 1H2 is drawn as a small circle to the right of the vertical line within the big circle. It has a square drawn around it to show that it has lost first the mass of 0.00436 amu, then a circle to show the 0.00100 amu is transferred to the second isotope of 1H2's neutron. It also has a small square under it to show that the negatron has been released from it. This negatron is connected to the neutron by a line to show that it comes from within the neutron. This negatron belongs to the positon when the neutron in the second isotope of 1H2 changes its neutron into a positon.

The neutron in the second isotope of 1H2 is also drawn as a small circle with a square drawn within the small circle to show that it becomes the positon in the isotope 2He3. The positon in the isotope 2He3 is also drawn this way to show that it came from the neutron in the second isotope of 1H2.

The neutron found in the isotope 2He3 is drawn as a small circle with another circle drawn around the small circle. There is a square drawn around both of them that indicates 0.00436 amu has been lost. The remaining mass is the single neutron found in the isotope 2He3 with 1.00037 amu's.

The lost mass of 0.00436 amu is made up of two sub-units. One sub-unit is 0.00434 amu and the other is a sub-unit of 0.00002 amu. There is a sub-unit of 0.00002 amu because when the isotope 1H3 is converted to the isotope 2He3, there is unit of 0.00002 amu lost.

Mass of 1H3	3.01605 amu's
Mass of 2He3	3.01603 amu's
Mass lost creating 2He3 from 1H3	0.00002 amu

The mass of the isotope 2H3 is equal to the total mass of all of its parts.

Mass of 1H1 intact within 2He3	1.00783 amu's
Mass of negatron in 2He3	0.00055 amu
Mass of positon in 2He3	1.00728 amu's
Mass of neutron in 2He3	1.00037 amu's
Mass of 2He3	3.01603 amu's

The mass of the isotope 2He3, the 1H1 that dropped out plus the real mass lost equals the mass of the two isotopes of 1H2.

Mass of 2He3	3.01603 amu's
Mass of 1H1 that dropped out	1.00783 amu's
Real mass lost	0.00436 amu
Mass of two isotopes of 1H2	4.02822 amu's

If the neutron in the first isotope of 1H2 can lose 0.00591 amu, then the proton can also lose this mass. This will be explained in the next step.

In this step, the proton and positon in the isotope 2He3 each have 1.00728 amu's. When this happens, the isotope 2He3 becomes stable. Any time an isotope has one or more proton or positon with less than this maximum mass, the isotope will be considered unstable whether it is radioactive or not. An unstable isotope will try to have its proton or positon regain the mass that will make it have 1.00728 amu's and become stable.

Sub-step 66-4-1b

2He3

FORMULA

1H2 + 1H2 = 2He3, one alpha 436 ray, one 1H1

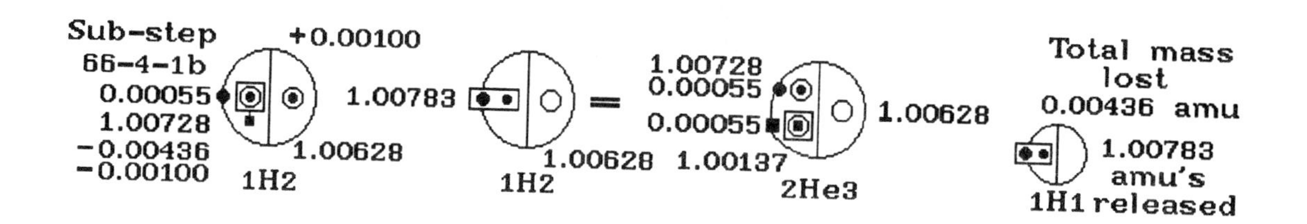

There are two isotopes of 1H2 with each having 2.01411 amu's creating the isotope 2He3 with 3.01603 amu's. There is one isotope of 1H1 that dropped out with 1.00783 amu's and 0.00436 amu lost plus 0.00100 amu that is transferred from the proton to the neutron in the first isotope of 1H2. When the 0.00100 is transferred, the proton in the first isotope of 1H2 releases a negatron that goes into an orbital with an electron already there.

Each isotope of 1H2 with 2.01411 amu's has one electron with 0.00055 amu, one proton with 1.00728 amu's and one neutron with 1.00628 amu's.

The isotope of 2He3 with 3.01603 amu's has one electron and one negatron with each having 0.00055 amu. There is one proton with 1.00728 amu's and one positon with 1.00137 amu's. The single neutron has 1.00628 amu's. The total mass lost is 0.00436 amu plus the isotope 1H1 that dropped out. There is one 436 alpha ray emitted which has an external magnetic field.

The first isotope of 1H2 with 2.01411 amu's loses 0.00436 amu from its proton and then transfers 0.00100 amu to its neutron making this neutron into the proton found in the isotope 2He3. The electron is not released because the 0.00100 amu is transferred and not lost. Instead, the proton releases a negatron from within itself. This negatron goes into the orbital with the electron already there.

The second isotope of 1H2 has its electron and proton released as an isotope of 1H1 with 1.00783 amu's. It is said to have dropped out. The neutron with 1.00628 amu's becomes the neutron in the isotope 2He3 with no loss of mass.

The proton in the first isotope of 1H2 in this step will change into a positon. The proton in the first isotope of 1H2 must be a proton because it has an electron in an orbital. It remains a proton until 0.00100 amu has been transferred to the neutron.

The proton in the first isotope 1H2 with 1.00728 amu's is drawn as a dot with small circle drawn around it and a square is drawn around both of them. This time the square around the dot and the small circle show that the lost mass is composed of a unit of 0.00436 amu. The 0.00436 amu is composed of a sub-unit of 0.00434 amu and a sub-unit of 0.00002 amu which does not release an electron from the proton. The small circle around the proton means that there is 0.00100 amu transferred that releases a negatron with 0.00055 amu. When this happens, the proton changes into a positon found in the isotope 2He3. This means that there is a negatron within the proton. This negatron is drawn as a square below the proton with a line connecting them. The 0.00100 amu transferred by this proton is regained by the neutron in the same isotope of 1H2 with 1.00628 amu's to become the proton with 1.00728 amu's in the isotope 2He3. Because this transferred mass of 0.00100 amu is regained, there is no lost mass of 0.00100 amu.

The proton in the first isotope of 1H2 loses 0.00434 amu and 0.00002 amu as a single unit of 0.00436 amu first. Then it loses a unit 0.00100 amu and releases a negatron with 0.00055 amu for a total lost and transferred mass of 0.00591 amu to become the positon in the isotope 2He3 with 1.00137 amu's. This means that the proton in the first isotope of 1H1 emits a 436 alpha ray. At this point in time the proton is still a proton. Because the 0.00100 amu is regained by the neutron in the first isotope of 1H2 changing it into a proton that causes the original proton to become a positon. There can not be two protons or two positons in this isotope. Since the transferred masses of 0.00055 amu and 0.00100 amu are not lost, the real mass lost is not 0.00591 amu but 0.00436 amu.

The second isotope of 1H2 with 2.01411 amu's has its electron and proton drawn with a rectangle around them to show that they are used as the isotope 1H1 with 1.00783 amu's within the second isotope 1H2 and that it drops out. When the isotope 1H1 is found intact within an isotope 1H2, it could drop out at some later time. Now is the time when the isotope 1H1 drops out of the second isotope of 1H2 and the mass of 1.00783 amu's is part of the total mass lost.

The neutron in the second isotope of 1H2 has 1.00628 amu's and it is used as is in the isotope 2He3 with no loss of mass. It is drawn as a small circle to the right of the vertical line in the second isotope 1H2 and it is drawn the same way in the isotope 2He3.

To find the amount of mass needed to create the isotope 2He3, add the masses of the two isotopes of 1H2 and then subtract the mass of the isotope 2He3 from this total.

Mass of first 1H2	2.01411 amu's
Mass of second 1H2	2.01411 amu's
Total mass needed to create 2He3	4.02822 amu's

Total mass needed to create 2He3	4.02822 amu's
Mass 2He3	3.01603 amu's
Total mass lost creating 2He3	1.01219 amu's

This mass lost creating the isotope 2He3 includes the isotope 1H1 that dropped out of the second isotope of 1H2. Subtract the mass of isotope 1H1 that dropped out from the total mass lost creating the isotope 2He3.

Total mass lost creating 2He3	1.01219 amu's
Mass of 1H1 that dropped out of second 1H2	1.00783 amu's
Mass lost creating 2He3	0.00436 amu

The mass of the electron and proton in the second isotope of 1H2 is used as is in the isotope 1H1. The isotope 2He3 is known to have an electron and a negatron therefore their mass is known. At this point, the isotope 1H1 is known to have dropped out of the second isotope of 1H2 taking with it the electron that would have been the negatron in the isotope 2He3. None the less, the isotope 2He3 must have a negatron. The neutron in the second isotope of 1H2 is known to have 1.00628 amu's and it is found as the neutron in the isotope 2He3. Therefore, all of the known masses of the isotope 2He3 are added together to find the mass of the positon which has its mass unknown. By subtracting the total known mass of the isotope 2He3 from the mass of the isotope 2He3, the remainder will be the mass of the positon and neutron.

Mass of electron and negatron in 2He3	0.00110 amu
Mass of proton in 2He3	1.00728 amu's
Mass of neutron in 2He3	1.00628 amu's
Known mass in 2He3	2.01466 amu's

Mass of 2He3	3.01603 amu's
Known mass in 2He3	2.01466 amu's
Mass of positon in 2He3	1.00137 amu's

When the neutron in the first isotope of 1H2 regains the 0.00100 amu to become the proton in the isotope 2He3, the original proton becomes a positon which must also have a negatron. This negatron must also come from the proton in the first isotope of 1H2. Now the proton in the first isotope of 1H2 has lost 0.00055 amu and 0.00100 amu plus the 0.00436 amu for a total lost mass of 0.00591 amu. Note that the 0.00436 amu did not release the electron.

Mass lost releasing negatron from proton in first 1H2	0.00100 amu
Mass of negatron released by first proton in 1H2	0.00055 amu
Mass lost not releasing electron by proton in first 1H2	0.00436 amu
Total mass lost by proton in first 1H2	0.00591 amu

This total mass lost by the proton in the isotope 1H2 is subtracted from this proton to find the positon in the isotope 2He3. This means that the proton in the first isotope of 1H2 becomes the positon found in the isotope 2He3.

Mass of first 1H2's proton	1.00728 amu's
Mass lost by it	0.00591 amu
Mass of positon in 2He3	1.00137 amu's

The negatron with 0.00055 amu that is released and the 0.00100 amu that released it is not lost when the isotope 2He3 is created. Therefore, their total mass can be subtracted from the total mass lost by the proton in the first isotope of 1H2 to find the real mass lost creating the isotope 2He3.

Total mass lost creating 2He3	0.00591 amu
Mass lost by proton in first 1H2	0.00155 amu
Real mass lost creating 2He3	0.00436 amu

This remaining mass lost by the first isotope 1H2's proton is lost as a 436 ray. The lost mass of 0.00436 amu is made up of two sub-units. One sub-unit is 0.00434 amu and the other is a sub-unit of 0.00002 amu.

There is a sub-unit of 0.00002 amu because when the isotope 1H3 is converted to the isotope 2He3, there is 0.00002 amu lost.

Mass of 1H3	3.01605 amu's
Mass of 2He3	3.01603 amu's
Mass lost creating 2He3 from 1H3	0.00002 amu

The mass of the isotope 2H3 is equal to the total mass of all of its parts.

Mass of proton in 2He3	1.00728 amu's
Mass of electron and negatron in 2He3	0.00110 amu
Mass of positon in 2He3	1.00137 amu's
Mass of neutron in 2He3	1.00628 amu's
Mass of 2He3	3.01603 amu's

The mass of the isotope 2He3, the 1H1 that dropped out plus the real mass lost equals the mass of the two isotopes of 1H2.

Mass of 2He3	3.01603 amu's
Mass of 1H1 that dropped out	1.00783 amu's
Real mass lost	0.00436 amu
Mass of two isotopes of 1H2	4.02822 amu's

Sub-step 66-4-1c

2He3

FORMULA

1H2 + 1H2 = 2He3, one alpha 436 ray, one 1H1

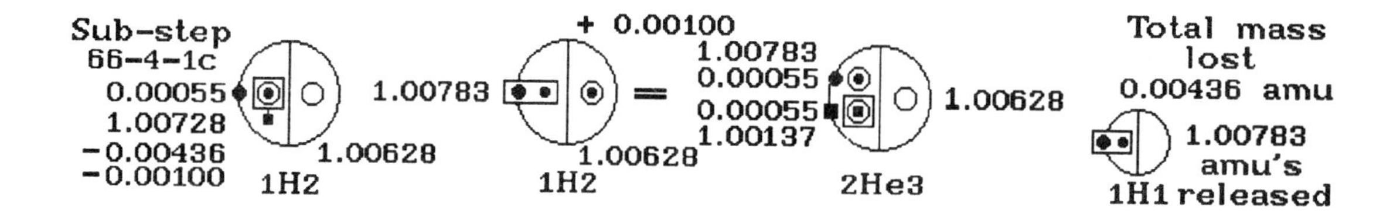

There are two isotopes of 1H2 with each having 2.01411 amu's creating the isotope 2He3 with 3.01603 amu's. The first isotope of 1H2 loses 0.00436 amu. The second isotope of 1H2 has an isotope of 1H1 drop out with 1.00783 amu's and plus 0.00100 amu that is transferred from the proton in the first isotope of 1H2 to the neutron in the second isotope of 1H2. When the 0.00100 is transferred, the proton in the first isotope of 1H2 releases a negatron that goes into an orbital with an electron already there.

Each isotope of 1H2 with 2.01411 amu's has one electron with 0.00055 amu, one proton with 1.00728 amu's and one neutron with 1.00628 amu's.

The isotope of 2He3 with 3.01603 amu's has one electron and one negatron with each having 0.00055 amu. There is one proton with 1.00728 amu's and one positon with 1.00137 amu's. The single neutron has 1.00628 amu's. The total mass lost is 0.00436 amu plus the isotope 1H1 that dropped out. There is one 436 alpha ray emitted which has an external magnetic field. There is a negatron with 0.00055 amu that came from the proton in the first isotope of 1H2.

The first isotope of 1H2 with 2.01411 amu's loses 0.00436 amu and then transfers 0.00100 amu to the neutron in the second isotope of 1H2 making this neutron into the proton found in the isotope 2He3. The electron is not released because the 0.00100 amu is transferred and not lost. Instead, the proton releases a negatron from within itself. This negatron goes into the orbital with the electron already there.

The neutron in the first isotope of 1H2 is the neutron in the isotope 2He3 with no loss of mass.

The second isotope of 1H2 has its electron and proton released as an isotope of 1H1 with 1.00783 amu's. It is said to have dropped out. The neutron with 1.00628 amu's receives 0.00100 amu from the proton in the first isotope of 1H2 to become the proton in the isotope 2He3 with 1.00728 amu's.

The proton in the first isotope of 1H2 in this step will change into a positon. The proton in the first isotope of 1H1 must be a proton because it has an electron in an orbital. It remains a proton until 0.00100 amu has been transferred to the neutron.

The proton in the first isotope 1H2 with 1.00728 amu's is drawn as a dot with small circle around it and a square is drawn around both of them. This time the square around the dot and small circle shows that 0.00436 amu has been lost. The small circle around the dot of the proton represents the 0.00100 amu that is transferred to the neutron in the second isotope of 1H2 which changes this neutron into the proton found in the isotope 2He3. The proton in the first isotope 1H2 is drawn the same way in the isotope 2He3 except that the dot changes into a small square. This is done to show that the positon in the isotope 2He3 is the remaining mass of the proton in the first isotope of 1H2.

The square shows around the dot and small circle of the proton in the first isotope of 1H2 is composed of a unit of 0.00436 amu that is composed of a sub-unit of 0.00434 amu and a sub-unit of 0.00002 amu which does not release an electron from the neutron. The small circle around the proton means that there is 0.00100 amu transferred that releases a negatron with 0.00055 amu which goes into an orbital with an electron already there. When this happens, the proton changes into a positon found in the isotope 2He3. This means that there is a negatron within the proton. This negatron is drawn as a square below the proton with a line connecting them. The 0.00100 amu transferred by this proton is regained by the neutron with 1.00628 amu's to become the proton with 1.00728 amu's in the isotope 2He3. Because this lost mass of 0.00100 amu is regained, there is no lost mass of 0.00100 amu.

The proton in the first isotope of 1H2 loses 0.00434 amu and 0.00002 amu as a single unit of 0.00436 amu first. Then it loses a unit 0.00100 amu and releases a negatron with 0.00055 amu for a total lost and transferred mass of 0.00591 amu to become the positon in the isotope 2He3 with 1.00137 amu's. This means that the proton in the first isotope of 1H1 emits a 436 alpha ray. At this point in time when the proton loses 0.00436 amu, the proton is still a proton. Because the 0.00100 amu is regained by the neutron in the second isotope of 1H2, it changes it into a proton that causes the original proton to become a positon. There can not be two protons or two positons in this isotope. Since the transferred masses of 0.00055 amu and 0.00100 amu are not lost, the real mass lost is not 0.00591 amu but 0.00436 amu

The second isotope of 1H2 with 2.01411 amu's has its electron and proton drawn with a rectangle around them to show that they are used as the isotope 1H1 with 1.00783 amu's within the second isotope 1H2 and that it drops out. When the isotope 1H1 is found intact within an isotope 1H2, it could drop out at some later time. Now is the time when the isotope 1H1 drops out of the second isotope of 1H2 and the mass of 1.00783 amu's is part of the total mass lost.

The neutron in the second isotope of 1H2 has 1.00628 amu's and it has 0.00100 amu transferred to it from the proton in the first isotope of 1H2. This neutron becomes the proton in the isotope 2He3 with 1.00728 amu's. It is drawn as a small circle with a dot in its center to the right of the vertical line in the second isotope 1H2 and it is drawn the same way in the isotope 2He3. The dot with a small circle shows that the transferred mass came from the proton in the first isotope of 1H2 and that it is this neutron that becomes the proton in the isotope 2He3.

To find the amount of mass needed to create the isotope 2He3, add the masses of the two isotopes of 1H2 and then subtract the mass of the isotope 2He3 from this total.

Mass of first 1H2	2.01411 amu's
Mass of second 1H2	2.01411 amu's
Total mass needed to create 2He3	4.02822 amu's

Total mass needed to create 2He3	4.02822 amu's
Mass 2He3	3.01603 amu's
Total mass lost creating 2He3	1.01219 amu's

This mass lost creating the isotope 2He3 includes the isotope 1H1 that dropped out of the second isotope of 1H2. Subtract the mass of isotope 1H1 that dropped out from the total mass lost creating the isotope 2He3.

Total mass lost creating 2He3	1.01219 amu's
Mass of 1H1 that dropped out of second 1H2	1.00783 amu's
Mass lost creating 2He3	0.00436 amu

The isotope 2He3 is known to have an electron and a negatron therefore their mass is known. At this point, the isotope 1H1 is known to have dropped out of the second isotope of 1H2 taking with it the electron that would have been the negatron in the isotope 2He3. None the less, the isotope 2He3 must have a negatron. The neutron in the second isotope of 1H2 is known to have 1.00628 amu's and it regains 0.00100 amu to become the proton in the isotope 2He3. The neutron in the first isotope of 1H2 becomes the neutron in the isotope 2He3 with no loss of mass. Therefore, all of the known masses of the isotope 2He3 can be added together to find the mass of the positon which has its mass unknown. By subtracting the total known mass of the isotope 2He3 from the mass of the isotope 2He3, the remainder will be the mass of the positon.

Mass of electron and negatron in 2He3	0.00110 amu
Mass of proton in 2He3	1.00728 amu's
Mass of neutron in 2He3	1.00628 amu's
Known mass in 2He3	2.01466 amu's

Mass of 2He3	3.01603 amu's
Known mass in 2He3	2.01466 amu's
Mass of positon in 2He3	1.00137 amu's

When the neutron in the first isotope of 1H2 regains the 0.00100 amu to become the proton in the isotope 2He3, the original proton becomes a positon which must also have a negatron. This negatron must also come from the proton in the first isotope of 1H2. Now the proton in the first isotope of 1H2 has lost 0.00055 amu and 0.00100 amu plus the 0.00436 amu for a total lost mass of 0.00591 amu. Note that the 0.00436 amu did not release the electron.

Mass lost releasing negatron from neutron in first 1H2	0.00100 amu
Mass of negatron released by first neutron in 1H2	0.00055 amu
Mass lost not releasing electron by neutron in first 1H2	0.00436 amu
Total mass lost by neutron in first 1H2	0.00591 amu

This total mass lost by the proton in the isotope 1H2 is subtracted from this proton to find the positon in the isotope 2He3. This means that the proton in the first isotope of 1H2 becomes the positon found in the isotope 2He3.

Mass of first 1H2's proton	1.00728 amu's
Mass lost by it	0.00591 amu
Mass of positon in 2He3	1.00137 amu's

The negatron with 0.00055 amu that is released and the 0.00100 amu that released it is not lost when the isotope 2He3 is created. Therefore, their total mass can be subtracted from the total mass lost by the proton in the first isotope of 1H2 to find the real mass lost creating the isotope 2He3.

Total mass lost creating 2He3	0.00591 amu
Mass lost by proton in first 1H2	0.00155 amu
Real mass lost creating 2He3	0.00436 amu

This remaining mass lost by the first isotope of 1H2's proton is lost as a 436 ray. It is composed of two sub-units. The first sub-unit is 0.00434 amu and the second sub-unit is 0.00002 amu.

There, is a sub-unit of 0.00002 amu because when the isotope 1H3 is converted to the isotope 2He3, there is 0.00002 amu lost. In the unit 0.00436 amu that is lost, it includes the sub-unit of 0.00002 amu.

Mass of 1H3	3.01605 amu's
Mass of 2He3	3.01603 amu's
Mass lost creating 2He3 from 1H3	0.00002 amu

The mass of the isotope 2H3 is equal to the total mass of all of its parts.

Mass of proton in 2He3	1.00728 amu's
Mass of electron and negatron in 2He3	0.00110 amu
Mass of positon in 2He3	1.00137 amu's
Mass of neutron in 2He3	1.00628 amu's
Mass of 2He3	3.01603 amu's

The mass of the isotope 2He3, the 1H1 that dropped out plus the real mass lost equals the mass of the two isotopes of 1H2.

Mass of 2He3	3.01603 amu's
Mass of 1H1 that dropped out	1.00783 amu's
Real mass lost	0.00436 amu
Mass of two isotopes of 1H2	4.02822 amu's

This completes all of the steps in STEP 4 using the formula 1H2 + 1H2 = 2He3.

All drawings in **STEP 4** using two isotopes of 1H2 to create the isotope 2He3

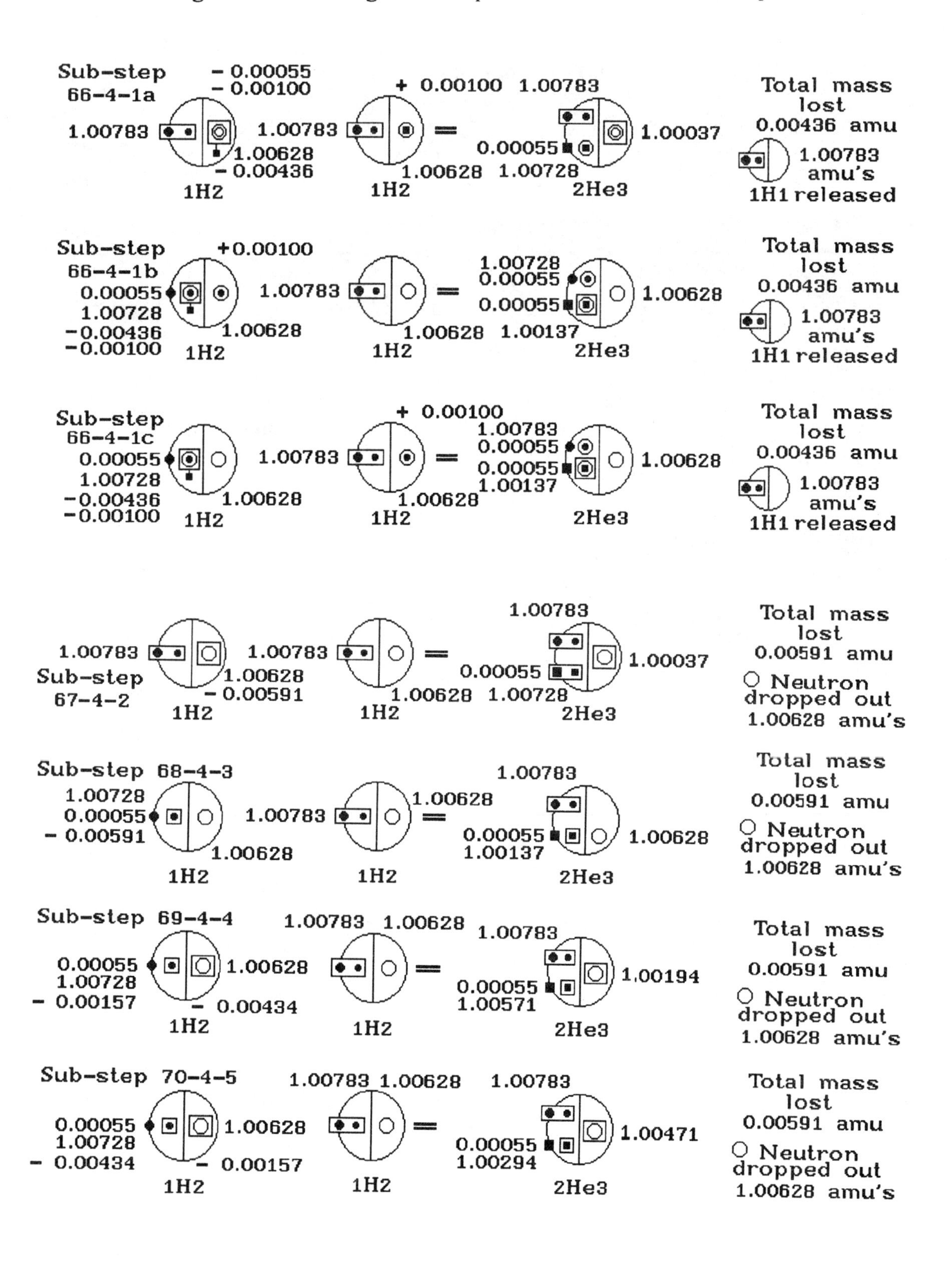

All drawings in **STEP 4** using two isotopes of 1H2 to create the isotope 2He3

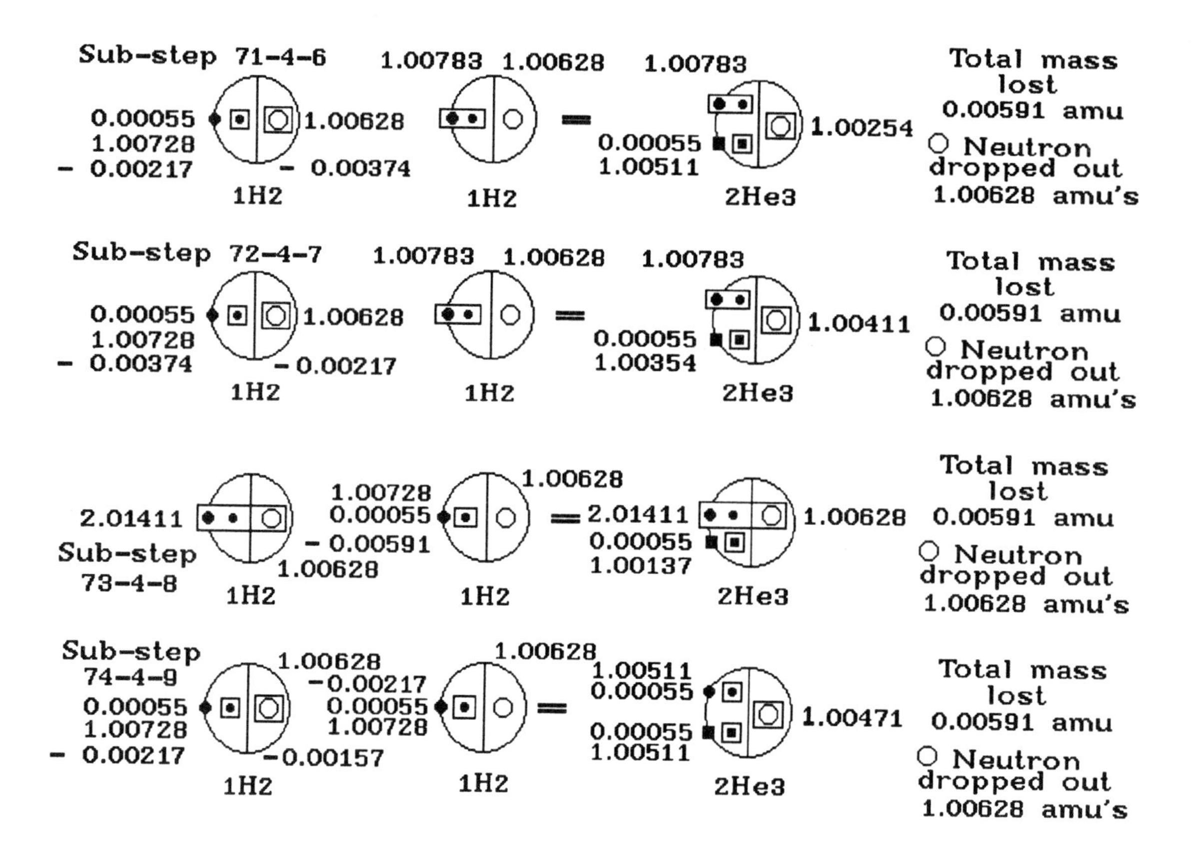

STEP 3

2He3

FORMULA

1H3 + 1H1 = 2He3

STEPS

3. 1H3 + 1H1 = 2He3

52-3-1	1H3 + 1H1 = 2He3, 591 ray, 1.00194 neutron
53-3-2	1H3 + 1H1 = 2He3, 591 ray, 1.00194 neutron
54-3-3	1H3 + 1H1 = 2He3, 591 ray, 1.00194 neutron
55-3-4	1H3 + 1H1 = 2He3, 157 ray, 1.00194 neutron
56-3-5	1H3 + 1H1 = 2He3, 157 ray, 434 ray, 1.00194 neutron
57-3-6	1H3 + 1H1 = 2He3, 434 ray, 157 ray, 1.00194 neutron
58-3-7	1H3 + 1H1 = 2He3, 217 ray, 374 ray, 1.00194 neutron
59-3-8	1H3 + 1H1 = 2He3, 374 ray, 217 ray, 1.00194 neutron
60-3-9	1H3 + 1H1 = 2He3, 374 ray, 217 ray, 1.00194 neutron
61-3-10	1H3 + 1H1 = 2He3, 374 ray, 217 ray, 1.00194 neutron
62-3-11	1H3 + 1H1 = 2He3, 374 ray, 1.00411 neutron
63-3-12	1H3 + 1H1 = 2He3, 374 ray, 1.00411 neutron
64-3-13	1H3 + 1H1 = 2He3, 157 ray, 1.00628 neutron
65-3-14	1H3 + 1H1 = 2He3, 217 ray, 157 ray, 217 ray, 1.00194 neutron

Sub-step 65-3-14

2He3

1H3 + 1H1 = 2He3, alpha 217 ray, gamma 157 ray, alpha 217 ray,
1.00194 neutron

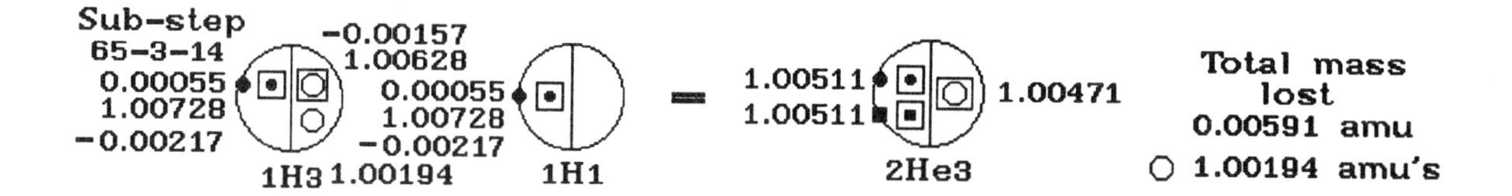

The isotope 1H3 with 3.01605 amu's has one electron with 0.00055 amu and one proton with 1.00728 amu's plus two neutrons creating the isotope 2He3 with 3.01603 amu's. The first neutron has 1.00628 amu's and the second neutron has 1.00194 amu's.

The isotope 1H1 with 1.00783 amu's has one electron with 0.00055 amu and one proton with 1.00728 amu's.

The isotope 2He3 with 3.01603 amu's has one electron with 0.00055 amu, one negatron with 0.00055 amu, one proton with 1.00511 amu's, one positon with 1.00511 amu's and one neutron with 1.00471 amu's.

The total mass lost is 0.00591 amu and it is emitted as two 217 rays and one 157 ray. Both 217 rays are emitted as an alpha rays which has an external magnetic field around each ray. The 157 ray is a gamma ray which does not have an external magnetic field. The second neutron in the isotope 1H3 drops out with 1.00194 amu's.

In this step and the next step, the isotope 1H3 has its electron drawn on the big circle as a dot with 0.00055 amu. The proton is drawn as a dot to the left of the vertical line within the big circle. This proton loses 0.00217 amu which does not release the electron. There is a square drawn around the dot to show that the 0.00217 amu has been lost. The electron and proton are drawn the same way in the isotope 2He3.

There are two neutrons in the isotope 1H3 and they are drawn one above the other to the right of the vertical line within the big circle. The top neutron is the first neutron with 1.00628 amu's and it loses 0.00157 amu to become the single neutron in the isotope 2He3 with 1.00471 amu's. It is drawn as a small circle with a square drawn around it to show that it has lost 0.00157 amu. It is drawn in the isotope 2He3 the same way as found in the isotope 1H3. The second neutron with 1.00194 amu's is drawn as a small circle below the first neutron and it is not needed in the isotope 2He3 so it drops out. This second neutron is drawn to the right of the equal sign with the value of 1.00194 amu's to show that it is the second neutron that dropped out.

The isotope 1H1 with 1.00783 amu's has one electron with 0.00055 amu and one proton with 1.00728 amu's. The electron is drawn as a dot on the big circle and as a square on the big circle in the isotope 2He3 to show that this electron becomes the negatron in the isotope 2He3. It does not lose any mass.

The proton with 1.00728 amu's in the isotope 1H1 loses 0.00217 amu to become the positon with 1.00511 amu's in the isotope 2He3. It has a square drawn around the dot in the isotope 1H1 to show that it has lost 0.00217 amu and it is drawn as a small square with a square drawn around the small square to the left of the vertical line within the big circle in the isotope 2He3 to show that it has become the positon in the isotope 2He3. The proton must be drawn as a square because the proton in the isotopes 1H3 and 2He3 are drawn as a dots. The electron in the isotopes 1H3 and 2He3 is drawn as a dot to show that it belongs to the proton. The negatron in the isotope 2He3 is drawn as a square to show that it belongs to the positon.

The isotope 2He3 with 3.001603 amu's has an electron and a negatron. The electron has 0.00055 amu and is drawn above the negatron as a dot while the negatron is drawn as a square. The negatron also has 0.00055 amu.

The isotope 2He3 has its proton drawn as a dot to show it comes from the isotope 1H3 while the positon is drawn as a square to show that it came from the isotope 1H1. The square around the dot shows that 0.00217 amu has been lost by the proton in the isotope 1H3 while the square drawn around the positon shows that 0.00217 amu has also been lost by the proton in the isotope 1H1.

There is a square drawn around the single neutron in the isotope 2He3 to show that this neutron is the neutron that came from the first neutron in the isotope 1H3. It has 1.00471 amu's because the first neutron in the isotope 1H3 with 1.00628 amu's lost 0.00157 amu.

The total mass lost shown to the right of the equal sign is 0.00591 amu. This value comes from adding all of the lost masses of the two mass units of 0.00217 amu and the single lost mass unit of 0.00157 amu. Note that the second neutron that dropped out is also shown here and its mass plus the total mass lost is the total mass lost by the isotopes 1H3 and 1H1. The second neutron that dropped out of the isotope 1H3 dropped out as mass while the other three masses are lost as rays which are three separate rays.

In this step, there are three parts that lost mass while in most cases it is only one or two parts that lose mass.

To find the mass that is lost creating the isotope 2He3, add the masses of the isotopes 1H3 and 1H1 and subtract the mass of the isotope 2He3 from their total.

Mass of 1H3	3.01605 amu's
Mass of 1H1	1.00783 amu's
Total mass of 1H3 and 1H1	4.02388 amu's

Total mass of 1H3 and 1H1	4.02388 amu's
Mass of 2He3	3.01603 amu's
Mass lost creating 2He3	1.00785 amu's

The mass lost creating the isotope 2He3 includes the mass of the second neutron that dropped out of the isotope 1H3. Subtract this neutron's mass from the total mass lost creating the isotope 2He3 to find the real mass lost.

Mass lost creating 2He3	1.00785 amu's
Mass of second neutron dropped out of 1H3	1.00194 amu's
Real mass lost creating 2He3	0.00591 amu

In this step, the isotope 1H3 has its proton lose 0.00217 amu to become the proton in the isotope 2He3. Subtract the lost mass from the mass of the proton to find the mass of the proton in the isotope 2He3.

Mass of proton in 1H3	1.00728 amu's
Mass lost by it	0.00217 amu
Mass of proton in 2He3	1.00511 amu's

The first neutron in the isotope 1H3 loses 0.00157 amu to become the single neutron in the isotope 2He3. Subtract the lost mass from the mass of the first neutron in the isotope 1H3 to find the mass of the single neutron in the isotope 2He3.

Mass of first neutron in 1H3	1.00628 amu's
Mass lost by it	0.00157 amu
Mass of single neutron in 2He3	1.00471 amu's

The electron is not lost or released and the second neutron is released or dropped out so the mass of the electron is found in the isotope 2He3 and the second neutron's mass is not. The electron does not lose any mass.

The isotope 1H1 has its proton lose the same mass as the proton in the isotope 1H3. Note that I called each proton in the isotopes 1H3 and 1H1 a proton. The drawing shows that one of them stays a proton while the other is a positon. It is the proton in the isotope 1H1 that becomes a positon. This proton loses 0.00217 amu to become the positon in the isotope 2He3 with 1.00511 amu's. To show how this happens, subtract the lost mass of 0.00217 amu from the proton in the isotope 1H1, the remainder will be the mass of the positon in the isotope 2He3.

Mass of proton in 1H1	1.00728 amu's
Mass lost by it	0.00217 amu
Mass of positon in 2He3	1.00511 amu's

The electron in the isotope 1H1 becomes the negatron in the isotope 2He3 with no loss of mass.

The proton in the isotope 1H3 and the proton in the isotope 1H1 each lost the same amount of mass of 0.00217 amu as a 217 alpha ray from each isotope. The first neutron in the isotope 1H3 lost 0.00157 amu as a 157 gamma ray. These lost masses make a total lost mass of 0.00591 amu. The lost mass of 0.00157 amu is composed of three sub-units. They are a sub-units of 0.00100 amu, 0.00055 amu and 0.00002 amu. They must be lost as a single ray in order to keep the first neutron from releasing an electron or negatron from within it.

By adding all of the now known masses of the isotope 2He3, the mass of the isotope 2He3 can be found.

Mass of electron and negatron in 2He3	0.00110 amu
Mass of proton in 2He3	1.00511 amu's
Mass of positon in 2He3	1.00511 amu's
Mass of single neutron in 2He3	1.00471 amu's
Mass of 2He3	3.01603 amu's

Almost all steps lose one or two units of mass while this step loses three units of mass. The next step will have the second neutron in the isotope lose the 0.00157 amu.

Sub-step 64-3-13

2He3

1H3 + 1H1 = 2He3, gamma 157 ray, 1.00628 neutron

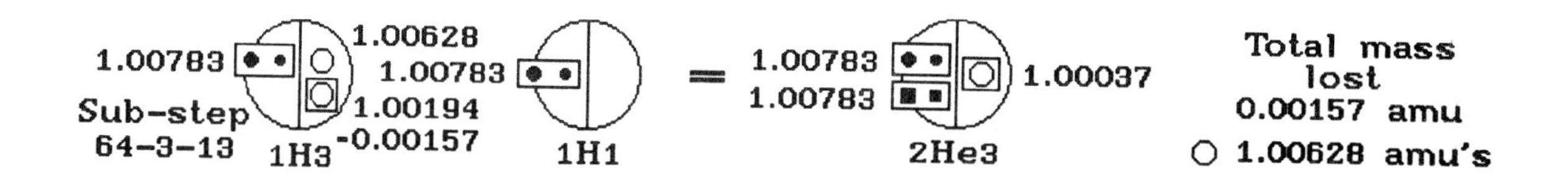

The isotope 1H3 with 3.01605 amu's has one electron with 0.00055 amu and one proton with 1.00728 amu's plus two neutrons creating the isotope 2He3 with 3.01603 amu's. The first neutron has 1.00628 amu's and the second neutron has 1.00194 amu's.

The isotope 1H1 with 1.00783 amu's has one electron with 0.00055 amu and one proton with 1.00728 amu's.

The isotope 2He3 with 3.01603 amu's has one electron with 0.00055 amu, one negatron with 0.00055 amu, one proton with 1.00728 amu's, one positon with 1.00728 amu's and one neutron with 1.00037 amu's.

The total mass lost is 0.00157 amu and it is emitted as one 157 gamma ray which does not have an external magnetic field. The first neutron in the isotope 1H3 drops out with 1.00628 amu's.

In this step, the isotope 1H3 has its electron and proton drawn with a rectangle around them to show that they are used as the isotope 1H1 within the isotope 1H3. The electron is drawn on the big circle as a dot with 0.00055 amu. The proton is drawn as a dot to the left of the vertical line within the big circle. When they are used as the isotope 1H1, they do not lose any mass.

There are two neutrons in the isotope 1H3 and they are drawn one above the other to the right of the vertical line within the big circle. The top neutron is the first neutron with 1.00628 amu's and it is not needed in the isotope 2He3 and it therefore drops out. The second neutron with 1.00194 amu's loses 0.00157 amu to become the single neutron in the isotope 2He3 with 1.00037 amu's. It is drawn with a square around it to show that it has lost 0.00157 amu. It is drawn in the isotope 2He3 the same way as found in the isotope 1H3.

The isotope 1H1 with 1.00783 amu's has one electron with 0.00055 amu and one proton with 1.00728 amu's. The electron is drawn as a dot on the big circle and as a square on the big circle in the isotope 2He3 to show that this electron becomes the negatron in the isotope 2He3. It does not lose any mass.

The proton with 1.00728 amu's in the isotope 1H1 is drawn as a dot to the left of the vertical line within the big circle and it is drawn as a square in the isotope 2He3 to show that the proton in the isotope 1H1 becomes a positon in the isotope 2He3. There is a rectangle drawn around the electron and proton in the isotope 1H1 and the negatron and positon in the isotope 2He3 to show that they are used as is which means they do not lose any mass and can drop out of the isotope 2He3 at some later time.

The isotope 2He3 with 3.001603 amu's has an electron and a negatron. The electron has 0.00055 amu and is drawn above the negatron as a dot while the negatron is drawn as a square. The negatron also has 0.00055 amu.

The proton in the isotope 2He3 is drawn as a dot above the positon to the left of the vertical line within the big circle. The positon is drawn as a square. There is a rectangle drawn around the electron and proton and one is also drawn around the negatron and positon to show that they are each used as the isotope 1H1 within the isotope 2He3.

There is one neutron in the isotope 2He3 with 1.00037 amu's. This neutron is drawn as a small circle with a square around it to show that it came from the second neutron in the isotope 1H3.

To find the mass that is lost creating the isotope 2He3, add the masses of the isotopes 1H3 and 1H1 then subtract the mass of the isotope 2He3 from their total.

Mass of 1H3	3.01605 amu's
Mass of 1H1	1.00783 amu's
Total mass of 1H3 and 1H1	4.02388 amu's

Total mass of 1H3 and 1H1	4.02388 amu's
Mass of 2He3	3.01603 amu's
Mass lost creating 2He3	1.00785 amu's

The mass lost creating the isotope 2He3 includes the mass of the first neutron that dropped out of the isotope 1H3. Subtract this neutron's mass from the total mass lost creating the isotope 2He3 to find the real mass lost.

Mass lost creating 2He3	1.00785 amu's
Mass of first neutron dropped out of 1H3	1.00628 amu's
Real mass lost creating 2He3	0.00157 amu

In this step, the isotope 1H3's proton does not lose any mass nor does the proton in the isotope 1H1. The first neutron with 1.00628 amu's drops out.

The second neutron with 1.00194 amu's in the isotope 1H3 loses 0.00157 amu to become the single neutron in the isotope 2He3. Subtract the lost mass from the mass of the second neutron in the isotope 1H3 to find the mass of the single neutron in the isotope 2He3.

Mass of second neutron in 1H3	1.00194 amu's
Mass lost by it	0.00157 amu
Mass of single neutron in 2He3	1.00037 amu's

The lost mass of 0.00157 amu is composed of three sub-units. They are sub-units of 0.00100 amu, 0.00055 amu and 0.00002 amu. In order for the electron or negatron to stay with their proton or positon, these three sub-units must be lost as a single unit of 0.00157 amu and emitted as a 157 ray. They must also be lost as a single unit and emitted as a ray in order to keep the second neutron from releasing an electron or negatron from within it.

By adding all of the now known masses of the isotope 2He3, the mass of the isotope 2He3 can be found.

Mass of 1H1 in 2He3 from 1H3	1.00783 amu's
Mass of 1H1 in 2He3	1.00783 amu's
Mass of single neutron in 2He3	1.00037 amu's
Mass of 2He3	3.01603 amu's

Sub-step 63-3-12

2He3

1H3 + 1H1 = 2He3, gamma 374 ray, 1.00411 neutron

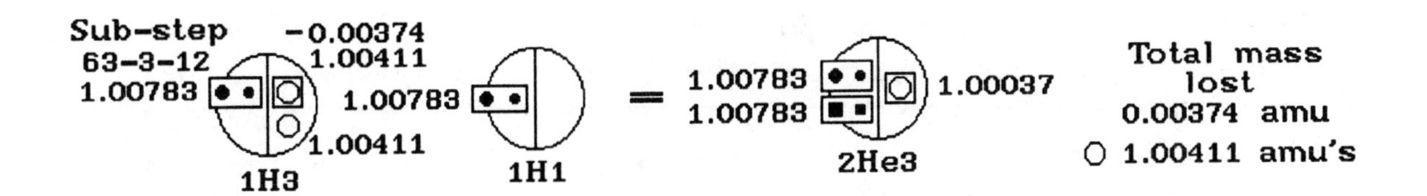

The isotope 1H3 with 3.01605 amu's has one electron with 0.00055 amu and one proton with 1.00728 amu's plus two neutrons creating the isotope 2He3 with 3.01603 amu's. Both neutrons have 1.00411 amu's.

The isotope 1H1 with 1.00783 amu's has one electron with 0.00055 amu and one proton with 1.00728 amu's.

The isotope 2He3 with 3.01603 amu's has one electron with 0.00055 amu, one negatron with 0.00055 amu, one proton with 1.00728 amu's, one positon with 1.00728 amu's and one neutron with 1.00037 amu's.

The total mass lost is 0.00374 amu and it is emitted as one 374 gamma ray which does not have an external magnetic field. There is one neutron that drops out with 1.00411 amu's.

The isotope 1H3 with 3.01605 amu's has its electron with 0.00055 amu drawn as a dot on the big circle and its proton with 1.00728 amu's drawn as a dot to the left of the vertical line within the big circle. There is a rectangle drawn around both of them to show that the isotope 1H1 is used in tact within the isotope 1H3 and 2He3. It can drop out of the isotope 2He3 at some later time. No mass is lost.

There are two neutrons in the isotope 1H3 and they are drawn as small circles with the first neutron with 1.00411 amu's drawn above the second neutron with the same mass of 1.00411 amu's to the right of the vertical line. The first neutron loses 0.00374 amu to become the single neutron found in the isotope 2He3 with 1.00037 amu's. It has a square drawn around the small circle to show that 0.00374 amu has been lost and it is drawn this same way in the isotope 2He3 to show that the single neutron came from the first neutron in the isotope 1H3. The second neutron with 1.00411 amu's is not needed so it drops out of the isotope 1H3 when the isotope 2He3 is created.

The isotope 1H1 has its electron with 0.00055 amu drawn as a dot on the big circle and its proton with 1.00728 amu's drawn as a dot to the left of the vertical line within the big circle. There is a rectangle drawn around both of them to show that the isotope 1H1 is used in tact within the isotope 2He3 and that it can drop out of the isotope 2He3 at some later time. No mass is lost.

The isotope 2He3 has its electron drawn as a dot on the big circle and the proton is drawn as a dot to the left of the vertical line within the big circle. They are enclosed within a rectangle to show that they are used as the isotope 1H1 within the isotope 2He3. The rectangle also show that the isotope of 1H1 can drop out at some later time.

There is a negatron and a positon in the isotope 2He3 with the negatron drawn as a square on the big circle. The negatron with 0.00055 amu was the electron found in the isotope 1H1. The positon with 1.00728 amu's is drawn as a square to the left of the vertical line within the big circle. It comes from the proton with 1.00728 amu's in the isotope 1H1. They are enclosed in a rectangle to show that the isotope 1H1 is found intact within the isotope 2He3 and it can drop out at some later time.

The single neutron in the isotope 2He3 has 1.00037 amu's is drawn as a small circle to the right of the vertical line in the big circle. It comes from the first neutron with 1.00411 amu's in the isotope 1H3.

To find the mass that is lost creating the isotope 2He3, add the masses of the isotopes 1H3 and 1H1 and subtract the mass of the isotope 2He3 from their total.

Mass of 1H3	3.01605 amu's
Mass of 1H1	1.00783 amu's
Total mass of 1H3 and 1H1	4.02388 amu's

Total mass of 1H3 and 1H1	4.02388 amu's
Mass of 2He3	3.01603 amu's
Mass lost creating 2He3	1.00785 amu's

The mass lost creating the isotope 2He3 includes the mass of the second neutron that dropped out of the isotope 1H3. Subtract this neutron's mass from the total mass lost creating the isotope 2He3 to find the real mass lost.

Mass lost creating 2He3	1.00785 amu's
Mass of second neutron dropped out of 1H3	1.00411 amu's
Real mass lost creating 2He3	0.00374 amu

The first neutron in the isotope 1H3 loses 0.00374 amu to become the single neutron in the isotope 2He3. Subtract the lost mass from the mass of the first neutron in the isotope 1H3 to find the mass of the single neutron in the isotope 2He3.

Mass of first neutron in 1H3	1.00411 amu's
Mass lost by it	0.00374 amu
Mass of single neutron in 2He3	1.00037 amu's

The lost mass of 0.00374 amu is composed of a sub-unit of 0.00217 amu and one of 0.00157 amu. The sub-unit of 0.00157 is composed of three sub-units. They are sub-units of 0.00100 amu, 0.00055 amu and 0.00002 amu. Because the unit of 0.00374 amu is emitted as a 374 gamma ray, all of these sub-units are part of the emitted ray. No electron or negatron can be released from within the first neutron in the isotope 1H3.

By adding all of the now known
masses of the isotope 2He3, the mass of the isotope 2He3 can be found.

Mass of 1H1 in 2He3 from 1H3	1.00783 amu's
Mass of 1H1 in 2He3	1.00783 amu's
Mass of single neutron in 2He3	1.00037 amu's
Mass of 2He3	3.01603 amu's

Sub-step 62-3-11

2He3

1H3 + 1H1 = 2He3, alpha 374 ray, 1.00411 neutron

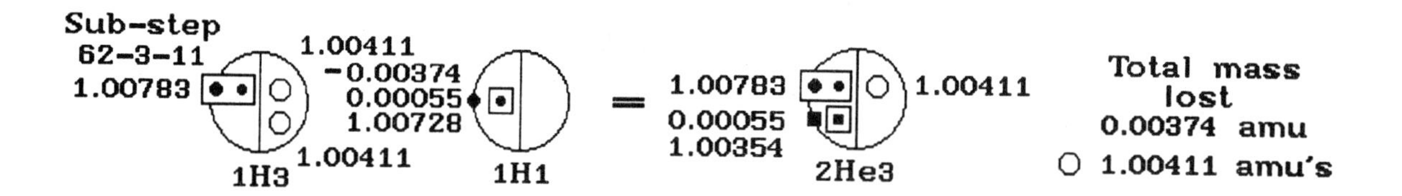

The isotope 1H3 with 3.01605 amu's has one electron with 0.00055 amu and one proton with 1.00728 amu's plus two neutrons creating the isotope 2He3 with 3.01603 amu's. Both neutrons have the same mass of 1.00411 amu's.

The isotope 1H1 with 1.00783 amu's has one electron with 0.00055 amu and one proton with 1.00728 amu's.

The isotope 2He3 with 3.01603 amu's has one electron with 0.00055 amu, one negatron with 0.00055 amu, one proton with 1.00728 amu's, one positon with 1.00354 amu's and one neutron with 1.00411 amu's.

The total mass lost of 0.00374 amu is emitted as alpha 374 ray from the proton in the isotope 1H1. The 374 alpha rays has an external magnetic field. The second neutron drops out with 1.00411 amu's from the isotope 1H3.

The isotope 1H3 has one electron with 0.00055 amu, one proton with 1.00728 amu's and two neutrons with 1.00411 amu's each. Note that there is no isotope of 1H2 found within the isotope 1H3 but there is an isotope of 1H1 plus the two neutrons. The electron and proton are enclosed within a rectangle to show that they are used as is with no loss of mass. The isotope 1H1 found in the isotopes 1H3 and 2He3 is intact and it can drop out of the isotope 2He3 at some later time. The two neutrons with 1.00411 amu's are held within the isotope 1H3 by a magnetic bond that is around them that comes from the proton and this magnetic field is around the entire isotope. This magnetic bond is weaker than the magnetic bond that is found between the proton and its two neutrons found in the isotope 1H3 when there is no isotope of 1H1 that can drop out at some later time. Therefore, these two neutrons can be separated and one leaves the isotope of 1H3. Note also that these two neutrons are not at full mass of 1.00628 amu's. This means that their internal magnetic field is weaker than that of a neutron with all of its mass. Remember that a stronger magnetic field will control a weaker magnetic field but it can not increase the strength of the weaker magnetic field.

The first neutron in the isotope 1H3 is found in the isotope 2He3 as the single neutron with 1.00411 amu's. The second neutron with 1.00411 amu's is not needed so it drops out of the isotope 1H3.

The isotope 1H1 has one electron that is drawn as a dot on the big circle while the proton is drawn as a dot to the left of the vertical line within the big circle. There is a square drawn around the proton to show that it has lost 0.00374 amu. The proton loses 0.00374 amu to become the positon in the isotope 2He3. This positon is drawn as a small square with a square drawn around it to show that it came from the proton in the isotope 1H1. The electron in the isotope 1H1 is drawn in the isotope 2He3 as a square on the big circle to show that it is a negatron that came from the isotope 1H1. The 0.00374 amu lost by the isotope 1H1's proton is the only mass lost creating the isotope 2He3 and it is emitted as a 374 alpha ray. In order for the proton in the isotope 1H1 to hold the second neutron in the isotope 1H3 by its magnetic field, the electron or negatron with a proton or positon must equal the mass of the isotope 1H2 when they gain a neutron. They must equal 2.001411 amu's and they do not.

To find the mass that is lost creating the isotope 2He3, add the masses of the isotopes 1H3 and 1H1 and subtract the mass of the isotope 2He3 from their total.

Mass of 1H3	3.01605 amu's
Mass of 1H1	1.00783 amu's
Total mass of 1H3 and 1H1	4.02388 amu's

Total mass of 1H3 and 1H1	4.02388 amu's
Mass of 2He3	3.01603 amu's
Mass lost creating 2He3	1.00785 amu's

The mass lost creating the isotope 2He3 includes the mass of the second neutron that dropped out of the isotope 1H3. Subtract this neutron's mass from the total mass lost creating the isotope 2He3 to find the real mass lost.

Mass lost creating 2He3	1.00785 amu's
Mass of second neutron dropped out of 1H3	1.00411 amu's
Real mass lost creating 2He3	0.00374 amu

In this step, the isotope 1H1 has its proton lose 0.00374 amu to become the positon in the isotope 2He3. Subtract the lost mass from the mass of the proton to find the mass of the positon in the isotope 2He3.

Mass of proton in 1H1	1.00728 amu's
Mass lost by it	0.00374 amu
Mass of positon in 2He3	1.00354 amu's

The electrons in the isotopes 1H3 and 1H1 are not lost or released and the second neutron is released or dropped out so the mass of the electron from the isotope 1H3 is found as an electron and the electron in the isotope 1H1 is found as a negatron in the isotope 2He3 while the second neutron's mass is not found in the isotope 2He3.

The proton in the isotope 1H3 is part of the isotope 1H1 found intact within the isotope 2He3 so it has the mass of 1.00728 amu's.

The proton in the isotope 1H1 lost the mass of 0.00374 amu for a total lost mass of 0.00374 amu. The lost mass of 0.00374 amu is composed of four sub-units. They are sub-units of 0.00217 amu, 0.00100 amu, 0.00055 amu and 0.00002 amu. In order for the electron or negatron to stay with their proton or positon, these four sub-units must be lost as a single unit of 0.00374 amu. They must also be lost as a single unit in order to keep the proton from releasing an electron or negatron from within any of it.

By adding all of the now known masses of the isotope 2He3, the mass of the isotope 2He3 can be found.

Mass of 1H1 intact in 2He3	1.00783 amu's
Mass of negatron in 2He3	0.00055 amu
Mass of positon in 2He3	1.00354 amu's
Mass of single neutron in 2He3	1.00411 amu's
Mass of 2He3	3.01603 amu's

Sub-step 61-3-10

2He3

1H3 + 1H1 = 2He3, alpha 374 ray, gamma 217 ray, 1.00194 neutron

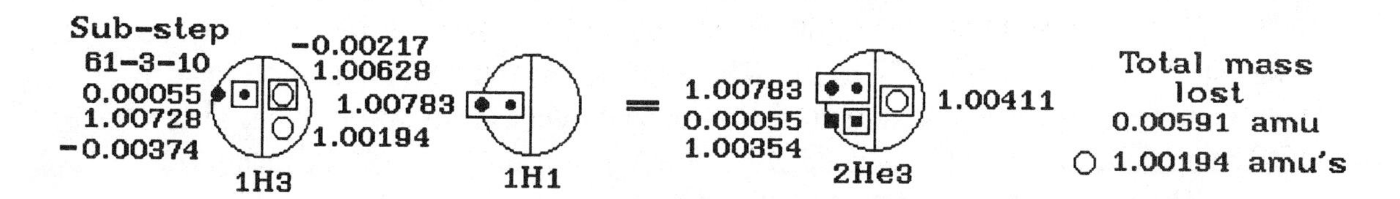

The isotope 1H3 with 3.01605 amu's has one electron with 0.00055 amu and one proton with 1.00728 amu's plus two neutrons creating the isotope 2He3 with 3.01603 amu's. The first neutron has 1.00628 amu's and the second neutron has 1.00194 amu's.

The isotope 1H1 with 1.00783 amu's has one electron with 0.00055 amu and one proton with 1.00728 amu's.

The isotope 2He3 with 3.01603 amu's has one electron with 0.00055 amu, one negatron with 0.00055 amu, one proton with 1.00728 amu's, one positon with 1.00354 amu's and one neutron with 1.00411 amu's.

The total mass lost is 0.00591 amu and it is emitted as one 374 alpha ray which has an external magnetic field and one 217 gamma ray which does not have an external magnetic field. The second neutron in the isotope 1H3 drops out with 1.00194 amu's.

In this step, the isotope 1H3 has its electron drawn on the big circle as a dot with 0.00055 amu. The proton is drawn as a dot to the left of the vertical line within the big circle. This proton loses 0.00374 amu which does not release the electron. There is a square drawn around the dot to show that the 0.00374 amu has been lost. This electron changes from a dot to a square on the big circle in the isotope 2He3 to show that the electron changes into a negatron. The proton in the isotope 1H3 is drawn as a dot with a square around it and this proton is drawn as a small square with another square drawn around the small square to show that the proton in the isotope 1H3 becomes the positon in the isotope 2He3.

There are two neutrons in the isotope 1H3 and they are drawn as small circles one above the other to the right of the vertical line within the big circle. The top neutron is the first neutron with 1.00628 amu's and it loses 0.00217 amu to become the single neutron in the isotope 2He3 with 1.00411 amu's. It is drawn with a square around it to show that it has lost 0.00217 amu. It is drawn in the isotope 2He3 the same way as found in the isotope 1H3. The second neutron with 1.00194 amu's is drawn as a small circle below the first neutron and it is not needed in the isotope 2He3 so it drops out.

The isotope 1H1 with 1.00783 amu's has one electron with 0.00055 amu and one proton with 1.00728 amu's. The electron is drawn as a dot on the big circle and as a square on the big circle in the isotope 2He3 to show that this electron becomes the negatron in the isotope 2He3. It does not lose any mass.

The proton with 1.00728 amu's in the isotope 1H1 does not lose any mass and it, along with its electron, has a rectangle drawn around them to show that they are used as the isotope 1H1 within the isotope 2He3. The rectangle shows that the isotope 1H1 can drop out of the isotope 2He3 at some later time.

The isotope 2He3 with 3.001603 amu's has an electron and a negatron. They are both drawn on the big circle. The electron has 0.00055 amu and is drawn above the negatron as a dot while the negatron is drawn as a square. The negatron also has 0.00055 amu.

The proton in the isotope 2He3 is drawn as a dot above the positon to the left of the vertical line within the big circle. The positon is drawn as a small square with another square drawn around it. This square around the small square shows that this proton was the proton found in the isotope 1H3.

There is one neutron with 1.00411 amu's in the isotope 2He3. This neutron is drawn as a small circle with a square around it to show that it was the first neutron in the isotope 1H3 because they are both drawn in the same way.

The second neutron in the isotope 1H3 that dropped out when the isotope 2He3 was created is drawn as a small circle to the right of the equal sign and the isotope 2He3 with the total amount of mass lost which is 0.00591 amu.

To find the mass that is lost creating the isotope 2He3, add the masses of the isotopes 1H3 and 1H1 and subtract the mass of the isotope 2He3 from their total.

Mass of 1H3	3.01605 amu's
Mass of 1H1	1.00783 amu's
Total mass of 1H3 and 1H1	4.02388 amu's

Total mass of 1H3 and 1H1	4.02388 amu's
Mass of 2He3	3.01603 amu's
Mass lost creating 2He3	1.00785 amu's

The mass lost creating the isotope 2He3 includes the mass of the second neutron that dropped out of the isotope 1H3. Subtract this neutron's mass from the total mass lost creating the isotope 2He3 to find the real mass lost.

Mass lost creating 2He3	1.00785 amu's
Mass of second neutron dropped out of 1H3	1.00194 amu's
Real mass lost creating 2He3	0.00591 amu

In this step, the isotope 1H3 has its proton lose 0.00374 amu to become the positon in the isotope 2He3. Subtract the lost mass from the mass of the proton to find the mass of the positon in the isotope 2He3.

Mass of proton in 1H3	1.00728 amu's
Mass lost by it	0.00374 amu
Mass of positon in 2He3	1.00354 amu's

If the total mass lost is 0.00591 amu and the proton in the isotope 1H3 lost 0.00374 amu, then the mass lost by the first neutron can be found by subtracting the lost mass of 0.00374 amu from the total mass lost creating the isotope 2He3.

Total mass lost creating 2He3	0.00591 amu
Mass lost by proton in 1H3	0.00374 amu
Mass lost by first neutron in 1H3	0.00217 amu

The first neutron in the isotope 1H3 loses 0.00217 amu to become the single neutron in the isotope 2He3. Subtract the lost mass from the mass of the first neutron in the isotope 1H3 to find the mass of the single neutron in the isotope 2He3.

Mass of first neutron in 1H3	1.00628 amu's
Mass lost by it	0.00217 amu
Mass of single neutron in 2He3	1.00411 amu's

The electrons are not lost or released. The second neutron is released or dropped out so the mass of the electron in the isotope 1H3 is found in the isotope 2He3 as an electron. The electron in the isotope 1H1 is found in the isotope 1H1 as the negatron. The electron and proton in the isotope 1H1 are used in the isotope 2He3 as the isotope 1H1 with 1.00783 amu's. The proton in the isotope 1H3 lost 0.00374 amu and the first neutron in the isotope 1H3 lost 0.00217 amu for a total lost mass of 0.00591 amu. The lost mass of 0.00374 amu is composed of four sub-units. They are sub-units of 0.00217 amu, 0.00100 amu, 0.00055 amu and 0.00002 amu. In order for the electron or negatron to stay with their proton or positon, these four sub-units must be lost as a single unit of 0.00374 amu. They must also be lost as a single unit in order for the proton not to release an electron or negatron from within it. The lost mass of 0.00217 amu is composed of a single unit. Only the unit of 0.00100 amu can release an electron or negatron so none can be released.

By adding all of the now known masses of the isotope 2He3, the mass of the isotope 2He3 can be found.

Mass of negatron in 2He3	0.00055 amu
Mass of positon in 2He3	1.00354 amu's
Mass of 1H1 in 2He3	1.00783 amu's
Mass of single neutron in 2He3	1.00411 amu's
Mass of 2He3	3.01603 amu's

Sub-step 60-3-9

2He3

1H3 + 1H1 = 2He3, gamma 374 ray, alpha 217 ray, 1.00194 neutron

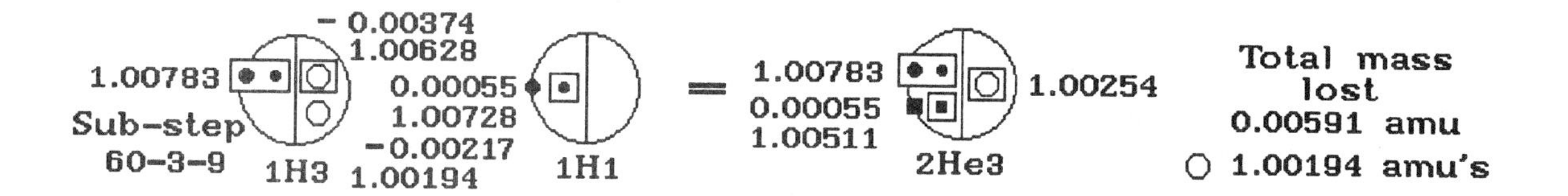

The isotope 1H3 with 3.01605 amu's has one electron with 0.00055 amu and one proton with 1.00728 amu's plus two neutrons creating the isotope 2He3 with 3.01603 amu's. The first neutron has 1.00628 amu's and the second neutron has 1.00194 amu's.

The isotope 1H1 with 1.00783 amu's has one electron with 0.00055 amu and one proton with 1.00728 amu's.

The isotope 2He3 with 3.01603 amu's has one electron with 0.00055 amu, one negatron with 0.00055 amu, one proton with 1.00728 amu's, one positon with 1.00511 amu's and one neutron with 1.00254 amu's.

The total mass lost is 0.00591 amu and it is emitted as one 374 gamma ray which does not have an external magnetic field and one 217 alpha ray which has an external magnetic field. The second neutron in the isotope 1H3 drops out with 1.00194 amu's.

The isotope 1H3 has 3.001605 amu's. Its electron is drawn on the big circle as a dot with 0.00055 amu. The proton with 1.00728 amu's is drawn as a dot also to the left of the vertical line within the big circle. This proton with its electron are enclosed in a rectangle to show that they are used as the isotope 1H1 within the isotope 2He3. The rectangle shows that the isotope 1H1 can drop out of the isotope 2He3 at a later time. No mass is lost.

There are two neutrons in the isotope 1H3 and they are drawn as small circles, one above the other, to the right of the vertical line within the big circle. The top neutron is the first neutron with 1.00628 amu's and it loses 0.00374 amu to become the single neutron in the isotope 2He3 with 1.00254 amu's. The second neutron with 1.00194 amu's is not needed in the isotope 2He3 so it drops out of the isotope 1H3.

The isotope 1H1 with 1.00783 amu's has one electron with 0.00055 amu and one proton with 1.00728 amu's. The electron is drawn as a dot on the big circle and as a square on the big circle in the isotope 2He3 to show that this electron becomes the negatron in the isotope 2He3. It does not lose any mass.

The proton with 1.00728 amu's in the isotope 1H1 loses 0.00217 amu to become the positon with 1.00511 amu's in the isotope 2He3. It has a square drawn around the dot in the isotope 1H1 to show that it has lost 0.00217 amu and it is drawn as a square to the left of the vertical line within the big circle to show that it has become the positon in the isotope 2He3.

The isotope 2He3 with 3.01603 amu's has an electron and a negatron. The electron has 0.00055 amu and is drawn above the negatron as a dot while the negatron is drawn as a square. The negatron also has 0.00055 amu.

The proton in the isotope 2He3 is drawn as a dot above the positon to the left of the vertical line within the big circle. The proton and its electron are enclosed within a rectangle to show that they are used as the isotope 1H1 within the isotope 2He3. This electron and proton was drawn the same way in the isotope 1H3. The positon is drawn as a small square with another square drawn around the small square to show that it was the proton in the isotope 1H1.

There is one neutron in the isotope 2He3 with 1.00254 amu's. This neutron is drawn as a small circle to the right of the vertical line within the big circle. There is a square drawn around it to show that it came from the first neutron in the isotope 1H3 because they are both drawn in the same way.

To find the mass that is lost creating the isotope 2He3, add the masses of the isotopes 1H3 and 1H1 and subtract the mass of the isotope 2He3 from their total.

Mass of 1H3	3.01605 amu's
Mass of 1H1	1.00783 amu's
Total mass of 1H3 and 1H1	4.02388 amu's

Total mass of 1H3 and 1H1	4.02388 amu's
Mass of 2He3	3.01603 amu's
Mass lost creating 2He3	1.00785 amu's

The mass lost creating the isotope 2He3 includes the mass of the second neutron that dropped out of the isotope 1H3. Subtract this neutron's mass from the total mass lost creating the isotope 2He3 to find the real mass lost.

Mass lost creating 2He3	1.00785 amu's
Mass of second neutron dropped out of 1H3	1.00194 amu's
Real mass lost creating 2He3	0.00591 amu

In this step, the isotope 1H1 has its proton lose 0.00217 amu to become the positon in the isotope 2He3. Subtract the lost mass from the mass of the proton to find the mass of the proton in the isotope 2He3.

Mass of proton in 1H1	1.00728 amu's
Mass lost by it	0.00217 amu
Mass of positon in 2He3	1.00511 amu's

If the proton in the isotope 1H1 lost 0.00217 amu, then the mass lost by the first neutron in the isotope 1H3 can be found by subtracting the lost mass of 0.00217 amu from the total mass lost creating the isotope 2He3.

Total mass lost creating 2He3	0.00591 amu
Mass lost by proton in 1H1	0.00217 amu
Mass lost by first neutron in 1H3	0.00374 amu

The first neutron in the isotope 1H3 loses 0.00374 amu to become the single neutron in the isotope 2He3. Subtract the lost mass from the mass of the first neutron in the isotope 1H3 to find the mass of the single neutron in the isotope 2He3.

Mass of first neutron in 1H3	1.00628 amu's
Mass lost by it	0.00374 amu
Mass of single neutron in 2He3	1.00254 amu's

The electrons are not lost or released and the second neutron is released or dropped out so the mass of the electron is found in the isotope 2He3 comes from the isotope 1H3 and the electron in the isotope 1H1 is found as the negatron in the isotope 2He3.

The first neutron in the isotope 1H3 lost 0.00374 amu and the proton in the isotope 1H1 lost 0.00217 amu for a total lost mass of 0.00591 amu. The lost mass of 0.00374 amu is composed of four sub-units. They are sub-units of 0.00217 amu, 0.00100 amu, 0.00055 amu and 0.00002 amu. In order for the electron or negatron to stay with their proton or positon, these four sub-units must be lost as a single unit of 0.00374 amu. They must also be lost as a single unit in order for the first neutron not to release an electron or negatron from within any of it.

The lost mass of 0.00217 amu is composed of a single unit. An electron or negatron can not be released because only 0.00100 amu can do this.

By adding all of the now known masses of the isotope 2He3, the mass of the isotope 2He3 can be found.

Mass of negatron in 2He3	0.00055 amu
Mass of positon in 2He3	1.00511 amu's
Mass of 1H1 in 2He3	1.00783 amu's
Mass of single neutron in 2He3	1.00254 amu's
Mass of 2He3	3.01603 amu's

Sub-step 59-3-8

2He3

1H3 + 1H1 = 2He3, alpha 374 ray, alpha 217 ray, 1.00194 neutron

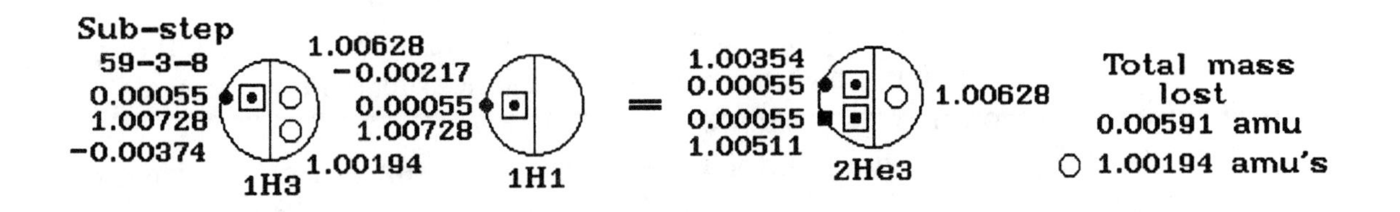

The isotope 1H3 with 3.01605 amu's has one electron with 0.00055 amu and one proton with 1.00728 amu's plus two neutrons creating the isotope 2He3 with 3.01603 amu's. The first neutron has 1.00628 amu's and the second neutron has 1.00194 amu's.

The isotope 1H1 with 1.00783 amu's has one electron with 0.00055 amu and one proton with 1.00728 amu's.

The isotope 2He3 with 3.01603 amu's has one electron with 0.00055 amu, one negatron with 0.00055 amu, one proton with 1.00354 amu's, one positon with 1.00511 amu's and one neutron with 1.00628 amu's.

The total mass lost is 0.00591 amu and it is emitted as one 374 ray and one 217 ray. Both of these rays are alpha rays and each has an external magnetic field. The second neutron in the isotope 1H3 drops out with 1.00194 amu's.

The isotope 1H3 with 3.01605 amu's has its electron drawn on the big circle as a dot with 0.00055 amu. The proton with 1.00728 amu's is drawn as a dot to the left of the vertical line within the big circle. This proton is enclosed in a square to show that it lost 0.00374 amu and it becomes the proton in the isotope 2He3 with 1.00354 amu's. It is drawn the same way in the isotope 2He3 to show that the proton in the isotope 2He3 came from the proton in the isotope 1H3. Note that this proton in the isotope 1H3 is drawn the same way as the proton in the isotope 1H1.

The proton in the isotope 1H1 with 1.00728 amu's loses 0.00217 amu to become the positon in the isotope 2He3 with 1.00511 amu's. It has a square drawn around it in the isotopes 1H1 and 2He3.

There are two neutrons in the isotope 1H3 and they are drawn as small circles, one above the other, to the right of the vertical line within the big circle. The top neutron is the first neutron with 1.00628 amu's and it becomes the single neutron in the isotope 2He3 with no loss of mass.

The second neutron with 1.00194 amu's in the isotope 1H3 is not needed in the isotope 2He3 so it drops out of the isotope 1H3 with no loss of mass. It is drawn as a small circle to the right of the equal sign and to the right of the isotope

2He3. The second neutron can not be used in the isotope 2He3 because it mass combined with the rest of the masses found in the isotope 2He3 will not equal the mass of 3.01603 amu's.

The isotope 1H1 with 1.00783 amu's has its electron drawn on the big circle as a dot with 0.00055 amu. The proton is drawn as a dot to the left of the vertical line within the big circle. This proton is drawn as a dot with a square drawn around it to show that it has lost 0.00217 amu. This electron is drawn as a square on the big circle to show that it is now a negatron with 0.00055 amu in the isotope 2He3. The proton is drawn as a square with another square drawn around the first square to show that is a positon with 1.00511 amu's.

The isotope 2He3 with 3.001603 amu's has an electron and a negatron. The electron has 0.00055 amu and is drawn above the negatron as a dot while the negatron is drawn as a square. The negatron also has 0.00055 amu.

The proton with 1.00354 amu's in the isotope 2He3 is drawn as a dot above the positon to the left of the vertical line within the big circle. The positon with 1.00511 amu's is drawn as a square with another square drawn around it to show that it is a positon that came from the isotope 1H1. These two parts have a square drawn around them even though they have different masses.

There is one neutron in the isotope 2He3 with 1.00628 amu's. This neutron is drawn as a small circle and its mass of 1.00628 shows that it came from the first neutron in the isotope 1H3 because they both have the same mass.

To find the mass that is lost creating the isotope 2He3, add the masses of the isotopes 1H3 and 1H1 and subtract the mass of the isotope 2He3 from their total.

Mass of 1H3	3.01605 amu's
Mass of 1H1	1.00783 amu's
Total mass of 1H3 and 1H1	4.02388 amu's

Total mass of 1H3 and 1H1	4.02388 amu's
Mass of 2He3	3.01603 amu's
Mass lost creating 2He3	1.00785 amu's

The mass lost creating the isotope 2He3 includes the mass of the first neutron that dropped out of the isotope 1H3. Subtract this neutron's mass from the total mass lost creating the isotope 2He3 to find the real mass lost.

Mass lost creating 2He3	1.00785 amu's
Mass of second neutron dropped out of 1H3	1.00194 amu's
Real mass lost creating 2He3	0.00591 amu

The proton in the isotope 1H3 loses 0.00374 amu to become the proton in the isotope 2He3. Subtract the lost mass from the mass of the proton in the isotope 1H3 to find the mass of the proton in the isotope 2He3.

Mass of proton in 1H3	1.00728 amu's
Mass lost by it	0.00374 amu
Mass of proton in 2He3	1.00354 amu's

If the proton in the isotope 1H3 lost 0.00374 amu, then this amount can be subtracted from the total mass lost creating the isotope 2He3 to find the mass lost by the proton in the isotope 1H1.

Total mass lost creating 2He3	0.00591 amu
Mass lost by proton in 1H3	0.00374 amu
Mass lost by proton in 1H1	0.00217 amu

The proton in the isotope 1H1 loses 0.00217 amu to become the positon in the isotope 2He3. Subtract this lost mass from the proton in the isotope 1H1 to find the mass of the positon in the isotope 2He3.

Mass of proton in 1H1	1.00728 amu's
Mass lost by it	0.00217 amu
Mass of positon in 2He3	1.00511 amu's

The electron and proton in the isotope 1H3 are used as the electron and proton in the isotope 2He3 while the electron and proton in the isotope 1H1 are used as the negatron and positon in the isotope 2He3. The first neutron with 1.00628 amu's in the isotope 1H3 is used as is in the isotope 2He3 and does not lose any mass. The second neutron is not used when the isotope 2He3 is created and it drops out with 1.00194 amu's.

The proton in the isotope 1H3 lost a unit of 0.00374 amu and the isotope 1H1 lost a unit of 0.00217 amu for a total lost mass of 0.00591 amu. The lost mass of 0.00374 amu is composed of two sub-units of 0.00217 amu and 0.00157 amu. The sub-unit of 0.00157 amu is composed of three sub-units. They are sub-units of 0.00100 amu, 0.00055 amu and 0.00002 amu. In order for the electron or negatron to stay with their proton or positon, these sub-units must be lost as single units of 0.00374 amu and 0.00217 amu. They must also be lost as a single unit in order for the proton and positon not to release an electron or negatron from within them. The lost mass of 0.00217 amu can not be broken down into sub-units.

By adding all of the now known masses of the isotope 2He3, the mass of the isotope 2He3 can be found.

Mass of electron and negatron in 2He3	0.00110 amu
Mass of proton in 2He3	1.00354 amu's
Mass of positon in 2He3	1.00511 amu's
Mass of single neutron in 2He3	1.00628 amu's
Mass of 2He3	3.01603 amu's

Sub-step 58-3-7

2He3

1H3 + 1H1 = 2He3, alpha 217 ray, alpha 374 ray, 1.00194 neutron

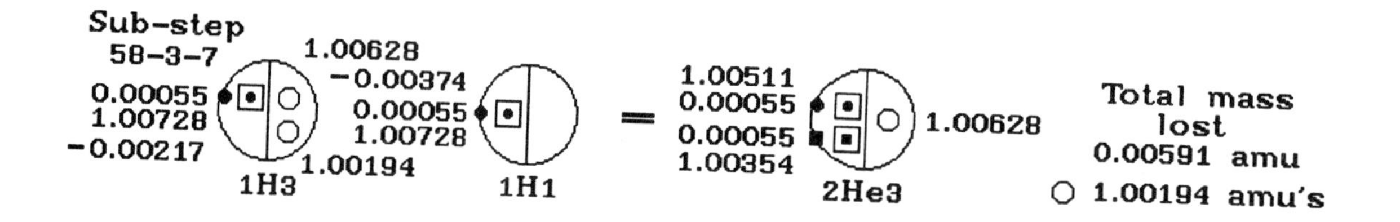

The isotope 1H3 with 3.01605 amu's has one electron with 0.00055 amu and one proton with 1.00728 amu's plus two neutrons creating the isotope 2He3 with 3.01603 amu's. The first neutron has 1.00628 amu's and the second neutron has 1.00194 amu's.

The isotope 1H1 with 1.00783 amu's has one electron with 0.00055 amu and one proton with 1.00728 amu's.

The isotope 2He3 with 3.01603 amu's has one electron with 0.00055 amu, one negatron with 0.00055 amu, one proton with 1.00511 amu's, one positon with 1.00354 amu's and one neutron with 1.00628 amu's.

The total mass lost is 0.00591 amu and it is emitted as one 217 ray and one 374 ray. Both of these rays are alpha rays and each has an external magnetic field. The second neutron in the isotope 1H3 drops out with 1.00194 amu's.

The isotope 1H3 with 3.01605 amu's has its electron drawn on the big circle as a dot with 0.00055 amu. The proton with 1.00728 amu's is drawn as a dot to the left of the vertical line within the big circle. This proton is enclosed in a square to show that it lost 0.00217 amu and it becomes the proton in the isotope 2He3 with 1.00511 amu's. It is drawn the same way in the isotope 2He3 to show that the proton in the isotope 2He3 came from the proton in the isotope 1H3.

There are two neutrons in the isotope 1H3 and they are drawn as small circles, one above the other, to the right of the vertical line within the big circle. The top neutron is the first neutron with 1.00628 amu's and it becomes the single neutron in the isotope 2He3 with no loss of mass.

The second neutron with 1.00194 amu's in the isotope 1H3 is not needed in the isotope 2He3 so it drops out of the isotope 1H3 with no loss of mass. It is drawn as a small circle to the right of the equal sign and to the right of the isotope 2He3.

The isotope 1H1 with 1.00783 amu's has its electron drawn on the big circle as a dot with 0.00055 amu. The proton is drawn as a dot also to the left of the vertical line within the big circle. This proton is drawn as a dot with a square drawn around it to show that it has lost 0.00374 amu. This electron in the isotope

2He3 is drawn as a square on the big circle to show that it is now a negatron with 0.00055 amu. This proton in the isotope 1H3 is drawn in the isotope 2He3 as a small square with a square drawn around it to show that is a positon with 1.00354 amu's.

The isotope 2He3 with 3.001603 amu's has an electron and a negatron. The electron has 0.00055 amu and it is drawn above the negatron as a dot while the negatron is drawn as a square. The negatron also has 0.00055 amu.

The proton with 1.00511 amu's in the isotope 2He3 is drawn as a dot above the positon to the left of the vertical line within the big circle. The positon with 1.00354 amu's is drawn as a square with another square drawn around it to show that it is a positon that came from the isotope 1H1.

There is one neutron in the isotope 2He3 with 1.00628 amu's. This neutron is drawn as a small circle to the right of the vertical line within the isotope 2He3 and its mass of 1.00628 shows that it came from the first neutron in the isotope 1H3 because they both have the same mass.

To find the mass that is lost creating the isotope 2He3, add the masses of the isotopes 1H3 and 1H1 and subtract the mass of the isotope 2He3 from their total.

Mass of 1H3	3.01605 amu's
Mass of 1H1	1.00783 amu's
Total mass of 1H3 and 1H1	4.02388 amu's

Total mass of 1H3 and 1H1	4.02388 amu's
Mass of 2He3	3.01603 amu's
Mass lost creating 2He3	1.00785 amu's

The mass lost creating the isotope 2He3 includes the mass of the second neutron that dropped out of the isotope 1H3. Subtract this neutron's mass from the total mass lost creating the isotope 2He3 to find the real mass lost.

Mass lost creating 2He3	1.00785 amu's
Mass of second neutron dropped out of 1H3	1.00194 amu's
Real mass lost creating 2He3	0.00591 amu

The proton in the isotope 1H3 loses 0.00217 amu to become the proton in the isotope 2He3. Subtract the lost mass from the mass of the proton in the isotope 1H3 to find the mass of the proton in the isotope 2He3.

Mass of proton in 1H3	1.00728 amu's
Mass lost by it	0.00217 amu
Mass of proton in 2He3	1.00511 amu's

Since the proton in the isotope 1H3 lost 0.00217 amu, the mass lost by the proton in the isotope 1H1 can be found by subtracting the 0.00217 amu from the total mass lost creating the isotope 2He3.

Total mass lost creating 2He3	0.00591 amu
Mass lost by proton in 1H3	0.00217 amu
Mass lost by proton in 1H1	0.00374 amu

The proton in the isotope 1H1 loses 0.00374 amu to become the positon in the isotope 2He3. Subtract this lost mass from the proton in the isotope 1H1 to find the mass of the positon in the isotope 2He3.

Mass of proton in 1H1	1.00728 amu's
Mass lost by it	0.00374 amu
Mass of positon in 2He3	1.00354 amu's

The electron and proton in the isotope 1H3 are used as the electron and proton in the isotope 2He3 while the electron and proton in the isotope 1H1 are used as the negatron and positon in the isotope 2He3. The first neutron with 1.00628 amu's in the isotope 1H3 is used as is in the isotope 2He3 and does not lose any mass. The second neutron is not used when the isotope 2He3 is created and it drops out with 1.00194 amu's.

The proton in the isotope 1H3 lost a unit 0.00217 amu and the isotope 1H1 lost a unit 0.00374 amu for a total lost mass of 0.00591 amu. The lost mass of 0.00217 amu can not be broken down into sub-units.

The lost mass of 0.00374 amu is composed of two sub-units. They are 0.00217 amu and 0.00157 amu. The lost mass of 0.00157 amu is composed of three sub-units. They are sub-units of 0.00100 amu, 0.00055 amu and 0.00002 amu. In order for the electron or negatron to stay with their proton or positon, these sub-units must be lost as single units of 0.00374 amu and 0.00217 amu. They each must also be lost as a single unit to keep each proton from releasing an electron or negatron from within either of them.

By adding all of the now known masses of the isotope 2He3, the mass of the isotope 2He3 can be found.

Mass of electron and negatron in 2He3	0.00110 amu
Mass of proton in 2He3	1.00511 amu's
Mass of positon in 2He3	1.00354 amu's
Mass of single neutron in 2He3	1.00628 amu's
Mass of 2He3	3.01603 amu's

Sub-step 57-3-6

2He3

1H3 + 1H1 = 2He3, alpha 434 ray, alpha 157 ray, 1.00194 neutron

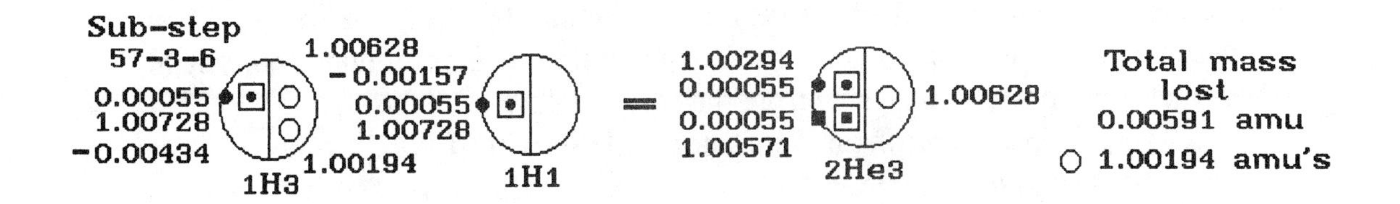

The isotope 1H3 with 3.01605 amu's has one electron with 0.00055 amu and one proton with 1.00728 amu's plus two neutrons creating the isotope 2He3 with 3.01603 amu's. The first neutron has 1.00628 amu's and the second neutron has 1.00194 amu's.

The isotope 1H1 with 1.00783 amu's has one electron with 0.00055 amu and one proton with 1.00728 amu's.

The isotope 2He3 with 3.01603 amu's has one electron with 0.00055 amu, one negatron with 0.00055 amu, one proton with 1.00294 amu's, one positon with 1.00571 amu's and one neutron with 1.00628 amu's.

The total mass lost is 0.00591 amu and it is emitted as one 434 ray and one 157 ray. Both rays are emitted as alpha rays and each has an external magnetic field. The second neutron in the isotope 1H3 drops out with 1.00194 amu's.

The isotope 1H3 with 3.01605 amu's has its electron drawn on the big circle as a dot with 0.00055 amu. The proton with 1.00728 amu's is drawn as a dot to the left of the vertical line within the big circle. This proton is enclosed in a square to show that it lost 0.00434 amu and it becomes the proton in the isotope 2He3 with 1.00294 amu's. It is drawn the same way in the isotope 2He3 to show that the proton in the isotope 2He3 came from the proton in the isotope 1H3.

There are two neutrons in the isotope 1H3 and they are drawn as small circles, one above the other, to the right of the vertical line within the big circle. The top neutron is the first neutron with 1.00628 amu's and it becomes the single neutron in the isotope 2He3 with no loss of mass.

The second neutron with 1.00194 amu's in the isotope 1H3 is not needed in the isotope 2He3 so it drops out of the isotope 1H3 with no loss of mass. It is drawn as a small circle to the right of the equal sign and to the right of the isotope 2He3.

The isotope 1H1 with 1.00783 amu's has its electron drawn on the big circle as a dot with 0.00055 amu. The proton is drawn as a dot to the left of the vertical line within the big circle. This proton is drawn as a dot with a square drawn around it to show that it has lost 0.00157 amu. This electron is drawn as a square on the big circle to show that it is now a negatron with 0.00055 amu in the isotope 2He3. The proton is drawn as a square with another square drawn around the first square to show that it is a positon with 1.00571 amu's in the isotope 2He3.

The isotope 2He3 with 3.001603 amu's has an electron and a negatron. The electron has 0.00055 amu and is drawn above the negatron as a dot while the negatron is drawn as a square. The negatron also has 0.00055 amu.

The proton with 1.00294 amu's in the isotope 2He3 is drawn as a dot above the positon to the left of the vertical line within the big circle. The positon with 1.00571 amu's is drawn as a square with another square drawn around it to show that it is a positon that came from the isotope 1H1.

There is one neutron in the isotope 2He3 with 1.00628 amu's. This neutron is drawn as a small circle and its mass of 1.00628 shows that it came from the first neutron in the isotope 1H3 because they both have the same mass.

To find the mass that is lost creating the isotope 2He3, add the masses of the isotopes 1H3 and 1H1 and subtract the mass of the isotope 2He3 from their total.

Mass of 1H3	3.01605 amu's
Mass of 1H1	1.00783 amu's
Total mass of 1H3 and 1H1	4.02388 amu's

Total mass of 1H3 and 1H1	4.02388 amu's
Mass of 2He3	3.01603 amu's
Mass lost creating 2He3	1.00785 amu's

The mass lost creating the isotope 2He3 includes the mass of the second neutron that dropped out of the isotope 1H3. Subtract this neutron's mass from the total mass lost creating the isotope 2He3 to find the real mass lost.

Mass lost creating 2He3	1.00785 amu's
Mass of second neutron dropped out of 1H3	1.00194 amu's
Real mass lost creating 2He3	0.00591 amu

The proton in the isotope 1H3 loses 0.00434 amu to become the proton in the isotope 2He3. Subtract the lost mass from the mass of the proton in the isotope 1H3 to find the mass of the proton in the isotope 2He3.

Mass of proton in 1H3	1.00728 amu's
Mass lost by it	0.00434 amu
Mass of proton in 2He3	1.00294 amu's

Since the proton in the isotope 1H3 lost 0.00434 amu, the mass lost by the proton in the isotope 1H1 can be found by subtracting this lost mass from the total mass lost creating the isotope 2He3.

Total mass lost creating 2He3	0.00591 amu
Mass lost by proton in 1H3	0.00434 amu
Mass lost by proton in 1H1	0.00157 amu

The proton in the isotope 1H1 loses 0.00157 amu to become the positon in the isotope 2He3. Subtract this lost mass from the proton in the isotope 1H1 to find the mass of the positon in the isotope 2He3.

Mass of proton in 1H1	1.00728 amu's
Mass lost by it	0.00157 amu
Mass of positon in 2He3	1.00571 amu's

The electron and proton in the isotope 1H3 are used as the electron and proton in the isotope 2He3 while the electron and proton in the isotope 1H1 are used as the negatron and positon in the isotope 2He3. The first neutron with 1.00628 amu's in the isotope 1H3 is used as is in the isotope 2He3 and does not lose any mass. The second neutron is not used when the isotope 2He3 is created and it drops out 1.00194 amu's.

The proton in the isotope 1H3 lost 0.00434 amu and the isotope 1H1 lost 0.00157 amu for a total lost mass of 0.00591 amu. The lost mass of 0.00434 amu is composed of two sub-units. They are both sub-units of 0.00217 amu. The lost mass of 0.00157 amu is composed of three sub-units. They are sub-units of 0.00100 amu, 0.00055 amu and 0.00002 amu. In order for the electron or negatron to stay with their proton or positon, these sub-units must be lost as single units of 0.00434 amu and 0.00157 amu. They must also be lost as a single unit to keep the protons in the isotopes 1H3 and 1H1 from releasing an electron or negatron from within either of them.

By adding all of the now known masses of the isotope 2He3, the mass of the isotope 2He3 can be found.

Mass of electron and negatron in 2He3	0.00110 amu
Mass of proton in 2He3	1.00294 amu's
Mass of positon in 2He3	1.00571 amu's
Mass of single neutron in 2He3	1.00628 amu's
Mass of 2He3	3.01603 amu's

Sub-step 56-3-5

2He3

1H3 + 1H1 = 2He3, alpha 157 ray, alpha 434 ray, 194 neutron

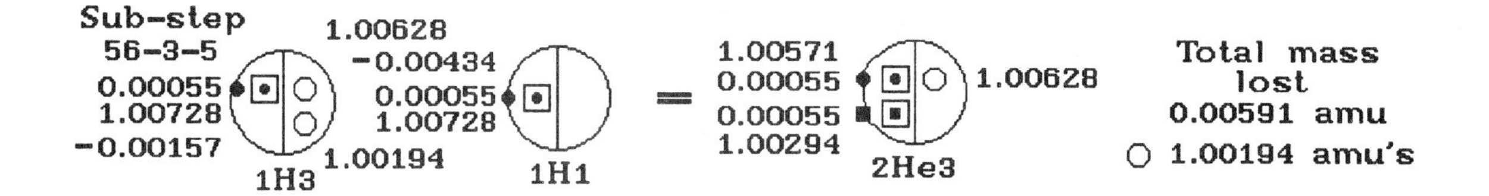

The isotope 1H3 with 3.01605 amu's has one electron with 0.00055 amu and one proton with 1.00728 amu's plus two neutrons creating the isotope 2He3 with 3.01603 amu's. The first neutron has 1.00628 amu's and the second neutron has 1.00194 amu's.

The isotope 1H1 with 1.00783 amu's has one electron with 0.00055 amu and one proton with 1.00728 amu's.

The isotope 2He3 with 3.01603 amu's has one electron with 0.00055 amu, one negatron with 0.00055 amu, one proton with 1.00571 amu's, one positon with 1.00294 amu's and one neutron with 1.00628 amu's.

The total mass lost is 0.00591 amu and it is emitted as one 157 ray and one 434 ray. Both rays are emitted as an alpha ray and each has an external magnetic field. The second neutron in the isotope 1H3 drops out with 1.00194 amu's.

The isotope 1H3 with 3.01605 amu's has its electron drawn on the big circle as a dot with 0.00055 amu. The proton with 1.00728 amu's is drawn as a dot to the left of the vertical line within the big circle. This proton is enclosed in a square to show that it lost 0.00157 amu and it becomes the proton in the isotope 2He3 with 1.00571 amu's. It is drawn the same way in the isotope 2He3 to show that the proton in the isotope 2He3 came from the proton in the isotope 1H3.

There are two neutrons in the isotope 1H3 and they are drawn as small circles, one above the other, to the right of the vertical line within the big circle. The top neutron is the first neutron with 1.00628 amu's and it becomes the single neutron in the isotope 2He3 with no loss of mass.

The second neutron with 1.00194 amu's in the isotope 1H3 is not needed in the isotope 2He3 so it drops out of the isotope 1H3 with no loss of mass. It is drawn as a small circle to the right of the equal sign and to the right of the isotope 2He3.

The isotope 1H1 with 1.00783 amu's has its electron drawn on the big circle as a dot with 0.00055 amu. The proton is drawn as a dot to the left of the vertical line within the big circle. This proton is drawn as a dot with a square drawn around it to show that it has lost 0.00434 amu. This electron is drawn as a square on the big circle to show that it is now a negatron with 0.00055 amu in the isotope 2He3. The proton in the isotope 1H1 is drawn in the isotope 2He3 as a small square with another square drawn around the small square to show that it is a positon with 1.00294 amu's.

The isotope 2He3 with 3.001603 amu's has an electron and a negatron. The electron has 0.00055 amu and is drawn above the negatron as a dot while the negatron is drawn as a square. The negatron also has 0.00055 amu.

The proton with 1.00571 amu's in the isotope 2He3 is drawn as a dot above the positon to the left of the vertical line within the big circle. The positon with 1.00294 amu's is drawn as a square with another square drawn around it to show that it is a positon that came from the isotope 1H1.

There is one neutron in the isotope 2He3 with 1.00628 amu's. This neutron is drawn as a small circle and its mass of 1.00628 amu's shows that it came from the first neutron in the isotope 1H3 because they both have the same mass.

To find the mass that is lost creating the isotope 2He3, add the masses of the isotopes 1H3 and 1H1 and subtract the mass of the isotope 2He3 from their total.

Mass of 1H3	3.01605 amu's
Mass of 1H1	1.00783 amu's
Total mass of 1H3 and 1H1	4.02388 amu's

Total mass of 1H3 and 1H1	4.02388 amu's
Mass of 2He3	3.01603 amu's
Mass lost creating 2He3	1.00785 amu's

The mass lost creating the isotope 2He3 includes the mass of the second neutron that dropped out of the isotope 1H3. Subtract this neutron's mass from the total mass lost creating the isotope 2He3 to find the real mass lost.

Mass lost creating 2He3	1.00785 amu's
Mass of second neutron dropped out of 1H3	1.00194 amu's
Real mass lost creating 2He3	0.00591 amu

The proton in the isotope 1H3 loses 0.00157 amu to become the proton in the isotope 2He3. Subtract the lost mass from the mass of the proton in the isotope 1H3 to find the mass of the proton in the isotope 2He3.

Mass of proton in 1H3	1.00728 amu's
Mass lost by it	0.00157 amu
Mass of proton in 2He3	1.00571 amu's

Since the proton in the isotope 1H3 lost 0.00157 amu, the mass lost by the proton in the isotope 1H1 can be found by subtracting this lost mass from the total mass creating the isotope 2He3.

Total lost mass creating 2He3	0.00591 amu
Mass lost by proton in 1H3	0.00157 amu
Mass lost by proton in 1H1	0.00434 amu

The proton in the isotope 1H1 loses 0.00434 amu to become the positon in the isotope 2He3. Subtract this lost mass from the proton in the isotope 1H1 to find the mass of the positon in the isotope 2He3.

Mass of proton in 1H1	1.00728 amu's
Mass lost by it	0.00434 amu
Mass of positon in 2He3	1.00294 amu's

The electron and proton in the isotope 1H3 are used as the electron and proton in the isotope 2He3 while the electron and proton in the isotope 1H1 are used as the negatron and positon in the isotope 2He3. The first neutron with 1.00628 amu's in the isotope 1H3 is used as is in the isotope 2He3 and does not lose any mass. The second neutron is not used when the isotope 2He3 is created and it drops out 1.00194 amu's.

The proton in the isotope 1H3 lost a unit of 0.00157 amu and the isotope 1H1 lost a unit of 0.00434 amu for a total lost mass of 0.00591 amu. The lost mass of 0.00157 amu is composed of three sub-units. They are a sub-units of 0.00100 amu, 0.00055 amu and 0.00002 amu. The lost mass of 0.00434 amu is composed of two sub-units. They are both sub-units of 0.00217 amu. The sub-unit of 0.00217 amu can not be broken down into sub-units.

In order for the electron or negatron to stay with their proton or positon, these sub-units must be lost as a single units of 0.00157 amu and 0.00434 amu. They must also be lost as a single unit in order for each proton in the isotopes 1H3 and 1H1 to keep from releasing an electron or negatron from within either of them.

By adding all of the now known masses of the isotope 2He3, the mass of the isotope 2He3 can be found.

Mass of electron and negatron in 2He3	0.00110 amu
Mass of proton in 2He3	1.00571 amu's
Mass of positon in 2He3	1.00294 amu's
Mass of single neutron in 2He3	1.00628 amu's
Mass of 2He3	3.01603 amu's

Sub-step 55-3-4

2He3

1H3 + 1H1 = 2He3, gamma 157 ray, 1.00628 neutron

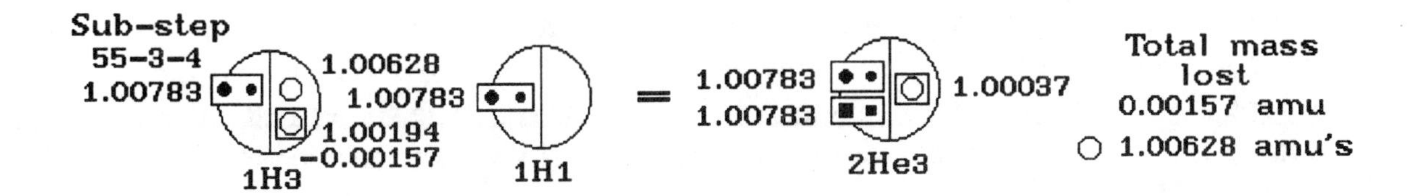

The isotope 1H3 with 3.01605 amu's has one electron with 0.00055 amu and one proton with 1.00728 amu's plus two neutrons creating the isotope 2He3 with 3.01603 amu's. The first neutron has 1.00628 amu's and the second neutron has 1.00194 amu's.

The isotope 1H1 with 1.00783 amu's has one electron with 0.00055 amu and one proton with 1.00728 amu's.

The isotope 2He3 with 3.01603 amu's has one electron with 0.00055 amu, one negatron with 0.00055 amu, one proton with 1.00728 amu's, one positon with 1.00728 amu's and one neutron with 1.00037 amu's.

The total mass lost is not 0.00591 amu as found in the previous steps. In this step it is 0.00157 amu. This lost mass is emitted as a 157 gamma ray which does not have an external magnetic field. The first neutron with 1.00628 amu's does not lose any mass and since it is not needed when the isotope 2He3 is created, it drops out with no loss of mass.

The isotope 1H3 with 3.01605 amu's has its electron drawn on the big circle as a dot with 0.00055 amu. The proton with 1.00728 amu's is drawn as a dot to the left of the vertical line within the big circle. The electron and proton are enclosed in a rectangle to show that they are used as the isotope 1H1 within the isotopes 1H3 and 2He3. The rectangle shows that this isotope of 1H1 can drop out of the isotope 2He3 at some later time.

There are two neutrons in the isotope 1H3 and they are drawn as small circles, one above the other, to the right of the vertical line within the big circle. The top neutron is the first neutron with 1.00628 amu's and it drops out of the isotope 1H3 because it is not needed in the isotope 2He3. The second neutron in the isotope 1H3 loses 0.00157 amu which is all of the mass that is lost. This second neutron becomes the single neutron in the isotope 2He3 with 1.00037 amu's.

The isotope 1H1 with 1.00783 amu's has its electron drawn on the big circle as a dot with 0.00055 amu. The proton is drawn as a dot to the left of the vertical line within the big circle. The electron and proton are drawn with a rectangle around both of them to show that the isotope 1H1 is used in the isotope 2He3 with no loss of mass. The rectangle shows that the isotope 1H1 can drop out of the isotope 2He3 at some later time.

The isotope 2He3 has its electron drawn as a dot on the big circle and the proton is drawn as a dot to the left of the vertical line within the big circle. They are drawn with a rectangle around them. They come from the isotope 1H3. The negatron is drawn below the electron on the big circle as a square. The positon is drawn below the proton as a square. The negatron and positon have a rectangle drawn around them to show that they are used as the isotope 1H1 within the isotope 2He3. They come from the isotope 1H1. The rectangle shows that either isotope 1H1 found within the isotope 2He3 can drop out at some later time as the isotope 1H1.

There is one neutron in the isotope 2He3 with 1.00037 amu's. This neutron is drawn as a small circle and its mass of 1.00037 shows that it came from the second neutron with 1.00194 amu's in the isotope 1H3. This is the only neutron that can lose 0.00157 amu and have the mass of 1.00037 amu's.

To find the mass that is lost creating the isotope 2He3, add the masses of the isotopes 1H3 and 1H1 and subtract the mass of the isotope 2He3 from their total.

Mass of 1H3	3.01605 amu's
Mass of 1H1	1.00783 amu's
Total mass of 1H3 and 1H1	4.02388 amu's

Total mass of 1H3 and 1H1	4.02388 amu's
Mass of 2He3	3.01603 amu's
Mass lost creating 2He3	1.00785 amu's

The mass lost creating the isotope 2He3 includes the mass of the first neutron that dropped out of the isotope 1H3. Subtract this neutron's mass from the total mass lost creating the isotope 2He3 to find the real mass lost.

Mass lost creating 2He3	1.00785 amu's
Mass of first neutron dropped out of 1H3	1.00628 amu's
Real mass lost creating 2He3	0.00157 amu

This real mass lost can now be subtracted from the mass of the second neutron in the isotope 1H3 to find the mass of the single neutron in the isotope 2He3.

Mass of second neutron in 1H3	1.00194 amu's
Mass lost by it	0.00157 amu
Mass of single neutron in 2He3	1.00037 amu's

The electron and proton in the isotope 1H3 are used as the isotope 1H1 in the isotope 2He3 while the electron and proton in the isotope 1H1 are used also as the isotope 1H1 as the negatron and positon in the isotope 2He3.

The first neutron with 1.00628 amu's in the isotope 1H3 drops out of the isotope 1H3. The second neutron in the isotope 1H3 lost 0.00157 amu for a total lost mass of 0.00157 amu and it is lost as a single unit. The lost mass of 0.00157 is composed of three sub-units. They are sub-units of 0.00100 amu, 0.00055 amu and 0.00002 amu.

By adding all of the now known masses of the isotope 2He3, the mass of the isotope 2He3 can be found.

Mass of 1H1 within 2He3 from 1H3	1.00783 amu's
Mass of 1H1 within 2He3 from 1H1	1.00783 amu's
Mass of single neutron in 2He3	1.00037 amu's
Mass of 2He3	3.01603 amu's

Sub-step 54-3-3

2He3

1H3 + 1H1 = 2He3, gamma 591 ray, 1.00194 neutron

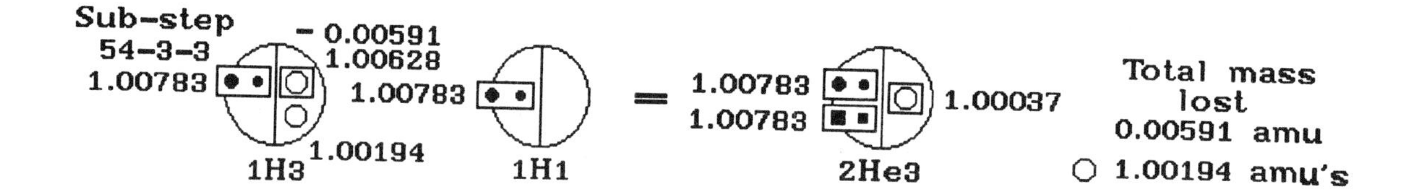

The isotope 1H3 with 3.01605 amu's has one electron with 0.00055 amu and one proton with 1.00728 amu's plus two neutrons creating the isotope 2He3 with 3.01603 amu's. The first neutron has 1.00628 amu's and the second neutron has 1.00194 amu's.

The isotope 1H1 with 1.00783 amu's has one electron with 0.00055 amu and one proton with 1.00728 amu's.

The isotope 2He3 with 3.01603 amu's has one electron with 0.00055 amu, one negatron with 0.00055 amu, one proton with 1.00728 amu's, one positon with 1.00728 amu's and one neutron with 1.00037 amu's.

The total mass lost is 0.00591 amu and it is emitted as one 591 gamma ray which does not have an external magnetic field. The second neutron in the isotope 1H3 drops out with 1.00194 amu's.

The isotope 1H3 with 3.01605 amu's has its electron drawn on the big circle as a dot with 0.00055 amu. The proton with 1.00728 amu's is drawn as a dot to the left of the vertical line within the big circle. The electron and proton are enclosed in a rectangle to show that they are used as the isotope 1H1 within the isotopes 1H3 and 2He3. The rectangle shows that this isotope of 1H1 can drop out of the isotope 2He3 at some later time.

There are two neutrons in the isotope 1H3 and they are drawn as small circles, one above the other, to the right of the vertical line within the big circle. The top neutron is the first neutron with 1.00628 amu's and it loses 0.00591 amu to become the single neutron in the isotope 2He3 with 1.00037 amu's. The 0.0591 amu is the only mass lost and it is emitted as a 591 gamma ray.

The second neutron with 1.00194 amu's in the isotope 1H3 is not needed in the isotope 2He3 so it drops out of the isotope 1H3 with no loss of mass. It is drawn as a small circle to the right of the equal sign and to the right of the isotope 2He3.

The isotope 1H1 with 1.00783 amu's has its electron drawn on the big circle as a dot with 0.00055 amu. The proton is drawn as a dot to the left of the vertical line within the big circle. The electron and proton are drawn with a rectangle around both of them to show that the isotope 1H1 is used in the isotope 2He3 with no loss of mass. The rectangle shows that the isotope 1H1 can drop out of the isotope 2He3 at some later time. This electron in the isotope 1H1 becomes the negatron in the isotope 2He3 and it is drawn as a square on the big circle in the isotope 2He3 to show that it is now a negatron with 0.00055 amu. The proton in the isotope 1H1 becomes the positon in the isotope 2He3 and it is drawn as a square to the left of the vertical line within the big circle.

There is one neutron with 1.00037 amu's in the isotope 2He3. This neutron is drawn as a small circle to the right of the big circle and its mass of 1.00037 amu's shows that it came from the first neutron in the isotope 1H3 because this is the only neutron that can lose 0.00591 amu and have the mass of 1.00037 amu's.

To find the mass that is lost creating the isotope 2He3, add the masses of the isotopes 1H3 and 1H1 and subtract the mass of the isotope 2He3 from their total.

Mass of 1H3	3.01605 amu's
Mass of 1H1	1.00783 amu's
Total mass of 1H3 and 1H1	4.02388 amu's

Total mass of 1H3 and 1H1	4.02388 amu's
Mass of 2He3	3.01603 amu's
Mass lost creating 2He3	1.00785 amu's

The mass lost creating the isotope 2He3 includes the mass of the second neutron that dropped out of the isotope 1H3. Subtract this neutron's mass from the total mass lost creating the isotope 2He3 to find the real mass lost.

Mass lost creating 2He3	1.00785 amu's
Mass of second neutron dropped out of 1H3	1.00194 amu's
Real mass lost creating 2He3	0.00591 amu

The only neutron in this step to lose mass is the first neutron in the isotope 1H3. Therefore, subtract the lost mass of 0.00591 amu from its mass to find the single neutron in the isotope 2He3.

Mass of first neutron in 1H3	1.00628 amu's
Mass lost by it	0.00591 amu
Mass of single neutron in 2He3	1.00037 amu's

The electron and proton in the isotope 1H3 are used as the isotope 1H1 in the isotope 2He3 while the electron and proton in the isotope 1H1 are used also as the isotope 1H1 as the negatron and positon in the isotope 2He3. The second neutron with 1.00194 amu's in the isotope 1H3 drops out of the isotope 1H3.

The first neutron in the isotope 1H3 lost 0.00591 amu for a total lost mass of 0.00591 amu and it is lost as a single unit. The lost mass of 0.00591 amu is composed of four sub-units. They are sub-units of 0.00434 amu, 0.00100 amu, 0.00055 amu and 0.00002 amu. The sub-unit of 0.00434 amu is composed of two sub-units of 0.00217 amu. The lost mass of 0.00591 amu must be lost as a single unit in order to keep the first neutron from releasing an electron or negatron from within it.

The reader is to keep in mind that in other steps the lost mass of 0.00591 amu could be composed of a sub-unit of 0.00374 amu and a sub-unit of 217 amu.

By adding all of the now known masses of the isotope 2He3, the mass of the isotope 2He3 can be found.

Mass of 1H1 within 2He3 from 1H3	1.00783 amu's
Mass of 1H1 within 2He3	1.00783 amu's
Mass of single neutron in 2He3	1.00037 amu's
Mass of 2He3	3.01603 amu's

Sub-step 53-3-2

2He3

1H3 + 1H1 = 2He3, alpha 591 ray, 194 neutron

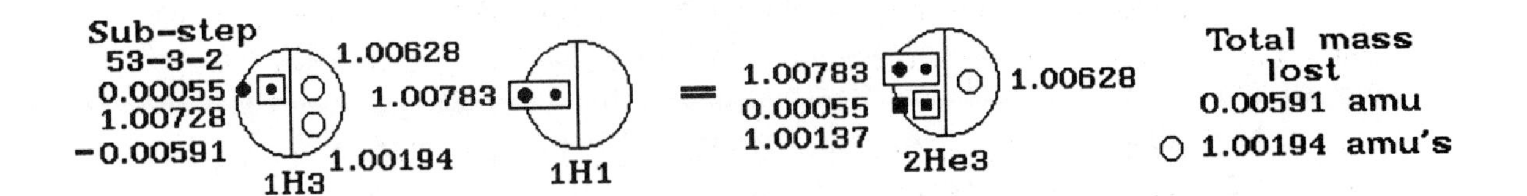

The isotope 1H3 with 3.01605 amu's has one electron with 0.00055 amu and one proton with 1.00728 amu's plus two neutrons creating the isotope 2He3 with 3.01603 amu's. The first neutron has 1.00628 amu's and the second neutron has 1.00194 amu's.

The isotope 1H1 with 1.00783 amu's has one electron with 0.00055 amu and one proton with 1.00728 amu's.

The isotope 2He3 with 3.01603 amu's has one electron with 0.00055 amu, one negatron with 0.00055 amu, one proton with 1.00728 amu's, one positon with 1.00137 amu's and one neutron with 1.00628 amu's.

The total mass lost is 0.00591 amu and it is emitted as one 591 alpha ray which has an external magnetic field. The second neutron in the isotope 1H3 drops out with 1.00194 amu's.

The isotope 1H3 has its electron drawn on the big circle as a dot with 0.00055 amu. The proton is drawn as a dot to the left of the vertical line within the big circle. This proton loses 0.00591 amu which does not release the electron. There is a square drawn around the dot to show that the 0.00591 amu has been lost. The electron is converted to a negatron and it is now drawn as a square on the big circle in the isotope 2He3. The proton with 1.00728 amu's in the isotope 1H3 is converted to a positon with 1.00137 amu's. It is drawn as a square to the right of the vertical line within the big circle in the isotope 2He3. There is a square drawn around the small square to show that this positon came from the proton in the isotope 1H3 because they are both drawn in the same way.

There are two neutrons in the isotope 1H3 and they are drawn as small circles one above the other to the right of the vertical line within the big circle. The top neutron is the first neutron with 1.00628 amu's and it is used as the single neutron in the isotope 2He3 with no loss of mass. The second neutron with 1.00194 amu's is drawn as a small circle below the first neutron and it is not needed in the isotope 2He3 so it drops out.

The isotope 1H1 with 1.00783 amu's has one electron with 0.00055 amu and one proton with 1.00728 amu's. The electron is drawn as a dot on the big circle and as a dot on the big circle in the isotope 2He3 to show that this electron stays an electron in the isotope 2He3. It does not lose any mass.

The proton with 1.00728 amu's in the isotope 1H1 does not lose any mass and it along with its electron has a rectangle drawn around them to show that they are used as the isotope 1H1 within the isotope 2He3. The rectangle shows that the isotope 1H1 can drop out of the isotope 2He3 at some later time.

The isotope 2He3 has one electron and one negatron drawn with the electron above the negatron on the big circle. Both have the same mass of 0.00055 amu. The proton is drawn above the positon and to the left of the vertical line within the big circle. The electron and proton have a rectangle drawn around them to show that they are used as the isotope 1H1 within the isotope 2He3. The rectangle shows that the isotope 1H1 within the isotope 2He3 can drop out at some later time. This isotope was the isotope of 1H1 used to create the isotope 2He3.

There is one neutron in the isotope 2He3 with 1.00628 amu's. This neutron is drawn as a small circle to the right of the vertical line within the big circle to show that it came from the first neutron in the isotope 1H3 because they both have the same mass and are drawn in the same way.

The second neutron with 1.00194 amu in the isotope 1H3 that dropped out when the isotope 2He3 was created is drawn as a small circle to the right of the equal sign and the isotope 2He3 with a statement of the total amount of mass lost which is 0.00591 amu.

To find the mass that is lost creating the isotope 2He3, add the masses of the isotopes 1H3 and 1H1 and subtract the mass of the isotope 2He3 from their total.

Mass of 1H3	3.01605 amu's
Mass of 1H1	1.00783 amu's
Total mass of 1H3 and 1H1	4.02388 amu's

Total mass of 1H3 and 1H1	4.02388 amu's
Mass of 2He3	3.01603 amu's
Mass lost creating 2He3	1.00785 amu's

The mass lost creating the isotope 2He3 includes the mass of the second neutron that dropped out of the isotope 1H3. Subtract this neutron's mass from the total mass lost creating the isotope 2He3 to find the real mass lost.

Mass lost creating 2He3	1.00785 amu's
Mass of second neutron dropped out of 1H3	1.00194 amu's
Real mass lost creating 2He3	0.00591 amu

In this step, the isotope 1H3 has its proton lose 0.00591 amu to become the positon in the isotope 2He3. Subtract the lost mass from the mass of the proton in the isotope 1H3 to find the mass of the positon in the isotope 2He3.

Mass of proton in 1H3	1.00728 amu's
Mass lost by it	0.00591 amu
Mass of positon in 2He3	1.00137 amu's

The electrons are not lost or released and the second neutron is released or dropped out so the mass of the electron in the isotope 1H3 is found in the isotope 2He3 as a negatron and the electron in the isotope 1H1 is found in the isotope 1H1 is the electron. The electron and proton in the isotope 1H1 are used in the isotope 2He3 as the isotope 1H1 with 1.00783 amu's.

The proton in the isotope 1H3 lost a single unit of 0.00591 amu for a total lost mass of 0.00591 amu. The lost mass of 0.00591 amu is composed of four sub-units. They are sub-units of 0.00434 amu, 0.00100 amu, 0.00055 amu and 0.00002 amu. The sub-unit of 0.00434 amu is composed of two sub-units of 0.00217 amu. In order for the electron that becomes a negatron to stay with its proton that becomes a positon, these sub-units must be lost as a single unit of 0.00591 amu.

The reader is to keep in mind that in other steps the lost mass of 0.00591 amu could be composed of a sub-unit of 0.00374 amu and a sub-unit of 217 amu.

By adding all of the now known masses of the isotope 2He3, the mass of the isotope 2He3 can be found.

Mass of negatron in 2He3	0.00055 amu
Mass of positon in 2He3	1.00137 amu's
Mass of 1H1 in 2He3	1.00783 amu's
Mass of single neutron in 2He3	1.00628 amu's
Mass of 2He3	3.01603 amu's

Sub-step 52-3-1

2He3

1H3 + 1H1 = 2He3, alpha 591 ray, 194 neutron

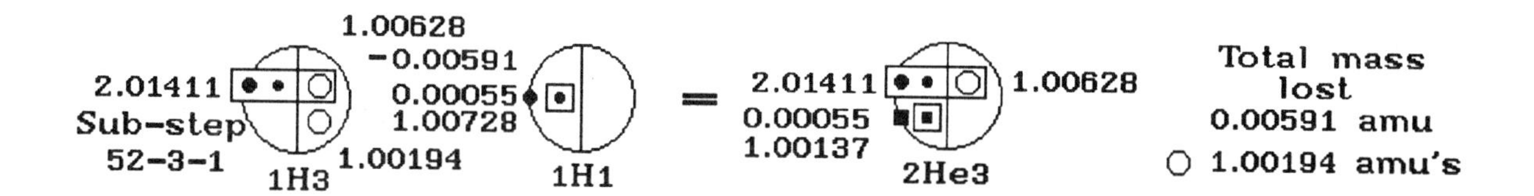

The isotope 1H3 with 3.01605 amu's has one electron with 0.00055 amu and one proton with 1.00728 amu's plus two neutrons creating the isotope 2He3 with 3.01603 amu's. The first neutron has 1.00628 amu's and the second neutron has 1.00194 amu's.

The isotope 1H1 with 1.00783 amu's has one electron with 0.00055 amu and one proton with 1.00728 amu's.

The isotope 2He3 with 3.01603 amu's has one electron with 0.00055 amu, one negatron with 0.00055 amu, one proton with 1.00728 amu's, one positon with 1.00137 amu's and one neutron with 1.00628 amu's.

The total mass lost is 0.00591 amu and it is emitted as one 591 alpha ray which has an external magnetic field. The second neutron in the isotope 1H3 drops out with 1.00194 amu's.

The isotope 1H3 with 3.01605 amu's has its electron drawn on the big circle as a dot with 0.00055 amu. The proton with 1.00728 amu's is drawn as a dot to the left of the vertical line within the big circle. The first neutron is drawn above the second neutron as a small circle. The electron, proton and first neutron all are enclosed within a rectangle to show that they are used as the isotope 1H2 with 2.01411 amu's within the isotopes 1H3 and 2He3. The rectangle shows that the isotope 1H2 can drop out of the isotope 2He3 at some later time and that no mass is lost.

There are two neutrons in the isotope 1H3 and they are drawn as small circles, one above the other, to the right of the vertical line within the big circle. The top neutron is the first neutron with 1.00628 amu's. The second neutron with 1.00194 amu's is not needed in the isotope 2He3 so it drops out of the isotope 1H3. It is drawn on the right side of the equal sign and to the right of the isotope of 2He3. The total mass lost is also shown here.

The isotope 1H1 with 1.00783 amu's has one electron with 0.00055 amu and one proton with 1.00728 amu's. The electron is drawn as a dot on the big circle and as a square on the big circle in the isotope 2He3 to show that this electron becomes the negatron in the isotope 2He3. It does not lose any mass.

The proton with 1.00728 amu's in the isotope 1H1 loses 0.00591 amu to become the positon with 1.00137 amu's in the isotope 2He3. It has a square drawn around the dot in the isotope 1H1 to show that it has lost 0.00591 amu and it is drawn as a square to the left of the vertical line within the big circle to show that it has become the positon in the isotope 2He3.

The isotope 2He3 with 3.001603 amu's has an electron and a negatron. The electron has 0.00055 amu and is drawn above the negatron as a dot while the negatron is drawn as a square on the big circle. The negatron also has 0.00055 amu.

The proton in the isotope 2He3 is drawn as a dot above the positon to the left of the vertical line within the big circle. The positon is drawn as a square with another square drawn around it to the left of the vertical line within the big circle.

There is one neutron in the isotope 2He3 with 1.00628 amu's. This neutron is drawn as a small circle to the right of the vertical line within the big circle. This shows that it came from the first neutron in the isotope 1H3 because they both have the same mass and they are drawn in the same way.

To find the mass that is lost creating the isotope 2He3, add the masses of the isotopes 1H3 and 1H1 and subtract the mass of the isotope 2He3 from their total.

Mass of 1H3	3.01605 amu's
Mass of 1H1	1.00783 amu's
Total mass of 1H3 and 1H1	4.02388 amu's

Total mass of 1H3 and 1H1	4.02388 amu's
Mass of 2He3	3.01603 amu's
Mass lost creating 2He3	1.00785 amu's

The mass lost creating the isotope 2He3 includes the mass of the second neutron that dropped out of the isotope 1H3. Subtract this neutron's mass from the total mass lost creating the isotope 2He3 to find the real mass lost.

Mass lost creating 2He3	1.00785 amu's
Mass of second neutron dropped out of 1H3	1.00194 amu's
Real mass lost creating 2He3	0.00591 amu

In this step, the isotope 1H1 has its proton lose 0.00591 amu to become the positon in the isotope 2He3. Subtract the lost mass from the mass of the proton to find the mass of the positon in the isotope 2He3.

Mass of proton in 1H1	1.00728 amu's
Mass lost by it	0.00591 amu
Mass of positon in 2He3	1.00137 amu's

The first neutron with 1.00628 amu's in the isotope 1H3 is used as the single neutron in the isotope 2He3 and does not lose any mass.

The second neutron with 1.00194 amu's drops out of the isotope 1H3 because it is not needed when the isotope 2He3 is created.

The electron in the isotope 1H3 is part of the isotope 1H2 found within the isotope 1H3 and 2He3. The electron in the isotope 1H1 has its proton lose 0.00591 amu which does not release an electron so this electron is found in the isotope 2He3 as a negatron. The proton in the isotope 1H3 is the proton found in the isotope 1H2 which is used as is in the isotope 1H3 and 2He3. The proton in the isotope 1H1 loses 0.00591 amu and becomes the positon in the isotope 2He3.

The proton in the isotope 1H1 lost 0.00591 amu for a total lost mass of 0.00591 amu and it is lost as a single unit. The lost mass of 0.00591 amu is composed of four sub-units. They are sub-units of 0.00434 amu, 0.00100 amu, 0.00055 amu and 0.00002 amu. The sub-unit of 0.00434 amu is composed of two sub-units of 0.00217 amu. In order for the electron or negatron to stay with its proton or positon, these sub-units must be lost as a single unit of 0.00591 amu. The reader is to keep in mind that in other steps the lost mass of 0.00591 amu could be composed of a sub-unit of 0.00374 amu and a sub-unit of 217 amu.

By adding all of the now known masses of the isotope 2He3, the mass of the isotope 2He3 can be found.

Mass of negatron in 2He3	0.00055 amu
Mass of positon in 2He3	1.00137 amu's
Mass of 1H2 in 2He3	2.01411 amu's
Mass of 2He3	3.01603 amu's

In these steps, the mass lost creating the isotope which includes the neutron that dropped out is 1.00785 amu's. Note that this is the mass of the isotope 1H1 plus the mass of 0.00002 amu which is lost as a sub-unit. In **STEP 1**, the isotope 1H1 is not used to create the isotope 2He3 while the 0.00002 amu is the only mass lost. In these steps, there is one neutron too many so one must drop out. The isotope 1H1 is not lost or caused to drop out. The remaining mass must be accounted for so the fact that 1.00785 amu's is needed to create the isotope 2He3 does not mean that the isotope 1H1 with 1.00783 amu's plus the 0.00002 amu is the lost mass under consideration.

All drawings in **STEP 3** creating the isotope 2He3 from isotopes of 1H3 and 1H1

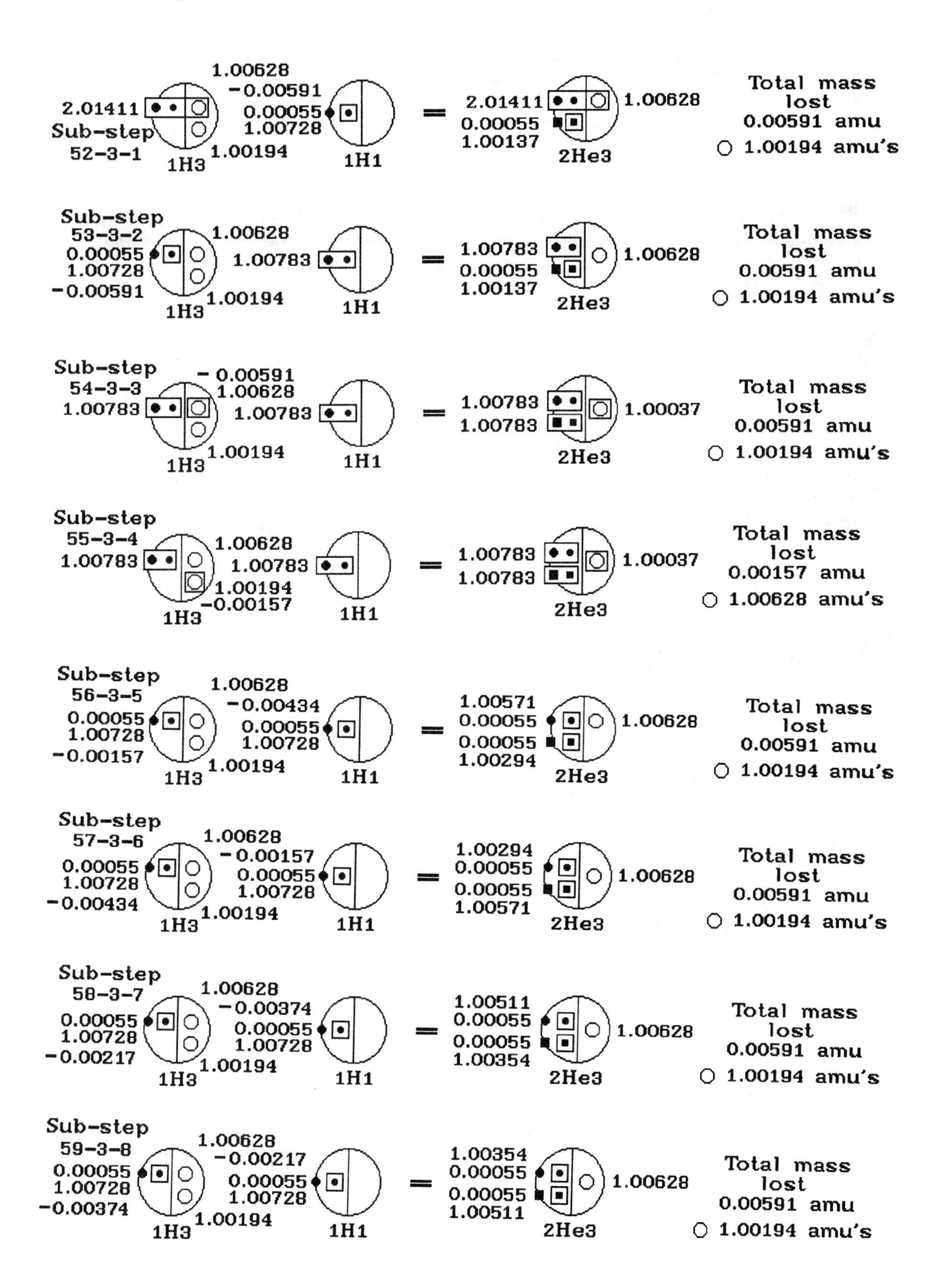

All drawings in **STEP 3** creating the isotope 2He3 from isotopes of 1H3 and 1H1

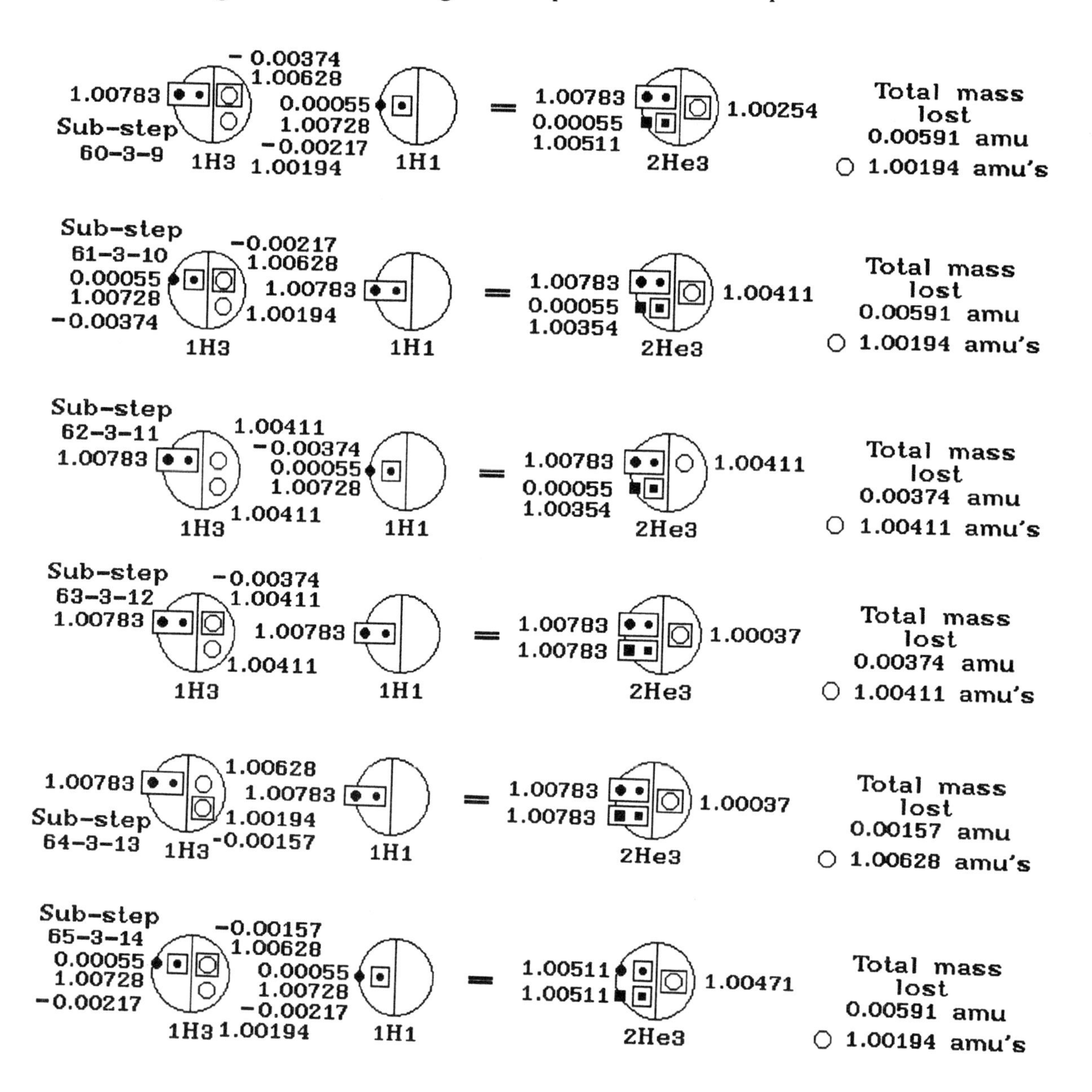

STEP 2

2He3

FORMULA

1H3 + 1H2 = 2He3

STEPS

1. 1H3 + 1H2 = 2He3

17-2-1 1H3 + 1H2 = 2He3, 591 ray, 1.00628/1.00194 neutrons

18-2-2 1H3 + 1H2 = 2He3, 591 ray, 1.00628/1.00194 neutrons

19-2-3 1H3 + 1H2 = 2He3, 591 ray, 1.00628/1.00194 neutrons

20-2-4 1H3 + 1H2 = 2He3, 591 ray, 1.00628/1.00194 neutrons

21-2-5 1H3 + 1H2 = 2He3, 591 ray, 1.00628/1.00194 neutrons

22-2-6 1H3 + 1H2 = 2He3, 591 ray, 1.00628/1.00194 neutrons

23-2-7 1H3 + 1H2 = 2He3, 434 ray, 157 ray, 1.00628/1.00194

24-2-8 1H3 + 1H2 = 2He3, 374 ray, 217 ray, 1.00628/1.00194

25-2-9 1H3 + 1H2 = 2He3, 434 ray, 157 ray, 1.00628/1.00194

26-2-10 1H3 + 1H2 = 2He3, 374 ray, 217 ray, 1.00628/1.00194

27-2-11 1H3 + 1H2 = 2He3, 157 ray, 434 ray, 1.00628/1.00194

28-2-12 1H3 + 1H2 = 2He3, 217 ray, 374 ray, 1.00628/1.00194

29-2-13 1H3 + 1H2 = 2He3, 157 ray, 434 ray, 1.00628/1.00194

30-2-14 1H3 + 1H2 = 2He3, 217 ray, 374 ray, 1.00628/1.00194

31-2-15 1H3 + 1H2 = 2He3, 434 ray, 157 ray, 1.00628/1.00194

32-2-16 1H3 + 1H2 = 2He3, 157 ray, 434 ray, 1.00628/1.00194

33-2-17 1H3 + 1H2 = 2He3, 157 ray, 434 ray, 1.00628/1.00194

34-2-18 1H3 + 1H2 = 2He3, 374 ray, 217 ray, 1.00628/1.00194

35-2-19 1H3 + 1H2 = 2He3, 217 ray, 374 ray, 1.00628/1.00194

36-2-20 1H3 + 1H2 = 2He3, 434 ray, 157 ray, 1.00628/1.00194

37-2-21 1H3 + 1H2 = 2He3, 591 ray, 1.00628/1.00194

38-2-22 1H3 + 1H2 = 2He3, 591 ray, 1.00628/1.00194

39-2-23 1H3 + 1H2 = 2He3, 157 ray, 1.00628/1.00628

40-2-24 1H3 + 1H2 = 2He3, 374 ray, 1.00411/1.00628

41-2-25 1H3 + 1H2 = 2He3, 157 ray, 217 ray, 1.00411/1.00628

42-2-26 1H3 + 1H2 = 2He3, 217 ray, 157 ray, 1.00411/1.00628

43-2-27 1H3 + 1H2 = 2He3, 591 ray, 1.00411/1.00411

STEP 2 1H3 + 1H2 = 2He3

44-2-28 1H3 + 1H2 = 2He3, 157 ray, 434 ray, 1.00411/1.00411
45-2-29 1H3 + 1H2 = 2He3, 217 ray, 374 ray, 1.00411/1.00411
46-2-30 1H3 + 1H2 = 2He3, 374 ray, 217 ray, 1.00411/1.00411
47-2-31 1H3 + 1H2 = 2He3, 434 ray, 157 ray, 1.00411/1.00411
48-2-32 1H3 + 1H2 = 2He3, 157 ray, 434 ray, 1.00411/1.00411
49-2-33 1H2 + 1H1 = 2He3, 217 ray, 374 ray, 1.00411/1.00411
50-2-34 1H2 + 1H2 = 2He3, 374 ray, 217 ray, 1.00411/1.00411
51-2-35 1H2 + 1H2 = 2He3, 434 ray, 157 ray, 1.00411/1.00411

Sub-step 51-2-35

2He3

1H3 + 1H2 = 2He3, alpha 434 ray, gamma 157 ray, 1.00411 neutron, 1.00411 neutron

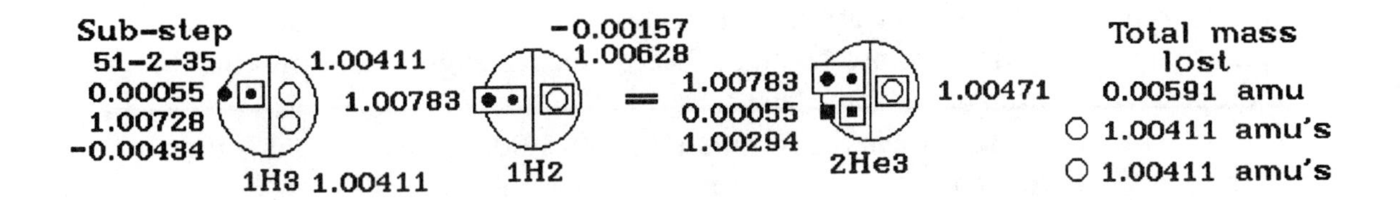

The isotope 1H3 with 3.01605 amu's has one electron with 0.00055 amu and one proton with 1.00728 amu's plus two neutrons. Both neutrons have 1.00411 amu's.

The isotope 1H2 with 2.01411 amu's has one electron with 0.00055 amu and one proton with 1.00728 amu's and one neutron with 1.00628 amu's.

The isotope 2He3 with 3.01603 amu's has one electron with 0.00055 amu, one negatron with 0.00055 amu, one proton with 1.00728 amu's, one positon with 1.00294 amu's and one neutron with 1.00471 amu's.

The total mass lost is 0.00591 amu and it is emitted as a 434 alpha ray and a 157 gamma ray. The 434 alpha ray which has an external magnetic field is emitted from the proton in the isotope 1H3 and the 157 gamma ray which does not have an external magnetic field is emitted from the single neutron in the isotope 1H2. There are two neutrons that drop out from the isotope 1H3 with each having 1.00411 amu's.

The isotope 1H3 with 3.01605 amu's has one electron with 0.00055 amu that is drawn as a dot on the big circle and it becomes the negatron with 0.00055 amu in the isotope 2He3 and is drawn as a square on the big circle.

The proton with 1.00728 amu's in the isotope 1H3 is drawn as a dot to the left of the vertical line within the big circle. There is a square drawn around the dot that is the proton in the isotope 1H3 and a square drawn around the small square of the positon in the isotope 2He3. These squares show that the proton in the isotope 1H3 lost 0.00434 amu to become the positon in the isotope 2He3 with 1.00294 amu's.

There are two neutrons in the isotope 1H3 that are drawn one above the other as small circles to the right of the vertical line within the big circle. The top neutron has 1.00411 amu's and is called the first neutron while the neutron below it is the second neutron and it has 1.00411 amu's. Neither of the two neutrons in the isotope 1H3 is needed when the isotope 2He3 is created so they drop out.

The isotope 1H2 has 2.01411 amu's. Its single electron is drawn as a dot on the big circle in the isotopes 1H2 and it is drawn as a dot on the big circle in the isotope 2He3 to show that the electron stays the electron with 0.00055 amu.

The proton in the isotope 1H2 has 1.00728 amu's and it is drawn as a dot to the left of the vertical line within the big circle in the isotopes 1H2 and it is drawn as a dot in the isotope 2He3 to show that the proton in the isotope 1H2 stays a proton in the isotope 2He3. There is a rectangle drawn around the electron and proton to show that they are used as the isotope 1H1 within the isotope 1H2 and around the electron and proton in the 2He3. The rectangle also shows that the isotope 1H1 can drop out at some later time.

There is one neutron in the isotope 1H2 with 1.00628 amu's and it is drawn as a small circle to the right of the vertical line within the big circle. This neutron loses 0.00157 amu to become the single neutron in the isotope 2He3 with 1.00471 amu's.

To find the mass that is lost creating the isotope 2He3, add the masses of the isotopes 1H3 and 1H2 and subtract the mass of the isotope 2He3 from their total.

Mass of 1H3	3.01605 amu's
Mass of 1H2	2.01411 amu's
Total mass of 1H3 and 1H2	5.03016 amu's

Total mass of 1H3 and 1H2	5.03016 amu's
Mass of 2He3	3.01603 amu's
Mass lost creating 2He3	2.01413 amu's

The mass lost creating the isotope 2He3 includes the mass of the two neutrons found in the isotope 1H3 that dropped out. Subtract their masses from the total mass lost creating the isotope 2He3 to find the real mass lost.

Mass of first neutron in 1H3	1.00411 amu's
Mass of second neutron in 1H3	1.00411 amu's
Mass of two neutrons dropping out of 1H3	2.00822 amu's

Mass lost creating 2He3	2.01413 amu's
Mass of two neutrons that dropped out of 1H3	2.00822 amu's
Real mass lost creating 2He3	0.00591 amu

The isotope 1H3 has its electron become the negatron and the proton becomes the positon in the isotope 2He3. To find the mass of the positon in the isotope 2He3, subtract the 0.00434 amu the proton lost.

Mass of proton in 1H3	1.00728 amu's
Mass lost by it	0.00434 amu
Mass of positon in 2He3	1.00294 amu's

The mass lost by the single neutron in the isotope 1H2 can be found by subtracting the mass the proton in the isotope 1H3 lost from the total mass lost creating the isotope 2He3.

Total mass lost creating 2He3	0.00591 amu
Mass lost by proton in 1H3	0.00434 amu
Mass of single neutron in 2He3	0.00157 amu

This single neutron with 1.00628 amu's in the isotope 1H2 loses 0.00157 amu to become the single neutron in the isotope 2He3. To find the mass of the single neutron in the isotope 2He3, subtract the mass the proton lost.

Mass of neutron in 1H2	1.00628 amu's
Mass lost by it	0.00157 amu
Mass of neutron in 2He3	1.00471 amu's

The electrons in the isotopes 1H3 and 1H2 are not lost or released because no 0.00100 amu has been transferred or emitted as a single unit and no 0.00002 amu has been lost creating the isotope 2He3.

The proton in the isotope 1H3 lost a unit of 0.00434 amu and the neutron in the isotope 1H2 lost a unit of 0.00157 amu for a total lost mass of 0.00591 amu. They are each lost as single units. The lost mass of 0.00434 is composed of a two sub-units. They are two sub-units of 0.00217 amu. The lost mass of 0.00157 amu is composed of three sub-units. They are sub-units of 0.00100 amu, 0.00055 amu and 0.00002 amu.

In order for the electron or negatron to stay with their proton or positon, these sub-units must be lost as a single unit of 0.00434 or 0.00157 amu. They must also be lost as single units in order for the proton, positon or neutron not to release an electron or negatron from within any of them.

By adding all of the now known masses of the isotope 2He3, the mass of the isotope 2He3 can be found.

Mass of 1H1 from 1H2 intact within 2He3	1.00783 amu's
Mass of positon 2He3	1.00294 amu's
Mass of negatron in 2He3	0.00055 amu
Mass of neutron in 2He3	1.00471 amu's
Mass of 2He3	3.01603 amu's

Sub-step 50-2-34

2He3

1H3 + 1H2 = 2He3, alpha 374 ray, gamma 217 ray, 1.00411 neutron, 1.00411 neutron

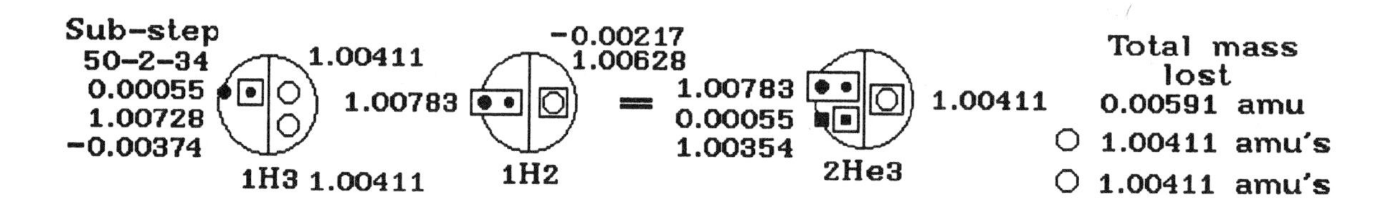

The isotope 1H3 with 3.01605 amu's has one electron with 0.00055 amu and one proton with 1.00728 amu's plus two neutrons. Both neutrons have 1.00411 amu's.

The isotope 1H2 with 2.01411 amu's has one electron with 0.00055 amu and one proton with 1.00728 amu's and one neutron with 1.00628 amu's.

The isotope 2He3 with 3.01603 amu's has one electron with 0.00055 amu, one negatron with 0.00055 amu, one proton with 1.00728 amu's, one positon with 1.00354 amu's and one neutron with 1.00411 amu's.

The total mass lost is 0.00591 amu and it is emitted as a 374 alpha ray and a 217 gamma ray. The 374 alpha ray has an external magnetic field is emitted from the proton in the isotope 1H3 and the 217 gamma ray does not have an external magnetic field is emitted from the single neutron in the isotope 1H2. There are two neutrons that drop out from the isotope 1H3 with each having 1.00411 amu's.

The isotope 1H3 with 3.01605 amu's has one electron with 0.00055 amu that is drawn as a dot on the big circle and it becomes the negatron with 0.00055 amu in the isotope 2He3 and is drawn as a square on the big circle.

The proton with 1.00728 amu's in the isotope 1H3 is drawn as a dot to the left of the vertical line within the big circle. There is a square drawn around the dot that is the proton in the isotope 1H3 and a square drawn around the small square of the positon in the isotope 2He3. These squares show that the proton with 1.00728 amu's in the isotope 1H3 lost 1.00374 amu to become the positon in the isotope 2He3 with 1.00354 amu's.

There are two neutrons in the isotope 1H3 that are drawn one above the other as small circles to the right of the vertical line within the big circle. The top neutron has 1.00411 amu's and is called the first neutron while the neutron below it is the second neutron and it has 1.00411 amu's. Neither of the two neutrons in the isotope 1H3 is needed when the isotope 2He3 is created so they drop out.

The isotope 1H2 has 2.01411 amu's. Its single electron is drawn as a dot on the big circle in the isotope 1H2 and it is drawn as a dot on the big circle in the isotope 2He3 to show that the electron stays the electron with 0.00055 amu.

The proton in the isotope 1H2 has 1.00728 amu's and it is drawn as a dot to the left of the vertical line within the big circle and it is drawn as a dot in the isotope 2He3 to show that the proton in the isotope 1H2 stays a proton in the isotope 2He3. There is a rectangle drawn around the electron and proton to show that they are used as the isotope 1H1 within the isotopes 1H2 and 2He3. The rectangle also shows that the isotope 1H1 can drop out at some later time.

There is one neutron in the isotope 1H2 with 1.00628 amu's and it is drawn as a small circle to the right of the vertical line within the big circle. This neutron loses 0.00217 amu to become the single neutron in the isotope 2He3 with 1.00411 amu's. There is a square drawn around the small circle in the isotope 1H2 and 2He3 to show that 0.00217 amu has been lost by the single neutron in the isotope 1H2 as it became the single neutron in the isotope 2He3.

To find the mass that is lost creating the isotope 2He3, add the masses of the isotopes 1H3 and 1H2 and subtract the mass of the isotope 2He3 from their total.

Mass of 1H3	3.01605 amu's
Mass of 1H2	2.01411 amu's
Total mass of 1H3 and 1H2	5.03016 amu's

Total mass of 1H3 and 1H2	5.03016 amu's
Mass of 2He3	3.01603 amu's
Mass lost creating 2He3	2.01413 amu's

The mass lost creating the isotope 2He3 includes the mass of the two neutrons found in the isotopes 1H3 that dropped out. Subtract their masses from the total mass lost creating the isotope 2He3 to find the real mass lost.

Mass of first neutron in 1H3	1.00411 amu's
Mass of second neutron in 1H3	1.00411 amu's
Mass of two neutrons dropping out of 1H3	2.00822 amu's

Mass lost creating 2He3	2.01413 amu's
Mass of two neutrons that dropped out of 1H3	2.00822 amu's
Real mass lost creating 2He3	0.00591 amu

The isotope 1H3 has its electron become the negatron and the proton becomes the positon in the isotope 2He3. To find the mass of the positon in the isotope 2He3, subtract the 0.00374 amu the proton lost.

Mass of proton in 1H3	1.00728 amu's
Mass lost by it	0.00374 amu
Mass of positon in 2He3	1.00354 amu's

Since the proton in the isotope 1H3 lost 0.00374 amu, the mass lost by the single neutron in the isotope 1H2 can be found by subtracting the proton's lost mass from the total mass lost creating the isotope 2He3.

Total mass lost creating 2He3	0.00591 amu
Mass lost by proton in 1H3	0.00374 amu
Mass lost by neutron in 1H2	0.00217 amu

This single neutron with 1.00628 amu's in the isotope 1H2 loses 0.00217 amu to become the single neutron in the isotope 2He3. To find the mass of the single neutron in the isotope 2He3, subtract the lost mass from it.

Mass of neutron in 1H2	1.00628 amu's
Mass lost by it	0.00217 amu
Mass of neutron in 2He3	1.00411 amu's

The electrons in the isotopes 1H3 and 1H2 are not lost or released because no 0.00100 amu has been transferred or emitted as a single unit and no 0.00002 amu has been lost creating the isotope 2He3.

The proton in the isotope 1H3 lost a unit of 0.00374 amu and the neutron in the isotope 1H2 lost a unit of 0.00217 amu for a total lost mass of 0.00591 amu. They are each lost as single units. The lost mass of 0.00217 amu does not have any sub-units. The lost mass of 0.00374 amu is composed of two sub-units. They are sub-units of 0.00217 amu and 0.00157 amu. The sub-unit of 0.00157 is composed of three sub-units. They are 0.00100 amu, 0.00055 amu and 0.00002 amu. Note that there are two mass units of 0.00217 amu. One is a single unit while the other is a sub-unit within a unit.

In order for the electron or negatron to stay with their proton or positon, these sub-units must be lost as a single unit of 0.00374 or 0.00217 amu. They must also be lost as single units in order for the proton, positon or neutron not to release an electron or negatron from within any of them.

By adding all of the now known masses of the isotope 2He3, the mass of the isotope 2He3 can be found.

Mass of 1H1 from 1H2 intact within 2He3	1.00783 amu's
Mass of positon 2He3	1.00354 amu's
Mass of negatron in 2He3	0.00055 amu
Mass of neutron in 2He3	1.00411 amu's
Mass of 2He3	3.01603 amu's

Sub-step 49-2-33

2He3

1H3 + 1H2 = 2He3, alpha 217 ray, gamma 374 ray, 1.00411 neutron, 1.00411 neutron

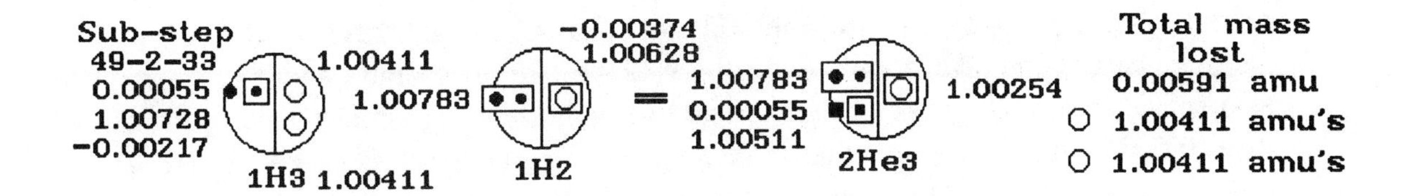

The isotope 1H3 with 3.01605 amu's has one electron with 0.00055 amu and one proton with 1.00728 amu's plus two neutrons. Both neutrons have 1.00411 amu's.

The isotope 1H2 with 2.01411 amu's has one electron with 0.00055 amu and one proton with 1.00728 amu's and one neutron with 1.00628 amu's.

The isotope 2He3 with 3.01603 amu's has one electron with 0.00055 amu, one negatron with 0.00055 amu, one proton with 1.00728 amu's, one positon with 1.00511 amu's and one neutron with 1.00254 amu's.

The total mass lost is 0.00591 amu and it is emitted as a 217 alpha ray and a 374 gamma ray. The 217 alpha ray which has an external magnetic field is emitted from the proton in the isotope 1H3 and the 374 gamma ray which has no external magnetic field is emitted from the single neutron in the isotope 1H2. There are two neutrons that drop out from the isotope 1H3 with each having 1.00411 amu's.

The isotope 1H3 with 3.01605 amu's has one electron with 0.00055 amu that is drawn as a dot on the big circle and it becomes the negatron with 0.00055 amu in the isotope 2He3 and is drawn as a square on the big circle.

The proton with 1.00728 amu's in the isotope 1H3 is drawn as a dot to the left of the vertical line within the big circle. There is a square drawn around the dot that is the proton in the isotope 1H3 and a square drawn around the small square of the positon in the isotope 2He3. These squares show that the proton in the isotope 1H3 lost 0.00217 amu to become the positon in the isotope 2He3 with 1.00511 amu's.

There are two neutrons in the isotope 1H3 that are drawn one above the other as small circles to the right of the vertical line within the big circle. The top neutron has 1.00411 amu's and is called the first neutron while the neutron below it is the second neutron and it has 1.00411 amu's. Neither of the two neutrons in the isotope 1H3 is needed when the isotope 2He3 is created so they drop out.

The isotope 1H2 has 2.01411 amu's. Its single electron is drawn as a dot on the big circle and it is drawn as a dot on the big circle in the isotope 2He3 to show that the electron stays the electron with 0.00055 amu.

The proton in the isotope 1H2 has 1.00728 amu's and it is drawn as a dot to the left of the vertical line within the big circle and it is drawn as a dot in the isotope 2He3 to show that the proton in the isotope 1H2 stays a proton in the isotope 2He3. There is a rectangle drawn around the electron and proton to show that they are used as the isotope 1H1 within the isotopes 1H2 and 2He3. The rectangle also shows that the isotope 1H1 can drop out at some later time.

There is one neutron in the isotope 1H2 with 1.00628 amu's and it is drawn as a small circle to the right of the vertical line within the big circle. This neutron loses 0.00374 amu to become the single neutron in the isotope 2He3 with 1.00254 amu's. There is a square drawn around the single neutron in the isotope 1H2 and the single neutron in the isotope 2He3 to show that the neutron in the isotope 1H2 lost 0.00374 amu and that it became the single neutron in the isotope 2He3.

To find the mass that is lost creating the isotope 2He3, add the masses of the isotopes 1H3 and 1H2 and subtract the mass of the isotope 2He3 from their total.

Mass of 1H3	3.01605 amu's
Mass of 1H2	2.01411 amu's
Total mass of 1H3 and 1H2	5.03016 amu's

Total mass of 1H3 and 1H2	5.03016 amu's
Mass of 2He3	3.01603 amu's
Mass lost creating 2He3	2.01413 amu's

The mass lost creating the isotope 2He3 includes the mass of the two neutrons found in the isotopes 1H3 that dropped out. Subtract their masses from the total mass lost creating the isotope 2He3 to find the real mass lost.

Mass of first neutron in 1H3	1.00411 amu's
Mass of second neutron in 1H3	1.00411 amu's
Mass of two neutrons dropping out of 1H3	2.00822 amu's

Mass lost creating 2He3	2.01413 amu's
Mass of two neutrons that dropped out of 1H3	2.00822 amu's
Real mass lost creating 2He3	0.00591 amu

The isotope 1H3 has its electron become the negatron and the proton becomes the positon in the isotope 2He3. To find the mass of the positon in the isotope 2He3, subtract the 0.00217 amu the proton lost.

Mass of proton in 1H3	1.00728 amu's
Mass lost by it	0.00217 amu
Mass of positon in 2He3	1.00511 amu's

Since the proton in the isotope 1H3 lost 0.00217 amu, the mass lost by the single neutron in the isotope 1H2 can be found by subtracting the lost mass of the proton from the total mass lost creating the isotope 2He3.

Total mass lost creating 2He3	0.00591 amu
Mass lost by proton in 1H3	0.00217 amu
Mass lost by neutron in 1H2	0.00374 amu

This single neutron with 1.00628 amu's in the isotope 1H2 loses 0.00374 amu to become the single neutron in the isotope 2He3. To find the mass of the single neutron in the isotope 2He3, subtract the lost mass from it.

Mass of neutron in 1H2	1.00628 amu's
Mass lost by it	0.00374 amu
Mass of neutron in 2He3	1.00254 amu's

The electrons in the isotopes 1H3 and 1H2 are not lost or released because no 0.00100 amu has been transferred or emitted as a single unit and no 0.00002 amu has been lost creating the isotope 2He3.

The proton in the isotope 1H3 lost a unit 0.00217 amu and the neutron in the isotope 1H2 lost a unit of 0.00374 amu for a total lost mass of 0.00591 amu. They are each lost as single units. The lost mass of 0.00217 does not have any sub-units. The lost mass of 0.00374 amu is composed of two sub-units. They are sub-units of 0.00217 amu and 0.00157 amu. The sub-unit of 0.00157 is composed of three sub-units. They are 0.00100 amu, 0.00055 amu and 0.00002 amu. Note that there are two mass of 0.00217 amu. One is a single unit while the other is a sub-unit within a unit.

In order for the electron or negatron to stay with their proton or positon, these sub-units must be lost as a single unit of 0.00217 or 0.00374 amu. They must also be lost as single units in order for the proton, positon or neutron to keep them from releasing an electron or negatron from within any of them.

By adding all of the now known masses of the isotope 2He3, the mass of the isotope 2He3 can be found.

Mass of 1H1 from 1H2 intact within 2He3	1.00783 amu's
Mass of positon 2He3	1.00511 amu's
Mass of negatron in 2He3	0.00055 amu
Mass of neutron in 2He3	1.00254 amu's
Mass of 2He3	3.01603 amu's

Sub-step 48-2-32

2He3

1H3 + 1H2 = 2He3, alpha 157 ray, gamma 434 ray, 1.00411 neutron, 1.00411 neutron

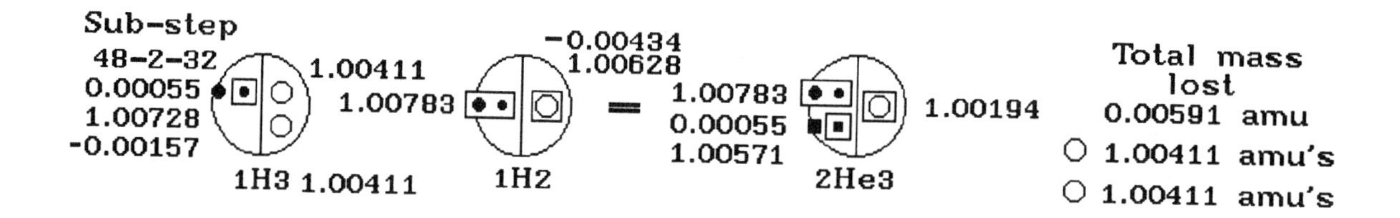

The isotope 1H3 with 3.01605 amu's has one electron with 0.00055 amu and one proton with 1.00728 amu's plus two neutrons. Both neutrons have 1.00411 amu's.

The isotope 1H2 with 2.01411 amu's has one electron with 0.00055 amu and one proton with 1.00728 amu's and one neutron with 1.00628 amu's.

The isotope 2He3 with 3.01603 amu's has one electron with 0.00055 amu, one negatron with 0.00055 amu, one proton with 1.00728 amu's, one positon with 1.00571 amu's and one neutron with 1.00194 amu's.

The total mass lost is 0.00591 amu and it is emitted as a 157 alpha ray and a 434 gamma ray. The 157 alpha ray which has an external magnetic field is emitted from the proton in the isotope 1H3 and the 434 gamma ray has no magnetic field is emitted from the single neutron in the isotope 1H2. There are two neutrons that drop out from the isotope 1H3 with each having 1.00411 amu's.

The isotope 1H3 with 3.01605 amu's has one electron with 0.00055 amu that is drawn as a dot on the big circle and it becomes the negatron with 0.00055 amu in the isotope 2He3 and is drawn as a square on the big circle.

The proton with 1.00728 amu's in the isotope 1H3 is drawn as a dot to the left of the vertical line within the big circle. There is a square drawn around the dot that is the proton in the isotope 1H3 and a square drawn around the small square of the positon in the isotope 2He3. These squares show that the proton in the isotope 1H3 lost 0.00157 amu to become the positon in the isotope 2He3 with 1.00571 amu's.

There are two neutrons in the isotope 1H3 that are drawn one above the other as small circles to the right of the vertical line within the big circle. The top neutron has 1.00411 amu's and is called the first neutron while the neutron below it is the second neutron and it has 1.00411 amu's. Neither of the two neutrons in the isotope 1H3 is needed when the isotope 2He3 is created so they drop out.

The isotope 1H2 has 2.01411 amu's. Its single electron is drawn as a dot on the big circle in the isotope 1H2 and it is drawn as a dot on the big circle in the isotope 2He3 to show that the electron stays the electron with 0.00055 amu.

The proton in the isotope 1H2 has 1.00728 amu's and it is drawn as a dot to the left of the vertical line within the big circle in the isotope 1H2 and it is drawn as a dot in the isotope 2He3 to show that the proton in the isotope 1H2 stays a proton in the isotope 2He3. There is a rectangle drawn around the electron and proton to show that they are used as the isotope 1H1 within the isotopes 1H2 and 2He3. The rectangle also shows that the isotope 1H1 can drop out of the isotope 2He3 at some later time.

There is one neutron in the isotope 1H2 with 1.00628 amu's and it is drawn as a small circle to the right of the vertical line within the big circle. This neutron loses 0.00434 amu to become the single neutron in the isotope 2He3 with 1.00194 amu's. There is a square drawn around the single neutron in the isotope 1H2 to show that it has lost 0.00434 amu and a square drawn around the single neutron in the isotope 2He3 to show that it came from the single neutron in the isotope 1H2.

To find the mass that is lost creating the isotope 2He3, add the masses of the isotopes 1H3 and 1H2 and subtract the mass of the isotope 2He3 from their total.

Mass of 1H3	3.01605 amu's
Mass of 1H2	2.01411 amu's
Total mass of 1H3 and 1H2	5.03016 amu's

Total mass of 1H3 and 1H2	5.03016 amu's
Mass of 2He3	3.01603 amu's
Mass lost creating 2He3	2.01413 amu's

The mass lost creating the isotope 2He3 includes the mass of the two neutrons found in the isotopes 1H3 that dropped out. Subtract their masses from the total mass lost creating the isotope 2He3 to find the real mass lost.

Mass of first neutron in 1H3	1.00411 amu's
Mass of second neutron in 1H3	1.00411 amu's
Mass of two neutrons dropping out of 1H3	2.00822 amu's

Mass lost creating 2He3	2.01413 amu's
Mass of two neutrons that dropped out of 1H3	2.00822 amu's
Real mass lost creating 2He3	0.00591 amu

The isotope 1H3 has its electron become the negatron and the proton becomes the positon in the isotope 2He3. To find the mass of the positon in the isotope 2He3, subtract the 0.00157 amu the proton lost.

Mass of proton in 1H3	1.00728 amu's
Mass lost by it	0.00157 amu
Mass of positon in 2He3	1.00571 amu's

Since the proton in the isotope 1H3 lost 0.00157 amu, the mass lost by the single neutron in the isotope 1H2 can be found by subtracting the mass lost by the proton from the total mass lost creating the isotope 2He3.

Total mass lost creating 2He3	0.00591 amu
Mass lost by proton in 1H3	0.00157 amu
Mass lost by neutron in 1H2	0.00434 amu

This single neutron with 1.00628 amu's in the isotope 1H2 loses 0.00434 amu to become the single neutron in the isotope 2He3. To find the mass of the single neutron in the isotope 2He3, subtract the lost mass from it.

Mass of neutron in 1H2	1.00628 amu's
Mass lost by it	0.00434 amu
Mass of neutron in 2He3	1.00194 amu's

The electrons in the isotopes 1H3 and 1H2 are not lost or released because no 0.00100 amu has been transferred or emitted as a single unit and no 0.00002 amu has been lost creating the isotope 2He3.

The proton in the isotope 1H3 lost a unit of 0.00157 amu and the neutron in the isotope 1H2 lost a unit of 0.00434 amu for a total lost mass of 0.00591 amu. They are each lost as single units. The lost mass of 0.00157 is composed of three sub-units. They are sub-units of 0.00100 amu, 0.00055 amu and 0.00002 amu. The 0.00434 amu is composed of two sub-units of 0.00217 amu. In order for the electron or negatron to stay with their proton or positon, these sub-units must be lost as a single unit of 0.00157 or 0.00434 amu. They must also be lost as single units in order for the proton, positon or neutron to keep their electron or negatron.

By adding all of the now known masses of the isotope 2He3, the mass of the isotope 2He3 can be found.

Mass of 1H1 from 1H2 intact within 2He3	1.00783 amu's
Mass of positon 2He3	1.00571 amu's
Mass of negatron in 2He3	0.00055 amu
Mass of neutron in 2He3	1.00194 amu's
Mass of 2He3	3.01603 amu's

Sub-step 47-2-31

2He3

1H3 + 1H2 = 2He3, alpha 434 ray, gamma 157 ray, 1.00411 neutron, 1.00411 neutron

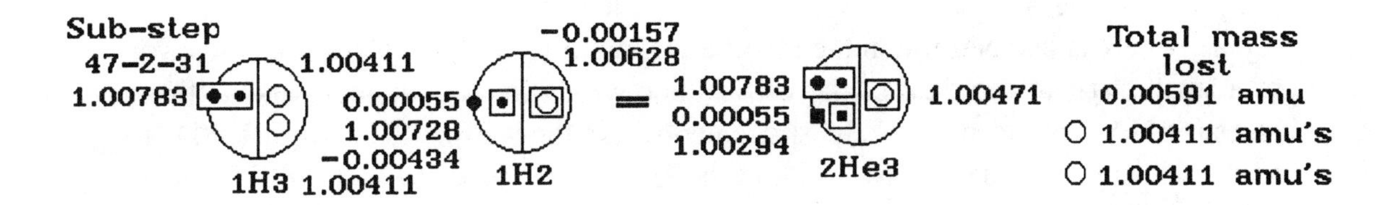

The isotope 1H3 with 3.01605 amu's has one electron with 0.00055 amu and one proton with 1.00728 amu's plus two neutrons. Both neutrons have 1.00411 amu's.

The isotope 1H2 with 2.01411 amu's has one electron with 0.00055 amu and one proton with 1.00728 amu's and one neutron with 1.00628 amu's.

The isotope 2He3 with 3.01603 amu's has one electron with 0.00055 amu, one negatron with 0.00055 amu, one proton with 1.00728 amu's, one positon with 1.00294 amu's and one neutron with 1.00471 amu's.

The total mass lost is 0.00591 amu and it is emitted as a 434 alpha ray which has an external magnetic field and a 157 gamma ray which has no external magnetic field. They are lost from the proton and a single neutron in the isotope 1H2. There are two neutrons that drop out from the isotope 1H3 with each having 1.00411 amu's.

The isotope 1H3 with 3.01605 amu's has one electron with 0.00055 amu that is drawn as a dot on the big circle and it becomes the electron with 0.00055 amu in the isotope 2He3 and is also drawn as a dot on the big circle.

The proton with 1.00728 amu's in the isotope 1H3 is drawn as a dot to the left of the vertical line within the big circle. The electron and proton have a rectangle drawn around them to show that they are used as the isotope 1H1 with 1.00783 amu's within the isotopes 1H3 and 2He3. The rectangle shows that the isotope 1H1 can drop out of the isotope 2He3 at some later time.

There are two neutrons in the isotope 1H3 that are drawn one above the other as small circles to the right of the vertical line within the big circle. The top neutron has 1.00411 amu's and is called the first neutron while the neutron below it is the second neutron and it has 1.00411 amu's. Neither of the two neutrons in the isotope 1H3 is needed when the isotope 2He3 is created so they drop out.

The isotope 1H2 has 2.01411 amu's. Its single electron is drawn as a dot on the big circle in the isotope 1H2 and it is drawn as a square on the big circle in the isotope 2He3 to show that the electron has become the negatron with 0.00055 amu.

The proton in the isotope 1H2 has 1.00728 amu's and it is drawn as a dot to the left of the vertical line within the big circle in the isotope 1H2 and it is drawn as a small square in the isotope 2He3 to show that the proton in the isotope 1H2 has become the positon in the isotope 2He3. The proton loses 0.00434 amu to become the positon in the isotope 2He3 with 1.00294 amu's. There is a square drawn around the proton to show that it has lost 0.00434 amu and there is a square drawn around the positon in the isotope 2He3 to show that this positon was the proton in the isotope 1H2.

There is one neutron in the isotope 1H2 with 1.00628 amu's and it is drawn as a small circle to the right of the vertical line within the big circle. This neutron loses 0.00157 amu to become the single neutron in the isotope 2He3 with 1.00471 amu's. There is a square drawn around the small circle to show that 0.00157 amu has been lost by the neutron. The single neutron in the isotope 2He3 is drawn in the same way to show that this neutron came from the neutron in the isotope 1H2.

To find the mass that is lost creating the isotope 2He3, add the masses of the isotopes 1H3 and 1H2 and subtract the mass of the isotope 2He3 from their total.

Mass of 1H3 3.01605 amu's
Mass of 1H2 2.01411 amu's
Total mass of 1H3 and 1H2 5.03016 amu's

Total mass of 1H3 and 1H2 5.03016 amu's
Mass of 2He3 3.01603 amu's
Mass lost creating 2He3 2.01413 amu's

The mass lost creating the isotope 2He3 includes the mass of the two neutrons found in the isotope 1H3 that dropped out. Subtract their masses from the total mass lost creating the isotope 2He3 to find the real lost mass.

Mass of first neutron in 1H3 1.00411 amu's
Mass of second neutron in 1H3 1.00411 amu's
Mass of two neutrons dropping out of 1H3 2.00822 amu's

Mass lost creating 2He3 2.01413 amu's
Mass of two neutrons that dropped out of 1H3 2.00822 amu's
Real mass lost creating 2He3 0.00591 amu

The isotope 1H3 has its electron and proton used as the isotope 1H1 with 1.00783 amu's within the isotopes 1H3 and 2He3. The electron and proton become the electron and proton in the isotope 2He3.

The isotope 1H2 has its electron become the negatron and the proton becomes the positon in the isotope 2He3. To find the mass of the positon in the isotope 2He3, subtract the 0.00434 amu the proton lost.

Mass of proton in 1H2	1.00728 amu's
Mass lost by it	0.00434 amu
Mass of positon in 2He3	1.00294 amu's

Since the proton in the isotope 1H2 lost 0.00434 amu, the mass lost by the single neutron can be found by subtracting the mass the proton lost from the total mass lost creating the isotope 2He3.

Total mass lost creating 2He3	0.00591 amu
Mass lost by proton in 1H2	0.00434 amu
Mass lost by neutron in 1H2	0.00157 amu

This single neutron with 1.00628 amu's in the isotope 1H2 loses 0.00157 amu to become the single neutron in the isotope 2He3. To find the mass of the single neutron in the isotope 2He3, subtract the lost mass from it.

Mass of neutron in 1H2	1.00628 amu's
Mass lost by it	0.00157 amu
Mass of neutron in 2He3	1.00471 amu's

The electrons in the isotopes 1H3 and 1H2 are not lost or released because no 0.00100 amu has been transferred or emitted as a single unit and no 0.00002 amu has been lost creating the isotope 2He3.

The proton lost a unit of 0.00434 amu and the neutron lost a unit of 0.00157 amu in the isotope 1H2 for a total lost mass of 0.00591 amu. They are each lost as single units. The lost mass of 0.00434 is composed of a two sub-units. They are both sub-units of 0.00217 amu. The lost mass of 0.00157 amu is composed of three sub-units. They are sub-units of 0.00100 amu, 0.00055 amu and 0.00002 amu.

In order for the electron or negatron to stay with their proton or positon, these sub-units must be lost as single units of 0.00434 amu and 0.00157 amu. They must also be lost as single units to keep the proton, positon or neutron from releasing an electron or negatron from within any of them.

By adding all of the now known masses of the isotope 2He3, the mass of the isotope 2He3 can be found.

Mass of 1H1 from 1H3 intact within 2He3	1.00783 amu's
Mass of positon 2He3	1.00294 amu's
Mass of negatron in 2He3	0.00055 amu
Mass of neutron in 2He3	1.00471 amu's
Mass of 2He3	3.01603 amu's

Sub-step 46-2-30

2He3

1H3 + 1H2 = 2He3, alpha 374 ray, gamma 217 ray, 1.00411 neutron, 1.00411 neutron

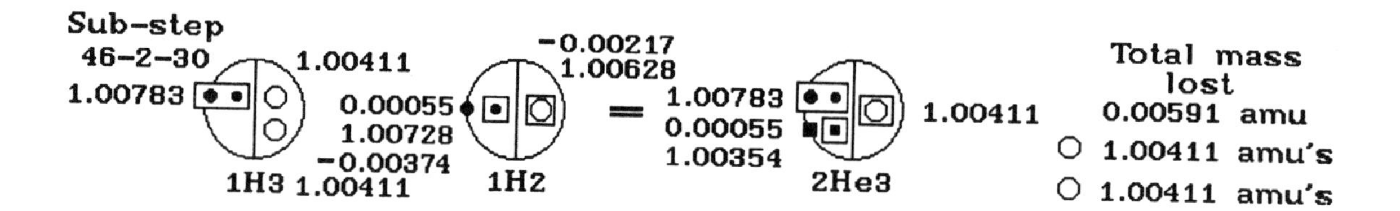

The isotope 1H3 with 3.01605 amu's has one electron with 0.00055 amu and one proton with 1.00728 amu's plus two neutrons. Both neutrons have 1.00411 amu's.

The isotope 1H2 with 2.01411 amu's has one electron with 0.00055 amu and one proton with 1.00728 amu's and one neutron with 1.00628 amu's.

The isotope 2He3 with 3.01603 amu's has one electron with 0.00055 amu, one negatron with 0.00055 amu, one proton with 1.00728 amu's, one positon with 1.00354 amu's and one neutron with 1.00411 amu's.

The total mass lost is 0.00591 amu and it is emitted as a 374 alpha ray which has an external magnetic field and a 217 gamma ray which has no external magnetic field. They are lost from the proton and single neutron in the isotope 1H2. There are two neutrons that drop out from the isotope 1H3 with each having 1.00411 amu's.

It should be noted that there are two neutrons with 1.00411 amu's each dropping out of the isotope 1H3 while another neutron with 1.00411 amu's is being created from the single neutron in the isotope 1H2 to become the single neutron in the isotope 2He3.

The isotope 1H3 with 3.01605 amu's has one electron with 0.00055 amu that is drawn as a dot on the big circle and it becomes the electron with 0.00055 amu in the isotope 2He3 and is also drawn as a dot on the big circle.

The proton with 1.00728 amu's in the isotope 1H3 is drawn as a dot to the left of the vertical line within the big circle. The electron and proton have a rectangle drawn around them to show that they are used as the isotope 1H1 with 1.00783 amu's within the isotopes 1H3 and 2He3. The rectangle shows that the isotope 1H1 can drop out of the isotope 2He3 at some later time.

There are two neutrons in the isotope 1H3 that are drawn one above the other as small circles to the right of the vertical line within the big circle. The top neutron has 1.00411 amu's and is called the first neutron while the neutron below it is the second neutron and it has 1.00411 amu's. Neither of the two neutrons in the isotope 1H3 is needed when the isotope 2He3 is created so they drop out.

The isotope 1H2 has 2.01411 amu's. Its single electron is drawn as a dot on the big circle in the isotope 1H2 and it is drawn as a square on the big circle in the isotope 2He3 to show that the electron has become the negatron with 0.00055 amu.

The proton in the isotope 1H2 has 1.00728 amu's and it is drawn as a dot to the left of the vertical line within the big circle in the isotope 1H2 and it is drawn as a small square in the isotope 2He3 to show that the proton in the isotope 1H2 has become the positon in the isotope 2He3. The proton loses 0.00374 amu to become the positon in the isotope 2He3 with 1.00354 amu's. There is a square drawn around the proton in the isotope 1H2 to show that the proton lost 0.00374 amu and there is a second square drawn around the positon in the isotope 2He3 to show that the positon came from the proton found in the isotope 1H2.

There is one neutron in the isotope 1H2 with 1.00628 amu's and it is drawn as a small circle to the right of the vertical line within the big circle. This neutron loses 0.00217 amu to become the single neutron in the isotope 2He3 with 1.00411 amu's. There is a square drawn around the neutron to show that the neutron has lost 0.00217 amu and there is a square drawn around the neutron in the isotope 2He3 to show that this neutron is the neutron found in the isotope 1H2.

To find the mass that is lost creating the isotope 2He3, add the masses of the isotopes 1H3 and 1H2 and subtract the mass of the isotope 2He3 from their total.

Mass of 1H3	3.01605 amu's
Mass of 1H2	2.01411 amu's
Total mass of 1H3 and 1H2	5.03016 amu's

Total mass of 1H3 and 1H2	5.03016 amu's
Mass of 2He3	3.01603 amu's
Mass lost creating 2He3	2.01413 amu's

The mass lost creating the isotope 2He3 includes the mass of the two neutrons found in the isotopes 1H3 that dropped out. Subtract their masses from the total mass lost creating the isotope 2He3 to find the real mass lost.

Mass of first neutron in 1H3	1.00411 amu's
Mass of second neutron in 1H3	1.00411 amu's
Mass of two neutrons dropping out of 1H3	2.00822 amu's

Mass lost creating 2He3	2.01413 amu's
Mass of two neutrons that dropped out of 1H3	2.00822 amu's
Real mass lost creating 2He3	0.00591 amu

The isotope 1H3 has its electron and proton used as the isotope 1H1 with 1.00783 amu's within the isotopes 1H3 and 2He3. The electron and proton become the electron and proton in the isotope 2He3.

The isotope 1H2 has its electron become the negatron and the proton becomes the positon in the isotope 2He3. To find the mass of the positon in the isotope 2He3, subtract the 0.00374 amu the proton lost.

Mass of proton in 1H2	1.00728 amu's
Mass lost by it	0.00374 amu
Mass of positon in 2He3	1.00354 amu's

Since the proton in the isotope 1H2 lost 0.00374 amu, the mass lost by the single neutron can be found by subtracting the mass the proton lost from the total mass lost creating the isotope 2He3.

Total mass lost creating 2He3	0.00591 amu
Mass lost by proton in 1H2	0.00374 amu
Mass lost by neutron in 1H2	0.00217 amu

This single neutron with 1.00628 amu's in the isotope 1H2 loses 0.00217 amu to become the single neutron in the isotope 2He3. To find the mass of the single neutron in the isotope 2He3, subtract the lost mass from it.

Mass of neutron in 1H2	1.00628 amu's
Mass lost by it	0.00217 amu
Mass of neutron in 2He3	1.00411 amu's

The electrons in the isotopes 1H3 and 1H2 are not lost or released because no 0.00100 amu has been transferred or emitted as a single unit and no 0.00002 amu has been lost creating the isotope 2He3.

The proton in the isotope 1H2 lost a unit of 0.00374 amu and the neutron lost a unit of 0.00217 amu for a total lost mass of 0.00591 amu. They are each lost as single units. The lost mass of 0.00374 amu is composed of two sub-units. They are sub-units of 0.00217 amu and 0.00157 amu. The 0.00217 amu is a single sub-unit while the 0.00157 amu is composed of three sub-units. They are sub-units of 0.00100 amu, 0.00055 amu and 0.00002 amu. There are two sub-units of 0.00217 amu shown and one is a single unit while the other belongs to a unit as a sub-unit.

In order for the electron or negatron to stay with their proton or positon, these sub-units must be lost as single units of 0.00374 amu and 0.00217 amu. They must also be lost as single units to keep the proton, positon or neutron from releasing an electron or negatron from within any of them.

By adding all of the now known masses of the isotope 2He3, the mass of the isotope 2He3 can be found.

Mass of 1H1 from 1H3 intact within 2He3	1.00783 amu's
Mass of positon 2He3	1.00354 amu's
Mass of negatron in 2He3	0.00055 amu
Mass of neutron in 2He3	1.00411 amu's
Mass of 2He3	3.01603 amu's

Sub-step 45-2-29

2He3

1H3 + 1H2 = 2He3, alpha 217 ray, gamma 374 ray, 1.00411 neutron, 1.00411 neutron

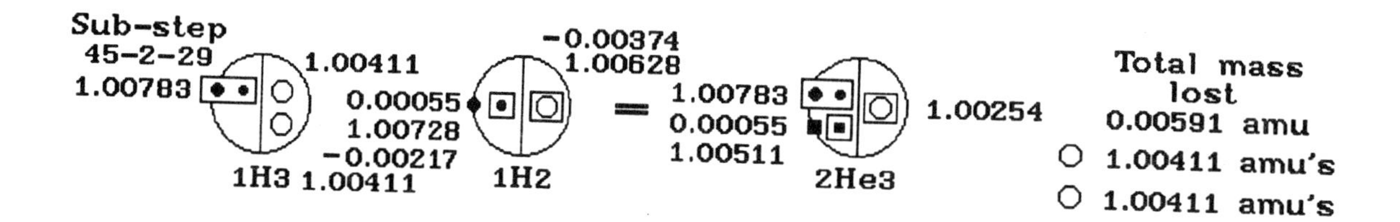

The isotope 1H3 with 3.01605 amu's has one electron with 0.00055 amu and one proton with 1.00728 amu's plus two neutrons. Both neutrons have 1.00411 amu's.

The isotope 1H2 with 2.01411 amu's has one electron with 0.00055 amu and one proton with 1.00728 amu's and one neutron with 1.00628 amu's.

The isotope 2He3 with 3.01603 amu's has one electron with 0.00055 amu, one negatron with 0.00055 amu, one proton with 1.00728 amu's, one positon with 1.00511 amu's and one neutron with 1.00254 amu's.

The total mass lost is 0.00591 amu and it is emitted as a 217 alpha ray which has an external magnetic field and a 374 gamma ray which has no external magnetic field. They are emitted from the proton and a single neutron in the isotope 1H2. There are two neutrons that drop out from the isotope 1H3 with each having 1.00411 amu's.

The isotope 1H3 with 3.01605 amu's has one electron with 0.00055 amu that is drawn as a dot on the big circle and it becomes the electron with 0.00055 amu in the isotope 2He3 and is also drawn as a dot on the big circle.

The proton with 1.00728 amu's in the isotope 1H3 is drawn as a dot to the left of the vertical line within the big circle. The electron and proton have a rectangle drawn around them to show that they are used as the isotope 1H1 with 1.00783 amu's within the isotope 1H3 and 2He3. The rectangle shows that the isotope 1H1 can drop out of the isotope 2He3 at some later time.

There are two neutrons in the isotope 1H3 that are drawn one above the other as small circles to the right of the vertical line within the big circle. The top neutron has 1.00411 amu's and is called the first neutron while the neutron below it is the second neutron and it has 1.00411 amu's. Neither of the two neutrons in the isotope 1H3 is needed when the isotope 2He3 is created so they drop out.

The isotope 1H2 has 2.01411 amu's. Its single electron is drawn as a dot on the big circle and it is drawn as a square on the big circle in the isotope 2He3 to show that the electron has become the negatron with 0.00055 amu.

The proton in the isotope 1H2 has 1.00728 amu's and it is drawn as a dot to the left of the vertical line within the big circle and it is drawn as a small square in the isotope 2He3 to show that the proton in the isotope 1H2 has become the positon in the isotope 2He3. The square also shows that the proton loses 0.00217 amu to become the positon in the isotope 2He3 with 1.00511 amu's. There is a square drawn around the proton to show that it has lost 0.00217 amu and there is a square drawn around the positon in the isotope 2He3 to show that it came from the proton in the isotope 1H2.

There is one neutron in the isotope 1H2 with 1.00628 amu's and it is drawn as a small circle to the right of the vertical line within the big circle. This neutron loses 0.00374 amu to become the single neutron in the isotope 2He3 with 1.00254 amu's. There is a square drawn around this neutron to show that it has lost 0.00374 amu and that it is the neutron in the isotope 2He3 which is drawn in the same way.

To find the mass that is lost creating the isotope 2He3, add the masses of the isotopes 1H3 and 1H2 and subtract the mass of the isotope 2He3 from their total.

Mass of 1H3	3.01605 amu's
Mass of 1H2	2.01411 amu's
Total mass of 1H3 and 1H2	5.03016 amu's

Total mass of 1H3 and 1H2	5.03016 amu's
Mass of 2He3	3.01603 amu's
Mass lost creating 2He3	2.01413 amu's

The mass lost creating the isotope 2He3 includes the mass of the two neutrons found in the isotopes 1H3 that dropped out. Subtract their masses from the total mass lost creating the isotope 2He3 to find the real mass lost.

Mass of first neutron in 1H3	1.00411 amu's
Mass of second neutron in 1H3	1.00411 amu's
Mass of two neutrons dropping out of 1H3	2.00822 amu's

Mass lost creating 2He3	2.01413 amu's
Mass of two neutrons that dropped out of 1H3	2.00822 amu's
Real mass lost creating 2He3	0.00591 amu

The isotope 1H3 has its electron and proton used as the isotope 1H1 with 1.00783 amu's within the isotopes 1H3 and 2He3. The electron and proton become the electron and proton in the isotope 2He3.

The isotope 1H2 has its electron become the negatron and the proton becomes the positon in the isotope 2He3. To find the mass of the positon in the isotope 2He3, subtract the 0.00217 amu the proton lost.

Mass of proton in 1H2	1.00728 amu's
Mass lost by it	0.00217 amu
Mass of positon in 2He3	1.00511 amu's

Since the proton in the isotope 1H2 lost 0.00217 amu, the mass lost by the single neutron can be found by subtracting the mass the proton lost from the total mass lost creating the isotope 2He3.

Total mass lost creating 2He3	0.00591 amu
Mass lost by proton in 1H2	0.00217 amu
Mass lost by neutron in 1H2	0.00374 amu

This single neutron with 1.00628 amu's in the isotope 1H2 loses 0.00374 amu to become the single neutron in the isotope 2He3. To find the mass of the single neutron in the isotope 2He3, subtract the lost mass from it.

Mass of neutron in 1H2	1.00628 amu's
Mass lost by it	0.00374 amu
Mass of neutron in 2He3	1.00254 amu's

The electrons in the isotopes 1H3 and 1H2 are not lost or released because no 0.00100 amu has been transferred or emitted as a single unit and no 0.00002 amu has been lost creating the isotope 2He3.

The proton lost a unit of 0.00217 amu and the neutron lost a unit of 0.00374 amu in the isotope 1H2 for a total lost mass of 0.00591 amu. They are each lost as single units. The lost mass of 0.00217 does not have any sub-units. The lost mass of 0.00374 amu is composed of two sub-units. They are sub-units of 0.00217 amu and 0.00157 amu. The 0.00217 amu is a single sub-unit while the 0.00157 amu is composed of three sub-units. They are sub-units of 0.00100 amu, 0.00055 amu and 0.00002 amu. There are two sub-units of 0.00217 amu shown and one is a single unit while the other belongs to a unit as a sub-unit. In order for the electron or negatron to stay with their proton or positon, these sub-units must be lost as single units of 0.00217 amu and 0.00374 amu. They must also be lost as single units to keep the proton, positon or neutron from releasing an electron or negatron from within any of them.

By adding all of the now known masses of the isotope 2He3, the mass of the isotope 2He3 can be found.

Mass of 1H1 from 1H3 intact within 2He3	1.00783 amu's
Mass of positon 2He3	1.00511 amu's
Mass of negatron in 2He3	0.00055 amu
Mass of neutron in 2He3	1.00254 amu's
Mass of 2He3	3.01603 amu's

Sub-step 44-2-28

2He3

1H3 + 1H2 = 2He3, alpha 157 ray, gamma 434 ray, 1.00411 neutron, 1.00411 neutron

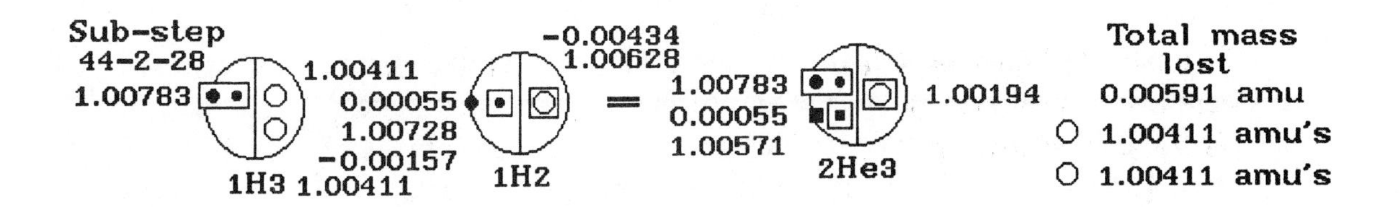

The isotope 1H3 with 3.01605 amu's has one electron with 0.00055 amu and one proton with 1.00728 amu's plus two neutrons. Both neutrons have 1.00411 amu's.

The isotope 1H2 with 2.01411 amu's has one electron with 0.00055 amu and one proton with 1.00728 amu's and one neutron with 1.00628 amu's.

The isotope 2He3 with 3.01603 amu's has one electron with 0.00055 amu, one negatron with 0.00055 amu, one proton with 1.00728 amu's, one positon with 1.00571 amu's and one neutron with 1.00194 amu's.

The total mass lost is 0.00591 amu and it is emitted as a 157 alpha ray which has an external magnetic field and a 434 gamma ray which has no external magnetic field. They are emitted from the proton and a single neutron in the isotope 1H2. There are two neutrons that drop out from the isotope 1H3 with each having 1.00411 amu's.

The isotope 1H3 with 3.01605 amu's has one electron with 0.00055 amu that is drawn as a dot on the big circle and it becomes the electron with 0.00055 amu in the isotope 2He3 and is also drawn as a dot on the big circle.

The proton with 1.00728 amu's in the isotope 1H3 is drawn as a dot to the left of the vertical line within the big circle. The electron and proton have a rectangle drawn around them to show that they are used as the isotope 1H1 with 1.00783 amu's within the isotopes 1H3 and 2He3. The rectangle shows that the isotope 1H1 can drop out of the isotope 2He3 at some later time.

There are two neutrons in the isotope 1H3 that are drawn one above the other as small circles to the right of the vertical line within the big circle. The top neutron has 1.00411 amu's and is called the first neutron while the neutron below it is the second neutron and it has 1.00411 amu's. Neither of the two neutrons in the isotope 1H3 is needed when the isotope 2He3 is created so they drop out.

The isotope 1H2 has 2.01411 amu's. Its single electron is drawn as a dot on the big circle in the isotope 1H2 and it is drawn as a square on the big circle in the isotope 2He3 to show that the electron has become the negatron with 0.00055 amu.

The proton in the isotope 1H2 has 1.00728 amu's and it is drawn as a dot to the left of the vertical line within the big circle in the isotope 1H2 and it is drawn as a small square in the isotope 2He3 to show that the proton in the isotope 1H2 has become the positon in the isotope 2He3. The proton loses 0.00157 amu to become the positon in the isotope 2He3 with 1.00571 amu's. The proton has a square drawn around it to show it has lost 0.00157 amu and this proton becomes the positon in the isotope 2He3 with a square drawn around the square that is the positon.

There is one neutron in the isotope 1H2 with 1.00628 amu's and it is drawn as a small circle to the right of the vertical line within the big circle. This neutron loses 0.00434 amu to become the single neutron in the isotope 2He3 with 1.00194 amu's. There is a square drawn around the neutron to show hat it has lost 0.00434 amu and it is drawn the same way in the isotope 2He3 to show that it was the neutron in the isotope 1H2.

To find the mass that is lost creating the isotope 2He3, add the masses of the isotopes 1H3 and 1H2 and subtract the mass of the isotope 2He3 from their total.

Mass of 1H3	3.01605 amu's
Mass of 1H2	2.01411 amu's
Total mass of 1H3 and 1H2	5.03016 amu's

Total mass of 1H3 and 1H2	5.03016 amu's
Mass of 2He3	3.01603 amu's
Mass lost creating 2He3	2.01413 amu's

The mass lost creating the isotope 2He3 includes the mass of the two neutrons found in the isotopes 1H3 that dropped out. Subtract their masses from the total mass lost creating the isotope 2He3 to find the real mass lost.

Mass of first neutron in 1H3	1.00411 amu's
Mass of second neutron in 1H3	1.00411 amu's
Mass of two neutrons dropping out of 1H3	2.00822 amu's

Mass lost creating 2He3	2.01413 amu's
Mass of two neutrons that dropped out of 1H3	2.00822 amu's
Real mass lost creating 2He3	0.00591 amu

The isotope 1H3 has its electron and proton used as the isotope 1H1 with 1.00783 amu's within the isotopes 1H3 and 2He3. The electron and proton become the electron and proton in the isotope 2He3.

The isotope 1H2 has its electron become the negatron and the proton becomes the positon in the isotope 2He3. To find the mass of the positon in the isotope 2He3, subtract the 0.00157 amu the proton lost.

Mass of proton in 1H2	1.00728 amu's
Mass lost by it	0.00157 amu
Mass of positon in 2He3	1.00571 amu's

Since the proton in the isotope 1H2 lost 0.00157 amu, the mass lost by the single neutron can be found by subtracting the mass the proton lost from the total mass lost creating the isotope 2He3.

Total mass lost creating the isotope 2He3	0.00591 amu
Mass lost by the proton in 1H2	0.00157 amu
Mass lost by neutron in 1H2	0.00434 amu

This single neutron with 1.00628 amu's in the isotope 1H2 loses 0.00434 amu to become the single neutron in the isotope 2He3. To find the mass of the single neutron in the isotope 2He3, subtract the lost mass from it.

Mass of neutron in 1H2	1.00628 amu's
Mass lost by it	0.00434 amu
Mass of neutron in 2He3	1.00194 amu's

The electrons in the isotopes 1H3 and 1H2 are not lost or released because no 0.00100 amu has been transferred or emitted as a single unit and no 0.00002 amu has been lost creating the isotope 2He3.

The proton lost a unit of 0.00157 amu and the neutron lost a unit of 0.00434 amu in the isotope 1H2 for a total lost mass of 0.00591 amu. They are each lost as single units. The lost mass of 0.00157 is composed of three sub-units. They are 0.00100 amu, 0.00055 amu and 0.00002 amu. The 0.00434 amu is composed of two sub-units of 0.00217 amu. In order for the electron or negatron to stay with their proton or positon, these sub-units must be lost as a single unit of 0.00157 amu or 0.00434 amu. They must also be lost as single units to keep the proton, positon or neutron from releasing an electron or negatron from within any of them.

By adding all of the now known masses of the isotope 2He3, the mass of the isotope 2He3 can be found.

Mass of 1H1 from 1H3 intact within 2He3	1.00783 amu's
Mass of positon 2He3	1.00571 amu's
Mass of negatron in 2He3	0.00055 amu
Mass of neutron in 2He3	1.00194 amu's
Mass of 2He3	3.01603 amu's

Sub-step 43-2-27

2He3

1H3 + 1H2 = 2He3, gamma 591 ray, 1.00411 neutron, 1.00411 neutron

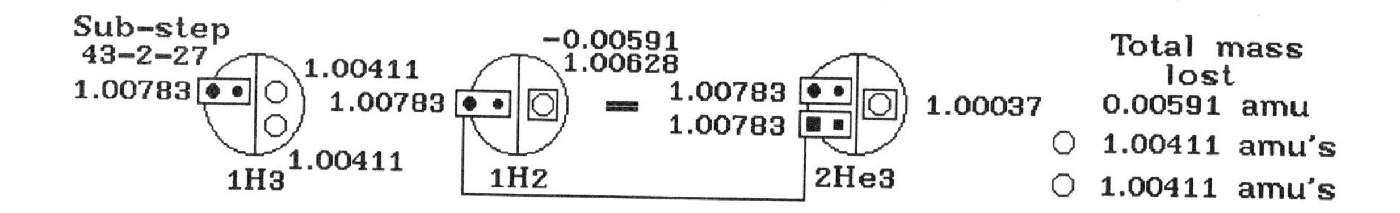

The isotope 1H3 with 3.01605 amu's has one electron with 0.00055 amu and one proton with 1.00728 amu's plus two neutrons. Both neutrons have 1.00411 amu's.

The isotope 1H2 with 2.01411 amu's has one electron with 0.00055 amu and one proton with 1.00728 amu's and one neutron with 1.00628 amu's.

The isotope 2He3 with 3.01603 amu's has one electron with 0.00055 amu, one negatron with 0.00055 amu, one proton with 1.00728 amu's, one positon with 1.00728 amu's and one neutron with 1.00037 amu's.

The total mass lost is 0.00591 amu and it is emitted as a 591 gamma ray which does not have an external magnetic field. It is emitted from the single neutron in the isotope 1H2. There are two neutrons that drop out from the isotope 1H3 with each having 1.00411 amu's.

The isotope 1H3 with 3.01605 amu's has one electron with 0.00055 amu that is drawn as a dot on the big circle and it becomes the electron with 0.00055 amu in the isotope 2He3 and is also drawn as a dot on the big circle.

The proton with 1.00728 amu's in the isotope 1H3 is drawn as a dot to the left of the vertical line within the big circle. The electron and proton have a rectangle drawn around them to show that they are used as the isotope 1H1 with 1.00783 amu's within the isotopes 1H3 and 2He3. The rectangle shows that the isotope 1H1 can drop out of the isotope 2He3 at some later time.

There are two neutrons in the isotope 1H3 that are drawn one above the other as small circles to the right of the vertical line within the big circle. The top neutron has 1.00411 amu's and is called the first neutron while the neutron below it is the second neutron and it has 1.00411 amu's. Neither of the two neutrons in the isotope 1H3 is needed when the isotope 2He3 is created so they drop out.

The isotope 1H2 has 2.01411 amu's. Its single electron is drawn as a dot on the big circle in the isotope 1H2 and it is drawn as a square on the big circle in the isotope 2He3 to show that the electron has become the negatron with 0.00055 amu.

The proton in the isotope 1H2 has 1.00728 amu's and it is drawn as a dot to the left of the vertical line within the big circle and it is drawn as a small square in the isotope 2He3 to show that the proton in the isotope 1H2 has become the positon in the isotope 2He3. There is a rectangle drawn around both of them in the isotopes 1H3 and 1H2 to show that they each become the isotope 1H1 with 1.00783 amu's within the isotope 2He3. The rectangle shows that either or both of the isotopes of 1H1 in the isotope 2He3 can drop out at some later time. There is a line drawn from the isotope of 1H1 in the isotope 1H2 to the isotope 1H1 in the isotope 2He3 to show which isotope of 1H1 came from the isotope 1H2.

There is a line drawn from the isotope 1H1 found intact within the isotope 1H2 to the isotope 1H1 found intact within the isotope 2He3 to show which isotope of 1H1 has the negatron and positon.

There is one neutron in the isotope 1H2 with 1.00628 amu's and it is drawn as a small circle to the right of the vertical line within the big circle. This neutron loses 0.00591 amu to become the single neutron in the isotope 2He3 with 1.00037 amu's. There is a square drawn around this neutron to show that it has lost 0.00591 amu and it is drawn the same way in the isotope 2He3 to show that the neutron in the isotope 2He3 came from the neutron in the isotope 1H2.

To find the mass that is lost creating the isotope 2He3, add the masses of the isotopes 1H3 and 1H2 and subtract the mass of the isotope 2He3 from their total.

Mass of 1H3	3.01605 amu's
Mass of 1H2	2.01411 amu's
Total mass of 1H3 and 1H2	5.03016 amu's

Total mass of 1H3 and 1H2	5.03016 amu's
Mass of 2He3	3.01603 amu's
Mass lost creating 2He3	2.01413 amu's

The mass lost creating the isotope 2He3 includes the mass of the two neutrons found in the isotope 1H3 that dropped out. Subtract their masses from the total mass lost creating the isotope 2He3 to find the real lost mass.

Mass of first neutron in 1H3	1.00411 amu's
Mass of second neutron in 1H3	1.00411 amu's
Mass of two neutrons dropping out of 1H3	2.00822 amu's

Mass lost creating 2He3	2.01413 amu's
Mass of two neutrons that dropped out of 1H3	2.00822 amu's
Real mass lost creating 2He3	0.00591 amu

The isotopes 1H3 and 1H2 have their electron and proton used as the isotope 1H1 with 1.00783 amu's within the isotope 2He3. The electron and proton in the isotope 1H3 become the electron and proton in the isotope 2He3. The isotope 1H2 has its electron and proton used as the isotope 1H1 with 1.00783 amu's and in the isotope 2He3 also. The electron becomes the negatron and the proton becomes the positon in the isotope 2He3.

This single neutron with 1.00628 amu's in the isotope 1H2 loses all of the mass that is lost to become the single neutron in the isotope 2He3. To find the mass of the single neutron in the isotope 2He3, subtract the lost mass from it.

Mass of neutron in 1H2	1.00628 amu's
Mass lost by it	0.00591 amu
Mass of neutron in 2He3	1.00037 amu's

The electrons in the isotopes 1H3 and 1H2 are not lost or released because no 0.00100 amu has been transferred or emitted as a single unit and no 0.00002 amu has been lost creating the isotope 2He3.

The neutron in the isotope 1H2 lost 0.00591 amu for a total lost mass of 0.00591 amu. It is lost as a single unit. The lost mass of 0.00591 is composed of two sub-units. They are sub-units of 0.00434 amu and 0.00157 amu. The 0.00434 amu is composed of two sub-units. They are sub-units of 0.00217 amu while the sub-unit of 0.00157 amu is composed of three sub-units of 0.00100 amu, 0.00055 amu and 0.00002 amu.

In order for the electron or negatron to stay with their proton or positon, these sub-units must be lost as a single unit of 0.00591 amu. It must be lost as single units to keep the neutron from releasing an electron or negatron from within it.

By adding all of the now known masses of the isotope 2He3, the mass of the isotope 2He3 can be found.

Mass of 1H1 from 1H3 intact within 2He3	1.00783 amu's
Mass of 1H1 from 1H2 intact within 2He3	1.00783 amu's
Mass of neutron in 2He3	1.00037 amu's
Mass of 2He3	3.01603 amu's

This step and the eight steps just given will lose 0.00591 amu. This step lost the entire 0.00591 amu while the remaining eight steps had a proton and a neutron lose this mass and the two neutrons in the isotope 1H3 always dropped out of the isotope 1H3. The first set of four steps lost this mass only from the proton and neutron in the isotope 1H2 while the last four steps lost it from the isotope 1H3's proton and the single neutron in the isotope 1H2. The electron and proton within the isotope 1H3 is always the isotope 1H1 in the first set of four steps while the isotope 1H2 has its electron and proton used as the isotope 1H1 within the isotope 1H2 in the last set of four steps. The single neutron in the isotope 1H2 always lost some of mass in all nine steps.

Sub-step 42-2-26

2He3

1H3 + 1H2 = 2He3, alpha 217 ray, gamma 157 ray, 1.00411 neutron, 1.00628 neutron

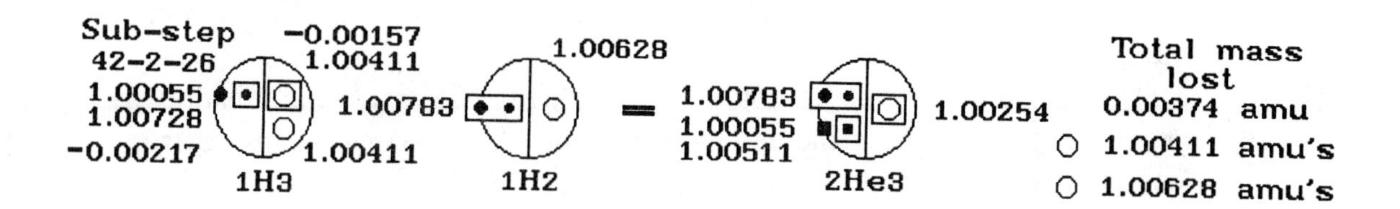

The isotope 1H3 with 3.01605 amu's has one electron with 0.00055 amu and one proton with 1.00728 amu's plus two neutrons. Both neutrons have 1.00411 amu's.

The isotope 1H2 with 2.01411 amu's has one electron with 0.00055 amu and one proton with 1.00728 amu's and one neutron with 1.00628 amu's.

The isotope 2He3 with 3.01603 amu's has one electron with 0.00055 amu, one negatron with 0.00055 amu, one proton with 1.00728 amu's, one positon with 1.00511 amu's and one neutron with 1.00254 amu's.

The total mass lost is 0.00374 amu and it is emitted as a 217 alpha ray which has an external magnetic field and a 157 gamma ray which has no external magnetic field. The 217 ray is emitted from the proton in the isotope 1H3. The 157 ray is emitted from one of the two neutrons in the isotope 1H3. There are two neutrons that drop out. The other neutron in the isotope 1H3 drops out with 1.00411 amu's. The second neutron to drop out comes from the isotope 1H2 with 1.00628 amu's.

The isotope 1H3 with 3.01605 amu's has one electron with 0.00055 amu that is drawn as a dot on the big circle and it becomes the negatron with 0.00055 amu in the isotope 2He3 and is drawn as a square on the big circle.

The proton with 1.00728 amu's in the isotope 1H3 is drawn as a dot to the left of the vertical line within the big circle. It has a square drawn around it to show that it has lost 0.00217 amu. It is drawn as a small square in the isotope 2He3 to show that it is now a positon. The positon has a square drawn around the small square to show that it was the proton in the isotope 1H3.

There are two neutrons in the isotope 1H3 that are drawn one above the other as small circles to the right of the vertical line within the big circle. The top neutron has 1.00411 amu's and is called the first neutron while the neutron below it is the second neutron and it has 1.00411 amu's. One of the two neutrons in the isotope 1H3 is not needed when the isotope 2He3 is created so it drops out. The other neutron loses 0.00157 amu to become the single neutron in the isotope 2He3 with 1.00254 amu's.

The isotope 1H2 has 2.01411 amu's. Its single electron is drawn as a dot on the big circle in the isotopes 1H2 and it is drawn as a dot on the big circle in the isotope 2He3 to show that the electron stays an electron with 0.00055 amu.

The proton in the isotope 1H2 has 1.00728 amu's and it is drawn as a dot to the left of the vertical line within the big circle and it is drawn as a dot within the isotope 2He3. The electron and proton have a rectangle drawn around both of them in the isotopes 1H2 and 2He3 to show that they are used as the isotope 1H1 within each of them. The isotope 1H1 can drop out of the isotope 2He3 at some later time.

There is one neutron in the isotope 1H2 with 1.00628 amu's and it is drawn as a small circle to the right of the vertical line within the big circle. This neutron is not needed when the isotope 2He3 is created so it drops out.

To find the mass that is lost creating the isotope 2He3, add the masses of the isotopes 1H3 and 1H2 and subtract the mass of the isotope 2He3 from their total.

Mass of 1H3	3.01605 amu's
Mass of 1H2	2.01411 amu's
Total mass of 1H3 and 1H2	5.03016 amu's

Total mass of 1H3 and 1H2	5.03016 amu's
Mass of 2He3	3.01603 amu's
Mass lost creating 2He3	2.01413 amu's

The mass lost creating the isotope 2He3 includes the mass of the neutron that dropped out of the isotope 1H3 and the neutron that dropped out of the isotope 1H2. Add their mass and subtract their masses from the total mass lost creating the isotope 2He3 to find the real mass lost.

Mass of neutron dropping out of 1H3	1.00411 amu's
Mass of neutron dropping out of 1H2	1.00628 amu's
Total mass of two neutrons dropping out	2.01039 amu's

Mass lost creating 2He3	2.01413 amu's
Mass of two neutrons that dropped out	2.01039 amu's
Real mass lost creating 2He3	0.00374 amu

Since the first neutron in the isotope 1H3 lost 0.00157 amu, the mass the proton loses can be found by subtracting the mass the first neutron in the isotope 1H3 lost from the total mass lost creating the isotope 2He3

Total mass lost creating the isotope 2He3	0.00374 amu
Mass lost by first neutron in 1H3	0.00157 amu
Mass lost by proton in 1H3	0.00217 amu

Before the mass of the two neutrons that dropped out can be subtracted from the mass lost creating the isotope 2He3, their mass must be found. Both neutrons in the isotope 1H3 have the same mass of 1.00411 amu's. Therefore, the top neutron is the first neutron and the neutron under it is called the second neutron. The first neutron is the neutron that loses 0.00157 amu and there is a square drawn around it to show this. This means that the second neutron with 1.00411 amu's in the isotope 1H3 is the neutron that drops out. The other neutron to drop out will come from the isotope 1H2 and since there is only one neutron in this isotope, this neutron must have 1.00628 amu's.

Mass of first neutrons in 1H3	1.00411 amu's
Mass lost by it	0.00157 amu
Mass of neutron in 2He3	1.00254 amu's

The isotope 1H3 has its electron used as an electron and it becomes the negatron in the isotope 2He3 with 0.00055 amu. The proton is a proton in the isotope 1H3 which loses 0.00217 amu to become the positon in the isotope 2He3. Subtract 0.00217 amu from the proton to find the mass of the positon in the isotope 2He3.

Mass of proton in 1H3	1.00728 amu's
Mass lost by it	0.00217 amu
Mass of positon in 2He3	1.00511 amu's

This single neutron with 1.00628 amu's in the isotope 1H2 does not lose any mass and it is not needed when the isotope 2He3 is created so it drops out.

The electrons in the isotopes 1H3 and 1H2 are not lost or released because no 0.00100 amu has been transferred or emitted as a single unit and no 0.00002 amu has been lost creating the isotope 2He3.

The first neutron in the isotope 1H3 lost a unit of 0.00157 amu and the proton in this same isotope loses a unit of 0.00217 amu for a total lost mass of 0.00374 amu. The lost mass of 0.00157 is composed of three sub-units. They are sub-units of 0.00100 amu, 0.00055 amu and 0.00002 amu. The unit of 0.00217 amu can not be broken down into any sub-units. In order for the electron or negatron to stay with their proton or positon, these sub-units must be lost as a single unit of 0.00157 amu and the unit of 0.00217 amu is already a unit. They must also be lost as single units to keep the proton, positon or neutron from releasing an electron or negatron from within either of them.

By adding all of the now known masses of the isotope 2He3, the mass of the isotope 2He3 can be found.

Mass of 1H1 from 1H2 intact within 2He3	1.00783 amu's
Mass of the positon in 2He3	1.00511 amu's
Mass of negatron in 2He3	0.00055 amu
Mass of neutron in 2He3	1.00254 amu's
Mass of 2He3	3.01603 amu's

Sub-step 41-2-25

2He3

1H3 + 1H2 = 2He3, alpha 157 ray, gamma 217 ray, 1.00411 neutron, 1.00628 neutron

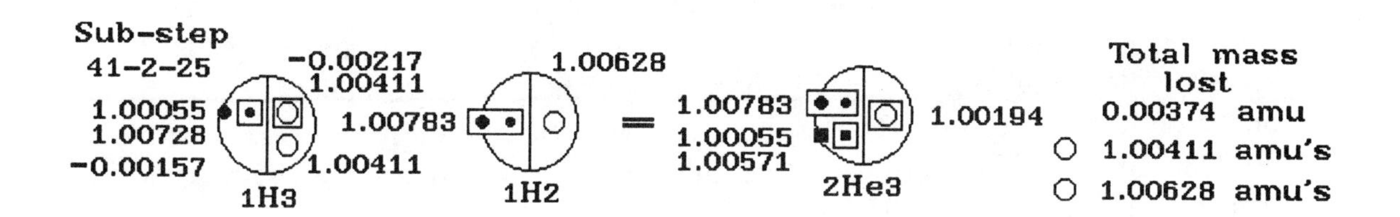

The isotope 1H3 with 3.01605 amu's has one electron with 0.00055 amu and one proton with 1.00728 amu's plus two neutrons. Both neutrons have 1.00411 amu's.

The isotope 1H2 with 2.01411 amu's has one electron with 0.00055 amu and one proton with 1.00728 amu's and one neutron with 1.00628 amu's.

The isotope 2He3 with 3.01603 amu's has one electron with 0.00055 amu, one negatron with 0.00055 amu, one proton with 1.00728 amu's, one positon with 1.00571 amu's and one neutron with 1.00194 amu's.

The total mass lost is 0.00374 amu and it is emitted as a 157 alpha ray which has an external magnetic field and a 217 gamma ray which has no external magnetic field. The 157 alpha ray is emitted from the proton in the isotope 1H3. The 217 gamma ray is emitted from one of the two neutrons found in this isotope. There are two neutrons that drop out. The other neutron in the isotope 1H3 drops out with 1.00411 amu's. The second neutron to drop out comes from the isotope 1H2 with 1.00628 amu's.

The isotope 1H3 with 3.01605 amu's has one electron with 0.00055 amu that is drawn as a dot on the big circle and it becomes the negatron with 0.00055 amu in the isotope 2He3 and is drawn as a square on the big circle.

The proton with 1.00728 amu's in the isotope 1H3 is drawn as a dot to the left of the vertical line within the big circle. There is a square drawn around it to show that it has lost 0.00157 amu. This proton becomes the positon in the isotope 2He3 and it is drawn as a square. This positon has another square drawn around it to show that it came from the proton in the isotope 1H3.

There are two neutrons in the isotope 1H3 that are drawn one above the other as small circles to the right of the vertical line within the big circle. The top neutron has 1.00411 amu's and is called the first neutron while the neutron below it is the second neutron and it has 1.00411 amu's. One of the two neutrons in the isotope 1H3 is not needed when the isotope 2He3 is created so it drops out. The other neutron loses 0.00217 amu to become the single neutron in the isotope 2He3 with 1.00194 amu's. To show that this proton lost 0.00217 amu, there is a square drawn around it. In the isotope 2He3, the neutron that has 1.00194 amu's has a square drawn around it to show that this neutron came from neutron in the isotope 1H3.

The isotope 1H2 has 2.01411 amu's. Its single electron is drawn as a dot on the big circle. It is also drawn as a dot on the big circle in the isotope 2He3 to show that the electron stays an electron with 0.00055 amu.

The proton in the isotope 1H2 has 1.00728 amu's and it is drawn as a dot to the left of the vertical line within the big circle in the isotopes 1H2 and it is drawn as a dot in the isotope 2He3 to show that the proton in the isotope 1H2 stays a proton in the isotope 2He3. There is a rectangle drawn around the electron and proton in the isotope 1H2 to show that they become the isotope 1H1 with 1.00783 amu's within the isotope 1H2 and 2He3. The rectangle shows that the isotope 1H1 in the isotope 2He3 can drop out at some later time.

There is one neutron in the isotope 1H2 with 1.00628 amu's and it is drawn as a small circle to the right of the vertical line within the big circle. This neutron is not needed when the isotope 2He3 is created so it drops out.

To find the mass that is lost creating the isotope 2He3, add the masses of the isotopes 1H3 and 1H2 and subtract the mass of the isotope 2He3 from their total.

Mass of 1H3	3.01605 amu's
Mass of 1H2	2.01411 amu's
Total mass of 1H3 and 1H2	5.03016 amu's

Total mass of 1H3 and 1H2	5.03016 amu's
Mass of 2He3	3.01603 amu's
Mass lost creating 2He3	2.01413 amu's

The mass lost creating the isotope 2He3 includes the mass of the two neutrons found in the isotopes 1H3 and 1H2 that dropped out. Subtract their masses from the total mass lost creating the isotope 2He3 to find the real mass lost.

Mass of first neutron dropping out of 1H3	1.00411 amu's
Mass of second neutron dropping out of 1H2	1.00628 amu's
Total mass of both neutrons dropping out	2.01039 amu's

Mass lost creating 2He3	2.01413 amu's
Mass of two neutrons that dropped out	2.01039 amu's
Real mass lost creating 2He3	0.00374 amu

Both neutrons in the isotope 1H3 have the same mass of 1.00411 amu's. The top neutron is called the first neutron and the neutron under it can be called the second neutron. It is the first neutron that loses 0.00217 amu to become the single neutron in the isotope 2He3 with 1.00194 amu's while the second neutron is not needed and it drops out with 1.00411 amu's. The other neutron to drop out will come from the isotope 1H2 and since there is only one neutron in this isotope, this neutron must have 1.00628 amu's.

Mass of one of two neutrons in 1H3	1.00411 amu's
Mass lost by it	0.00217 amu
Mass of neutron in 2He3	1.00194 amu's

Since the first neutron in the isotope 1H3 lost 0.00217 amu, the mass of the proton found in this isotope can be found by subtracting the mass it lost from the total mass lost creating the isotope 2He3.

Total mass lost creating 2He3	0.00374 amu
Mass lost by first neutron in 1H3	0.00217 amu
Mass lost by proton in 1H3	0.00157 amu

The isotope 1H3 has electron used as an electron in the isotope 1H3 and it becomes the negatron in the isotope 2He3 with 0.00055 amu. The proton in the isotope 1H3 loses 0.00157 amu to become the positon in the isotope 2He3. Subtract 0.00157 amu from the proton to find the mass of the positon in the isotope 2He3.

Mass of proton in 1H3	1.00728 amu's
Mass lost by it	0.00157 amu
Mass of positon in 2He3	1.00571 amu's

This single neutron with 1.00628 amu's in the isotope 1H2 does not lose any mass and it is not needed when the isotope 2He3 is created so it drops out.

The electrons in the isotopes 1H3 and 1H2 are not lost or released because no 0.00100 amu has been transferred or emitted as a single unit and no 0.00002 amu has been lost creating the isotope 2He3.

One of the two neutrons in the isotope 1H3 lost a unit 0.00217 amu and the proton in this same isotope loses a unit of 0.00157 amu for a total lost mass of 0.00374 amu. They are each lost as a single unit. The lost mass of 0.00157 is composed of three sub-units. They are 0.00100 amu, 0.00055 amu and 0.00002 amu. The unit of 0.00217 amu can not be broken down into any sub-units. In order for the electron or negatron to stay with their proton or positon, these sub-units must be lost as single units of 0.00157 amu and 0.00217 amu. They must also be lost as single units in order to keep the proton, positon or neutron from releasing an electron or negatron from within either of them.

By adding all of the now known masses of the isotope 2He3, the mass of the isotope 2He3 can be found.

Mass of 1H1 from 1H2 intact within 2He3	1.00783 amu's
Mass of the positon in 2He3	1.00571 amu's
Mass of negatron in 2He3	0.00055 amu
Mass of neutron in 2He3	1.00194 amu's
Mass of 2He3	3.01603 amu's

Sub-step 40-2-24

2He3

1H3 + 1H2 = 2He3, gamma 374 ray, 1.00411 neutron, 1.00628 neutron

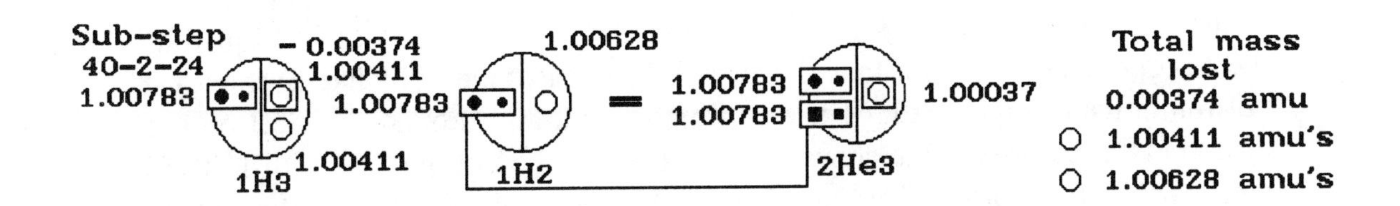

The isotope 1H3 with 3.01605 amu's has one electron with 0.00055 amu and one proton with 1.00728 amu's plus two neutrons. Both neutrons have 1.00411 amu's.

The isotope 1H2 with 2.01411 amu's has one electron with 0.00055 amu and one proton with 1.00728 amu's and one neutron with 1.00628 amu's.

The isotope 2He3 with 3.01603 amu's has one electron with 0.00055 amu, one negatron with 0.00055 amu, one proton with 1.00728 amu's, one positon with 1.00728 amu's and one neutron with 1.00037 amu's.

The total mass lost is 0.00374 amu and it is emitted as a 374 gamma ray which has no external magnetic field. It is lost from one of the two neutrons in the isotope 1H3. There are two neutrons that drop out. The first neutron to drop out has 1.00411 amu's and it drops out of the isotope 1H3. The second neutron to drop out comes from the isotope 1H2 with 1.00628 amu's.

The isotope 1H3 with 3.01605 amu's has one electron with 0.00055 amu that is drawn as a dot on the big circle and it becomes the electron with 0.00055 amu in the isotope 2He3 and is also drawn as a dot on the big circle.

The proton with 1.00728 amu's is drawn as a dot to the left of the vertical line within the big circle. The electron and proton have a rectangle drawn around them to show that they are used as the isotope 1H1 with 1.00783 amu's within the isotope 1H3 and 2He3. The rectangle shows that the isotope 1H1 can drop out of the isotope 2He3 at some later time.

There are two neutrons in the isotope 1H3 that are drawn one above the other as small circles to the right of the vertical line within the big circle. The top neutron has 1.00411 amu's and is called the first neutron while the neutron below it is the second neutron and it has 1.00411 amu's. One of the two neutrons in the isotope 1H3 is not needed when the isotope 2He3 is created so it drops out. The other neutron loses 0.00374 amu to become the single neutron in the isotope 2He3 with 1.00037 amu's. The first neutron loses the 0.00374 amu and there is a square drawn around the small circle to show that it has lost 0.00374 amu. There is a square drawn around the neutron in the isotope 2He3 to show that this neutron was the first neutron in the isotope 1H3.

The isotope 1H2 has 2.01411 amu's. Its single electron is drawn as a dot on the big circle in the isotope 1H2 and it is drawn as a square on the big circle in the isotope 2He3 to show that the electron has become the negatron with 0.00055 amu.

The proton in the isotope 1H2 has 1.00728 amu's and it is drawn as a dot to the left of the vertical line within the big circle. It is drawn as a small square in the isotope 2He3 to show that the proton in the isotope 1H2 has become the positon in the isotope 2He3. There is a rectangle drawn around the electron and proton in the isotope 1H2 to show that they become the isotope 1H1 with 1.00783 amu's within the isotopes 1H2 and 2He3. The rectangle shows that the isotope 1H1 in the isotope 2He3 can drop out at some later time.

There is a line drawn from the isotope 1H1 found intact within the isotope 1H2 to the isotope 1H1 found intact within the isotope 2He3 to show which isotope of 1H1 has the negatron and positon.

There is one neutron in the isotope 1H2 with 1.00628 amu's and it is drawn as a small circle to the right of the vertical line within the big circle. This neutron is not needed when the isotope 2He3 is created so it drops out.

To find the mass that is lost creating the isotope 2He3, add the masses of the isotopes 1H3 and 1H2 and subtract the mass of the isotope 2He3 from their total.

Mass of 1H3	3.01605 amu's
Mass of 1H2	2.01411 amu's
Total mass of 1H3 and 1H2	5.03016 amu's

Total mass of 1H3 and 1H2	5.03016 amu's
Mass of 2He3	3.01603 amu's
Mass lost creating 2He3	2.01413 amu's

Both neutrons in the isotope 1H3 have the same mass of 1.00411 amu's. The first neutron to drop out comes from the isotope 1H3. In this step, one of the two neutrons in the isotope 1H3 will lose 0.00374 amu to become the single neutron in the isotope 2He3 with 1.00037 amu's while the other neutron is not needed and it drops out with 1.00411 amu's. The other neutron to drop out will come from the isotope 1H2 and since there is only one neutron in this isotope, this neutron must have 1.00628 amu's.

Mass of first neutron dropping out from 1H3	1.00411 amu's
Mass of neutron dropping out from 1H2	1.00628 amu's
Mass of two neutrons that dropped out	2.01039 amu's

The mass lost creating the isotope 2He3 includes the mass of the two neutrons found in the isotopes 1H3 and 1H2 that dropped out. Subtract their masses from the total mass lost creating the isotope 2He3 to find the real mass lost.

Mass lost creating 2He3	2.01413 amu's
Mass of two neutrons that dropped out	2.01039 amu's
Real mass lost creating 2He3	0.00374 amu

Mass of one of two neutrons in 1H3	1.00411 amu's
Mass lost by it	0.00374 amu
Mass of neutron in 2He3	1.00037 amu's

The isotope 1H3 has electron and proton used as the isotope 1H1 with 1.00783 amu's within the isotopes 1H3 and 2He3. The electron and proton become the electron and proton in the isotope 2He3.

The isotope 1H2 has electron and proton used as the isotope 1H1 with 1.00783 amu's and in the isotope 2He3 also. The electron becomes the negatron and the proton becomes the positon in the isotope 2He3.

This single neutron with 1.00628 amu's in the isotope 1H2 does not lose any mass and it is not needed when the isotope 2He3 is created so it drops out.

The electrons in the isotopes 1H3 and 1H2 are not lost or released because no 0.00100 amu has been transferred or emitted as a single unit and no 0.00002 amu has been lost creating the isotope 2He3.

The first neutron in the isotope 1H3 lost a unit of 0.00374 amu for a total lost mass of 0.00374 amu. It is lost as a single unit. The lost mass of 0.00374 is composed of four sub-units. They are sub-units of 0.00217 amu, 0.00100 amu, 0.00055 amu and 0.00002 amu. The lost mass of 0.00374 amu must be lost as a single unit in order to keep the neutron from releasing an electron or negatron from within it.

By adding all of the now known masses of the isotope 2He3, the mass of the isotope 2He3 can be found.

Mass of 1H1 from 1H3 intact within 2He3	1.00783 amu's
Mass of 1H1 from 1H2 intact within 2He3	1.00783 amu's
Mass of neutron in 2He3	1.00037 amu's
Mass of 2He3	3.01603 amu's

The lost mass of 0.00374 amu is lost as 0.00157 amu and 0.00217 amu in two of the steps and one step has it lost as a single unit of 0.00374 amu from the isotope 1H3. The isotope 1H2 is used in the same way in all three steps.

Sub-step 39-2-23

2He3

1H3 + 1H2 = 2He3, gamma 157 ray, 1.00628 neutron, 1.00628 neutron

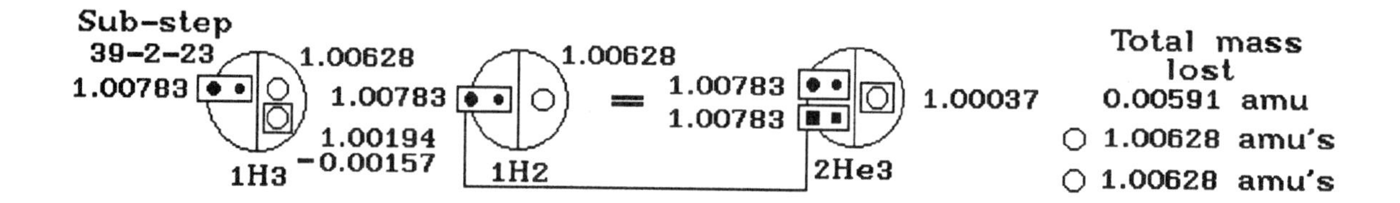

The isotope 1H3 with 3.01605 amu's has one electron with 0.00055 amu and one proton with 1.00728 amu's plus two neutrons. The first neutron has 1.00628 amu's and the second neutron has 1.00194 amu's.

The isotope 1H2 with 2.01411 amu's has one electron with 0.00055 amu and one proton with 1.00728 amu's and one neutron with 1.00628 amu's.

The isotope 2He3 with 3.01603 amu's has one electron with 0.00055 amu, one negatron with 0.00055 amu, one proton with 1.00728 amu's, one positon with 1.00728 amu's and one neutron with 1.00037 amu's.

The total mass lost is 0.00157 amu and it is emitted as a 157 gamma ray which does not have an external magnetic field. The 157 gamma ray is emitted from the second neutron in the isotope 1H3. There are two neutrons that drop out. The first neutron to drop out has 1.00628 amu's and it drops out of the isotope 1H3. The second neutron to drop out comes from the isotope 1H2 with 1.00628 amu's.

The isotope 1H3 with 3.01605 amu's has one electron with 0.00055 amu that is drawn as a dot on the big circle and it becomes the electron with 0.00055 amu in the isotope 2He3 and is also drawn as a dot on the big circle.

The proton with 1.00728 amu's in the isotope 1H3 is drawn as a dot to the left of the vertical line within the big circle. The electron and proton have a rectangle drawn around them to show that they are used as the isotope 1H1 with 1.00783 amu's within the isotope 1H3 and 2He3. The rectangle shows that the isotope 1H1 can drop out of the isotope 2He3 at some later time.

There is a line drawn from the isotope 1H1 found intact within the isotope 1H2 to the isotope 1H1 found intact within the isotope 2He3 to show which isotope of 1H1 has the negatron and positon.

There are two neutrons in the isotope 1H3 that are drawn one above the other as small circles to the right of the vertical line within the big circle. The top neutron has 1.00628 amu's and is called the first neutron while the neutron below it is the second neutron and it has 1.00194 amu's. The first neutron is not needed when the isotope 2He3 is created so it drops out. The second neutron loses

0.00157 amu to become the single neutron in the isotope 2He3 with 1.00037 amu's.

The isotope 1H2 has 2.01411 amu's. Its single electron is drawn as a dot on the big circle in the isotopes 1H2 and it is drawn as a square on the big circle in the isotope 2He3 to show that the electron has become the negatron with 0.00055 amu.

The proton in the isotope 1H2 has 1.00728 amu's and it is drawn as a dot to the left of the vertical line within the big circle and it is drawn as a small square in the isotope 2He3 to show that the proton in the isotope 1H2 has become the positon in the isotope 2He3. There is a rectangle drawn around both of them in the isotopes 1H2 and 2He3 to show that they are used as the isotope 1H1 with 1.00783 amu's. The rectangle shows that the isotope 1H1 in the isotope 2He3 can drop out at some later time.

There is a line drawn from the isotope 1H1 found intact within the isotope 1H2 to the isotope 1H1 found intact within the isotope 2He3 to show which isotope of 1H1 has the negatron and positon.

There is one neutron in the isotope 1H2 with 1.00628 amu's and it is drawn as a small circle to the right of the vertical line within the big circle. This neutron is not needed when the isotope 2He3 is created so it drops out. This means that both neutrons that drop out have the same mass of 1.00628 amu's.

To find the mass that is lost creating the isotope 2He3, add the masses of the isotopes 1H3 and 1H2 and subtract the mass of the isotope 2He3 from their total.

Mass of 1H3	3.01605 amu's
Mass of 1H2	2.01411 amu's
Total mass of 1H3 and 1H2	5.03016 amu's

Total mass of 1H3 and 1H2	5.03016 amu's
Mass of 2He3	3.01603 amu's
Mass lost creating 2He3	2.01413 amu's

The mass lost creating the isotope 2He3 includes the mass of the two neutrons found in the isotopes 1H3 and 1H2 that dropped out. Subtract their masses from the total mass lost creating the isotope 2He3 to find the real mass lost.

Mass of first neutron in 1H3 that dropped out	1.00628 amu's
Mass of neutron in 1H2 that dropped out	1.00628 amu's
Mass of two neutrons that dropped out	2.01256 amu's

Mass lost creating 2He3	2.01413 amu's
Mass of two neutrons that dropped out	2.01256 amu's
Real mass lost creating 2He3	0.00157 amu

The mass that is lost is lost by the second neutron in the isotope 1H3. Subtract this lost mass from it to find the mass of the single neutron found in the isotope 2He3.

Mass of second neutron in 1H3	1.00194 amu's
Mass lost by it	0.00157 amu
Mass of neutron in 2He3	1.00037 amu's

The isotope 1H3 has electron and proton used as the isotope 1H1 with 1.00783 amu's within the isotopes 1H3 and 2He3. The electron and proton become the electron and proton in the isotope 2He3.

The isotope 1H2 has electron and proton used as the isotope 1H1 with 1.00783 amu's and in the isotopes 1H1 and 2He3 also. The electron becomes the negatron and the proton becomes the positon in the isotope 2He3.

The electrons in the isotopes 1H3 and 1H2 are not lost or released because no 0.00100 amu has been transferred or emitted as a single unit and no 0.00002 amu has been lost creating the isotope 2He3.

The second neutron in the isotope 1H3 lost a single unit of 0.00157 amu. The lost mass of 0.00157 is composed of three sub-units. They are sub-units of 0.00100 amu, 0.00055 amu and 0.00002 amu. The lost mass of 0.00157 amu must be lost as single units to keep the neutron from releasing an electron or negatron from within it.

By adding all of the now known masses of the isotope 2He3, the mass of the isotope 2He3 can be found.

Mass of 1H1 from 1H3 intact within 2He3	1.00783 amu's
Mass of 1H1 from 1H2 intact within 2He3	1.00783 amu's
Mass of neutron in 2He3	1.00037 amu's
Mass of 2He3	3.01603 amu's

Sub-step 38-2-22

2He3

1H3 + 1H2 = 2He3, gamma 591 ray, 1.00628 neutron, 1.00194 neutron

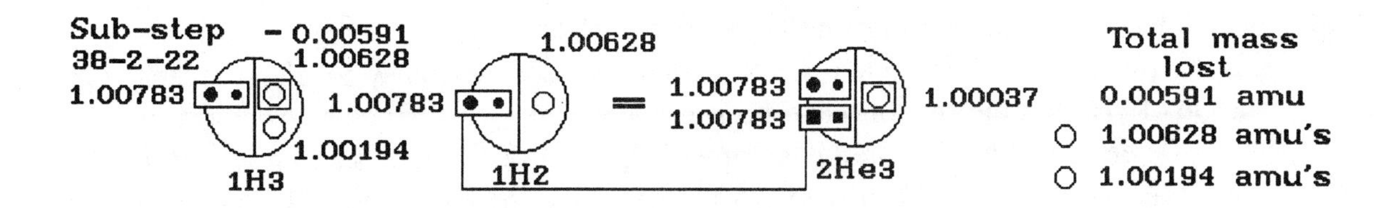

The isotope 1H3 with 3.01605 amu's has one electron with 0.00055 amu and one proton with 1.00728 amu's plus two neutrons. The first neutron has 1.00628 amu's and the second neutron has 1.00194 amu's.

The isotope 1H2 with 2.01411 amu's has one electron with 0.00055 amu and one proton with 1.00728 amu's and one neutron with 1.00628 amu's.

The isotope 2He3 with 3.01603 amu's has one electron with 0.00055 amu, one negatron with 0.00055 amu, one proton with 1.00728 amu's, one positon with 1.00728 amu's and one neutron with 1.00037 amu's.

The total mass lost is 0.00591 amu and it is emitted as a 591 gamma ray which does not have an external magnetic field. It is lost from the first neutron in the isotope 1H3. There are two neutrons that drop out. The first neutron to drop out has 1.00628 amu's and it drops out of the isotope 1H2. The second neutron to drop out comes from the isotope 1H3 with 1.00194 amu's.

The isotope 1H3 with 3.01605 amu's has one electron with 0.00055 amu that is drawn as a dot on the big circle and it becomes the electron with 0.00055 amu in the isotope 2He3 and is also drawn as a dot on the big circle.

The proton with 1.00728 amu's in the isotope 1H3 is drawn as a dot to the left of the vertical line within the big circle. The electron and proton have a rectangle drawn around them to show that they are used as the isotope 1H1 with 1.00783 amu's within the isotopes 1H3 and 2He3. The rectangle shows that the isotope 1H1 can drop out of the isotope 2He3 at some later time.

There are two neutrons in the isotope 1H3 that are drawn one above the other as small circles to the right of the vertical line within the big circle. The top neutron has 1.00628 amu's and is called the first neutron while the neutron below it is the second neutron and it has 1.00194 amu's. The first neutron is needed when the isotope 2He3 is created. The second neutron is not needed so it drops out.

The isotope 1H2 has 2.01411 amu's. Its single electron is drawn as a dot on the big circle in the isotopes 1H2 and it is drawn as a square on the big circle in the isotope 2He3 to show that the electron has become the negatron with 0.00055 amu.

The proton in the isotope 1H2 has 1.00728 amu's and it is drawn as a dot to the left of the vertical line within the big circle in the isotope 1H2 and it is drawn as a small square in the isotope 2He3 to show that the proton in the isotope 1H2 has become the positon in the isotope 2He3. There is a rectangle drawn around the electron and proton in the isotope 1H2 to show that they become the isotope 1H1 with 1.00783 amu's within the isotopes 1H2 and 2He3. The rectangle shows that the isotope 1H1 in the isotope 2He3 can drop out at some later time.

There is a line drawn from the isotope 1H1 found intact within the isotope 1H2 to the isotope 1H1 found intact within the isotope 2He3 to show which isotope of 1H1 has the negatron and positon.

There is one neutron in the isotope 1H2 with 1.00628 amu's and it is drawn as a small circle to the right of the vertical line within the big circle. This neutron is not needed so it drops out.

To find the mass that is lost creating the isotope 2He3, add the masses of the isotopes 1H3 and 1H2 and subtract the mass of the isotope 2He3 from their total.

Mass of 1H3	3.01605 amu's
Mass of 1H2	2.01411 amu's
Total mass of 1H3 and 1H2	5.03016 amu's

Total mass of 1H3 and 1H2	5.03016 amu's
Mass of 2He3	3.01603 amu's
Mass lost creating 2He3	2.01413 amu's

The mass lost creating the isotope 2He3 includes the mass of the two neutrons found in the isotopes 1H3 and 1H2 that dropped out. Subtract their masses from the total mass lost creating the isotope 2He3 to find the real mass lost.

Mass of neutron in 1H2	1.00628 amu's
Mass of second neutron in 1H3	1.00194 amu's
Mass of two neutrons that dropped out	2.00822 amu's

Mass lost creating 2He3	2.01413 amu's
Mass of two neutrons that dropped out	2.00822 amu's
Real mass lost creating 2He3	0.00591 amu

The isotope 1H3 has its electron and proton used as the isotope 1H1 with 1.00783 amu's within the isotopes 1H3 and 2He3. The electron and proton become the electron and proton in the isotope 2He3.

The isotope 1H2 has its electron and proton used as the isotope 1H1 with 1.00783 amu's and in the isotope 2He3 also. The electron becomes the negatron and the proton becomes the positon in the isotope 2He3.

This first neutron in the isotope 1H3 loses 0.00591 amu. Subtract this mass from the first neutron to find the mass of the single neutron it becomes in the isotope 2He3.

Mass of first neutron in 1H3	1.00628 amu's
Mass lost by it	0.00591 amu
Mass of neutron in 2He3	1.00037 amu's

The electrons in the isotopes 1H3 and 1H2 are not lost or released because no 0.00100 amu has been transferred or emitted as a single unit and no 0.00002 amu has been lost creating the isotope 2He3.

The first neutron in the isotope 1H3 lost a unit of 0.00591 amu for a total lost mass of 0.00591 amu and it is lost as a single unit. The lost mass of 0.00591 is composed of four sub-units. They are sub-units of 0.00434 amu, 0.00100 amu, 0.00055 amu and 0.00002 amu. Note that the last three sub-units could be a single sub-unit of 0.00157 amu. The sub-unit of 0.00434 amu is composed of two sub-units of 0.00217 amu. The lost mass of 591 amu must also be lost as a single unit to keep the neutron from releasing an electron or negatron from within it.

The unit of 0.00591 amu could be composed of two other sub-units. They are sub-units of 0.00374 amu and 0.00217 amu.

By adding all of the now known masses of the isotope 2He3, the mass of the isotope 2He3 can be found.

Mass of 1H1 from 1H3 intact within 2He3	1.00783 amu's
Mass of 1H1 from 1H2 intact within 2He3	1.00783 amu's
Mass of neutron in 2He3	1.00037 amu's
Mass of 2He3	3.01603 amu's

Sub-step 37-2-21

2He3

1H3 + 1H2 = 2He3, gamma 591 ray, 1.00628 neutron, 1.00194 neutron

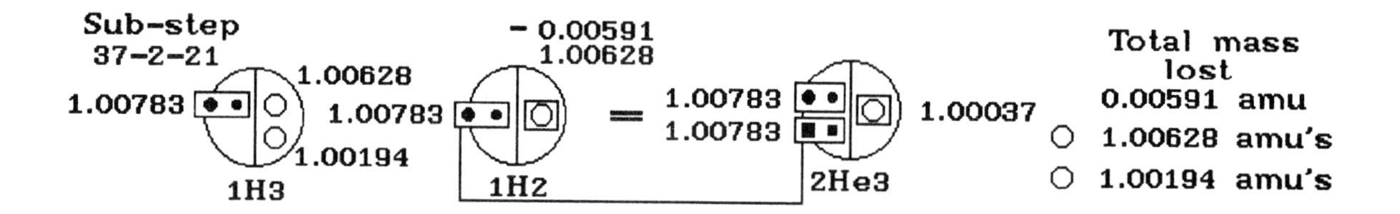

The isotope 1H3 with 3.01605 amu's has one electron with 0.00055 amu and one proton with 1.00728 amu's plus two neutrons. The first neutron has 1.00628 amu's and the second neutron has 1.00194 amu's.

The isotope 1H2 with 2.01411 amu's has one electron with 0.00055 amu and one proton with 1.00728 amu's and one neutron with 1.00628 amu's.

The isotope 2He3 with 3.01603 amu's has one electron with 0.00055 amu, one negatron with 0.00055 amu, one proton with 1.00728 amu's, one positon with 1.00728 amu's and one neutron with 1.00037 amu's.

The total mass lost is 0.00591 amu and it is emitted as a 591 gamma ray which does not have an external magnetic field. There are two neutrons that drop out from the isotope 1H3. The first neutron has 1.00628 amu's and the second one has 1.00194 amu's.

The isotope 1H3 with 3.01605 amu's has one electron with 0.00055 amu that is drawn as a dot on the big circle and it becomes the electron with 0.00055 amu in the isotope 2He3 and is also drawn as a dot on the big circle.

The proton with 1.00728 amu's in the isotope 1H3 is drawn as a dot to the left of the vertical line within the big circle. The electron and proton have a rectangle drawn around them to show that they are used as the isotope 1H1 with 1.00783 amu's within the isotopes 1H3 and 2He3. The rectangle shows that the isotope 1H1 can drop out of the isotope 2He3 at some later time.

There are two neutrons in the isotope 1H3 that are drawn one above the other as small circles to the right of the vertical line within the big circle. The top neutron has 1.00628 amu's and it is called the first neutron while the neutron below it is the second neutron and it has 1.00194 amu's. Neither of these neutrons are needed when the isotope 2He3 is created so they both drop out.

The isotope 1H2 has 2.01411 amu's. Its single electron is drawn as a dot on the big circle in the isotopes 1H2 and it is drawn as a square on the big circle in the isotope 2He3 to show that the electron has become the negatron with 0.00055 amu.

The proton in the isotope 1H2 has 1.00728 amu's and it is drawn as a dot to the left of the vertical line within the big circle and it is drawn as a small square in the isotope 2He3 to show that the proton in the isotope 1H2 has become the positon in the isotope 2He3. There is a rectangle drawn around both of them in the isotope 1H2 to show that they become the isotope 1H1 within the isotopes 1H2 and 2He3. The rectangle shows that the isotope 1H1 in the isotope 2He3 can drop out at some later time.

There is a line drawn from the isotope 1H1 found intact within the isotope 1H2 to the isotope 1H1 found intact within the isotope 2He3 to show which isotope of 1H1 has the negatron and positon.

There is one neutron in the isotope 1H2 with 1.00628 amu's and it is drawn as a small circle to the right of the vertical line within the big circle. There is a square drawn around it to show that it has lost 0.00591 amu in the isotope 1H2 and it is drawn this same way in the isotope 2He3.

To find the mass that is lost creating the isotope 2He3, add the masses of the isotopes 1H3 and 1H2 and subtract the mass of the isotope 2He3 from their total.

Mass of 1H3	3.01605 amu's
Mass of 1H2	2.01411 amu's
Total mass of 1H3 and 1H2	5.03016 amu's

Total mass of 1H3 and 1H2	5.03016 amu's
Mass of 2He3	3.01603 amu's
Mass lost creating 2He3	2.01413 amu's

The mass lost creating the isotope 2He3 includes the mass of the two neutrons found in the isotope 1H3 that dropped out. Subtract their masses from the total mass lost creating the isotope 2He3 to find the real mass lost.

Mass of first neutron in 1H3	1.00628 amu's
Mass of second neutron in 1H3	1.00194 amu's
Mass of two neutrons that dropped out of 1H3	2.00822 amu's

Mass lost creating 2He3	2.01413 amu's
Mass of two neutrons that dropped out of 1H3	2.00822 amu's
Real mass lost creating 2He3	0.00591 amu

The isotope 1H3 has its electron and proton used as the isotope 1H1 with 1.00783 amu's within the isotopes 1H3 and 2He3. The electron and proton become the electron and proton in the isotope 2He3.

The isotope 1H2 has its electron and proton used as the isotope 1H1 with 1.00783 amu's and in the isotope 2He3 also. The electron becomes the negatron and the proton becomes the positon.

This single neutron in the isotope 1H2 loses 0.00591 amu so subtract this mass from the single neutron to find the mass of the single neutron it becomes in the isotope 2He3.

Mass of neutron in 1H2	1.00628 amu's
Mass lost by it	0.00591 amu
Mass of neutron in 2He3	1.00037 amu's

The electrons in the isotopes 1H3 and 1H2 are not lost or released because no 0.00100 amu has been transferred or emitted as a single unit and no 0.00002 amu has been lost creating the isotope 2He3.

The single neutron in the isotope 1H2 lost a unit of 0.00591 amu for a total lost mass of 0.00591 amu and it is lost as a single unit. The lost mass of 0.00591 is composed of four sub-units. They are sub-units of 0.00434 amu, 0.00100 amu, 0.00055 amu and 0.00002 amu. The last three sub-units can make a sub-unit of 0.00157 amu. The sub-unit of 0.00434 amu is composed of two sub-units of 0.00217 amu. The lost mass of 0.00591 amu must be lost as a single unit to keep the neutron from releasing an electron or negatron from within it.

The lost mass of 0.00591 amu can also be lost as two sub-units. One is a sub-unit of 374 amu and the other is a sub-unit of 0.00217 amu.

By adding all of the now known masses of the isotope 2He3, the mass of the isotope 2He3 can be found.

Mass of 1H1 from 1H3 intact within 2He3	1.00783 amu's
Mass of 1H1 from 1H2 intact within 2He3	1.00783 amu's
Mass of neutron in 2He3	1.00037 amu's
Mass of 2He3	3.01603 amu's

Sub-step 36-2-20

2He3

1H3 + 1H2 = 2He3, alpha 434 ray, gamma 157 ray, 1.00628 neutron,
1.00194 neutron

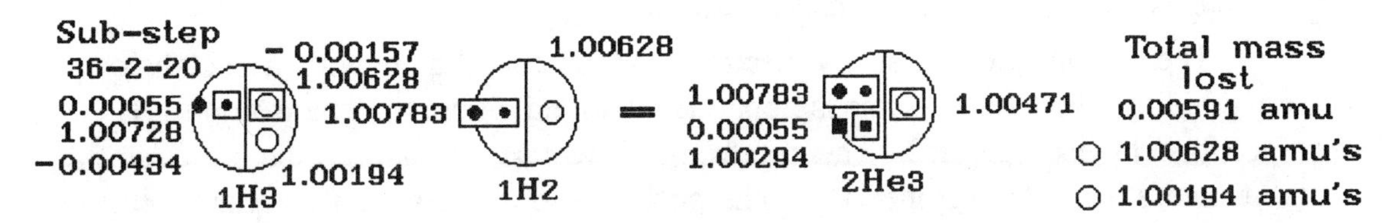

The isotope 1H3 with 3.01605 amu's has one electron with 0.00055 amu and one proton with 1.00728 amu's plus two neutrons. The first neutron has 1.00628 amu's and the second neutron has 1.00194 amu's.

The isotope 1H2 with 2.01411 amu's has one electron with 0.00055 amu and one proton with 1.00728 amu's and one neutron with 1.00628 amu's.

The isotope 2He3 with 3.01603 amu's has one electron with 0.00055 amu, one negatron with 0.00055 amu, one proton with 1.00728 amu's, one positon with 1.00294 amu's and one neutron with 1.00471 amu's.

The total mass lost is 0.00591 amu and it is emitted as a 434 alpha ray which has an external magnetic field and a 157 gamma ray which has no external magnetic field. There are two neutrons that drop out of the isotope 1H3. The first neutron has 1.00628 amu's and the second neutron in the isotope 1H3 has 1.00194 amu's.

The isotope 1H3 with 3.01605 amu's has one electron with 0.00055 amu. The electron is drawn as a dot on the big circle in the isotope 1H3 and it becomes the negatron with 0.00055 amu in the isotope 2He3 and is drawn as a square on the big circle.

The proton with 1.00728 amu's in the isotope 1H3 is drawn as a dot to the left of the vertical line within the big circle. There is a square drawn around it to show that it loses 0.00434 amu. It is drawn as a small square with a square around it in the isotope 2He3 to show that the proton in the isotope 1H3 is the positon with 1.00294 amu's in the isotope 2He3.

There are two neutrons in the isotope 1H3 that are drawn one above the other as small circles to the right of the vertical line within the big circle. The top neutron has 1.00628 amu's and is called the first neutron while the neutron below it is the second neutron and it has 1.00194 amu's. The first neutron becomes the single neutron with 1.00471 amu's found in the isotope 2He3. The second neutron in the isotope 1H3 is not needed when the isotope 2He3 is created so it drops out.

The isotope 1H2 has 2.01411 amu's. Its single electron is drawn as a dot on the big circle in the isotopes 1H2 and 2He3. The proton has 1.00728 amu's and it is drawn as a dot to the left of the vertical line within the big circle in the isotopes 1H2 and 2He3.

The electron and proton in the isotope 1H2 have a rectangle drawn around both of them to show that they are used as the isotope 1H1 with 1.00783 amu's in the isotopes 1H2 and 2He3. The rectangle in the isotope 2He3 shows that the isotope 1H1 can drop out of the isotope 2He3 at a later time.

There is one neutron in the isotope 1H2 with 1.00628 amu's and it is drawn as a small circle to the right of the vertical line within the big circle. This neutron is not needed when the isotope 2He3 is created so it drops out.

The isotope 2He3 has its electron and proton drawn with a rectangle around both of them to show the isotope 1H1 is found within it. The isotope 1H1 comes from the isotope 1H2. The negatron is drawn as a square on the big circle and it comes from the isotope 1H3. The positon is drawn as a small square with another square drawn around it to show that it came from the isotope 1H3. The single neutron is drawn as a small circle to the right of the vertical line within the big circle. This neutron has a square drawn around it to show it came from the neutron in the isotope 1H3.

To find the mass that is lost creating the isotope 2He3, add the masses of the isotopes 1H3 and 1H2 and subtract the mass of the isotope 2He3 from their total.

Mass of 1H3	3.01605 amu's
Mass of 1H2	2.01411 amu's
Total mass of 1H3 and 1H2	5.03016 amu's

Total mass of 1H3 and 1H2	5.03016 amu's
Mass of 2He3	3.01603 amu's
Mass lost creating 2He3	2.01413 amu's

The mass lost creating the isotope 2He3 includes the mass of the second neutron found in the isotope 1H3 that dropped out and the single neutron found in the isotope 1H2. Subtract their masses from the total mass lost creating the isotope 2He3 to find the real mass lost.

Mass of neutron in 1H2 that dropped out	1.00628 amu's
Mass of second neutron in 1H3 that dropped out	1.00194 amu's
Mass of two neutrons that dropped out	2.00822 amu's

Mass lost creating 2He3	2.01413 amu's
Mass of two neutrons that dropped out	2.00822 amu's
Real mass lost creating 2He3	0.00591 amu

The isotope 1H2 has its electron and proton used as the isotope 1H1 within the isotopes 1H2 and 2He3. Its neutron with 1.00628 amu's drops out. The electron and proton become the electron and proton in the isotope 2He3. This means that the only proton that can become the positon in the isotope 2He3 must come from the isotope 1H3. This proton loses 0.00434 amu so subtract this mass from the proton to find the mass of the positon it becomes in the isotope 2He3.

Mass of proton in 1H3	1.00728 amu's
Mass lost by it	0.00434 amu
Mass of positon in 2He3	1.00294 amu's

Since the proton in the isotope 1H3 lost 0.00434 amu, the mass the first neutron loses can be found by subtracting the mass the proton lost from the total mass lost creating the isotope 2He3.

Total mass lost creating 2He3	0.00591 amu
Mass lost by proton in 1H3	0.00434 amu
Mass lost by the first neutron in 1H3	0.00157 amu

The first neutron with 1.00628 amu's in the isotope 1H3 loses 0.00157 amu to become the single neutron in the isotope 2He3. Subtract the lost mass from the mass of the first neutron in the isotope 1H3 to find the mass of the single neutron in the isotope 2He3.

Mass of first neutron in 1H3	1.00628 amu's
Mass lost by it	0.00157 amu
Mass of neutron in 2He3	1.00471 amu's

The electrons in the isotopes 1H3 and 1H2 are not lost or released because no 0.00100 amu has been transferred or emitted as a single unit and no 0.00002 amu has been lost creating the isotope 2He3.

The proton in the isotope 1H3 lost a unit of 0.00434 amu and the first neutron lost a unit of 0.00157 amu for a total lost mass of 0.00591 amu. They are each lost as a single unit. The unit of 0.00434 is composed of two sub-units of 0.00217 amu. The unit of 0.00157 amu is composed of three sub-units. They are sub-units of 0.00100 amu, 0.00055 amu and 0.00002 amu. In order for the electron or negatron to stay with their proton or positon, these sub-units must be lost as single units of 0.00434 amu and 0.00157 amu. They must also be lost as single units in order for the proton, positon or neutron from releasing an electron or negatron from within any of them.

By adding all of the now known masses of the isotope 2He3, the mass of the isotope 2He3 can be found.

Mass of 1H1 intact within 2He3	1.00783 amu's
Mass of neutron in 2He3	1.00471 amu's
Mass of negatron in 2He3	0.00055 amu
Mass of positon in 2He3	1.00294 amu's
Mass of 2He3	3.01603 amu's

Sub-step 35-2-19

2He3

1H3 + 1H2 = 2He3, alpha 217 ray, gamma 374 ray, 1.00628 neutron, 1.00194 neutron

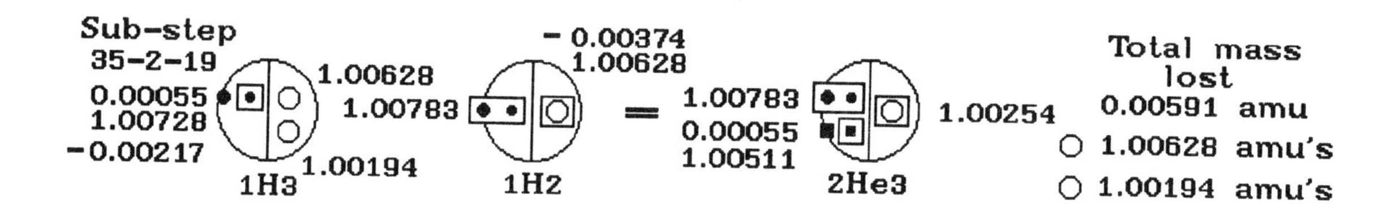

The isotope 1H3 with 3.01605 amu's has one electron with 0.00055 amu and one proton with 1.00728 amu's plus two neutrons. The first neutron has 1.00628 amu's and the second neutron has 1.00194 amu's.

The isotope 1H2 with 2.01411 amu's has one electron with 0.00055 amu and one proton with 1.00728 amu's and one neutron with 1.00628 amu's.

The isotope 2He3 with 3.01603 amu's has one electron with 0.00055 amu, one negatron with 0.00055 amu, one proton with 1.00728 amu's, one positon with 1.00511 amu's and one neutron with 1.00254 amu's.

The total mass lost is 0.00591 amu and it is emitted as a 217 alpha ray which has an external magnetic field and a 374 gamma ray which has no external magnetic field. There are two neutrons that drop out from the isotope 1H3. The first neutron has 1.00628 amu's and the second one has 1.00194 amu's.

The isotope 1H3 with 3.01605 amu's has one electron with 0.00055 amu. The electron in the isotope 1H3 is drawn as a dot on the big circle and it becomes the negatron with 0.00055 amu which is drawn as a square on the big circle in the isotope 2He3.

The proton with 1.00728 amu's in the isotope 1H3 is drawn as a dot to the left of the vertical line within the big circle. There is a square drawn around it to show that it loses 0.00217 amu. It is drawn as a small square with a square around it in the isotope 2He3 to show that the proton in the isotope 1H3 is the positon with 1.00511 amu's in the isotope 2He3.

There are two neutrons in the isotope 1H3 that are drawn one above the other as small circles to the right of the vertical line within the big circle. The top neutron has 1.00628 amu's and is called the first neutron while the neutron below it is the second neutron and it has 1.00194 amu's. Neither of these neutrons are needed when the isotope 2He3 is created so they both drop out.

The isotope 1H2 has 2.01411 amu's. Its single electron is drawn as a dot on the big circle in the isotopes 1H2 and 2He3. The proton in the isotope 1H2 has 1.00728 amu's and it is drawn as a dot to the left of the vertical line within the big circle in the isotopes 1H2 and 2He3.

The electron and proton in the isotope 1H2 has a rectangle drawn around both of them to show that they are used as the isotope 1H1 with 1.00783 amu's in the isotopes 1H2 and 2He3. The rectangle in the isotope 2He3 shows that the isotope 1H1 can drop out of the isotope 2He3 at a later time.

There is one neutron in the isotope 1H2 with 1.00628 amu's and it is drawn as a small circle to the right of the vertical line within the big circle. The single neutron in the isotope 1H2 loses 0.00374 amu to become the single neutron in the isotope 2He3 with 1.00254 amu's. There is a square drawn around the small circle to show that 0.00374 amu has been lost.

The isotope 2He3 has its electron and proton drawn with a rectangle around both to show that the isotope 1H1 is found intact within it. The negatron is drawn as a square on the big circle and it comes from the isotope 1H3 where it is drawn as a dot. The positon is drawn as a small square with another square drawn around it to show that it came from the isotope 1H3. The single neutron in the isotope 2He3 is drawn as a small circle with a square drawn around it. The square shows that this neutron came from the isotope 1H2.

To find the mass that is lost creating the isotope 2He3, add the masses of the isotopes 1H3 and 1H2 and subtract the mass of the isotope 2He3 from their total.

Mass of 1H3	3.01605 amu's
Mass of 1H2	2.01411 amu's
Total mass of 1H3 and 1H2	5.03016 amu's

Total mass of 1H3 and 1H2	5.03016 amu's
Mass of 2He3	3.01603 amu's
Mass lost creating 2He3	2.01413 amu's

The mass lost creating the isotope 2He3 includes the mass of the two neutrons found in the isotope 1H3 that dropped out. Subtract their masses from the total mass lost creating the isotope 2He3 to find the real mass lost.

Mass of first neutron in 1H3	1.00628 amu's
Mass of second neutron in 1H3	1.00194 amu's
Mass of two neutrons that dropped out of 1H3	2.00822 amu's

Mass lost creating 2He3	2.01413 amu's
Mass of two neutrons that dropped out of 1H3	2.00822 amu's
Real mass lost creating 2He3	0.00591 amu

The isotope 1H2 has its electron and proton used as the isotope 1H1 within the isotopes 1H2 and 2He3. Its neutron with 1.00628 amu's becomes the single neutron in the isotope 2He3 when it loses 0.00374 amu. The electron and proton become the electron and proton in the isotope 2He3. This means that the only proton that can become the positon in the isotope 2He3 must come from the isotope 1H3. This proton loses 0.00217 amu so subtract this mass from the proton to find the mass of the positon it becomes in the isotope 2He3.

Mass of proton in 1H3	1.00728 amu's
Mass lost by it	0.00217 amu
Mass of positon in 2He3	1.00511 amu's

Since the proton in the isotope 1H3 lost 0.00217 amu, the mass lost by the single neutron in the isotope 1H2 can be found by subtracting the mass the proton lost from the total mass lost creating the isotope 2He3.

Total mass lost creating the isotope 2He3	0.00591 amu
Mass lost by proton in 1H3	0.00217 amu
Mass lost by neutron in 1H2	0.00374 amu

The single neutron in the isotope 1H2 loses 0.00374 amu to become the single neutron in the isotope 2He3. To find the mass of the single neutron in the isotope 2He3, subtract the lost mass from the single neutron in the isotope 1H2.

Mass of neutron in 1H2	1.00628 amu's
Mass lost by it	0.00374 amu
Mass of neutron in 2He3	1.00254 amu's

The electrons in the isotopes 1H3 and 1H2 are not lost or released because no 0.00100 amu has been transferred or emitted as a single unit and no 0.00002 amu has been lost creating the isotope 2He3.

The proton in the isotope 1H3 lost a unit of 0.00217 amu and the single neutron in the isotope 1H2 lost a unit of 0.00374 amu for a total lost mass of 0.00591 amu and they are each lost as a single unit. The lost mass of 0.00217 is composed of a single unit. It does not have any sub-units. The lost mass of 0.00374 amu is composed of four sub-units. They are sub-units of 0.00217 amu, 0.00100 amu, 0.00055 amu and 0.00002 amu. Note that there are two masses of 0.00217 amu. One is lost as a single unit while the other is lost as sub-unit of the lost mass of 0.00374 amu. In order for the electron or negatron to stay with their proton or positon, these sub-units must be lost as single units of 0.00217 and 0.00374 amu. They must also be lost as a single units to keep the proton, positon or neutron from releasing an electron or negatron from within any of them.

By adding all of the now known masses of the isotope 2He3, the mass of the isotope 2He3 can be found.

Mass of 1H1 intact within 2He3	1.00783 amu's
Mass of neutron in 2He3	1.00254 amu's
Mass of negatron in 2He3	0.00055 amu
Mass of positon in 2He3	1.00511 amu's
Mass of 2He3	3.01603 amu's

Sub-step 34-2-18

2He3

1H3 + 1H2 = 2He3, alpha 374 ray, gamma 217 ray, 1.00628 neutron, 1.00194 neutron

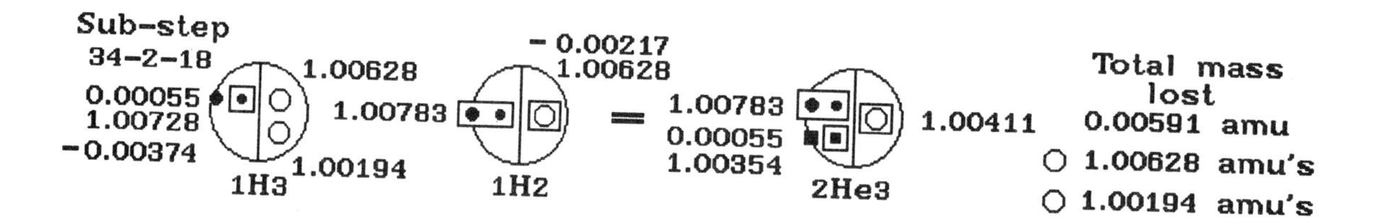

The isotope 1H3 with 3.01605 amu's has one electron with 0.00055 amu and one proton with 1.00728 amu's plus two neutrons. The first neutron has 1.00628 amu's and the second neutron has 1.00194 amu's.

The isotope 1H2 with 2.01411 amu's has one electron with 0.00055 amu and one proton with 1.00728 amu's and one neutron with 1.00628 amu's.

The isotope 2He3 with 3.01603 amu's has one electron with 0.00055 amu, one negatron with 0.00055 amu, one proton with 1.00728 amu's, one positon with 1.00354 amu's and one neutron with 1.00411 amu's.

The total mass lost is 0.00591 amu and it is emitted as a 374 alpha ray which has an external magnetic field and a 217 gamma ray which has no external magnetic field. There are two neutrons that drop out from the isotope 1H3. The first neutron has 1.00628 amu's and the second one has 1.00194 amu's.

The isotope 1H3 with 3.01605 amu's has one electron with 0.00055 amu that is drawn as a dot on the big circle and it becomes the negatron with 0.00055 amu in the isotope 2He3 and is drawn as a square on the big circle.

The proton with 1.00728 amu's in the isotope 1H3 is drawn as a dot to the left of the vertical line within the big circle. There is a square drawn around it to show that it loses 0.00374 amu. It is drawn as a square with a square around it in the isotope 2He3 to show that the proton in the isotope 1H3 is the positon with 1.00354 amu's in the isotope 2He3.

There are two neutrons in the isotope 1H3 that are drawn one above the other as small circles to the right of the vertical line within the big circle. The top neutron has 1.00628 amu's and is called the first neutron while the neutron below it is the second neutron and it has 1.00194 amu's. Neither of these neutrons are needed when the isotope 2He3 is created so they both drop out.

The isotope 1H2 has 2.01411 amu's. Its single electron is drawn as a dot on the big circle in the isotopes 1H2 and 2He3. The proton in the isotope 1H2 has 1.00728 amu's and it is drawn as a dot to the left of the vertical line within the big circle in the isotopes 1H2 and 2He3.

The electron and proton in the isotope 1H2 have a rectangle drawn around both of them to show that they are used as the isotope 1H1 with 1.00783 amu's in the isotope 2He3. They are drawn the same way in the isotope 2He3. The rectangle in the isotope 2He3 shows that the isotope 1H1 can drop out of the isotope 2He3 at a later time.

There is one neutron in the isotope 1H2 with 1.00628 amu's and it is drawn as a small circle to the right of the vertical line within the big circle. The single neutron in the isotope 1H2 loses 0.00217 amu to become the single neutron in the isotope 2He3 with 1.00411 amu's. The neutron has a square drawn around the small circle to show that it has lost 0.00217 amu to become the single neutron in the isotope 2He3 with 1.00411 amu's. The single neutron in the isotope 2He3 is drawn the same way as the neutron in the isotope 1H2.

The isotope 2He3 has 3.01603 amu's. It has one electron and one negatron with each having 0.00055 amu. The electron is drawn as a dot on the big circle. The negatron is drawn as a square on the big circle below the electron. There is one proton and one positon with the proton drawn above the positon. The proton is drawn as a dot with 1.00728 amu's. There is a rectangle around it and its electron to show that there is an isotope of 1H1 within the isotopes 1H2 and 2He3. The rectangle shows that the isotope 1H1 can drop out of the isotope 2He3 at some later time.

The positon with 1.00354 amu's is drawn as a small square below the proton. It has a square drawn around the small square to show that the positon came from the proton in the isotope 1H3. The proton and positon are both drawn to the left of the vertical line within the big circle. The single neutron with 1.00411 amu's is drawn as a small circle to the right of the vertical line within the big circle. It has a square drawn around it to show that it came from the isotope 1H2. They are drawn alike.

To find the mass that is lost creating the isotope 2He3, add the masses of the isotopes 1H3 and 1H2 and subtract the mass of the isotope 2He3 from their total.

Mass of 1H3	3.01605 amu's
Mass of 1H2	2.01411 amu's
Total mass of 1H3 and 1H2	5.03016 amu's

Total mass of 1H3 and 1H2	5.03016 amu's
Mass of 2He3	3.01603 amu's
Mass lost creating 2He3	2.01413 amu's

The mass lost creating the isotope 2He3 includes the mass of the two neutrons found in the isotope 1H3 that dropped out. Subtract their masses from the total mass lost creating the isotope 2He3 to find the real mass lost.

Mass of first neutron in 1H3	1.00628 amu's
Mass of second neutron in 1H3	1.00194 amu's
Mass of two neutrons that dropped out of 1H3	2.00822 amu's

Mass lost creating 2He3	2.01413 amu's
Mass of two neutrons that dropped out of 1H3	2.00822 amu's
Real mass lost creating 2He3	0.00591 amu

The isotope 1H2 has its electron and proton used as the isotope 1H1 within the isotopes 1H2 and 2He3. Its neutron with 1.00628 amu's becomes the single neutron in the isotope 2He3 when it loses 0.00217 amu. The electron and proton become the electron and proton in the isotope 2He3. This means that the only proton that can become the positon in the isotope 2He3 must come from the isotope 1H3. This proton loses 0.00374 amu. Subtract this mass from the proton to find the mass of the positon it becomes in the isotope 2He3.

Mass of proton in 1H3	1.00728 amu's
Mass lost by it	0.00374 amu
Mass of positon in 2He3	1.00354 amu's

Since the proton in the isotope 1H3 lost 0.00374 amu, the mass the single neutron in the isotope 1H2 loses can be found by subtracting the mass the proton lost from the total mass lost creating the isotope 2He3.

Total mass lost creating 2He3	0.00591 amu
Mass lost by proton in 1H3	0.00374 amu
Mass lost by neutron in 1H2	0.00217 amu

The single neutron in the isotope 1H2 loses 0.00217 amu to become the single neutron in the isotope 2He3. To find the mass of the single neutron in the isotope 2He3, subtract the lost mass from the single neutron in the isotope 1H2.

Mass of neutron in 1H2	1.00628 amu's
Mass lost by it	0.00217 amu
Mass of neutron in 2He3	1.00411 amu's

The electrons in the isotopes 1H3 and 1H2 are not lost or released because no 0.00100 amu has been transferred or emitted as a single unit and no 0.00002 amu has been lost creating the isotope 2He3.

The proton in the isotope 1H3 lost a unit of 0.00374 amu and the single neutron in the isotope 1H2 lost a unit of 0.00217 amu for a total lost mass of 0.00591 amu. They are each lost as a single unit. The lost mass of 0.00374 amu is composed of four sub-units. They are sub-units of 0.00217 amu, 0.00100 amu, 0.00055 amu and 0.00002 amu. The lost mass of 0.00217 is composed of a single unit. Note that there are two masses of 0.00217 amu. One is lost as a single unit while the other is lost as a sub-unit of the lost mass of 0.00374 amu. In order for the electron or negatron to stay with their proton or positon, these sub-units must be lost as single units of 0.00374 and 0.00217 amu. They must also be lost as a single units to keep the proton, positon or neutron from releasing an electron or negatron from within any of them.

By adding all of the now known masses of the isotope 2He3, the mass of the isotope 2He3 can be found.

Mass of 1H1 intact within 2He3	1.00783 amu's
Mass of neutron in 2He3	1.00411 amu's
Mass of negatron in 2He3	0.00055 amu
Mass of positon in 2He3	1.00354 amu's
Mass of 2He3	3.01603 amu's

Sub-step 33-2-17

2He3

1H3 + 1H2 = 2He3, alpha 157 ray, gamma 434 ray, 1.00628 neutron, 1.00194 neutron

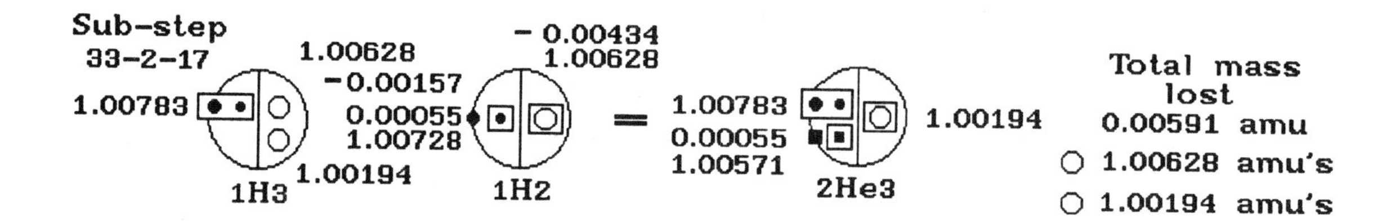

The isotope 1H3 with 3.01605 amu's has one electron with 0.00055 amu and one proton with 1.00728 amu's plus two neutrons. The first neutron has 1.00628 amu's and the second neutron has 1.00194 amu's.

The isotope 1H2 with 2.01411 amu's has one electron with 0.00055 amu and one proton with 1.00728 amu's and one neutron with 1.00628 amu's.

The isotope 2He3 with 3.01603 amu's has one electron with 0.00055 amu, one negatron with 0.00055 amu, one proton with 1.00728 amu's, one positon with 1.00571 amu's and one neutron with 1.00194 amu's.

The total mass lost is 0.00591 amu and it is emitted as a 157 alpha ray which has an external magnetic field and a 434 gamma ray which has no external magnetic field. There are two neutrons that drop out of the isotope 1H3. The first neutron drops out of the isotope 1H3 has 1.00628 amu's and the second isotope to drop out has 1.00194 amu's.

The isotope 1H3 with 3.01605 amu's has one electron with 0.00055 amu that is drawn as a dot on the big circle and it becomes the electron with 0.00055 amu in the isotope 2He3 and is also drawn as a dot on the big circle.

The proton with 1.00728 amu's in the isotope 1H3 is drawn as a dot to the left of the vertical line within the big circle. It is drawn as a dot in the isotope 2He3. The electron and proton have a rectangle drawn around them to show that they are used as the isotope 1H1 with 1.00783 amu's within the isotopes 1H3 and 2He3. The rectangle also shows that the isotope 1H1 can drop out of the isotope 2He3 at some later time.

There are two neutrons in the isotope 1H3 that are drawn one above the other as small circles to the right of the vertical line within the big circle. The top neutron has 1.00628 amu's and is called the first neutron while the neutron below it is the second neutron and it has 1.00194 amu's. Neither of these neutrons are needed when the isotope 2He3 is created so they both drop out.

The isotope 1H2 has 2.01411 amu's. Its single electron is drawn as a dot on the big circle in the isotopes 1H2 and it is drawn as a square on the big circle in the isotope 2He3 to show that the electron has become the negatron with 0.00055 amu.

The proton in the isotope 1H2 has 1.00728 amu's and it is drawn as a dot to the left of the vertical line within the big circle and it is drawn as a small square in the isotope 2He3 to show that the proton in the isotope 1H2 has become the positon in the isotope 2He3. There is a square drawn around the dot in the isotope 1H2 and a square around the small square in the isotope 2He3 to show that the proton in the isotope 1H2 has lost 0.00157 amu to become the positon in the isotope 2He3 with 1.00571 amu's.

There is one neutron in the isotope 1H2 with 1.00628 amu's and it is drawn as a small circle to the right of the vertical line within the big circle. There is a square drawn around the neutron in the isotopes 1H2 and 2He3 to show that this neutron has lost 0.00434 amu in the isotope 1H2 to become the single neutron in the isotope 2He3 with 1.00194 amu's.

The isotope 2He3 has 3.01603 amu's. It has one electron and one negatron with each having 0.00055 amu. The electron is drawn as a dot on the big circle. The negatron is drawn as a square on the big circle below the electron. There is one proton and one positon with the proton drawn above the positon. The proton is drawn as a dot with 1.00728 amu's. There is a rectangle around it and its electron to show that there is an isotope of 1H1 within the isotope 2He3 and 1H3. The rectangle shows that the isotope 1H1 can drop out of the isotope 2He3 at some later time. The positon with 1.00354 amu's is drawn as a small square below the proton. It has a square drawn around the small square to show that the positon came from the proton in the isotope 1H2. The proton and positon are both drawn to the left of the vertical line within the big circle. The single neutron with 1.00411 amu's is drawn as a small circle to the right of the vertical line within the big circle. It has a square drawn around it to show that it came from the isotope 1H2. They are drawn alike.

It should be noted that there is a neutron in the isotope 1H3 that drops out with 1.00194 amu's and there is a neutron in the isotope 1H2 that changes into this same mass. This happens because it is the isotope 1H2 that loses the lost mass forcing the neutron in the isotope 1H3 to drop out.

To find the mass that is lost creating the isotope 2He3, add the masses of the isotopes 1H3 and 1H2 and subtract the mass of the isotope 2He3 from their total.

Mass of 1H3	3.01605 amu's
Mass of 1H2	2.01411 amu's
Total mass of 1H3 and 1H2	5.03016 amu's

Total mass of 1H3 and 1H2	5.03016 amu's
Mass of 2He3	3.01603 amu's
Mass lost creating 2He3	2.01413 amu's

The mass lost creating the isotope 2He3 includes the mass of the two neutrons that dropped out of the isotope 1H3. Subtract their masses from the total mass lost creating the isotope 2He3 to find the real mass lost.

Mass of first neutron in 1H3	1.00628 amu's
Mass of second neutron in 1H3	1.00194 amu's
Mass of two neutrons that dropped out of 1H3	2.00822 amu's

Mass lost creating 2He3	2.01413 amu's
Mass of two neutrons that dropped out of 1H3	2.00822 amu's
Real mass lost creating 2He3	0.00591 amu

The isotope 1H3 has its electron and proton used as the isotope 1H1 within the isotopes 1H3 and 2He3. The electron and proton become the electron and proton in the isotope 2He3. This means that the only proton that can become the positon in the isotope 2He3 must come from the isotope 1H2. This proton loses 0.00157 amu. Subtract this lost mass from the proton to find the mass of the positon it becomes in the isotope 2He3.

Mass of proton in 1H2	1.00728 amu's
Mass lost by it	0.00157 amu
Mass of positon in 2He3	1.00571 amu's

Since the proton in the isotope 1H2 lost 0.00157 amu, the mass by the single neutron in the isotope 1H2 can be found by subtracting the mass the proton lost from the total mass lost creating the isotope 2He3.

Total mass lost creating 2He3	0.00591 amu
Mass lost by proton in 1H2	0.00157 amu
Mass lost by neutron in 1H2	0.00434 amu

The single neutron in the isotope 1H2 loses 0.00434 amu to become the single neutron in the isotope 2He3. Subtract this lost mass from the single neutron in the isotope 1H2 to find the mass of the single neutron in the isotope 2He3.

Mass of neutron in 1H2	1.00628 amu's
Mass lost by it	0.00434 amu
Mass of neutron in 2He3	1.00194 amu's

The electrons in the isotopes 1H3 and 1H2 are not lost or released because no 0.00100 amu has been transferred or emitted as a single unit and no 0.00002 amu has been lost creating the isotope 2He3.

The proton in the isotope 1H2 lost a unit of 0.00157 amu and the single neutron lost a unit of 0.00434 amu for a total lost mass of 0.00591 amu. They are each lost as a single unit. The lost mass of 0.00157 is composed of three sub-units. They are sub-units of 0.00100 amu, 0.00055 amu and 0.00002 amu. The sub-unit of 0.00434 amu is composed of two sub-units of 0.00217 amu. In order for the electron or negatron to stay with their proton or positon, these sub-units must be lost as a single unit of 0.00157 amu and 0.00434 amu. They must also be lost as a single unit to keep the proton, positon or neutron from releasing an electron or negatron from within any of them.

By adding all of the now known masses of the isotope 2He3, the mass of the isotope 2He3 can be found.

Mass of 1H1 intact within 2He3	1.00783 amu's
Mass of neutron in 2He3	1.00194 amu's
Mass of negatron in 2He3	0.00055 amu
Mass of positon in 2He3	1.00571 amu's
Mass of 2He3	3.01603 amu's

Sub-step 32-2-16

2He3

1H3 + 1H2 = 2He3, alpha 157 ray, gamma 434 ray, 1.00628 neutron, 1.00194 neutron

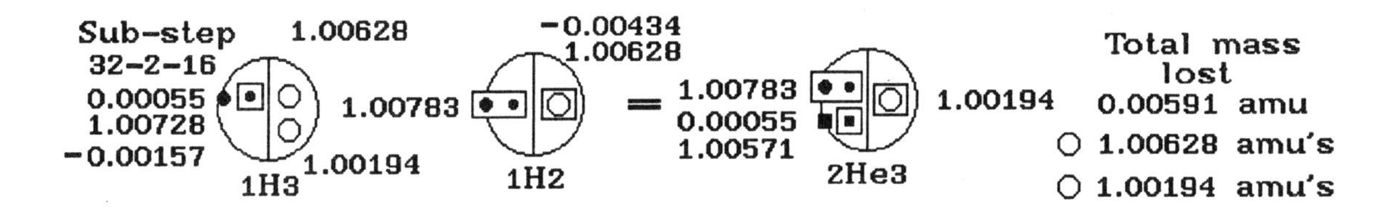

The isotope 1H3 with 3.01605 amu's has one electron with 0.00055 amu and one proton with 1.00728 amu's plus two neutrons. The first neutron has 1.00628 amu's and the second neutron has 1.00194 amu's.

The isotope 1H2 with 2.01411 amu's has one electron with 0.00055 amu and one proton with 1.00728 amu's and one neutron with 1.00628 amu's.

The isotope 2He3 with 3.01603 amu's has one electron with 0.00055 amu, one negatron with 0.00055 amu, one proton with 1.00728 amu's, one positon with 1.00571 amu's and one neutron with 1.00194 amu's.

The total mass lost is 0.00591 amu and it is emitted as a 157 alpha ray which has an external magnetic field and a 434 gamma ray which does not have an external magnetic field. There are two neutrons that drop out from the isotope 1H3. The first neutron has 1.00628 amu's and the second one has 1.00194 amu's.

The isotope 1H3 with 3.01605 amu's has one electron with 0.00055 amu that is drawn as a dot on the big circle and it becomes the negatron with 0.00055 amu in the isotope 2He3 and is drawn as a square on the big circle.

The proton with 1.00728 amu's in the isotope 1H3 is drawn as a dot to the left of the vertical line within the big circle. There is a square drawn around it to show that it loses 0.00157 amu. It is drawn as a square with a square around it in the isotope 2He3 to show that the proton in the isotope 1H3 is the positon with 1.00571 amu's in the isotope 2He3.

There are two neutrons in the isotope 1H3 that are drawn one above the other as small circles to the right of the vertical line within the big circle. The top neutron has 1.00628 amu's and is called the first neutron while the neutron below it is the second neutron and it has 1.00194 amu's. Neither of these neutrons are needed when the isotope 2He3 is created so they both drop out.

The isotope 1H2 has 2.01411 amu's. Its single electron is drawn as a dot on the big circle in the isotopes 1H2 and 2He3. The proton in the isotope 1H2 has 1.00728 amu's and it is drawn as a dot to the left of the vertical line within the big circle in the isotopes 1H2 and 2He3.

The electron and proton in the isotope 1H2 has a rectangle drawn around both of them to show that they are used as the isotope 1H1 with 1.00783 amu's in the isotope 2He3. They are drawn the same way in the isotope 2He3. The rectangle in the isotope 2He3 shows that the isotope 1H1 can drop out of the isotope 2He3 at a later time.

There is one neutron in the isotope 1H2 with 1.00628 amu's and it is drawn as a small circle to the right of the vertical line within the big circle. The single neutron in the isotope 1H2 loses 0.00434 amu to become the single neutron in the isotope 2He3 with 1.00194 amu's.

The isotope 2He3 has 3.01603 amu's. It has one electron and one negatron with each having 0.00055 amu. The electron is drawn as a dot on the big circle. The negatron is drawn as a square on the big circle below the electron. There is one proton and one positon with the proton drawn above the positon. The proton is drawn as a dot with 1.00728 amu's. There is a rectangle around it and its electron to show that there is an isotope of 1H1 within the isotope 2He3. The rectangle shows that the isotope 1H1 can drop out at some later time. The positon with 1.00571 amu's is drawn as a small square below the proton. It has a square drawn around the small square to show that the positon came from the proton in the isotope 1H3. The proton and positon are both drawn to the left of the vertical line within the big circle. The single neutron with 1.00194 amu's is drawn as a small circle to the right of the vertical line within the big circle. It has a square drawn around it to show that it came from the isotope 1H2. They are drawn alike.

It should be noted that there is a neutron in the isotope 1H3 that drops out with 1.00194 amu's and there is a neutron in the isotope 1H2 that changes into this same mass. This happens because it is the isotope 1H2 that loses the lost mass forcing the neutron in the isotope 1H3 to drop out.

To find the mass that is lost creating the isotope 2He3, add the masses of the isotopes 1H3 and 1H2 and subtract the mass of the isotope 2He3 from their total.

Mass of 1H3	3.01605 amu's
Mass of 1H2	2.01411 amu's
Total mass of 1H3 and 1H2	5.03016 amu's

Total mass of 1H3 and 1H2	5.03016 amu's
Mass of 2He3	3.01603 amu's
Mass lost creating 2He3	2.01413 amu's

The mass lost creating the isotope 2He3 includes the mass of the two neutrons found in the isotope 1H3 that dropped out. Subtract their masses from the total mass lost creating the isotope 2He3 to find the real mass lost.

Mass of first neutron in 1H3	1.00628 amu's
Mass of second neutron in 1H3	1.00194 amu's
Mass of two neutrons that dropped out of 1H3	2.00822 amu's

Mass lost creating 2He3	2.01413 amu's
Mass of two neutrons that dropped out of 1H3	2.00822 amu's
Real mass lost creating 2He3	0.00591 amu

The isotope 1H2 has its electron and proton used as the isotope 1H1 within the isotopes 1H2 and 2He3. Its neutron with 1.00628 amu's becomes the single neutron in the isotope 2He3 when it loses 0.00434 amu. The electron and proton become the electron and proton in the isotope 2He3. This means that the only proton that can become the positon in the isotope 2He3 must come from the isotope 1H3. This proton loses 0.00157 amu so subtract this mass from the proton to find the mass of the positon it becomes in the isotope 2He3.

Mass of proton in 1H3	1.00728 amu's
Mass lost by it	0.00157 amu
Mass of positon in 2He3	1.00571 amu's

Since the proton in the isotope 1H3 lost 0.00157 amu, the mass lost by the neutron in the isotope 1H2 can be found by subtracting the mass the proton lost from the total mass lost creating the isotope 2He3.

Total mass lost creating 2He3	0.00591 amu
Mass lost by proton in 1H3	0.00157 amu
Mass lost by neutron in 1H2	0.00434 amu

The single neutron in the isotope 1H2 loses 0.00434 amu to become the single neutron in the isotope 2He3. To find the mass of the single neutron in the isotope 2He3, subtract the lost mass from the single neutron in the isotope 1H2.

Mass of neutron in 1H2	1.00628 amu's
Mass lost by it	0.00434 amu
Mass of neutron in 2He3	1.00194 amu's

The electrons in the isotopes 1H3 and 1H2 are not lost or released because no 0.00100 amu has been transferred or emitted as a single unit and no 0.00002 amu has been lost creating the isotope 2He3.

The proton in the isotope 1H3 lost a unit of 0.00157 amu and the single neutron in the isotope 1H2 lost a unit of 0.00434 amu for a total lost mass of 0.00591 amu and they are each lost as a single unit. The lost mass of 0.00157 amu is composed of 0.00100 amu, 0.00055 amu and 0.00002 amu. The lost mass of 0.00434 is composed of two sub-units. They are sub-units of 0.00217 amu. In order for the electron or negatron to stay with their proton or positon, these sub-units must be lost as single units of 0.00434 and 0.00157 amu. They must also be lost as single units in to keep the proton, positon or neutron from releasing an electron or negatron from within any of them.

By adding all of the now known masses of the isotope 2He3, the mass of the isotope 2He3 can be found.

Mass of 1H1 intact within 2He3	1.00783 amu's
Mass of neutron in 2He3	1.00194 amu's
Mass of negatron in 2He3	0.00055 amu
Mass of positon in 2He3	1.00571 amu's
Mass of 2He3	3.01603 amu's

Sub-step 31-2-15

2He3

1H3 + 1H2 = 2He3, alpha 434 ray, gamma 157 ray, 1.00628 neutron, 1.00194 neutron

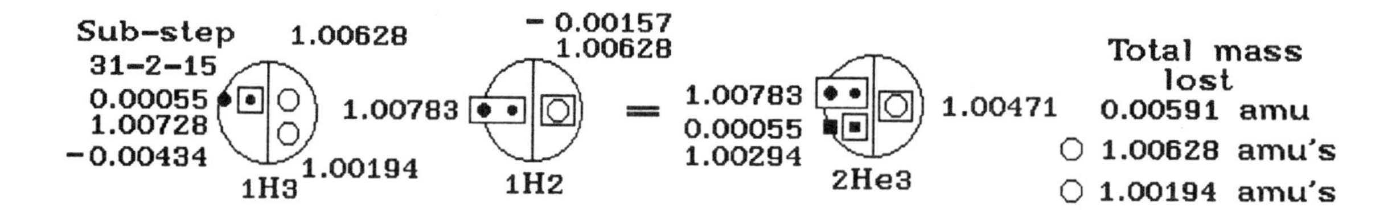

The isotope 1H3 with 3.01605 amu's has one electron with 0.00055 amu and one proton with 1.00728 amu's plus two neutrons. The first neutron has 1.00628 amu's and the second neutron has 1.00194 amu's.

The isotope 1H2 with 2.01411 amu's has one electron with 0.00055 amu and one proton with 1.00728 amu's and one neutron with 1.00628 amu's.

The isotope 2He3 with 3.01603 amu's has one electron with 0.00055 amu, one negatron with 0.00055 amu, one proton with 1.00728 amu's, one positon with 1.00294 amu's and one neutron with 1.00471 amu's.

The total mass lost is 0.00591 amu and it is emitted as a 434 alpha ray which has an external magnetic field and a 157 gamma ray which has no external magnetic field. There are two neutrons that drop out from the isotope 1H3. The first neutron has 1.00628 amu's and the second one has 1.00194 amu's.

The isotope 2He3 has 3.01603 amu's. It has one electron and one negatron with each having 0.00055 amu. The electron is drawn as a dot on the big circle. The negatron is drawn as a square on the big circle below the electron. There is one proton and one positon with the proton drawn above the positon. The proton is drawn as a dot with 1.00728 amu's. There is a rectangle around it and its electron to show that there is an isotope of 1H1 within the isotope 2He3. The rectangle shows that the isotope 1H1 can drop out at some later time.

The positon with 1.00294 amu's is drawn as a small square below the proton. It has a square drawn around the small square to show that the positon came from the proton in the isotope 1H3. The proton and positon are both drawn to the left of the vertical line within the big circle.

The single neutron with 1.00471 amu's is drawn as a small circle to the right of the vertical line within the big circle. It has a square drawn around it to show that it came from the isotope 1H2. They are drawn alike.

The isotope 1H3 with 3.01605 amu's has one electron with 0.00055 amu that is drawn as a dot on the big circle and it becomes the negatron with 0.00055 amu in the isotope 2He3 and is drawn as a square on the big circle.

The proton with 1.00728 amu's in the isotope 1H3 is drawn as a dot to the left of the vertical line within the big circle. There is a square drawn around it to show that it loses 0.00434 amu. It is drawn as a small square with a square around it in the isotope 2He3 to show that the proton in the isotope 1H3 is the positon with 1.00294 amu's in the isotope 2He3.

There are two neutrons in the isotope 1H3 that are drawn one above the other as small circles to the right of the vertical line within the big circle. The top neutron has 1.00628 amu's and is called the first neutron while the neutron below it is the second neutron and it has 1.00194 amu's. Neither of these neutrons are needed when the isotope 2He3 is created so they both drop out.

The isotope 1H2 has 2.01411 amu's. Its single electron is drawn as a dot on the big circle in the isotopes 1H2 and 2He3. The proton in the isotope 1H2 has 1.00728 amu's and it is drawn as a dot to the left of the vertical line within the big circle in the isotopes 1H2 and 2He3.

The electron and proton in the isotope 1H2 has a rectangle drawn around both of them to show that they are used as the isotope 1H1 with 1.00783 amu's in the isotope 2He3. They are drawn the same way in the isotope 2He3. The rectangle in the isotope 2He3 shows that the isotope 1H1 can drop out of the isotope 2He3 at a later time.

There is one neutron in the isotope 1H2 with 1.00628 amu's and it is drawn as a small circle to the right of the vertical line within the big circle. There is a square drawn around the neutron to show that it has lost 0.00157 amu to become the single neutron found in the isotope 2He3.

The isotope 2He3 with 3.01603 amu's has one electron with 0.00055 amu drawn as a dot and a negatron with 0.00055 amu drawn as a square. Both are drawn on the big circle. There is a proton with 1.00628 amu's drawn as a dot and the electron and proton is drawn with a rectangle drawn around the electron and proton to show that they come from the isotope 1H2. There is square drawn around small square that is a positon to show that it came from the isotope 1H3. There is one neutron with 1.00471 amu's in the isotope 2He3 with a square drawn around it to show that it cames from the single neutron in the isotope 1H2.

To find the mass that is lost creating the isotope 2He3, add the masses of the isotopes 1H3 and 1H2 and subtract the mass of the isotope 2He3 from their total.

Mass of 1H3	3.01605 amu's
Mass of 1H2	2.01411 amu's
Total mass of 1H3 and 1H2	5.03016 amu's

Total mass of 1H3 and 1H2	5.03016 amu's
Mass of 2He3	3.01603 amu's
Mass lost creating 2He3	2.01413 amu's

The mass lost creating the isotope 2He3 includes the mass of the two neutrons found in the isotope 1H3 that dropped out. Subtract their masses from the total mass lost creating the isotope 2He3 to find the real mass lost.

Mass of first neutron in 1H3	1.00628 amu's
Mass of second neutron in 1H3	1.00194 amu's
Mass of two neutrons that dropped out of 1H3	2.00822 amu's

Mass lost creating 2He3	2.01413 amu's
Mass of two neutrons that dropped out of 1H3	2.00822 amu's
Real mass lost creating 2He3	0.00591 amu

The isotope 1H2 has its electron and proton used as the isotope 1H1 within the isotopes 1H2 and 2He3. Its neutron with 1.00628 amu's becomes the single neutron in the isotope 2He3. The electron and proton become the electron and proton in the isotope 2He3. This means that the only proton that can become the positon in the isotope 2He3 must come from the isotope 1H3. This proton loses 0.00434 amu so subtract this mass from the proton to find the mass of the positon it becomes in the isotope 2He3.

Mass of proton in 1H3	1.00728 amu's
Mass lost by it	0.00434 amu
Mass of positon in 2He3	1.00294 amu's

Since the proton in the isotope 1H3 lost 0.00434 amu, the mass lost by the single neutron in the isotope 1H2 can be found by subtracting the mass the proton lost from the total mass lost creating the isotope 2He3.

Total mass lost creating 2He3	0.00591 amu
Mass lost by proton in 1H3	0.00434 amu
Mass lost by neutron in 1H2	0.00157 amu

The single neutron in the isotope 1H2 loses 0.00157 amu to become the single neutron in the isotope 2He3. To find the mass of the single neutron in the isotope 2He3, subtract the lost mass from the single neutron in the isotope 1H2.

Mass of neutron in 1H2	1.00628 amu's
Mass lost by it	0.00157 amu
Mass of neutron in 2He3	1.00471 amu's

The electrons in the isotopes 1H3 and 1H2 are not lost or released because no 0.00100 amu has been transferred or emitted as a single unit and no 0.00002 amu has been lost creating the isotope 2He3.

The proton in the isotope 1H3 lost a unit of 0.00434 amu and the single neutron in the isotope 1H2 lost a unit of 0.00157 amu for a total lost mass of

0.00591 amu and they are each lost as single unit. The lost mass of 0.00434 is composed of two sub-units. They are sub-units of 0.00217 amu. The lost mass of 0.00157 amu is composed of three sub-units. They are sub-units of 0.00100 amu, 0.00055 amu and 0.00002 amu. In order for the electron or negatron to stay with their proton or positon, these sub-units must be lost as single units of 0.00434 and 0.00157 amu. They must also be lost as single units to keep the proton, positon or neutron from releasing an electron or negatron from within any of them.

By adding all of the now known masses of the isotope 2He3, the mass of the isotope 2He3 can be found.

Mass of 1H1 intact within 2He3	1.00783 amu's
Mass of neutron in 2He3	1.00471 amu's
Mass of negatron in 2He3	0.00055 amu
Mass of positon in 2He3	1.00294 amu's
Mass of 2He3	3.01603 amu's

Sub-step 30-2-14

2He3

1H3 + 1H2 = 2He3, alpha 217 ray, alpha 374 ray, 1.00628 neutron, 1.00194 neutron

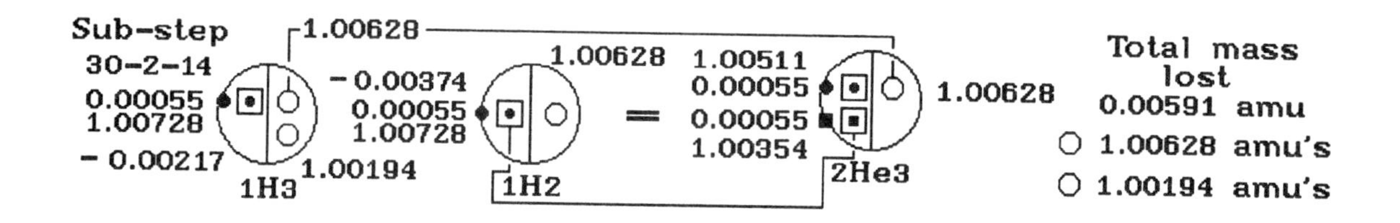

The isotope 1H3 with 3.01605 amu's has one electron with 0.00055 amu and one proton with 1.00728 amu's plus two neutrons. The first neutron has 1.00628 amu's and the second neutron has 1.00194 amu's.

The isotope 1H2 with 2.01411 amu's has one electron with 0.00055 amu and one proton with 1.00728 amu's and one neutron with 1.00628 amu's.

The isotope 2He3 with 3.01603 amu's has one electron with 0.00055 amu, one negatron with 0.00055 amu, one proton with 1.00511 amu's, one positon with 1.00354 amu's and one neutron with 1.00628 amu's.

The total mass lost is 0.00591 amu and it is emitted as a 217 alpha ray and a 374 alpha ray. Each alpha ray has an external magnetic field. There are two neutrons that drop out. The first neutron comes from the isotope 1H2 with 1.00628 amu's and the second one comes from the isotope 1H3 with 1.00194 amu's.

The isotope 1H3 with 3.01605 amu's has one electron with 0.00055 amu that is drawn as a dot on the big circle and it stays the electron with 0.00055 amu in the isotope 2He3 and is drawn as a dot on the big circle.

The proton with 1.00728 amu's in the isotope 1H3 is drawn as a dot to the left of the vertical line within the big circle. It is drawn as a dot in the isotope 2He3 to show that the proton in the isotope 1H3 is the proton in the isotope 2He3. The proton has a square drawn around it to show that it has lost 0.00217 amu to become the proton with 1.00511 amu's in the isotope 2He3.

There are two neutrons in the isotope 1H3 that are drawn one above the other as small circles to the right of the vertical line within the big circle. The top neutron has 1.00628 amu's and is called the first neutron while the neutron below it is the second neutron and it has 1.00194 amu's. The first neutron becomes the single neutron in the isotope 2He3 when it is created. The second neutron is not needed when the isotope 2He3 is created so it drops out.

Because there are two neutron with 1.00628 amu's in two different isotopes, there is a line drawn from the first neutron in the isotope 1H3 to the single neutron in the isotope 2He3 to show that it is this neutron that becomes the neutron in the isotope 2He3.

The isotope 1H2 has 2.01411 amu's. Its single electron is drawn as a dot on the big circle in the isotopes 1H2 and it is drawn as a square on the big circle in the isotope 2He3 to show that the electron in the isotope 1H2 is the negatron in the isotope 2He3 with 0.00055 amu.

The proton in the isotopes 1H2 has 1.00728 amu's and it is drawn as a dot to the left of the vertical line within the big circle and it is drawn as a small square in the isotope 2He3 to show that this proton in the isotope 1H2 is the positon in the isotope 2He3. There is a square drawn around the proton in the isotope 1H2 and a square around the positon in the isotope 2He3 to show that the proton in the isotope 1H2 has lost 0.00374 amu to become the positon in the isotope 2He3 with 1.00354 amu's.

The isotope 1H3 and 1H2 each have their electron and proton drawn as a dot with each proton losing mass. Each proton in the isotopes of 1H3 and 1H2 has a square drawn around it to show each proton has lost some mass. A line has been drawn from the proton in the isotope 1H2 to the positon in the isotope 2He3 to show which proton became the positon in the isotope 2He3.

There is one neutron in the isotope 1H2 with 1.00628 amu's and it is drawn as a small circle to the right of the vertical line within the big circle. This neutron is not needed when the isotope 2He3 is created so it drops out.

The isotope 2He3 with 3.01603 amu's has one electron with 0.00055 amu drawn as a dot and a negatron with 0.00055 amu drawn as a square. Both are drawn on the big circle. There is a proton with 1.00511 amu's drawn as a dot and a positon 1.00354 amu's drawn as a small square with a square drawn around both of them. There is a line drawn from the positon to the proton in the isotope 1H2 to show that the positon was the proton in the isotope 1H2. The proton is drawn above the positon. There is one neutron with 1.00628 amu's in the isotope 2He3 and it comes from the first neutron in the isotope 1H3. There is a line drawn between them to show this.

To find the mass that is lost creating the isotope 2He3, add the masses of the isotopes 1H3 and 1H2 and subtract the mass of the isotope 2He3 from their total.

Mass of 1H3	3.01605 amu's
Mass of 1H2	2.01411 amu's
Total mass of 1H3 and 1H2	5.03016 amu's

Total mass of 1H3 and 1H2	5.03016 amu's
Mass of 2He3	3.01603 amu's
Mass lost creating 2He3	2.01413 amu's

The mass lost creating the isotope 2He3 includes the mass of the two neutrons found in the isotopes 1H3 and 1H2 that dropped out from each isotope of 1H3 and 1H2. Subtract their masses from the total mass lost creating the isotope 2He3 to find the real mass lost.

Mass of neutron in 1H2	1.00628 amu's
Mass of second neutron in 1H3	1.00194 amu's
Mass of two neutrons that dropped out of 1H3	2.00822 amu's

Mass lost creating 2He3	2.01413 amu's
Mass of two neutrons that dropped out of 1H3	2.00822 amu's
Real mass lost creating 2He3	0.00591 amu

The proton in the isotope 1H3 loses 0.00217 amu to become the proton in the isotope 2He3. To find the mass of the proton in the isotope 2He3, subtract the lost mass from the proton in the isotope 1H3.

Mass of proton in 1H3	1.00728 amu's
Mass lost by it	0.00217 amu
Mass of proton in 2He3	1.00511 amu's

Since the proton in the isotope 1H3 lost 0.00217 amu, the mass lost by the proton in the isotope 1H2 can be found by subtracting the mass the proton lost in the isotope 1H3 from the total mass lost creating the isotope 2He3.

Total mass lost creating 2He3	0.00591 amu
Mass lost by proton in 1H3	0.00217 amu
Mass lost by proton in 1H2	0.00374 amu

The proton in the isotope 1H2 loses 0.00374 amu to become the positon in the isotope 2He3. Subtract the lost mass of 0.00374 from this proton to find the mass of the positon in the isotope 2He3.

Mass of proton in 1H2	1.00728 amu's
Mass lost by it	0.00374 amu
Mass of positon in 2He3	1.00354 amu's

The electrons in the isotopes 1H3 and 1H2 are not lost or released because no 0.00100 amu has been transferred or emitted as a single unit and no 0.00002 amu has been lost creating the isotope 2He3.

The proton in the isotope 1H3 lost a unit of 0.00217 amu and the proton in the isotope 1H2 lost a unit of 0.00374 amu for a total lost mass of 0.00591 amu. The 0.00217 amu and the 0.00374 amu are each lost as a single unit. The unit of 0.00217 amu does not have any sub-units. The lost mass of 0.00374 amu is composed of four sub-units. They are sub-units of 0.00217 amu, 0.00100 amu, 0.00055 amu and 0.00002 amu. Note that there are two masses of 0.00217 amu. One is a unit and the other is a sub-unit of the unit 0.00374 amu. In order for the electron or negatron to stay with their proton and positon, these sub-units must be lost as a single unit of 0.00217 amu or 0.00374 amu. They must also be lost as single units to keep the proton or positon from releasing an electron or negatron from within either of them.

By adding all of the now known masses of the isotope 2He3, the mass of the isotope 2He3 can be found.

Mass of electron and negatron in 2He3	0.00110 amu
Mass of proton in 2He3	1.00511 amu's
Mass of positon in 2He3	1.00354 amu's
Mass of neutron in 2He3	1.00628 amu's
Mass of 2He3	3.01603 amu's

Sub-step 29-2-13

2He3

1H3 + 1H2 = 2He3, alpha 157 ray, alpha 434 ray, 1.00628 neutron, 1.00194 neutron

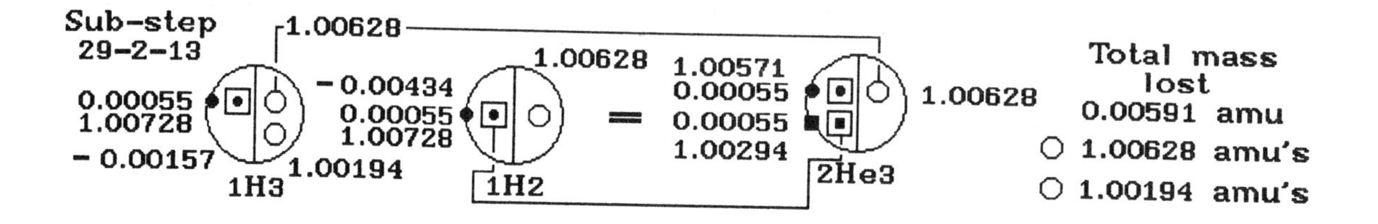

The isotope 1H3 with 3.01605 amu's has one electron with 0.00055 amu and one proton with 1.00728 amu's plus two neutrons. The first neutron has 1.00628 amu's and the second neutron has 1.00194 amu's.

The isotope 1H2 with 2.01411 amu's has one electron with 0.00055 amu and one proton with 1.00728 amu's and one neutron with 1.00628 amu's.

The isotope 2He3 with 3.01603 amu's has one electron with 0.00055 amu, one negatron with 0.00055 amu, one proton with 1.00571 amu's, one positon with 1.00294 amu's and one neutron with 1.00628 amu's.

The total mass lost is 0.00591 amu and it is emitted as a 157 alpha ray and a 434 alpha ray. Each alpha ray has an external magnetic field. There are two neutrons that drop out. The first neutron comes from the isotope 1H2 has 1.00628 amu's and the second one comes from the isotope 1H3 with 1.00194 amu's.

The isotope 1H3 with 3.01605 amu's has one electron with 0.00055 amu that is drawn as a dot on the big circle and it stays the electron with 0.00055 amu in the isotope 2He3 and is drawn as a dot on the big circle.

The proton with 1.00728 amu's in the isotope 1H3 is drawn as a dot to the left of the vertical line within the big circle. It is drawn as a dot in the isotope 2He3 to show that the proton in the isotope 1H3 is the proton in the isotope 2He3. The proton has a square drawn around it to show that it has lost 0.00157 amu to become the proton with 1.00571 amu's in the isotope 2He3.

There are two neutrons in the isotope 1H3 that are drawn one above the other as small circles to the right of the vertical line within the big circle. The top neutron has 1.00628 amu's and is called the first neutron while the neutron below it is the second neutron and it has 1.00194 amu's. The first of these neutrons becomes the single neutron when the isotope 2He3 is created while the second neutron is not needed and it drops out.

Because there are two neutrons with 1.00628 amu's in two different isotopes, there is a line drawn from the first neutron in the isotope 1H3 to the single neutron in the isotope 2He3 to show that it is this neutron that becomes the neutron in the isotope 2He3.

The isotope 1H2 has 2.01411 amu's. Its single electron is drawn as a dot on the big circle in the isotope 1H2 and it is drawn as a square on the big circle in the isotope 2He3 to show that the electron in the isotope 1H2 is the negatron in the isotope 2He3 with 0.00055 amu.

The proton in the isotope 1H2 has 1.00728 amu's and it is drawn as a dot to the left of the vertical line within the big circle and it is drawn as a small square in the isotope 2He3 to show that the proton in the isotope 1H2 is the positon in the isotope 2He3. There is a square drawn around the proton in the isotope 1H2 and a square around the positon in the isotope 2He3 to show that the proton in the isotope 1H2 has lost 0.00434 amu to become the positon in the isotope 2He3 with 1.00294 amu's.

The isotopes 1H3 and 1H2 each have their electron and proton drawn as a dot with each proton losing mass. Each proton in the isotopes of 1H3 and 1H2 has a square drawn around it to show each proton has lost some mass. A line has been drawn from the proton in the isotope 1H2 to the positon in the isotope 2He3 to show which proton became the positon in the isotope 2He3.

There is one neutron in the isotope 1H2 with 1.00628 amu's and it is drawn as a small circle to the right of the vertical line within the big circle. This neutron is not needed when the isotope 2He3 is created so it drops out.

The isotope 2He3 with 3.01603 amu's has an electron and a negatron drawn on the big circle. They both have 0.00055 amu. The electron is drawn as a dot while the negatron is drawn as a square. There is a proton with 1.00571 amu's and a positon with 1.00294 amu's drawn within the big circle and to the left of the vertical line. The proton is drawn above the positon. The single neutron with 1.00628 amu's is drawn as a small circle to the right of the vertical line within the big circle.

There is a line drawn from the positon in the isotope 2He3 to the proton in the isotope 1H2 to show where the positon came from.

There is a line drawn from the single neutron to the first neutron in the isotope 1H3 to show where the single neutron in the isotope 2He3 came from.

To find the mass that is lost creating the isotope 2He3, add the masses of the isotopes 1H3 and 1H2 and subtract the mass of the isotope 2He3 from their total.

Mass of 1H3	3.01605 amu's
Mass of 1H2	2.01411 amu's
Total mass of 1H3 and 1H2	5.03016 amu's

Total mass of 1H3 and 1H2	5.03016 amu's
Mass of 2He3	3.01603 amu's
Mass lost creating 2He3	2.01413 amu's

The mass lost creating the isotope 2He3 includes the mass of the two neutrons found in the isotopes 1H3 and 1H2 that dropped out. Subtract their masses from the total mass lost creating the isotope 2He3 to find the real mass lost.

Mass of neutron in 1H2	1.00628 amu's
Mass of second neutron in 1H3	1.00194 amu's
Mass of two neutrons that dropped out of 1H3	2.00822 amu's

Mass lost creating 2He3	2.01413 amu's
Mass of two neutrons that dropped out of 1H3	2.00822 amu's
Real mass lost creating 2He3	0.00591 amu

The proton in the isotope 1H3 loses 0.00157 amu to become the proton in the isotope 2He3. To find the mass of the proton in the isotope 2He3, subtract the lost mass from the proton in the isotope 1H3.

Mass of proton in 1H3	1.00728 amu's
Mass lost by it	0.00157 amu
Mass of proton in 2He3	1.00571 amu's

Since the proton in the isotope 1H3 lost 0.00157 amu, the mass lost by the proton in the isotope 1H3 can be found by subtracting the mass the proton lost in the isotope 1H3 from the total mass lost creating the isotope 2He3.

Total mass lost creating 2He3	0.00591 amu
Mass lost by proton in 1H3	0.00157 amu
Mass lost by proton in 1H2	0.00434 amu

The proton in the isotope 1H2 loses a of 0.00434 amu to become the positon in the isotope 2He3. Subtract the lost mass of 0.00434 from this proton to find the mass of the positon in the isotope 2He3.

Mass of proton in 1H2	1.00728 amu's
Mass lost by it	0.00434 amu
Mass of positon in 2He3	1.00294 amu's

The electrons in the isotopes 1H3 and 1H2 are not lost or released because no 0.00100 amu has been transferred or emitted as a single unit and no 0.00002 amu has been lost creating the isotope 2He3.

The proton in the isotope 1H3 lost a unit of 0.00157 amu and the proton in the isotope 1H2 lost a unit of 0.00434 amu for a total lost mass of 0.00591 amu. The 0.00157 amu and the 0.00434 amu are each lost as single units. The unit of 0.00157 amu is composed of three sub-units. They are sub-units of 0.00100 amu,

0.00055 amu and 0.00002 amu. The lost mass of 0.00434 is composed of two sub-units. They are sub-units of 0.00217 amu. In order for the electron or negatron to stay with their proton and positon, these sub-units must be lost as a single unit of 0.00157 amu and 0.00434 amu. They must also be lost as single units to keep the proton or positon from releasing an electron or negatron from within either of them.

By adding all of the now known masses of the isotope 2He3, the mass of the isotope 2He3 can be found.

Mass of electron and negatron in 2He3	0.00110 amu
Mass of proton in 2He3	1.00571 amu's
Mass of positon in 2He3	1.00294 amu's
Mass of neutron in 2He3	1.00628 amu's
Mass of 2He3	3.01603 amu's

Sub-step 28-2-12

2He3

1H3 + 1H2 = 2He3, alpha 217 ray, alpha 374 ray, 1.00628 neutron, 1.00194 neutron

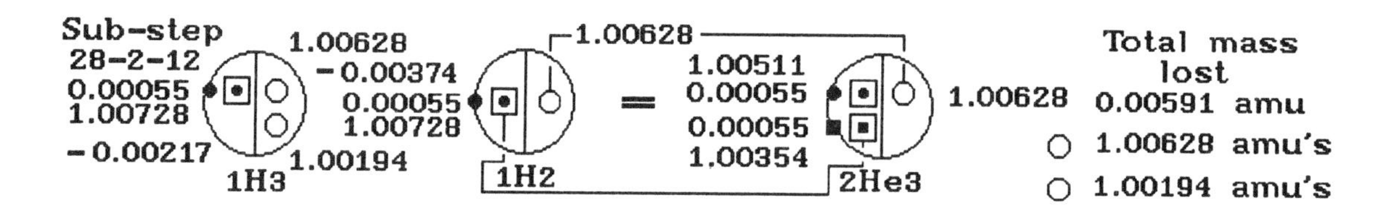

The isotope 1H3 with 3.01605 amu's has one electron with 0.00055 amu and one proton with 1.00728 amu's plus two neutrons. The first neutron has 1.00628 amu's and the second neutron has 1.00194 amu's.

The isotope 1H2 with 2.01411 amu's has one electron with 0.00055 amu and one proton with 1.00728 amu's and one neutron with 1.00628 amu's.

The isotope 2He3 with 3.01603 amu's has one electron with 0.00055 amu, one negatron with 0.00055 amu, one proton with 1.00511 amu's, one positon with 1.00354 amu's and one neutron with 1.00628 amu's.

The total mass lost is 0.00591 amu and it is emitted as a 217 alpha ray and a 374 alpha ray. Each alpha ray has an external magnetic field. There are two neutrons that drop out. They both come from the isotope 1H3 and the first neutron has 1.00628 amu's while the second neutron has 1.00194 amu's.

The isotope 1H3 with 3.01605 amu's has one electron with 0.00055 amu that is drawn as a dot on the big circle and it stays an electron with 0.00055 amu in the isotope 2He3 and is drawn as a dot on the big circle.

The proton with 1.00728 amu's in the isotope 1H3 is drawn as a dot to the left of the vertical line within the big circle. It is drawn as a dot in the isotope 2He3 to show that the proton in the isotope 1H3 is the proton in the isotope 2He3. The proton has a square drawn around it to show that it has lost 0.00217 amu to become the proton with 1.00511 amu's in the isotope 2He3.

There are two neutrons in the isotope 1H3 that are drawn one above the other as small circles to the right of the vertical line within the big circle. The top neutron has 1.00628 amu's and is called the first neutron while the neutron below it is the second neutron and it has 1.00194 amu's. Neither of these neutrons are needed when the isotope 2He3 is created so they both drop out.

The isotope 1H2 has 2.01411 amu's. Its single electron is drawn as a dot on the big circle in the isotopes 1H2 and it is drawn as a square on the big circle in the isotope 2He3 to show that the electron in the isotope 1H2 is the negatron in the isotope 2He3 with 0.00055 amu.

The proton in the isotope 1H2 has 1.00728 amu's and it is drawn as a dot to the left of the vertical line within the big circle and it is drawn as a small square in the isotope 2He3 to show that the proton in the isotope 1H2 becomes the positon in the isotope 2He3. There is a square drawn around the dot in the isotope 1H2 and a square around the positon in the isotope 2He3 to show that the proton in the isotope 1H2 has lost 0.00374 amu to become the positon in the isotope 2He3 with 1.00354 amu's. There is a line drawn from the proton in the isotope 1H2 to the positon in the isotope 2He3 to show which proton becomes the positon.

There is one neutron in the isotope 1H2 with 1.00628 amu's and it is drawn as a small circle to the right of the vertical line within the big circle. This neutron becomes the single neutron in the isotope 2He3. There is a line drawn from the neutron in the isotope 1H2 to the neutron in the isotope 2He3 because there also is a neutron in the isotope 1H3 with 1.00628 amu's.

The isotope 2He3 has an electron drawn as a dot and a negatron drawn as a square. They both have 0.00055 amu and are both drawn on the big circle with the electron above the negatron. There is a proton with 1.00511 amu's drawn as a dot and a positon with 1.00354 amu's drawn as a square. They are both drawn to the left of the vertical line within the big circle. The proton is above the positon. There is a single neutron with 1.00628 amu's drawn as a small circle within the big circle and to the right of the vertical line.

To find the mass that is lost creating the isotope 2He3, add the masses of the isotopes 1H3 and 1H2 and subtract the mass of the isotope 2He3 from their total.

Mass of 1H3	3.01605 amu's
Mass of 1H2	2.01411 amu's
Total mass of 1H3 and 1H2	5.03016 amu's

Total mass of 1H3 and 1H2	5.03016 amu's
Mass of 2He3	3.01603 amu's
Mass lost creating 2He3	2.01413 amu's

The mass lost creating the isotope 2He3 includes the mass of the two neutrons found in the isotope 1H3 that dropped out. Subtract their masses from the total mass lost creating the isotope 2He3 to find the real mass lost.

Mass of first neutron in 1H3	1.00628 amu's
Mass of second neutron in 1H3	1.00194 amu's
Mass of two neutrons that dropped out of 1H3	2.00822 amu's

Mass lost creating 2He3	2.01413 amu's
Mass of two neutrons that dropped out of 1H3	2.00822 amu's
Real mass lost creating 2He3	0.00591 amu

The proton in the isotope 1H3 loses 0.00217 amu to become the proton in the isotope 2He3. To find the mass of the proton in the isotope 2He3, subtract the lost mass in the isotope 1H3 from the proton.

Mass of proton in 1H3	1.00728 amu's
Mass lost by it	0.00217 amu
Mass of proton in 2He3	1.00511 amu's

Since the proton in the isotope lost 0.00217 amu, the mass lost by the proton in the isotope 1H2 can be found by subtracting the mass lost by the proton in the isotope 1H3 from the total mass lost creating the isotope 2He3.

Total mass lost creating 2He3	0.00591 amu
Mass lost by proton in 1H3	0.00217 amu
Mass lost by proton in 1H2	0.00374 amu

The proton in the isotope 1H2 loses 0.00374 amu to become the positon in the isotope 2He3. Subtract this lost mass to find the mass of the positon in the isotope 2He3.

Mass of proton in 1H2	1.00728 amu's
Mass lost by it	0.00374 ams
Mass of positon in 2He3	1.00354 amu's

The electrons in the isotopes 1H3 and 1H2 are not lost or released because no 0.00100 amu has been transferred or emitted as a single unit and no 0.00002 amu has been lost creating the isotope 2He3.

The proton in the isotope 1H3 lost a unit of 0.00217 amu and the proton in the isotope 1H2 lost a unit of 0.00374 amu for a total lost mass of 0.00591 amu. The 0.00217 amu and the 0.00374 amu are each lost as single units. The lost mass of 0.00217 amu does not have any sub-units. The lost mass of 0.00374 amu is composed of four sub-units. They are sub-units of 0.00217 amu, 0.00100 amu, 0.00055 amu and 0.00002 amu. In order for the electron or negatron to stay with their proton or positon, these sub-units must be lost as singles unit of 0.00217 amu or 0.00374 amu. They must also be lost as single units to keep the proton or positon from releasing an electron or negatron from within either of them.

By adding all of the now known masses of the isotope 2He3, the mass of the isotope 2He3 can be found.

Mass of electron and negatron in 2He3	0.00110 amu
Mass of proton in 2He3	1.00511 amu's
Mass of positon in 2He3	1.00354 amu's
Mass of neutron in 2He3	1.00628 amu's
Mass of 2He3	3.01603 amu's

Sub-step 27-2-11

2He3

1H3 + 1H2 = 2He3, alpha 157 ray, alpha 434 ray, 1.00628 neutron, 1.00194 neutron

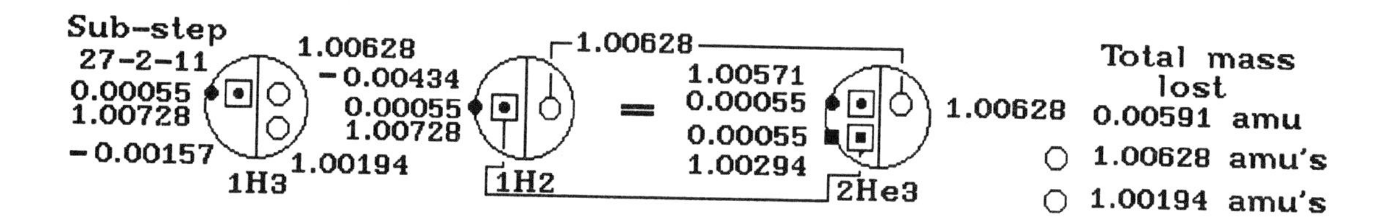

The isotope 1H3 with 3.01605 amu's has one electron with 0.00055 amu and one proton with 1.00728 amu's plus two neutrons. The first neutron has 1.00628 amu's and the second neutron has 1.00194 amu's.

The isotope 1H2 with 2.01411 amu's has one electron with 0.00055 amu and one proton with 1.00728 amu's and one neutron with 1.00628 amu's.

The isotope 2He3 with 3.01603 amu's has one electron with 0.00055 amu, one negatron with 0.00055 amu, one proton with 1.00571 amu's, one positon with 1.00294 amu's and one neutron with 1.00628 amu's.

The total mass lost is 0.00591 amu and it is emitted as a 157 alpha ray and a 434 alpha ray. Each alpha ray has an external magnetic field. There are two neutrons that drop out from the isotope 1H3. The first neutron has 1.00628 amu's and the second one has 1.00194 amu's.

The isotope 1H3 with 3.01605 amu's has one electron with 0.00055 amu that is drawn as a dot on the big circle and it stays the electron with 0.00055 amu in the isotope 2He3 and is drawn as a dot on the big circle.

The proton with 1.00728 amu's in the isotope 1H3 is drawn as a dot to the left of the vertical line within the big circle. It is drawn as a dot in the isotope 2He3 to show that the proton in the isotope 1H3 is the proton in the isotope 2He3. The proton has a square drawn around it to show that it has lost 0.00157 amu to become the proton with 1.00571 amu's in the isotope 2He3.

There are two neutrons in the isotope 1H3 that are drawn one above the other as small circles to the right of the vertical line within the big circle. The top neutron has 1.00628 amu's and is called the first neutron while the neutron below it is the second neutron and it has 1.00194 amu's. Neither of these neutrons are needed when the isotope 2He3 is created so they both drop out.

The isotope 1H2 has 2.01411 amu's. Its single electron is drawn as a dot on the big circle in the isotope 1H2 and it is drawn as a square on the big circle in the isotope 2He3 to show that the electron in the isotope 1H2 has become the negatron in the isotope 2He3 with 0.00055 amu.

The proton in the isotope 1H2 has 1.00728 amu's and it is drawn as a dot to the left of the vertical line within the big circle and it is drawn as a small square in the isotope 2He3 to show that the proton in the isotope 1H2 has become the positon in the isotope 2He3. There is a square drawn around the proton in the isotope 1H2 and a square around the positon in the isotope 2He3 to show that the proton with 1.00728 amu's in the isotope 1H2 has lost 0.00434 amu to become the positon in the isotope 2He3 with 1.00294 amu's.

The isotopes 1H3 and 1H2 each have their electron and proton drawn as a dot with each proton losing mass. Each proton in the isotopes of 1H3 and 1H2 has a square drawn around it to show each proton has lost some mass. A line has been drawn from the proton in the isotope 1H2 to the positon in the isotope 2He3 to show which proton became the positon in the isotope 2He3.

There is one neutron in the isotope 1H2 with 1.00628 amu's and it is drawn as a small circle to the right of the vertical line within the big circle. This neutron is the single neutron in the isotope 2He3. There is a line drawn from the neutron in the isotope 1H2 to the single neutron in the isotope 2He3 because the isotope 1H3 also has a neutron with 1.00628 amu's. The line shows that the neutron in the isotope 1H2 is the neutron in the isotope 2He3.

The isotope 2He3 with 3.01603 amu's has an electron and a negatron drawn on the big circle with the electron drawn over the negatron. The electron is drawn as a dot while the negatron is drawn as a square. They both have 0.00055 amu. There is a proton drawn as a dot over the positon which is drawn as a square. They are both drawn to the left of the vertical line within the big circle. They each have a square drawn around them to show they come from a proton found in the isotopes 1H3 and 1H2. There is a line drawn from the positon with 1.00571 amu's to the proton with 1.00728 amu's to show where the positon came from. The single neutron in the isotope 2He3 is drawn as a small circle to the right of the vertical line within the big circle. There is a line drawn from this neutron with 1.00628 amu's to the neutron with 1.00628 amu's in the isotope 1H2 to show where the neutron in the isotope 2He3 came from. This line is needed because there are two neutrons with 1.00628 amu's. One is found in the isotope 1H3 while the other one is found in the isotope 1H2.

To find the mass that is lost creating the isotope 2He3, add the masses of the isotopes 1H3 and 1H2 and subtract the mass of the isotope 2He3 from their total.

Mass of 1H3	3.01605 amu's
Mass of 1H2	2.01411 amu's
Total mass of 1H3 and 1H2	5.03016 amu's

Total mass of 1H3 and 1H2	5.03016 amu's
Mass of 2He3	3.01603 amu's
Mass lost creating 2He3	2.01413 amu's

The mass lost creating the isotope 2He3 includes the mass of the two neutrons found in the isotope 1H3 that dropped out. Subtract their masses from the total mass lost creating the isotope 2He3 to find the real mass lost.

Mass of first neutron in 1H3	1.00628 amu's
Mass of second neutron in 1H3	1.00194 amu's
Mass of two neutrons that dropped out of 1H3	2.00822 amu's

Mass lost creating 2He3	2.01413 amu's
Mass of two neutrons that dropped out of 1H3	2.00822 amu's
Real mass lost creating 2He3	0.00591 amu

The proton in the isotope 1H3 loses 0.00157 amu to become the proton in the isotope 2He3. To find the mass of the proton in the isotope 2He3, subtract the lost mass from the proton in the isotope 1H3.

Mass of proton in 1H3	1.00728 amu's
Mass lost by it	0.00157 amu
Mass of proton in 2He3	1.00571 amu's

The proton in the isotope 1H2 lost 0.00434 amu to become the positon in the isotope 2He3. Subtract this lost mass from the proton in the isotope 1H2 to find the mass of the positon.

Mass of proton in 1H2	1.00728 amu's
Mass lost by it	0.00434 amu
Mass of positon in 2He3	1.00294 amu's

The electrons in the isotopes 1H3 and 1H2 are not lost or released because no 0.00100 amu has been transferred or emitted as a single unit and no 0.00002 amu has been lost creating the isotope 2He3.

The proton in the isotope 1H3 lost a unit of 0.00157 amu and the proton in the isotope 1H2 lost a unit of 0.00434 amu for a total lost mass of 0.00591 amu. The 0.00434 amu and the 0.00157 amu are each lost as single units. The lost mass of 0.00157 is composed of three sub-units. They are sub-units of 0.00100 amu, 0.00055 amu and 0.00002 amu. The unit of 0.00434 amu is composed of two sub-units of 0.00217 amu. In order for the electron or negatron to stay with their proton or positon, these sub-units must be lost as single units of 0.00157 amu and 0.00434 amu. They must also be lost as single units to keep the proton or positon from releasing an electron or negatron from within either of them.

By adding all of the now known masses of the isotope 2He3, the mass of the isotope 2He3 can be found.

Mass of electron and negatron in 2He3	0.00110 amu
Mass of proton in 2He3	1.00571 amu's
Mass of positon in 2He3	1.00294 amu's
Mass of neutron in 2He3	1.00628 amu's
Mass of 2He3	3.01603 amu's

Sub-step 26-2-10

2He3

1H3 + 1H2 = 2He3, alpha 374 ray, alpha 217 ray, 1.00628 neutron, 1.00194 neutron

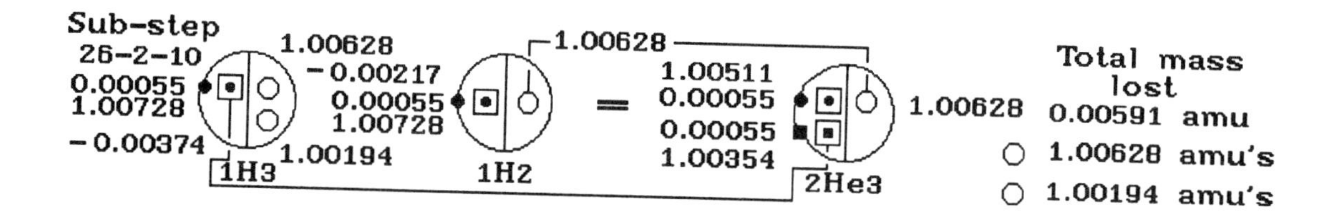

The isotope 1H3 with 3.01605 amu's has one electron with 0.00055 amu and one proton with 1.00728 amu's plus two neutrons. The first neutron has 1.00628 amu's and the second neutron has 1.00194 amu's.

The isotope 1H2 with 2.01411 amu's has one electron with 0.00055 amu and one proton with 1.00728 amu's and one neutron with 1.00628 amu's.

The isotope 2He3 with 3.01603 amu's has one electron with 0.00055 amu, one negatron with 0.00055 amu, one proton with 1.00511 amu's, one positon with 1.00354 amu's and one neutron with 1.00628 amu's.

The total mass lost is 0.00591 amu and it is emitted as a 374 alpha ray and a 217 alpha ray. Each alpha ray has an external magnetic field. There are two neutrons that drop out of the isotope 1H3. The first neutron to drop out has 1.00628 amu's and the second neutron to drops out with 1.00194 amu's.

The isotope 1H3 with 3.01605 amu's has one electron with 0.00055 amu that is drawn as a dot on the big circle and it becomes the negatron with 0.00055 amu in the isotope 2He3 and is drawn as a square on the big circle.

The proton with 1.00728 amu's in the isotope 1H3 is drawn as a dot to the left of the vertical line within the big circle. It is drawn as a small square in the isotope 2He3 to show that the proton in the isotope 1H3 has become the positon in the isotope 2He3. The proton has a square drawn around it to show that it has lost 0.00374 amu to become the positon with 1.00354 amu's in the isotope 2He3. There is a line drawn from the proton in the isotope 1H3 to the positon in the isotope 2He3 to show which proton is the positon.

There are two neutrons in the isotope 1H3 that are drawn one above the other as small circles to the right of the vertical line within the big circle. The top neutron has 1.00628 amu's and is called the first neutron while the neutron below it is the second neutron and it has 1.00194 amu's. Neither of these neutrons are needed when the isotope 2He3 is created so they both drop out.

The isotope 1H2 has 2.01411 amu's. Its single electron is drawn as a dot on the big circle in the isotope 1H2 and it is drawn as a dot on the big circle in the isotope 2He3 to show that the electron in the isotope 1H2 is the electron in the isotope 2He3 with 0.00055 amu.

The proton in the isotope 1H2 has 1.00728 amu's and it is drawn as a dot to the left of the vertical line within the big circle and it is drawn as a dot in the isotope 2He3 to show that the proton in the isotope 1H2 is the proton in the isotope 2He3. There is a square drawn around the proton in the isotope 1H2 and a square around the proton in the isotope 2He3 to show that the proton in the isotope 1H2 has lost 0.00217 amu to become the proton in the isotope 2He3 with 1.00511 amu's.

There is one neutron in the isotope 1H2 with 1.00628 amu's and it is drawn as a small circle to the right of the vertical line within the big circle. This neutron becomes the single neutron in the isotope 2He3. There is a line drawn from the neutron in the isotope 1H2 to the neutron in the isotope 2He3 because there also is a neutron in the isotope 1H3 with 1.00628 amu's.

The isotope 2He3 has 3.01603 amu's. It has an electron and a negatron drawn on the big circle. The electron is drawn above the negatron. They both have 0.00055 amu. There is a proton and a positon drawn to the left of the vertical line within the big circle. The proton is drawn above the positon. There is a line drawn from the positon with 1.0354 amu's in the isotope 2He3 to the proton with 1.00728 amu's in the isotope 1H3 to show which proton is the positon. The single neutron in the isotope 2He3 with 1.00628 amu's is drawn as a small circle to the right of the vertical line. There is a line drawn from the single neutron in the isotope 2He3 to the single neutron in the isotope 1H2. This line is needed because there are two neutrons with the mass of 1.00628 amu's. The first neutron in the isotope 1H3 has 1.00628 amu's and the single neutron in the isotope 1H2 has 1.00628 amu's.

To find the mass that is lost creating the isotope 2He3, add the masses of the isotopes 1H3 and 1H2 and subtract the mass of the isotope 2He3 from their total.

Mass of 1H3	3.01605 amu's
Mass of 1H2	2.01411 amu's
Total mass of 1H3 and 1H2	5.03016 amu's

Total mass of 1H3 and 1H2	5.03016 amu's
Mass of 2He3	3.01603 amu's
Mass lost creating 2He3	2.01413 amu's

The mass lost creating the isotope 2He3 includes the mass of the two neutrons found in the isotope 1H3 that dropped out. Subtract their masses from the total mass lost creating the isotope 2He3 to find the real mass lost.

Mass of first neutron in 1H3	1.00628 amu's
Mass of second neutron in 1H3	1.00194 amu's
Mass of two neutrons that dropped out of 1H3	2.00822 amu's

Mass lost creating 2He3	2.01413 amu's
Mass of two neutrons that dropped out of 1H3	2.00822 amu's
Real mass lost creating 2He3	0.00591 amu

The proton in the isotope 1H3 loses 0.00374 amu to become the positon in the isotope 2He3. To find the mass of the positon in the isotope 2He3, subtract the lost mass in the isotope 1H3 from the proton.

Mass of proton in 1H3	1.00728 amu's
Mass lost by it	0.00374 amu
Mass of positon in 2He3	1.00354 amu's

Since the proton in the isotope 1H3 lost 0.00374 amu, the mass lost by the proton in the isotope 1H2 can be found by subtracting the mass lost by the proton in the isotope 1H3 from the total mass lost creating the isotope 2He3.

Total mass lost creating 2He3	0.00591 amu
Mass lost by proton in 1H3	0.00374 amu
Mass lost by proton in 1H2	0.00217 amu

The proton in the isotope 1H2 lost 0.00217 amu to become the proton in the isotope 2He3. Subtract the lost mass from this proton in the isotope 1H2 to find the mass of the proton in the isotope 2He3.

Mass of proton in 1H2	1.00728 amu's
Mass lost by it	0.00217 amu
Mass of proton in 2He3	1.00511 amu's

The electrons in the isotopes 1H3 and 1H2 are not lost or released because no 0.00100 amu has been transferred or emitted as a single unit and no 0.00002 amu has been lost creating the isotope 2He3.

The proton in the isotope 1H3 lost a unit of 0.00374 amu and the proton in the isotope 1H2 lost a unit of 0.00217 amu for a total lost mass of 0.00591 amu. The 0.00374 amu and the 0.00217 amu are each lost as single units. The lost mass of 0.00374 is composed of four sub-units. They are sub-units of 0.00217 amu, 0.00100 amu, 0.00055 amu and 0.00002 amu. The unit of 0.00217 amu does not have any sub-units. In order for the electron or negatron to stay with their proton or positon, these sub-units must be lost as single units of 0.00374 amu and 0.00217 amu. They must also be lost as single units to keep the proton or positon from releasing an electron or negatron from within either of them.

By adding all of the now known masses of the isotope 2He3, the mass of the isotope 2He3 can be found.

Mass of electron and negatron in 2He3	0.00110 amu
Mass of proton in 2He3	1.00511 amu's
Mass of positon in 2He3	1.00354 amu's
Mass of neutron in 2He3	1.00628 amu's
Mass of 2He3	3.01603 amu's

Sub-step 25-2-9

2He3

1H3 + 1H2 = 2He3, alpha 434 ray, alpha 157 ray, 1.00628 neutron, 1.00194 neutron

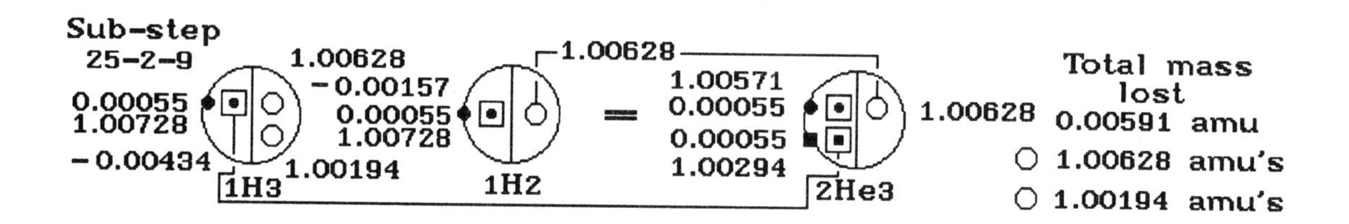

The isotope 1H3 with 3.01605 amu's has one electron with 0.00055 amu and one proton with 1.00728 amu's plus two neutrons. The first neutron has 1.00628 amu's and the second neutron has 1.00194 amu's.

The isotope 1H2 with 2.01411 amu's has one electron with 0.00055 amu and one proton with 1.00728 amu's and one neutron with 1.00628 amu's.

The isotope 2He3 with 3.01603 amu's has one electron with 0.00055 amu, one negatron with 0.00055 amu, one proton with 1.00571 amu's, one positon with 1.00294 amu's and one neutron with 1.00628 amu's.

The total mass lost is 0.00591 amu and it is emitted as a 434 alpha ray and a 157 alpha ray. Each alpha ray has an external magnetic field. There are two neutrons that drop out from the isotope 1H3. The first neutron has 1.00628 amu's and the second one has 1.00194 amu's.

The isotope 1H3 with 3.01605 amu's has one electron with 0.00055 amu that is drawn as a dot on the big circle and it becomes the negatron with 0.00055 amu in the isotope 2He3 and is drawn as a square on the big circle.

The proton with 1.00728 amu's in the isotope 1H3 is drawn as a dot to the left of the vertical line within the big circle. It is drawn as a small square in the isotope 2He3 to show that the proton in the isotope 1H3 has become the positon in the isotope 2He3. The proton has a square drawn around it to show that it has lost 0.00434 amu to become the positon with 1.00294 amu's in the isotope 2He3.

There are two neutrons in the isotope 1H3 that are drawn one above the other as small circles to the right of the vertical line within the big circle. The top neutron has 1.00628 amu's and is called the first neutron while the neutron below it is the second neutron and it has 1.00194 amu's. Neither of these neutrons are needed when the isotope 2He3 is created so they both drop out.

The isotope 1H2 has 2.01411 amu's. Its single electron is drawn as a dot on the big circle in the isotope 1H2 and it is drawn as a dot on the big circle in the isotope 2He3 to show that the electron in the isotope 1H2 is the electron in the isotope 2He3 with 0.00055 amu.

The proton in the isotope 1H2 has 1.00728 amu's and it is drawn as a dot to the left of the vertical line within the big circle and it is drawn as a dot in the isotope 2He3 to show that the proton in the isotope 1H2 is the proton in the isotope 2He3. There is a square drawn around the proton in the isotope 1H2 and a square around the proton in the isotope 2He3 to show that the proton in the isotope 1H2 has lost 0.00157 amu to become the proton in the isotope 2He3 with 1.00571 amu's.

The isotopes 1H3 and 1H2 each have their electron and proton drawn as a dot with each proton losing mass. Each proton in the isotopes of 1H3 and 1H2 has a square drawn around it to show each proton has lost some mass. A line has been drawn from the proton in the isotope 1H3 to the positon in the isotope 2He3 to show which proton became the positon in the isotope 2He3.

There is one neutron in the isotope 1H2 with 1.00628 amu's and it is drawn as a small circle to the right of the vertical line within the big circle. This neutron is the single neutron in the isotope 2He3. There is a line drawn from the neutron in the isotope 1H2 to the single neutron in the isotope 2He3 because the isotope 1H3 also has a neutron with 1.00628 amu's. The line shows that the neutron in the isotope 1H2 is the neutron in the isotope 2He3.

The isotope 2He3 has 3.01603 amu's. It has one electron and one negatron with the electron drawn above the negatron on the big circle. They both have 0.00055 amu. There is one proton and one positon drawn with the proton over the positon to the left of the vertical line within the big circle. The proton has 1.00571 amu's and the positon has 1.00294 amu. There is a line drawn from the positon to the proton in the isotope 1H3 to show which proton is the positon in the isotope 2He3. The single neutron in the isotope 2He3 is drawn as a small circle to the right of the vertical line within the big circle. It has 1.00628 amu's. There is a line drawn from the neutron in the isotope 2He3 to the single neutron in the isotope 1H2 to show which neutron is found in the isotope 2He3. This is done because there are two neutron with 1.00628 amu's. There is one in the isotope 1H3 and one in the isotope 1H2.

To find the mass that is lost creating the isotope 2He3, add the masses of the isotopes 1H3 and 1H2 and subtract the mass of the isotope 2He3 from their total.

Mass of 1H3	3.01605 amu's
Mass of 1H2	2.01411 amu's
Total mass of 1H3 and 1H2	5.03016 amu's

Total mass of 1H3 and 1H2	5.03016 amu's
Mass of 2He3	3.01603 amu's
Mass lost creating 2He3	2.01413 amu's

The mass lost creating the isotope 2He3 includes the mass of the two neutrons found in the isotope 1H3 that dropped out. Subtract their masses from the total mass lost creating the isotope 2He3 to find the real mass lost.

Mass of first neutron in 1H3	1.00628 amu's
Mass of second neutron in 1H3	1.00194 amu's
Mass of two neutrons that dropped out of 1H3	2.00822 amu's

Mass lost creating 2He3	2.01413 amu's
Mass of two neutrons that dropped out of 1H3	2.00822 amu's
Real mass lost creating 2He3	0.00591 amu

The proton in the isotope 1H3 loses 0.00434 amu to become the positon in the isotope 2He3. To find the mass of the positon in the isotope 2He3, subtract the lost mass of the proton in the isotope 1H3 from the proton in the isotope 1H3.

Mass of proton in 1H3	1.00728 amu's
Mass lost by it	0.00434 amu
Mass of positon in 2He3	1.00294 amu's

Since the proton in the isotope 1H3 lost 0.00434 amu, the mass lost by the proton in the isotope 1H2 can be found by subtracting the mass the proton in the isotope 1H3 lost from the total mass lost creating the isotope 2He3.

Total mass lost creating 2He3	0.00591 amu
Mass lost by proton in 1H3	0.00434 amu
Mass lost by proton in 1H2	0.00157 amu

The proton in the isotope 1H2 lost 0.00157 amu to become the proton in the isotope 2He3. Subtract this lost mass to find the mass of the proton in the isotope 2He3.

Mass of proton in 1H2	1.00728 amu's
Mass lost by it	0.00157 amu
Mass of proton in 2He3	1.00571 amu's

The electrons in the isotopes 1H3 and 1H2 are not lost or released because no 0.00100 amu has been transferred or emitted as a single unit and no 0.00002 amu has been lost creating the isotope 2He3.

The proton in the isotope 1H3 lost a unit of 0.00434 amu and the proton in the isotope 1H2 lost a unit of 0.00157 amu for a total lost mass of 0.00591 amu. The 0.00434 amu and the 0.00157 amu are each lost as single units. The lost mass of 0.00434 is composed of two sub-units. They are sub-units of 0.00217 amu. The unit of 0.00157 amu is composed of three sub-units. They are sub-units of 0.00100 amu, 0.00055 amu and 0.00002 amu. In order for the electron or negatron to stay with their proton or positon, these sub-units must be lost as single units of 0.00434 amu and 0.00157 amu. They must also be lost as single units to keep the proton or positon from releasing an electron or negatron from within either of them.

By adding all of the now known masses of the isotope 2He3, the mass of the isotope 2He3 can be found.

Mass of electron and negatron in 2He3	0.00110 amu
Mass of proton in 2He3	1.00571 amu's
Mass of positon in 2He3	1.00294 amu's
Mass of neutron in 2He3	1.00628 amu's
Mass of 2He3	3.01603 amu's

Sub-step 24-2-8

2He3

1H3 + 1H2 = 2He3, alpha 374 ray, alpha 217 ray, 1.00628 neutron, 1.00194 neutron

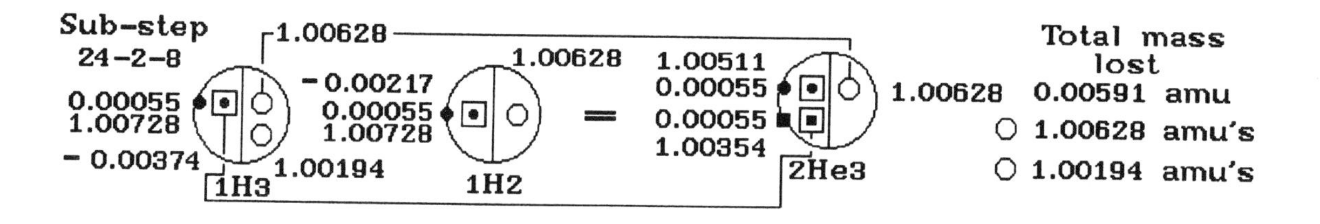

The isotope 1H3 with 3.01605 amu's has one electron with 0.00055 amu and one proton with 1.00728 amu's plus two neutrons. The first neutron has 1.00628 amu's and the second neutron has 1.00194 amu's.

The isotope 1H2 with 2.01411 amu's has one electron with 0.00055 amu and one proton with 1.00728 amu's and one neutron with 1.00628 amu's.

The isotope 2He3 with 3.01603 amu's has one electron with 0.00055 amu, one negatron with 0.00055 amu, one proton with 1.00511 amu's, one positon with 1.00354 amu's and one neutron with 1.00628 amu's.

The total mass lost is 0.00591 amu and it is emitted as a 434 alpha ray and a 157 alpha ray. Each alpha ray has an external magnetic field. There are two neutrons that drop out from the isotope 1H3. The neutron in the isotope 1H2 drops out with 1.00628 amu's and the second neutron in the isotope 1H2 drops out with 1.00194 amu's.

The isotope 1H3 with 3.01605 amu's has one electron with 0.00055 amu that is drawn as a dot on the big circle and it becomes the negatron with 0.00055 amu in the isotope 2He3 and is drawn as a square on the big circle.

The proton with 1.00728 amu's in the isotope 1H3 is drawn as a dot to the left of the vertical line within the big circle. It is drawn as a small square in the isotope 2He3 to show that the proton in the isotope 1H3 has become the positon in the isotope 2He3. The proton has a square drawn around it to show that it has lost 0.00374 amu to become the positon with 1.00354 amu's in the isotope 2He3.

There are two neutrons in the isotope 1H3 that are drawn one above the other as small circles to the right of the vertical line within the big circle. The top neutron has 1.00628 amu's and is called the first neutron while the neutron below it is the second neutron and it has 1.00194 amu's. The first neutron with 1.00628 amu's is the single neutron in the isotope 2He3 while the second neutron with 1.00194 amu's is not needed when the isotope 2He3 is created so it drops out. There is a line drawn from the first neutron in the isotope 1H3 to the single neutron in the isotope 2He3 to show that it is the first neutron in the isotope 1H3 that becomes the single neutron in the isotope 2He3. This line is drawn because there are two neutrons with 1.00628 amu's.

The isotope 1H2 has 2.01411 amu's. Its single electron is drawn as a dot on the big circle in the isotope 1H2 and it is drawn as a dot on the big circle in the isotope 2He3 to show that the electron in the isotope 1H2 is the electron in the isotope 2He3 with 0.00055 amu.

The proton in the isotope 1H2 has 1.00728 amu's and it is drawn as a dot to the left of the vertical line within the big circle and it is drawn as a dot in the isotope 2He3 to show that the proton in the isotope 1H2 is the proton in the isotope 2He3. There is a square drawn around the dot in the isotope 1H2 and a square around the dot in the isotope 2He3 to show that the proton in the isotope 1H2 has lost 0.00217 amu to become the proton in the isotope 2He3 with 1.00511 amu's.

The isotopes 1H3 and 1H2 each have their electron and proton drawn as a dot with each proton losing mass. Each proton in the isotopes of 1H3 and 1H2 has a square drawn around it to show each proton has lost some mass. A line has been drawn from the proton in the isotope 1H3 to the positon in the isotope 2He3 to show which proton became the positon in the isotope 2He3.

There is one neutron in the isotope 1H2 with 1.00628 amu's and it is drawn as a small circle to the right of the vertical line within the big circle. This neutron is not needed when the isotope 2He3 is created so it drops out.

The isotope 2He3 has 3.01603 amu's. It has one electron and one negatron drawn on the big circle. The electron is drawn as a dot to show that it is the electron in the isotope 1H2. The negatron is drawn as a square to show that it is the electron in the isotope 1H3. They both have 0.00055 amu.

The proton in the isotope 2He3 is drawn as a dot to show that it comes from the isotope 1H2. There is a square drawn around it to show that also comes from the isotope of 1H2. The positon is drawn as a small square below the proton. There is a line drawn from the positon in the isotope 2He3 to the proton in the isotope 1H3 to show which proton became the positon.

The single neutron in the isotope 2He3 is drawn as a small circle to the right of the vertical line within the big circle. There is a line drawn from the neutron in the isotope 2He3 to the first neutron in the isotope 1H3 to show which neutron is the neutron in the isotope 2He3. This is needed because there are two neutron with 1.00628 amu's. The first neutron in the isotope 1H3 and the single neutron in the isotope 1H2 each have 1.00628 amu's.

To find the mass that is lost creating the isotope 2He3, add the masses of the isotopes 1H3 and 1H2 and subtract the mass of the isotope 2He3 from their total.

Mass of 1H3	3.01605 amu's
Mass of 1H2	2.01411 amu's
Total mass of 1H3 and 1H2	5.03016 amu's

Total mass of 1H3 and 1H2	5.03016 amu's
Mass of 2He3	3.01603 amu's
Mass lost creating 2He3	2.01413 amu's

The mass lost creating the isotope 2He3 includes the mass of the two neutrons found in the isotope 1H3 that dropped out. Subtract their masses from the total mass lost creating the isotope 2He3 to find the real mass lost.

Mass of first neutron in 1H3	1.00628 amu's
Mass of second neutron in 1H3	1.00194 amu's
Mass of two neutrons that dropped out of 1H3	2.00822 amu's

Mass lost creating 2He3	2.01413 amu's
Mass of two neutrons that dropped out of 1H3	2.00822 amu's
Real mass lost creating 2He3	0.00591 amu

The proton in the isotope 1H3 loses 0.00374 amu to become the positon in the isotope 2He3. To find the mass of the positon in the isotope 2He3, subtract the lost mass of the proton in the isotope 1H3 from the proton in the isotope 1H3.

Mass of proton in 1H3	1.00728 amu's
Mass lost by it	0.00374 amu
Mass of positon in 2He3	1.00354 amu's

Since the proton in the isotope 1H3 lost 0.00374 amu, the mass of the proton in the isotope 1H2 can be found by subtracting the mass the proton lost in the isotope 1H3 from the total mass lost creating the isotope 2He3.

Total mass lost creating the 2He3	0.00591 amu
Mass lost by the proton in 1H3	0.00374 amu
Mass lost by proton in 1H2	0.00217 amu

The proton in the isotope 1H2 lost 0.00217 amu to become the proton in the isotope 2He3. Subtract this lost mass to find the mass of the proton in the isotope 2He3.

Mass of proton in 1H2	1.00728 amu's
Mass lost by it	0.00217 amu
Mass of proton in 2He3	1.00511 amu's

The electrons in the isotopes 1H3 and 1H2 are not lost or released because no 0.00100 amu has been transferred or emitted as a single unit and no 0.00002 amu has been lost creating the isotope 2He3.

The proton in the isotope 1H3 lost a unit of 0.00374 amu and the proton in the isotope 1H2 lost a unit of 0.00217 amu for a total lost mass of 0.00591 amu.

The 0.00374 amu and the 0.00217 amu are each lost as single units. The lost mass of 0.00374 is composed of four sub-units. They are sub-units of 0.00217 amu, 0.00100 amu, 0.00055 amu and 0.00002 amu. The unit of 0.00217 amu does not have any sub-units. In order for the electron or negatron to stay with their proton or positon, these sub-units must be lost as single units of 0.00374 amu and 0.00217 amu. It must also be lost as single units to keep the proton, positon or neutron from releasing an electron or negatron from within any of them.

By adding all of the now known masses of the isotope 2He3, the mass of the isotope 2He3 can be found.

Mass of electron and negatron in 2He3	0.00110 amu
Mass of proton in 2He3	1.00511 amu's
Mass of positon in 2He3	1.00354 amu's
Mass of neutron in 2He3	1.00628 amu's
Mass of 2He3	3.01603 amu's

Sub-step 23-2-7

2He3

1H3 + 1H2 = 2He3, alpha 434 ray, alpha 157 ray, 1.00628 neutron, 1.00194 neutron

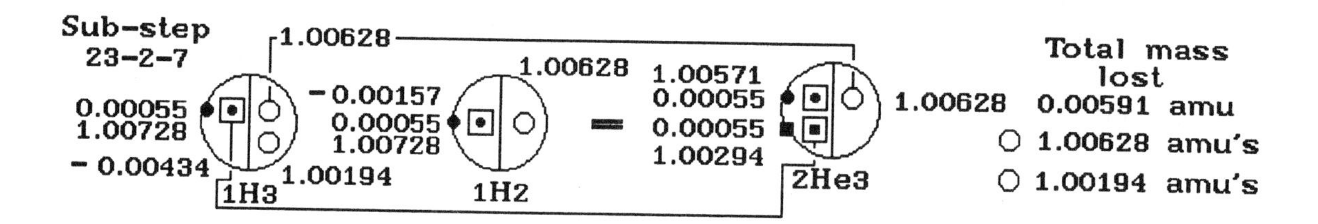

The isotope 1H3 with 3.01605 amu's has one electron with 0.00055 amu and one proton with 1.00728 amu's plus two neutrons. The first neutron has 1.00628 amu's and the second neutron has 1.00194 amu's.

The isotope 1H2 with 2.01411 amu's has one electron with 0.00055 amu and one proton with 1.00728 amu's and one neutron with 1.00628 amu's.

The isotope 2He3 with 3.01603 amu's has one electron with 0.00055 amu, one negatron with 0.00055 amu, one proton with 1.00571 amu's, one positon with 1.00294 amu's and one neutron with 1.00628 amu's.

The total mass lost is 0.00591 amu and it is emitted as a 434 alpha ray and a 157 alpha ray. Each alpha ray has an external magnetic field. There are two neutrons that drop out. The isotope 1H2 has its neutron drop out with 1.00628 amu's and the isotope 1H3 has its second neutron drop out with 1.00194 amu's.

The isotope 1H3 with 3.01605 amu's has one electron with 0.00055 amu that is drawn as a dot on the big circle and it becomes the negatron with 0.00055 amu in the isotope 2He3 and is drawn as a square on the big circle.

The proton with 1.00728 amu's in the isotope 1H3 is drawn as a dot to the left of the vertical line within the big circle. It is drawn as a small square in the isotope 2He3 to show that the proton in the isotope 1H3 has become the positon in the isotope 2He3. The proton has a square drawn around it to show that it has lost 0.00434 amu to become the positon with 1.00294 amu's in the isotope 2He3.

There are two neutrons in the isotope 1H3 that are drawn one above the other as small circles to the right of the vertical line within the big circle. The top neutron has 1.00628 amu's and is called the first neutron while the neutron below it is the second neutron and it has 1.00194 amu's. The first neutron with 1.00628 amu's is the single neutron in the isotope 2He3 while the second neutron with 1.00194 amu's is not needed when the isotope 2He3 is created so it drops out. There is a line drawn from the first neutron in the isotope 1H3 to the single neutron in the isotope 2He3 to show that it is the first neutron in the isotope 1H3 that becomes the single neutron in the isotope 2He3. This line is drawn because there are two neutrons with 1.00628 amu's.

The isotope 1H2 has 2.01411 amu's. Its single electron is drawn as a dot on the big circle in the isotope 1H2 and it is also drawn as a dot on the big circle in the isotope 2He3 to show that the electron in the isotope 1H2 is the electron in the isotope 2He3 with 0.00055 amu.

The proton in the isotope 1H2 has 1.00728 amu's and it is drawn as a dot to the left of the vertical line within the big circle and it is drawn as a dot in the isotope 2He3 to show that the proton in the isotope 1H2 is the proton in the isotope 2He3. There is a square drawn around the dot in the isotope 1H2 and a square around the dot in the isotope 2He3 to show that the proton in the isotope 1H2 has lost 0.00157 amu to become the proton in the isotope 2He3 with 1.00571 amu's.

The isotopes 1H3 and 1H2 each have their electron and proton drawn as a dot with each proton losing mass. Each proton in the isotopes of 1H3 and 1H2 has a square drawn around it to show each proton has lost some mass. A line has been drawn from the proton in the isotope 1H3 to the positon in the isotope 2He3 to show which proton became the positon in the isotope 2He3.

There is one neutron in the isotope 1H2 with 1.00628 amu's and it is drawn as a small circle to the right of the vertical line within the big circle. This neutron is not needed when the isotope 2He3 is created so it drops out.

To find the mass that is lost creating the isotope 2He3, add the masses of the isotopes 1H3 and 1H2 and subtract the mass of the isotope 2He3 from their total.

Mass of 1H3	3.01605 amu's
Mass of 1H2	2.01411 amu's
Total mass of 1H3 and 1H2	5.03016 amu's

Total mass of 1H3 and 1H2	5.03016 amu's
Mass of 2He3	3.01603 amu's
Mass lost creating 2He3	2.01413 amu's

The mass lost creating the isotope 2He3 includes the mass of the two neutrons found in the isotope 1H3 that dropped out. Subtract their masses from the total mass lost creating the isotope 2He3 to find the real mass lost.

Mass of neutron in 1H2	1.00628 amu's
Mass of second neutron in 1H3	1.00194 amu's
Mass of two neutrons that dropped out of 1H3	2.00822 amu's

Mass lost creating 2He3	2.01413 amu's
Mass of two neutrons that dropped out of 1H3	2.00822 amu's
Real mass lost creating 2He3	0.00591 amu

The proton in the isotope 1H3 loses 0.00434 amu to become the positon in the isotope 2He3. To find the mass of the positon in the isotope 2He3, subtract the lost mass of the proton in the isotope 1H3 from the proton in the isotope 1H3.

Mass of proton in 1H3	1.00728 amu's
Mass lost by it	0.00434 amu
Mass of positon in 2He3	1.00294 amu's

Since the proton in the isotope 1H3 lost 0.00434 amu, the mass of the positon in the isotope 2He3 can be found by subtracting the lost mass of the proton in the isotope 1H3 from the total mass lost creating the isotope 2He3.

Total mass lost creating the isotope 2He3	0.00591 amu
Mass lost by the proton in 1H3	0.00434 amu
Mass lost by proton in 1H2	0.00157 amu

The proton in the isotope 1H2 lost 0.00157 amu to become the proton in the isotope 2He3. Subtract the mass lost to find the mass of the proton in the isotope 2He3.

Mass of proton in 1H2	1.00728 amu's
Mass lost by it	0.00157 amu
Mass of proton in 2He3	1.00571 amu's

The electrons in the isotopes 1H3 and 1H2 are not lost or released because no 0.00100 amu has been transferred or emitted as a single unit and no 0.00002 amu has been lost creating the isotope 2He3.

The proton in the isotope 1H3 lost a unit of 0.00434 amu and the proton in the isotope 1H2 lost a unit of 0.00157 amu for a total lost mass of 0.00591 amu. The 0.00434 amu and the 0.00157 amu are each lost as single units. The lost mass of 0.00434 is composed of two sub-units. They are sub-units of 0.00217 amu. The unit of 0.00157 amu is composed of three sub-units. They are sub-units of 0.00100 amu, 0.00055 amu and 0.00002 amu. In order for the electron or negatron to stay with their proton or positon, these sub-units must be lost as single units of 0.00434 amu and 0.00157 amu. They must also be lost as single units to keep the proton, positon or neutron from releasing an electron or negatron from within any of them.

By adding all of the now known masses of the isotope 2He3, the mass of the isotope 2He3 can be found.

Mass of electron and negatron in 2He3	0.00110 amu
Mass of proton in 2He3	1.00571 amu's
Mass of positon in 2He3	1.00294 amu's
Mass of neutron in 2He3	1.00628 amu's
Mass of 2He3	3.01603 amu's

Sub-step 22-2-6

2He3

1H3 + 1H2 = 2He3, alpha 591 ray, 1.00628 neutron, 1.00194 neutron

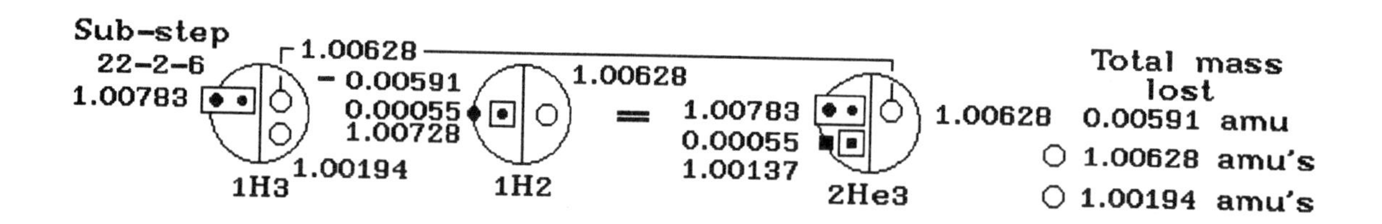

The isotope 1H3 with 3.01605 amu's has one electron with 0.00055 amu and one proton with 1.00728 amu's plus two neutrons. The first neutron has 1.00628 amu's and the second neutron has 1.00194 amu's.

The isotope 1H2 with 2.01411 amu's has one electron with 0.00055 amu and one proton with 1.00728 amu's and one neutron with 1.00628 amu's.

The isotope 2He3 with 3.01603 amu's has one electron with 0.00055 amu, one negatron with 0.00055 amu, one proton with 1.00728 amu's, one positon with 1.00137 amu's and one neutron with 1.00628 amu's.

The total mass lost is 0.00591 amu and it is emitted as a 591 alpha ray which has an external magnetic field. There are two neutrons that drop out. The first neutron drops out of the isotope 1H2 has 1.00628 amu's and the second isotope to drop out comes from the isotope 1H3 with 1.00194 amu's.

The isotope 1H3 with 3.01605 amu's has one electron with 0.00055 amu that is drawn as a dot on the big circle and it becomes the electron with 0.00055 amu in the isotope 2He3 and is drawn as a dot on the big circle.

The proton with 1.00728 amu's in the isotope 1H3 is drawn as a dot to the left of the vertical line within the big circle. It is drawn as a dot in the isotope 2He3. The electron and proton have a rectangle drawn around them to show that they are used as the isotope 1H1 within the isotopes 1H3 and 2He3. The rectangle shows that the isotope 1H1 can drop out of the isotope 2He3 at some later time.

There are two neutrons in the isotope 1H3 that are drawn one above the other as small circles to the right of the vertical line within the big circle. The top neutron has 1.00628 amu's and is called the first neutron while the neutron below it is the second neutron and it has 1.00194 amu's. The first neutron with 1.00628 amu's is the single neutron in the isotope 2He3 while the second neutron with 1.00194 amu's is not needed when the isotope 2He3 is created so it drops out. There is a line drawn from the first neutron in the isotope 1H3 to the single neutron in the isotope 2He3 to show that it is the first neutron in the isotope 1H3 that becomes the single neutron in the isotope 2He3. This line is drawn because there are two neutrons with 1.00628 amu's.

The isotope 1H2 has 2.01411 amu's. Its single electron is drawn as a dot on the big circle in the isotope 1H2 and it is drawn as a square on the big circle in the isotope 2He3 to show that the electron has become the negatron with 0.00055 amu. The proton in the isotope 1H2 has 1.00728 amu's and it is drawn as a dot to the left of the vertical line within the big circle in the isotopes 1H2 and it is drawn as a small square in the isotope 2He3 to show that the proton in the isotope 1H2 has become the positon in the isotope 2He3. There is a square drawn around the dot in the isotope 1H2 and a square around the square in the isotope 2He3 to show that the proton in the isotope 1H2 has lost 0.00591 amu to become the positon in the isotope 2He3 with 1.00137 amu's.

There is one neutron in the isotope 1H2 with 1.00628 amu's and it is drawn as a small circle to the right of the vertical line within the big circle. This neutron is not needed when the isotope 2He3 is created so it drops out.

The isotope 2He3 has 3.01603 amu's. It has one electron and one negatron drawn on the big circle. The electron is drawn as a dot in the isotope 2He3 on the big circle while the negatron is drawn as a square on the big circle. There is a proton and a positon in the isotope 2He3. The proton is drawn as a dot on the right side of the vertical line within the big circle. There is a rectangle drawn around the electron and proton in the isotope 2He3 and 1H3 to show that the isotope 1H1 is found intact within both of them and that the isotope of 1H1 can drop out of the isotope 2He3 at some later time. The single neutron is drawn as a small circle to the right of the vertical line within the big circle. There is a line drawn from the single neutron in the isotope 2He3 to the first neutron in the isotope 1H3. This line is needed because there are two neutrons with the mass of 1.00628 amu's. The first neutron in the isotope 1H3 and the single neutron in the isotope 1H2 have 1.00628 amu's.

To find the mass that is lost creating the isotope 2He3, add the masses of the isotopes 1H3 and 1H2 and subtract the mass of the isotope 2He3 from their total.

Mass of 1H3	3.01605 amu's
Mass of 1H2	2.01411 amu's
Total mass of 1H3 and 1H2	5.03016 amu's

Total mass of 1H3 and 1H2	5.03016 amu's
Mass of 2He3	3.01603 amu's
Mass lost creating 2He3	2.01413 amu's

The mass lost creating the isotope 2He3 includes the mass of the two neutrons that dropped out. The single neutron in the isotope 1H2 and the second neutron in the isotope 1H3 dropped out when the isotope 2He3 was created because they were not needed. Subtract their masses from the total mass lost creating the isotope 2He3 to find the real mass lost.

Mass of neutron in 1H2	1.00628 amu's
Mass of second neutron in 1H3	1.00194 amu's
Mass of neutrons that dropped out of 1H3 and 1H2	2.00822 amu's

Mass lost creating 2He3	2.01413 amu's
Mass of neutrons that dropped out of 1H3 and 1H2	2.00822 amu's
Real mass lost creating 2He3	0.00591 amu

The isotope 1H3 has its electron and proton used as the isotope 1H1 within the isotopes 1H3 and 2He3. The electron and proton become the electron and proton in the isotope 2He3. This means that the only proton that can become the positon in the isotope 2He3 must come from the isotope 1H2. This proton loses 0.00591 amu so subtract this mass from the proton to find the mass of the positon it becomes in the isotope 2He3.

Mass of proton in 1H2	1.00728 amu's
Mass lost by it	0.00591 amu
Mass of positon in 2He3	1.00137 amu's

The electrons in the isotopes 1H3 and 1H2 are not lost or released because no 0.00100 amu has been transferred or emitted as a single unit and no 0.00002 amu has been lost creating the isotope 2He3.

The proton in the isotope 1H3 lost a unit 0.00591 amu for a total lost mass of 0.00591 amu and it is lost as a single unit. The lost mass of 0.00591 is composed of four sub-units. They are sub-units of 0.00434 amu, 0.00100 amu, 0.00055 amu and 0.00002 amu. The last three sub-units could be sub-units of 0.00157 amu. The sub-unit of 0.00434 amu is composed of two sub-units of 0.00217 amu. In order for the electron or negatron to stay with its proton, these sub-units must be lost as a single unit of 0.00591 amu. It must also be lost as a single unit to keep the proton from releasing an electron or negatron from within the proton. The lost mass of 0.591 amu can be composed of sub-units of 0.00374 amu and 0.00217 amu.

By adding all of the now known masses of the isotope 2He3, the mass of the isotope 2He3 can be found.

Mass of 1H1 intact within 2He3	1.00783 amu's
Mass of neutron in 2He3	1.00628 amu's
Mass of negatron in 2He3	0.00055 amu
Mass of positon in 2He3	1.00137 amu's
Mass of 2He3	3.01603 amu's

Sub-step 21-2-5

2He3

1H3 + 1H2 = 2He3, alpha 591 ray, 1.00628 neutron, 1.00194 neutron

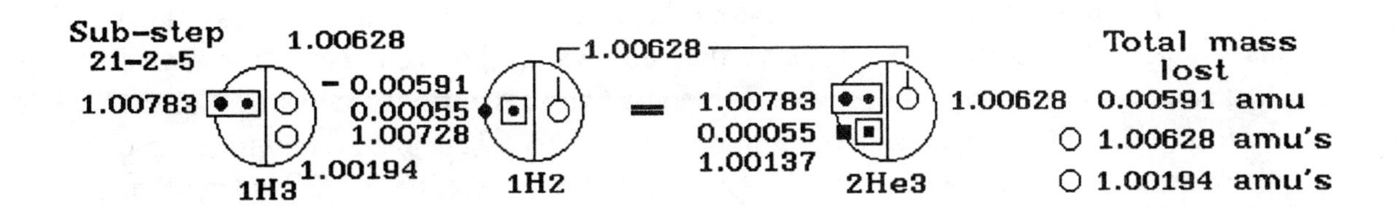

The isotope 1H3 with 3.01605 amu's has one electron with 0.00055 amu and one proton with 1.00728 amu's plus two neutrons. The first neutron has 1.00628 amu's and the second neutron has 1.00194 amu's.

The isotope 1H2 with 2.01411 amu's has one electron with 0.00055 amu and one proton with 1.00728 amu's and one neutron with 1.00628 amu's.

The isotope 2He3 with 3.01603 amu's has one electron with 0.00055 amu, one negatron with 0.00055 amu, one proton with 1.00728 amu's, one positon with 1.00137 amu's and one neutron with 1.00628 amu's.

The total mass lost is 0.00591 amu and it is emitted as a 591 alpha ray which has an external magnetic field. There are two neutrons that drop out from the isotope 1H3. The first neutron has 1.00628 amu's and the second one has 1.00194 amu's.

The isotope 1H3 with 3.01605 amu's has one electron with 0.00055 amu that is drawn as a dot on the big circle and it becomes the electron with 0.00055 amu in the isotope 2He3 and is drawn as a dot on the big circle.

The proton with 1.00728 amu's in the isotope 1H3 is drawn as a dot to the left of the vertical line within the big circle. It is drawn as a dot in the isotope 2He3. The electron and proton have a rectangle drawn around them to show that they are used as the isotope 1H1 within the isotopes 1H3 and 2He3. The rectangle shows that the isotope 1H1 can drop out of the isotope 2He3 at some later time.

There are two neutrons in the isotope 1H3 that are drawn one above the other as small circles to the right of the vertical line within the big circle. The top neutron has 1.00628 amu's and is called the first neutron while the neutron below it is the second neutron and it has 1.00194 amu's. Neither of these neutrons are needed when the isotope 2He3 is created so they both drop out.

The isotope 1H2 has 2.01411 amu's. Its single electron is drawn as a dot on the big circle in the isotope 1H2 and it is drawn as a square on the big circle in the isotope 2He3 to show that the electron has become the negatron with 0.00055 amu.

The proton in the isotope 1H2 has 1.00728 amu's and it is drawn as a dot to the left of the vertical line within the big circle in the isotope 1H2 and it is drawn as a small square in the isotope 2He3 to show that the proton in the isotope 1H2 has become the positon in the isotope 2He3. There is a square drawn around the dot in the isotope 1H2 and a square around the square in the isotope 2He3 to show that the proton in the isotope 1H2 has lost 0.00591 amu to become the positon in the isotope 2He3 with 1.00137 amu's.

There is one neutron in the isotope 1H2 with 1.00628 amu's and it is drawn as a small circle to the right of the vertical line within the big circle. It is drawn this same way in the isotope 2He3. There is a line drawn from the neutron in the isotope 1H2 to the single neutron in the isotope 2He3. This line is drawn because there are two neutrons with 1.00628 amu's and it is the isotope 1H2's neutron that is used in the isotope 2He3.

The isotope 2He3 has 3.01603 amu's. It has one electron and one negatron drawn on the big circle. The electron is drawn as a dot on the big circle in the isotope 1H3 and 2He3. The negatron is drawn as a dot on the big circle in the isotope 1H3 and as a square on the big circle in the isotope 2He3. The proton in the isotope 2He3 is drawn as dot on the left side of the vertical line within the isotope 2He3. The electron and proton has a rectangle drawn around them to show that the isotope 1H1 is used as is within the isotope 2He3. It is drawn this same way in the isotope 1H3 to show that the isotope of 1H1 is used with each of them. The rectangle shows that the isotope 1H1 found within the isotope 2He3 can drop out at some later time. The single neutron in the isotope 2He3 is drawn as a small circle to the right of the vertical line within the big circle. There is a line drawn from the neutron in the isotope 2He3 to the single neutron in the isotope 1H2. The line is needed because there are two neutrons with the mass of 1.00628 amu's. One is found in the isotope 1H3 and the other is found in the isotope 1H2.

To find the mass that is lost creating the isotope 2He3, add the masses of the isotopes 1H3 and 1H2 and subtract the mass of the isotope 2He3 from their total.

Mass of 1H3	3.01605 amu's
Mass of 1H2	2.01411 amu's
Total mass of 1H3 and 1H2	5.03016 amu's

Total mass of 1H3 and 1H2	5.03016 amu's
Mass of 2He3	3.01603 amu's
Mass lost creating 2He3	2.01413 amu's

The mass lost creating the isotope 2He3 includes the mass of the two neutrons found in the isotope 1H3 that dropped out. Subtract their masses from the total mass lost creating the isotope 2He3 to find the real mass lost.

Mass of first neutron in 1H3	1.00628 amu's
Mass of second neutron in 1H3	1.00194 amu's
Mass of two neutrons that dropped out of 1H3	2.00822 amu's

Mass lost creating 2He3	2.01413 amu's
Mass of two neutrons that dropped out of 1H3	2.00822 amu's
Real mass lost creating 2He3	0.00591 amu

The isotope 1H3 has its electron and proton used as the isotope 1H1 within the isotopes 1H3 and 2He3. The electron and proton become the electron and proton in the isotope 2He3. This means that the only proton that can become the positon in the isotope 2He3 must come from the isotope 1H2. This proton loses 0.00591 amu so subtract this mass from the proton to find the mass of the positon it becomes in the isotope 2He3.

Mass of proton in 1H2	1.00728 amu's
Mass lost by it	0.00591 amu
Mass of positon in 2He3	1.00137 amu's

The electrons in the isotopes 1H3 and 1H2 are not lost or released because no 0.00100 amu has been transferred or emitted as a single unit and no 0.00002 amu has been lost creating the isotope 2He3.

The proton in the isotope 1H3 lost a unit of 0.00591 amu for a total lost mass of 0.00591 amu and it is lost as a single unit. The lost mass of 0.00591 is composed of four sub-units. They are sub-units of 0.00434 amu, 0.00100 amu, 0.00055 amu and 0.00002 amu. The last three sub-units could be a sub-unit of 0.00157 amu. The sub-unit of 0.00434 amu is composed of two sub-units of 0.00217 amu. In order for the electron or negatron to stay with its proton, these sub-units must be lost as a single unit of 0.00591 amu. It must also be lost as a single unit to keep the proton from releasing an electron or negatron from within the proton. The lost mass of 0.00591 could also be composed of sub-units of 0.00374 amu and 0.00217 amu.

By adding all of the now known masses of the isotope 2He3, the mass of the isotope 2He3 can be found.

Mass of 1H1 intact within 2He3	1.00783 amu's
Mass of neutron in 2He3	1.00628 amu's
Mass of negatron in 2He3	0.00055 amu
Mass of positon in 2He3	1.00137 amu's
Mass of 2He3	3.01603 amu's

Sub-step 20-2-4

2He3

1H3 + 1H2 = 2He3, alpha 591 ray, 1.00628 neutron, 1.00194 neutron

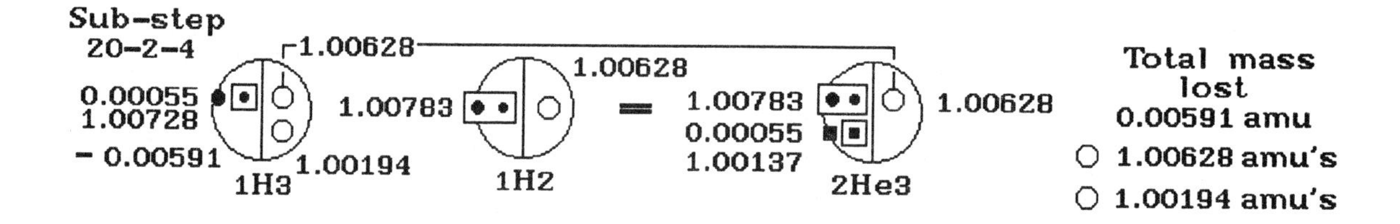

The isotope 1H3 with 3.01605 amu's has one electron with 0.00055 amu and one proton with 1.00728 amu's plus two neutrons. The first neutron has 1.00628 amu's and the second neutron has 1.00194 amu's.

The isotope 1H2 with 2.01411 amu's has one electron with 0.00055 amu and one proton with 1.00728 amu's and one neutron with 1.00628 amu's.

The isotope 2He3 with 3.01603 amu's has one electron with 0.00055 amu, one negatron with 0.00055 amu, one proton with 1.00728 amu's, one positon with 1.00137 amu's and one neutron with 1.00628 amu's.

The total mass lost is 0.00591 amu and it is emitted as a 591 alpha ray which has an external magnetic field. There are two neutrons that drop out. The single neutron in the isotope 1H2 drops out with 1.00628 amu's and the second neutron in the isotope 1H3 drops out with 1.00194 amu's.

The isotope 1H3 with 3.01605 amu's has one electron with 0.00055 amu that is drawn as a dot on the big circle and it becomes the negatron with 0.00055 amu in the isotope 2He3 and is drawn as a square on the big circle.

The proton with 1.00728 amu's in the isotope 1H3 is drawn as a dot to the left of the vertical line within the big circle. There is a square drawn around it to show that it loses 0.00591 amu. It is drawn as a small square with a square around it in the isotope 2He3 to show that the proton in the isotope 1H3 is the positon in the isotope 2He3.

There are two neutrons in the isotope 1H3 that are drawn one above the other as small circles to the right of the vertical line within the big circle. The top neutron has 1.00628 amu's and is called the first neutron while the neutron below it is the second neutron and it has 1.00194 amu's. The first neutron becomes the single neutron found in the isotope 2He3. There is a line drawn from the first neutron in the isotope 1H3 to the single neutron in the isotope 2He3 to show that the first neutron in the isotope 1H3 is the single neutron in the isotope 2He3. This line is needed because the single neutron in the isotope 1H2 also has 1.00628 amu's. The second neutron in the isotope 1H3 is not needed when the isotope 2He3 is created so it drops out.

The isotope 1H2 has 2.01411 amu's. Its single electron is drawn as a dot on the big circle in the isotopes 1H2 and 2He3. The proton has 1.00728 amu's and it is drawn as a dot to the left of the vertical line within the big circle in the isotopes 1H2 and 2He3.

The electron and proton in the isotope 1H2 has a rectangle drawn around both of them to show that they are used as the isotope 1H1 in the isotope 2He3. They are drawn the same way in the isotope 2He3. The rectangle in the isotope 2He3 shows that the isotope 1H1 can drop out of the isotope 2He3 at a later time.

There is one neutron in the isotope 1H2 with 1.00628 amu's and it is drawn as a small circle to the right of the vertical line within the big circle. This neutron is not needed when the isotope 2He3 is created so it drops out.

The isotope 2He3 has 3.01603 amu's. It has one electron and one negatron drawn on the big circle. They both have 0.00055 amu. There is one proton and one positon drawn to the left of the vertical line within the big circle. The proton is drawn as a dot to show that it comes from the isotope 1H2 and the positon is drawn as a small square with another square drawn around it to show that it is the proton in the isotope 1H3. The negatron and positon have a rectangle drawn around them to show that they came from the isotope 1H2. The rectangle also shows that the isotope 1H1 can drop out of the isotope 2He3 at some later time.

The single neutron in the isotope 2He3 has 1.00628 amu's is drawn as a small circle to the right of the vertical line. It comes from the first neutron in the isotope 1H3. There is a line drawn from the single neutron in the isotope 2He3 to the first neutron in the isotope 1H3 to show where the single neutron in the isotope 2He3 came from. This line is needed because there are two isotopes with 1.00628 amu's. One is found in the isotope 1H3 and the other is found in the isotope 1H2.

To find the mass that is lost creating the isotope 2He3, add the masses of the isotopes 1H3 and 1H2 and subtract the mass of the isotope 2He3 from their total.

Mass of 1H3	3.01605 amu's
Mass of 1H2	2.01411 amu's
Total mass of 1H3 and 1H2	5.03016 amu's

Total mass of 1H3 and 1H2	5.03016 amu's
Mass of 2He3	3.01603 amu's
Mass lost creating 2He3	2.01413 amu's

The mass lost creating the isotope 2He3 includes the mass of the neutron found in the isotope 1H3 that dropped out and the mass of the neutron found in the isotope 1H2 that dropped out. Subtract their masses from the total mass lost creating the isotope 2He3 to find the real mass lost.

Mass of neutron in 1H2	1.00628 amu's
Mass of second neutron in 1H3	1.00194 amu's
Mass of two neutrons that dropped out of 1H3	2.00822 amu's

Mass lost creating 2He3	2.01413 amu's
Mass of two neutrons that dropped out of 1H3	2.00822 amu's
Real mass lost creating 2He3	0.00591 amu

The isotope 1H2 has its electron and proton used as the isotope 1H1 within the isotopes 1H2 and 2He3. Its neutron with 1.00628 amu's becomes the single neutron in the isotope 2He3. The electron and proton become the electron and proton in the isotope 2He3. This means that the only proton that can become the positon in the isotope 2He3 must come from the isotope 1H3. This proton loses 0.00591 amu so subtract this mass from the proton in the isotope 1H3 to find the mass of the positon it becomes in the isotope 2He3.

Mass of proton in 1H3	1.00728 amu's
Mass lost by it	0.00591 amu
Mass of positon in 2He3	1.00137 amu's

The electrons in the isotopes 1H3 and 1H2 are not lost or released because no 0.00100 amu has been transferred or emitted as a single unit and no 0.00002 amu has been lost creating the isotope 2He3.

The proton in the isotope 1H3 lost a unit of 0.00591 amu for a total lost mass of 0.00591 amu and it is lost as a single unit. The lost mass of 0.00591 is composed of four sub-units. They are sub-units of 0.00434 amu, 0.00100 amu, 0.00055 amu and 0.00002 amu. The last three sub-units could be a sub-unit of 0.00157 amu. The sub-unit of 0.00434 amu is composed of two sub-units of 0.00217 amu. In order for the electron or negatron to stay with its proton, these sub-units must be lost as a single unit of 0.00591 amu. It must also be lost as a single unit to keep the proton from releasing an electron or negatron from within the proton. The lost mass of 0.00591 amu could also be composed of a sub-unit of 0.00374 amu and a sub-unit of 0.00217 amu.

By adding all of the now known masses of the isotope 2He3, the mass of the isotope 2He3 can be found.

Mass of 1H1 intact within 2He3	1.00783 amu's
Mass of neutron in 2He3	1.00628 amu's
Mass of negatron in 2He3	0.00055 amu
Mass of positon in 2He3	1.00137 amu's
Mass of 2He3	3.01603 amu's

Sub-step 19-2-3

2He3

1H3 + 1H2 = 2He3, alpha 591 ray, 1.00628 neutron, 1.00194 neutron

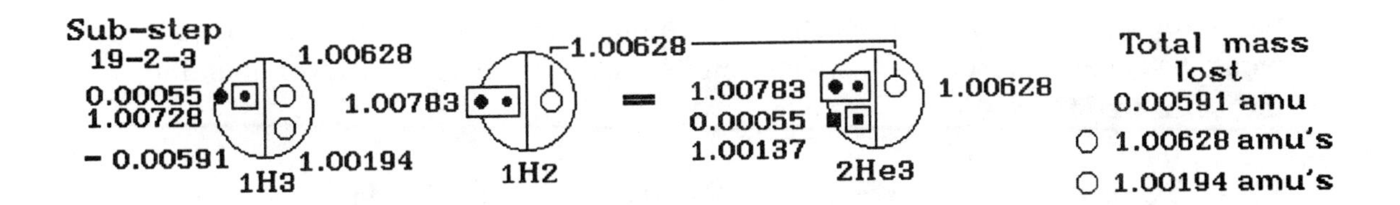

The isotope 1H3 with 3.01605 amu's has one electron with 0.00055 amu and one proton with 1.00728 amu's plus two neutrons. The first neutron has 1.00628 amu's and the second neutron has 1.00194 amu's.

The isotope 1H2 with 2.01411 amu's has one electron with 0.00055 amu and one proton with 1.00728 amu's and one neutron with 1.00628 amu's.

The isotope 2He3 with 3.01603 amu's has one electron with 0.00055 amu, one negatron with 0.00055 amu, one proton with 1.00728 amu's, one positon with 1.00137 amu's and one neutron with 1.00628 amu's.

The total mass lost is 0.00591 amu and it is emitted as a 591 alpha ray which has an external magnetic field. There are two neutrons that drop out from the isotope 1H3. The first neutron has 1.00628 amu's and the second one has 1.00194 amu's.

The isotope 1H3 with 3.01605 amu's has one electron with 0.00055 amu that is drawn as a dot on the big circle and it becomes the negatron with 0.00055 amu in the isotope 2He3 and is drawn as a square on the big circle.

The proton with 1.00728 amu's in the isotope 1H3 is drawn as a dot to the left of the vertical line within the big circle. There is a square drawn around it to show that it loses 0.00591 amu. It is drawn as a small square with a square around it in the isotope 2He3 to show that the proton in the isotope 1H3 is the positon in the isotope 2He3.

There are two neutrons in the isotope 1H3 that are drawn one above the other as small circles to the right of the vertical line within the big circle. The top neutron has 1.00628 amu's and is called the first neutron while the neutron below it is the second neutron and it has 1.00194 amu's. Neither of these neutrons are needed when the isotope 2He3 is created so they both drop out.

The isotope 1H2 has 2.01411 amu's. Its single electron with 0.00055 amu is drawn as a dot on the big circle in the isotopes 1H2 and 2He3. The proton in the isotope 1H2 has 1.00728 amu's and it is drawn as a dot to the left of the vertical line within the big circle in the isotopes 1H2 and 2He3.

The electron and proton in the isotope 1H2 has a rectangle drawn around both of them to show that they are used as the isotope 1H1 in the isotope 2He3. They are drawn the same way in the isotope 2He3. The rectangle in the isotope 2He3 shows that the isotope 1H1 can drop out of the isotope 2He3 at a later time.

The single neutron in the isotope 1H2 becomes the single neutron in the isotope 2He3 with 1.00628 amu's. There is a line drawn from the neutron in the isotope 2He3 to the neutron in the isotope 1H2 to show which neutron is used in the isotope 2He3. The first neutron in the isotope 1H3 has the same mass so there would be some confusion as to which isotope provided the single neutron found in the isotope 2He3.

To find the mass that is lost creating the isotope 2He3, add the masses of the isotopes 1H3 and 1H2 and subtract the mass of the isotope 2He3 from their total.

Mass of 1H3	3.01605 amu's
Mass of 1H2	2.01411 amu's
Total mass of 1H3 and 1H2	5.03016 amu's

Total mass of 1H3 and 1H2	5.03016 amu's
Mass of 2He3	3.01603 amu's
Mass lost creating 2He3	2.01413 amu's

The mass lost creating the isotope 2He3 includes the mass of the two neutrons found in the isotope 1H3 that dropped out. Subtract their masses from the total mass lost creating the isotope 2He3 to find the real mass lost.

Mass of first neutron in 1H3	1.00628 amu's
Mass of second neutron in 1H3	1.00194 amu's
Mass of two neutrons that dropped out of 1H3	2.00822 amu's

Mass lost creating 2He3	2.01413 amu's
Mass of two neutrons that dropped out of 1H3	2.00822 amu's
Real mass lost creating 2He3	0.00591 amu

The isotope 1H2 has its electron and proton used as the isotope 1H1 within the isotopes 1H2 and 2He3. Its neutron with 1.00628 amu's becomes the single neutron in the isotope 2He3. The electron and proton become the electron and proton in the isotope 2He3. This means that the only proton that can become the positon in the isotope 2He3 must come from the isotope 1H3. This proton loses 0.00591 amu so subtract this mass from the proton found in the isotope 1H3 to find the mass of the positon it becomes in the isotope 2He3.

Mass of proton in 1H3	1.00728 amu's
Mass lost by it	0.00591 amu
Mass of positon in 2He3	1.00137 amu's

The electrons in the isotopes 1H3 and 1H2 are not lost or released because no 0.00100 amu has been transferred or emitted as a single unit and no 0.00002 amu has been lost creating the isotope 2He3.

The proton in the isotope 1H3 lost a unit of 0.00591 amu for a total lost mass of 0.00591 amu and it is lost as a single unit. The lost mass of 0.00591 is composed of four sub-units. They are sub-units of 0.00434 amu, 0.00100 amu, 0.00055 amu and 0.00002 amu. The last three sub-unit could be a sub-unit of 0.00157 amu. The sub-unit of 0.00434 amu is composed of two sub-units of 0.00217 amu. In order for the electron or negatron to stay with its proton, these sub-units must be lost as a single unit of 0.00591 amu. It must also be lost as a single unit to keep the proton from releasing an electron or negatron from within the proton. The lost mass of 0.00591 could also be composed to sub-units of 0.00374 amu and a sub-unit of 0.00217 amu.

By adding all of the now known masses of the isotope 2He3, the mass of the isotope 2He3 can be found.

Mass of 1H1 intact within 2He3	1.00783 amu's
Mass of neutron in 2He3	1.00628 amu's
Mass of negatron in 2He3	0.00055 amu
Mass of positon in 2He3	1.00137 amu's
Mass of 2He3	3.01603 amu's

Sub-step 18-2-2

2He3

1H3 + 1H2 = 2He3, alpha 591 ray, 1.00628 neutron, 1.00194 neutron

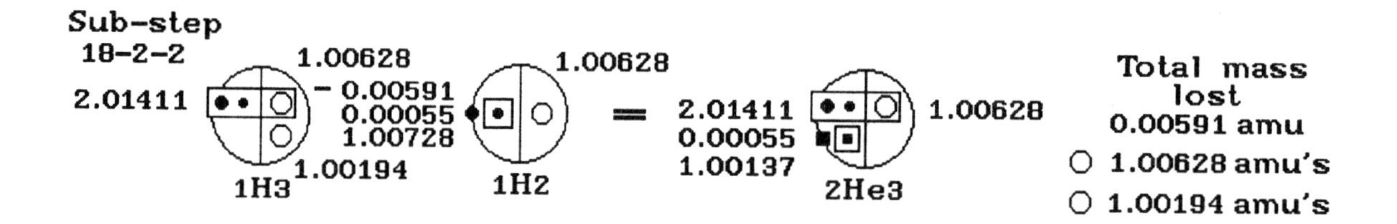

The isotope 1H3 with 3.01605 amu's has one electron with 0.00055 amu and one proton with 1.00728 amu's plus two neutrons. The first neutron has 1.00628 amu's and the second neutron has 1.00194 amu's.

The isotope 1H2 with 2.01411 amu's has one electron with 0.00055 amu and one proton with 1.00728 amu's and one neutron with 1.00628 amu's.

The isotope 2He3 with 3.01603 amu's has one electron with 0.00055 amu, one negatron with 0.00055 amu, one proton with 1.00728 amu's, one positon with 1.00137 amu's and one neutron with 1.00628 amu's.

The total mass lost is 0.00591 amu and it is emitted as a 591 alpha ray which has an external magnetic field. There are two neutrons that drop out. The first neutron to drop out comes from the isotope 1H2 has 1.00628 amu's and the second neutron to drop out comes from the isotope 1H3 with has 1.00194 amu's.

The isotope 1H3 with 3.01605 amu's has one electron with 0.00055 amu that is drawn as a dot on the big circle. The proton with 1.00728 amu's in the isotope 1H3 is drawn as a dot to the left of the vertical line within the big circle.

There are two neutrons in the isotope 1H3 that are drawn one above the other as small circles to the right of the vertical line within the big circle. The top neutron has 1.00628 amu's and is called the first neutron while the neutron below it is the second neutron and it has 1.00194 amu's. The first neutron with 1.00628 amu's is part of the isotope 1H2 found within the isotope 1H3. The electron and proton with the first neutron make the isotope 1H2 with 2.01411 amu's found within the isotope 1H3 and 2He3. They each have a rectangle drawn around them to show that they are the isotope 1H2 and this rectangle indicates that the isotope 1H2 found within the isotope 2He3 can drop out at some later time.

The second neutron with 1.00194 amu's in the isotope 1H3 is not needed when the isotope 2He3 is created so it drops out.

The single neutron in the isotope 1H2 is not needed either when the isotope 2He3 is created so it also drops out. This means that the two neutrons that drop out come from two different isotopes.

The electron in the isotope 1H2 (not the isotope 1H2 in the isotope 1H3) has 0.00055 amu and it is drawn as a dot on the big circle and as a square in the isotope 2He3 on its big circle. The proton in the isotope 1H2 is drawn as a dot within the big circle and to the left of the vertical line. This proton with 1.00728 amu's loses 0.00591 amu. To show this, a square is drawn around the dot that is the proton. It becomes the positon in the isotope 2He3 with 1.00137 amu's and it is drawn as a small square with a square drawn around it to show that this positon comes from the proton in the isotope 1H2 that is not part of the isotope 1H3.

The isotope 2He3 has an electron, proton and first neutron used as the isotope 1H2 within it with 2.01411 amu's. They have a rectangle drawn around them to show that the are the isotope 1H2 with 2.01411 amu's. This isotope of 1H2 comes from the isotope of 1H3. The negatron with 0.00055 amu and positon with 1.00137 amu's come from the isotope 1H2 that is not part of the isotope 1H3. The negatron is drawn as a square on the big circle and the positon is drawn as a small square to the left of the vertical line within the big circle. There is a square drawn around the positon to show that the positon comes from the proton in the isotope 1H2. The single neutron in the isotope 2He3 is the neutron that is part of the isotope 1H2 found intact within the isotope 2He3 with 1.00628 amu's. It is drawn as a small circle to the right of the vertical line within the big circle and within the rectangle of the isotope 1H2 within the isotope 2He3.

To find the mass that is lost creating the isotope 2He3, add the masses of the isotopes 1H3 and 1H2 and subtract the mass of the isotope 2He3 from their total.

Mass of 1H3	3.01605 amu's
Mass of 1H2	2.01411 amu's
Total mass of 1H3 and 1H2	5.03016 amu's

Total mass of 1H3 and 1H2	5.03016 amu's
Mass of 2He3	3.01603 amu's
Mass lost creating 2He3	2.01413 amu's

The mass lost creating the isotope 2He3 includes the mass of the two neutrons that dropped out. Subtract their masses from the total mass lost creating the isotope 2He3 to find the real mass lost.

Mass of neutron in 1H2	1.00628 amu's
Mass of second neutron in 1H3	1.00194 amu's
Mass of two neutrons that dropped out	2.00822 amu's

Mass lost creating 2He3	2.01413 amu's
Mass of two neutrons that dropped out	2.00822 amu's
Real mass lost creating 2He3	0.00591 amu

The isotope 1H2 found within the isotope 1H3 is used as the isotope 1H2 within the isotope 2He3 so its neutron with 1.00628 amu's becomes the single neutron in the isotope 2He3. The electron and proton become the electron and proton in the isotope 2He3. This means that the only proton that can become the positon in the isotope 2He3 must come from the isotope 1H2 that is by itself. This proton loses 0.00591 amu so subtract this mass from the proton in the isotope 1H2 that is by itself to find the mass of the positon it becomes in the isotope 2He3.

Mass of proton in 1H2 that is by itself	1.00728 amu's
Mass lost by it	0.00591 amu
Mass of positon in 2He3	1.00137 amu's

The electrons in the isotopes 1H3 and 1H2 are not lost or released because no 0.00100 amu has been transferred or emitted as a single unit and no 0.00002 amu has been lost creating the isotope 2He3.

The proton in the isotope 1H2 lost a unit of 0.00591 amu for a total lost mass of 0.00591 amu and it is lost as a single unit. The lost mass of 0.00591 is composed of four sub-units. They are sub-units of 0.00434 amu, 0.00100 amu, 0.00055 amu and 0.00002 amu. The last three sub-units could be a sub-unit of 0.00157 amu. The sub-unit of 0.00434 amu is composed of two sub-units of 0.00217 amu. In order for the electron or negatron to stay with its proton, these sub-units must be lost as a single unit of 0.00591 amu. It must also be lost as a single unit to keep the proton from releasing an electron or negatron from within the proton. The lost mass of 0.00591 could also be made up of sub-units of 0.00374 amu and 0.00217 amu.

By adding all of the now known masses of the isotope 2He3, the mass of the isotope 2He3 can be found.

Mass of 1H2 intact within 2He3	2.01411 amu's
Mass of negatron in 2He3	0.00055 amu
Mass of positon in 2He3	1.00137 amu's
Mass of 2He3	3.01603 amu's

Sub-step 17-2-1

2He3

1H3 + 1H2 = 2He3, alpha 591 ray, 1.00628 neutron, 1.00194 neutron

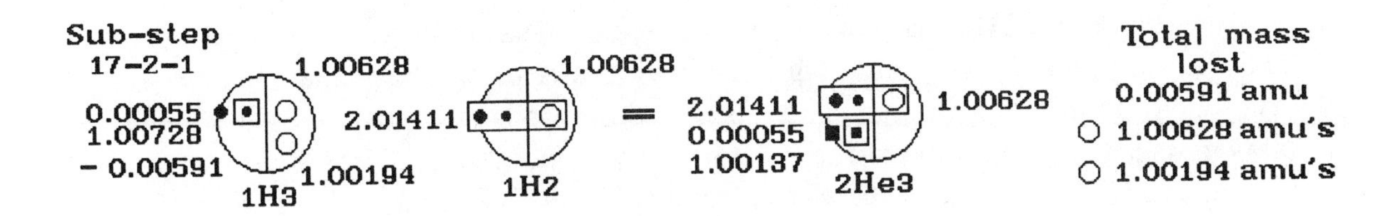

The isotope 1H3 with 3.01605 amu's has one electron with 0.00055 amu and one proton with 1.00728 amu's plus two neutrons. The first neutron has 1.00628 amu's and the second neutron has 1.00194 amu's.

The isotope 1H2 with 2.01411 amu's has one electron with 0.00055 amu and one proton with 1.00728 amu's and one neutron with 1.00628 amu's.

The isotope 2He3 with 3.01603 amu's has one electron with 0.00055 amu, one negatron with 0.00055 amu, one proton with 1.00728 amu's, one positon with 1.00137 amu's and one neutron with 1.00628 amu's.

The total mass lost is 0.00591 amu and it is emitted as a 591 alpha ray which has an external magnetic field. There are two neutrons that drop out from the isotope 1H3. The first neutron has 1.00628 amu's and the second one has 1.00194 amu's.

The isotope 1H3 with 3.01605 amu's has one electron with 0.00055 amu that is drawn as a dot on the big circle and it becomes the negatron with 0.00055 amu in the isotope 2He3 and is drawn as a square on the big circle.

The proton with 1.00728 amu's in the isotope 1H3 is drawn as a dot to the left of the vertical line within the big circle. There is a square drawn around it to show that it loses 0.00591 amu. It is drawn as a square with a square around it in the isotope 2He3 to show that the proton in the isotope 1H3 is the positon in the isotope 2He3.

There are two neutrons in the isotope 1H3 that are drawn one above the other as small circles to the right of the vertical line within the big circle. The top neutron has 1.00628 amu's and is called the first neutron while the neutron below it is the second neutron and it has 1.00194 amu's. Neither of these neutrons are needed when the isotope 2He3 is created so they both drop out.

The isotope 1H2 has 2.01411 amu's and it is used as is in the isotope 2He3. Its single electron is drawn as a dot on the big circle in the isotopes 1H2 and 2He3.

The proton in the isotope 1H2 has 1.00728 amu's and it is drawn as a dot to the left of the vertical line within the big circle in the isotopes 1H2 and 2He3.

There is one neutron in the isotope 1H2 with 1.00628 amu's and it is drawn as a small circle to the right of the vertical line within the big circle.

The electron, proton and neutron in the isotope 1H2 have a rectangle drawn around all of these parts to show that they are used as is in the isotope 2He3. They are drawn the same way in the isotope 2He3. The rectangle in the isotope 2He3 shows that the isotope 1H2 can drop out of the isotope 2He3 at a later time.

The isotope 2He3 has 3.01603 amu's. It has one electron and one negatron and they are both drawn on the big circle with the electron above the negatron. They both have 0.00055 amu. There is one proton and one positon which are drawn with the proton above the positon. The proton with 1.00728 amu's is drawn as a dot to the left of the vertical line within the big circle and it is part of the isotope 1H2 found within the isotope 2He3. The positon with 1.00137 amu's is drawn as a small square to the left of the vertical line within the big circle and it has a square drawn around it to show that it came from the proton in the isotope 1H3. The single neutron is drawn as a small circle to the right of the vertical line with in the big circle and it is part of the isotope 1H2 found within the isotope 2He3. There is a rectangle drawn around all of the parts of the isotope 1H2 in the isotope 2He3 show that they are intact within the isotope 2He3 and it is drawn this same way for the isotope 1H2 that is by itself. The rectangle drawn around the 1H2 found within the isotope 2He3 shows that this isotope can drop out of the isotope 2He3 at some later time.

To find the mass that is lost creating the isotope 2He3, add the masses of the isotopes 1H3 and 1H2 and subtract the mass of the isotope 2He3 from their total.

Mass of 1H3	3.01605 amu's
Mass of 1H2	2.01411 amu's
Total mass of 1H3 and 1H2	5.03016 amu's

Total mass of 1H3 and 1H2	5.03016 amu's
Mass of 2He3	3.01603 amu's
Mass lost creating 2He3	2.01413 amu's

The mass lost creating the isotope 2He3 includes the mass of the two neutrons found in the isotope 1H3 that dropped out. Subtract their masses from the total mass lost creating the isotope 2He3 to find the real mass lost.

Mass of first neutron in 1H3	1.00628 amu's
Mass of second neutron in 1H3	1.00194 amu's
Mass of two neutrons that dropped out of 1H3	2.00822 amu's

Mass lost creating 2He3	2.01413 amu's
Mass of two neutrons that dropped out of 1H3	2.00822 amu's
Real mass lost creating 2He3	0.00591 amu

The isotope 1H2 is used as the isotope 1H2 within the isotope 2He3 so its neutron with 1.00628 amu's becomes the single neutron in the isotope 2He3. The electron and proton become the electron and proton in the isotope 2He3. This means that the only proton that can become the positon in the isotope 2He3 must come from the isotope 1H3. This proton loses 0.00591 amu so subtract this mass from the proton in the isotope 1H3 to find the mass of the positon it becomes in the isotope 2He3.

Mass of proton in 1H3	1.00728 amu's
Mass lost by it	0.00591 amu
Mass of positon in 2He3	1.00137 amu's

The electrons in the isotopes 1H3 and 1H2 are not lost or released because no 0.00100 amu has been transferred or emitted as a single unit and no 0.00002 amu has been lost creating the isotope 2He3.

The proton in the isotope 1H3 lost a unit of 0.00591 amu for a total lost mass of 0.00591 amu and it is lost as a single unit. The lost mass of 0.00591 is composed of four sub-units. They are sub-units of 0.00434 amu, 0.00100 amu, 0.00055 amu and 0.00002 amu. The last three sub-units could be a sub-unit of 0.00157 amu. The sub-unit of 0.00434 amu is composed of two sub-units of 0.00217 amu. In order for the electron or negatron to stay with its proton, these sub-units must be lost as a single unit of 0.00591 amu. It must also be lost as a single unit to keep the proton from releasing an electron or negatron from within the proton. The lost mass of 0.00591 can also be composed of units of 0.00374 amu and 0.00217 amu.

By adding all of the now known masses of the isotope 2He3, the mass of the isotope 2He3 can be found.

Mass of 1H2 intact within 2He3	2.01411 amu's
Mass of negatron in 2He3	0.00055 amu
Mass of positon in 2He3	1.00137 amu's
Mass of 2He3	3.01603 amu's

This completes all of the steps in **STEP 2**. The next pages will show all of the steps with drawings only.

All drawings in **STEP 2** creating the isotope 2He3 from isotopes of 1H3 and 1H2

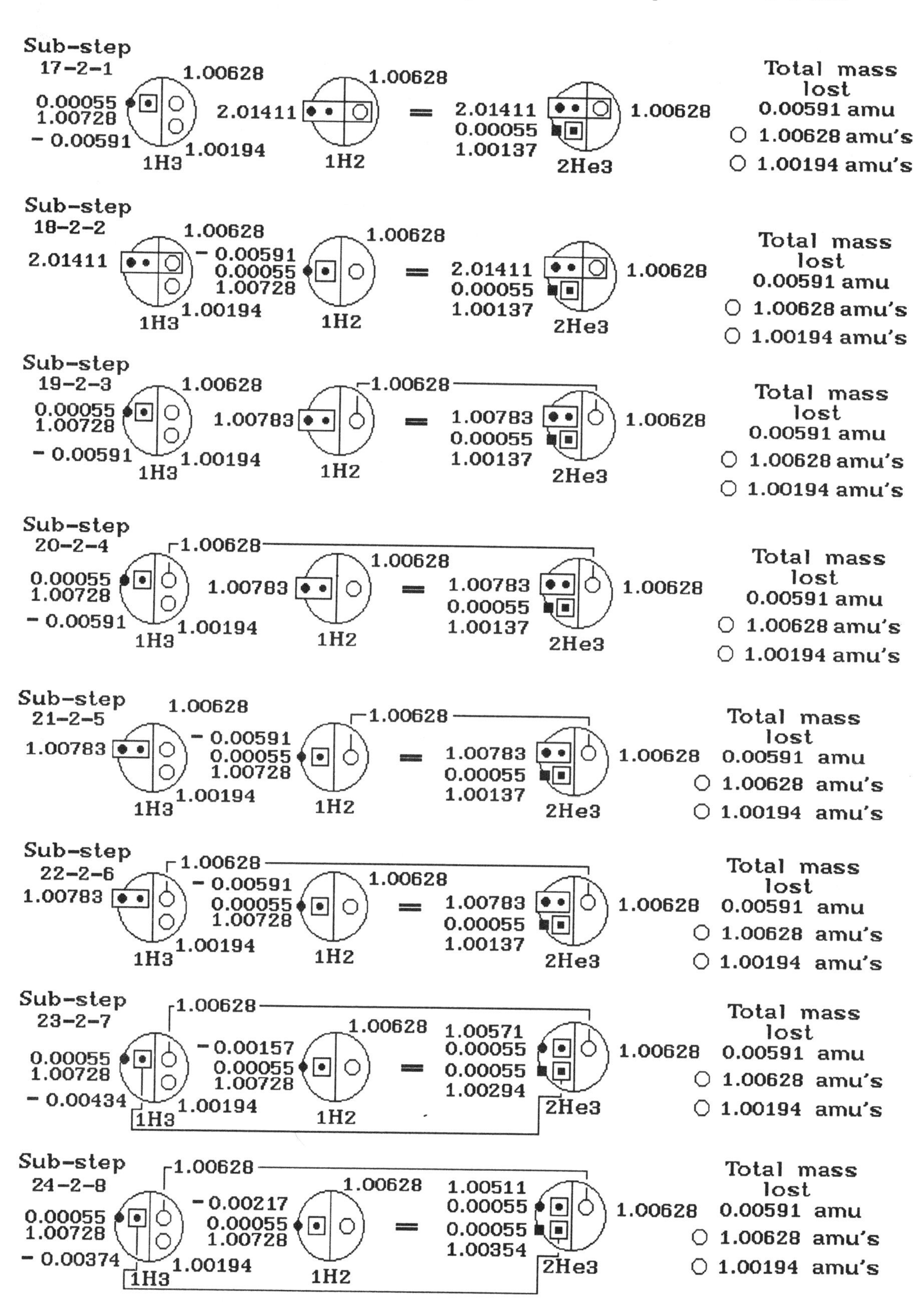

All drawings in **STEP 2** creating the isotope 2He3 from isotopes of 1H3 and 1H2

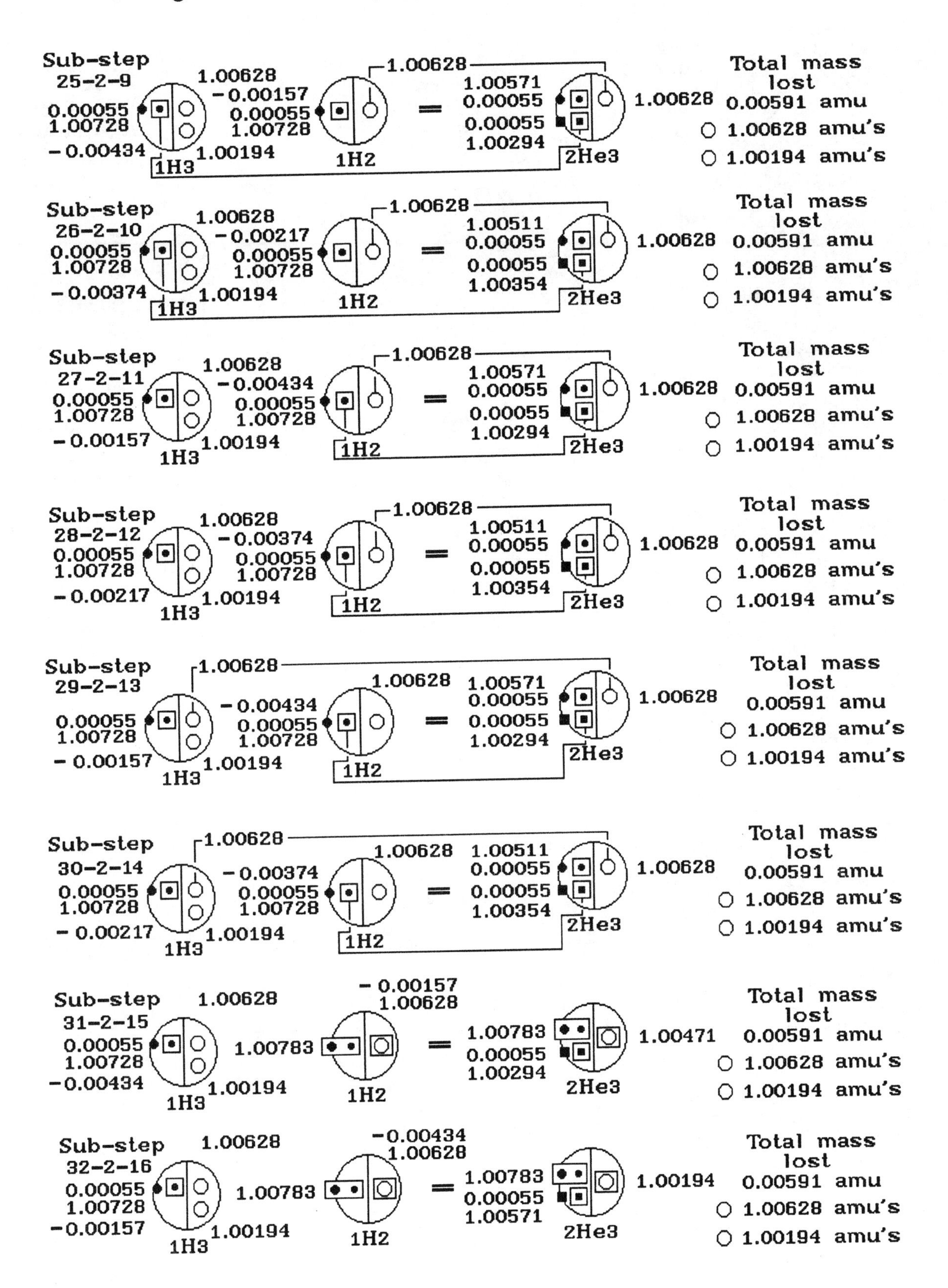

All drawings in **STEP 2** creating the isotope 2He3 from isotopes of 1H3 and 1H2

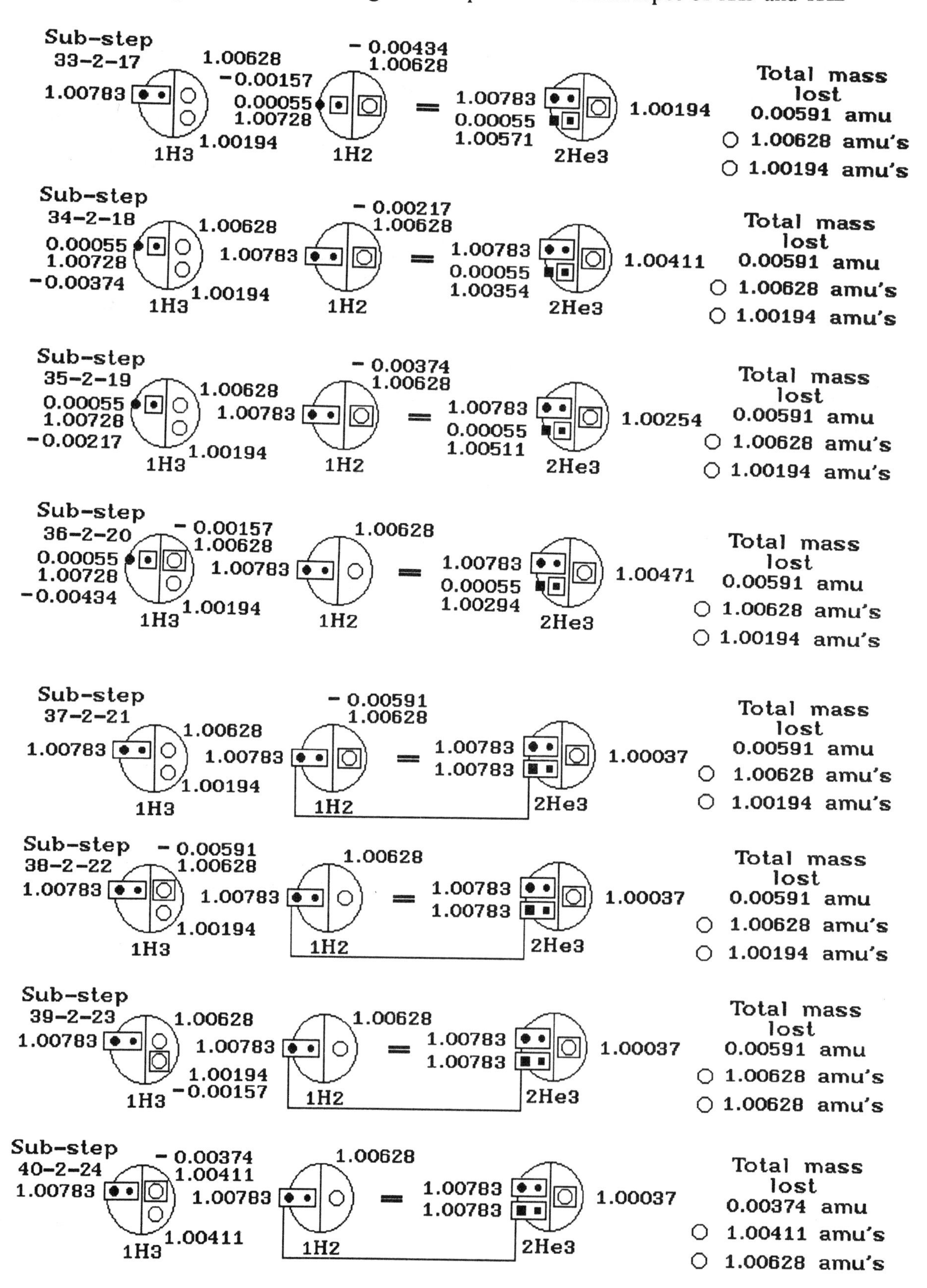

All drawings in **STEP 2** creating the isotope 2He3 from isotopes of 1H3 and 1H2

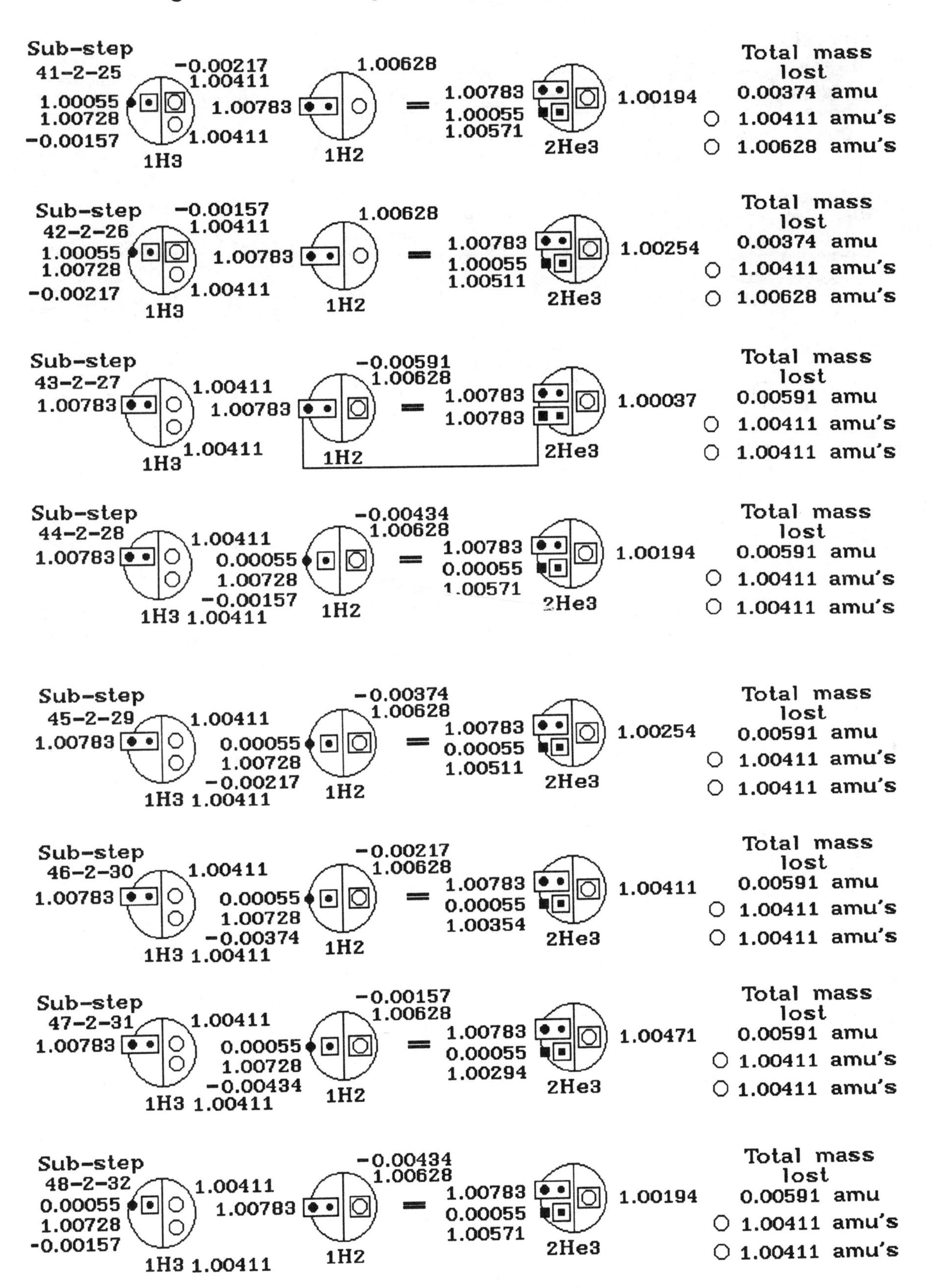

All drawings in **STEP 2** creating the isotope 2He3 from isotopes of 1H3 and 1H2

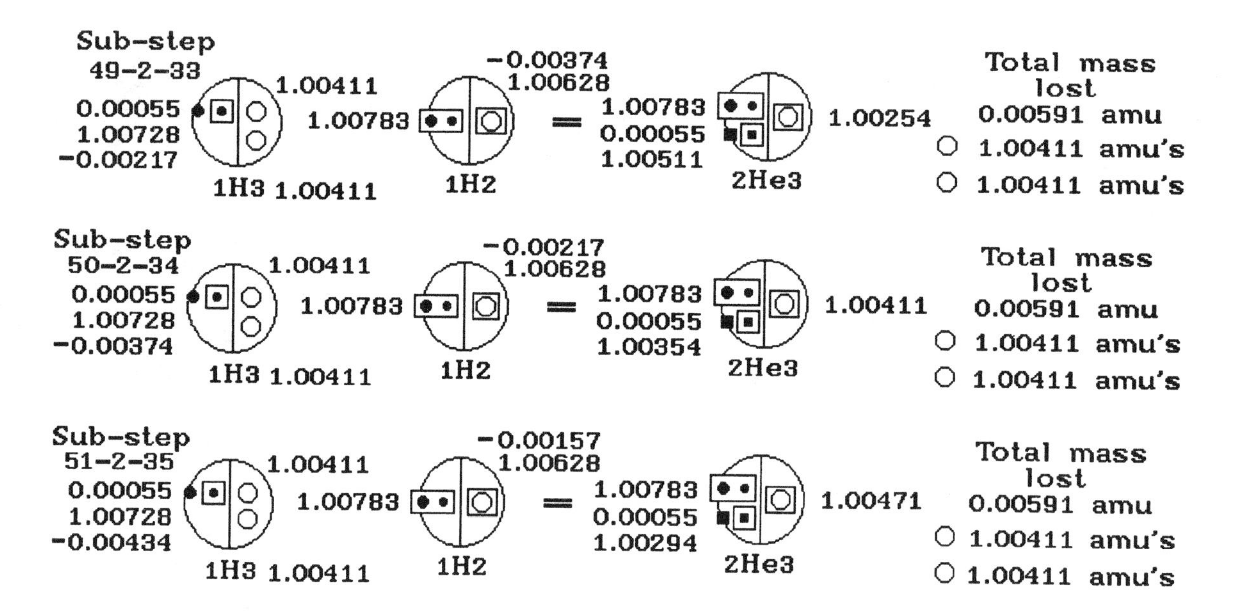

STEP 1

2He3

FORMULA

1H3 = 2He3

STEPS

1. 1H3 = 2He3

1-1-1	1H3 = 2He3, #2 ray, transfer 0.00100 amu, neg. orbit
2-1-2	1H3 = 2He3, #2 ray, transfer 0.00100 amu, 0.00434 amu, neg. orbit
3-1-3	1H3 = 2He3, #2 ray, transfer 0.00100 amu, 0.00217 amu, neg. orbit
4-1-4	1H3 = 2He3, #2 ray, transfer 0.00100 amu, neg. orbit
5-1-5a	1H3 = 2He3, #2 ray, transfer 0.00100 amu, 0.00434 amu, neg. orbit
5-1-5b	2He3 to 2He3
6-1-6	1H3 = 2He3, #2 ray, transfer 0.00100 amu, 0.00434 amu, neg. orbit
7-1-7	1H3 = 2He3, #2 ray, transfer 2(0.00100 amu), neg. orbit
8-1-8	1H3 = 2He3, #2 ray, transfer 0.00100 amu, 0.00317 amu, neg. orbit
9-1-9	1H3 = 2He3, #2 ray, transfer 2(0.00100 amu), 0.00434 amu, neg. orbit
10-1-10	1H3 = 2He3, #2 ray, transfer 0.00100 amu, 0.00534 amu, neg. orbit
11-1-11	1H3 = 2He3, #2 ray, neg. orbit
12-1-12	1H3 = 2He3, #2 ray, transfer 0.00100 amu, neg. orbit
13-1-13	1H3 = 2He3, #2 ray, transfer 0.00100 amu, 0.00217 amu, neg. orbit
14-1-14	1H3 = 2He3, #2 ray, transfer 0.00217/0.00317/0.00100 amu, neg.orbit
15-1-15	1H3 = 2He3, #2 ray, transfer 0.00100 amu, neg. orbit
16-1-16	1H3 = 2He3, #2 ray, transfer 0.00100 amu, 0.00217 amu, neg. orbit

Sub-step 16-1-16

2He3

1H3 = 2He3, #2 alpha ray, transferred 0.00100 amu, 0.00217 amu, neg. orbit

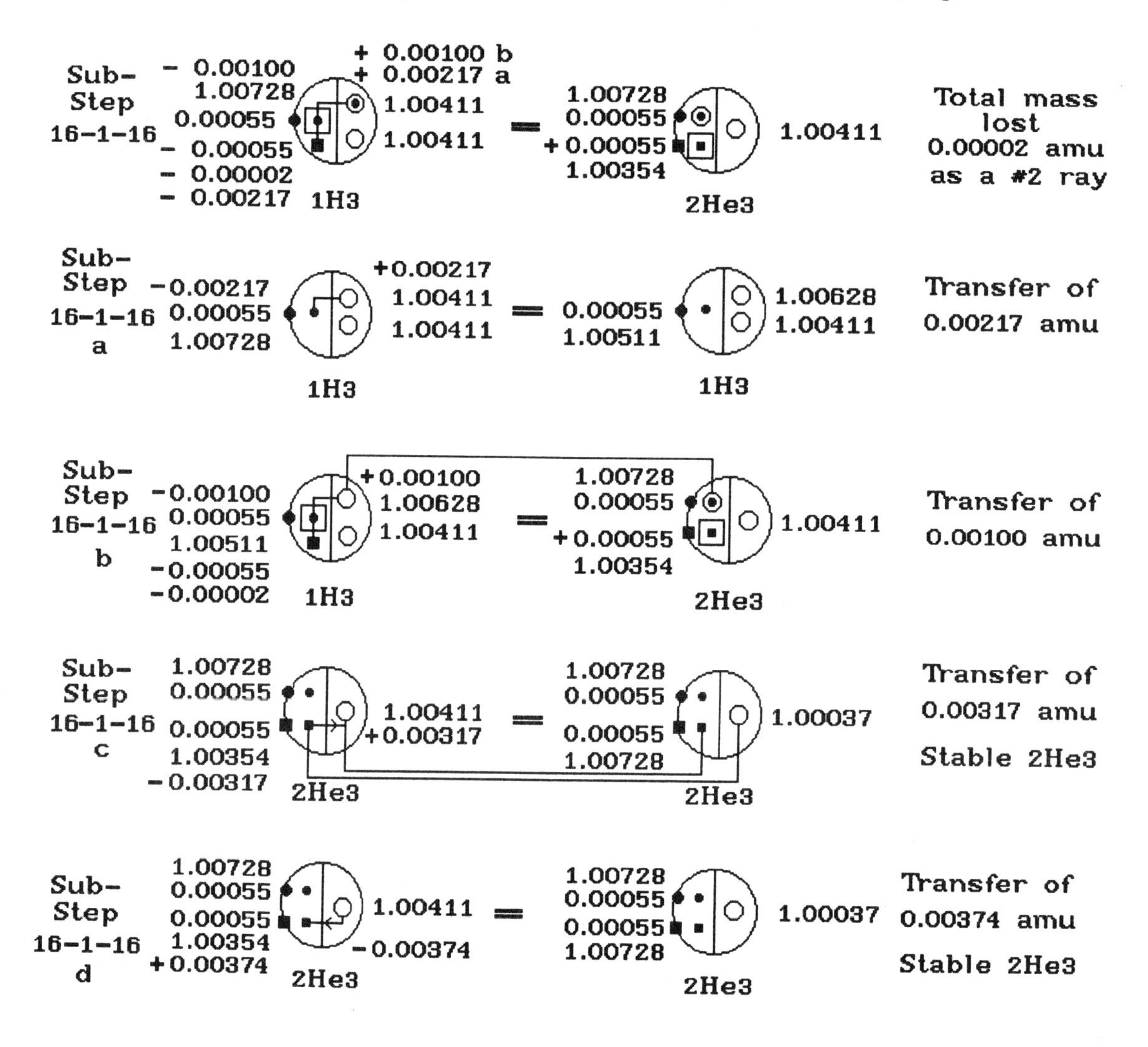

The isotope 1H3 with 3.01605 amu's has one electron with 0.00055 amu and one proton with 1.00728 amu's plus two neutrons. Both neutrons have the same mass of 1.00411 amu's.

The isotope 2He3 with 3.01603 amu's has one electron with 0.00055 amu, one negatron with 0.00055 amu, one proton with 1.00728 amu's, one positon with 1.00354 amu's and one neutron with 1.00411 amu's.

The total mass lost is 0.00002 amu and it is emitted as a #2 alpha ray which has an external magnetic field.

The drawings for this step are drawn with the first drawing showing the isotope 1H3 creating the isotope 2He3 directly. Under this drawing are two more drawings that explain in more detail how the first drawing creates the isotope 2He3. The write-up gives the way to read the first drawing because it will be noted that it has the first neutron with a dot in the center of the small circle while the other two drawings (a and b) do not. Because of this there may be some confusion when a drawing is made that goes directly from the isotope 1H3 to the isotope 2He3.

There is another drawing, sub-step 16-1-16c that shows how the new isotope of 2He3 then changes from the isotope 2He3 in sub-step 16-1-16b to its final stable form. This may be considered a non-radioactive decay.

The last drawing of sub-step 16-1-16d shows how the isotope 2He3 becomes stable.

The isotope 1H3 with 3.01605 amu's has its electron drawn on the big circle as a dot and it is drawn this same way in the isotope 2He3. The proton in the isotope 1H3 is drawn as a dot with a square drawn around it. In the isotope 1H3, it is a proton and in the isotope 2He3, it is a positon which means that the dot in the isotope 1H3 changes to a small square in the isotope 2He3. The square drawn around the dot does not mean that the proton with 1.00728 amu's has transferred a single unit of 0.00217 amu to the first neutron that has 1.00411 amu's. This transferred mass is lost by the proton in the isotope 1H3 which now has 1.00511 amu's. Because the lost mass of the proton is transferred to the first neutron with 1.00411 amu's, this mass is not lost by the isotope 1H3 as this isotope becomes the isotope 2He3. When this happens, the electron is not released. The electron that belongs to the proton becomes the electron in the isotope 2He3 with 0.00055 amu and it is drawn as a dot on the big circle. The value of 1.00511 amu's is the value of a proton which has 0.00100 amu more than a neutron. The first neutron with 1.00411 amu's gains the transferred mass of 0.00217 amu to become a neutron with 1.00628 amu's.

The proton with 1.00511 amu's in the isotope of 1H3 then transfers 0.00100 amu to the new first neutron with 1.00628 amu's to make the first neutron in the isotope 1H3 into the proton in the isotope 2He3 with 1.00728 amu's. The first neutron in the isotope 1H3 is drawn as a dot with a small circle around it and it is drawn this same way in the isotope 2He3 to show that the first neutron in the isotope 1H3 becomes the proton in the isotope 2He3. At this point, the drawing of this proton is a dot as shown in drawing 16-1-16a and the neutron is drawn as a small circle with a line drawn from the proton to the first neutron to show that 0.00217 amu has been transferred. There is no square drawn around them.

When the 0.00100 amu is transferred to the isotope 2He3, as shown in drawing 16-1-16b, the proton with 1.00511 amu's in the isotope 1H3 releases a negatron with 0.00055 amu from within itself and loses 0.00002 amu to become the positon with 1.00354 amu's. The 0.00100 amu causes the negatron to go into the orbital that already has an electron in it. It does not release the negatron. The 0.00002 amu that is lost as a #2 alpha ray is the lost mass that releases the negatron causing the isotope 2He3 to be created. This means that the transferred mass of 0.00100 amu did not cause the isotope 2He3 to be created.

This negatron is drawn in sub-step 16-1-16b as a small square below the proton. The small square is connected to the proton by a line to show that the small square comes from it. There is no arrow drawn for the electron or negatron because they are not released completely from the isotope. There is a line connecting the proton to the first neutron in the isotope 1H3 to show that it is the neutron that regains the mass released by the proton. In the isotope 2He3 there is a small circle drawn around the dot that is the proton found in the isotope 1H3 to show that this proton has transferred 0.00100 amu to the first neutron. When this is done, the proton in the isotope 1H3 is drawn as a dot with a square around it and because this proton is the positon in the isotope 2He3, it is drawn the same way except this proton is drawn as a small square. This means that there is now a proton and a positon in the isotope 2He3 with an electron and a negatron. Because the first neutron becomes a proton, the first neutron disappears which leaves the second neutron as the single neutron in the isotope 2He3 with 1.00411 amu's. It is drawn as a small circle to the right of the vertical line within the big circle in the isotopes 1H3 and 2He3.

Now that the first neutron is a proton in the isotope 2He3, the proton in the isotope 1H3 that transferred 0.00217 amu and 0.00100 amu and released, a negatron loses 0.00002 amu. The 0.00002 amu is the excess mass that must be lost as a ray to change the isotope 1H3 in the isotope 2He3. The loss of 0.00002 amu takes place at the same time the 0.00100 amu is transferred. The square drawn around the dot that is the proton in the isotope 1H3 represents the 0.00002 amu that is lost. It does not represent any mass that is transferred.

In previous steps the 0.00002 amu was included with the units that were lost or transferred and in these cases, all of the mass that the square represents was lost or transferred. This may seem confusing. Just remember that in the previous steps the 0.00002 amu was not a single unit by itself whereas when the isotope 1H3 is by itself causing the isotope 2He3 to be created, it is lost as a single unit. When the 0.00002 amu is part of a unit, it was called a sub-unit.

Now that the isotope 2He3 has been created with the masses shown, this new isotope of 2He3 then changes into its final form. This final form will have one electron and one negatron with each having 0.00055 amu, a proton and a positon with each having 1.00728 amu's and a single neutron with 1.00037 amu's.

The new isotope of 2He3 with 3.01603 amu's has its single neutron transfer a single unit of 0.00317 amu to the positon that has 1.00354 amu's to make this positon into a positon with 1.00728 amu's. The single neutron does not transfer any mass to the proton because it already has 1.00728 amu's. This leaves the single neutron with a mass of 1.00037 amu's. This final form of the isotope 2He3 still has 3.01603 amu's. This is sub-step 16-1-16c.

The positon has a line drawn from it to the single neutron to show that the positon has transferred 0.00317 amu. The single neutron is drawn as a small circle on the right side of the vertical line within both isotopes of 2He3. The single neutron in the first isotope of 2He3 has 1.00411 amu's and it has 1.00037 amu's in the second isotope of 2He3.

Both isotopes of 2He3 have their electron and negatron drawn with the electron above the negatron on the big circle. The electron is drawn as a dot and the negatron is drawn as a square. The proton and positon are both drawn to the left of the vertical line within the big circle. The proton is drawn as a dot and the positon is drawn as a small square.

There is a line with an arrow drawn on it pointing to the single neutron in the isotope 2He3 on the left side of equal sign to show that there is 0.00327 amu transferred to the single neutron in this isotope.

There is a line drawn from the positon in the isotope 2He3 on the left side of the equal sign to the single neutron in the other isotope of 2He3 to show that this positon becomes the single neutron in the isotope of 2He3. There is another line drawn from the single neutron in the isotope 2He3 on the left side of the equal sign to the positon in the other isotope of 2He3 to show that the single neutron becomes the positon in the other isotope of 2He3.

The above is one way the final isotope of 2He3 can be created from a single isotope of 1H3 but if the single neutron in the non-final isotope of 2He3 transfers 0.00374 amu to the positon, then this positon has 1.00728 amu's and the single neutron has 1.00037 amu's in the final isotope of 2He3.

It may be asked why would the non-final isotope of 2He3 switch its positon and neutron when it can go directly to the final form when it transfers 0.00374 amu to the positon, and the positon stays a positon, and the single neutron stays a neutron? The answer is found when the isotope 2He3 becomes part of an molecule that must transfer either 0.00317 amu or 0.00374 amu in order for it to become part of the molecule.

In the drawing of the sub-step 16-1-16d the non-final isotope of 2He3 has its electron drawn as a dot on the big circle and the negatron drawn as a square below the dot. The proton is drawn as a dot to the left of the vertical line within the big circle and the positon is drawn as a small square below it. The single neutron is drawn as small circle to the right of the vertical line within the big circle. There is a line drawn from the single neutron to the positon with an arrow pointing toward the positon to show that the single neutron does the transferring.

The final isotope of 2He3 is drawn the same way as the non-final isotope of 2He3 without the lines.

This transfer of mass involves radioactivity because there is no ray emitted until the 0.00002 amu has been emitted as a #2 alpha ray and the negatron goes into its orbital. All of this transferring of mass is an internal movement of mass and does not require any outside energy to make it happen.

To find the mass that is lost creating the isotope 2He3, subtract the isotope 2He3 from the isotope 1H3.

Mass of 1H3	3.01605 amu's
Mass of 2He3	3.01603 amu's
Mass lost creating 2He3	0.00002 amu

In this step I have stated that the proton has 1.00728 amu's. The proton comes from the first neutron in the isotope 1H3. This neutron has 1.00411 amu's. To find the mass this neutron must regain to become a proton with 1.00728 amu's, subtract the mass of the neutron from the proton.

Mass of proton in 2He3	1.00728 amu's
Mass of neutron in 1H3	1.00411 amu's
Mass neutron regains becoming proton	0.00317 amu

The mass that the neutron regains from the proton in the isotope 1H3 is a two step process. The 0.00317 amu is not transferred to the first neutron in the isotope 1H3 as a single unit. The 0.00217 amu is transferred to the first neutron making this neutron have 1.00628 amu's and the proton now is a proton with 1.00511 amu's. These two masses plus the electron and second neutron will still make an isotope of 1H3.

Mass new first neutron in 1H3	1.00411 amu's
Mass transferred not releasing negatron from 1H3's proton	0.00217 amu
Mass of new proton	1.00628 amu's

Mass of proton in 1H3	1.00728 amu's
Mass transferred not releasing negatron from 1H3	0.00217 amu
Mass of new proton in 1H3	1.00511 amu's

Mass of electron in 1H3	0.00055 amu
Mass of new proton in 1H3	1.00511 amu's
Mass of new first neutron in 1H3	1.00628 amu's
Mass of second neutron in 1H3	1.00411 amu's
Mass of 1H3	3.01605 amu's

The new proton with 1.00511 amu's in the isotope 1H3 transfers 0.00100 amu to the first neutron to make this neutron the proton found in the isotope 2He3 with 1.00728 amu's. The first neutron disappears. When this happens, the new proton loses 0.00002 amu as a #2 alpha ray and releases from within itself a negatron with 0.00055 amu. The 0.00002 amu is the only mass lost creating the isotope 2He3.

Now add the transferred mass, the mass of the negatron released and the real mass lost and subtract this transferred, released and lost mass from the mass of the new proton in the isotope 1H3. This remaining mass of the new proton in the isotope 1H3 will be the mass of the positon in the isotope 2He3.

Mass transferred from new proton in 1H3	0.00100 amu
Mass of released negatron from new proton in 1H3	0.00055 amu
Real mass lost from new proton in 1H3	0.00002 amu
Total transferred, released and lost mass from new proton in 1H3	0.00157 amu

Mass of new proton in 1H3	1.00511 amu's
Total transferred, released and lost mass from new proton in 1H3	0.00157 amu
Mass of positon in 2He3	1.00354 amu's

This mass of the positon in the isotope 2He3 can not come from the electron or the second neutron in the isotope 1H3 because they do not lose any mass. The only place the positon can come from is the new proton in the isotope 1H3.

The 0.00002 amu is the only mass that is really lost so it must be lost as a single unit. Add this lost mass, the mass transferred that does not release the negatron, the mass that is lost releasing the negatron and the negatron itself to find the mass lost by the original proton in the isotope 1H3.

Real mass lost by new proton in 1H3	0.00002 amu
Mass not causing negatron from original proton in 1H3	0.00217 amu
Mass lost causing orbiting negatron from new proton in 1H3	0.00100 amu
Mass of negatron released from new proton in 1H3	0.00055 amu
Total mass lost by original proton in 1H3	0.00374 amu

By subtracting the mass of the negatron and the real mass lost, the remaining mass will be the mass the first neutron regains to become the proton in the isotope 2He3.

Total mass transferred and lost by original proton in 1H3	0.00374 amu
Mass of negatron and real mass lost	0.00057 amu
Mass regained by first neutron in 1H3	0.00317 amu

Now add this mass to the first neutron in the isotope 1H3 to get the mass of the new proton in the isotope 2He3.

Mass of first neutron in 1H3	1.00411 amu's
Mass making it into a new proton in 2He3	0.00317 amu
Mass of new proton in 2He3	1.00728 amu's

Now all of the masses of the isotope 2He3 are known. Add them to find the mass of the isotope 2He3. This is shown in the drawing of sub-step 16-1-16b.

Mass of electron and negatron in 2He3	0.00110 amu
Mass of proton in 2He3	1.00354 amu's
Mass of positon in 2He3	1.00728 amu's
Mass of neutron in 2He3	1.00411 amu's
Mass of 2He3	3.01603 amu's

The isotope 2He3 now changes into its final form by transferring 0.00317 amu to the single neutron with 1.00411 amu's from the positon with 1.00354 amu's in the non-final isotope of 2He3 to make the positon the single neutron in the final isotope 2He3. When this is done, the neutron has 1.00728 amu's which makes it the positon and the positon that transferred the 0.00317 amu becomes the single neutron with 1.00037 amu's. The proton already has 1.00728 amu's in the non-final and the final isotope of 2He3. This is shown in the drawing of sub-step 16-1-16c.

Mass of positon in non-final 2He3	1.00354 amu's
Mass transferred from it to neutron in non-final 2He3	0.00317 amu
Mass of new neutron in final 2He3	1.00037 amu's

Mass of neutron in non-final 2He3	1.00411 amu's
Total mass transferred to it from non-final positon 2He3	0.00317 amu
Mass of new positon in final 2He3	1.00728 amu's

In the drawing of sub-step 16-1-16c, the positon and neutron trade places in the non-final isotope of 2He3 to make the final isotope of 2He3. However it is possible for the single neutron in the non-final isotope of 2He3 to transfer 0.00374 amu to the positon and thus make the positon have 1.00728 amu's and the single neutron to have 1.00037 amu's.

The isotope 2He3 now changes into its final form by transferring 0.00374 amu from the single neutron with 1.00411 amu's to the positon with 1.00354 amu's in the non-final isotope of 2He3. When this is done, the proton and positon both have 1.00728 amu's and the single neutron has 1.00037 amu's in the final isotope of 2He3. This is shown in the drawing of sub-step 16-1-16d.

Mass of positon in non-final 2He3	1.00354 amu's
Mass transferred to it from non-final 2He3	0.00374 amu
Mass of positon in final 2He3	1.00728 amu's

Mass of neutron in non-final 2He3	1.00411 amu's
Total mass transferred to final 2He3	0.00374 amu
Mass of neutron in final 2He3	1.00037 amu's

By adding these final masses, the mass of the final isotope of 2He3 will be found. This final isotope of 2He3 is the result of a radioactive decay because no mass was lost and no ray was emitted until the 0.00002 amu was emitted as a #2 alpha ray and the negatron was released from the proton in the isotope 1H3. This movement of the internal parts did not require any outside energy.

This final isotope of 2He3 is stable because the proton and positon each has 1.00728 amu's and the single neutron has 1.00037 amu's. This is the maximum mass they can have.

Mass of electron and negatron in final 2He3	0.00110 amu
Mass of proton in final 2He3	1.00728 amu's
Mass of positon in final 2He3	1.00728 amu's
Mass of neutron in final 2He3	1.00037 amu's
Mass of final 2He3	3.01603 amu's

Sub-step 15-1-15

2He3

1H3 = 2He3, #2 alpha ray, transferred 0.00100 amu, neg. orbit

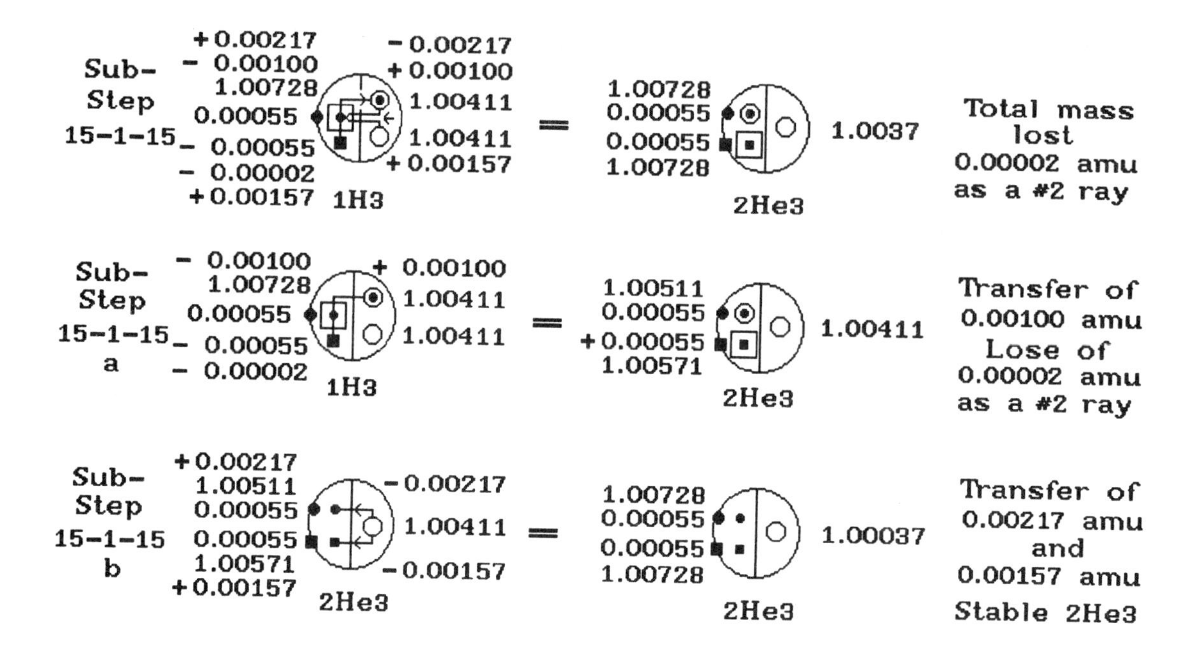

The isotope 1H3 with 3.01605 amu's has one electron with 0.00055 amu and one proton with 1.00728 amu's plus two neutrons. Both neutrons have the same mass of 1.00411 amu's.

The isotope 2He3 with 3.01603 amu's has one electron with 0.00055 amu, one negatron with 0.00055 amu, one proton with 1.00511 amu's, one positon with 1.00571 amu's and one neutron with 1.00411 amu's.

The total mass lost is 0.00002 amu and it is emitted as a #2 alpha ray.

The isotope 1H3 with 3.01605 amu's has its electron drawn on the big circle as a dot and it is drawn this same way in the isotope 2He3. The proton in the isotope 1H3 is drawn as a dot with a square drawn around it. In the isotope 1H3, it is a proton and in the isotope 2He3, it is a positon which means that the dot in the isotope 1H3 changes to a square in the isotope 2He3. The line drawn from the dot to the first neutron means that the proton with 1.00728 amu's has transferred 0.00100 amu as a single unit to the first neutron that has 1.00411 amu's. The 0.00100 amu does not release the negatron from the proton. It does cause the negatron to go into an orbit with the electron that is already there.

There is a square drawn around the proton in the isotope 1H3 to show that 0.00002 amu has been lost as a single unit. This lost mass of 0.00002 amu causes the release of the negatron and the creation of the isotope 2He3. This negatron goes into an orbital with the electron already there. They each have 0.00055 amu. (Do not write the amu at the end of this sentence as amu's.)

The proton with 1.00728 amu's transfers 0.00100 amu to the first neutron with 1.00411 amu's to make the first neutron the proton in the isotope 2He3 with 1.00511 amu's. When the first neutron becomes a proton, the first neutron disappears and the mass that has gained the 0.00100 amu now appears as the proton in the isotope 2He3. The first neutron in the isotope 1H3 is drawn as a small circle with a dot in its middle and the proton it becomes is drawn the same way in the isotope 2He3. This shows that the first neutron in the isotope 1H3 becomes the proton in isotope 2He3. When the 0.00100 amu is transferred, it causes a negatron with 0.00055 amu from within the proton to go into an orbital with the electron. When the 0.00100 amu is transferred and the negatron is released, there is 0.00002 amu lost creating the isotope 2He3. When all of this mass is lost by the proton with 1.00728 amu's in the isotope 1H3, this proton becomes the positon in the isotope 2He3 with 1.00571 amu's. The drawing of the proton in the isotope 1H3 is drawn the same way in the isotope 2He3 except the dot changes into a small square to show that the proton in the isotope 1H3 becomes the positon in the isotope 2He3. The negatron is drawn as a small square below the proton with a line connecting the two in the isotope 1H3 and as a square on the big circle in the isotope 2He3.

There is a line drawn from the proton to the first neutron to show that the mass that is transferred is transferred from the proton to the first neutron.

The second neutron in the isotope 1H3 with 1.00411 amu's becomes the single neutron in the isotope 2He3 with no loss of mass. This is shown in the drawing of sub-step 15-1-15a.

Now that the isotope 2He3 has been created with the masses shown, this new isotope of 2He3 then changes into its final form. This final form will have one electron and one negatron with each having 0.00055 amu, a proton and a positon with each having 1.00728 amu's and a single neutron with 1.00037 amu's. This is shown in the drawing of sub-step 15-1-15b.

The new isotope of 2He3 with 3.01603 amu's has its single neutron transfer a single unit of 0.00217 amu to the proton that has 1.00511 amu's to make this proton into a proton with 1.00728 amu's. The single neutron will also transfer 0.00157 amu to the positon that has 1.00571 amu's to make the positon have 1.00728 amu's. This leaves the single neutron with a mass of 1.00037 amu's. This final form of the isotope 2He3 still has 3.01603 amu's.

The single neutron has a line drawn from the single neutron to the proton to show that it has transferred 0.00217 amu. There is also a line drawn from the single neutron to the positon to show that the single neutron has transferred 0.00157 amu. The single neutron is drawn as a small circle on the right side of the vertical line within both isotopes of 2He3. The single neutron in the first isotope of 2He3 has 1.00411 amu's and it has 1.00037 amu's in the second isotope of 2He3.

The electron and negatron are drawn with the electron above the negatron on the big circle. The electron is drawn as a dot and the negatron is drawn as a square. The proton and positon are both drawn to the left of the vertical line within the big circle. The proton is drawn as a dot and the positon is drawn as a small square.

This transfer of mass involves radioactivity because all of this transferring of mass is an internal movement of mass and does not require any outside energy to make it happen.

To find the mass that is lost creating the isotope 2He3, subtract the isotope 2He3 from the isotope 1H3.

Mass of 1H3	3.01605 amu's
Mass of 2He3	3.01603 amu's
Mass lost creating 2He3	0.00002 amu

The electron with 0.00055 amu in the isotope 1H3 does not lose any mass and becomes the electron found in the isotope 2He3. The isotope 2He3 must have a negatron with 0.00055 amu so the masses of the electron and negatron are known.

The second neutron in the isotope 1H3 is used as the single neutron in the isotope 2He3 with 1.00411 amu's.

The masses of the proton and positon in the isotope 2He3 are unknown and must now be found.

The first neutron in the isotope 1H3 is converted to the proton in the isotope 2He3 by regaining 0.00100 amu. When this happens, the proton must release a negatron and lose 0.00002 amu. Therefore, the proton in the isotope 1H3 does not become a neutron in the isotope 2He3. Because of this, the proton in the isotope 1H3 must transfer 0.00100 amu, release 0.00055 amu and lose 0.00002 amu for a total lost mass of 0.00157 amu. Subtract this total lost mass from the proton in the isotope 1H3 and the remaining mass becomes the positon in the isotope 2He3. The neutron in isotope 1H3 is 1.00411 amu's. (Sub-step 15-1-15a)

Mass transferred from proton in 1H3	0.00100 amu
Mass of released negatron from proton in 1H3	0.00055 amu
Mass lost from proton releasing negatron in 1H3	0.00002 amu
Total mass lost by proton in 1H3	0.00157 amu

Mass of proton in 1H3	1.00728 amu's
Total mass lost by proton in 1H3	0.00157 amu
Mass of positon in 1H3	1.00571 amu's

The first neutron in the isotope 1H3 has 1.00411 amu's and it regains 0.00100 amu from the proton. Add this regained mass to the first neutron in the isotope 1H3 to find the mass of the proton in the isotope 2He3.

Mass of first neutron in 1H3	1.00411 amu's
Regained mass from proton in 1H3	0.00100 amu
Mass of proton in 2He3	1.00511 amu's

Now all of the masses of the isotope 2He3 are known. Adding these masses will give the mass of the isotope 2He3. (Sub-step 15-1-15a)

Mass of electron and negatron in 2He3	0.00110 amu
Mass of proton in 2He3	1.00511 amu's
Mass of positon in 2He3	1.00571 amu's
Mass of neutron in 2He3	1.00411 amu's
Mass of 2He3	3.01603 amu's

The isotope 2He3 now changes into its final form by transferring 0.00217 amu from the single neutron with 1.00411 amu's to the proton and 0.00157 amu to the positon. When this is done, the proton and positon both have 1.00728 amu's and the single neutron has 1.00037 amu's. (Sub-step 15-1-15b)

Mass of proton in non-final 2He3	1.00511 amu's
Mass transferred to it from non-final 2He3	0.00217 amu
Mass of proton in final 2He3	1.00728 amu's

Mass of positon in non-final 2He3	1.00571 amu's
Mass transferred to it from non-final 2He3	0.00157 amu
Mass of positon in final 2He3	1.00728 amu's

Mass transferred to proton in final 2He3	0.00217 amu
Mass transferred to positon in final 2He3	0.00157 amu
Total mass transferred in final 2He3	0.00374 amu

Mass of neutron in non-final 2He3	1.00411 amu's
Total mass transferred from neutron in non-final 2He3	0.00374 amu
Mass of neutron in final 2He3	1.00037 amu's

By adding these final masses, the mass of the final isotope of 2He3 will be found. This final isotope of 2He3 is the result of a non-radioactive decay because no mass was lost and no ray was emitted. (Sub-step 15-1-15b)

Mass of electron and negatron in final 2He3	0.00110 amu
Mass of proton in final 2He3	1.00728 amu's
Mass of positon in final 2He3	1.00728 amu's
Mass of neutron in final 2He3	1.00037 amu's
Mass of final 2He3	3.01603 amu's

Sub-step 14-1-14

2He3

1H3 = 2He3, #2 gamma ray, transfer 0.00217 amu, 0.00317 amu, 0.00100 amu, negatron orbit

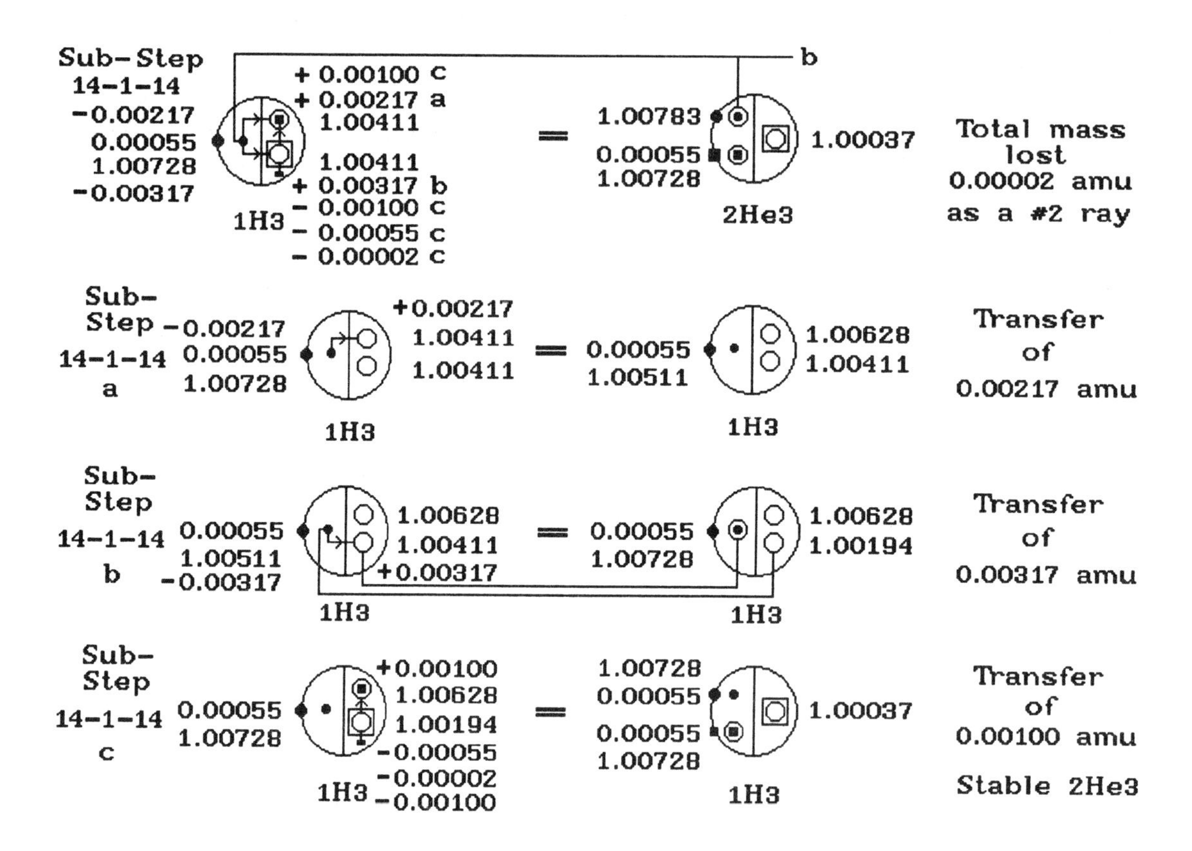

The isotope 1H3 with 3.01605 amu's has one electron with 0.00055 amu and one proton with 1.00728 amu's plus two neutrons. Both neutrons have the same mass of 1.00411 amu's.

The isotope 2He3 with 3.01603 amu's has one electron with 0.00055 amu, one negatron with 0.00055 amu, one proton with 1.00728 amu's, one positon with 1.00728 amu's and one neutron with 1.00037 amu's. The total mass lost is 0.00002 amu and it is emitted as a #2 gamma ray which has no external magnetic field.

The proton in the isotope 1H3 first transfers 0.00217 amu to the first neutron and then 0.00317 amu to the second neutron.

If the proton in the isotope 1H3 has 1.00728 amu's and it transfers 0.00317 amu as a single unit to the first neutron, which has 1.00411 amu's, then this first neutron would have 1.00728 amu's to become the proton again. This means that the 0.00317 amu can not be transferred as a single unit first.

Because of the above, the proton must transfer 0.00217 amu to the first neutron with 1.00411 amu's shown in sub-step 14-1-14a. When this happens, the proton with 1.00728 amu's stays a proton with 1.0511 amu's and the first neutron that received the transferred mass stays a neutron with 1.00628 amu's. The second neutron still has 1.00411 amu's. Now the proton can transfer 0.00317 amu to the second neutron which becomes the proton in the isotope 1H3 with 1.00728 amu's and the second neutron with 1.00411 amu then is no longer in the isotope 1H3 shown in sub-step 14-1-14b. The proton that had 0.00511 amu's loses 0.00317 amu to become a new second neutron with 1.00194 amu's. The proton that had 0.00511 amu vanishes and a new second neutron with 1.00194 amu's appears in the isotope 1H3. This new second neutron in sub-step 14-1-14c then transfers 0.00100 amu to the first neutron with 1.00628 amu's making the first neutron receiving this mass into the proton with 1.00728 amu's. This means that the first neutron vanishes and a proton with 1.00728 amu's appears. When the new second neutron with 1.00194 amu's transferred the 0.00100 amu, it caused the negatron to go into an orbital with the electron already there. It did not release the negatron from within itself. This new second neutron has now lost 0.00155 amu to have 1.00039 amu's remaining in the isotope 1H3. It loses 0.00002 amu at the same time the 0.00100 amu is transferred to become the single neutron with 1.00037 amu's in the isotope 2He3. The loss of 0.00002 amu causes the negatron to be released from within the second neutron in the isotope 1H3. When this is done, the isotope 2He3 is created. The 0.00002 amu is the only mass lost and it is emitted as a #2 gamma ray. When the 0.00002 amu is emitted as a gamma ray, the radioactivity stops and the isotope 2He3 becomes stable.

Now let us look at the magnetic field as these transfers take place.

The single isotope of 1H3 is radioactive and it is going to change into the isotope 2He3 that is not radioactive. The electron and proton have external magnetic fields while the two neutrons have internal magnetic fields. The isotope 1H3 has its electron and proton providing the magnetic field that is around the entire isotope. It is this fact that holds the two neutrons within the isotope 1H3.

It is known that the isotope 1H3 is not stable and will change into an isotope that has more protons. The final isotope will have all of its proton and positon with 1.00728 amu's while the neutrons may not have the maximum mass of 1.00628 amu's.

Radioactivity occurs when an isotope does not have all of the protons and positons it can have to become stable with its protons and positons all having 1.00728 amu's. There are two neutrons in the isotope 1H3 with each having 1.00411 amu's and one proton that has 1.00728 amu's. Note that the proton in the isotope 1H3 already has 1.00728 amu's and the isotope 1H3 is not stable. There are two neutrons and each neutron must have a proton or positon. It must have the proton and positon found in the isotope 2He3 with the proton and positon each having 1.00728 amu's to be a stable isotope. It is believed today that the isotope 2He3 is stable. With the steps just given, there are isotopes of 2He3 that must convert to a final stable form that is not radioactive.

When the second neutron changes the first neutron into a positon, it must release a negatron to go into an orbital with the electron that is already there. Because any neutron has its magnetic field within itself, it becomes important to know what happens to the magnetic field when a proton or positon changes into a neutron or when a neutron changes into a proton or positon. It also becomes important to know what happens to the electron or negatron when this happens.

When the 0.00217 amu is transferred to the first neutron in the isotope 1H3, it carried with it, the magnetic field that belonged to it. This is sub-step 14-1-14a. This means that the proton lost some of its magnetic field because the 0.00217 amu went to the first neutron with 1.00411 amu's which stayed a first neutron with 1.00628 amu's and the magnetic field around the mass of 0.00217 amu became part of the internal magnetic field on the first neutron. At this point in time the magnetic field around the entire isotope of 1H3 is decreased.

The proton in sub-step 14-1-14b now has 1.00511 amu's and it transfers 0.00317 amu to the second neutron with 1.00411 amu's making the second neutron have 1.00728 amu's in the isotope 1H3. When it does this, the proton with 1.00511 amu's becomes the new second neutron with 1.00194 amu's in the isotope 1H3. This means that the proton that had 1.00511 amu's becomes a second neutron in the isotope 1H3 and its magnetic field becomes internal and thus the magnetic field around the entire isotope of 1H3 is decreased by 1.00511 amu's. The old second neutron with 1.00411 amu's regained 0.00317 amu to become the proton in the isotope 1H3. This means that the old second neutron has its internal magnetic field become external and the magnetic field around the entire isotope is increased by 1.00728 amu's The result of these changes is that the new proton regains 0.00217 amu which increases its magnetic field from that of a proton with 1.00511 amu's to a proton with 1.00728 amu's.

The new second neutron in sub-step 14-1-14c with 1.00194 amu's in the isotope 1H3 then transfers 0.00100 amu to the first neutron that has 1.00628 amu's to make it a positon with 1.00728 amu's in the isotope 2He3. This then decreases the new second neutron's internal magnetic field by 0.00100 amu. Because the first neutron becomes a positon, this 0.00100 amu is added to the overall magnetic field. The new second neutron then releases a negatron with 0.00055 amu. This decreases the internal magnetic field of the new second neutron by 0.00055 amu and increases the overall magnetic field by 0.00055 amu. At this time the new second neutron loses 0.00002 amu to make the isotope 2He3 have 3.01603 amu's. This lost 0.00002 amu decreases the new second neutron in the isotope 1H3 by 0.00002 amu and the new second neutron in the isotope 1H3 becomes the single neutron in the isotope 2He3. This results in the isotope 1H3 changing into an isotope of 2He3 with a decrease in its internal magnetic field equal to the mass of 0.00002 amu. Because this lost mass comes from the second neutron and its magnetic field is internal, there is no loss of its magnetic field equal to this lost mass found around the entire isotope of 2He3. This means that the real lost magnetic field comes from the second neutron and not from the proton in the isotope 1H3. The negatron that is released from the second neutron in the isotope 1H3 carries with it an external magnetic field which is added to the total magnetic field of the isotope 2He3 just created. Since the old first neutron in the isotope 1H3 in sub-step 14-1-14a becomes the positon in the isotope 2He3 and the new first neutron vanishes, the isotope 2He3 gains an external magnetic field of 1.00728 amu's plus the negatron's magnetic field for a total gain of 1.00783 amu's in sub-step 14-1-14c. This is the same as the magnetic field around the isotope 1H1 except there is no isotope of 1H1 found within the isotopes 1H3 or 2He3. This means the orbital of the electron and negatron in this case is different than when the isotope of 1H1 is found within the isotopes 1H3 and 2He3. Remember that when a weaker magnetic field becomes part of a larger magnetic field, the larger magnetic field controls the weaker magnetic field and the overall magnetic field does not increase. In this step, when the positon comes into existence in the isotope 2He3 with a negatron going into an orbit with the electron, the magnetic field is not increased because of the negatron.

Remember that the 0.00002 amu is emitted as a #2 gamma ray which is pure energy and pure energy is magnetism. This magnetism is added to the magnetism of the cosmos or to all of the magnetism found around everything. This word "everything" means all empty space.

When the proton with 1.00728 amu's in the isotope 1H3 transfers 0.00217 amu, it loses 0.00217 amu, then 0.00317 amu is transferred from the proton with 1.00511 amu's for a total lost mass of 0.00534 amu. The reader might think that the proton in the isotope 1H3 should lose 0.00217 amu to each neutron then 0.00100 amu with the release of the negatron and the loss of 0.00002 amu to create the isotope 2He3. Since this does not happen, the loss of 0.00534 amu is part of the positon that comes from the new first neutron in the isotope 1H3 and the 0.00534 amu is therefore not lost.

The electron that belongs to the proton becomes the electron in the isotope 2He3 with 0.00055 amu and it is drawn as a dot on the big circle. In the real world, the electron is never released because the isotope that is the isotope 1H3 becomes the isotope that is the isotope 2He3. There is one isotope that is converted to a new isotope which means that the electron never leaves it.

Because the isotope of 1H3 has only one proton and the isotope 2He3 has a proton and a positon, the magnetic field of the isotope 2He3 is twice as strong as the isotope 1H3.

It should be pointed out that in the isotope 1H3 there is one proton working against the mass of two equal neutrons which means that the work done by the proton to overcome the mass of the two neutrons is greater than in the isotope 2He3 where there are two masses working to overcome the mass of a single neutron. The result of this is that the isotope 2He3's magnetic field can hold the single neutron better than the proton in the isotope 1H3 can hold its two neutrons. This results is the proton in the isotope 1H3 is holding its spinning electron at a slower rate on its axis than the proton or positon spinning on their axis. The magnetic spin is the cause of a voltage found on an isotope thus when the isotope 2He3 has each proton's and positon's magnetic spin increased because they each have only one neutron slowing them down. This increases the voltage but does not increase the magnetic field around any of the parts. However, this decreases the number of photons the proton or positon can receive and keep so the magnetic field will decrease if it can not keep a normal amount of photons because the photon adds to the magnetic field. The result of this is that a new voltage is established for the isotope 2He3 and since all isotopes with different masses for their proton, positon and neutron will end up with the masses found for them in this step will change the position of the electron and negatron as they orbit the nucleus.

In the drawing of sub-step 14-1-14, the isotope 1H3 has its electron drawn as a dot on the big circle with the negatron drawn below it as a square. The proton drawn as a dot on the left side of the vertical line within the big circle. The first neutron is drawn as a small circle with the second neutron drawn the same way under it. They are both drawn to the right of the vertical line within the big circle. There is a line drawn from the proton to each of the neutrons with an arrow drawn on the lines pointing to each neutron. The line to the first neutron shows that this neutron gained 0.00217 amu from the proton. The line to the second neutron shows that this neutron gained 0.00317 amu.

There is a line drawn from the second neutron to the first neutron with an arrow drawn on it to show that the second neutron transferred 0.00100 amu to the first neutron.

The first neutron has a small square drawn in the center of the small circle to show that this neutron becomes the positon in the isotope 2He3. The second neutron has a square drawn around the small circle to show that it has lost 0.00002 amu. There is a small square drawn below the second neutron with a line connecting them. The line shows that the second neutron released a negatron from within itself. When the 0.00002 amu is lost as a #2 gamma ray, the isotope 2He3 is created.

The isotope 2He3 has its electron drawn as a dot and the negatron drawn as a square below it on the big circle just like the isotope 1H3. The proton is drawn as a dot with a small circle drawn around it to show that this proton came from the second neutron in the isotope 1H3. The positon is drawn below the proton as a small square with a small circle drawn around it to show that the positon came from the first neutron. The single neutron is drawn as a small circle with a square drawn around it. The square is drawn around this neutron to show that the lost mass of 0.00002 amu came from it.

There is a line drawn from the proton in the isotope 1H3 to the proton in the isotope 2He3 to show that after the 0.00317 amu has been transferred, the proton in the isotope 1H3 becomes the proton in the isotope 2He3. This line has the letter b put at its end to show that the reader should look at sub-step 14-1-14b to understand what has happened.

The total mass lost is given to the right of the isotope 2He3.

In order for the creation of the isotope 2He3 by an isotope of 1H3 to be understood, there are three drawings drawn below the sub-step 14-1-14.

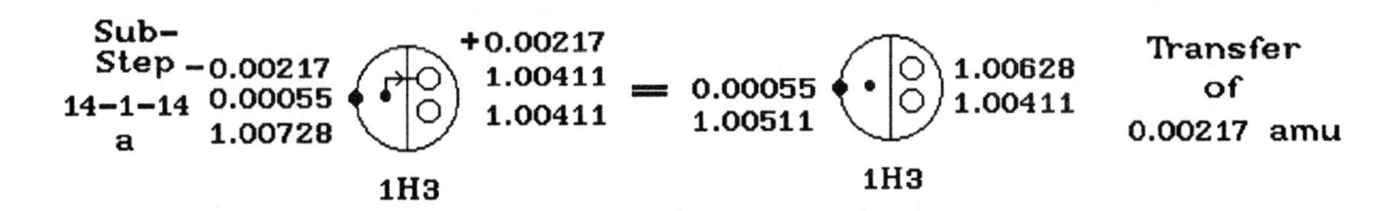

The drawing of sub-step 14-1-14a has the electron drawn as a dot on the big circle. The proton is drawn as a dot to the left of the vertical line within the big circle. There is a line with an arrow pointing to the first neutron that is drawn as a small circle. The small circle has nothing in its middle. The line shows that 0.00217 amu has been transferred from the proton to the first neutron.

After the 0.00217 amu has been transferred, the isotope 1H3 is still an isotope of 1H3 as shown by the isotope 1H3 drawn to the right of the equal sign. The electron is still a dot on the big circle and the proton is drawn as a dot to the left of the vertical line within the big circle with no line connecting the proton to the first neutron as is found in the other isotope of 1H3. The second neutron is drawn as a small circle below the first neutron. The proton now has 0.00511 amu, the first neutron has 1.00628 amu's and the second neutron has 1.00411 amu's.

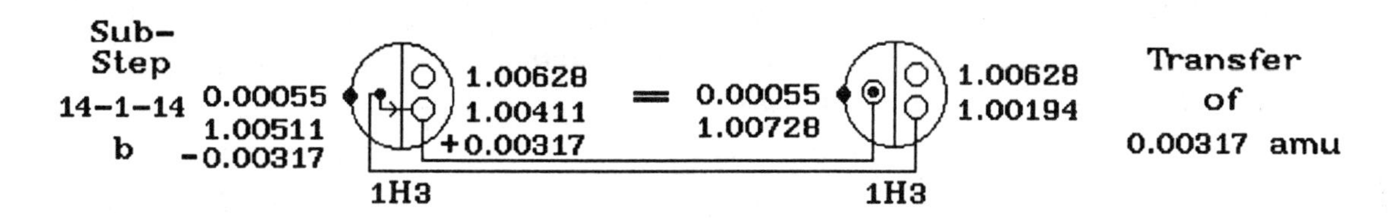

The drawing of sub-step 14-1-14b has the isotope 1H3 found on the left side of the equal sign drawn the same as the isotope 1H3 found on the right side of sub-step 14-1-14a. Now the proton has a line with an arrow pointing toward the second neutron to show that the proton has transferred 0.00317 amu to the second neutron making the second neutron into the proton found in the other isotope of 1H3. There is a line drawn from the second neutron on the left side of the equal sign to the proton in the isotope 1H3 on the right side of the equal sign to show that this neutron becomes the proton in the other isotope of 1H3. This proton in the isotope found on the right side of the equal sign is drawn as a dot with a small circle drawn around it to show that it came from the second neutron found in the isotope 1H3 on the left side of the equal sign.

There is a line drawn from the proton found in the isotope on the left side of the equal sign to the second neutron in the isotope 1H3 found on the right side of the equal sign to show that this proton becomes the second neutron in the other isotope of 1H3. Both neutrons in each of the isotopes are drawn as small circles with nothing in their centers. The first neutron is drawn above the second neutron. The proton on the left side of the equal sign has 1.00511 amu's and the neutron it becomes in the other isotope of 1H3 has 1.00194 amu's. The first neutron in the isotope 1H3 on the left side of the equal sign has 1.00628 amu's and it also has this same mass in the other isotope of 1H3. The second neutron in the isotope 1H3 found on the left side of the equal sign has 1.00411 amu's and the proton it becomes in the other isotope of 1H3 has 1.00728 amu's. All electrons have 0.00055 amu because they never lose any mass.

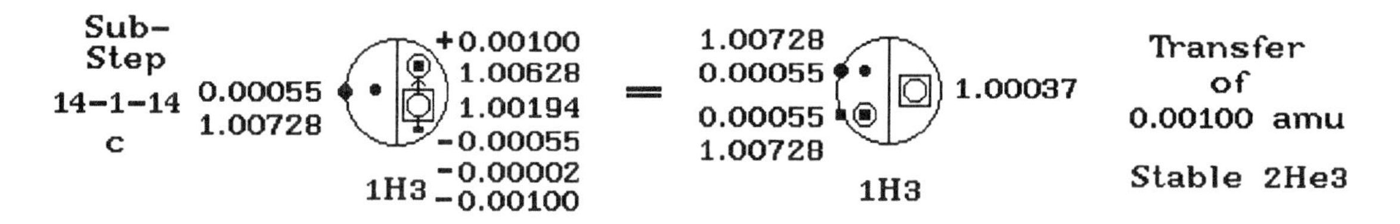

In sub-step 14-1-14c the electron is drawn as a dot on the big circle in the isotope 1H3 found on the left side of the equal sign and it is drawn this same way in the isotope 2He3 which is drawn on the right side of the equal sign. The electron in the isotope 2He3 is drawn above the negatron on the big circle.

The proton in the isotope 1H3 on the left side of the equal sign is drawn as a dot on the left side of the vertical line within the big circle and it is drawn the same way in the isotope 2He3. The first neutron in the isotope 1H3 on the left side of the equal sign is drawn as a small circle with a small square in its middle and it is drawn this same way in the isotope 2He3 to show that the first neutron in the isotope 1H3 becomes the positon in the isotope 2He3. The second neutron in the isotope 1H3 is drawn as a small circle with nothing in its middle and it has a square drawn around it. It is drawn this same way in the isotope 2He3 to show that the single neutron in the isotope 2He3 came from the second neutron in the isotope 1H3. There is a line with an arrow on it pointing towards the first neutron in the isotope 1H3 to show that the second neutron has transferred 0.00100 amu to the first neutron.

There is a small square that is the negatron which is released from the second neutron in the isotope 1H3. This negatron is drawn below the second neutron with a connecting line to show that the negatron came from the second neutron. It comes from within the second neutron and has 0.00055 amu.

The proton in the isotope 1H3 has 1.00728 amu's in the isotopes 1H3 and 2He3. The first neutron in the isotope 1H3 has 1.00628 amu's in the isotope 1H3 and it becomes the positon in the isotope 2He3 with 1.00728 amu's. The second neutron in the isotope 1H3 has 1.00194 amu's and the single neutron it becomes has 1.00037 amu's. The electron and negatron each have 0.00055 amu.

The total mass lost is shown on the right side of the isotope 2He3.

To find the mass that is lost creating the isotope 2He3, subtract the isotope 2He3 from the isotope 1H3.

Mass of 1H3	3.01605 amu's
Mass of 2He3	3.01603 amu's
Mass lost creating 2He3	0.00002 amu

This 0.00002 amu is lost by the second neutron in the isotope 1H3 as a single unit that becomes a #2 ray. This is the only mass that is really lost.

In the isotope 2He3, the only thing known is the electron. Since this isotope must have a negatron, its mass is known. Electrons and negatrons have the same mass and they never lose any of it.

This step shows both neutrons in the isotope 1H3 have 1.00411 amu's each. It looks like both neutrons regain the same amount of mass because the proton and positon each have 1.00728 amu's in the isotope 2He3. It also looks like the first neutron in the isotope 1H3 regains this mass and becomes a proton, keeping its electron in the isotope 2He3. If each neutron has the same mass of 1.00411 amu's and they become a proton and positon with the same mass of 1.00728 amu's, then each must regain the same amount of mass which should mean that the proton must transfer 0.00217 amu and 0.00100 amu twice with the electron and a negatron released to go into an orbital because of the two 0.00100 amu that are transferred but since they are only transferred, no electron is released. The negatron is released but it is released from the second neutron and not from within the proton to go with the electron in the isotope 2He3. The proton and positon must each have 1.00728 amu's if the single neutron has 1.00037 amu's. By having the 0.00217 amu regained first by the first neutron, the proton stays a proton and the first neutron stay a neutron. When the second neutron regains 0.00317 amu and it has 1.00411 amu's, this will make a neutron with 1.00628 amu's.

Mass of proton in 1H3	1.00728 amu's
Mass transferred by it	0.00217 amu
New mass of proton in 1H3	1.00511 amu's

Mass of first neutron in 1H3	1.00411 amu's
Mass regained by first neutron from proton in 1H3	0.00217 amu
Mass of first neutron in 1H3	1.00628 amu's

Now that the proton in the isotope 1H3 has 1.00511 amu's, it can transfer 0.00317 amu to the second neutron. By doing this, the second neutron in the isotope 1H3 becomes the positon with 1.00728 amu's in the isotope 2He3.

Mass of second neutron in 1H3	1.00411 amu's
Mass regained by it from proton in 1H3	0.00317 amu
Mass of positon in 2He3	1.00728 amu's

When the new proton with 1.00511 amu's transfers 0.00317 amu, it becomes a new neutron in the isotope 1H3 with 1.00194 amu's. This new neutron then transfers 0.00100 amu to the first neutron which has 1.00628 amu's. This 0.00100 amu is therefore lost by the new second neutron. The new first neutron regains the 0.00100 amu to become the proton in the isotope 2He3.

Mass of new proton in 1H3	1.00511 amu's
Mass transferred to second neutron in 1H3	0.00317 amu
Mass of new second neutron in 2He3	1.00194 amu's

Mass of new second neutron in 1H3	1.00194 amu's
Mass transferred to first neutron in 1H3	0.00100 amu
New mass of new second neutron in 1H3	1.00094 amu's

Mass of new first neutron in 1H3	1.00628 amu's
Mass transferred to first neutron from second neutron in 1H3	0.00100 amu
Mass of proton in 2He3	1.00728 amu's

This new proton now has the same mass as the positon. The new second neutron in the isotope 1H3 now has 1.00094 amu's which must release a negatron from within itself. The negatron has 0.00055 amu. Subtract this mass from the new second neutron. The remaining mass will have 0.00002 amu more than the single neutron in the isotope 2He3 can have and it must therefore be lost. Subtract it from the remaining mass of the new second neutron and when it is lost, the isotope 2He3 will be created. This new second neutron must lose 0.00002 amu to become the single neutron in the isotope 2He3. This 0.00002 amu is the only mass lost and it is emitted as a #2 gamma ray. All of the other masses are transferred between the proton and its neutrons and between the two neutrons.

Remaining mass of new second neutron in 1H3	1.00094 amu's
Mass of released negatron in 1H3	0.00055 amu
Mass now found for new second neutron in 1H3	1.00039 amu's

Remaining mass of now found for new second neutron in 1H3	1.00039 amu's
Mass of lost creating 2He3	0.00002 amu
Mass of single neutron in 2He3	1.00037 amu's

Now all the masses of the isotope 2He3 are known. Adding all of these masses will give the mass of the isotope 2He3.

Mass of electron and negatron in 2He3	0.00110 amu
Mass of proton and positon in 2He3	2.01456 amu's
Mass of neutron in 2He3	1.00037 amu's
Mass of 2He3	3.01603 amu's

Sub-step 13-1-13

2He3

1H3 = 2He3, #2 gamma ray, transferred 0.00100 amu, 0.00217 amu, neg. orbit

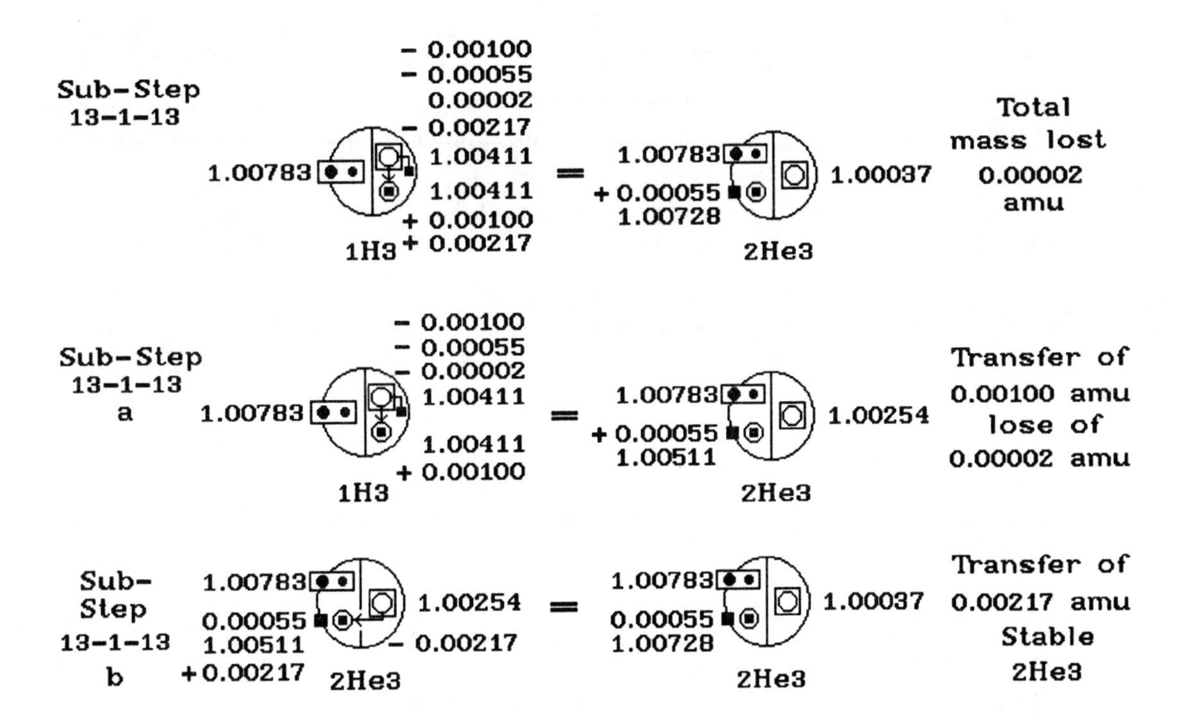

The isotope 1H3 with 3.01605 amu's has one electron with 0.00055 amu and one proton with 1.00728 amu's plus two neutrons. Both neutrons have the same mass of 1.00411 amu's.

The isotope 2He3 with 3.01603 amu's has one electron with 0.00055 amu, one negatron with 0.00055 amu, one proton with 1.00728 amu's, one positon with 1.00728 amu's and one neutron with 1.00037 amu's.

The total mass lost is 0.00002 amu and it is emitted as a #2 ray which has no external magnetic field.

The isotope 1H3 with 3.01605 amu's has its electron with 0.00055 amu and proton with 1.00728 amu's each drawn as a dot with a rectangle drawn around both in the isotopes 1H3 and 2He3. The rectangle means that they are used as the isotope 1H1 with 1.00783 amu's and that the isotope 1H1 can drop out of the isotope 2He3 at some later time.

The first neutron with 1.00411 amu's in the isotope 1H3 is drawn as a small circle with nothing drawn in its middle to the right of the vertical line within the big circle. There is a line drawn from the first neutron to the second neutron with an arrow drawn on the line pointing toward the second neutron to show that the first neutron has transferred 0.00100 amu and 0.00217 amu. When this is done, the first neutron with 1.00411 amu's becomes the single neutron with 1.00037 amu's in the isotope 2He3. There is a square drawn around the first neutron to show that 0.00002 amu has been lost. There a small square drawn to the left and slightly below the first neutron with a line connecting it to this neutron. The small square is the negatron that is released from within the first neutron with 0.00055 amu.

The second neutron with 1.00411 amu's becomes a positon with 1.00728 amu's in the isotope 2He3 when it regains 0.00100 amu and 0.00217 amu. It is drawn as a small circle below the first neutron and to the right of the vertical line within the big circle. There is a line connecting it to the first neutron with an arrow drawn on the line pointing at the second neutron to show that the transferred masses are transferred to the second neutron in the isotope 1H3. Note that there is no positon in the isotope 1H3 and there is one in the isotope 2He3. This positon in the isotope 2He3 was the second neutron in the isotope 1H3 and to show this, there is a small square drawn in the middle of the small circle of the second neutron in the isotope 1H3 and it is drawn this same way in the isotope 2He3 to show that it is the neutron that becomes a positon.

When the first neutron transferred 0.00100 amu, it did not release a negatron from within itself because the 0.00002 amu did this. However it did cause the negatron to go into an orbital with the electron that is already there. The reason the negatron does not leave the isotope 1H3 completely is that when the positon comes into existence, it gains an external magnetic field which holds the negatron in an orbital. This holding of the negatron in an orbit is part of the reason the isotope 2He3 is created because the isotope 2He3 must have a negatron to go with its positon. The second neutron in the isotope 1H3 must disappear when the positon appears in the isotope 2He3.

In previous steps, the 0.00100 amu is emitted as a 100 ray and the electron is released completely from the isotope. Its mass is part of the total mass lost creating the new isotope. In this step the 0.00100 amu is transferred to another neutron and it is therefore not lost or emitted as a ray. There is a negatron with 0.00055 amu that is released but this time it is released from within the neutron and stays with the newly created isotope of 2He3 in an orbital with the electron. This means that no mass has been lost.

The 0.00002 amu that is lost from the first neutron is not regained by the second neutron and it is the real mass that is lost creating the isotope 2He3. It is emitted as a #2 gamma ray. The square drawn around the first neutron shows it loses the 0.00002 amu and it does not include any transferred mass between the two neutrons in the isotope 1H3.

The isotope 2He3 with 3.01603 amu's has its electron and proton drawn with a rectangle around them which gives them a mass of 1.00783 amu's. This rectangle shows that the electron and proton form the isotope 1H1 within the isotope 2He3 and since it is drawn the same way in the isotope 1H3, it shows that it came from the isotope 1H3.

There is a square drawn on the big circle in the isotope 2He3 to show that the negatron with 0.00055 amu came from within the first neutron in the isotope

1H3. There is a small square drawn to the left of the vertical line within the big circle. The positon has 1.00728 amu's. This small square has a circle drawn around it to show that it came from the second neutron in the isotope 1H3.

The single neutron with 1.00037 amu's in the isotope 2He3 is drawn as a small circle to the right of the vertical line within the big circle. It has a square drawn around it to show that the 0.00002 amu is lost when it came into existence. The reader will want to know which part of the isotope 2He3 lost the real mass lost creating the isotope 2He3. This is why the square is drawn around the single neutron in the isotope 2He3.

To find the mass that is lost creating the isotope 2He3, subtract the isotope 2He3 from the isotope 1H3.

Mass of 1H3	3.01605 amu's
Mass of 2He3	3.01603 amu's
Mass lost creating 2He3	0.00002 amu

This 0.00002 amu is lost by the first neutron in the isotope 1H3. It is lost as a single unit as a #2 gamma ray. This is the only mass that is really lost. All of the other masses are transferred by the first neutron except for the negatron which goes into an orbital with the electron the isotope 2He3.

The known mass in the isotope 2He3 comes from the isotope 1H1 that is intact within the isotope 2He3 and the fact that the isotope 2He3 must have a negatron with 0.00055 amu. By adding these two masses and subtracting their total from the mass of the isotope 2He3, the remaining mass will be the mass of the positon and the neutron found in the isotope 2He3.

Mass of 1H1 within 2He3	1.00783 amu's
Mass of negatron in 2He3	0.00055 amu
Total known mass in 2He3	1.00838 amu's

Mass of 2He3	3.01603 amu's
Total known mass in 2He3	1.00838 amu's
Mass of positon and neutron in 2He3	2.00765 amu's

The first neutron in the isotope 1H3 is known to lose the mass that will release the negatron. It is also known that the second neutron regains this transferred mass. Therefore the first neutron must transfer more mass than the 0.00100 amu and it must release the negatron at the same time. The first neutron will lose 0.00155 amu plus the 0.00002 amu that is really lost creating the isotope 2He3. By subtracting the 0.00157 amu from the first neutron, the remaining mass still will not make the positon in the isotope 2He3.

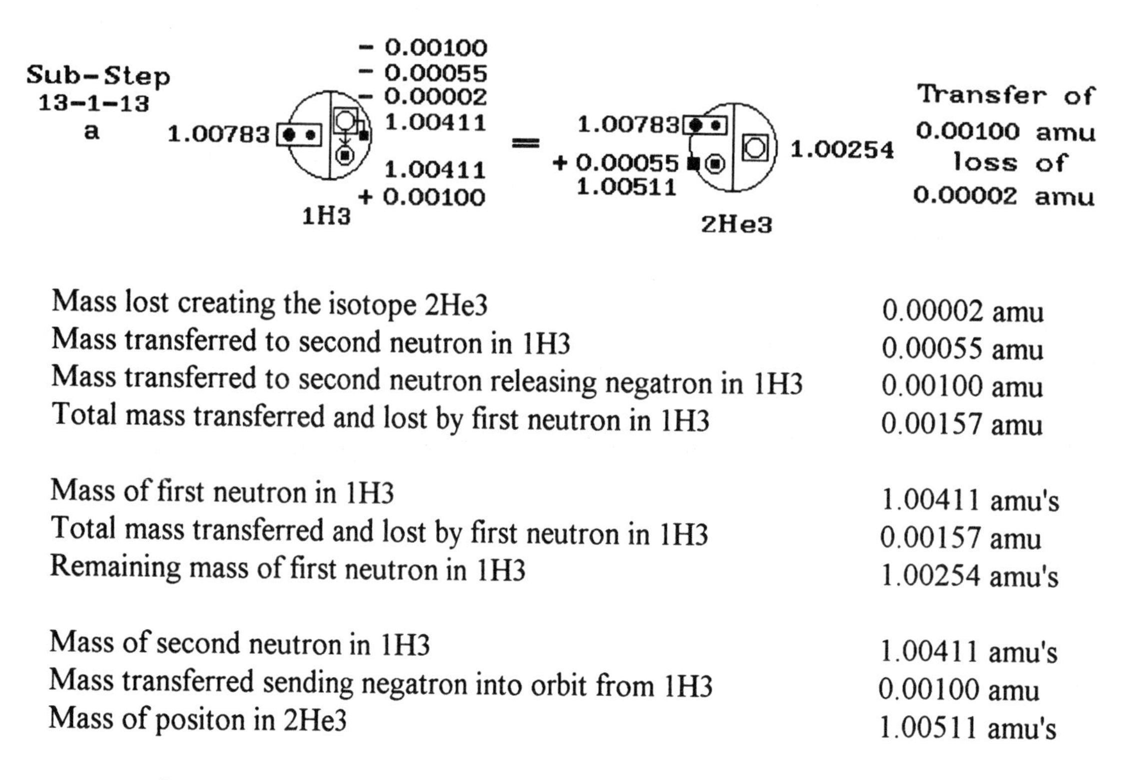

Mass lost creating the isotope 2He3	0.00002 amu
Mass transferred to second neutron in 1H3	0.00055 amu
Mass transferred to second neutron releasing negatron in 1H3	0.00100 amu
Total mass transferred and lost by first neutron in 1H3	0.00157 amu

Mass of first neutron in 1H3	1.00411 amu's
Total mass transferred and lost by first neutron in 1H3	0.00157 amu
Remaining mass of first neutron in 1H3	1.00254 amu's

Mass of second neutron in 1H3	1.00411 amu's
Mass transferred sending negatron into orbit from 1H3	0.00100 amu
Mass of positon in 2He3	1.00511 amu's

The remaining mass of the first neutron in the isotope 1H3 is the mass of a neutron. Since the transferred mass of 0.00100 amu goes to the second neutron in the isotope 1H3, this second neutron would have 1.00511 amu's and this also is the mass of a proton or positon. Because there already is a proton in the isotope 2He3, this mass must be the mass of the positon. The second neutron at this point in time becomes the positon and the first neutron becomes the single neutron in the isotope 2He3. Their total mass equals the mass of the isotope 2He3.

Mass of 1H1 in 2He3	1.00783 amu's
Mass of negatron in 2He3	0.00055 amu
Mass of positon in 2He3	1.00511 amu's
Mass of neutron in 2He3	1.00254 amu's
Mass of 2He3	3.01603 amu's

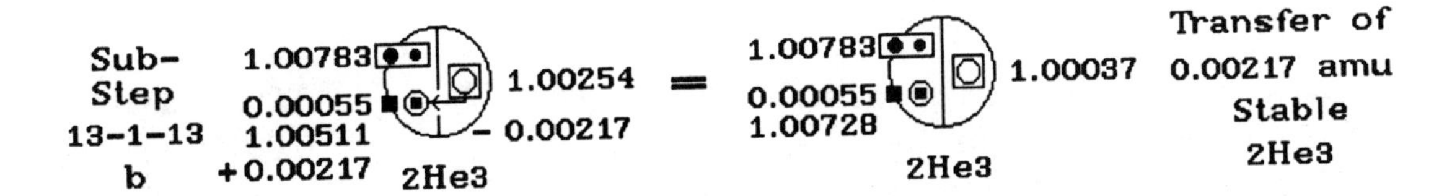

In this step, this isotope of 2He3 does not stay as it is. More mass is transferred to make a final isotope of 2He3. This means that the positon must gain more mass and the single neutron must lose this mass to the positon. For this to happen, the only mass the positon can gain will make it into a normal positon with 1.00728 amu's. Subtract the mass of the positon from the mass of a normal neutron to find the mass the positon must gain. Then subtract this gained mass by the positon from the mass of the single neutron in the isotope 2He3.

Mass of normal positon	1.00728 amu's
Mass of positon in 2He3	1.00511 amu's
Mass gained by positon in 2He3	0.00217 amu

Mass of neutron in 2He3	1.00254 amu's
Mass transferred to positon in 2He3	0.00217 amu
Mass of neutron in 2He3	1.00037 amu

A normal positon is 1.00728 amu's. In this step, the positon is a normal positon. By subtracting its mass from the mass that make up the positon and neutron in the isotope 2He3, then the remaining mass will be the neutron in the isotope 2He3.

Mass of positon and neutron in 2He3	2.00765 amu's
Mass of positon in 2He3	1.00728 amu's
Mass of neutron in 2He3	1.00037 amu's

The neutron in the isotope 2He3 came from the first neutron in the isotope 1H3. This first neutron has 1.00411 amu's so by subtracting the mass of the neutron in the isotope 2He3 from the first neutron in the isotope 1H3, the mass lost by the first neutron in the isotope 1H3 will be known.

Mass of first neutron in 1H3	1.00411 amu's
Mass of neutron in 2He3	1.00037 amu's
Total mass lost by first neutron in 1H3	0.00374 amu

The first neutron in the isotope 1H3 must transfer the masses that will not release a negatron and the mass that is lost creating the isotope 2He3. By adding the mass of the negatron that is released plus the 0.00100 amu that is transferred and then subtracting their total from the mass lost by the neutron in the isotope

2He3, the remaining mass that is transferred which also does not release the negatron will be known.

Mass of negatron released from first neutron in 1H3	0.00055 amu
Mass transferred from first neutron in 1H3	0.00100 amu
Total known mass lost by first neutron in 1H3	0.00155 amu

Total mass lost by first neutron in 2He3	0.00374 amu
Total known mass lost by first neutron in 1H3	0.00155 amu
Mass transferred not releasing negatron within neutron in 1H3	0.00219 amu

The mass transferred which does not release the negatron from within the first neutron in the isotope 1H3 includes the real mass lost creating the isotope 2He3. Subtract this real mass lost from the mass transferred to the second neutron in the isotope 1H3 to find the mass that is transferred that does not send the negatron in an orbital.

Mass transferred not releasing negatron within neutron in 1H3	0.00219 amu
Real mass lost creating 2He3	0.00002 amu
Mass transferred not releasing negatron within neutron in 1H3	0.00217 amu

Now all the masses of the mass lost by the first neutron in the isotope 1H3 are known. Add all of these parts and subtract their total from the first neutron in the isotope 1H3 to find the mass of this neutron.

Mass lost creating 2He3 by first neutron in 1H3	0.00002 amu
Mass transferred not releasing negatron by first neutron in 1H3	0.00217 amu
Mass of negatron by first neutron in 1H3	0.00055 amu
Mass transferred by first neutron in 1H3	0.00100 amu
Total mass lost and transferred by first neutron in 1H3	0.00374 amu

Mass of first neutron in 1H3	1.00411 amu's
Total mass lost and transferred by first neutron in 1H3	0.00374 amu
Mass of neutron in 2He3	1.00037 amu's

This mass transferred by the first neutron in the isotope 1H3 has the 0.00217 amu and the 0.00100 amu regained by the second neutron to make this second neutron into the positon in the isotope 2He3.

Mass not releasing negatron regained by second neutron	0.00217 amu
Mass releasing negatron regained by second neutron	0.00100 amu
Total mass regained by second neutron in 1H3	0.00317 amu

Mass of second neutron in 1H3	1.00411 amu's
Mass regained by it	0.00317 amu
Mass of positon in 2He3	1.00728 amu's

All of the masses making up the isotope 2He3 are now known. Add all of the known masses to find the mass of 2He3.

Mass of 1H1 in 2He3	1.00783 amu's
Mass of negatron in 2He3	0.00055 amu
Mass of positon in 2He3	1.00728 amu's
Mass of neutron in 2He3	1.00037 amu's
Mass of 2He3	3.01603 amu's

This step shows the electron and proton used as the isotope 1H1 with the first neutron with 1.00411 amu's transferring 0.00100 amu to the second neutron in the isotope 1H3 and then transferring 0.00217 amu to it. The next step will show what happens when the 0.00217 amu is transferred first instead of the 0.00100 amu.

Sub-step 12-1-12

2He3

1H3 = 2He3, #2 gamma ray, transfer 0.00217 amu, 0.00100 amu, neg. orbit

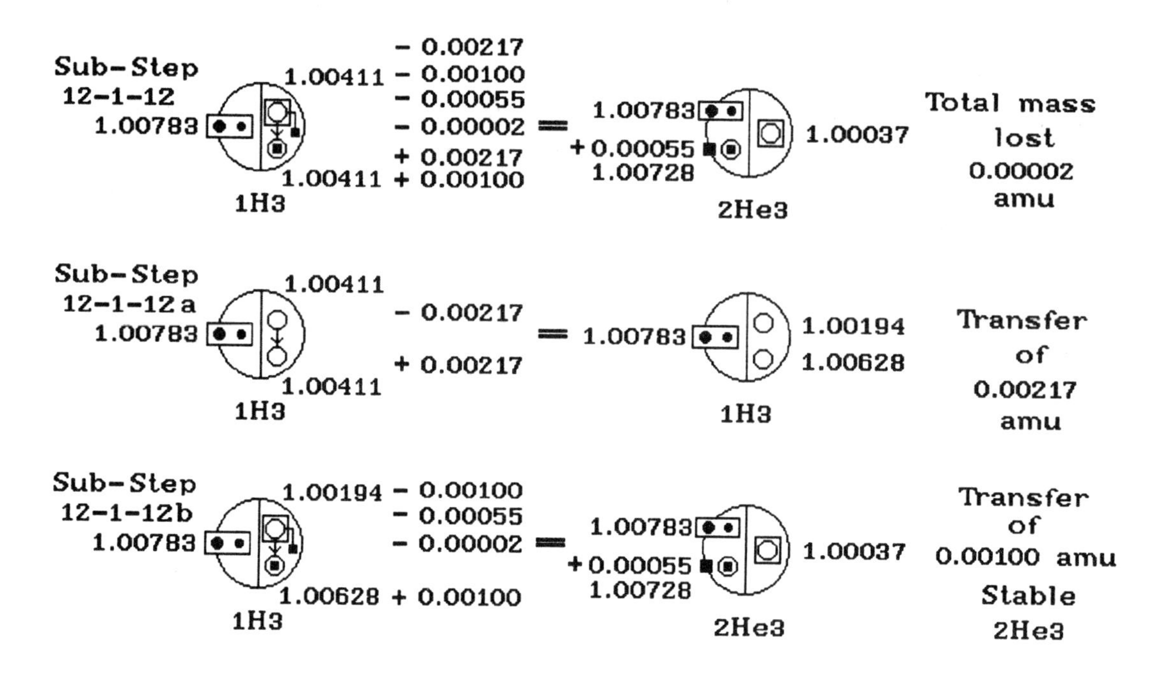

The isotope 1H3 with 3.01605 amu's has one electron with 0.00055 amu and one proton with 1.00728 amu's plus two neutrons. Both neutrons have the same mass of 1.00411 amu's.

The isotope 2He3 with 3.01603 amu's has one electron with 0.00055 amu, one negatron with 0.00055 amu, one proton with 1.00728 amu's, one positon with 1.00728 amu's and one neutron with 1.00037 amu's.

The total mass lost is 0.00002 amu and it is emitted as a #2 gamma ray which has no external magnetic field.

The isotope 1H3 with 3.01605 amu's has its electron with 0.00055 amu and proton with 1.00728 amu's each drawn as a dot with a rectangle drawn around both in the isotopes 1H3 and 2He3. The rectangle means that they are used as the isotope 1H1 with 1.00783 amu's and that the isotope 1H1 can drop out of the isotope 2He3 at some later time.

The first neutron with 1.00411 amu's in the isotope 1H3 is drawn as a small circle with nothing in its middle, to the right of the vertical line within the big circle. There is a line drawn from the first neutron to the second neutron with an arrow drawn on the line pointing toward the second neutron to show that the first neutron has transferred 0.00217 amu and 0.00100 amu. When this is done, the first neutron with 1.00411 amu's becomes the single neutron with 1.00037 amu's in the isotope 2He3. There is a square drawn around the first neutron to show that 0.00002 amu has been lost. There a small square drawn to the left and slightly below the first neutron with a line connecting it to this neutron. The small square is the negatron that is released from within the first neutron with 0.00055 amu.

The second neutron with 1.00411 amu's becomes a positon with 1.00728 amu's in the isotope 2He3 when it regains 0.00217 amu and 0.00100 amu. It is drawn as a small circle below the first neutron and to the right of the vertical line within the big circle. There is a line connecting it to the first neutron with an arrow drawn on the line pointing at the second neutron to show that the transferred masses are transferred to the second neutron in the isotope 1H3. Note that there is no positon in the isotope 1H3 and there is one in the isotope 2He3. This positon in the isotope 2He3 was the second neutron in the isotope 1H3 and to show this, there is a small square drawn in the middle of the small circle of the second neutron in the isotope 1H3 and it is drawn this same way in the isotope 2He3 to show that it is the neutron that becomes a positon.

When the first neutron transferred 0.00100 amu, it did not release a negatron from within itself because the 0.00002 amu did this. However, it did cause the negatron to go into an orbital with the electron that is already there. The reason the negatron does not leave the isotope 1H3 completely is that when the positon comes into existence, it gains an external magnetic field which holds the negatron in an orbital. This holding of the negatron in an orbit is part of the reason the isotope 2He3 is created because the isotope 2He3 must have a negatron to go with its positon. The second neutron in the isotope 1H3 must disappear when the positon appears in the isotope 2He3.

In previous steps, the 0.00100 amu is emitted as a 100 ray and the electron is released completely from the isotope. Its mass is part of the total mass lost creating the new isotope. In this step, the 0.00100 amu is transferred to another neutron and it is therefore not lost or emitted as a ray. There is a negatron with 0.00055 amu that is released but this time it is released from within the neutron and stays with the newly created isotope of 2He3 in an orbital with the electron. This means that no mass has been lost.

The 0.00002 amu that is lost from the first neutron is not regained by the second neutron and it is the real mass that is lost creating the isotope 2He3. It is emitted as a #2 gamma ray. The square drawn around the first neutron shows it loses the 0.00002 amu and it does not include any transferred mass between the two neutrons in the isotope 1H3.

The isotope 2He3 with 3.01603 amu's has its electron and proton drawn with a rectangle around them which gives them a mass of 1.00783 amu's. This rectangle shows that the electron and proton form the isotope 1H1 within the isotope 2He3 and since it is drawn the same way in the isotope 1H3, it shows that it came from the isotope 1H3.

There is a square drawn on the big circle in the isotope 2He3 to show that the negatron with 0.00055 amu came from within the first neutron in the isotope

1H3. There is a small square drawn to the left of the vertical line within the big circle. This small square is the positon with 1.00728 amu's. This small square has a circle drawn around it to show that it came from the second neutron in the isotope 1H3.

The single neutron with 1.00037 amu's in the isotope 2He3 is drawn as a small circle to the right of the vertical line within the big circle. It has a square drawn around it to show that the 0.00002 amu is lost when it came into existence. The reader will want to know which part of the isotope 2He3 lost the real mass that is lost creating the isotope 2He3. This is why the square is drawn around the single neutron in the isotope 2He3.

To find the mass that is lost creating the isotope 2He3, subtract the isotope 2He3 from the isotope 1H3.

Mass of 1H3	3.01605 amu's
Mass of 2He3	3.01603 amu's
Mass lost creating 2He3	0.00002 amu

This 0.00002 amu is lost by the first neutron in the isotope 1H3. It is lost as a single unit as a #2 gamma ray. This is the only mass that is really lost. All of the other masses are transferred by the first neutron except for the negatron which goes into an orbital with the electron in the isotope 2He3.

It is known that the stable isotope of 2He3 has an electron and a negatron with 0.00110 amu. There is one proton and one positon with each having 1.00728 amu's and one neutron with 1.00037 amu's. Since it is known that one isotope of 1H3 will become these parts in the isotope 2He3, the mass that is transferred can be found by noting that the single neutron in the isotope 2He3 comes from one of the two equal neutrons in the isotope 1H3. By subtracting the mass of the single neutron in the isotope 2He3 from one of the two neutrons found in the isotope 1H3, the mass that is transferred plus the lost will be found.

Mass of first neutron in 1H3	1.00411 amu's
Mass of single neutron in 2He3	1.00037 amu's
Mass transferred plus lost mass creating 2He3	0.00374 amu

The first neutron in the isotope 1H3 is known to lose the transferred mass that will release the negatron. It is also known that the second neutron regains this transferred mass. Since the 0.00100 amu causes the negatron to leave the first neutron and this would make an a positon of 0.00511 amu which can not happen in this step, there must be more mass transferred to the second neutron. The additional mass transferred must be transferred first.
Therefore, the first neutron must transfer more mass than the 0.00100 amu and it must release the negatron at the same time. The first neutron will lose 0.00155 amu plus the 0.00002 amu. The 0.00002 amu is the only mass that is really lost

creating the isotope 2He3. By subtracting the 0.00157 amu from the first neutron, the remaining mass will make the first neutron have 1.00254 amu's which happens because these masses are lost first. I showed this in the last step. In this step, the first transferred mass is 0.00217 amu. This 0.00217 amu must be found.

Mass lost creating the isotope 2He3	0.00002 amu
Mass of released negatron from first neutron in 1H3	0.00055 amu
Mass transferred to second neutron in 1H3	0.00100 amu
Known mass transferred and lost by first neutron in 1H3	0.00157 amu

With the total mass transferred, released and lost by the first neutron in the isotope 1H3 known, the mass transferred first in this step from the first neutron can be found by subtracting the known mass that is transferred, lost and released from the total known mass transferred from the first neutron.

Total know mass transferred creating 2He3	0.00374 amu
Known mass transferred, released and lost by first neutron in 1H3	0.00157 amu
Additional mass transferred first by first neutron in 1H3	0.00217 amu

With the knowledge that the first neutron transferred 0.00217 amu first to the second neutron, the steps used to create the isotope 2He3 can now be found. By subtracting the 0.00217 amu from the first neutron in the isotope 1H3, the remaining mass will still be a neutron.

Mass of first neutron in 1H3	1.00411 amu's
Mass transferred to second neutron in 1H3	0.00217 amu
Remaining mass of first neutron in 1H3	1.00194 amu's

This remaining mass of the first neutron in the isotope 1H3 then loses 0.00002 amu and releases a negatron from within itself. At this time there is 0.00100 amu transferred to the second neutron that causes the negatron to go into an orbital that has an electron already in it. The 0.00100 amu is transferred and not lost. The fact that the 0.00100 is transferred, not lost prevents the negatron from leaving the isotope of 1H3. Subtracting this lost, released and transferred mass from the remaining mass will give the mass of the single neutron found in the isotope 2He3.

Remaining mass of first neutron in 1H3	1.00194 amu's
Mass lost, released and transferred from first neutron in 1H3	0.00157 amu
Mass of single neutron in 2He3	1.00037 amu's

Since the first neutron becomes the single neutron in the isotope 2He3, the second neutron must become the positon in the isotope 2He3 because the isotope 1H1 that is intact within the isotope 1H3 and it is also intact within the isotope 2He3. Therefore, the proton in the isotope 2He3 must have 1.00728 amu's. The second neutron in the isotope 1H3 must receive 0.00217 amu and 0.00100 amu from the first neutron. The second neutron has 1.00411 amu's. By adding the transferred mass from the first neutron, the mass of the positon in the isotope 2He3 can be found.

Mass transferred first to second neutron in 1H3	0.00217 amu
Mass transferred next to second neutron in 1H3	0.00100 amu
Total mass transferred to second neutron in 1H3	0.00317 amu

Mass of second neutron in 1H3	1.00411 amu's
Total mass transferred to second neutron in 1H3	0.00317 amu
Mass of positon in 2He3	1.00728 amu's

The electron and negatron each have 0.00055 amu for a total of 0.00110 amu in the isotope 2He3. The isotope 1H1 found intact within the isotope 2He3 has 1.00783 amu's. All of the masses of the isotope 2He3 are now known and it is also now known how they got there. Adding these masses will give the isotope 2He3.

Mass of 1H1 in 2He3	1.00783 amu's
Mass of negatron in 2He3	0.00055 amu
Mass of positon in 2He3	1.00728 amu's
Mass of neutron in 2He3	1.00037 amu's
Mass of 2He3	3.01603 amu's

In this step with the electron and proton used as the isotope 1H1 with the first neutron having 1.00411 amu's, transferring 0.00217 amu to the second neutron in the isotope 1H3 and then transferring 0.00100 amu to it. The next step will show what happens when the isotope 1H2 is found intact with the isotope 1H3 and 2He3. This means that the second neutron in the isotope 1H3 loses all of the mass that is lost and no mass can be transferred.

Sub-step 11-1-11

2He3

1H3 = 2He3, #2 gamma ray, neg. orbit, transfer 0.00591 amu

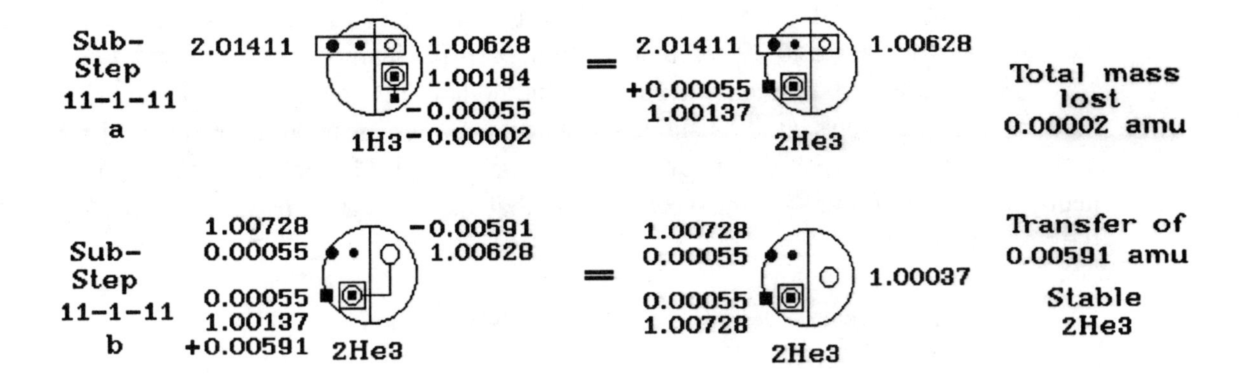

The isotope 1H3 with 3.01605 amu's has one electron with 0.00055 amu and one proton with 1.00728 amu's plus two neutrons. The first neutron has 1.00628 amu's and the second neutron has 1.00194 amu's.

The isotope 2He3 with 3.01603 amu's has one electron with 0.00055 amu, one negatron with 0.00055 amu, one proton with 1.00728 amu's, one positon with 1.00137 amu's and one neutron with 1.00628 amu's.

The total mass lost is 0.00002 amu and it is emitted as a #2 gamma ray which has no external magnetic field.

The isotope 1H3 in sub-step 11-1-11a with 3.01605 amu's has its electron with 0.00055 amu and the proton with 1.00728 amu's each drawn as a dot. The first neutron with 1.00628 amu's drawn as a small circle. There is a rectangle drawn around them in the isotopes 1H3 and 2He3. The rectangle means that they are used as the isotope 1H2 with 2.01411 amu's within the isotopes 1H3 and 2He3. The isotope 1H2 can drop out of the isotope 2He3 at some later time.

The second neutron with 1.00194 amu's drawn as a small circle. The middle part of the small circle has a square drawn in it. It is drawn this same way as the positon with 1.00137 amu's in the isotope 2He3 to show that the second neutron in the isotope 1H3 has become the positon in the isotope 2He3. There is also 0.00002 amu lost by the second neutron in the isotope 1H3 which is the only mass that is lost creating the isotope 2He3. The second neutron in the isotope 1H3 releases a negatron with 0.00055 amu from within itself. The negatron is drawn as a square below the second neutron and as a square on the big circle in the isotope 2He3. There is a square drawn around the small circle that has a small square within it to show that 0.00002 amu has been lost.

In all of the steps where 0.00002 amu has been lost creating a new isotope, a positon has been the result. This means that when a positon is the result of 0.00002 amu being lost, there is a negatron released from within a proton or a neutron. It is not released from an orbital. It must combine with a positon to create a new isotope. The negatron is drawn as a square on the big circle in the isotope 2He3.

In sub-step 11-1-11b, the first isotope of 2He3 found on the left side of the equal sign is drawn the same as the isotope 2He3 in sub-step 11-1-11a except there is a line drawn from the single neutron to the positon to show that this neutron has transferred 0.00591 amu to the positon, making the positon have 1.00728 amu's and the single neutron now has 1.00037 amu's. The isotope 2He3 now has 3.01603 amu's and it is stable. The rectangle drawn in sub-step 11-1-11a around the isotope 1H2 is not drawn in sub-step 11-1-11b because the single neutron transfers 0.00591 amu which means that there is no isotope 1H2 within the isotope 2He3.

To find the mass that is lost creating the isotope 2He3, subtract the isotope 2He3 from the isotope 1H3. (Sub-step 11-1-11a.)

Mass of 1H3	3.01605 amu's
Mass of 2He3	3.01603 amu's
Mass lost creating 2He3	0.00002 amu

This 0.00002 amu is lost by the isotope 1H3. It is lost as a single unit as a #2 gamma ray. This is the only mass that is really lost.

The mass of the single neutron in the isotope 2He3 is part of the isotope 1H2 that is intact within the isotope 2He3. This means that this neutron has 1.00628 amu's in both the isotopes 1H3 and 2He3.

The mass of the second neutron in the isotope 1H3 can be found by subtracting the isotope 1H2 from the isotope 1H3.

Mass of 1H3	3.01605 amu's
Mass of 1H2	2.01411 amu's
Mass of second neutron in 1H3	1.00194 amu's

The mass of the second neutron in the isotope 1H3 is known to be 1.00194 amu's. There is only one neutron in the isotope 2He3 so the lost mass must come from the second neutron in the isotope 1H3. The second neutron will become the positon in the isotope 2He3. This second neutron in the isotope 1H3 will release a negatron from within itself as the needed negatron in the isotope 2He3. The negatron will go into an orbital with the electron and both have the same mass of 0.00055 amu.

To find the mass of the positon, add the mass of the isotope 1H2 that is intact within the isotope 2He3 and the negatron. Then subtract this total from the isotope 2He3. The mass that remains must be the positon.

Mass of 1H2 in 2He3	2.01411 amu's
Mass of negatron in 2He3	0.00055 amu
Total known mass in 2He3	2.01466 amu's

Mass of 2He3	3.01603 amu's
Total known mass of 2He3	2.01466 amu's
Mass of positon in 2He3	1.00137 amu's

Now that the mass of the positon is known, the mass lost by the second neutron in the isotope 1H3 can be found by subtracting the mass of the positon from it.

Mass of second neutron in 1H3	1.00194 amu's
Mass of positon in 2He3	1.00137 amu's
Mass lost by second neutron in1H3	0.00057 amu

The lost mass by the second neutron includes the mass of the negatron that goes into an orbital and the mass that split it from the second neutron. There is a 0.00100 amu in the second neutron and it stays as part of the neutron that now becomes a positon because no mass needs to be transferred. This means that if the second neutron is to become a positon, it must keep the 0.00100 amu and lose the 0.00055 amu as a negatron. Because the negatron is part of the isotope 2He3, its mass is not lost and it is not part of the 0.00002 amu that is really lost. By subtracting the mass of the negatron from the mass lost by the second neutron in 1H3, the remainder will be the real mass lost.

Mass lost by second neutron in 1H3	0.00057 amu
Mass of negatron in 2He3	0.00055 amu
Mass lost releasing negatron from second neutron in 1H3	0.00002 amu

The 0.00002 amu released the negatron from the isotope 1H3's second neutron because it is the only lost mass that could do this in this step.

Now all of the masses of the isotope 2He3 are known. By adding these parts, the mass of the isotope 2He3 will be found.

Mass of 1H2 in 2He3	2.01411 amu's
Mass of negatron in 2He3	0.00055 amu
Mass of positon in 2He3	1.00137 amu's
Mass of 2He3	3.01603 amu's

This isotope is not the final isotope of 2He3. The final isotope of 2He3 must have a proton and a positon with each having 1.00728 amu's. This means the single neutron will have 1.00037 amu's and the electron and negatron will each have 0.00055 amu for a total mass of 3.01603 amu's. (Sub-step 11-1-11b)

The isotope 1H3 has the isotope 1H2 found within it and the isotope 1H3 is radioactive. Because the isotope 1H3 is radioactive, it must change into a different isotope within a given time. In this step, this means that the single neutron found the isotope 2He3 in Sub-step 11-1-11a must change because the proton must have its mass remain as 1.00728 amu's.

In Sub-step 11-1-11b, the single neutron in the first isotope of 2He3 has its neutron with 1.00628 amu's and it transfers 0.00591 amu to the positon. The positon has 1.00137 amu's and when it receives the 0.00591 amu, it becomes a positon with 1.00728 amu's and the neutron that transferred this mass becomes a neutron with 1.00037 amu's.

Mass of neutron in first 2He3	1.00628 amu's
Mass transferred to positon in first 2He3	0.00591 amu
Mass of neutron in second 2He3	1.00037 amu's

Mass of positon in first 2He3	1.00137 amu's
Mass transferred from neutron in first 2He3	0.00591 amu
Mass of positon in second 2He3	1.00728 amu's

The electron and proton have a total mass of 1.00783 amu's which is not an isotope of 1H1. Now all of the masses of the stable isotope of 2He3 are known. Add them to find the mass of the isotope 2He3 that is stable.

Mass of electron and proton stable 2He3	1.00783 amu's
Mass of negatron in stable 2He3	0.00055 amu
Mass of positon in stable 2He3	1.00728 amu's
Mass of neutron in stable 2He3	1.00037 amu's
Mass of stable 2He3	3.01603 amu's

Sub-step 10-1-10

2He3

1H3 = 2He3, #2 alpha ray, transfer 0.00100 amu, 0.00534 amu, neg. orbit

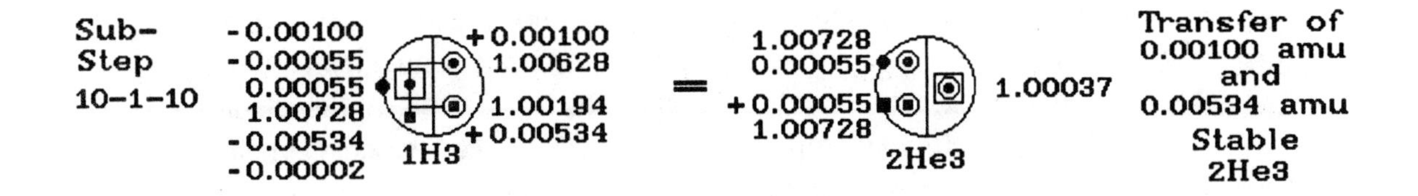

The isotope 1H3 with 3.01605 amu's has one electron with 0.00055 amu and one proton with 1.00728 amu's plus two neutrons. The first neutron has 1.00628 amu's and the second neutron has 1.00194 amu's.

The isotope 2He3 with 3.01603 amu's has one electron with 0.00055 amu, one negatron with 0.00055 amu, one proton with 1.00728 amu's, one positon with 1.00728 amu's and one neutron with 1.00037 amu's.

The total mass lost is 0.00002 amu and it is emitted as a #2 alpha ray which has an external magnetic field.

The isotope 1H3 with 3.01605 amu's has its electron drawn as a dot on the big circle in the isotopes 1H3 and 2He3. The proton with 1.00728 amu's is drawn as a dot with a small circle drawn around it to show that 0.00100 amu has been transferred to the first neutron causing the negatron with 0.00055 amu to go into an orbital with the electron already there. At the time the 0.00100 amu has been transferred and the 0.00055 amu of the negatron has been released, the proton loses 0.00002 amu to create the isotope 2He3. Note that the electron that is with the proton is not released. There is a line drawn from the proton to the first neutron to show that the 0.00100 amu has been transferred. There is a small square drawn below the proton with a line connecting them. This small square is the negatron released from the proton that is drawn as a square on the big circle in the isotope 2He3.

There is a square drawn around the dot that is the proton and a small circle which shows that the proton will come from the first neutron. This square shows that 0.00002 amu has been lost as a single unit by the proton in the isotope 1H3. This square drawn around the proton drawn as a dot in the isotope 1H3 and as the single neutron as a dot with a small circle in the isotope 2He3 does not include any mass that has been transferred.

There is a line drawn from the proton to the second neutron which has 1.00194 amu's to show that the proton has transferred 0.00534 amu to the second neutron making this second neutron into the positon with 1.00728 amu's. The released negatron with 0.00055 amu belongs to the positon.

When this transferring of mass by the proton in the isotope 1H3 is complete and the 0.00002 amu has been lost, the remaining mass becomes the single neutron in the isotope 2He3. The single neutron comes from the proton in the isotope 1H3 which is the reason the dot is drawn in the center of the small circle. The square drawn around them represents the 0.00002 amu because the reader may want to know it caused the single neutron to come into existence.

Now look at what happens to the first and second neutrons. The first neutron with 1.00628 amu's receives the 0.00100 amu and becomes the proton with 1.00728 amu's as soon as the original proton releases the negatron. Since the electron was not released when the 0.00100 amu was transferred, the electron then belongs to the proton that came from the first neutron. The second neutron with 1.00194 amu's then receives 0.00534 amu as a single unit to become the positon with 1.00728 amu's and the negatron that was released from the original proton now belongs to the positon. At the same time this is happening, the original proton loses 0.00002 amu as a #2 alpha ray creating the isotope 2He3. The remaining mass of the original proton is now 1.00037 amu's. The original proton is now the single neutron in the isotope 2He3. Note that there now is an electron and a negatron in orbit around the nucleus of the isotope 2He3.

If the proton in the isotope 1H3 has 1.00728 amu's and it transfers 0.00100 amu to the first neutron, then this first neutron would have 1.00728 amu's to become the proton again. The proton that transferred the 0.00100 amu must then release a negatron and thereby lose 0.00055 amu. The negatron goes into an orbital with the other electron that is already there. This proton has now transferred and released 0.00155 amu. This proton is no longer the original proton nor is it the positon in the isotope 2He3. It now has 1.00573 amu's. It now transfers 0.00534 amu to the second neutron that has 1.00194 amu's to make this second neutron into the positon with 1.00728 amu's and at the same time lose 0.00002 amu to create the isotope 2He3. The remaining mass of the original neutron now becomes the single neutron with 1.00037 amu's in the isotope 2He3.

This means that the first neutron with 1.00628 amu's vanishes and a proton appears with 1.00728 amu's and the remaining mass of the original proton and the second neutron trade places with the original proton now becoming the single neutron with 1.00037 amu's in the isotope 2He3 and the second neutron becoming the positon with 1.00728 amu's.

In the above steps the 0.00002 amu that must be lost when the isotope 1H3 changes into the isotope 2He3. By adding the masses of the parts before and after this mass has been transferred and no mass lost, then the isotope 2He3 would have the mass of the isotope 1H3 which is 3.01605 amu's. Since this can not happen, then the 0.00002 amu must be lost when the negatron is released from the proton and the 0.00534 amu is transferred to the second neutron.

The isotope 2He3 now has an electron and negatron with each having 0.00055 amu, a proton and positon with each having the mass of 1.00728 amu's plus one neutron with 1.00037 amu's. Their total mass equals the mass of 3.01603 amu's which is the mass of the isotope 2He3. All of the masses that are transferred, released and lost occur at the same time. There is no sequence of events.

To find the mass that is lost creating the isotope 2He3, subtract the isotope 2He3 from the isotope 1H3.

Mass of 1H3	3.01605 amu's
Mass of 2He3	3.01603 amu's
Mass lost creating 2He3	0.00002 amu

The proton with 1.00728 amu's transfers 0.00100 amu to the first neutron which has 1.00628 amu's making this first neutron into a proton with 1.00728 amu's. At the same time this transfer of 0.00100 amu takes place, the negatron is released from the proton and the 0.00002 amu is lost. This is a total lost mass of 0.00157 amu from the proton in the isotope 1H3. Also at this same time, the 0.00534 amu is transferred to the second neutron making the second neutron into the positon with 1.00728 amu's in the isotope 2He3. Adding this transferred mass to the 0.00157 amu means that 0.00691 amu has been transferred, released and lost by the proton in the isotope 1H3. By subtracting this 0.00691 amu from the proton, the remaining mass is the single neutron found in the isotope 2He3.

Mass of first neutron in 1H3	1.00628 amu's
Mass transferred to it from proton in 1H3	0.00100 amu
Mass of proton in 2He3	1.00728 amu's

Mass of second neutron in 1H3	1.00194 amu's
Mass transferred to it from proton in 1H3	0.00534 amu
Mass of positon in 2He3	1.00728 amu's

Mass transferred to first neutron from proton in 1H3	0.00100 amu
Mass of negatron released from proton in 1H3	0.00055 amu
Mass transferred to second neutron	0.00534 amu
Mass really lost creating 2He3	0.00002 amu
Total mass lost by proton in 1H3	0.00691 amu

Mass of proton in 1H3	1.00728 amu's
Total mass lost by proton in 1H3	0.00691 amu
Mass of neutron in 2He3	1.00037 amu's

The electron that is with the original proton stays with the new proton that comes from the first neutron in the isotope 1H3. The way this happens is understood when the original neutron transferred 0.00100 amu to the first neutron, the first neutron becomes the new proton. This means that the electron is released and regained or to put another way, the electron never is released because the transfer of mass from the original proton to the first neutron is an internal transfer and therefore no mass has been lost and the electron is not released. This is shown by drawing the electron as a dot on the big circle in the isotopes 1H3 and 2He3.

The negatron comes from within the proton because the proton transfers 0.00100 amu to the first neutron in the isotope 1H3 and at the same time loses 0.00002 amu.

The single unit of 0.00534 amu is composed of two sub-units, one sub-unit of 0.00434 amu and another sub-unit of 0.00100 amu. Because it is a single unit of 0.00534 amu, the negatron can not be released by it. Also the lost mass of 0.00002 amu is lost because the isotope of 1H3 cannot stay an isotope of 1H3. The 0.00100 amu is transferred because the creation of the isotope of 1H3 became radioactive when the neutron 1.00194 amu's came into being and not because the 0.00534 amu is transferred. The transferring and losing of mass enables the any isotope to have protons and positons to have their maximum mass of 1.00728 amu's thus making the isotope 2He3 stable.

Now all of the masses that make up the stable isotope of 2He3 are known.

Mass of electron and negatron in 2He3	0.00110 amu
Mass of proton in 2He3	1.00728 amu's
Mass of positon in 2He3	1.00728 amu's
Mass of neutron in 2He3	1.00037 amu's
Mass of 2He3	3.01603 amu's

Sub-step 9-1-9

2He3

1H3 = 2He3, #2 alpha ray, transfer 0.00434 amu, 2(0.00100 amu), neg. orbit

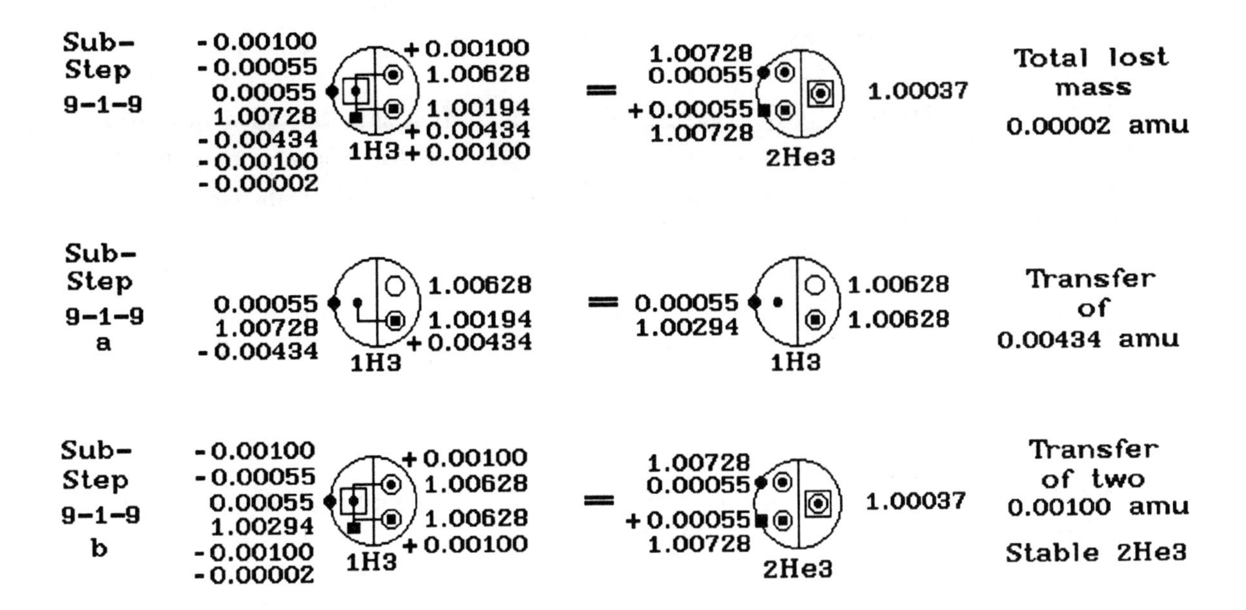

The isotope 1H3 with 3.01605 amu's has one electron with 0.00055 amu and one proton with 1.00728 amu's plus two neutrons. The first neutron has 1.00628 amu's and the second neutron has 1.00194 amu's.

The isotope 2He3 with 3.01603 amu's has one electron with 0.00055 amu, one negatron with 0.00055 amu, one proton with 1.00728 amu's, one positon with 1.00728 amu's and one neutron with 1.00037 amu's.

The total mass lost is 0.00002 amu and it is emitted as a #2 alpha ray which has an external magnetic field.

In sub-step 10-1-10, there is a mass of 0.00534 amu transferred to the second neutron in the isotope 1H3. In this step, the 0.00534 amu is broken down into two units, one unit of 0.00100 amu and another unit of 0.00434 amu.

Before the first neutron in the isotope 1H3 can receive 0.00100 amu, the proton transfers 0.00434 amu to the second neutron which has 1.00194 amu's.

If the proton transfers 0.00434 amu to the second neutron, the second neutron with 1.00194 amu's stays a neutron with 1.00628 amu's and the proton stays a proton with 1.00294 amu's.

Now both neutrons in the isotope 1H3 have 1.00628 amu's and they become the proton and positon in the isotope 2He3. To do this, they each must regain 0.00100 amu. The proton in the isotope 1H3 now has 1.00294 amu's and it can transfer 0.00100 amu to each neutron making them into a proton and positon.

This is an internal transfer of mass and the proton still has its electron which is in an orbital around the nucleus.

The second neutron in the isotope 1H3 regains 0.00100 amu from the proton to become the positon in the isotope 2He3. This positon needs a negatron to go into an orbital with the electron which means that the proton must have a negatron within itself which it now releases. Both the electron and negatron have 0.00055 amu. When the negatron is released, the proton has lost two units of 0.00100 amu and a negatron. The isotope 1H3 loses 0.00002 amu at this time to become the isotope 2He3.

The first neutron in the isotope 1H3 is drawn as a small circle with a dot drawn in its middle. The proton in the isotope 2He3 is drawn the same way to show that the first neutron in the isotope 1H3 becomes the proton in the isotope 2He3.

The second neutron in the isotope 1H3 is drawn as a small circle with a square in its middle to show that this neutron becomes the positon in the isotope 2He3. The second neutron in the isotope 1H3 and the positon in the isotope 2He3 are drawn the same way showing that the second neutron in the isotope 1H3 became the positon in the isotope 2He3.

There is a line drawn from the proton in the isotope 1H3 to the second neutron to show that 0.00434 amu has been transferred (Step 9-1-9a). This line also shows that 0.00100 amu has been transferred next.

After the 0.00434 amu has been transferred, there are two units of 0.00100 amu transferred from the proton to each of the two neutrons. This is shown by drawing a line from the proton to each neutron. There already is a line drawn from the proton to the second neutron.

In Sub-step 11-1-11b, there is a square drawn around the proton to show that the proton in the isotope 1H3 has lost 0.00002 amu at the time the two 0.00100 amu are transferred. This square around the proton does not include any mass that has been transferred.

In the isotope 2He3 there is a dot drawn in the small circle that shows the proton in the isotope 2He3 came from the first neutron in the isotope 1H3. There is a square drawn within a small circle that show the positon came from the second neutron. (Sub-step 9-1-9b)

The original proton in the isotope 1H3 is drawn as a dot with a square around it and the single neutron in the isotope 2He3 is drawn as a dot within a small circle and it has a square drawn around both of them in the isotope 2He3. It is drawn this way to show that after the proton in the isotope 1H3 has transferred, released and lost its masses, the remaining mass is the single neutron in the isotope 2He3.

The electron that is with the original proton stays with the new proton that comes from the first neutron in the isotope 1H3. The way this happens is understood when the original neutron transferred 0.00100 amu to the first neutron, the first neutron becomes the new proton. This means that the electron is released and regained or to put it another way, the electron never is released because the transfer of mass from the original proton to the first neutron is an internal transfer and therefore no mass has been lost and the electron is not released. This is shown by drawing the electron as a dot on the big circle in the isotopes 1H3 and 2He3.

The negatron comes from within the original proton because the original proton transfers 0.00100 amu to the second neutron in the isotope 1H3 and at the same time loses 0.00002 amu. The other 0.00100 amu does not have the 0.00002 amu lost with it and it therefore can not release a negatron but it can keep the electron that is with the original proton. The 0.00100 amu is transferred at the same time the negatron is released and the 0.00002 amu has been lost.

When the proton in the isotope 1H3 transferred 0.00434 amu to the second neutron, this proton became a proton with 1.00294 amu's in the isotope 1H3. This makes a new isotope of 2He3 that is unstable.

The transferring of the two 0.00100 amu makes a proton and a positon in the isotope 2He3 with the positon needing a negatron. The negatron is released from the new proton with 1.00294 amu's to go into an orbital with the electron. It is drawn as a square below the proton with a line connecting them to show that the negatron came from the proton in the isotope 1H3. It is also drawn on the big circle in the isotope 2He3 as a square. The remaining mass of the new proton now is 1.00039 amu's which has 0.00002 amu more than the single neutron in the isotope 2He3 can have and this 0.00002 amu is then emitted as a #2 alpha ray. When this happens, the isotope 2He3 is created with 1.01603 amu's. The single neutron in the isotope 2He3 now has 1.00037 amu's. This make a stable isotope of 2He3.

To find the mass that is lost creating the isotope 2He3, subtract the isotope 2He3 from the isotope 1H3.

Mass of 1H3	3.01605 amu's
Mass of 2He3	3.01603 amu's
Mass lost creating 2He3	0.00002 amu

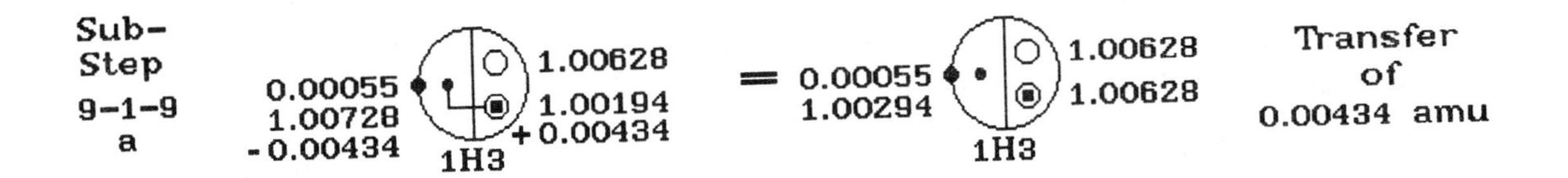

The proton with 1.00728 amu's transfers 0.00434 amu to the second neutron. The proton stays a proton and the second neutron stays a neutron. The drawing that shows this is sub-step 9-1-9a.

Mass of proton in 1H3	1.00728 amu's
Mass transferred from proton in 1H3	0.00434 amu
Mass of new proton in 1H3	1.00294 amu's

Mass of second neutron in 1H3	1.00194 amu's
Mass transferred to it from proton in 1H3	0.00434 amu
Mass of new second neutron in 1H3	1.00628 amu's

Now there is a new isotope of 1H3 with an electron having 0.00055 amu, a proton with 1.00294 amu's and two neutrons with 1.00628 amu's for a total mass of 3.01605 amu's.

Mass of electron in new 1H3	0.00055 amu
Mass of new proton in new 1H3	1.00294 amu's
Mass of first neutron in new 1H3	1.00628 amu's
Mass of new second neutron in new 1H3	1.00628 amu's
Mass of new 1H3	3.01605 amu's

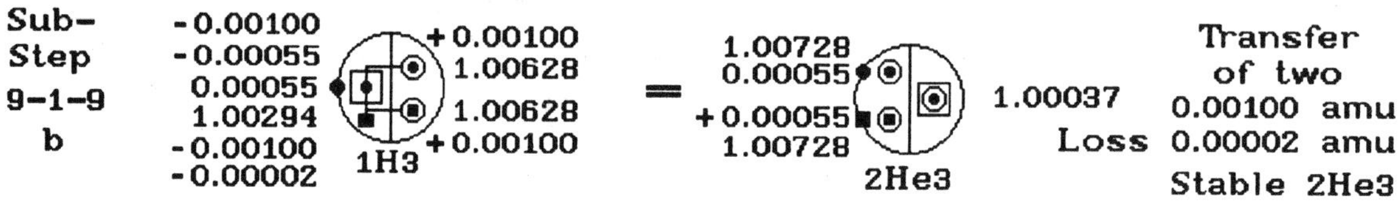

Sub-Step 9-1-9 b

-0.00100
-0.00055
0.00055
1.00294
-0.00100
-0.00002
+0.00100
1.00628
1.00628
+0.00100
1H3

=

1.00728
0.00055
+0.00055
1.00728
2He3

1.00037

Transfer of two 0.00100 amu
Loss 0.00002 amu
Stable 2He3

The new proton with 1.00294 amu's then transfers 0.00100 amu to the first neutron to make this neutron the proton in the isotope 2He3 with 1.00728 amu's. The remaining mass of the proton also transfers 0.00100 amu to the second neutron and releases a negatron from within itself. When the 0.00100 amu is transferred, the negatron is released to go into an orbital with the electron already there. At the time the 0.00100 amu is transferred to the second neutron, 0.00002 amu is lost. The remaining mass of the proton is 1.00037 amu's that is the single neutron in the isotope 2He3. Sub-step 9-1-9b drawing shows this.

Mass of new proton in 1H3	1.00294 amu's
Mass transferred from proton to first neutron in 1H3	0.00100 amu
Mass remaining of new proton in 1H3	1.00194 amu's

Mass remaining of new proton in 1H3	1.00194 amu's
Mass transferred to second neutron in 1H3	0.00100 amu
Mass remaining of new proton in 2He3	1.00094 amu's

Mass remaining of new proton in the isotope 1H3 also lost 0.00002 amu when the negatron was released. Subtract this mass from the remaining mass of the proton to find the remaining mass of the new proton that now becomes the single neutron in the isotope of 2He3.

Remaining mass of proton in 1H3	1.00094 amu's
Mass of negatron released by remaining mass of proton in 1H3	0.00055 amu
Remaining mass of proton in 1H3 changing into neutron in 2He3	1.00039 amu's

Remaining mass of proton in 1H3 changing into neutron in 2He3	1.00039 amu's
Mass lost creating 2He3	0.00002 amu
Mass of single neutron in 2He3	1.00037 amu's

All of these masses that are transferred, released and lost are transferred, released and lost in a sequence from the proton found in the isotope 1H3 causing the first neutron to become the new proton, the second neutron becoming the positon and the proton becoming the single neutron in the isotope 2He3.

The proton in the isotope 1H3 transferred 0.00100 amu to the first neutron and it also transferred 0.00100 amu and 0.00434 amu to the second neutron and lost 0.00002 amu with the release of a negatron which has 0.00055 amu. The 0.00100 amu and 0.00434 amu are each lost as single units of mass. Each transferred mass of 0.00100 amu and the 0.00434 amu are transferred internally within the isotope 1H3 and therefore no ray is emitted.

When the masses of 0.00434 amu and 0.00100 amu are transferred, the negatron with 0.00055 amu is released and 0.00002 amu is lost from the original proton to the second neutron plus the 0.00100 amu transferred to the first neutron are all added together, they equal 0.00691 amu. By subtracting this total mass from the original proton in the isotope 1H3, the mass of the single neutron in the isotope 2He3 will be found.

Mass transferred to first neutron	0.00100 amu
Mass transferred to second neutron	0.00434 amu
Mass transferred to second neutron	0.00100 amu
Mass of negatron released from proton	0.00055 amu
Mass really lost creating 2He3	0.00002 amu
Total mass lost by proton in 1H3	0.00691 amu

Mass of proton in 1H3	1.00728 amu's
Total mass lost by proton in 1H3	0.00691 amu
Mass of neutron in 2He3	1.00037 amu's

Now all of the masses that make up the stable isotope of 2He3 are known.

Mass of electron and negatron in 2He3	0.00110 amu
Mass of proton in 2He3	1.00728 amu's
Mass of positon in 2He3	1.00728 amu's
Mass of neutron in 2He3	1.00037 amu's
Stable mass of 2He3	3.01603 amu's

Sub-step 8-1-8

2He3

1H3 = 2He3, #2 alpha ray, transfer 0.00100 amu, 0.00317 amu, 0.00217 amu, negatron orbit

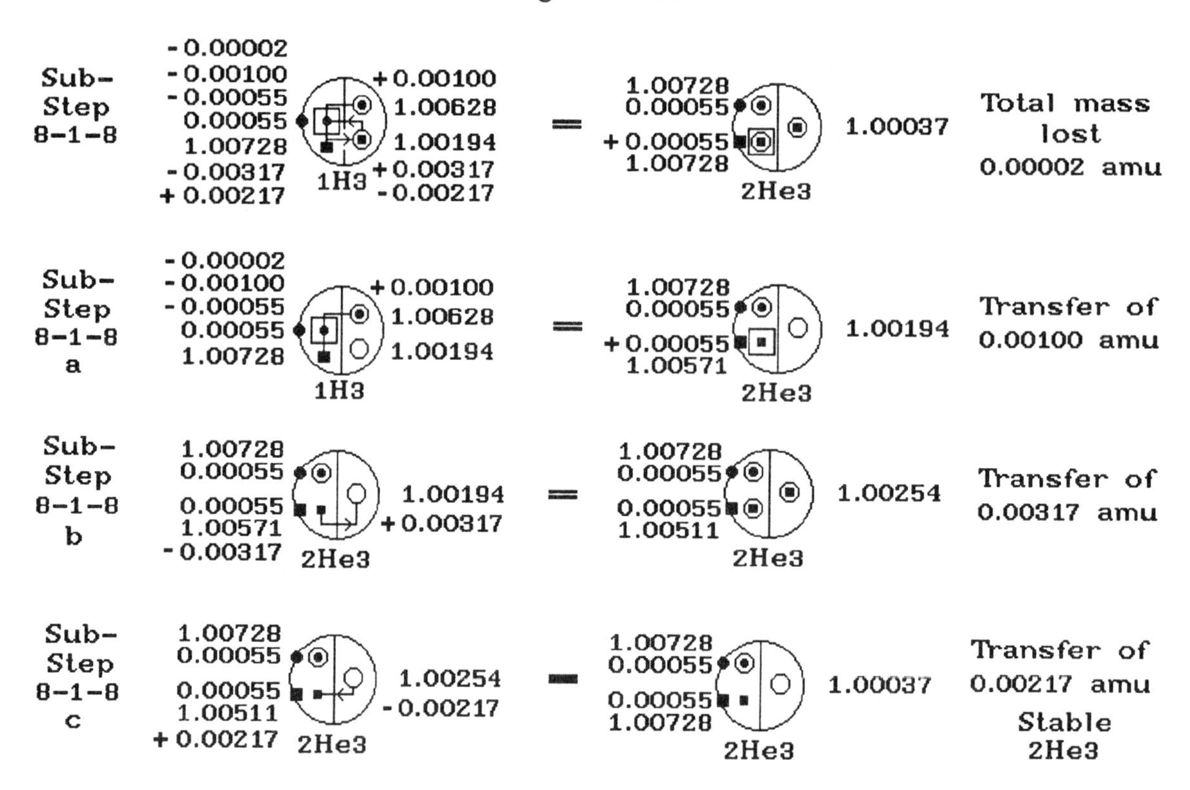

The isotope 1H3 with 3.01605 amu's has one electron with 0.00055 amu and one proton with 1.00728 amu's plus two neutrons. The first neutron has 1.00628 amu's and the second one has 1.00194 amu's.

The isotope 2He3 with 3.01603 amu's has one electron with 0.00055 amu, one negatron with 0.00055 amu, one proton with 1.00728 amu's, one positon with 1.00728 amu's and one neutron with 1.00037 amu's.

The total mass lost is 0.00002 amu and it is emitted as a #2 alpha ray which has an external magnetic field.

The isotope 1H3 has its electron drawn as a dot on the big circle. The proton is drawn as a dot on the left side of the vertical line within the big circle. The proton has a square drawn around the dot.

The proton in the isotope 1H3 has 1.00728 amu's and it transfers 0.00100 amu to the first neutron. This first neutron would have 1.00728 amu's to become the proton again but before the first neutron can replace the proton, the proton will lose 0.00002 amu as a #2 ray and release its negatron. The negatron has 0.00055 amu and it goes into an orbit with the electron already there. These sets of events will make the proton have 1.00571 amu's which now becomes the positon in the isotope 2He3. The electron that was with the proton before it lost the 0.00100 amu stays an electron with the proton that was the first neutron. The second neutron at this point in time stays as it is and becomes the single neutron in the newly created isotope of 2He3 with 1.00194 amu's. The isotope 2He3 now has a proton with 1.00728 amu's and a positon with 1.00571 amu's, an electron and a negatron with 0.00055 amu each plus one neutron with 1.00194 amu's. Their total mass equals the mass of 3.01603 amu's which is the mass of the isotope 2He3. The drawing of sub-step 8-1-8a shows how this is done.

The positon in the isotope 2He3 now has 1.00571 amu's and it transfers 0.00317 amu to the single neutron with 1.00194 amu's. The positon now has 1.00511 amu's and the single neutron has 1.00254 amu's. The positon and the single neutron trade places. These masses again make an isotope of 2He3. This is shown in the drawing of sub-step 8-1-8b.

Now the positon has 1.00511 amu's and it receives from the single neutron with 1.00254 amu's, the mass that will make it into a positon with 1.00728 amu's. This means that the single neutron transfers 0.00217 amu to the positon with 1.00511 amu's. The single neutron now has 1.00037 amu's. The first isotope of 2He3 in the drawing of sub-step 8-1-8c has become the second isotope of 2He3 with an electron and negatron having a total of 0.00110 amu, a proton and a positon each having 1.00728 amu's and a neutron that has 1.00037 amu's. This makes a stable isotope of 3.01603 amu's.

The electron, in the isotopes 1H3 and 2He3, is drawn as a dot on the big circle and the negatron is drawn as a square below the dot on the big circle of the isotope 2He3. There is a square drawn below the proton in the isotope 1H3 that has a line connecting them. This square is the negatron that dropped out of the proton and the line shows this. This is shown in the drawings of sub-step 8-1-8 and sub-step 8-1-8a.

The first drawing of sub-step 8-1-8 shows that the proton in the isotope 1H3 has transferred 0.00100 amu to the first neutron in the isotope 1H3. The proton is drawn as a dot on the left side of the vertical line. There is a line drawn from the proton to the first neutron showing that the 0.00100 amu has been transferred from the proton to the first neutron. There is a square drawn around the proton showing that the 0.00002 amu has been lost as #2 alpha ray. There is another line drawn from the proton to the second neutron to show that 0.00317 amu has been transferred to the second neutron. This line has an arrow drawn on it pointing toward the second neutron.

The first neutron in the isotope 1H3 is drawn as a small circle with a dot in its middle to show that it has regained 0.00100 amu in the isotope 1H3 and because this neutron becomes the proton in the isotope 2He3, it is drawn the same way in the isotope 2He3. The plus in front of the 0.00100 amu in the isotope 1H3 means that the first neutron has regained this mass.

The second neutron in the isotope 1H3 is drawn the same way except the dot is now drawn as a square in the middle of the small circle to show that it is the positon in the isotope 2He3 and that the positon comes from the isotope 1H3's second neutron.

In the first drawing for the isotope 2He3, the electron is drawn as a dot with the negatron drawn below it as a square on the big circle.

The proton is drawn as a dot with a small circle around it. The small circle indicates that it was the first neutron in the isotope 1H3.

The positon is drawn below the proton as a small square with a small circle around the small square to show that at one point in time, the positon was a neutron. There is a square drawn around the small square that is the positon and small circle to show that 0.00002 amu has been lost by the proton. Remember that when the 0.00002 amu was lost, it was part of a proton and not a positon.

The single neutron is drawn as a small circle with a square in its middle to show that at one point in time the single neutron in the isotope 2He3 changes into a positon.

The drawings of sub-steps 8-1-8a, 8-1-8b and 8-1-8c are drawn to show how to understand the first drawing of sub-step 8-1-8.

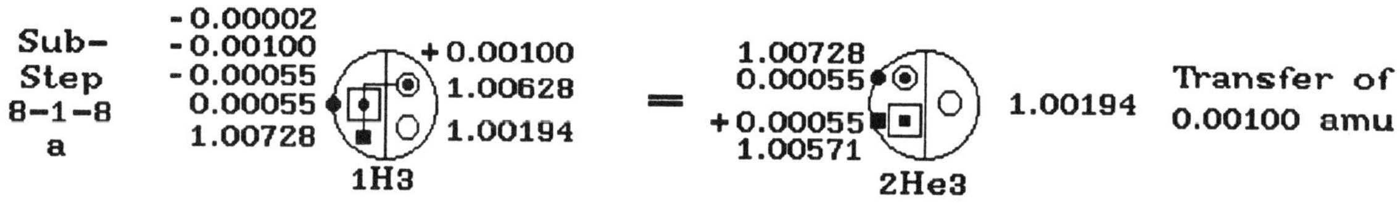

Sub-Step 8-1-8 a

-0.00002
-0.00100
-0.00055
0.00055
1.00728
+0.00100
1.00628
1.00194
1H3

=

1.00728
0.00055
+0.00055
1.00571
1.00194
2He3

Transfer of 0.00100 amu

The drawing of sub-step 8-1-8a has the isotope 1H3's electron drawn as a dot on the big circle and it is drawn this same way in the isotope 2He3 to show that the electron in the isotope 1H3 stays an electron in the isotope 2He3.

The drawing of sub-step 8-1-8a shows the loss of 0.00002 amu as a square drawn around the dot that is the proton in the isotope 1H3. The loss of 0.00002 amu causes a negatron to be released from within the proton. The negatron goes into an orbital that already has an electron in it. There is a line drawn between the proton and the negatron to show that the negatron came from the proton.

This proton has a line drawn from it to the first neutron to show that 0.00100 amu has been transferred to the first neutron. The first neutron is drawn as a small circle with a dot in its middle. Since the first neutron has 1.00628 amu's and it receives 0.00100 amu from the proton, it becomes the proton in the isotope 2He3. This is why the dot is in the middle of the first neutron and it is drawn this same way in the isotope 2He3's proton.

The isotope 2He3 has its electron drawn above the negatron on the big circle. The negatron is drawn as a square.

The proton is drawn as a dot with a small circle around it. It is drawn this way to show that it came from the first neutron in the isotope 1H3 because they are both drawn alike.

The positon is drawn as a small square with another square drawn around it. This square around the small square is drawn in the isotope 2He3 to indicate which part of the isotope is the result of the loss of 0.0002 amu. The reader is to understand that it is at this point in time that the proton in the isotope 1H3 emitted a #2 alpha ray as it becomes a positon in the isotope 2He3. In the remaining steps, this square around the small square is not drawn because the transferred masses are transferred without emitting a ray.

The single neutron with 1.00194 amu's is drawn as a small circle with nothing in its middle. The fact that there is nothing in the middle shows that the second neutron in the isotope 1H3 is used as is with no loss of mass in the isotope 2He3.

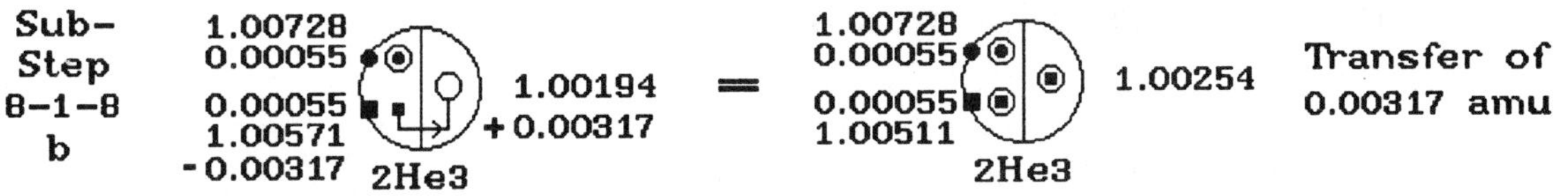

Sub-Step 8-1-8 b

1.00728
0.00055
0.00055
1.00571
-0.00317
2He3
1.00194
+0.00317
=
1.00728
0.00055
0.00055
1.00511
2He3
1.00254
Transfer of 0.00317 amu

The drawings of sub-step 8-1-8b is drawn because in sub-step 8-1-8a has an isotope of 2He3 that is not the final stable step. The isotope 2He3 that is on the left side of the equal sign will be called the first isotope of 2He3 and the other isotope of 2He3 will be call the second isotope of 2He3.

In the drawing of sub-step 8-1-8b, the first isotope of 2He3 is drawn like the isotope of 2He3 in sub-step 8-1-8a with the same masses for it parts. There is no square drawn around the positon. There is a line drawn with an arrow on it pointing to the single neutron from the positon to show that 0.00317 amu has been transferred to the single neutron in the second isotope of 2He3. The positon with 1.00571 amu's transfers 0.00317 amu to the single neutron with 1.00194 amu's. When it does this, the positon's remaining mass is 1.00254 amu's. This changes the positon into the single neutron in the isotope 2He3. The single neutron with 1.00194 amu's receives this transferred mass and becomes a positon with 1.00511 amu's. This means that the positon and neutron trade places. This is shown in the second isotope 2He3 by having the positon drawn as a small square with a small circle drawn around it and the single neutron is drawn the same way.

All of these masses will make the isotope 2He3 as shown in the second isotope of 2He3. This second isotope of 2He3 is still not the final stable isotope of 2He3.

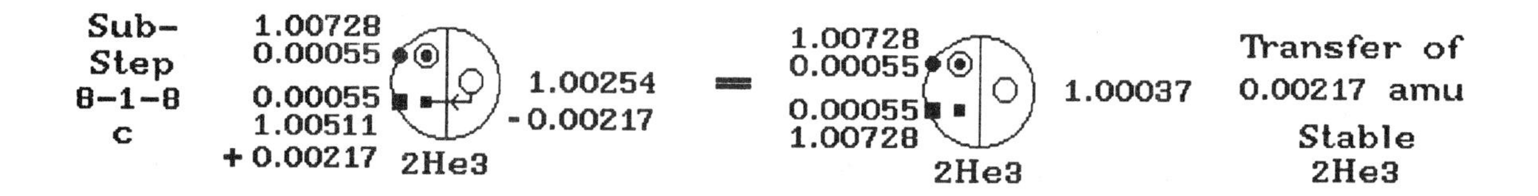

Sub-step 8-1-8c has its first isotope of 2He3 drawn the same way as the second isotope of 2He3 found in sub-step 8-1-8b. The first isotope of 2He3 has the same masses as found in sub-step 8-1-8b. Now there is a line with an arrow on it pointing from the single neutron to the positon in the first isotope of 2He3. The single neutron in the first isotope of 2He3 transfers 0.00217 amu to the positon. This makes the single neutron in the first isotope of 2He3 change its mass from 1.00254 amu's to 1.00037 amu's to become the single neutron in the second isotope of 2He3. The positon with 1.00511 amu's in the first isotope of 2He3 changes its mass to become the positon in the second isotope of 2He3 with 1.00728 amu's.

This makes the electron drawn as a dot on the big circle with the negatron drawn below it. They both have 0.00055 amu. The proton is drawn as a dot with a small circle around it with 1.00728 amu's. The positon is drawn as a small square only with 1.00728 amu's. The single neutron is drawn as a small circle to the right of the vertical line with nothing in its middle. It has 1.00037 amu's. Because the proton and positon now have their maximum mass of 1.00728 amu's, this is the final stable isotope of 2He3.

It should be pointed out that all of these internal transfers of masses are needed because the transferred mass of 0.00317 amu carried with it a sub-unit of 0.00100 amu to the isotope of 2He3 and left it there when the neutron then transferred 0.00217 amu to the positon to create a stable isotope of 2He3.

I believe the above happens when a molecule needs a stable isotope of 2He3 and it has an isotope of 1H3 as shown in sub-step 8-1-8a. When this happens, the molecule needs an isotope of 2He3 with a proton having 1.00571 amu's. The molecule then needs an isotope of 2He3 with a proton having 1.00511 amu's. After the molecule receives this isotope of 2He3, it needs changed again. This last change makes the molecule stable as shown in the last sub-step of 8-1-8c. Thus the molecule ends up with a stable isotope of 2He3 and the changes it has been going through stops.

In the above, the needs of the molecule are met by using positons with different masses. It could need the positon with 1.00571 amu's and then need a stable isotope of 2He3 but can not find one so it causes the isotope of 2He3 with the positon of 1.00571 amu's to change to a positon with 1.00511 amu's, causing it to release its sub-unit of 0.00100 amu and then the isotope of 2He3 with 1.00511 amu's can transfer 0.00217 amu to the positon and the molecule can then use this stable isotope that results with a proton and positon having 1.00728 amu's each.

The sub-unit of 0.00100 amu does not leave the isotope of 2He3. It remains as part of the single neutron in the isotope 2He3 within the molecule.

The molecule can tell which positon it wants by the way the proton controls the magnetic field of the electron and the positon controls the negatron's magnetic field. The balancing of the isotope is done by the way the single neutron is used mechanically as the proton and positon spin on their axis within the isotope of 2He3. I suggest that the reader look at the DNA molecule.

To find the mass that is lost creating the isotope 2He3, subtract the isotope 2He3 from the isotope 1H3.

Mass of 1H3	3.01605 amu's
Mass of 2He3	3.01603 amu's
Mass lost creating 2He3	0.00002 amu

The first neutron has 0.00100 amu transferred to it as a single unit with 0.00055 amu released as a negatron. There is 0.00002 amu lost creating the isotope 2He3. Now all of these masses can be added together to find the total mass transferred, released and lost by the proton in the isotope 1H3.

Mass transferred to first neutron in 1H3	0.00100 amu
Mass of negatron released by proton in 1H3	0.00055 amu
Mass lost creating 2He3	0.00002 amu
Total mass lost by proton creating 2He3	0.00157 amu

Mass of proton in 1H3	1.00728 amu's
Mass transferred, released, lost by proton in 1H3	0.00157 amu
Mass of positon in 2He3	1.00571 amu's

Mass of first neutron in 1H3	1.00628 amu's
Mass transferred to it from proton in 1H3	0.00100 amu
Mass of proton in 2He3	1.00728 amu's

The second neutron in the isotope 1H3 is used as is in the isotope 2He3 in Sub-step 8-1-8a so the isotope 2He3 now has all of its parts.

Mass of electron and negatron in 2He3	0.00110 amu
Mass of proton in 2He3	1.00728 amu's
Mass of positon in 2He3	1.00571 amu's
Mass of neutron in 2He3	1.00194 amu's
Mass of 2He3	3.01603 amu's

Because the positon has 1.00571 amu's, this isotope of 2He3 is unstable. This means the positon must have 1.00728 amu's to make the isotope 2He3 stable. It can not do this in this step unless the single neutron has 1.00037 amu's. This single neutron in the isotope 2He3 has 1.00194 amu's. Before it can become the single neutron with 1.00037 amu's in the isotope 2He3, it must regain 0.00317 amu from the positon that has 1.00571 amu's. When the neutron regains this mass, it stays a neutron with 1.00254 amu. The remaining mass of the positon is 1.00511 amu's.

Mass of positon in 2He3	1.00571 amu's
Mass transferred to neutron in 2He3	0.00317 amu
Mass of positon in 2He3	1.00511 amu's

Mass of neutron in 2He3	1.00194 amu's
Mass transferred from positon in 2He3	0.00317 amu's
Mass of neutron in 2He3	1.00254 amu's

It will no be noted that the proton in Sub-step 8-1-8 has lost 0.00474 amu before it becomes part of a stable isotope of 2He3 as the single neutron. Sub-step 8-1-8b shows this.

Mass transferred to first neutron	0.00100 amu
Mass transferred to the second neutron	0.00317 amu
Mass of negatron released from proton	0.00055 amu
Mass really lost creating 2He3	0.00002 amu
Total mass transferred, released and lost by proton in 1H3	0.00474 amu

Mass of proton in 1H3	1.00728 amu's
Total mass transferred, released and lost by proton in 1H3	0.00474 amu
Mass of neutron in 2He3	1.00254 amu's

The neutron in the isotope 2He3 in Sub-step 8-1-8c now transfers back to the positon enough mass to make the positon have its maximum mass of 1.00728 amu's. When this happens, the neutron remains a neutron with 1.00037 amu's.

Mass of neutron in 2He3	1.00254 amu's
Mass transferred back to positon in 2He3	0.00217 amu
Mass of neutron in 2He3	1.00037 amu

Mass of positon in 2He3	1.00511 amu's
Mass transferred to it from neutron in 2He3	0.00217 amu
Mass of positon in 2He3	1.00728 amu's

All of the masses of the stable isotope of 2He3 found in sub-step 8-1-8c are now known. Add them to find the mass of the stable isotope of 2He3.

Mass of electron and negatron in stable 2He3	0.00110 amu
Mass of proton in stable 2He3	1.00728 amu's
Mass of positon in stable 2He3	1.00728 amu's
Mass of neutron in stable 2He3	1.00037 amu's
Mass of stable 2He3	3.01603 amu's

Sub-step 7-1-7

2He3

1H3 = 2He3, #2 alpha ray, transfer 0.00100 amu, 0.00217 amu, 0.00374 amu, negatron orbit

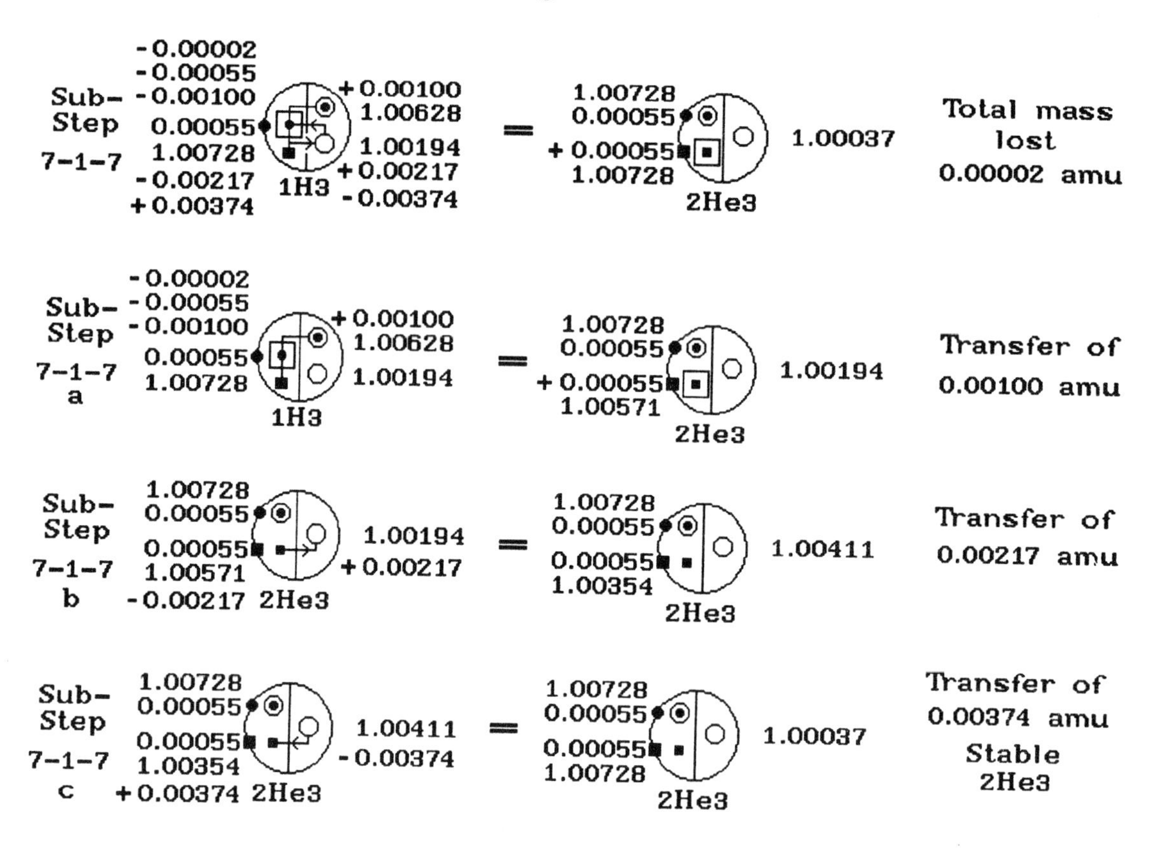

The isotope 1H3 with 3.01605 amu's has one electron with 0.00055 amu and one proton with 1.00728 amu's plus two neutrons. The first neutron has 1.00628 amu's and the second neutron has 1.00194 amu's.

The isotope 2He3 with 3.01603 amu's has one electron with 0.00055 amu, one negatron with 0.00055 amu, one proton with 1.00728 amu's, one positon with 1.00728 amu's and one neutron with 1.00037 amu's.

The total mass lost is 0.00002 amu and it is emitted as a #2 alpha ray.

The isotope 1H3 with 3.01605 amu's has its electron with 0.00055 amu drawn as a dot on the big circle while the proton with 1.00728 amu's is drawn as a dot with a square around it to the left of the vertical line within the big circle. The square drawn around the proton means that the proton with 1.00728 amu's has lost a single unit of 0.00002 amu as a #2 ray. The loss of 0.00002 amu releases a negatron from within the proton to go into an orbital that has an electron in it. The negatron is drawn as a square below the proton with a line connecting them to show that the negatron comes from within the proton. The electron and negatron each have 0.00055 amu.

The proton then transfers 0.00100 amu to the first neutron. It is the loss of 0.00100 amu that is transferred to the first neutron that causes the negatron to go into an orbital with the electron that is already there. The proton has now lost 0.00002 amu, 0.00055 amu and 0.00100 amu for a total lost mass of 0.00157 amu. When this mass is lost by the proton with 1.00728 amu's, it becomes a positon with 1.00571 amu's in the isotope 2He3.

There is a line drawn from the proton to the first neutron to show that the proton has transferred 0.00100 amu to the first neutron in the isotope 1H3. The first neutron is drawn as a small circle with a dot in its middle. The first neutron has 1.00628 amu's and when it receives the 0.00100 amu from the proton, it becomes the proton found in the isotope 2He3. The dot with a small circle drawn around it in the isotopes 1H3 and 2He3 is drawn this way to show that the proton in the isotope 2He3 comes from the first neutron in the isotope 1H3. The first neutron in the isotope 1H3 has 1.00628 amu's and it becomes the proton in the isotope 2He3 with 1.00728 amu's. It may look like the proton with 1.00728 amu's and the first neutron that now has 1.00728 amu's trade places but they do not. The first neutron in the isotope 1H3 vanishes and the proton that was the first neutron now is found in the isotope 2He3.

The second neutron in the isotope 1H3 with 1.00194 amu's is drawn as a small circle with nothing in it's middle to the right of the vertical line within the big circle. It is drawn this way in the isotopes 1H3 and 2He3. At this point in time nothing has been transferred to it. This is shown in the drawing of sub-step 7-1-7a. The isotope 1H3 has now become an isotope of 2He3 with 3.01603 amu's. This new isotope of 2He3 is not stable. It will change into another unstable isotope before it becomes a stable isotope of 2He3.

In the drawing of sub-step 7-1-7b, the isotope 2He3 now has its positon with 1.00571 amu's transfer 0.00217 amu to the second neutron. The second neutron receives 0.00217 amu to become the single neutron with 1.00411 amu's in the isotope 2He3. It is drawn as a small circle with nothing in its middle and it is drawn the same way in the isotopes 1H3 and 2He3. There is a line with an arrow pointing to the single neutron. The line is drawn from the positon with 1.00571 amu's to the single neutron with 1.00194 amu's in the first isotope 2He3 to show that 0.00217 amu has been transferred. The reader is to keep in mind that the loss of the transferred 0.00217 amu does not cause a negatron to be released from within the positon. The positon with 1.00571 amu's has now lost 0.00217 amu to become a new positon with 1.00354 amu's in the second isotope of 2He3. The single neutron in the first isotope with 1.00194 amu's is now the single neutron with 1.00411 amu's in the second isotope of 2He3. This means that the first isotope of 2He3 has changed into a different isotope of 2He3 as shown in the drawing of sub-step 7-1-7b. This new isotope of 2He3 is unstable and must change into a stable isotope of 2He3.

The second unstable isotope of 2He3 in the drawing of sub-step 7-1-7b becomes the first isotope of 2He3 in the drawing of sub-step 7-1-7c. Now the first isotope of 2He3 has the single neutron with 1.00411 amu's transfer 0.00374 amu to the positon that has 1.00354 amu's. There is a line with an arrow drawn on it pointing to the positon. The line is drawn from the single neutron to the positon in the first isotope of 2He3 to show that the positon receives 0.00374 amu from the single neutron. The single neutron with 1.00411 amu's becomes the single neutron with 1.00037 amu's in the stable isotope of 2He3. The positon with 1.00354 amu's becomes a positon with 1.00728 amu's in the stable isotope of 2He3. Now the proton and the positon each have 1.00728 amu's. This is the maximum they can have and this makes a stable isotope of 2He3.

There are four drawings which show how the radioactive isotope of 1H3 changes into a nonradioactive isotope of 2He3. The first drawing puts all of the events into one drawing. The other three drawing brake this first drawing down into the sub-steps that let the reader see how it is done.

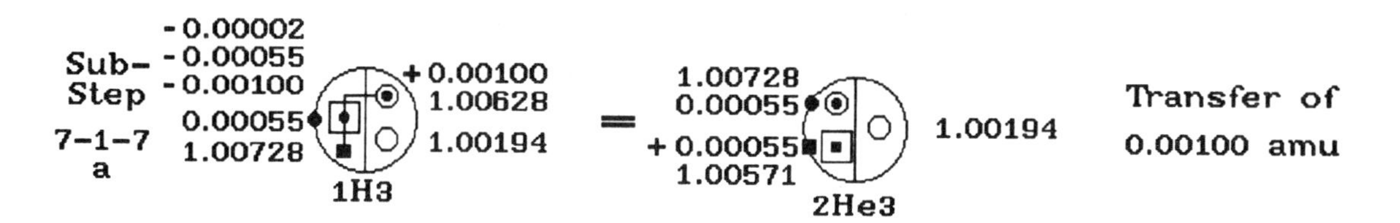

In sub-step 7-1-7a, the proton with 1.00728 amu's in the isotope 1H3 transfers 0.00100 amu to the first neutron which has 1.00628 amu's. At the same time, the proton loses 0.00002 amu and releases a negatron with 0.00055 amu. There is a line drawn from the proton to the first neutron showing that the proton transferred the 0.00100 amu to the first neutron. The proton has an electron that is in an orbital. The electron is drawn as a dot on the big circle while the proton is drawn as a dot to the left of a vertical line within the big circle. There is a square drawn around the dot that is the proton to show that 0.00002 amu has been lost by the proton. There is a small square drawn below the proton that is the negatron that is released from within the proton. This negatron goes into the same orbital that has the electron in it. There is a line drawn between the proton and the negatron to show that the negatron came from within the proton.

The first neutron in the isotope 1H3 is drawn as a small circle to the right of the vertical line within the big circle. This first neutron has a dot within the small circle to show that it becomes the proton in the isotope 2He3. It is drawn the same way in the isotope 2He3 to show that the proton came from the first neutron in the isotope 1H3. When the first neutron regained the 0.00100 amu, it became the proton in the isotope 2He3 which means that the first neutron disappears from the left side of the vertical line within the big circle and the proton appears on the left side of the vertical line. The proton that lost the 0.00100 amu also lost 0.00002 amu and 0.00055 amu for a total lost mass of 0.00157 amu. This makes the proton have a mass of 1.00571 amu's. It becomes the positon in the isotope 2He3. The negatron is drawn as a square on the big circle below the dot that is the electron. The negatron belongs to this positon. The positon is drawn as a small square with a square drawn around it to show that it is the mass that lost the 0.00002 amu.

The second neutron with 1.00194 amu's in the isotope 1H3 is drawn as a small circle to the right of the vertical line within the big circle below the first neutron. At this point in time, the second neutron becomes the single neutron in the isotope 2He3 with no loss of mass. In sub-step 7-1-7b it will receive some mass from the mass that remains of the proton and stay a neutron.

Sub-Step 7-1-7 b
1.00728
0.00055
0.00055
1.00571
-0.00217 2He3
1.00194
+0.00217
=
1.00728
0.00055
0.00055
1.00354
2He3
1.00411
Transfer of 0.00217 amu

The isotope of 2He3 created from the isotope 1H3 in the sub-step 7-1-7a becomes the first isotope of 2He3 in the drawing of sub-step 7-1-7b. The isotope to the right of the equal sign becomes the second isotope of 2He3.

The positon in the first isotope 2He3 with 1.00571 amu's then transfers 0.00217 amu to the single neutron that has 1.00194 amu's to make this neutron into the single neutron with 1.00411 amu's and the positon that had 1.00571 amu's stays the positon in the second isotope of 2He3 with 1.00354 amu's. The electron is still drawn as a dot on the big circle with the negatron drawn as a square below it. They both have 0.00055 amu. The proton is drawn as a dot with a small circle around it and now the square that was around the proton is not drawn. This square around the proton in the isotope 2He3 is not drawn because the proton in the isotope 1H3 becomes the positon in the isotope 2He3. It never becomes a neutron. Both of these isotopes are unstable and Sub-step 7-1-7c shows how they become stable.

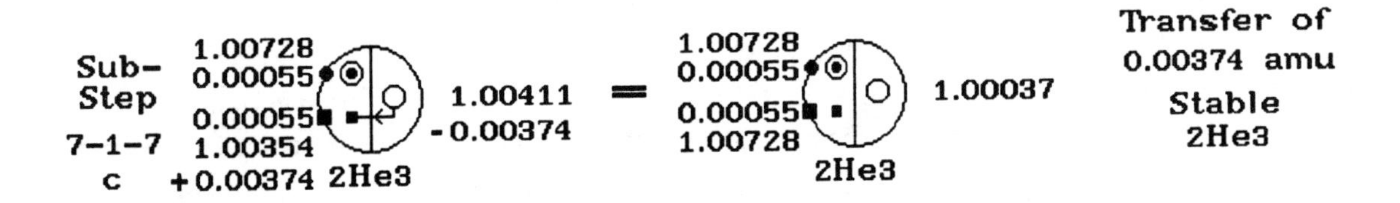

The isotope 2He3 now has a proton with 1.00728 amu's and a positon with 1.00354 amu's, an electron and a negatron with 0.00055 amu each plus one neutron with 1.00411 amu's. Their total mass equals the mass of 3.01603 amu's which is the mass of the isotope 2He3.

The first isotope of 2He3 in drawing of sub-step 7-1-7c is not stable. Its single neutron with 1.00411 amu's transfers 0.00374 amu to the positon which has 1.00354 amu's. When the positon receives the transferred mass of 0.00374 amu, it becomes the positon with 1.00728 amu's in the stable isotope of 2He3 which is the second isotope of 2He3 in the drawing of sub-step 7-1-7c. The single neutron changes from a neutron with 1.00411 amu's to a neutron with 1.00037 amu's.

The electron and proton is drawn the same way as shown in sub-step 7-1-7b. The negatron is still a square on the big circle below the electron. The positon is drawn as a small square again but this time there is a line connecting it to the single neutron in the first isotope of 2He3. There is an arrow drawn on the line pointing to the positon to show that the single neutron with 1.00411 amu's has transferred 0.00374 amu to the positon. The single neutron is drawn as a small circle to the right of the vertical line within the big circle. When it transfers 0.00374 amu it becomes the single neutron found in the second isotope of 2He3 with 1.00037 amu's. The positon with 1.00354 amu's becomes a positon with 1.00728 amu's. Because the proton and positon in the second isotope of 2He3 now both have 1.00728 amu's, the second isotope of 2He3 becomes stable. There is no way for them to gain more mass which means that the single neutron must have the mass of 1.00037 amu's.

To find the mass that is lost creating the isotope 2He3, subtract the isotope 2He3 from the isotope 1H3.

Mass of 1H3	3.01605 amu's
Mass of 2He3	3.01603 amu's
Mass lost creating 2He3	0.00002 amu

This 0.00002 amu is lost by the proton in the isotope 1H3 along with a single unit of 0.00100 amu, plus the negatron with 0.00055 amu that goes into an orbital with the electron. This 0.00002 amu is lost as a single unit as a #2 alpha ray. This is the only mass that is really lost.

In the isotope 2He3, the electron has 0.00055 amu. Since this isotope must have a negatron, its mass is known. Electrons and negatrons have the same mass and they never lose any of it. The first neutron regains 0.00100 amu to become the proton in the isotope 2He3. The 0.00100 amu can be added to the 1.00628 amu's to have the mass of the proton in the isotope 2He3. At the same time the 0.00100 amu is transferred to the first neutron, the proton loses 0.00002 amu as a #2 alpha ray which causes a negatron to drop out of the proton. (Drawing sub-step 7-1-7a)

Mass of proton in 1H3	1.00728 amu's
Mass lost creating 2He3 releasing negatron	0.00002 amu
Remaining mass of proton in 1H3	1.00726 amu's

Remaining mass of proton in 1H3	1.00726 amu's
Mass of released negatron in 1H3	0.00055 amu
Remaining mass of proton in 1H3	1.00671 amu's

Remaining mass of proton in 1H3	1.00671 amu's
Mass transferred to first neutron in 1H3	0.00100 amu
Mass of positon in 2He3	1.00571 amu's

Mass of first neutron in 1H3	1.00628 amu's
Mass regained from proton in 1H3	0.00100 amu
Mass of proton in 2He3	1.00728 amu's

In this step, the second neutron in the isotope 1H3 is known to have regained 0.00217 amu to become the single neutron in the isotope 2He3. By adding this mass to the mass of the second neutron, the mass of the single neutron in the isotope 2He3 will be found. (Drawing sub-step 7-1-7b)

Mass of second neutron in 1H3	1.00194 amu's
Mass regained by second neutron in 1H3	0.00217 amu
Mass of single neutron in 2He3	1.00411 amu's

The proton in the isotope 1H3 becomes a positon in the isotope 2He3. It loses by transfer, 0.00100 amu and 0.00217 amu with the release of a negatron with 0.00055 amu and the loss of 0.00002 amu. Adding all of these masses and subtracting their total from the mass of the proton, the remaining mass will be the positon in the isotope 2He3. (Drawings of sub-step 7-1-7b-c)

Mass transferred to first neutron in 1H3	0.00100 amu
Mass transferred to second neutron in 1H3	0.00217 amu
Mass of negatron released from proton in 1H3	0.00055 amu
Mass lost creating 2He3	0.00002 amu
Total mass lost by proton in 1H3	0.00374 amu

Mass of proton in 1H3	1.00728 amu's
Total mass lost by proton in 1H3	0.00374 amu
Mass of positon in 2He3	1.00354 amu's

Now all of the masses of the isotope 2He3 are known. This isotope of 2He3 is also unstable and it will change into a stable isotope of 2He3.

Mass of electron and negatron in 2He3	0.00110 amu
Mass of proton in 2He3	1.00728 amu's
Mass of positon in 2He3	1.00354 amu's
Mass of neutron in 2He3	1.00411 amu's
Mass of 2He3	3.01603 amu's

This unstable isotope of 2He3 has an electron and a negatron with 0.00055 amu each, a proton with 1.00728 amu's, a positon with 1.00354 amu's and a single neutron with 1.00411 amu's for a total mass of 3.01603 amu's. For this isotope of

2He3 to become a stable isotope, the positon must gain enough mass to make a positon with its maximum mass of 1.00728 amu's. To do this it must gain 0.00374 amu. The only place this mass can come from is the single neutron with 1.00411 amu's. The single neutron in the first isotope of 2He3 loses 0.00374 amu to become the single neutron in the second isotope of 2He3 with 1.00037 amu's. When the positon regains this transferred mass, it becomes a positon with 1.00728 amu's in the second isotope of 2He3. Now the proton and positon each have the same mass of 1.00728 amu's which is the maximum mass they can have. This makes the second isotope of 2He3 stable as shown in drawing of sub-step 7-1-7c.

Mass of neutron in first 2He3	1.00411 amu's
Mass transferred to positon in first 2He3	0.00374 amu
Mass of neutron in stable isotope of 2He3	1.00037 amu's

Mass of positon in first 2He3	1.00354 amu's
Mass regained by positon in first 2He3	0.00374 amu
Mass of positon in stable isotope of 2He3	1.00728 amu's

Mass of electron and negatron in stable 2He3	0.00110 amu
Mass of proton in stable 2He3	1.00728 amu's
Mass of positon in stable 2He3	1.00728 amu's
Mass of neutron in stable 2He3	1.00037 amu's
Mass of stable 2He3	3.01603 amu's

Sub-step 6-1-6

2He3

1H3 = 2He3, #2 alpha ray, transfer 2(0.00100 amu), 0.00434, neg. orbit

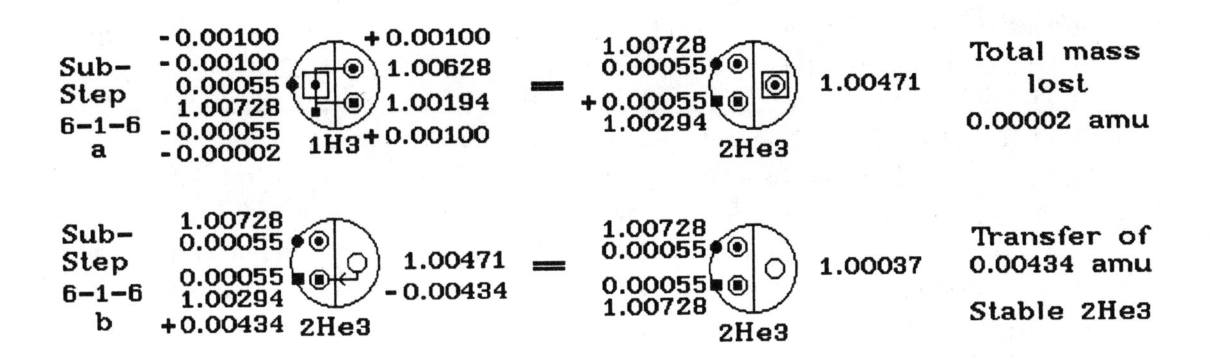

The isotope 1H3 with 3.01605 amu's has one electron with 0.00055 amu and one proton with 1.00728 amu's plus two neutrons. The first neutron has 1.00628 amu's and the second one has 1.00194 amu's.

The isotope 2He3 with 3.01603 amu's has one electron with 0.00055 amu, one negatron with 0.00055 amu, one proton with 1.00728 amu's, one positon with 1.00728 amu's and one neutron with 1.00037 amu's.

The total mass lost is 0.00002 amu and it is emitted as a #2 alpha ray which has an external magnetic field.

The isotope 1H3 has its electron and proton each drawn as a dot with the proton having a square drawn around the dot. This proton becomes the single neutron in the isotope 2He3 after it has transferred the mass it must lose to both of the neutrons in the isotope 1H3.

The proton with 1.00728 amu's transfers 0.00100 amu to the second neutron that has 1.00194 amu's to make this neutron into a positon with 1.00294 amu's. When it does this, it loses 0.00002 amu which causes the negatron to be released. The remaining mass of the proton then transfers 0.00100 amu to the first neutron making this first neutron into the proton with 1.00728 amu's found in the isotope 2He3. The first neutron vanishes when the proton appears in the isotope 2He3. Now the remaining mass of the positon is 1.00471 amu's and this remaining mass is now the single neutron in the isotope 2He3.

The proton in the isotope 1H3 loses 0.00002 amu. It is drawn as a dot with a square around the proton. The square around the dot that is the proton represents the loss of 0.00002 amu only. It does not represent any transferred mass.

The electron in the isotopes 1H3 and 2He3 is drawn as a dot on the big circle and the negatron is drawn as a square on the big circle in the isotope 2He3.

There is a square drawn below the proton in the isotope 1H3 that has a line connecting them. This square is the negatron that dropped out of the proton and the line shows this.

The first neutron in the isotope 1H3 is drawn as a small circle with a dot drawn in its center to show that it has regained 0.00100 amu because this neutron becomes the proton in the isotope 2He3. It is drawn the same way in the isotope 2He3. The plus in front of the 0.00100 amu in the isotope 1H3 means that the first neutron has regained this mass.

The second neutron in the isotope 1H3 is drawn as a small circle with a square drawn in its middle to show that it is the positon in the isotope 2He3. The positon in the isotope 2He3 is drawn the same way.

All of the above takes place because the 0.00100 amu was transferred first to the second neutron and then another 0.00100 amu was transferred to the first neutron. Note that each of the transferred 0.00100 amu had a different result. The first transferred 0.00100 amu released a negatron to go into an orbital. The second 0.00100 amu transferred changed the first neutron into a proton which kept its electron. Do not say that the neutron kept its electron because a neutron does not have an electron or negatron in an orbital.

The square around the small circle with a dot in its middle in the isotope 2He3 shows that 0.00002 amu has been lost. The dot in the center of the small circle represents the proton. This neutron came from the proton found in the isotope 1H3. Because no mass has been lost, the electron belonging to the proton is not released. It already is in an orbital around the nucleus of the isotope 2He3.

There is a line drawn to each of the two neutrons in the isotope 1H3 from the proton to show that they have received 0.00100 amu from the proton.

There also is 0.00002 amu lost from the proton in the isotope 1H3 which is emitted as a #2 alpha ray. This is the real mass lost creating the isotope 2He3.

The isotope of 2He3 found in Sub-step 6-1-6a is not stable. It then transfers 0.00434 amu from the single neutron that has 1.00471 amu's to the positon that has 1.00294 amu's. When this is done, the positon has 1.00728 amu's and the single neutron has 1.00037 amu's. There is an electron and a negatron with 0.00055 amu each plus a proton with 1.00728 amu's to go with the positon and single neutron to make a stable isotope of 2He3. Sub-step 6-1-6b shows this.

To find the mass that is lost creating the isotope 2He3, subtract the isotope 2He3 from the isotope 1H3.

Mass of 1H3	3.01605 amu's
Mass of 2He3	3.01603 amu's
Mass lost creating 2He3	0.00002 amu

The mass the second neutron regains from the proton in the isotope 1H3 is 0.00100 amu which will release the negatron from within the proton. Add it to the second neutron to find the mass of the positon it becomes in the isotope 2He3.

Mass of second neutron in 1H3	1.00194 amu's
Mass regained by it	0.00100 amu
Mass of positon in 2He3	1.00294 amu's

The mass of the first neutron is 1.00628 amu's in the isotope 1H3 and it regains 0.00100 amu to become the proton in the isotope 2He3.

Mass of first neutron in 1H3	1.00628 amu's
Mass regained by first neutron in 1H3	0.00100 amu
Mass of proton in 2He3	1.00728 amu's

The original proton in the isotope 1H3 transferred 0.00100 amu to the first neutron and it also transferred 0.00100 amu to the second neutron and lost 0.00002 amu with the release of a negatron with 0.00055 amu. By adding all of these masses and subtracting their total from the mass of the proton, the mass of the single neutron in the isotope 2He3 will be known.

Mass transferred from proton to first neutron in 1H3	0.00100 amu
Mass transferred from proton to second neutron in 1H3	0.00100 amu
Mass of negatron released from proton in 1H3	0.00055 amu
Mass really lost creating 2He3	0.00002 amu
Total mass lost by proton in 1H3	0.00257 amu

Mass of proton in 1H3	1.00728 amu's
Total mass lost by proton in 1H3	0.00257 amu
Mass of neutron in 2He3	1.00471 amu's

Now all of the masses of the isotope 2He3 are known. Add them to find the mass of the isotope 2He3. (Sub-step 6-1-6a)

Mass of electron and negatron in 2He3	0.00110 amu
Mass of proton in 2He3	1.00728 amu's
Mass of positon in 2He3	1.00294 amu's
Mass of neutron in 2He3	1.00471 amu's
Mass of 2He3	3.01603 amu's

Because the isotope 2He3 is not stable, it's single neutron transfers 0.00434 amu to the positon. Subtract the mass lost by the single neutron to find the final mass of the single neutron in the stable isotope 2He3. Sub-step 6-1-6b shows this.

Mass of neutron in first 2He3	1.00471 amu's
Mass transferred to positon in first 2He3	0.00434 amu
Mass of final neutron in stable 2He3	1.00037 amu's

The positon regains the mass that is transferred to it from the single neutron in the first isotope of 2He3. Add this regained mass to the mass of the positon to find the final mass of the positon in the isotope 2He3.

Mass of positon in first 2He3	1.00294 amu's
Mass regained by it	0.00434 amu
Mass of positon in stable 2He3	1.00728 amu's

The first isotope of 2He3 is unstable. It becomes stable when the proton and positon each have 1.00728 amu's. It can not be radioactive because no ray has been emitted.

Mass of electron and negatron in stable 2He3	0.00110 amu
Mass of proton in stable 2He3	1.00728 amu's
Mass of positon in stable 2He3	1.00728 amu's
Mass of neutron in stable 2He3	1.00037 amu's
Mass of stable 2He3	3.01603 amu's

Sub-step 5-1-5

2He3

1H3 = 2He3, #2 alpha ray, transfer 0.00100 amu, 0.00434 amu, 0.00591 amu, neg. orbit

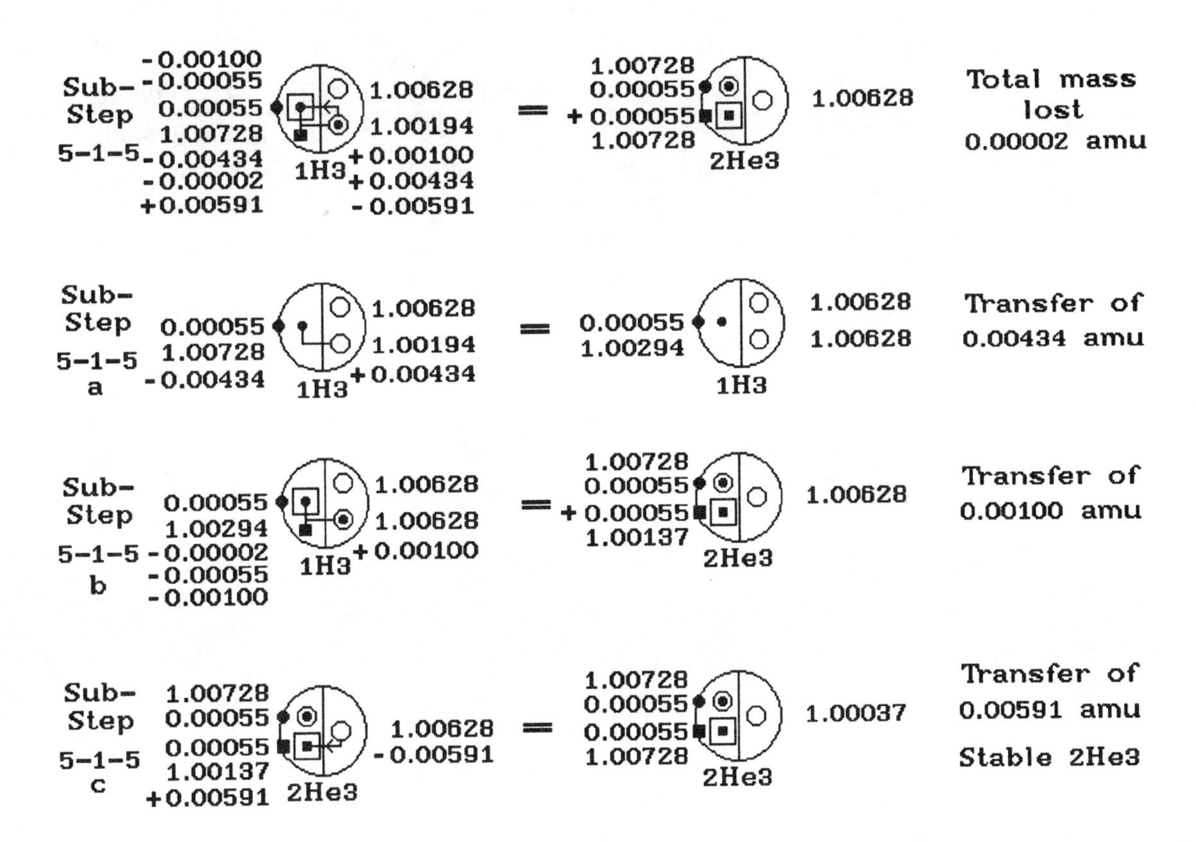

The isotope 1H3 with 3.01605 amu's has one electron with 0.00055 amu and one proton with 1.00728 amu's plus two neutrons. The first neutron has 1.00628 amu's and the second neutron has 1.00194 amu's.

The isotope 2He3 with 3.01603 amu's has one electron with 0.00055 amu, one negatron with 0.00055 amu, one proton with 1.00728 amu's, one positon with 1.00137 amu's and one neutron with 1.00628 amu's.

The total mass lost is 0.00002 amu and it is emitted as a #2 alpha ray.

The isotope 1H3 with 3.01605 amu's has its electron drawn as a dot above the square that is the negatron on the big circle. The electron and negatron each have 0.00055 amu.

The proton with 1.00728 amu's is drawn as a dot with a square drawn around it to show that the proton has lost 0.00002 amu as a #2 alpha ray. There is a line drawn from the proton to the second neutron. This line shows that 0.00434 amu has been transferred to the second neutron. After this 0.00434 amu has been transferred, the proton loses 0.00002 amu which releases a negatron from within itself. At this same time, the proton transfers 0.00100 amu to the new second neutron to make this new neutron into the proton in the isotope 2He3. The remainder of the proton becomes the positon in the isotope 2He3. The line showing the proton has transferred 0.00434 amu also shows that the 0.00100 amu has been transferred to the second neutron. The second neutron is drawn as a small circle with a dot in its middle. Since this second neutron becomes the proton in the isotope 2He3, it is drawn the same way in the isotope 2He3.

The isotope 2He3 created by the isotope 1H3 is not stable. To become stable, the single neutron with 1.00628 amu's transfers 0.00591 amu to the positon that has 1.00137 amu's to make it have 1.00728 amu's. Now the proton and positon each have 1.00728 amu's. This makes an isotope of 2He3 with it neutron having 1.00037 amu's. They go with the electron and negatron to make a stable isotope of 2He3. All of this is shown in drawing of Sub-step 5-1-5.

Sub-Step 5-1-5 a

0.00055 1.00728 -0.00434 | 1H3 | 1.00628 1.00194 +0.00434 = 0.00055 1.00294 | 1H3 | 1.00628 1.00628 — Transfer of 0.00434 amu

The drawing of Sub-step 5-1-5a shows that the 0.00434 amu has been transferred from the proton to the second neutron in the isotope 1H3. At this point in time the line drawn between the proton and the second neutron only shows that 0.00434 amu has been transferred. There is no square drawn around the proton because the lost mass of 0.00002 amu has not been lost. Because the proton with 1.00728 amu's loses by transferring 0.00434 amu and remains a proton in the isotope 1H3 with 1.00294 amu's, it is drawn as a dot in the isotopes 1H3 and 1H3 to the left of the vertical line within the big circle. The second neutron with 1.00194 amu's regains the transferred mass of 0.00434 amu to remain the second neutron with 1.00628 amu's in the isotope 1H3. The first neutron with 1.00628 amu's is drawn as a small circle in the isotope 1H3 which shows that it never lost any mass. Later on in sub-step 5-1-5c it will transfer some of its mass and become a positon in the isotope 2He3.

Sub-Step 5-1-5 b

0.00055 1.00294 -0.00002 -0.00055 -0.00100 | 1H3 | 1.00628 1.00628 +0.00100 = 1.00728 0.00055 +0.00055 1.00137 | 2He3 | 1.00628 — Transfer of 0.00100 amu

In the drawing of sub-step 5-1-5b, the proton with 1.00294 amu's in the isotope 1H3 transfers 0.00100 amu to the second neutron that has 1.00628 amu's to make this second neutron into the proton found in the isotope 2He3 with 1.00728 amu's. There is a line drawn between the proton and the second neutron to show this. The second neutron with 1.00628 amu's in the isotope 1H3 is drawn as a small circle that is drawn to the left of the vertical line within the big circle has a dot in its middle to show that this second neutron in the isotope 1H3 becomes a proton in the isotope 2He3. It is drawn this same way in the isotope 2He3 as a proton with 1.00728 amu's which is now drawn on the right side of the vertical line within the big circle. The second neutron in the isotope 1H3 vanishes when the proton shows up in the isotope 2He3.

The first neutron with 1.00628 amu's has not lost any mass at this point in time in the isotopes 1H3 and 2He3 so it is still drawn as a small circle to the left of the vertical line within the big circle. At the same time the 0.00100 amu was transferred to the second neutron in the isotope 1H3, the 0.00002 amu was lost and the negatron with 0.00055 amu was released from within the proton that has 1.00294 amu's. The negatron is drawn as a small square below the proton. There is a line drawn between the proton and the negatron to show that the negatron came from within the proton. This proton is drawn as a dot with a square drawn around it to show that 0.00002 amu has been lost. This proton that has 1.00294 amu's has now lost 0.00157 amu to become the positon found in the isotope 2He3 with 1.00137 amu's. This proton in the isotope 1H3 is drawn as a dot which is then drawn as a small square in the isotope 2He3 to show that a proton in the isotope 1H3 has become a positon in the isotope 2He3. The square around the proton in the isotope 1H3 is also drawn around the positon in the isotope 2He3 to show that this positon came from the proton in the isotope 1H3.

When the proton in the isotope 1H3 lost 0.00002 amu, the 0.00002 amu was emitted as a #2 alpha ray. At this time the isotope 2He3 was created. This isotope 2He3 is unstable.

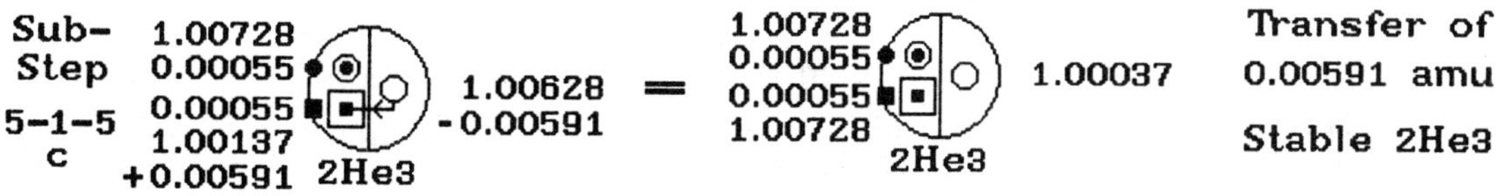

Sub-Step 5-1-5 c
1.00728
0.00055
0.00055
1.00137
+0.00591
2He3
1.00628
-0.00591
=
1.00728
0.00055
0.00055
1.00728
2He3
1.00037
Transfer of 0.00591 amu
Stable 2He3

The isotope with 3.01603 amu's in drawing of sub-step 5-1-5b is not stable because its positon do not have 1.00728 amu's. For this isotope of 2He3 to be come a stable isotope with the same mass of 3.01603 amu's, the single neutron must transfer 0.00591 amu to the positon. When this happens, the positon regains 0.00591 amu to become a positon with 1.00728 amu's and the single neutron becomes the single neutron in the stable isotope of 2He3 with 1.00037 amu's. This is shown in drawing of sub-step 5-1-5c.

The isotope of 2He3 in sub-step 5-1-5b is the first isotope of 2He3 in the drawing of sub-step 5-1-5c. There is a line with an arrow on it pointing toward the single neutron in the first isotope of 2He3 to show that single neutron has transferred 0.00591 amu to the positon. This positon has 1.00137 amu's and receives 0.00591 amu to become the positon with 1.00728 amu's in the stable isotope of 2He3. The stable isotope of 2He3 is the second isotope found in drawing of sub-step 5-1-5c. The single neutron in the drawing of sub-step 5-1-5b found in the isotope of 2He3 has 1.00628 amu's and when it transfers the 0.00591 amu, it becomes the single neutron in the stable isotope of 2He3 with 1.00037 amu's. Note that the square around the positon is still drawn because the reader should know which mass found in the isotope 2He3 lost the real mass that is lost (0.00002 amu).

To find the mass that is lost creating the isotope 2He3, subtract the isotope 2He3 from the isotope 1H3.

Mass of 1H3	3.01605 amu's
Mass of 2He3	3.01603 amu's
Mass lost creating 2He3	0.00002 amu

The electron with 0.00055 amu in the isotope 1H3 does not lose any mass and becomes the electron found in the isotope 2He3. The isotope 2He3 must have a negatron with 0.00055 amu so the masses of the electron and negatron are known. At this point in time, the first neutron in the isotope 1H3 is used as the single neutron in the isotope 2He3 with 1.00628 amu's. The masses of the proton and the positon in the isotope 2He3 are unknown.

The proton with 1.00728 amu's in the isotope 1H3 transfers 0.00434 amu to the second neutron with 1.00194 amu's. When this 0.00434 amu has been transferred, the proton has 1.00294 amu's remaining which is the mass of a positon. This is shown in the drawing of sub-step 5-1-5a.

Mass of proton in 1H3	1.00728 amu's
Mass transferred to second neutron in 1H3	0.00434 amu
Remaining mass of proton in 1H3	1.00294 amu's

This positon with 1.00294 amu's then transfers 0.00100 amu to the second neutron which now has 1.00628 amu's. When it receives the 0.00100 amu, it becomes the proton with 1.00728 amu's. When the positon with 1.00294 amu's transfers the 0.00100 amu, it releases a negatron with 0.00055 amu from within itself and at the same time loses 0.00002 amu that is emitted as a #2 alpha ray. This means that the positon with 1.00294 amu's loses 0.00157 amu to become the positon in the isotope 2He3 with 1.00137 amu's. This is shown in the drawing of sub-step 5-1-5b.

Mass lost by proton creating 2He3	0.00002 amu
Mass of negatron released from proton in 1H3	0.00055 amu
Mass transferred to second neutron in 1H3	0.00100 amu
Total mass lost by proton in 1H3	0.00157 amu
Mass of proton in 1H3	1.00294 amu's
Mass lost by it	0.00157 amu
Mass of positon in 2He3	1.00137 amu's

The electron that was with the proton when it had 1.00728 amu's in the isotope 1H3 was not released when the 0.00434 amu was transferred because this transferred mass will not release an electron. When the 0.00100 amu was transferred, it caused the proton which is now a positon to release a negatron from within itself and therefore the proton that became a positon could not release the electron. This means that the electron was never released from the isotope 1H3 and became the electron in the isotope 2He3 when the second neutron became the proton with 1.00728 amu's.

The proton with 1.00728 amu's in the isotope 1H3 must transfer 0.00434 amu, 0.00100 amu and lose 0.00055 amu. Add to this, the lost mass of 0.00002 amu and subtract their total mass from the proton in the isotope 1H3 and the remaining mass is the positon in the isotope 2He3.

Mass transferred to second neutron in 2He3	0.00434 amu
Mass transferred to second neutron in 2He3	0.00100 amu
Mass of negatron released from proton in 1H3	0.00055 amu
Mass lost creating 2He3	0.00002 amu
Total mass lost by proton in 1H3	0.00591 amu

Mass of proton in 1H3	1.00728 amu's
Total mass lost by proton in 1H3	0.00591 amu
Mass of positon in 2He3	1.00137 amu's

The second neutron with 1.00194 amu's regained 0.00434 amu and 0.00100 amu to become the proton in the isotope 2He3 with 1.00728 amu's.

Mass regained by second neutron in 1H3	0.00434 amu
Mass regained by second neutron in 1H3	0.00100 amu
Total mass regained by second neutron in 1H3	0.00534 amu

Mass of second neutron in 1H3	1.00194 amu's
Total mass regained by second neutron in 1H3	0.00534 amu
Mass of proton in 2He3	1.00728 amu's

The first neutron with 1.00628 amu's in the isotope 1H3 is used as is and does not lose any mass to become the single neutron in the isotope 2He3.

Now all of the masses of the isotope 2He3 are known. Adding these masses will give the mass of the isotope 2He3. (Drawing sub-step 5-1-5b)

Mass of electron and negatron in 2He3	0.00110 amu
Mass of proton in 2He3	1.00137 amu's
Mass of positon in 2He3	1.00728 amu's
Mass of neutron in 2He3	1.00628 amu's
Mass of 3He3	3.01603 amu's

This isotope of 2He3 is not stable. It changes into a stable isotope of 2He3 by having the single neutron in the first isotope of 2He3 transfer 0.00591 amu to the positon which has 1.00137 amu's. This is shown by drawing a line with an arrow on it pointing to the positon from the single neutron. When this is done, the single neutron with 1.00628 amu's becomes the single neutron with 1.00037 amu's in the stable isotope of 2He3. The positon which regains this transferred 0.00591 amu becomes a positon with 1.00728 amu's. Now there is a proton and a positon with each having 1.00728 amu's which makes this isotope of 2He3 stable. (Drawing sub-step 5-1-5c)

Mass of neutron in 2He3	1.00628 amu's
Mass transferred to positon in 2He3	0.00591 amu
Mass of neutron in stable 2He3	1.00037 amu's

Mass of positon in 2He3	1.00137 amu's
Mass transferred to it from neutron in 2He3	0.00591 amu
Mass of positon in stable 2He3	1.00728 amu's

Mass of electron and negatron in stable 2He3	0.00110 amu
Mass of proton in stable 2He3	1.00728 amu's
Mass of positon in stable 2He3	1.00728 amu's
Mass of neutron in stable 2He3	1.00037 amu's
Mass of stable 2He3	3.01603 amu's

Sub-step 4-1-4

2He3

1H3 = 2He3, #2 alpha ray, transfer 0.00100 amu, 0.00157 amu, neg. orbit

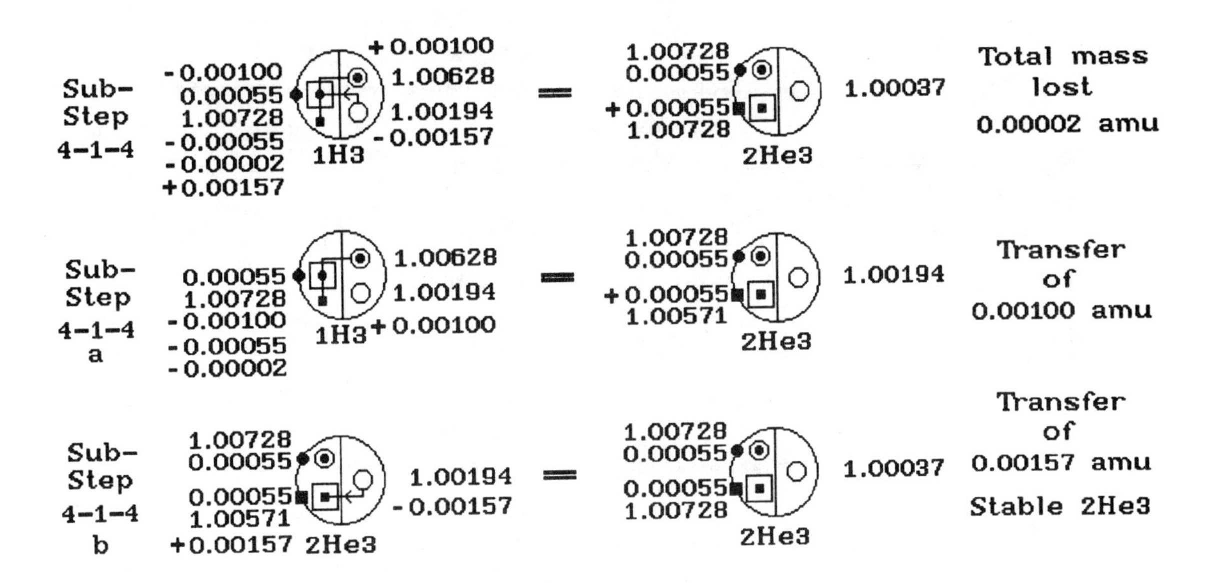

The isotope 1H3 with 3.01605 amu's has one electron with 0.00055 amu and one proton with 1.00728 amu's plus two neutrons. The first neutron has 1.00628 amu's and the second neutron has 1.00194 amu's.

The isotope 2He3 with 3.01603 amu's has one electron with 0.00055 amu, one negatron with 0.00055 amu, one proton with 1.00728 amu's, one positon with 1.00728 amu's and one neutron with 1.00037 amu's.

The total mass lost is 0.00002 amu and it is emitted as a #2 alpha ray which has an external magnetic field.

The isotope 1H3 with 3.01605 amu's has an electron with 0.00055 amu. The electron is drawn as a dot on the big circle.

The proton with 1.00728 amu's is drawn as a dot with a square drawn around it to show that it has lost 0.00002 amu. The proton releases a negatron with 0.00055 amu from within itself when the 0.00002 amu is lost as a #2 alpha ray. At this same time the proton transfers 0.00100 amu to the first neutron with 1.00628 amu's to make the first neutron the proton with 1.00728 amu's. When the 0.00100 amu is transferred, it causes the negatron with 0.00055 amu to go into an orbital with the electron already there. This negatron is drawn as a square below the proton with a line connecting them both.

To show that the 0.00100 amu has been transferred, there is a line drawn from the proton to the first neutron. After the 0.00002 amu has bee lost and the 0.00100 amu transferred with the negatron released from within the proton, the proton in the isotope 1H3 becomes a positon with 1.00571 amu's in the isotope 2He3. The proton with its square around it in the isotope 1H3 is drawn the same way in the isotope 2He3 to show that the positon in the isotope 2He3 came from the proton in the isotope 1H3.

The first neutron with 1.00628 amu's in the isotope 1H3 regains 0.00100 amu from the proton. This happens at the same time the 0.00002 amu and the negatron is released. The first neutron changes into the proton found in the isotope 2He3 only after the proton has lost 0.00157 amu and becomes the positon with 1.00571 amu's. At this time, the first neutron with 1.00628 amu's changes into a proton with 1.00728 amu's in the isotope 2He3. The first neutron is drawn as a small circle with a dot in its middle. It is drawn on the right side of the vertical line within the big circle. To show that the first neutron becomes the proton in the isotope 2He3, the proton in the isotope 2He3 is drawn the same way as the first neutron in the isotope 1H3. Note that the electron that belonged to the proton in the isotope 1H3 still belongs to the proton in the isotope 2He3. The reason the electron stays with the proton is because this is an internal exchange of masses with no ray being emitted. The ray that is emitted is not an internal exchange of mass but the transfer of 0.00100 amu is therefore the negatron that is released from the proton is not the cause for a ray.

There is a line drawn from the proton to the first neutron in the isotope 1H3 to show that 0.00100 amu has been transferred. There is another line with an arrow pointing at the proton drawn from this neutron to the proton to show that 0.00157 amu has been transferred from the neutron to the proton. This proton will later on become a positon. The above is shown in the drawing of sub-step 4-1-4.

The drawing sub-step 4-1-4 shows all that happens to the isotope 1H3 when it becomes the isotope 2He3. Below this drawing are two more drawings that will explain the drawing of sub-step 4-1-4.

The isotope of 1H3 has its proton drawn as a dot to the left of the vertical line within the big circle. This dot has a square drawn around it to show that 0.00002 amu has been lost. There is a line drawn from this proton to the first neutron to show that 0.00100 amu has been transferred. The first neutron with 1.00628 amu's in the isotope 1H3 disappears and the proton with 1.00728 amu's in the isotope 2He3 appears. To show this, the first neutron in the isotope 1H3 is drawn as a small circle with a dot in its middle on the right side of the vertical line within the big circle while the proton it becomes is drawn the same way except it is drawn on the left side of the vertical line within the big circle in the isotope 2He3. At this point in time, the second neutron in the isotope 1H3 with 1.00194 amu's becomes the single neutron in the isotope 2He3 with no loss of mass. It is drawn as a small circle on the right side of the vertical line in the big circle and it is drawn the same way in the isotope 2He3. The drawing of sub-step 4-1-4a shows how this is done.

This isotope with 3.01603 amu's is not stable because its proton and positon do not have 1.00728 amu's each. For this isotope of 2He3 to become a stable isotope with the same mass, the single neutron must transfer 0.00157 amu to the positon. When this happens, the positon regains 0.00157 amu to become a positon with 1.00728 amu's and the single neutron becomes the single neutron in the stable isotope of 2He3 with 1.00037 amu's. This is shown in the drawing of sub-step 4-1-4b by drawing a line from the single neutron to the positon with an arrow drawn on the line pointing toward the positon.

To find the mass that is lost creating the isotope 2He3, subtract the isotope 2He3 from the isotope 1H3.

Mass of 1H3	3.01605 amu's
Mass of 2He3	3.01603 amu's
Mass lost creating 2He3	0.00002 amu

The electron with 0.00055 amu in the isotope 1H3 does not lose any mass and becomes the electron found in the isotope 2He3. The isotope 2He3 must have a negatron with 0.00055 amu so the masses of the electron and negatron are known. The second neutron in the isotope 1H3 is used as the single neutron in the isotope 2He3 with 1.00194 amu's. The masses of the proton and positon in the isotope 2He3 are unknown.

The proton with 1.00728 amu's in the isotope 1H3 transfers 0.00100 amu to the first neutron and when it does this, it releases a negatron with 0.00055 amu from within itself. The negatron goes into the orbital with the electron. This proton also loses 0.00002 amu which is the only mass that is lost and it is emitted as a #2 alpha ray. By adding all of the masses that are transferred, released and lost, the total mass lost by the proton in the isotope 1H3 will be found. Subtract this total from the mass of the proton in the isotope 1H3 to find the mass of the positon in the isotope 2He3. (Drawing sub-step 4-1-4a)

Mass transferred to first neutron in 2He3	0.00100 amu
Mass of negatron released from 1H3's proton	0.00055 amu
Mass lost creating 2He3	0.00002 amu
Total mass lost by proton in 1H3	0.00157 amu

Mass of proton in 1H3	1.00728 amu's
Total mass lost by proton in 1H3	0.00157 amu
Mass of positon in 2He3	1.00571 amu's

The first neutron with 1.00628 amu's regains 0.00100 amu to become the proton in the isotope 2He3 with 1.00728 amu's.

Mass of first neutron in 1H3	1.00628 amu's
Total mass regained by first neutron in 1H3	0.00100 amu
Mass of proton in 2He3	1.00728 amu's

The second neutron with 1.00194 amu's in the isotope 1H3 is used as is and does not lose any mass to become the single neutron in the isotope 2He3.

Now all of the masses of the isotope 2He3 are known. Adding these masses will give the mass of the isotope 2He3. (Drawing sub-step 4-1-4a)

Mass of electron and negatron in 2He3	0.00110 amu
Mass of proton in 2He3	1.00728 amu's
Mass of positon in 2He3	1.00571 amu's
Mass of neutron in 2He3	1.00194 amu's
Mass of 3He3	3.01603 amu's

The first isotope of 2He3 in the drawing of sub-step 4-1-4a given above is not stable and if it is to become stable, its single neutron with 1.00194 amu's must transfer 0.00157 amu to the positon. When this happens, the single neutron in the unstable isotope of 2He3 has 1.00037 amu's and the positon with 1.00571 amu's has 1.00728 amu's. The stable isotope of 2He3 now has a proton and a positon with 1.00728 amu's each. This is the reason the second isotope of 2He3 becomes stable. (Drawing sub-step 4-1-4b)

Mass of neutron in 2He3	1.00194 amu's
Mass transferred to positon in 2He3	0.00157 amu
Mass of neutron in stable 2He3	1.00037 amu's

Mass of positon in 2He3	1.00571 amu's
Mass transferred to it from neutron in 2He3	0.00157 amu
Mass of positon in stable 2He3	1.00728 amu's

Mass of electron and negatron in stable 2He3	0.00110 amu
Mass of proton in stable 2He3	1.00728 amu's
Mass of positon in stable 2He3	1.00728 amu's
Mass of neutron in stable 2He3	1.00037 amu's
Mass of stable 2He3	3.01603 amu's

Sub-step 3-1-3

2He3

1H3 = 2He3, #2 gamma ray, transferred 0.00100 amu, neg. orbit

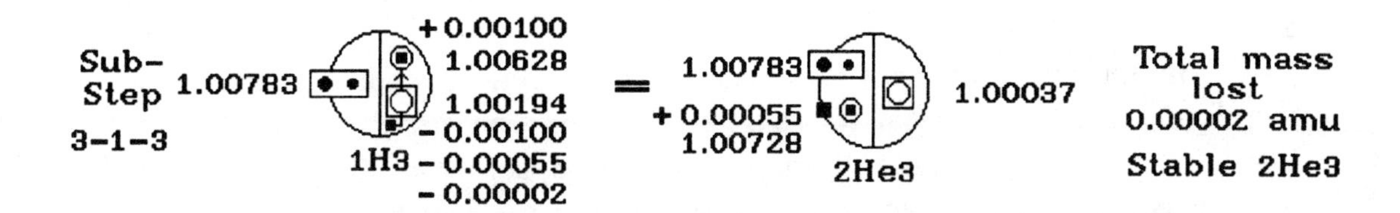

The isotope 1H3 with 3.01605 amu's has one electron with 0.00055 amu and one proton with 1.00728 amu's plus two neutrons. The first neutron has 1.00628 amu's and the second neutron has 1.00194 amu's.

The isotope 2He3 with 3.01603 amu's has one electron with 0.00055 amu, one negatron with 0.00055 amu, one proton with 1.00728 amu's, one positon with 1.00728 amu's and one neutron with 1.00037 amu's.

The total mass lost is 0.00002 amu and it is emitted as a #2 gamma ray which does not have an external magnetic field.

The isotope 1H3 has its electron drawn as a dot on the big circle and proton is drawn as a dot to the left of the vertical line within the big circle. There is a rectangle drawn around both in the isotopes 1H3 and 2He3. The rectangle means that they are used as the isotope 1H1 with 1.00783 amu's and the isotope 1H1 can drop out at some later time from this isotope 2He3.

The first neutron is drawn as a small circle with a small square in its middle to show that the first neutron in the isotope 1H3 becomes the positon in the isotope 2He3. The first neutron in the isotope 1H3 is drawn on the right side of the vertical line within the big circle while the positon is drawn on the left side below the proton as a square with a small circle drawn around it to show that it comes from the first neutron in the isotope 1H3.

The second neutron is drawn as a small circle with nothing in the middle. There is a square drawn around the small circle to show that 0.00002 amu has been lost creating the isotope 2He3. When this mass is lost, the second neutron releases a negatron from within itself. This negatron is drawn as a square below the second neutron in the isotope 1H3 and as a square on the big circle below the electron. There is a line drawn between the second neutron and the negatron to show that the negatron comes from within the second neutron.

There is a line drawn between the second neutron and the first neutron with an arrow pointing toward the first neutron. The arrow shows that the second neutron transfers 0.00100 amu to the first neutron.

When the first neutron in the isotope 1H3 with 1.00628 amu's receives 0.00100 amu from the second neutron which has 1.00194 amu's, it becomes a positon with 1.00728 amu's in the isotope 2He3. The second neutron in the isotope 1H3 then releases a negatron with 0.00055 amu from within itself.

This means that the second neutron with 1.00194 amu's in the isotope 1H3 transfers 0.00100 amu, releases one negatron and loses 0.00002 amu. After this mass is gone, it becomes the single neutron in the isotope 2He3 with 1.00037 amu's.

In this step, the 0.00100 amu is transferred to another neutron and therefore it is not lost nor emitted as a ray. There is a negatron with 0.00055 amu that is released and it stays with the newly created isotope of 2He3 in an orbital with the electron. This means that none of this mass has been lost.

The 0.00002 amu that is lost from the second neutron is not regained by the first neutron and it is the real mass that is lost creating the isotope 2He3. It is emitted as a #2 gamma ray. It is the loss of 0.00002 amu that causes the negatron to drop out of the second neutron and the 0.00100 amu is the reason the negatron goes in an orbital with the electron. This last statement may seem strange because in one case, the lost mass causes a negatron to drop out and in the second case, the 0.00100 amu cause the negatron to go into an orbital. If the 0.00002 amu does not cause the negatron to leave the second neutron, the 0.00100 amu could not make it go into an orbital. Remember that when 0.00100 amu is lost, not transferred, the negatron would leave the isotope completely whereas just transferring the 0.00100 amu only lets the negatron go into an orbital.

To find the mass that is lost creating the isotope 2He3, subtract the isotope 2He3 from the isotope 1H3.

Mass of 1H3	3.01605 amu's
Mass of 2He3	3.01603 amu's
Mass lost creating 2He3	0.00002 amu

This 0.00002 amu is lost by the second neutron in the isotope 1H3 It is lost as a single unit as a #2 gamma ray. This is the only mass that is really lost. The 0.00100 amu transferred by the second neutron is regained by the first neutron except for the negatron which is regained by the isotope 2He3 and goes into the orbital with the electron.

The known mass in the isotope 2He3 comes from the isotope 1H1 that is intact within it and the fact that the isotope 2He3 must have a negatron with 0.00055 amu. By adding these two masses and subtracting their total from the mass of the isotope 2He3, the remaining mass will be the mass of the positon and the neutron found in the isotope 2He3.

Mass of 1H1 within 2He3	1.00783 amu's
Mass of negatron in 2He3	0.00055 amu
Total known mass in 2He3	1.00838 amu's

Mass of 2He3	3.01603 amu's
Total known mass in 3He3	1.00838 amu's
Mass of positon and neutron in 2He3	2.00765 amu's

The single neutron in the isotope 2He3 was the second neutron in the isotope 1H3. This second neutron has 1.00194 amu's. By subtracting the mass of the single neutron in the isotope 2He3 from the second neutron in the isotope 1H3, the mass lost by the second neutron in the isotope 1H3 will be known.

Mass of second neutron in 1H3	1.00194 amu's
Mass of neutron in 2He3	1.00037 amu's
Mass lost by second neutron in 1H3	0.00157 amu

This mass lost by the second neutron includes the mass of the negatron that was released from within the second neutron and the mass that it transferred to the first neutron in the isotope 1H3. Subtracting this total from the mass lost by the second neutron in the isotope 1H3 will give the mass that is really lost creating the isotope 2He3.

Mass of negatron released from second neutron in 1H3	0.00055 amu
Mass releasing negatron from second neutron in 1H3	0.00100 amu
Total known mass lost by second neutron in 1H3	0.00155 amu

Total mass lost by second neutron in 1H3	0.00157 amu
Total known mass lost by second neutron in 1H3	0.00155 amu
Mass remaining that is lost by second neutron in 1H3	0.00002 amu

The mass of the first neutron is 1.00628 amu's and it regains 0.00100 amu from the second neutron which has 1.00194 amu's. Add this regained mass to the mass of the first neutron in the isotope 1H3. When this is done, the first neutron with 1.00628 amu's has 1.00728 amu's which is the mass of a proton or a positon. The first neutron in the isotope 1H3 becomes the positon in the isotope 2He3 because there already is a proton in the isotope 2He3 with 1.00728 amu's.

Mass of first neutron in 1H3	1.00628 amu's
Mass transferred to it from second neutron in 1H3	0.00100 amu
Mass of positon in 2He3	1.00728 amu's

Now all of the masses lost by the second neutron in the isotope 1H3 are known. Add all of these masses and subtract their total from the second neutron in the isotope 1H3 to find the mass of this neutron.

Mass lost creating 2He3 by first neutron	0.00002 amu
Mass of negatron by first neutron	0.00055 amu
Mass transferred releasing negatron by second neutron	0.00100 amu
Total mass lost by second neutron in 1H3	0.00157 amu

Mass of second neutron in 1H3	1.00194 amu's
Total mass lost by second neutron in 1H3	0.00157 amu
Mass of neutron in 2He3	1.00037 amu's

The isotope of 1H1 is used intact within the isotopes of 1H3 and 2He3. This means that the proton in the isotope 2He3 has 1.00728 amu's. Now the proton and positon both have 1.00728 amu's which make a stable isotope of 2He3.

Now all of the masses of the isotope 2He3 are known. Add them to find the mass of the stable isotope of 2He3.

Mass of 1H1 in 2He3	1.00783 amu's
Mass of negatron in 2He3	0.00055 amu
Mass of positon in 2He3	1.00728 amu's
Mass of neutron in 2He3	1.00037 amu's
Mass of stable 2He3	3.01603 amu's

Sub-step 2-1-2

2He3

1H3 = 2He3, #2 gamma ray, transferred 0.00100 amu, 0.00434 amu, neg. orbit

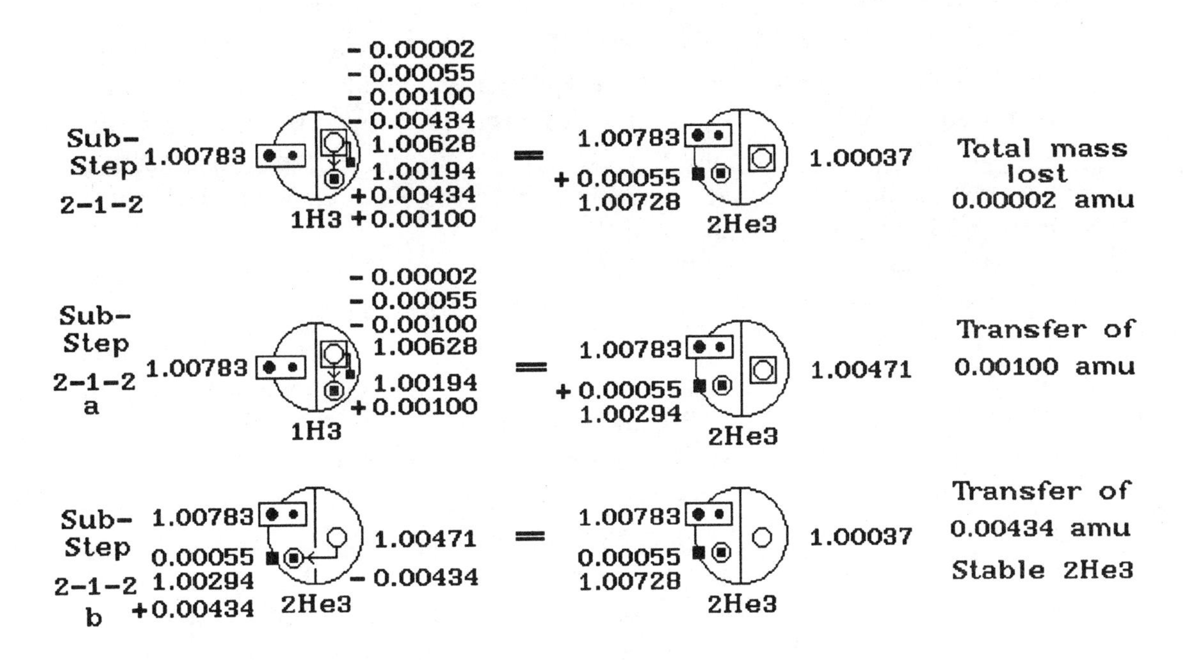

The isotope 1H3 with 3.01605 amu's has one electron with 0.00055 amu and one proton with 1.00728 amu's plus two neutrons. The first neutron has 1.00628 amu's and the second neutron has 1.00194 amu's.

The isotope 2He3 with 3.01603 amu's has one electron with 0.00055 amu, one negatron with 0.00055 amu, one proton with 1.00728 amu's, one positon with 1.00728 amu's and one neutron with 1.00037 amu's.

The total mass lost is 0.00002 amu and it is emitted as a #2 gamma ray which does not have an external magnetic field.

The isotope 1H3 has its electron drawn as a dot on the big circle and proton is drawn as a dot to the left of the vertical line within the big circle. There is a rectangle drawn around both in the isotopes 1H3 and 2He3. The rectangle means that they are used as the isotope 1H1 with 1.00783 amu's and the isotope 1H1 can drop out at some later time from this isotope 2He3.

The first neutron is drawn as a small circle with nothing in its middle. There is nothing in the middle because the first neutron after it loses 0.00002 amu, transfers the 0.00100 amu and releases a negatron from within itself, at which time it becomes the single neutron in the isotope 2He3. The first neutron in the isotope 1H3 and the single neutron in the isotope 2He3 are drawn on the right side of the vertical line within the big circle. There is a square drawn around each neutron.

There is a small square drawn below the first neutron with a line connecting them. This small square is the negatron that drops out of the first neutron and the negatron is drawn as a square on the big circle in the isotope 2He3. If the line connecting the negatron to the first neutron was not drawn, then the negatron would have been found by itself in the nucleus of the isotope 1H3 and this does not happen.

The second neutron in the isotope 1H3 becomes a positon in the isotope 2He3. It is drawn as a small circle with a square in its middle to show that it is the second neutron that becomes a positon. This positon is drawn the same way in the isotope 2He3 as the second neutron in the isotope 1H3 except the second neutron is drawn on the right side of the vertical line within the big circle while the positon is on the left side of the vertical line.

If the first neutron in the isotope 1H3 with 1.00628 amu's transfers 0.00434 amu to the second neutron it will have 1.00194 amu's which is the same mass as the second neutron. The first neutron can not transfer 0.00434 amu first but it can lose 0.00002 amu and transfer 0.00100 amu before it loses 0.00434 amu. The loss of 0.00002 amu as a #2 gamma ray causes the negatron to be released from within the first neutron and at the same time it transfers 0.00100 amu to the second neutron that does cause the negatron to go into an orbital with the electron that is already there. This is shown in the drawing of sub-step 2-1-2a.

After the above mass is gone, the first neutron has 1.00471 amu's. It then transfers 0.00434 amu to the second neutron. When this is done, the first neutron becomes the single neutron in the isotope 2He3 with 1.00037 amu's. This is shown in the drawing of sub-step 2-1-2b.

The second neutron with 1.00194 amu's in the isotope 1H3 regains the 0.00100 amu to become a positon with 1.00294 amu's in the isotope 2He3. When this is done, the isotope 2He3 will be created. This isotope of 2He3 in not stable until the single neutron transfers 0.00434 amu to the positon. When this happens the single neutron has 1.00037 amu's and the positon has 1.00728 amu's. Now there is a proton with 1.00728 amu's and a positon with 1.00728 amu's, making the final isotope of 2He3 stable. This is shown in the drawing of sub-step 2-1-2b.

In Sub-step 2-1-2 there is a line drawn from the single neutron in the isotope 2He3 to the positon with an arrow drown on it pointing toward the positon. This is drawn this way to show that the 0.00434 amu is transferred from the single neutron with 1.00471 amu's to the positon with 1.00294 amu's. This creates the final stable isotope of 2He3 that has an isotope of 1H1 within itself, a negatron, a positon and a single neutron. These masses will give the isotope 2He3 that is stable and not radioactive.

To find the mass that is lost creating the isotope 2He3, subtract the isotope 2He3 from the isotope 1H3.

Mass of 1H3	3.01605 amu's
Mass of 2He3	3.01603 amu's
Mass lost creating 2He3	0.00002 amu

This 0.00002 amu is lost by the first neutron in the isotope 1H3. It is lost as a single unit as a #2 gamma ray. This is the only mass that is really lost. When the 0.00002 amu is lost, a negatron is released from within the first neutron. At this same time, the 0.00100 amu is transferred to the second neutron. Add their masses and subtract their total mass from the mass of the first neutron to find the mass of the single neutron found in the isotope 2He3.

Mass lost by first neutron in 1H3	0.00002 amu
Mass of negatron released by first neutron in 1H3	0.00055 amu
Mass transferred to second neutron in 1H3	0.00100 amu
Mass lost by first neutron in 1H3	0.00157 amu

Mass of first neutron in 1H3	1.00628 amu's
Mass lost by it	0.00157 amu
Mass of neutron in 2He3	1.00471 amu's

This 0.00100 amu is regained by the second neutron which makes this second neutron into the positon in the isotope 2He3. The second neutron disappears from the isotope 1H3 and shows up as a positon in the isotope 2He3. Add the transferred mass to the second neutron to show that its new mass is the mass of the positon in the isotope 2He3.

Mass of second neutron in 1H3	1.00194 amu's
Mass regained by it	0.00100 amu
Mass of positon in 2He3	1.00294 amu's

The isotope 1H1 is used as is in the isotopes 1H3 and 2He3. This means that there is a proton in the isotopes 1H3 and 2He3 that has 1.00728 amu's. The mass of the electron and proton makes an isotope of 1H1 with 1.00783 amu's. All of the masses are now known for the unstable isotope of 2He3.

Mass of 1H1 in unstable 2He3	1.00783 amu's
Mass of negatron in unstable 2He3	0.00055 amu
Mass of positon in unstable 2He3	1.00294 amu's
Mass of neutron in unstable 2He3	1.00471 amu's
Mass of unstable 2He3	3.01603 amu's

To make this unstable isotope of 2He3 stable, the proton and positon must both have 1.00728 amu's. This is done by transferring 0.00434 amu from the single neutron to the positon. When this is done the single neutron will have 1.00037 amu's and the positon will have 1.00728 amu's.

Mass of single neutron in unstable 2He3	1.00471 amu's
Mass transferred to positon in unstable 2He3	0.00434 amu
Mass of neutron in stable 2He3	1.00037 amu's
Mass of positon in unstable 2He3	1.00294 amu's
Mass transferred from neutron in 2He3	0.00434 amu
Mass of positon in stable 2He3	1.00728 amu's

All of the masses of the stable isotope of 2He3 are now known.

Mass of 1H1 intact within the stable 2He3	1.00783 amu's
Mass of negatron in stable 2He3	0.00055 amu
Mass of positon in stable 2He3	1.00728 amu's
Mass of neutron in stable 2He3	1.00037 amu's
Mass of stable 2He3	3.01603 amu's

Sub-step 1-1-1

2He3

1H3 = 2He3, #2 gamma ray, transfer 0.00157 amu, neg. orbit

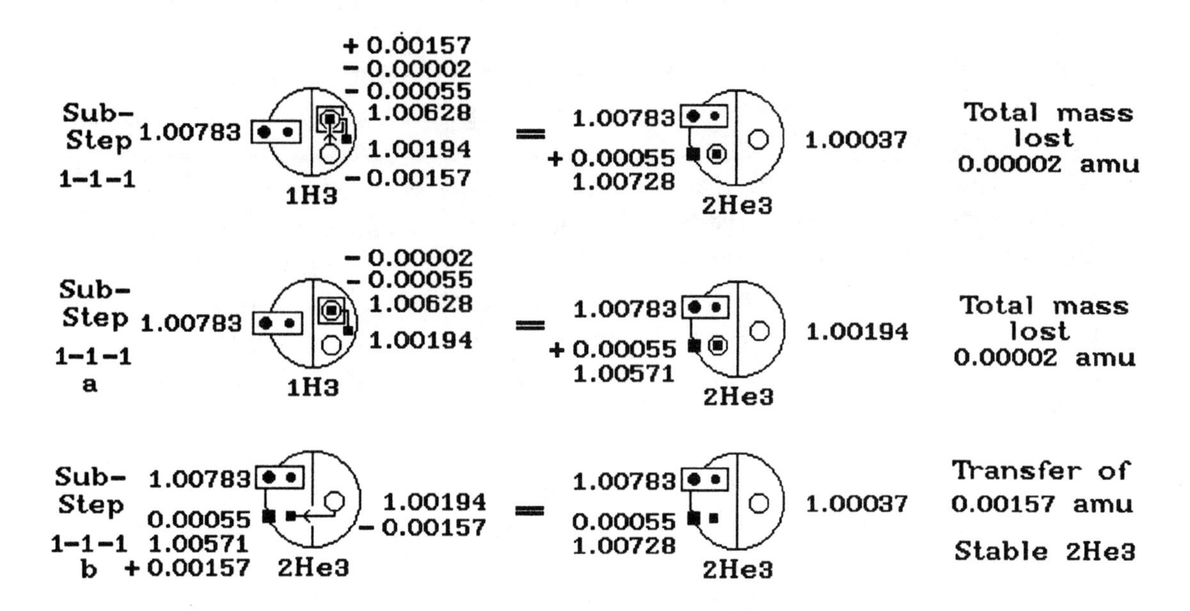

The isotope 1H3 with 3.01605 amu's has one electron with 0.00055 amu and one proton with 1.00728 amu's plus two neutrons. The first neutron has 1.00628 amu's and the second neutron has 1.00194 amu's.

The isotope 2He3 with 3.01603 amu's has one electron with 0.00055 amu, one negatron with 0.00055 amu, one proton with 1.00728 amu's, one positon with 1.00728 amu's and one neutron with 1.00037 amu's.

The total mass lost is 0.00002 amu and it is emitted as a #2 gamma ray which does not have an external magnetic field.

The isotope 1H3 has its electron drawn as a dot on the big circle and the proton is drawn as a dot to the left of the vertical line within the big circle. There is a rectangle drawn around both in the isotopes 1H3 and 2He3. The rectangle means that they are used as the isotope 1H1 with 1.00783 amu's and the isotope 1H1 can drop out at some later time from this isotope 2He3.

The first neutron is drawn as a small circle with a small square drawn in its middle area. This first neutron is drawn on the right side of the vertical line within the big circle. There is a square drawn around the small circle with a square in its middle to show that 0.00002 amu has been lost as a #2 gamma ray. When this 0.00002 amu has been lost, the negatron is released from the first neutron to go into an orbital that has an electron already in it. There is a line drawn from the first neutron to the negatron to show that the first neutron released the negatron.

When the first neutron with 1.00628 amu's in the isotope 1H3 loses 0.00002 amu and 0.00055 amu, it becomes the positon with 1.00571 amu's in the isotope 2He3. The first neutron vanishes when the positon appears in the isotope 2He3.

The second neutron has 1.00194 amu's and it is used as is with no loss of mass in the isotope 2He3. It is drawn as a small circle with nothing in its middle, to the right of the vertical line within the big circle below the first neutron in the isotope 1H3. When it becomes the single neutron in the isotope 2He3, it is drawn the same way as the second neutron in the isotope 1H3 except that there is no other neutron above it.

The isotope of 2He3 that is created by the loss of 0.00002 amu and the negatron that is released by the first neutron is unstable and it becomes stable by transferring 0.00157 amu to the positon that has 1.00571 amu's. This change the single neutron in the unstable isotope of 2He3 into the single neutron found in the stable isotope of 2He3 with 1.00037 amu's. The positon in the unstable isotope of 2He3 changes into a positon with 1.00728 amu's. This makes the stable isotope have a proton and a positon with each having 1.00728 amu's.

What I am about to say regarding the drawing of the Sub-step 1-1-1 is important because it most likely will be missed by the reader.

I want to direct the attention of the reader to the drawing of Sub-step 1-1-1 where there is a line drawn from the first neutron to the negatron. Note that this line does not go to the square in the middle of the small circle. It is the first neutron that releases the negatron and not the positon. The small square represents the positon and therefore the line showing the release of the negatron can not be connected to it. This is shown is the drawing of Sub-step 1-1-1a.

Now look at the line drawn in Sub-step 1-1-1 from the second neutron to the first neutron. While this line is drawn to the first neutron, it is connected to the small square in the middle of the small circle. This line has an arrow on it to show that 0.00157 amu has been transferred. This means that the second neutron does not transfer the 0.00157 amu to the first neutron therefore the line drawn from the second neutron must connect to the positon that is the small square. This is shown in the drawing of Sub-step 1-1-1b.

In Sub-step 1-1-1b, the stable isotope of 2He3 has an isotope of 1H1 within it having 1.00783 amu's, a negatron with 0.00055 amu and positon having 1.00728 amu's with a single neutron that has 1.00037 amu's. Their combined mass creates the mass of a stable isotope of 2He3.

To find the mass that is lost creating the isotope 2He3, subtract the isotope 2He3 from the isotope 1H3.

Mass of 1H3	3.01605 amu's
Mass of 2He3	3.01603 amu's
Mass lost creating 2He3	0.00002 amu

This 0.00002 amu is lost by the first neutron in the isotope 1H3. It is lost as a single unit as a #2 gamma ray. This is the only mass that is really lost. When the 0.00002 amu is lost, a negatron is released from within the first neutron. Add the real mass that is lost and the mass of the negatron then subtract their total lost mass from the first neutron to find the mass of the single neutron in the unstable isotope of 2He3.

Mass lost creating 2He3	0.00002 amu
Mass of negatron released by first neutron in 2He3	0.00055 amu
Mass lost by first neutron in 2He3	0.00057 amu

Mass of first neutron in 1H3	1.00628 amu's
Mass lost by it	0.00057 amu
Mass of positon in 2He3	1.00571 amu's

The isotope 1H1 is found intact within the isotopes 1H3 and 2He3 with 1.00783 amu's. This means that there is a proton in the isotope 2He3 with 1.00728 amu's. All of these parts make an unstable isotope of 2He3.

Mass of 1H1 in unstable 2He3	1.00783 amu's
Mass of negatron in unstable 2He3	0.00055 amu
Mass of positon in unstable 2He3	1.00571 amu's
Mass of neutron in unstable 2He3	1.00194 amu's
Mass of unstable 2He3	3.01603 amu's

To become stable the single neutron with 1.00194 amu's in the unstable isotope of 2He3 transfers 0.00157 amu to the positon that has 1.00571 amu's making it have 1.00728 amu's.

Mass of neutron in unstable 2He3	1.00194 amu's
Mass transferred to positon in unstable 2He3	0.00157 amu
Mass of neutron in stable 2He3	1.00037 amu's

Mass of positon in unstable 2He3	1.00571 amu's
Mass transferred to it from neutron in unstable 2He3	0.00157 amu
Mass of positon in stable 2He3	1.00728 amu's

Now add all of the masses that make up the stable isotope of 2He3.

Mass of 1H1 in stable 2He3	1.00783 amu's
Mass of negatron in stable 2He3	0.00055 amu
Mass of positon in stable 2He3	1.00728 amu's
Mass of neutron in stable 2He3	1.00037 amu's
Mass of stable 2He3	3.01603 amu's

This completes all of the steps using one isotope of 1H3 to create the isotope 2He3 with the loss of 0.00002 amu plus the release of a negatron.

I will now show all of the drawings showing how this is done in **STEP 1.** All drawings in **STEP 1** using the isotope 1H3 to create the isotope 2He3

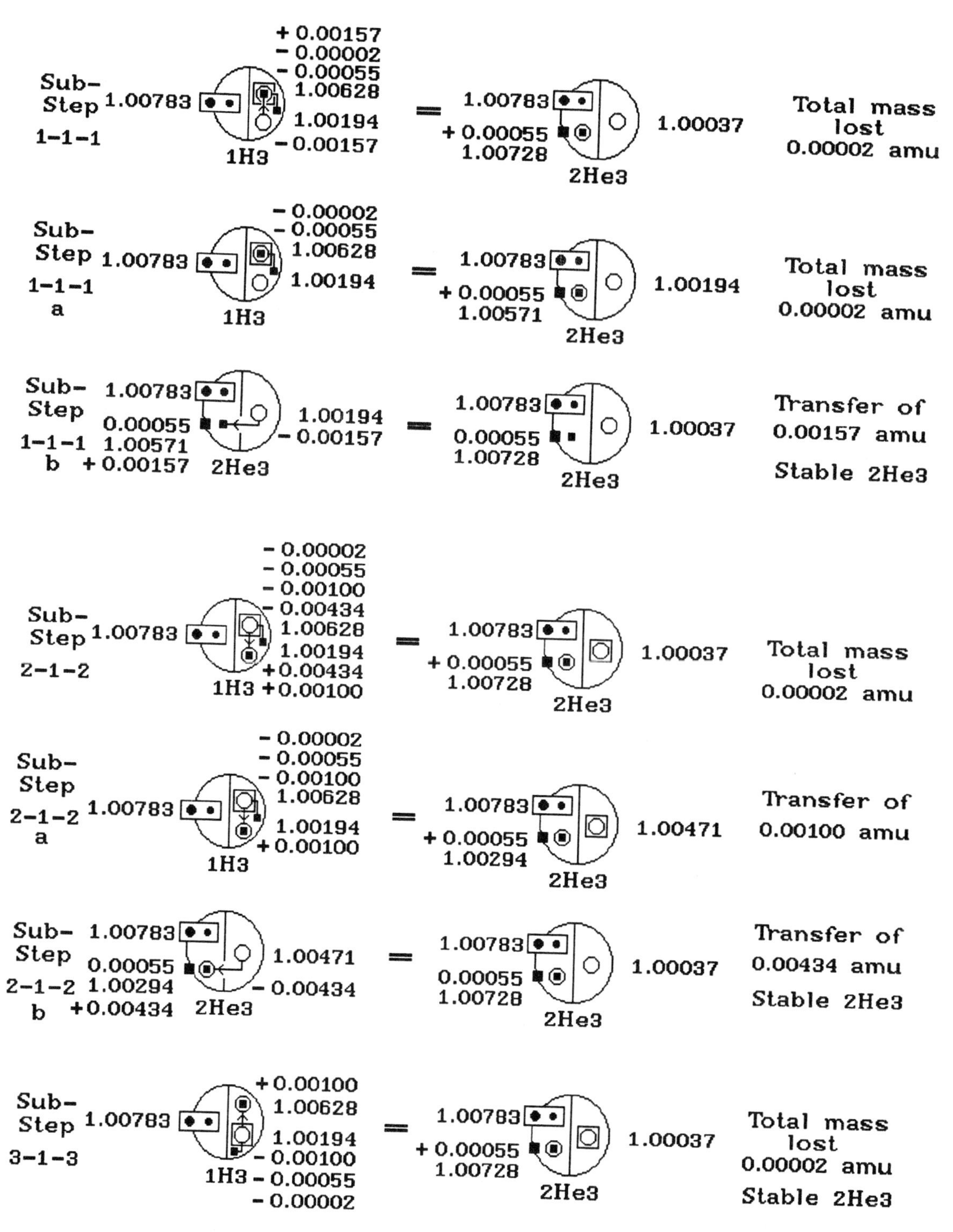

All drawings in **STEP 1** using the isotope 1H3 to create the isotope 2He3

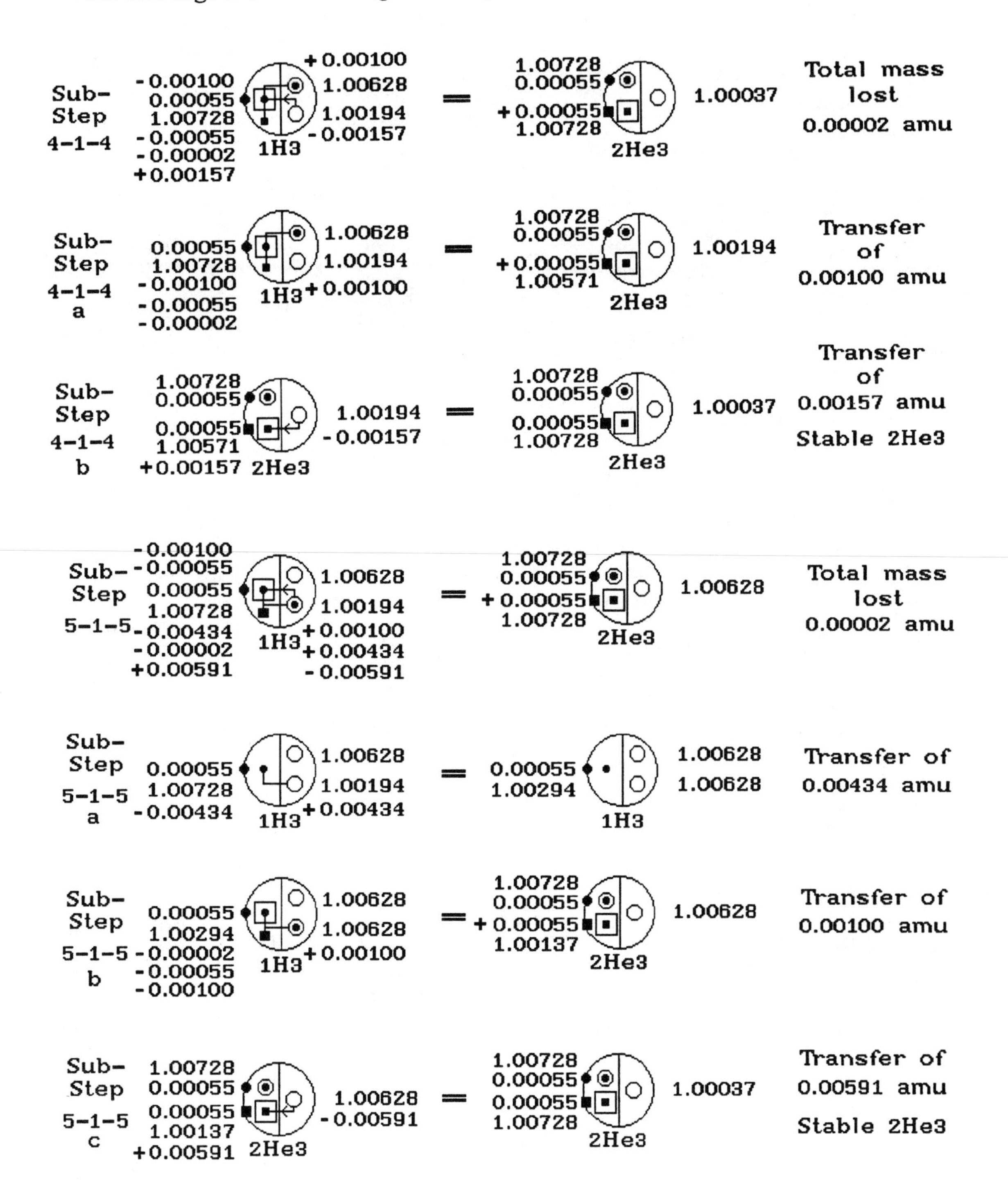

All drawings in **STEP 1** using the isotope 1H3 to create the isotope 2He3

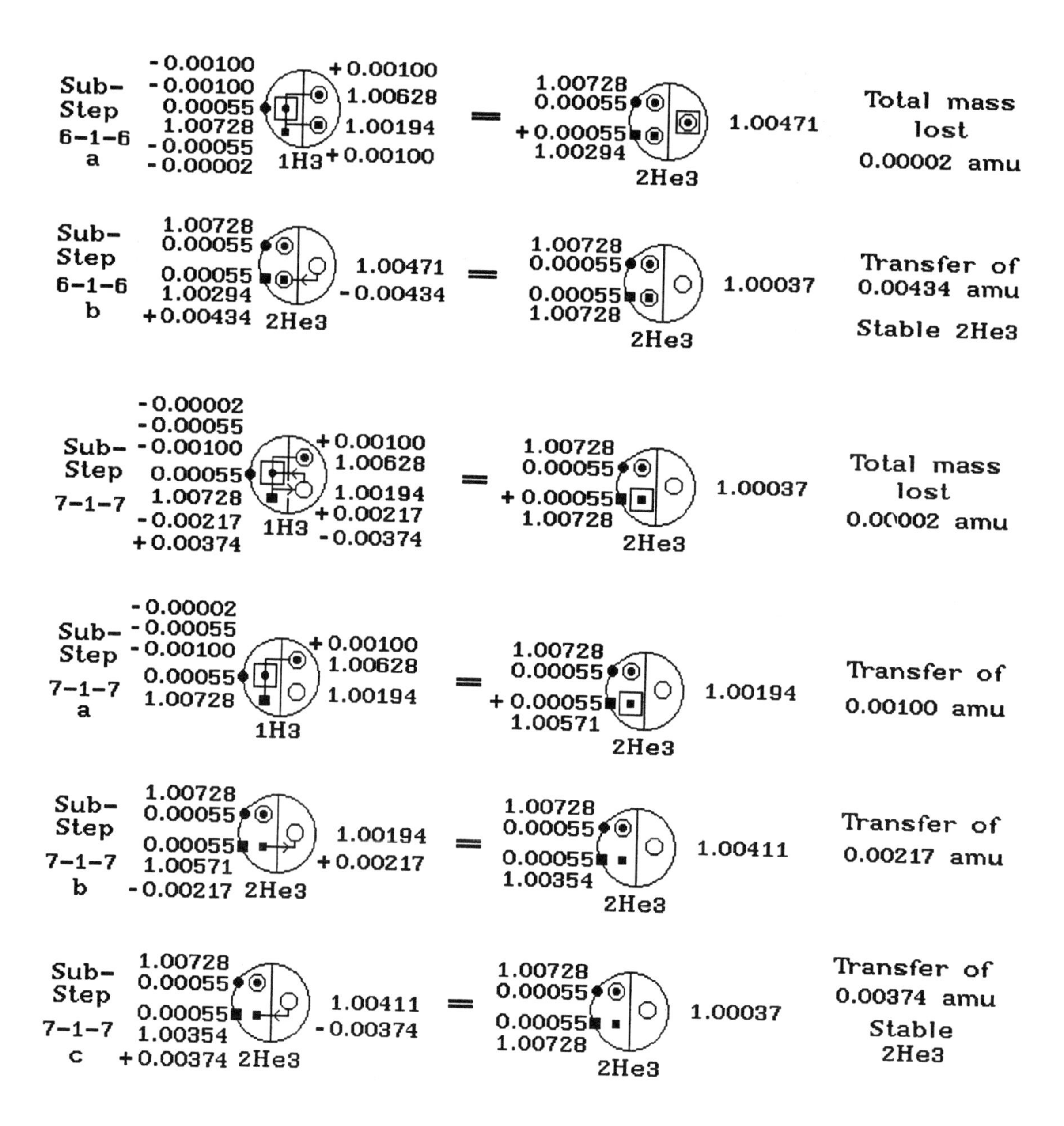

All drawings in **STEP 1** using the isotope 1H3 to create the isotope 2He3

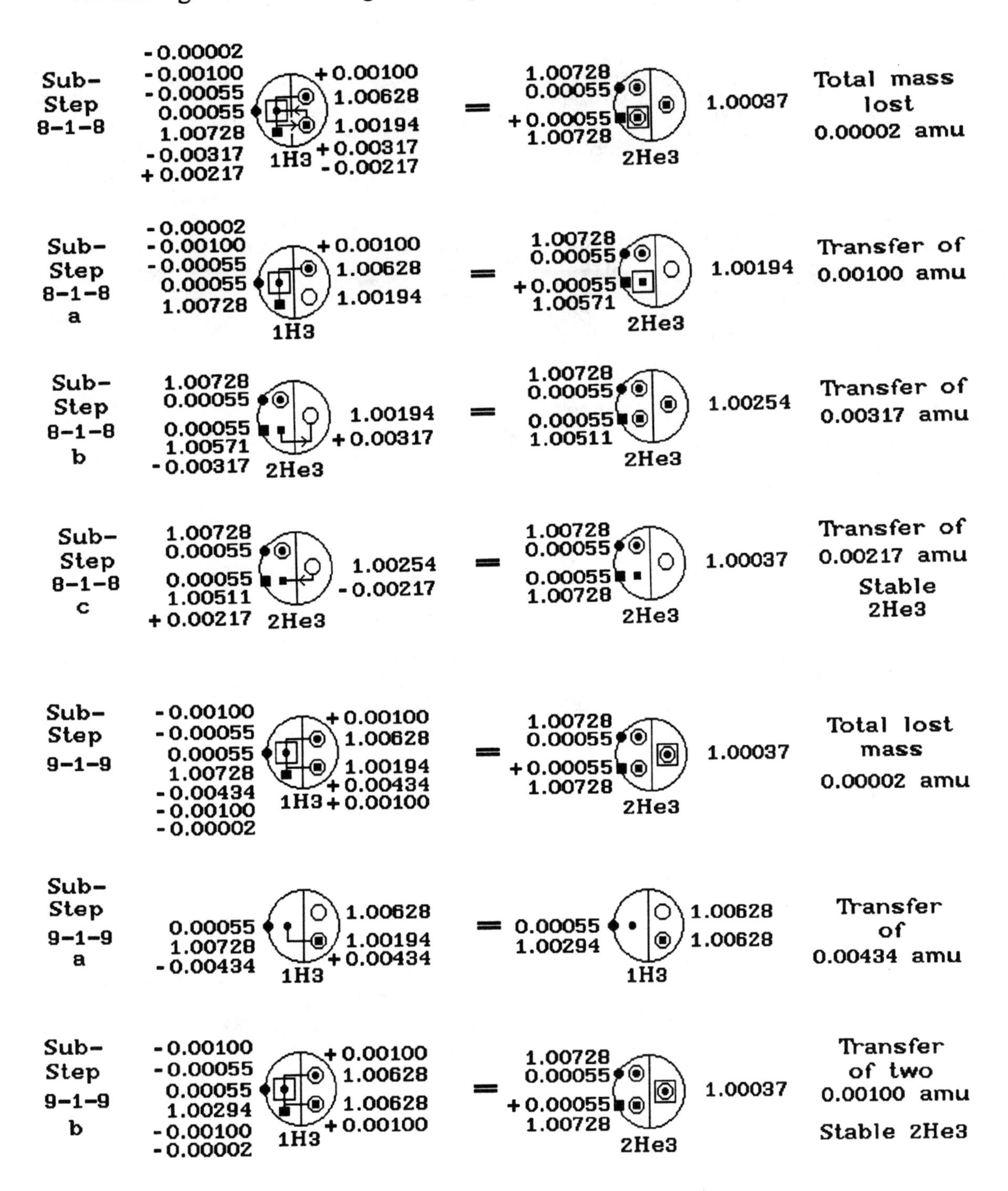

All drawings in **STEP 1** using the isotope 1H3 to create the isotope 2He3

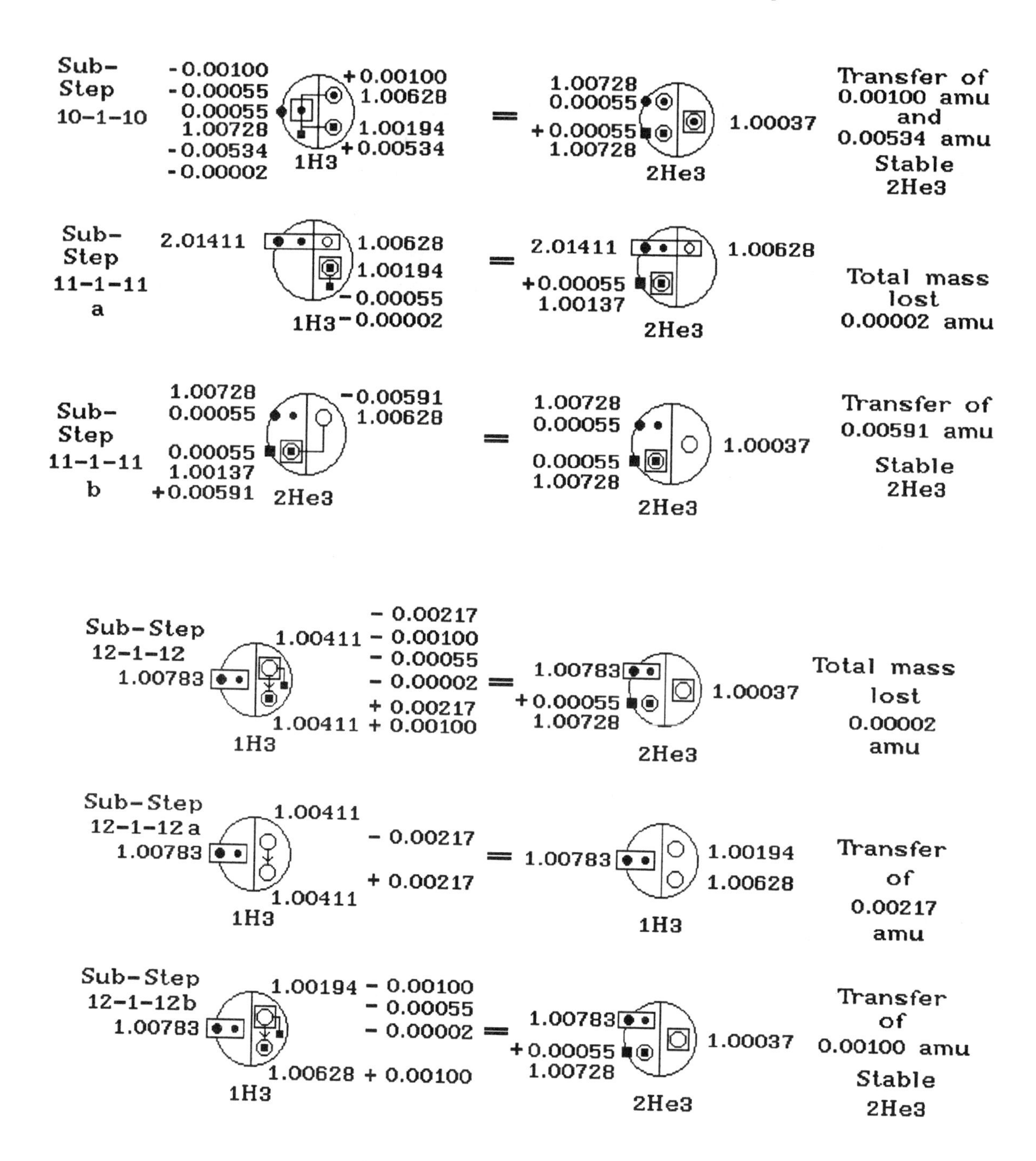

All drawings in **STEP 1** using the isotope 1H3 to create the isotope 2He3

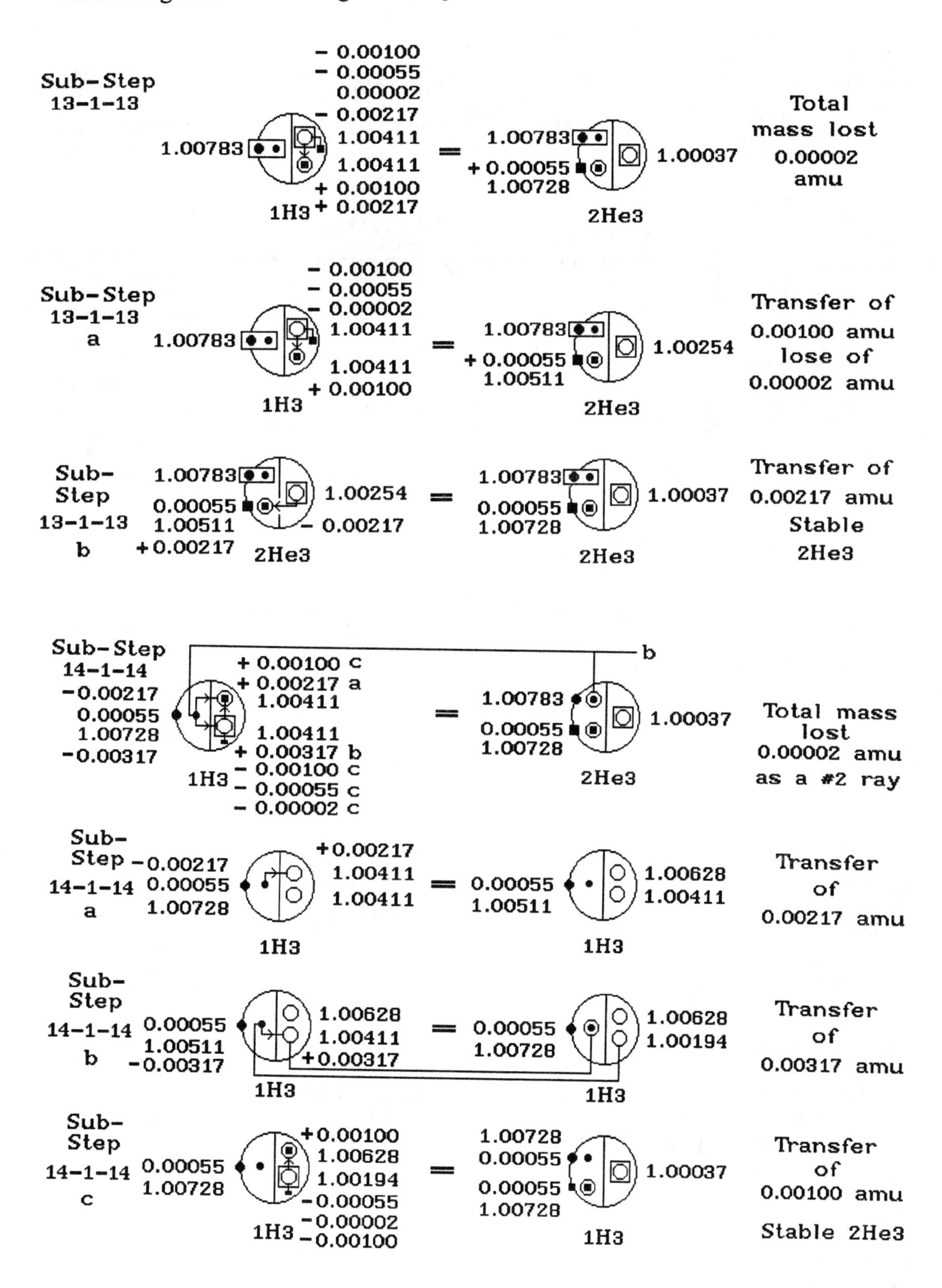

All drawings in **STEP 1** using the isotope 1H3 to create the isotope 2He3

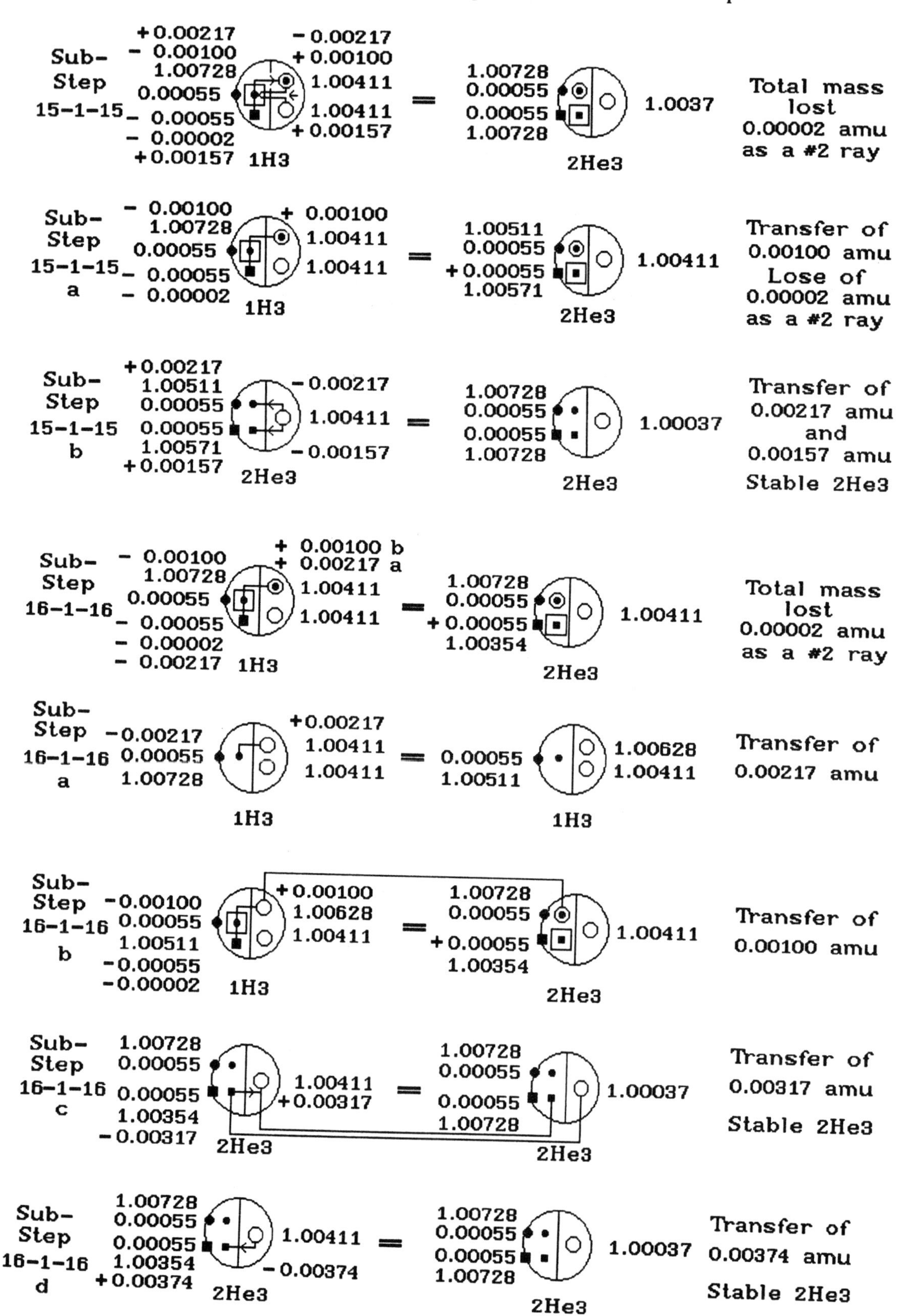

DRAWINGS OF ALL THE STEPS CREATING

THE ISOTOPE 2He3

FORMULAS

STEPS

1. 1H3 = 2He3
2. 1H3 + 1H2 = 2He3
3. 1H3 + 1H1 = 2He3
4. 1H2 + 1H2 = 2He3
5. 1H2 + 1H1 = 2He3
6. 1H1 + 1H1 + 1H1 = 2He3

   or

   2(1H1) + 1H1 = 2He3

DRAWINGS OF ALL THE STEPS CREATING THE ISOTOPE 2He3

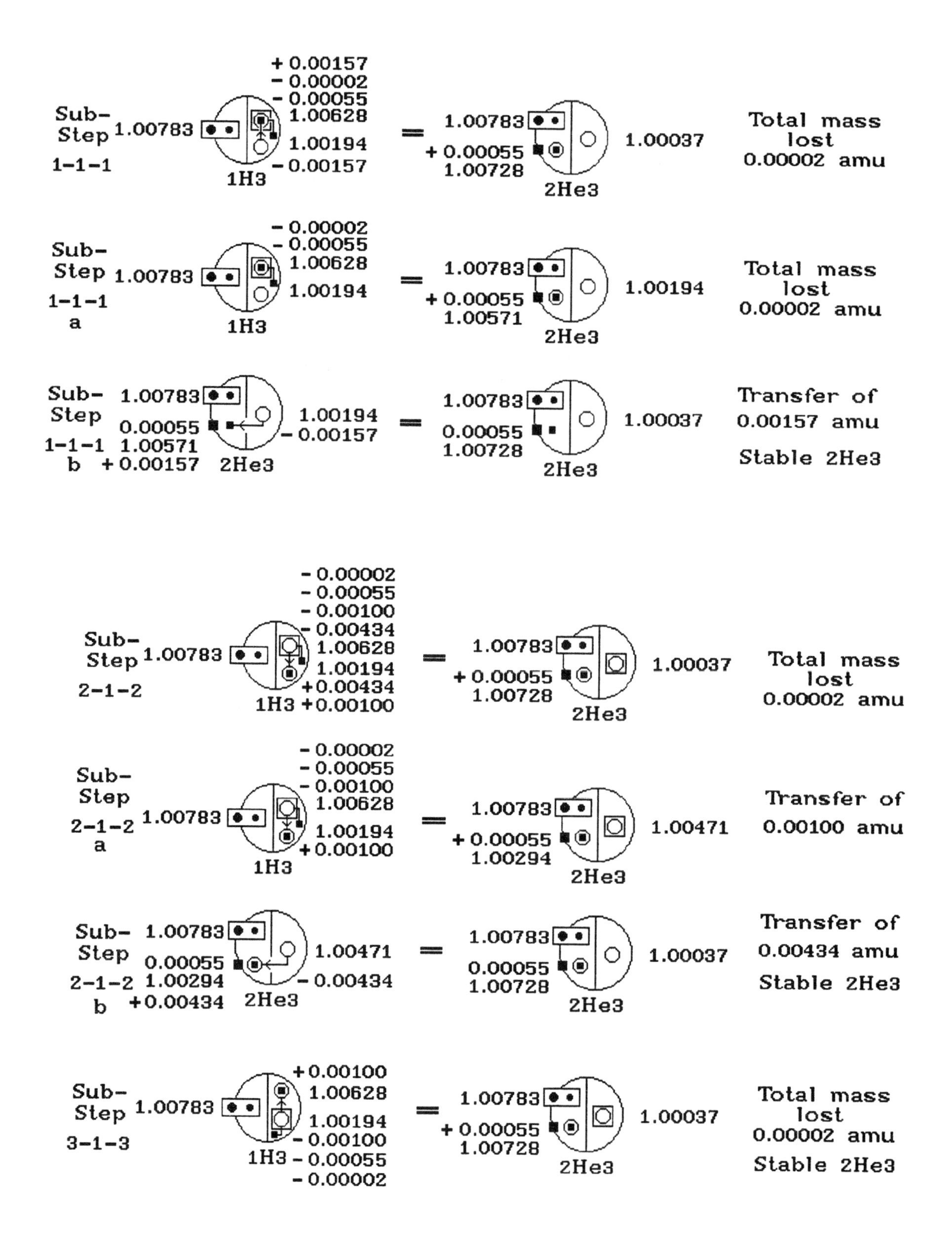

DRAWINGS OF ALL THE STEPS CREATING THE ISOTOPE 2He3

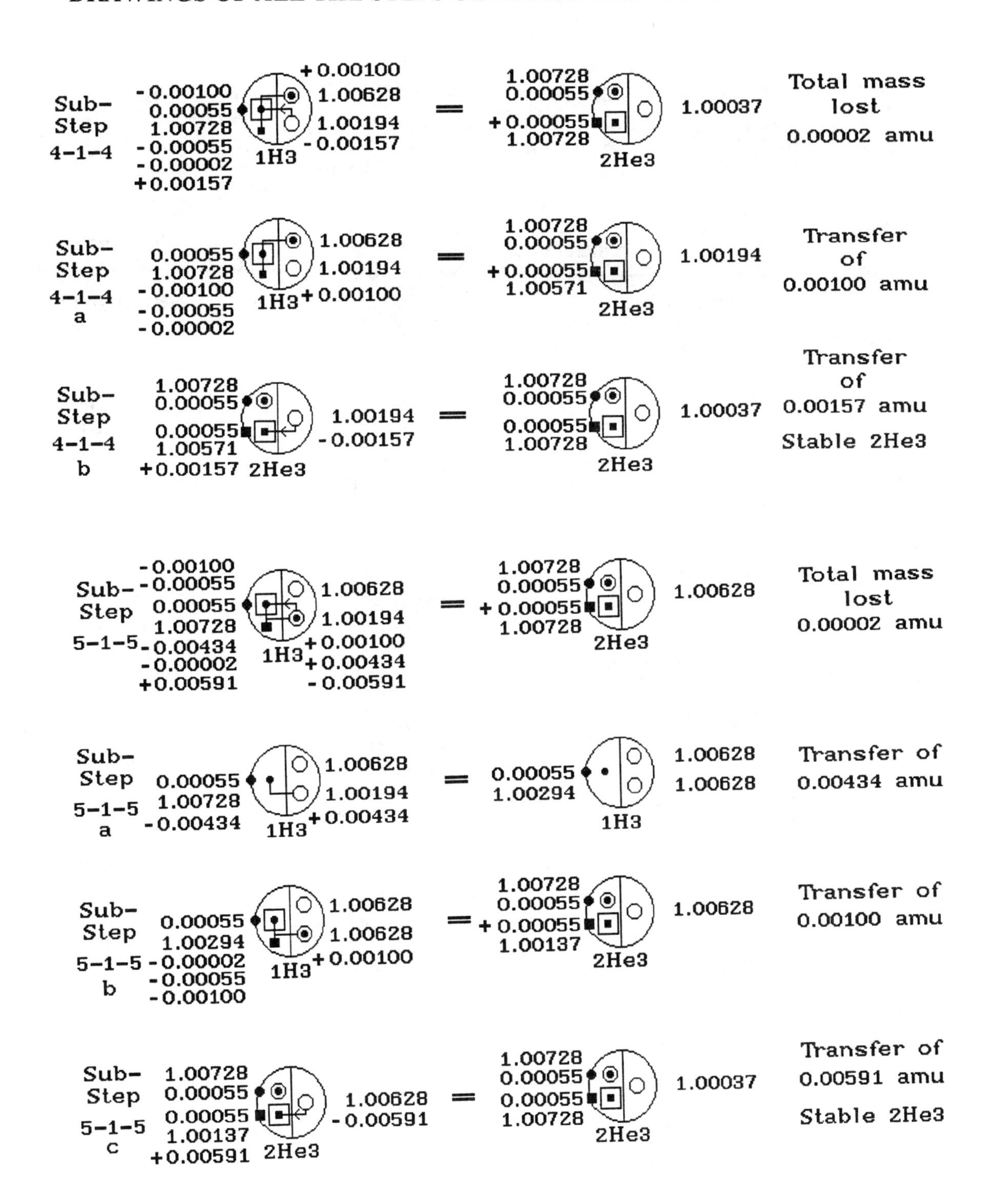

DRAWINGS OF ALL THE STEPS CREATING THE ISOTOPE 2He3

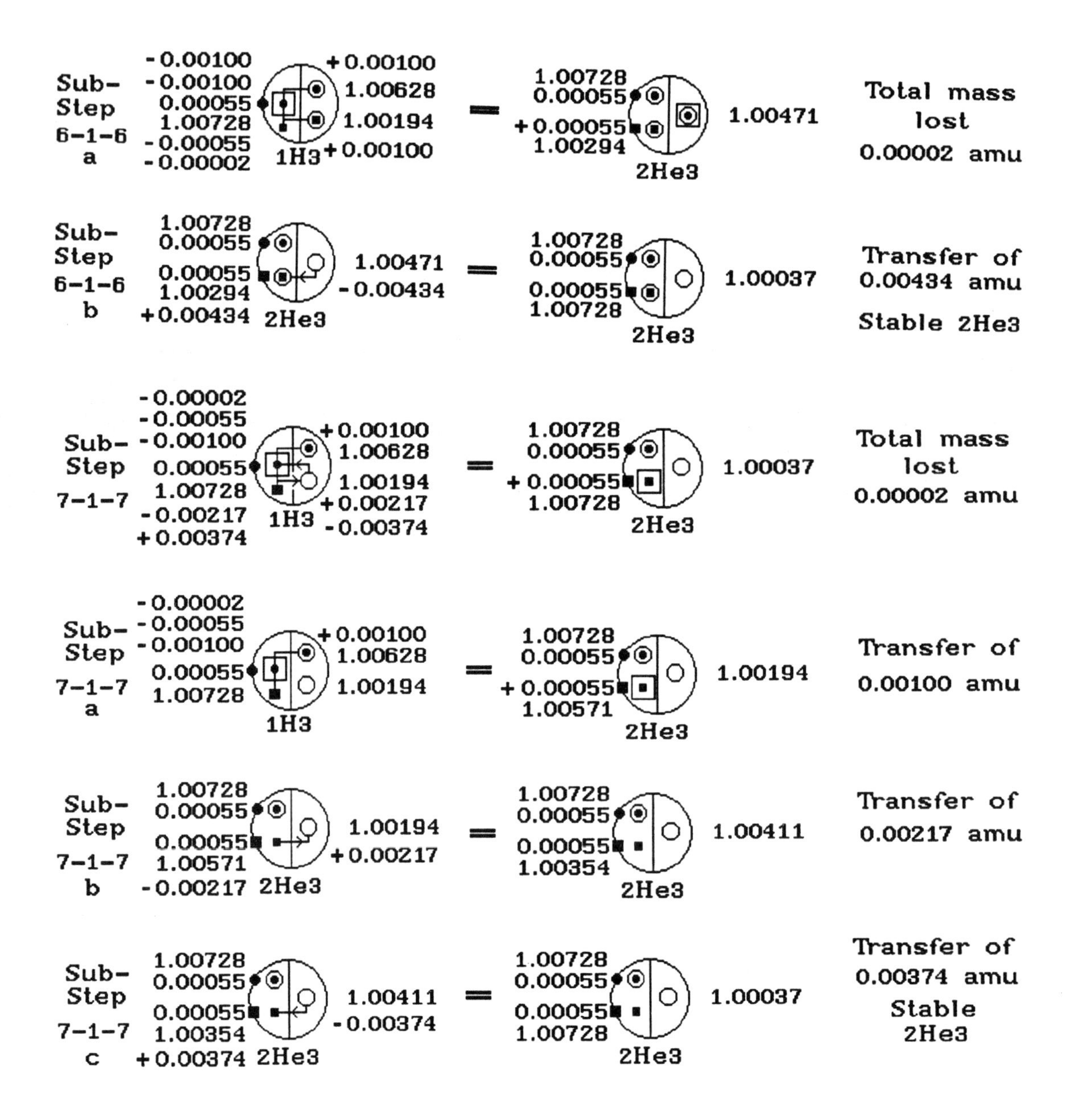

DRAWINGS OF ALL THE STEPS CREATING THE ISOTOPE 2He3

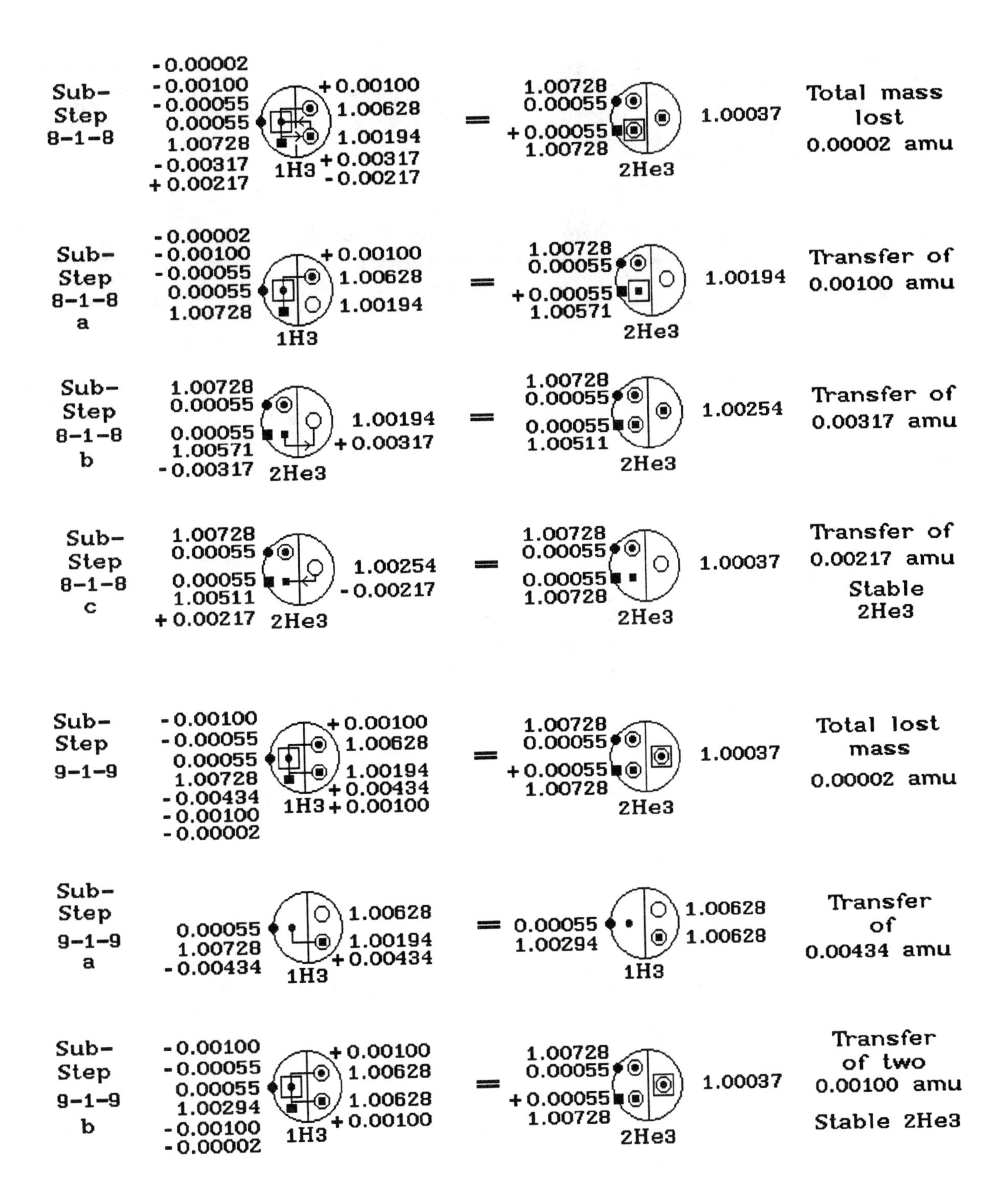

DRAWINGS OF ALL THE STEPS CREATING THE ISOTOPE 2He3

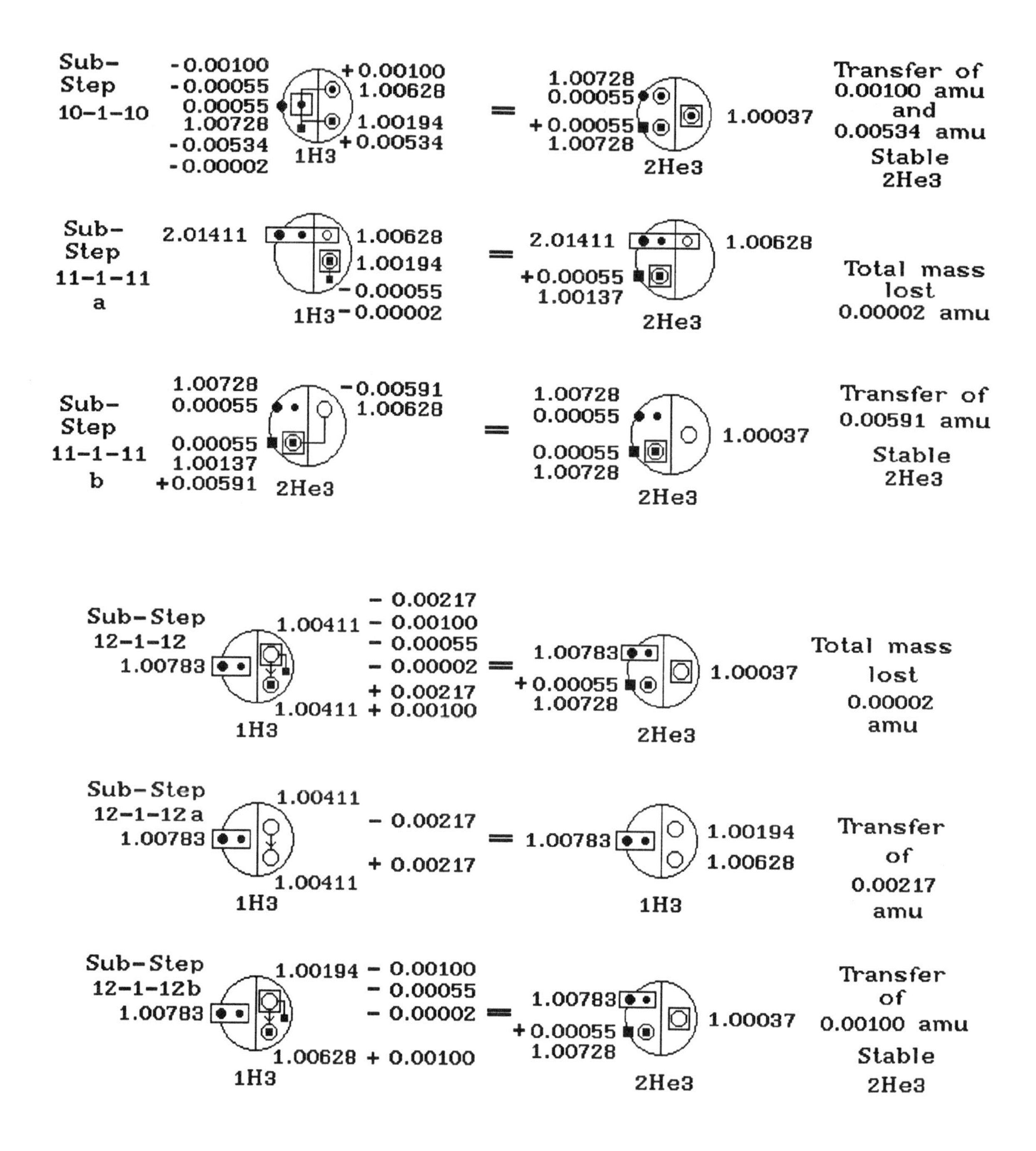

DRAWINGS OF ALL THE STEPS CREATING THE ISOTOPE 2He3

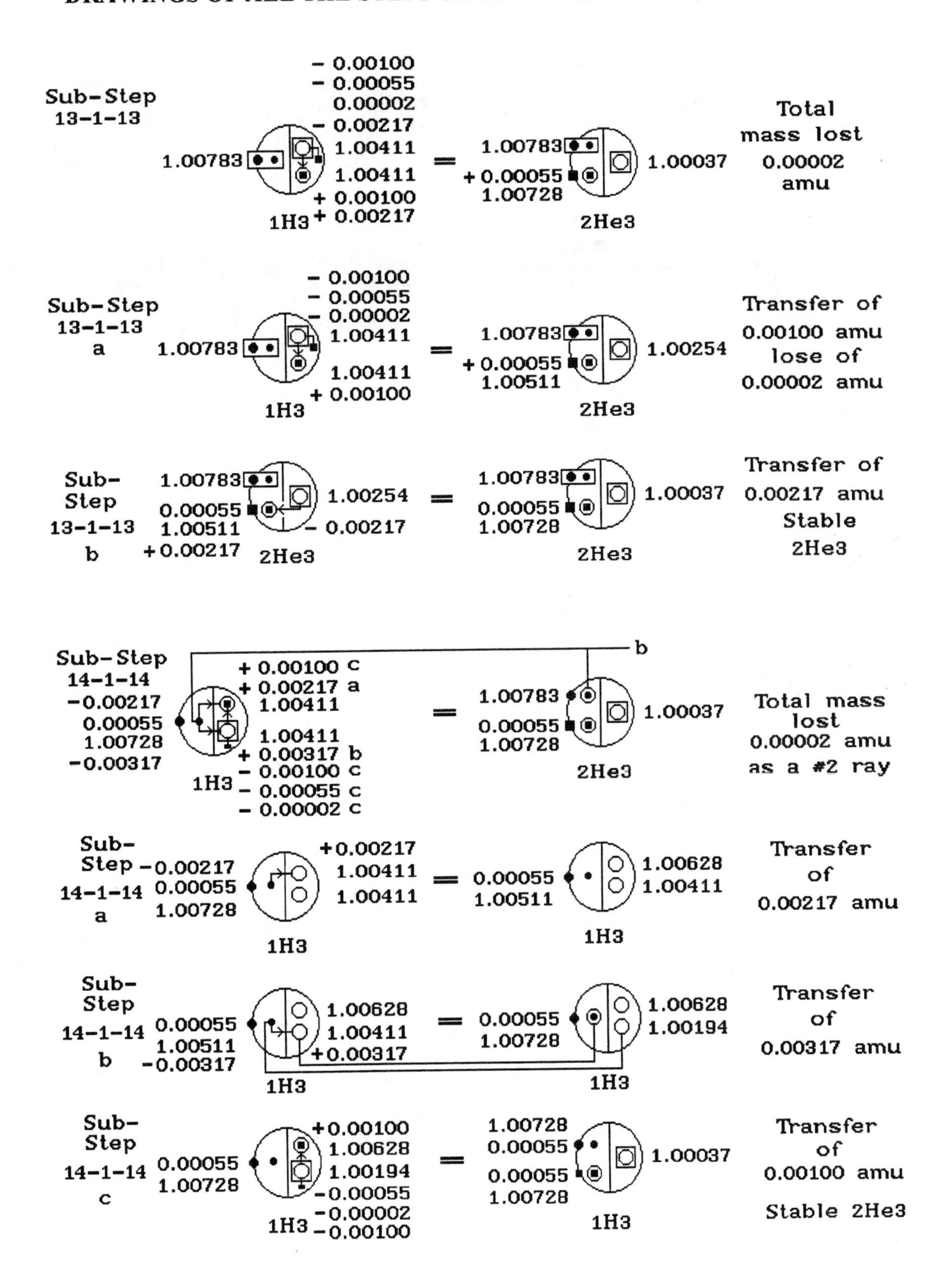

DRAWINGS OF ALL THE STEPS CREATING THE ISOTOPE 2He3

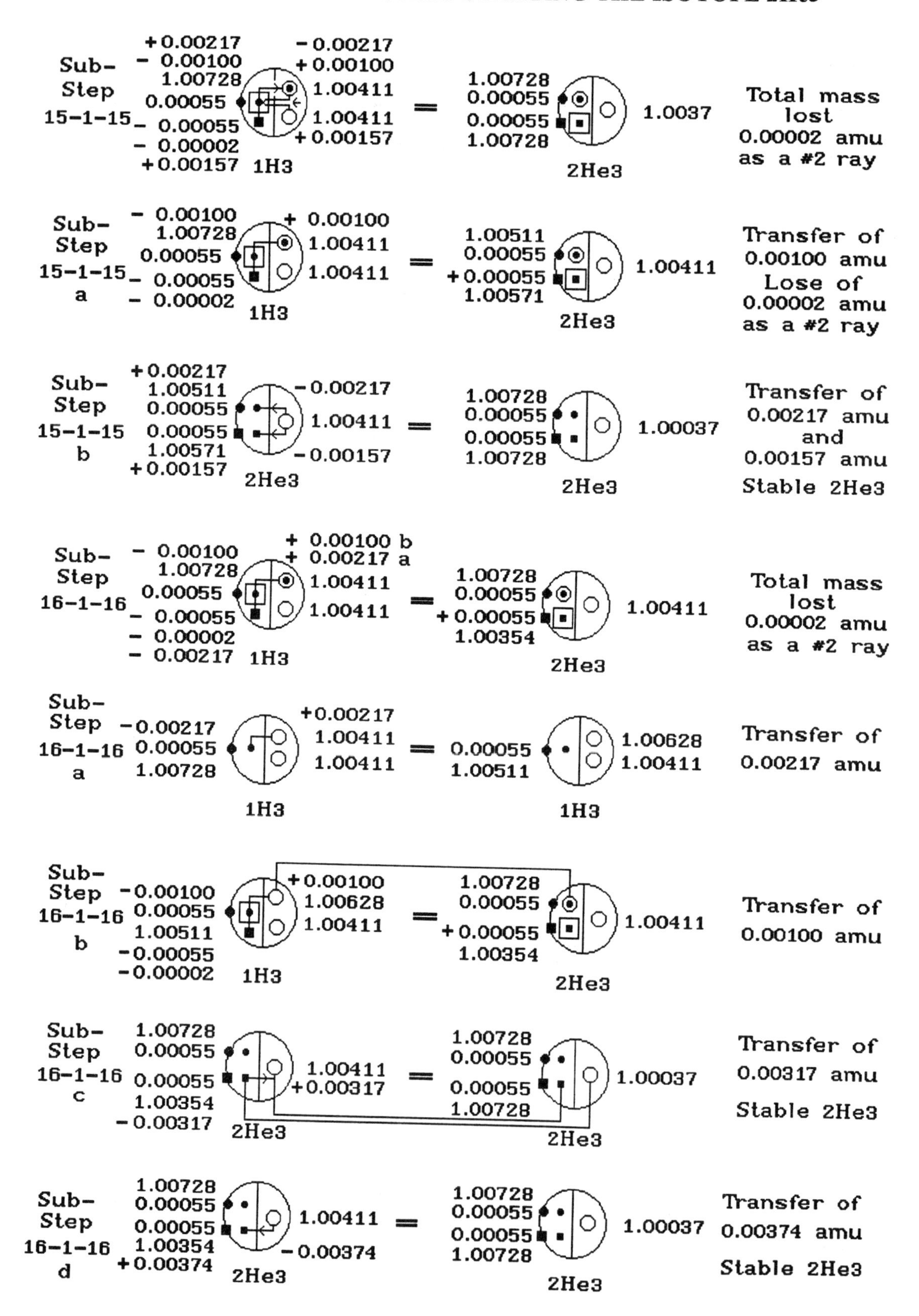

DRAWINGS OF ALL THE STEPS CREATING THE ISOTOPE 2He3

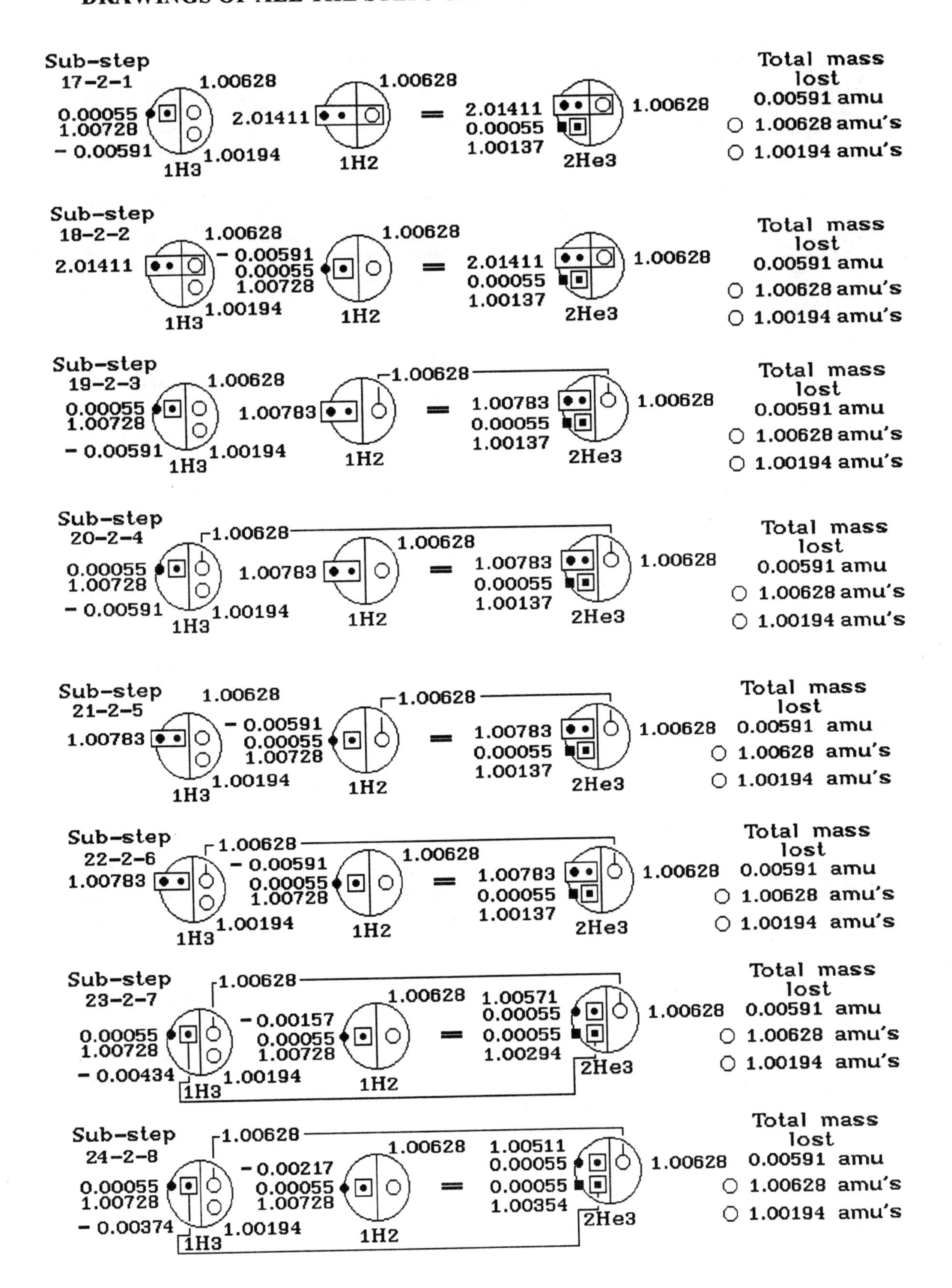

DRAWINGS OF ALL THE STEPS CREATING THE ISOTOPE 2He3

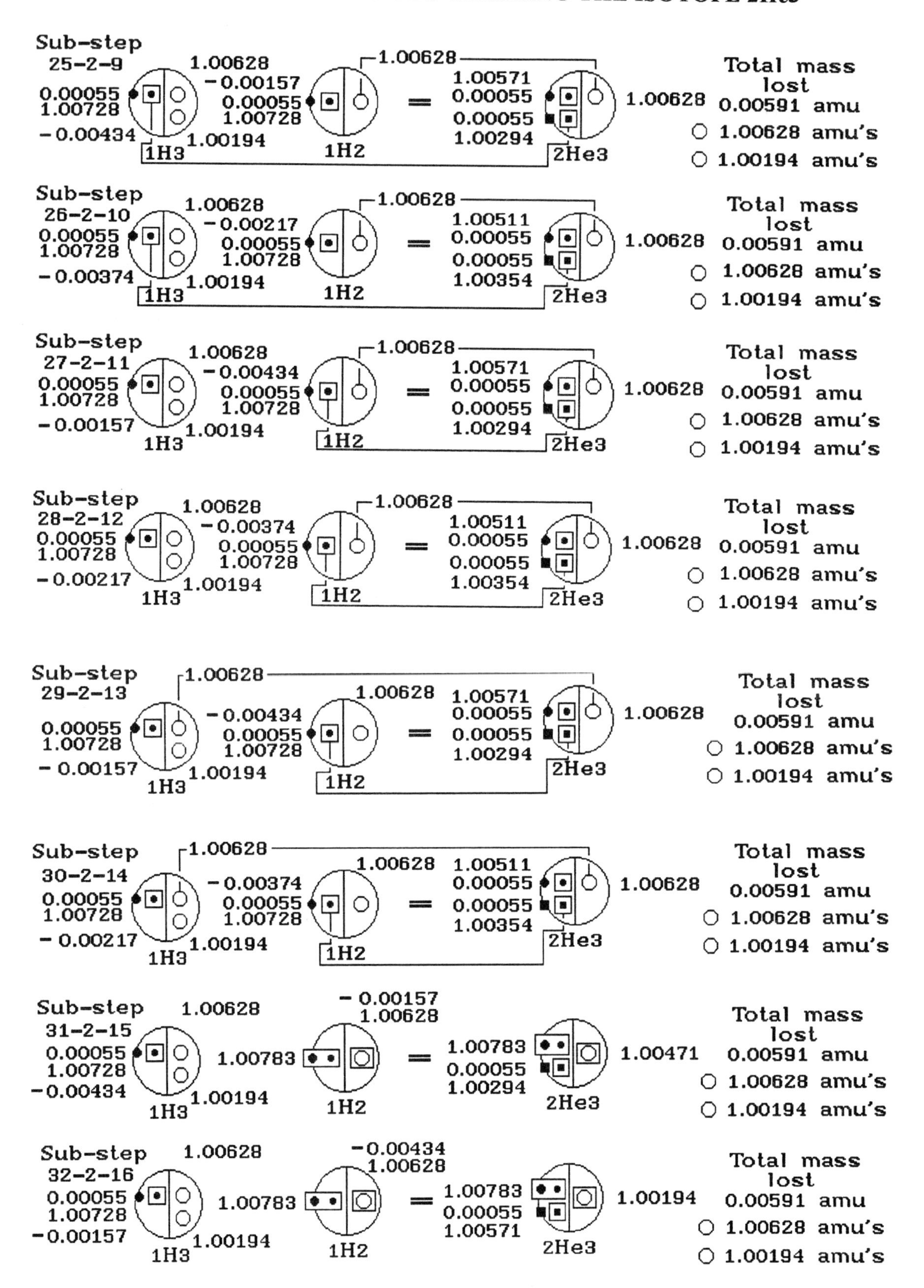

DRAWINGS OF ALL THE STEPS CREATING THE ISOTOPE 2He3

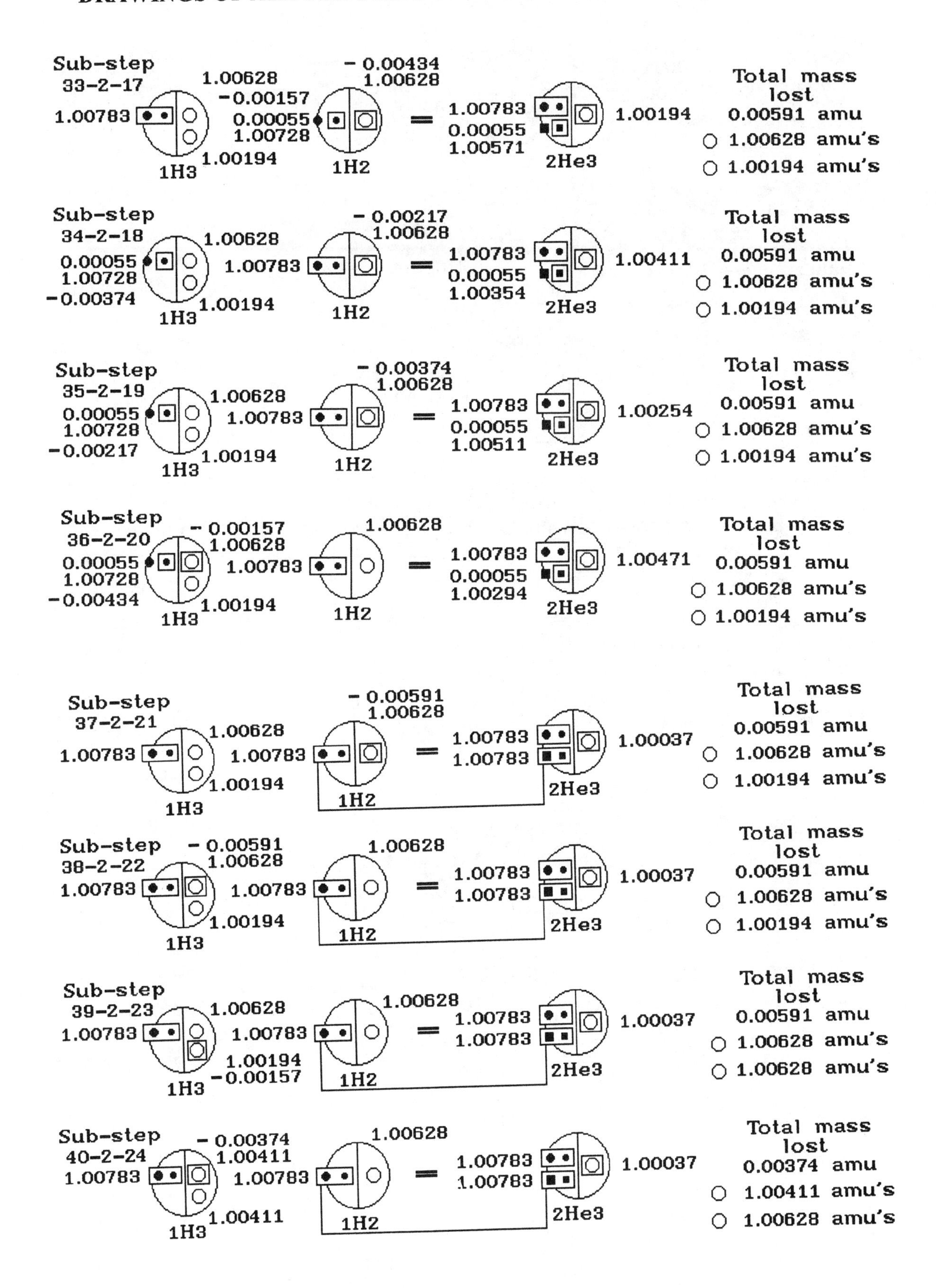

DRAWINGS OF ALL THE STEPS CREATING THE ISOTOPE 2He3

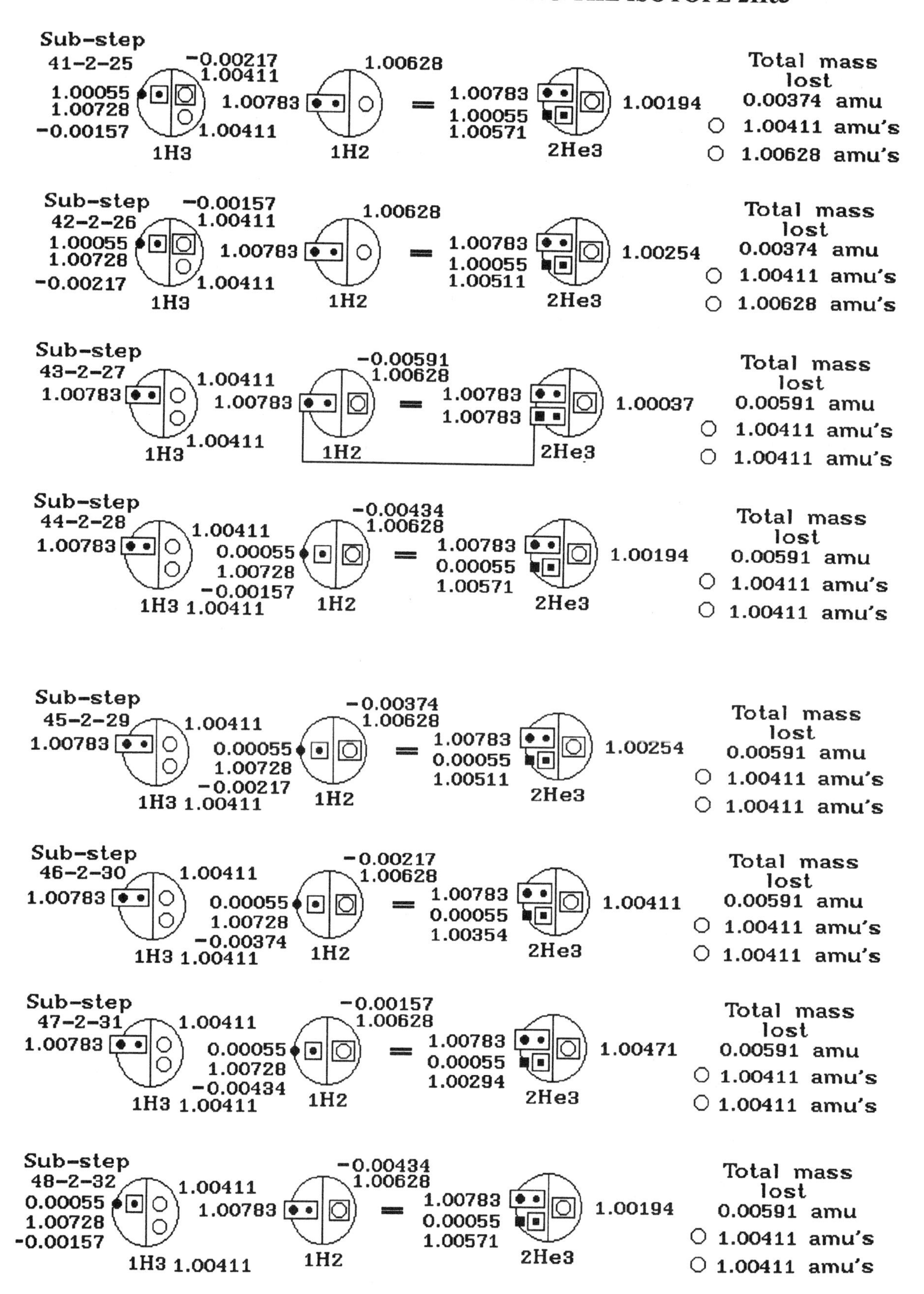

DRAWINGS OF ALL THE STEPS CREATING THE ISOTOPE 2He3

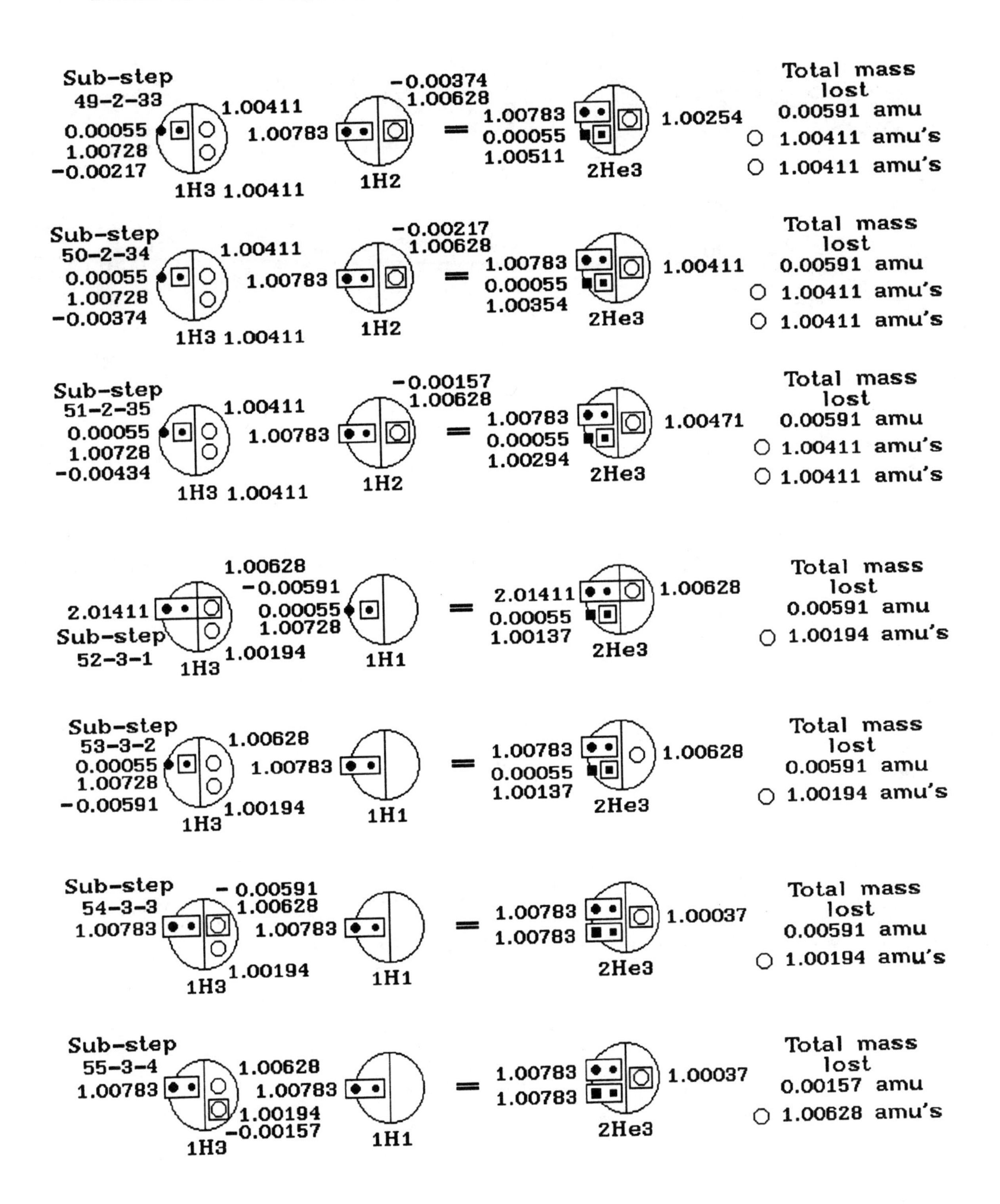

DRAWINGS OF ALL THE STEPS CREATING THE ISOTOPE 2He3

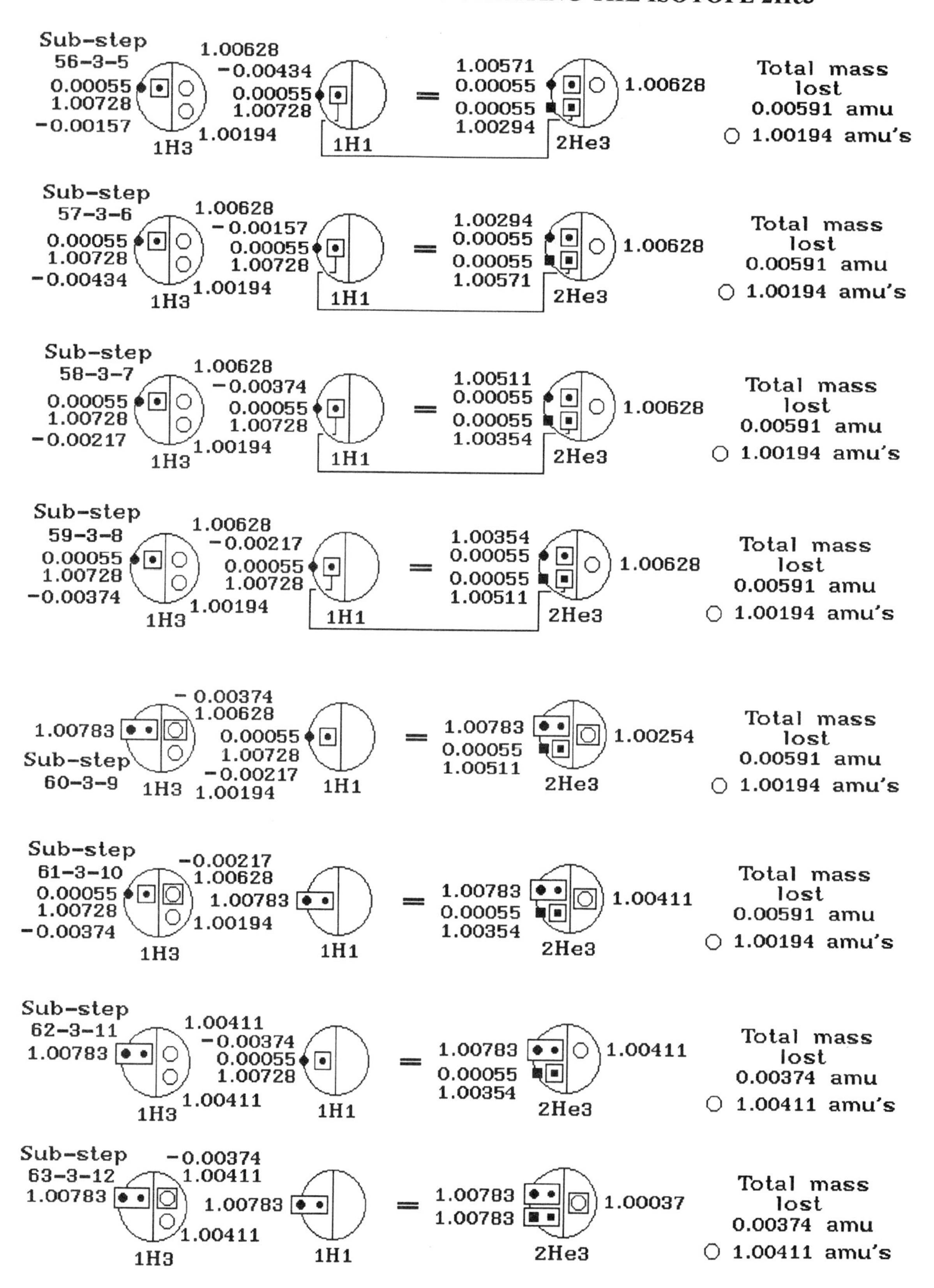

DRAWINGS OF ALL THE STEPS CREATING THE ISOTOPE 2He3

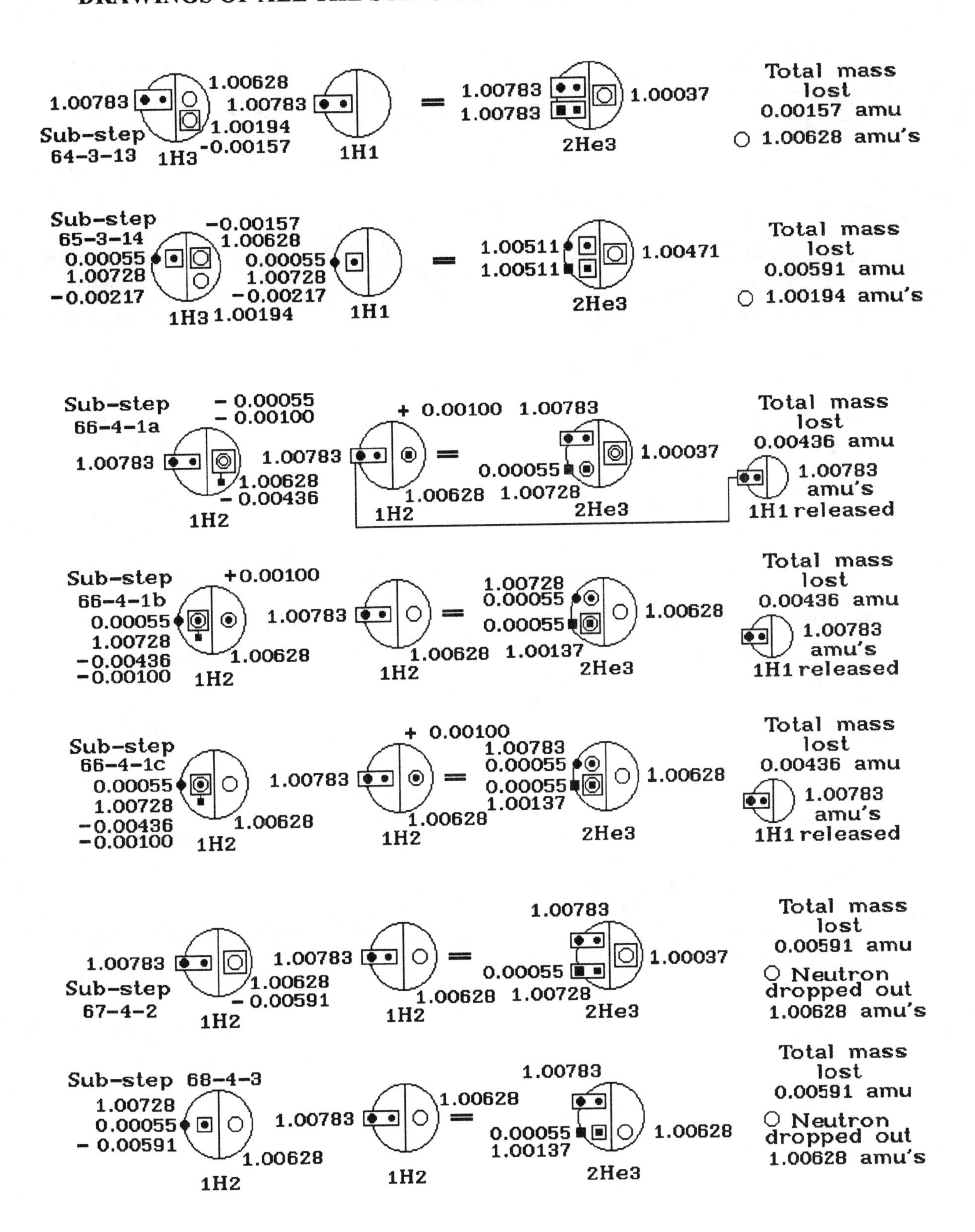

DRAWINGS OF ALL THE STEPS CREATING THE ISOTOPE 2He3

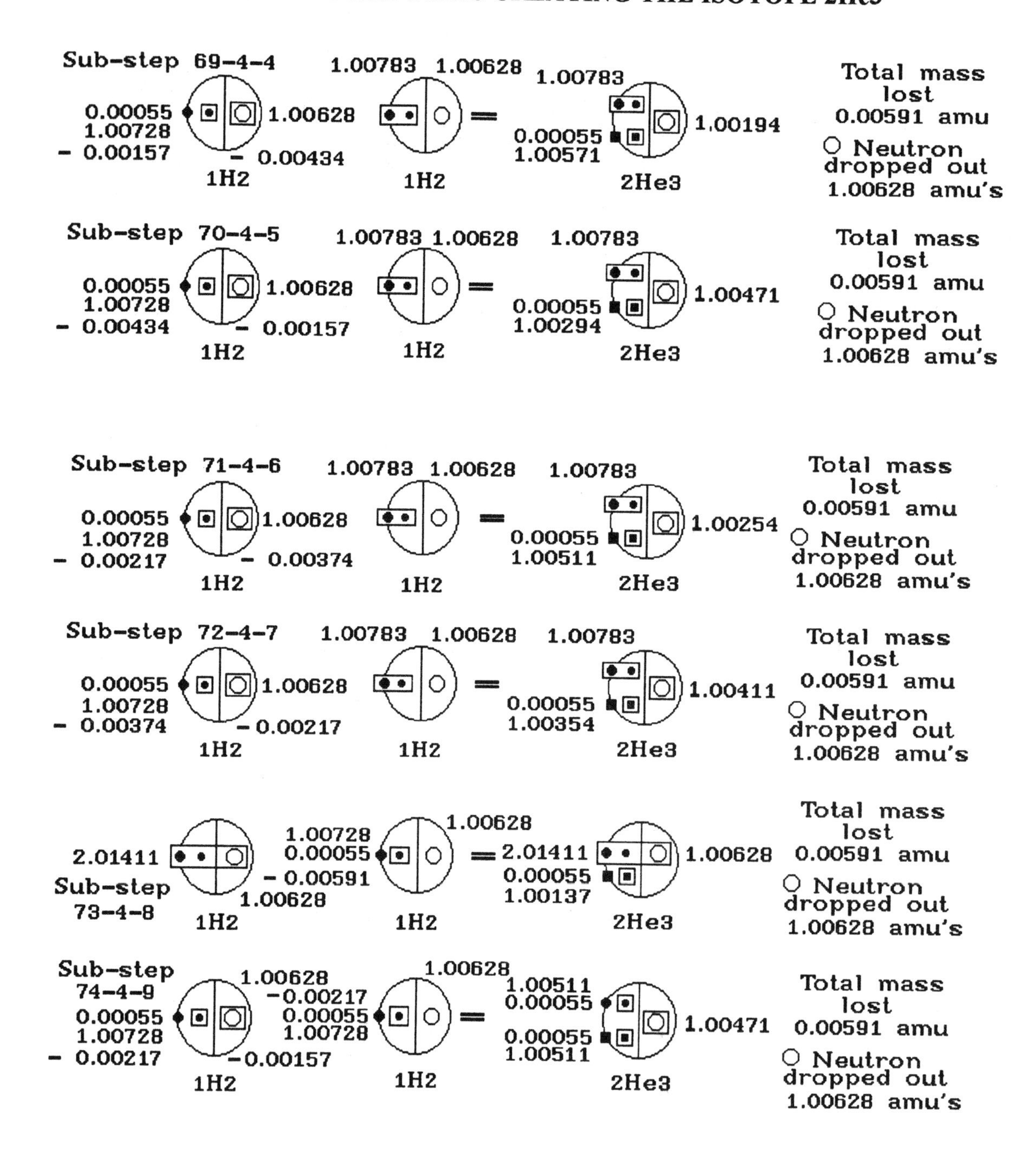

DRAWINGS OF ALL THE STEPS CREATING THE ISOTOPE 2He3

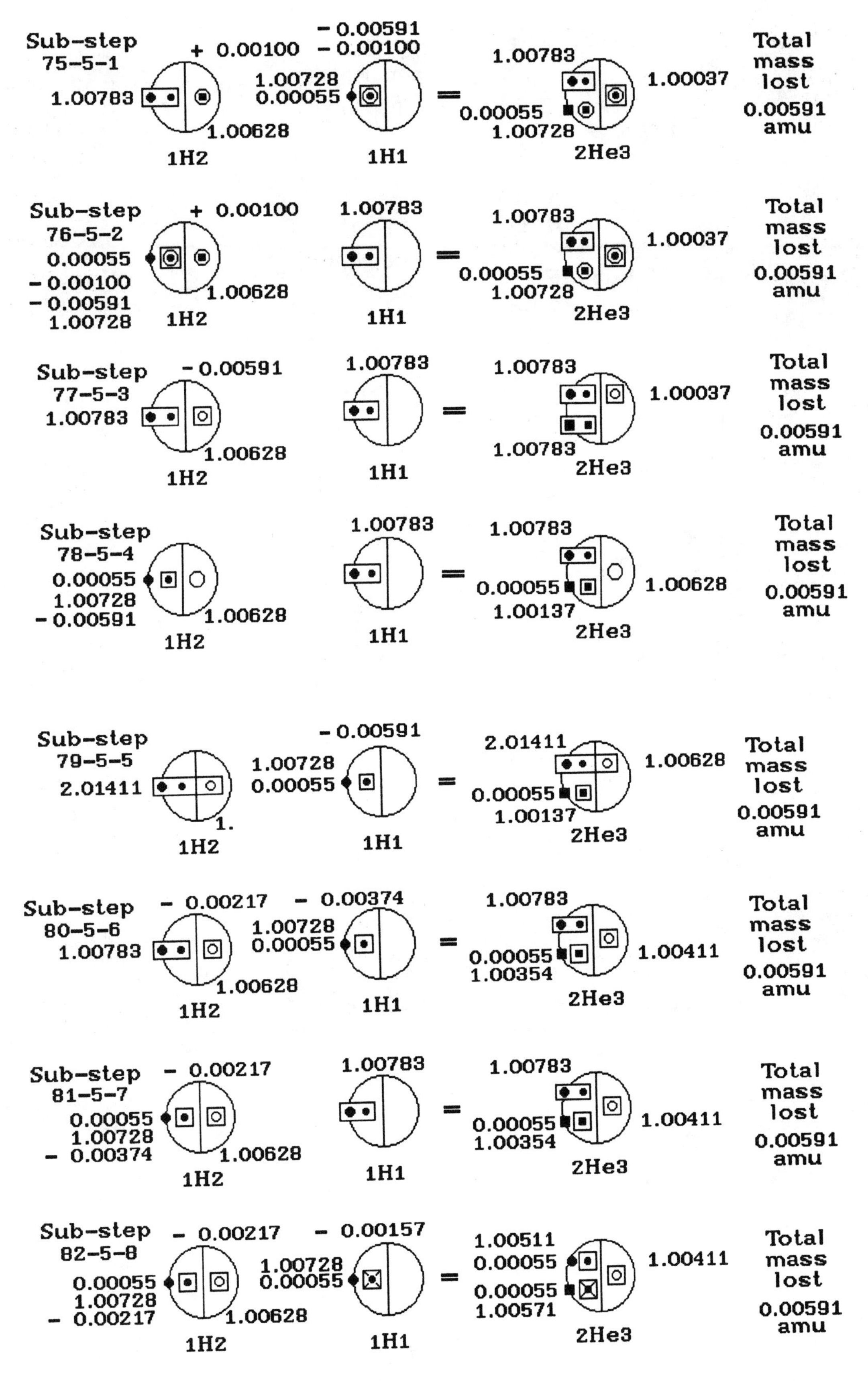

DRAWINGS OF ALL THE STEPS CREATING THE ISOTOPE 2He3

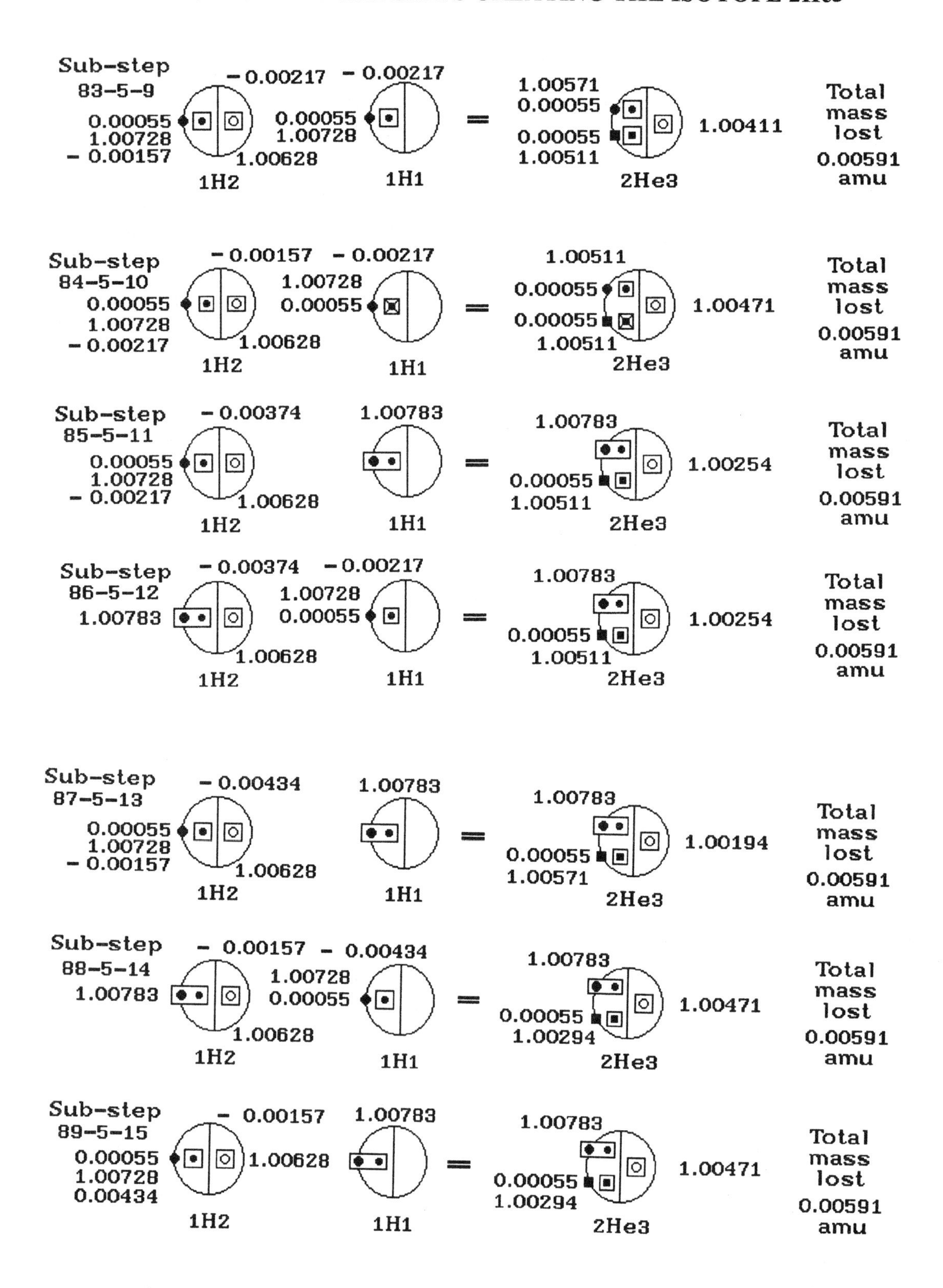

DRAWINGS OF ALL THE STEPS CREATING THE ISOTOPE 2He3

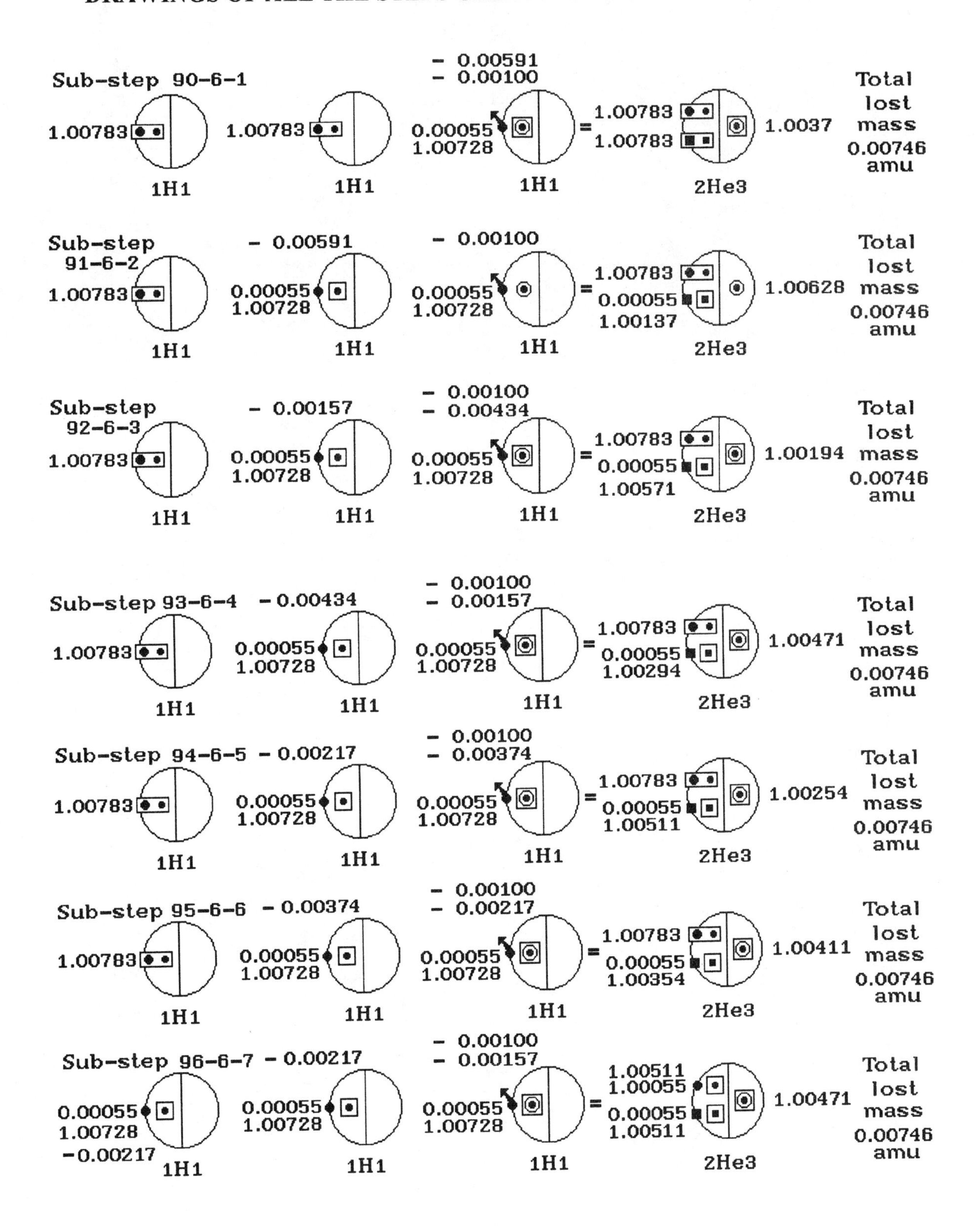

Notes

PART 5

BACK TO THE ORIGINAL NEUTRON

Now that all of the **STEPS** have been given, let's go back to the Original Neutron.

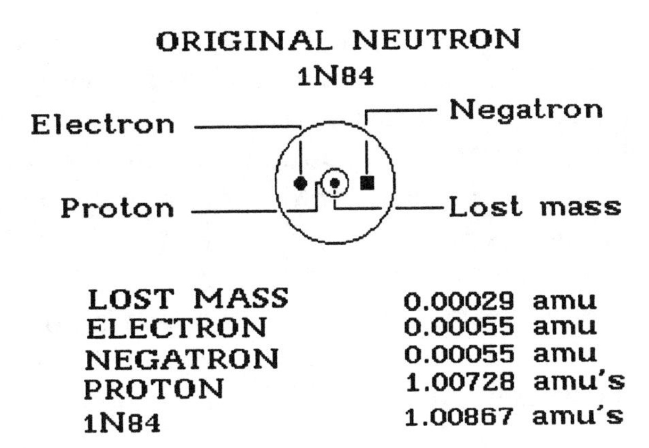

LOST MASS	0.00029 amu
ELECTRON	0.00055 amu
NEGATRON	0.00055 amu
PROTON	1.00728 amu's
1N84	1.00867 amu's

The Original Neutron loses 0.00084 amu.

From the drawing of the Original Neutron, the Original Neutron has a limit to its magnetic field. This magnetic field will not combine with the magnetic field found around an electron, negatron, proton or positon. Found within this magnetic field is a positron that has more mass than a normal proton. The normal proton has a magnetic field that is external and it will combine with other magnetic fields except when it is within the positron. The amount of mass the positron has more than a proton is the mass that is lost when the Original Neutron becomes part of an isotope and either an electron or negatron. The Original Neutron is not an isotope so when it becomes part of an isotope, it must lose a certain amount of mass that is returned to pure energy which is magnetism. This mass will always include 0.00029 amu and the mass of either an electron or negatron. This lost mass of 0.00029 amu will include the electron and negatron when the Original Neutron becomes a normal neutron within an isotope just created. If the Original Neutron does not become a normal neutron in a just created isotope, then the isotope 1H1 is created with the loss of either an electron or negatron from the positron and it loses the 0.00029 amu.

There is an electron and a negatron that is part of the positron. One or both of these masses must be lost when the Original Neutron becomes a normal neutron in an isotope being created. An electron and a negatron never lose any mass so they will always have 0.00055 amu. By adding the mass of the electron and negatron to the lost mass of 0.00029 amu, their sum will then become a constant that the Original Neutron will always lose when it becomes part of an isotope as a normal neutron. A normal neutron is any neutron found within an isotope that is not the Original Neutron. This new constant is 0.00139 amu.

It will be noted that only the proton, positon and neutron can lose mass as well as the Original Neutron. Also the reader is to note that in all of the **STEPS** given, the proton or positon is never released but an electron, negatron and a neutron can be. This becomes important when the Original Neutron changes into an isotope.

In Step 01N139 there are two isotopes of 1H1 used as is with one Original Neutron that loses 0.00830 amu to create the isotope 2He3. When the single Original Neutron lost the 0.00830 amu, it became the single normal neutron in the isotope 2He3 with 1.00037 amu's. At this point in time, I will use the method given in the chemistry book "Principles of Modern Chemistry by David W. Oxtoby and Norman H. Nachtrieb, copyright 1987 as shown on pages 374 and 375.

Below are two drawings showing the creation of the isotopes 2He3 and 2He4. The first drawing of Step 01N139 shows how two isotopes of 1H1 and a single Original Neutron can create the isotope 2He3 with the loss of 0.00830 amu. Drawing 02N139 shows how two isotopes of 1H1 can combine with two Original Neutrons to create the isotope 2He4 with the loss of 0.03040 amu. The mass of each electron and a negatron is 0.00055 amu, each proton and positon is 1.00728 amu's and the normal neutron having no more than 1.00628 amu's. In the case of the isotope 2He3, its single neutron has 1.00037 amu's and the isotope 2He4 has each neutron having the mass of 0.99347 amu. The Original Neutron will have 1.00867 amu's. The isotope of 2He3 has 3.01603 amu's and the isotope 2He4 has 4.00260 amu's.

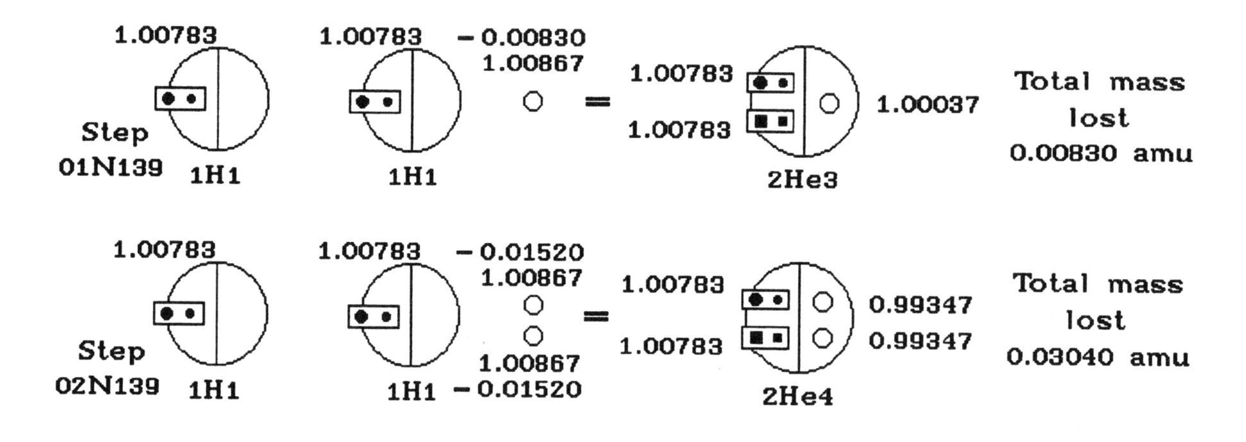

USING THE ORIGINAL NEUTRON TO CREATE THE ISOTOPE 2He3

Changing mass	=	4.00260 - 2(1.00783) - 2(1.00867)
2He4	=	4.00260 - 2.01566 - 2.01734
	=	4.00260 - 4.03300
	=	- 0.03040 amu

Using this format for the isotope 2He3, the following is given.

Changing mass	=	3.01603 - 2(1.00783) - 1.00867
2He3	=	3.01603 - 2.01566 - 1.00867
	=	3.01603 - 3.02433
	=	- 0.00830 amu

In the chemistry books, two Original Neutrons with each having 1.00867 amu's are combined with two isotopes of 1H1 to create the new isotope 2He4. The chemistry book mentioned above does not state that the two isotopes of 1H1 will combine with one Original Neutron to create the isotope 2He3. When two isotopes of 1H1 combine with two Original Neutrons to create the isotope 2He4, 0.03040 amu is lost. Each Original Neutron loses 0.001520 amu. When two isotopes of 1H1 are combined with one Original Neutron to create the isotope of 2He3, there is 0.00830 amu lost. Because only one Original Neutron is used to create the isotope 2He3, the entire 0.00830 amu is lost as a single unit by the Original Neutron.

Since the two isotopes of 1H1 are used as is, their parts do not lose any mass. This means that the single Original Neutron must lose all of the mass that is lost. To find the mass of the single neutron in the isotope 2He3, add the masses of the two isotopes of 1H1 and subtract their total mass from the mass of the isotope of 2He3.

Mass of 1H1 intact in 2He3	1.00783 amu's
Mass of 1H1 intact in 2He3	1.00783 amu's
Total mass of the two 1H1's in 2He3	2.01566 amu's

Mass of 2He3	3.01603 amu's
Total mass of the two 1H1's in 2He3	2.01566 amu's
Mass of single neutron in 2He3	1.00037 amu's

This single neutron in the isotope 2He3 came from the single Original Neutron which has 1.00867 amu's. To find the mass lost by the Original Neutron, subtract the mass of the single neutron in the isotope 2He3 from the mass of the Original Neutron.

Mass of Original Neutron	1.00867 amu's
Mass of single neutron in 2He3	1.00037 amu's
Mass lost by Original Neutron	0.00830 amu

The Original neutron loses 0.00830 amu to become the single neutron in the isotope 2He3.

Original Neutron	1.00867 amu's
Mass lost by it	0.00830 amu
Mass of single neutron in 2He3	1.00037 amu's

The last two proofs are given to let the reader see why the isotope of 2He3 has a single neutron with 1.00037 amu's.

The lost mass of 0.00830 amu is composed of different sub-units. To find these different sub-units, subtract the constant of 0.00139 amu. The remainder will also be composed of other sub-units.

Lost mass by Original Neutron creating 2He3	0.00830 amu
Constant in positron	0.00139 amu
Remainder of lost mass by Original Neutron	0.00691 amu

This remainder of lost mass by the Original Neutron creating the isotope 2He3 is composed of a sub-unit of 0.00591 amu and one sub-unit of 0.00100 amu. The 0.00591 amu is composed of the sub-units of 0.00434 amu and 0.00157 amu. The sub-unit of 0.00434 is composed of two sub-units of 0.00217 amu while the sub-unit of 0.00157 amu is composed of a sub-unit of 0.00100 amu, 0.00055 amu and one sub-unit of 0.00002 amu. The sub-unit of 0.00591 amu could also be composed of two sub-units. One sub-unit of 0.00374 amu and one sub-unit of 0.00217 amu. The constant of 0.00139 amu is composed of a sub-unit of 0.00110 amu and a sub-unit of 0.00029 amu. The sub-unit of 0.00110 amu is composed of two sub-units of 0.00055 amu.

There is a proton in each of the isotopes of 1H1 and each proton has 1.00728 amu's. Their magnetic fields are therefore equal. Remember that the larger magnetic field controls the smaller magnetic field. In this case, the two magnetic fields are equal so neither controls the other. Also remember that combining two magnetic fields does not increase their magnetic field for the isotope they are in. This means that the magnetic field around the isotope 2He3 has the strength of a single proton. However, the two protons become a proton and a positon in the isotope 2He3 with only a single neutron to control their spin. When the isotope 2He3 is compared to the isotope of 2He4, the spin of the proton and positon in the isotope 2He4 is balanced by two neutrons of equal mass. This gives more stability to the isotope 2He4.

USING THE ORIGINAL NEUTRON TO CREATE THE ISOTOPE 2He4

In the chemistry books, two Original Neutrons with each having 1.00867 amu's are combined with two isotopes of 1H1 to create the new isotope 2He4. These books do not state that the two isotopes of 1H1 will combine with one Original Neutron to create the isotope 2He3. When two isotopes of 1H1 combine with two Original Neutrons to create the isotope 2He4, 0.03040 amu is lost. Each Original Neutron loses 0.001520 amu. When two isotopes of 1H1 are combined with one Original Neutron to create the isotope of 2He3, there is 0.00830 amu lost. Because only one Original Neutron is used to create the isotope 2He3, the entire 0.00830 amu is lost as a single unit.

In the chemistry book "PRINCIPLES OF MODERN CHEMISTRY" by David W. Oxtoby and Norman H. Nachtrieb at the University of Chicago, copyright 1987, on page 374 and 375 is a description of how to calculate the binding energies of nuclei by means of the first law of thermodynamics through the use of Einstein's relationship and accurate mass determinations with a mass spectrometer. This book then states that a correction is needed due to the differences in the binding energies of electrons (which of course are present both in the Hydrogen atoms and in the atom being formed). The book then states that the mass of the electron is very small and need not concern any one when calculating the binding energies at this point in the books write-up. The failure to include the mass of the electrons leads to some very bad answers. In the case of two isotopes of Hydrogen (1H1) which does not have a neutron, the book gives a formula that uses a changing mass equal to the mass of Helium four (2He4) minus the mass of two Hydrogen one isotopes of 1H1, mass of 1.007825 amu's each and two neutrons with the mass of 1.008665 amu's each. The isotope of Helium four (2He4) has the mass of 4.002603 amu's. The mass of Helium (2He4) is 4.00260 amu's, the Hydrogen one (1H1) isotope is 1.00783 amu's and the neutrons will each have the mass of the Original Neutron of 1.00867 amu's. These masses are used because they are the ones used in all of the **STEPS** to this point.

Changing mass 2He4 = 4.00260 - 2(1.00783) - 2(1.00867)
= 4.00260 - 2.01566 - 2.01734
= 4.00260 - 4.03300
= - 0.03040 amu

Mass of two 1H1's	2.01566 amu's
Mass of two Original Neutrons	2.01734 amu's
Total mass of two 1H1's and two Original Neutrons	4.03300 amu's

Total mass of two 1H1's and two Original Neutrons	4.03300 amu's
Mass of 2He4	3.00260 amu's
Difference between them	1.03040 amu's

Note that the mass of the two Hydrogen isotopes 2(1H1) and the two neutrons is greater than the mass of the isotope of Helium four (2He4). This greater mass must then be the binding mass for the isotope of Helium four (2He4). It seems that the logic used is that a greater mass will create a lesser mass with the excess mass. This just is not possible. What is really happening is that the greater mass creates something with less mass by losing the excess mass. Because, in this case, we are dealing with two isotopes of Hydrogen one, 2(1H1), and two neutrons that now combine to form the isotope of Helium four (2He4). The two neutrons are Original Neutrons and in the formula they are by themselves. They are not at this point in time bonded to the two Hydrogen atoms. The book states that when they are bonded, the binding mass is - 0.030377 amu.

Changing mass = 4.002603 - 2(1.007825) - 2(1.008665) = - 0.030377 amu

Changing mass = 4.00260 - 2(1.00783) - 2(1.00867) = - 0.03040 amu
2He4 = 4.00260 - 2.01566 - 2.01734 = - 0.03040 amu
= 4.00260 - 4.03300 = - 0.03040 amu

Changing mass = 2.01566 + 2.01734 = 4.00260 + 0.03040 amu

The combining of two Hydrogen One isotopes with two Original Neutrons gives more mass than the single isotope of Helium Four by 0.03040 amu. If two wooden sticks had glue on one of their ends, the sticks would be glued to each other if I combined them with the glue. Now look at the last formula. The Helium Four must have the electron and proton found in each of the Hydrogen One isotopes plus two neutrons which have the mass of the Original Neutron. When these isotopes combine, the electrons and protons keep their mass. This means that the two Original Neutrons lose the lost mass. If the combining of the four isotopes make one new isotope with the loss of 0.03040 amu, then the lost mass can not be called the bonding or binding mass. This is true because the two isotopes and two neutrons were separate parts. When they combined to make the isotope of Helium Four, they lost mass which can not be called a glue. It is true they stuck together but this is because they lost mass.

The two Original Neutrons do not have external magnetic fields while the two isotopes of Hydrogen One do. This means that the two isotopes of Hydrogen One combined their magnetic field around the two Original Neutrons after the two Original Neutrons lost the 0.03040 amu to create the isotope of Helium Four. Each Original Neutron must lose half of the 0.03040 amu to change into the two neutrons with 0.99347 amu found in the newly created isotope of Helium Four. This means that the terms "bonding or binding mass" is not the correct way to state this. The loss of 0.03040 amu decreases the total mass but does not decrease the magnetic field of each of the two isotopes of Hydrogen One. By changing the Original Neutrons into regular neutrons, they can become part of the isotope of Helium Four. For the isotope of Helium Four to be created, the two isotopes of Hydrogen One must be found within the isotope of 2He4. The trouble with them combining is that their combining must lose some lost mass which does not happen. This means that if the isotope of Helium Four is to be created, the two isotopes of Hydrogen One must be found within the isotope of 2He4 as individual isotopes of 1H1 with an external magnetic field around both isotopes of 1H1 and the two normal neutrons. When this happens, they cause each of the two Original Neutrons to lose 0.001520 amu to become the neutrons found in the isotope of Hydrogen Four. When the two Original Neutrons become neutrons with 0.99347 amu, they are found within the magnetic field of the isotope of Helium Four that has two isotopes of Hydrogen One intact within it. This is shown in the drawing below.

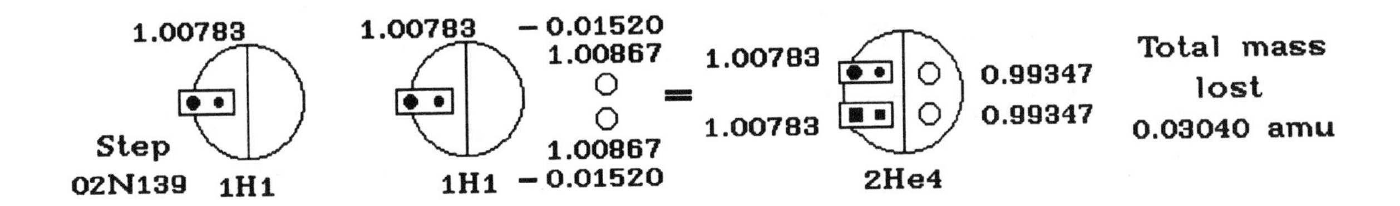

Since the isotope of Helium four (2He4) has an electron and a negatron, these two masses must come from the two isotopes of Hydrogen One 2(1H1). The isotope of Helium Four (2He4) also has a proton and a positon which must also come from the two isotopes of Hydrogen One 2(1H1). There are two neutrons in the isotope Helium Four (2He4) so these two neutrons must come from the two neutrons which are Original Neutrons. However, if all of the masses of the two Hydrogen isotopes 2(1H1) and the two neutrons are added together, they will have more mass than the mass of the Helium Four (2He4) isotope they form. This means that the two neutrons must lose the lost mass that is lost creating the isotope Helium Four (2He4). This lost mass is the so called binding mass as given in the chemistry books. They say it is 0.030377 amu which is rounded off to 0.003040 amu. If the two neutrons lost this mass, then each neutron lost half of it. Each neutron then lost 0.01520 amu. If this lost mass of 0.01520 amu is subtracted from the mass of one of the neutrons, then the remaining mass will be the mass of one of the two neutrons found in the Helium Four isotope (2He4).

0.5(0.03040) = 0.01520 amu

Original Neutron	1.00867 amu's
Lost mass by it	0.01520 amu
Mass of one of the neutrons in 2He4	0.99347 amu

With this mass of one of the neutrons in the isotope 2He4, it can be doubled to find the mass of both of the neutrons found in the isotope 2He4.

Mass of first neutron in 2He4	0.99347 amu
Mass of second neutron in 2He4	0.99347 amu
Mass of both neutrons in 2He4	1.98694 amu's

The mass of each of the two isotopes of 1H1 is 1.00783 amu's and there are two of these in the isotope 2He4. Their combined mass is 2.01566 amu's and by adding this combined mass to the mass of the two neutrons, the mass of the isotope 2He4 will be found.

Mass of two 1H1's in 2He4	2.01566 amu's
Mass of two neutrons in 2He4	1.98694 amu's
Mass of 2He4	4.00260 amu's

Now that the mass of the two neutrons found in the isotope 2He4 is known, the mass of one of the two neutrons can be subtracted from the mass of the Original Neutron. The remaining mass will be the mass lost by the Original neutron creating one of the two neutrons in the isotope 2He4.

Mass of the Original Neutron	1.00867 amu's
Mass of one of the neutrons in 2He4	0.99347 amu
Mass lost by Original Neutron	0.01520 amu

This mass lost by the Original Neutron is composed of three sub-units. I have shown that the Original Neutron loses 0.00029 amu plus an electron and a negatron. This means that the masses of the electron and the negatron are lost by the Original Neutron which is 0.00110 amu. This 0.00110 amu is added to the 0.00029 amu of lost mass to make a total of 0.00139 amu lost by the Original Neutron. By subtracting this lost mass from the lost mass of 0.01520 amu, the remaining mass will be the mass that the Original Neutron lost which did not release either an electron or a negatron.

Mass of electron and negatron in Original Neutron	0.00110 amu
Mass lost by Original Neutron's positron	0.00029 amu
Mass lost by Original Neutron	0.00139 amu

Mass lost by Original Neutron	0.01520 amu
Mass constant of Original Neutron	0.00139 amu
Mass lost not releasing electron and negatron	0.01381 amu

0.00690 amu	
0.00002 amu	
0.00100 amu	The mass lost that does not release either an electron or a negatron is composed of five sub-units.
0.00155 amu	
0.00434 amu	
0.01381 amu	

There are other neutrons in the isotope 2He4 with different masses other than 0.99457 amu. If each one of these neutrons is subtracted from the Original Neutron and then subtract the new constant of 0.00139 amu from this remainder of the two neutrons, the mass that will not release an electron or negatron will be found.

Mass of Original Neutron	1.00867 amu's
Mass of neutron in 2He4	0.99347 amu
Mass lost by Original Neutron	0.01520 amu

Mass lost by Original Neutron	0.01520 amu
Mass constant of Original Neutron	0.00139 amu
Mass lost not releasing electron or negatron from Original Neutron	0.01381 amu

Mass of Original Neutron	1.00867 amu's
Mass of neutron in 2He4	1.00194 amu's
Mass lost by Original Neutron	0.00673 amu

Mass lost by Original Neutron	0.00673 amu
Mass constant of Original Neutron	0.00139 amu
Mass lost not releasing electron or negatron from Original Neutron	0.00534 amu

Mass of Original Neutron	1.00867 amu's
Mass of neutron in 2He4	1.00411 amu's
Mass lost by Original Neutron	0.00456 amu

Mass lost by Original Neutron	0.00456 amu
Mass constant of Original Neutron	0.00139 amu
Mass lost not releasing electron or negatron from Original Neutron	0.00317 amu

Mass of Original Neutron	1.00867 amu's
Mass of neutron in 2He4	1.00628 amu's
Mass lost by Original Neutron	0.00239 amu

Mass lost by Original Neutron	0.00239 amu
Mass constant of Original Neutron	0.00139 amu
Mass lost releasing electron or negatron from Original Neutron	0.00100 amu

The way to understand what is happening that ends up with the Original Neutron losing 0.00100 amu is that the sequence of events starts with the Original Neutron losing 0.00139 amu. This releases the negatron completely from the positron and changes the positron into a proton that has an electron in an orbital around the proton. This is an isotope of 1H1. The positron is now a proton or positon with 1.00728 amu's which then loses the 0.00100 amu and releases the electron from its orbital around the proton. The proton now becomes a neutron with 1.00628 amu's. The Original Neutron went from an Original Neutron to an isotope of 1H1 which then lost its electron and changed into a neutron that is not an Original Neutron.

The lost mass of 0.01381 amu includes a sub-unit of 0.00100 amu. The lost masses of 0.00534 amu and 0.00317 amu both include this sub-unit of 0.00100 amu while the last use of the constant ends up with this lost mass only. When the lost mass includes this 0.00100 amu, no electron or negatron is released from the neutron and if it is transferred, no ray is emitted.

Now change from the neutron found in the isotope 2He4 to its proton or positon to find the mass that will not release an electron or negatron.

Mass of Original Neutron	1.00867 amu's
Mass of proton or positon in 2He4	1.00294 amu's
Mass lost by Original Neutron	0.00573 amu

Mass lost by Original Neutron	0.00573 amu
Mass constant of Original Neutron	0.00139 amu
Mass lost not releasing electron or negatron from Original Neutron	0.00434 amu

Mass of Original Neutron	1.00867 amu's
Mass of proton or positon in 2He4	1.00511 amu's
Mass lost by Original Neutron	0.00356 amu

Mass lost by Original Neutron	0.00573 amu
Mass constant of Original Neutron	0.00139 amu
Mass lost not releasing electron or negatron from Original Neutron	0.00217 amu

Mass of Original Neutron	1.00867 amu's
Mass of proton or positon in 2He4	1.00728 amu's
Mass lost by Original Neutron	0.00139 amu

Mass lost by Original Neutron	0.00139 amu
Mass constant of Original Neutron	0.00139 amu
Mass lost releasing electron or negatron from Original Neutron	0.00000 amu

This last line of reasoning shows that the Original Neutron does not lose any additional mass when the electron, negatron and lost mass are lost by it when the positron becomes a proton or positon in the isotope 1H1. The chemistry book did not believe that the electron's mass of 0.00055 amu was worth being concerned about but the above math shows that it must be accounted for. However, the mass given in the formula for the isotope of 1H1 was 1.00783 amu's and this mass includes the mass of the electron. This means that the mass of the electron was taken under consideration.

With the above information about the mass of the Original Neutron, the following statement will be made by me for the first time ever in atomic structure and in the world of chemistry. It is my belief that this statement will be the greatest break through in chemistry ever made to this date, June 28, 1995.

The atomic mass of the Original Neutron that is given in most chemistry books is 1.00867 amu's which is rounded off to five places to the right of the decimal point. The part of this number that is 0.00867 amu is composed of ten (10) sub-units. The remainder will be 1.00000 amu. The 1.00000 amu's and 0.00867 amu are called sub-units of 1.00867 amu's. Thus the Original Neutron is composed of two sub-units and the sub-unit of 0.00867 amu is composed of ten (10) more sub-units.

1.	0.00055 amu	1.00867 amu's
2.	0.00055 amu	0.00867 amu
3.	0.00055 amu	1.00000 amu's
4.	0.00055 amu	
5.	0.00055 amu	
6.	0.00029 amu	
7.	0.00029 amu	
8.	0.00100 amu	
9.	0.00217 amu	
10.	0.00217 amu	

	0.00867 amu	

These ten (10) sub-units make up seven (7) sub-units. These seven (7) sub-units make up four (4) sub-units. These four (4) sub-units make up three (3) sub-units. One of these last three sub-units will change as different isotopes are created.

0.00055 amu	0.00110 amu	0.00139 amu	0.00139 amu
0.00055 amu	0.00029 amu	0.00139 amu	0.00673 amu
0.00055 amu	0.00110 amu	0.00534 amu	0.00055 amu
0.00055 amu	0.00029 amu	0.00055 amu	--------
0.00055 amu	0.00434 amu	---------	0.00867 amu
0.00029 amu	0.00100 amu	0.00867 amu	
0.00029 amu	0.00055 amu		
0.00217 amu	---------		
0.00217 amu	0.00867 amu		
0.00100 amu			

0.00867 amu			

Now that the two sub-units of the Original Neutron are known, what about the other protons, positons and neutrons found in isotopes? What units and sub-units are they made up of? If the sub-unit of 1.00000 amu's is treated as a single unit at this point, then it is the second sub-unit that needs to be explained as a unit. There are protons or neutrons made up of 1.00728 amu's, 1.00511 amu's and 1.00294 amu's. There are neutrons of 1.00628 amu's, 1.00411 amu's and 1.00194 amu's. All of these protons, positons and neutrons have two sub-units in them which if used as a unit, they have two units.

Proton or positon

1.00728 amu's	1.00511 amu's	1.00294 amu's
0.00139 amu	0.00139 amu	0.00139 amu
0.00534 amu	0.00317 amu	0.00100 amu
0.00055 amu	0.00055 amu	0.00055 amu
--------	---------	---------
0.00728 amu	0.00511 amu	0.00294 amu

Neutrons

1.00628 amu's	1.00411 amu's	1.00194 amu's
0.00139 amu	0.00139 amu	0.00139 amu
0.00434 amu	0.00217 amu	0.00055 amu
0.00055 amu	0.00055 amu	---------
---------	--------	
0.00628 amu	0.00411 amu	0.00194 amu

All of the above dealt with neutrons with more mass than 1.00000 amu's. The isotope 2He4 has two neutrons that are less than this amount. How do I find the units and sub-units that are used starting from the Original Neutron's mass?

The mass of each of the neutrons in the isotope 2He4 is 0.99347 amu. This amount must be subtracted from the mass of the Original Neutron's mass of 1.00867 amu's to find the first sub-unit.

Mass of Original Neutron	1.00867 amu's
Mass of first or second neutron in 2He4	0.99347 amu
Mass lost by Original Neutron	0.01520 amu

Next I must know the mass that is less than 1.00000 amu's. This sub-unit will then be subtracted from the mass lost by the Original Neutron to find the mass that will give the unit of 0.00867 amu.

First unit of 1.00867 amu's	1.00000 amu's
Mass of first or second neutron in 2He4	0.99347 amu
Mass less than 1.00000 amu's	0.00653 amu

Mass lost by Original Neutron	0.01520 amu
Mass less than 1.00000 amu's	0.00653 amu
Second unit of 1.00867 amu's	0.00867 amu

Now I know that the Original Neutron lost an electron, negatron and the lost mass. This is the sub-unit of 0.00139 amu. Subtract it from the second unit of 1.00867 amu's to find the sub-unit 0.00728 amu which is the sub-unit of a proton or positon.

Second unit of 1.00867 amu's	0.00867 amu
Constant of 1.00867 amu's	0.00139 amu
Mass of sub-unit of proton or neutron	0.00728 amu

The sub-unit of 0.00728 amu is composed of three (3) sub-units.

0.00139 amu
0.00534 amu
0.00055 amu

0.00728 amu

Note that the sub-unit 0.00139 amu has already been used once and it is used again in the sub-unit 0.00728 amu.

The sub-unit of 0.01520 amu is the amount of mass the Original Neutron lost creating or becoming one of the neutrons in the isotope 2He4 with 0.99347 amu. Subtract from this sub-unit the lost mass of 0.00137 amu to find the sub-unit that is found using two isotopes of 1H1 and two Original Neutrons creating the isotope 2He4.

Mass lost using two Original Neutrons	0.01520 amu
Constant of Original Neutron	0.00139 amu
Mass lost from 2(1H1) and 2 Original neutrons	0.01381 amu

The sub-unit of 0.01381 amu is composed of five sub-units. These sub-units differ from the sub-units found in the second unit of the Original Neutron because they involve the use of two isotopes of 1H1. This shows that the sub-units found in the second unit of 1.00867 amu's is related to the use of the two isotopes of 1H1 used in creating the isotope 2He4.

0.00690 amu
0.00002 amu
0.00100 amu
0.00155 amu
0.00434 amu

0.01381 amu

CHANGED-TRANSFERRED-RELEASED-LOST

The above four words have lead to a great deal of confusion when used to describe what has happened to mass. I intend here to call the attention of the reader to make a special effort to pay attention to the way these words are used within a sentence.

When a new isotope is created, it may be a new isotope of the same atom or it may be an isotope of a totally different atom. Sometimes mass is "changed" into a "ray" and some times it is "transferred" from one isotope to another isotope or from one isotope's internal parts. The end results is that a new isotope is created. In every case some mass will be "lost".

The term "lost" has many ways it can be used and the reader is to pay close attention to this. An isotope will have an internal part that is either a proton, positon or a neutron that will have some of its mass changed into a ray. When this happens, the mass is changed from mass into pure energy that is magnetism. Note that the meaning of the word "lost" is that it is not mass but pure energy which is no longer part of the mass it belonged to. The mass has "changed" its form.

The term "changed" as given above is a change in form but it could be a change in place. Mass could be in a proton, positon or neutron and "transferred" to a different place. Mass can be moved or "transferred" from proton to proton, from proton to positon, from positon to positon, from proton to neutron, from positon to neutron or from neutron to neutron. In this "transfer" of mass from one place to another, the place it is "transferred" from must "lose" the mass that is "transferred". This is called "lost mass" that means that the isotope that had the transferred mass does not have it any more. It has been "transferred" from one place to another but it is not "released". It does show up in the isotope being created. No mass has been truly "lost" creating a new isotope.

In the case of an electron or negatron that is "released" from an isotope, it leaves the isotope it was a part of and it no longer belongs to that isotope. It is also possible for an electron or negatron to be "released" from within a proton, positon or neutron and be found as part of the new isotope being created. In this case, the electron or negatron goes in to an orbital that has an electron or negatron already there. The reader is to note which type of "release" has taken place.

Sometimes there is a part that is not needed in the new isotope being created. When this happens, the part that is not needed is "released". It is no longer part of anything used to create the new isotope and it is not found in the newly created isotope. It is not emitted as a ray. It is "lost mass" from the total mass needed to create the new isotope.

I want to call special attention to the proton that has some of its mass removed from it and transferred to a neutron. There is also some mass that is not transferred because it has been changed into pure energy. The mass that was transferred was lost from the proton and the mass that was changed into pure energy was also lost from the proton. Here is a situation where the reader can become confused. I am calling the mass that was transferred as lost mass and the real mass lost I am also calling lost mass. Note how I use the words "lost and transferred". I also said that the mass was "transferred and lost". When I said this, both masses are lost from the proton but one of the lost masses has been transferred and not released from the new isotope being created while the other has been changed into a ray and it is therefore the real mass that is lost creating the new isotope.

RADIOACTIVITY

When the isotopes 1H2 and 1H1 combine to create the isotope 1H3 with the loss of 0.00100 amu and the release of an electron from either the isotope 1H2 or 1H1, the isotope 1H3 thus created is said to be radioactive. It also happens when three isotopes of 1H1 combine to create the isotope 1H3 with the loss of 0.00100 from two of the isotopes of 1H1 and the release of an electron from each. Thus radioactivity comes into existence when a proton or positon changes into a neutron with the release of an electron or negatron that leaves the isotope that loses 0.00100 amu. In the isotope 1H3 there is one neutron more than the proton found in the nucleus. The proton has 1.00728 amu's and the first neutron has 1.00628 amu's with the second neutron having 1.00194 amu's. There is one electron with 0.00055 amu. The isotope of 1H2 has one electron with 0.00055 amu, one proton with 1.00728 amu's and one neutron with 1.00628 amu's to make an isotope with 2.01411 amu's. Note that this isotope has a proton with 1.00728 amu's and a neutron with 1.00628 amu's and it is not radioactive. This shows that the second neutron in the isotope 1H3 is not stable. It changes into a positon, the positon has 1.00137 amu's which in the isotope 2He3 is unstable.

The first neutron has its maximum mass of 1.00628 amu's and the second neutron does not. The second neutron with 1.00194 amu's needs 0.00434 amu to become a neutron with 1.00628 amu's. If the second neutron does not regain this mass, it transfers mass to the first neutron and becomes a positon with the release of a negatron from within itself and loses 0.00002 amu.

It is also possible for three isotopes of 1H1 to combine to create the isotope of 1H3 with a proton having 1.00728 amu's and two neutrons with each neutron having 1.00411 amu's. If the isotope 1H3 is to be converted to a stable isotope of 2He3, there are several ways it can do this.

Now that I have given the idea that a neutron needs to be 1.00628 amu's. The reader is to note that when an isotope has its proton or positon with 1.00728 amu's and there is no other proton or positon with less mass, the isotope should be stable and the word "stable" means that the isotope will need outside energy to change. Radioactivity needs an internal change to become stable.

There is no way for an isotope of 1H3 to have two neutrons with their maximum mass of 1.00628 amu's and have its proton with 1.00728 amu's. When the proton has less than its maximum, it will try to regain the mass needed to have its maximum mass. Thus when the neutrons in the isotope 1H3 have their maximum mass and the proton does not, some of the mass in the neutrons will be transferred to the proton to make it have its maximum mass. If the isotope of 1H3 can not transfer mass to the proton then it will combine with some other isotope to change into a new isotope.

In Step 3-1-3 using a single isotope of 1H3 to create the isotope 2He3, the isotope 1H3 has a proton with its maximum mass of 1.00728 amu's and the first neutron has its maximum mass of 1.00628 amu's with the second neutron having 1.00194 amu's. Because the isotope of 1H3 is by itself and does not combine with any other isotope in this step, the second neutron cannot transfer any mass to the proton because it has its maximum mass. This second neutron will then transfer to the first neutron enough mass to make the first neutron become a positon with the release of a negatron from within itself that goes into an orbital with an electron already there. The mass that is transferred that releases the negatron from within the second neutron will always be 0.00100 amu. At the time the second neutron transfers the mass to the first neutron, it loses 0.00002 amu. The remaining mass of the second neutron will become the single neutron in the isotope 2He3 with 1.00037 amu's. When this is done, the isotope 2He3 is created and it is not radioactive.

The statements made above indicates that the transfer of 0.00100 amu is needed to release a negatron and lose 0.00002 amu. However, this is not the case in Step 11-1-11 where the isotope 1H3 is by itself and it changes into the isotope 2He3. In this step, there is an isotope of 1H2 found within the isotope 1H3. The fact that the isotope 1H2 is intact within the isotope 1H3 prevents the second neutron from transferring 0.00100 amu to the first neutron. Within a given time, the isotope of 1H3 must change into the isotope 2He3. It does this by losing 0.00002 amu and releasing a negatron to go into an orbital with the electron already there. When the second neutron does this, it becomes a positon with

1.00137 amu. Because the positon in the isotope 2He3 does not equal 1.00728 amu's, this isotope of 2He3 is not stable. It becomes stable by having the single neutron with 1.00628 amu's transferring 0.00591 amu to the positon making it have the required maximum mass of 1.00728 amu's. The single neutron now has 1.00037 amu's which is the required mass if the proton and positon are to each have 1.00728 amu's. This isotope of 2He3 is stable.

The above two steps use an isotope of 1H3 that is by itself. In other steps, this isotope combines with other isotopes to create new isotopes. If the other isotopes have an electron that can become a negatron and a proton that can become a positon, then mass is lost which include the 0.00002 amu and the 0.00002 amu is not lost as a single unit. No negatron is released from within any mass.

These steps where the isotope 1H3 is used involves radioactivity because the isotope 1H3 is radioactive. It will be noted that after the isotope 2He3 is created, there is no radioactivity. This means that the isotope 2He3 does not have the ability to change into another isotope without some outside source causing it. After all the transferring of mass from an isotope of 1H3 that is by itself, the only real mass lost is 0.00002 amu. It is the movement of the internal parts until it is possible for the 0.00002 amu to be lost that must be the reason the isotope 1H3 stops being radioactive and becomes stable. However, the movement of internal parts is not always the reason for radioactivity to stop. It only causes one of the parts making up the isotope 1H3 to lose 0.00002 amu. In Step 11-1-11, the isotope 1H2 is drawn with a rectangle around the electron, proton and the first neutron within the isotopes 1H3 and 2He3. The second neutron in the isotope 1H3 then loses only 0.00002 amu and releases a negatron from within itself. When this happens, the second neutron becomes the positon in the isotope 2He3 with no other mass lost. The 0.00002 amu is emitted as a #2 ray. Where the radioactive isotope of 1H3 is the only isotope that is converted to the isotope 2He3, it is the loss of 0.00002 amu that causes radioactivity to stop.

It was the loss of 0.00100 by a proton or neutron in the isotope 1H1 or 1H2 that caused the second neutron in the isotope 1H3 to come into existence and at this time radioactivity also came into existence. Now 0.00002 amu is lost creating the isotope 2He3 and radioactivity stops. The loss of 0.00100 amu released an electron completely from the isotope or if the 0.00002 amu is lost and the negatron is released and the 0.00100 amu is transferred to a proton, positon or another neutron, then it causes the negatron to go into an orbital with an electron that is already there. At the time the isotope 1H3 was created, one of its parts has 0.00002 amu too much for the isotope to stay an isotope of 1H3. When the isotope 2He3 is created from the single isotope of 1H3, this 0.00002 amu is lost, stopping radioactivity and the isotope 2He3 stays as it is.

The reader is to understand that today's scientists as of 1996 do not know that this is how radioactivity is created or how it is stopped until they read it here. They know of Alpha, Beta and Gamma Rays that are emitted by radioactive isotopes which are unusually measured in Megavolts. However they do not know of the lost masses of 0.00002, 0.00591, 0.00434, 0.00217 amu and many other masses that are lost by protons and neutrons. All of these lost masses are called #2 ray, 591 ray, 434 ray and 217 ray. These rays are all called Gamma Rays. The lost mass of 0.00100 amu is known by todays scientist as a Beta Ray. Since I created the words negatron, positon and called the neutrons as first or second neutron, they could not know these names. It could be that when a proton or positon emits a ray, it is the Alpha Ray and when a neutron emits a ray it is a Gamma Ray. Todays scientist talk about the duel relationship of mass and a ray. They can not tell when one is mass or a ray. When the electron is released at the same time 0.00100 amu is lost as a 100 ray, a mass is emitted and at the same time a ray is emitted. This is a duel relationship. They also talk about whether an electron is emitted from within the nucleus of an isotope and here, I show that there is either an electron or a negatron released from within a proton, positon or a neutron in an isotope.

ISOTOPES MAGNETIC FIELDS

The magnetic fields of the isotopes 1H1, 1H2 and 1H3 are all caused by one proton and not the neutrons. The magnetic field of a proton is external. The electron has an external magnetic field which works with the proton's magnetic field so it plays a part of the over all magnetic field found around the entire isotope. The neutrons have their magnetic field found internally so the magnetic fields of the neutrons are important to their bonding but play only a small part in the overall magnetic field of an isotope.

The magnetic fields of the isotopes 2He3, 2He4 and 2He5 have two protons. If each proton and positon has 1.00728 amu's, each proton and positon combine their magnetic fields but no more than the value of one of them. The neutrons increase the mass of an isotope but they do not increase their magnetic field. When looking at the magnetic field of an isotope of 2He3, 2He4 and 2He5, the proton and positon makes it look like there are two magnetic fields because each proton and positon has a magnetic field. However, they combine their magnetic fields to form only one magnetic field around an isotope. Keep this in mind as the proton and positon hold the neutrons within the isotope's magnetic field.

There is one more thing that should be considered. I know that an electrical arc is composed of two arcs that form one arc. They are wound around each other and the arc is therefore composed of two magnetic fields wound around each other that have one magnetic field. This is important when there is one proton and one positon. Their single magnetic field should be composed of two magnetic fields wound around each other as the single magnetic field found around the isotope they are in. This also means that the spin of the proton and positon cause the arcs to have their magnetic fields spinning within the single arc. This allows an outside magnetic field to pull the magnetic fields apart from each other.

A magnetic field goes outward from the center of the mass that has it. The farther you get from the center, the weaker the magnet field gets. These two statements are important when one tries to understand a magnetic field that goes outward from a proton or positon. The greater a mass is concentrated, the stronger the magnetic field. This means that the closer protons, positons and neutrons are packed within the nucleus, the greater is the magnetic field found around the isotope. If there is a lot of mass found in the nucleus and it is packed so that its parts are as close as they can get, the overall magnetic field will be much stronger than any nucleus with less mass and still packed as close as the masses can get. Each mass has a graviton in its center which keeps the mass concentrated. However, there is a graviton in the center of a nucleus that keeps all of its parts together as an isotope. Thus when there is a lot of mass, the graviton must be stronger to keep these parts together as an isotope.

THE RULES FOR UNDERSTANDING THE DRAWINGS

In all drawings, there are some rules or reasons for making the drawings the way they are drawn. The electron that stays an electron is always drawn as a dot on the big circle. When there is a negatron, it will always be drawn as a square on the big circle with an electron already there. The negatron can come from within the proton, positon or any neutron found within an isotope. It may already be in an orbital with an electron. The proton is always drawn as a dot to the left of the vertical line within the big circle. A proton can stay a proton in the creation of a new isotope or it can become a positon. The positon is always drawn as a small square. If there are two neutrons, the first neutron will always be drawn as the top neutron as a small circle to the right of the vertical line within the big circle and it will be called the first neutron if there are only two neutrons. Other isotopes have three or more neutrons and the last two neutrons will have a neutron called a first and a second neutron. The statement of "last two neutrons" comes from the fact that the math used to find the lost mass and tell where this lost mass comes from will assume that it is the last two electrons in the isotopes under consideration. The second neutron will be drawn below the first neutron as a small circle just like the first neutron. The neutron can also have a negatron drop out and go into an orbital with the electron already there. The reader is to be aware that it may look like the negatron will come from a neutron and think that this neutron will become the positon in the isotope being created. Do not assume this. In Step 14-1-14, the second neutron releases a negatron from within it because it transfers 0.00100 amu to the first neutron. This means that the first neutron becomes the positon and the second neutron becomes the single neutron in the isotope 2He3.

The square drawn around any proton, positon or neutron means that some mass is really lost and it is always lost as a ray. A rectangle drawn around two or more parts of an isotope always indicates that these parts form an isotope that can drop out at some later time. These parts are always found within the isotope that has the rectangle drawn in it.

There will always be a line drawn from the negatron to the proton, positon or neutron that released it. There will always be a line drawn from the proton, positon or neutron that releases, loses or transfers some of its mass to another proton, positon or neutron. When there may be some confusion as to which part within an isotope is the part in the new isotope being created, a line will be drawn between them to the correct part to prevent this confusion. Because there will also be some confusion as to the direction the transferred mass goes, an arrow will be drawn on the line between the two masses.

The isotopes used in creating the new isotope will always be drawn on the left side of an equal sign and the isotope being created along with any mass that drops out will be drawn on the right side of the equal sign. Generally, but not always, the total mass lost will be shown on the right side of the equal sign and to the right of the neutron or isotope that dropped out.

Every sub-step has its write-up and drawing given in such a way that they will stand alone. This is done so that the reader can take one sub-step and put it with any other sub-step to show some line of reasoning this new combination of sub-steps will show. The way I set up my steps or sub-steps may not show the reader what the reader sees and by rearranging them in a new way, a different understanding will come to be understood.

From all of the material given in this book, the reader should now be able to understand how all of the atoms with their isotopes have been created from the Original Neutron. The reader will also now know what the Original Neutron is made up of. I believe this is the first time this has ever been done. The uses of bonding or binding is now fully explained. The electron and negatron has been shown to be needed even if their mass is small.

The use of orbitals by electrons and negatrons has been fully explained when their Shells and sub-shells are used. The use of photons is explained and how frequencies and quanta are used by these orbitals. (Open and closed doors)

Now the reader knows that all space is filled with magnetism and that all pure energy is magnetism and that this magnetism can create electrical current within a wire in only three ways.

I have shown how time came into existence and how it will end. The rules of the cosmos is explained and from there, all the methods and rules that govern the atoms, their isotopes and the parts of these isotopes are explained.

I want the reader to remember the balloon and the battleship given at the beginning of this book.

PART 6

DRAWINGS OF ALL THE STEPS

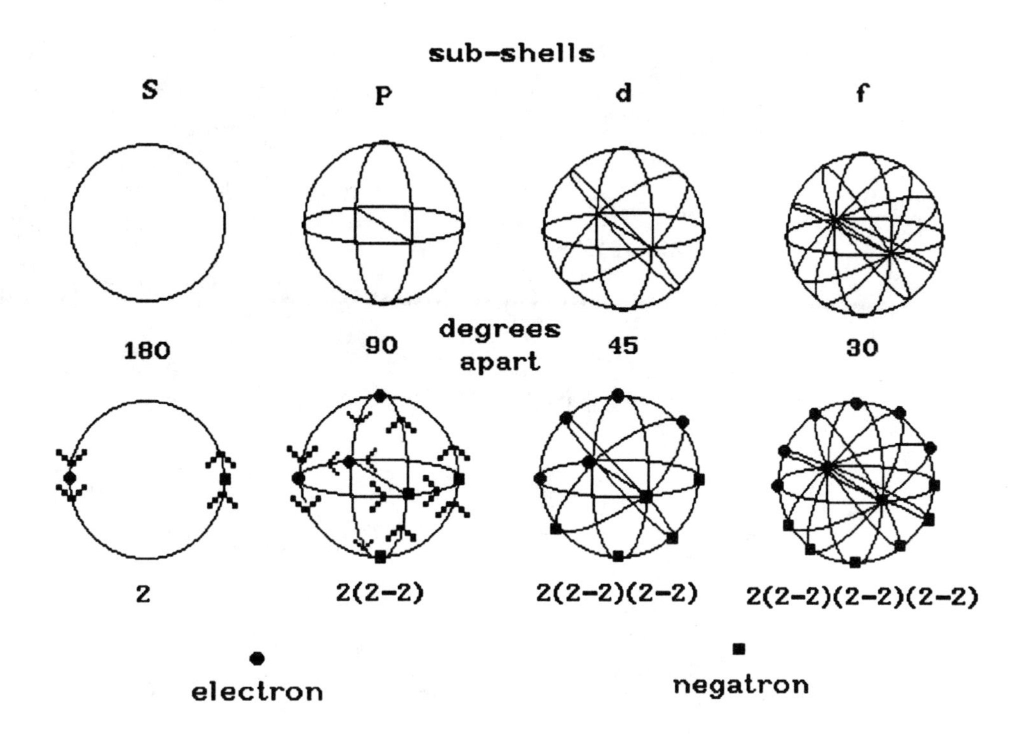

SUB-SHELL ORBITALS

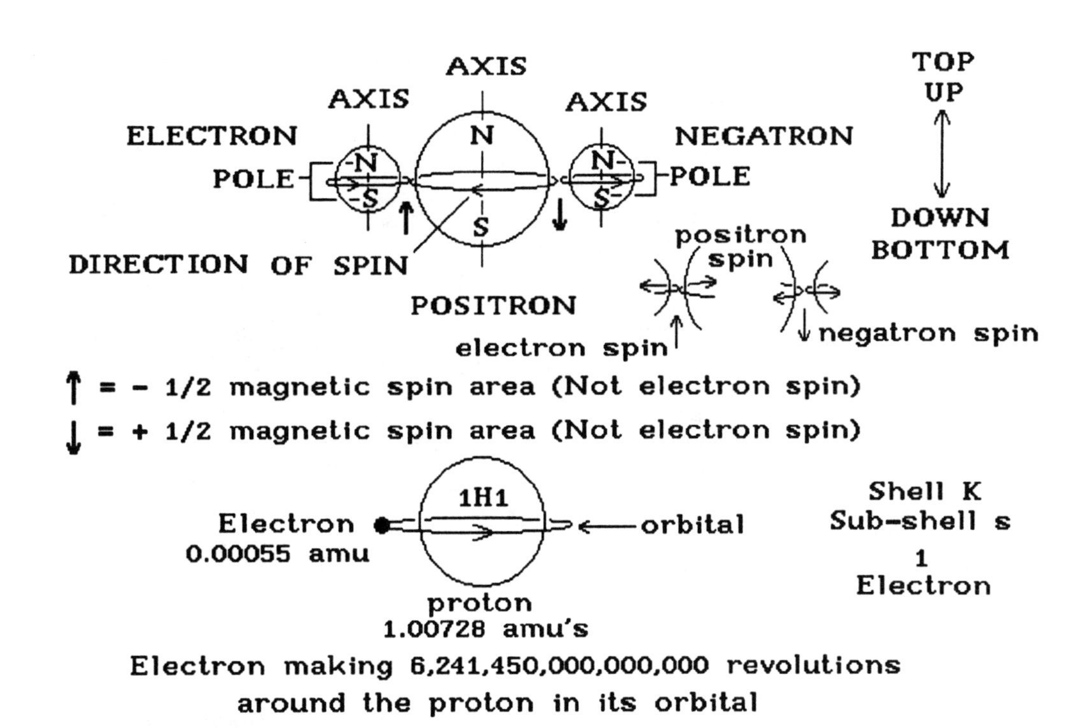

DRAWINGS OF ALL THE STEPS

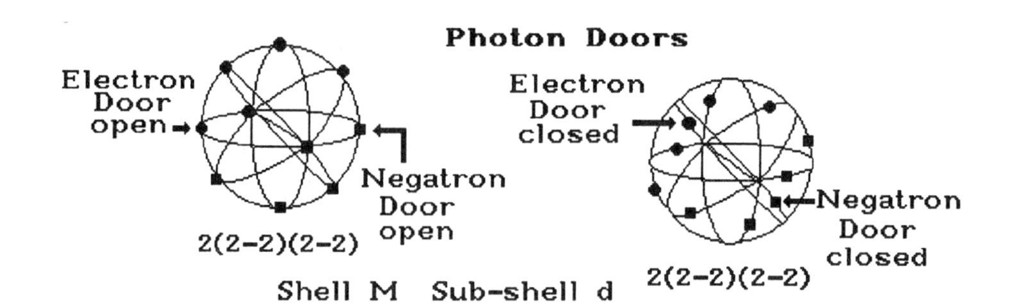

MAGNETIC BOND

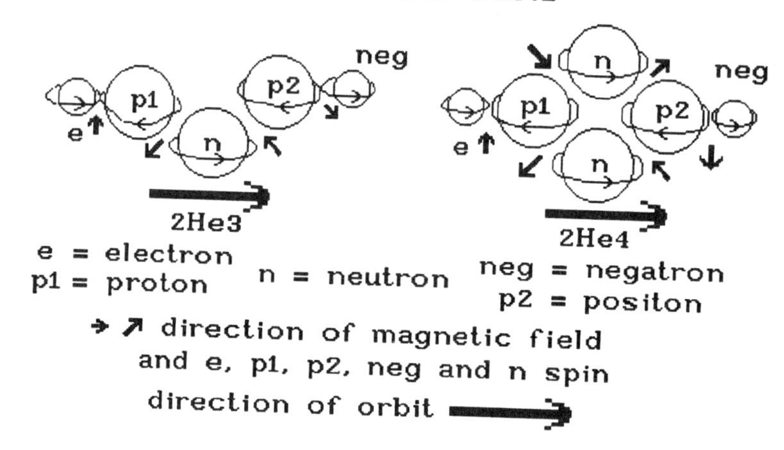

INTERNAL CONSTRUCTION OF 1H2

Isotope 1H1 is internal part of isotope 1H2.

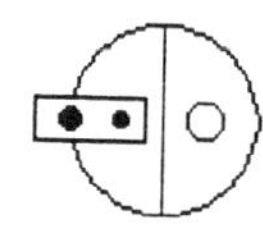

Electron and proton in 1H2 are separate from neutron.

Isotope 1H2

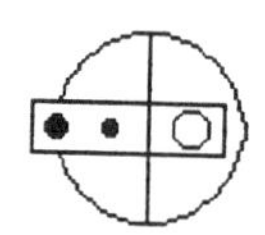

It takes all parts of 1H2 to make 1H2.

INTERNAL CONSTRUCTION OF 1H3

Isotope 1H3

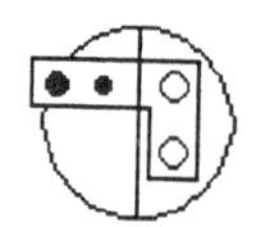

It takes all parts of 1H3 to make 1H3.

Isotope 1H2 is internal part of isotope 1H3.

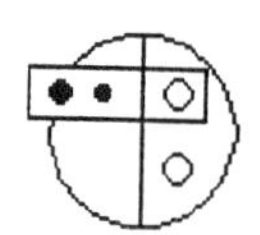

Electron, proton and neutron in 1H3 are intact as 1H2 with the second neutron by itself.

DRAWINGS OF ALL THE STEPS

MESON

ORIGINAL NEUTRON

1.00867 amu's

1N84

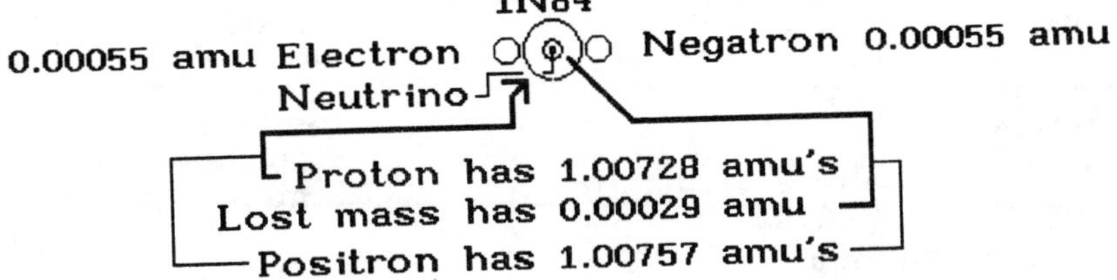

ORIGINAL NEUTRON

1N84

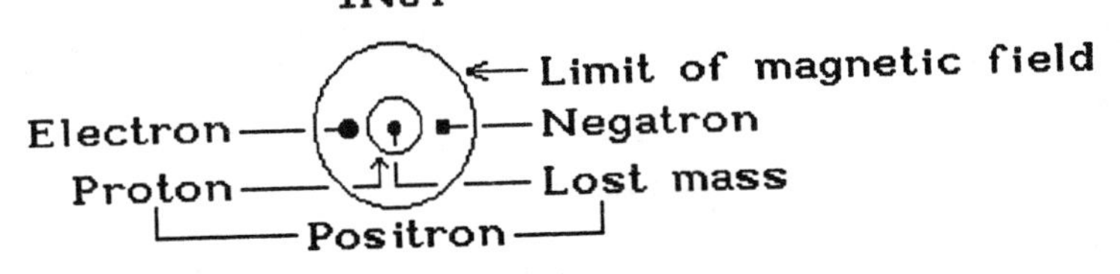

Lost mass	0.00029 amu
Electron	0.00055 amu
Negatron	0.00055 amu
Proton	1.00728 amu's
Positron	1.00757 amu's
1N84	1.00867 amu's

HYDROGEN 1

1H1

1.00783 amu's

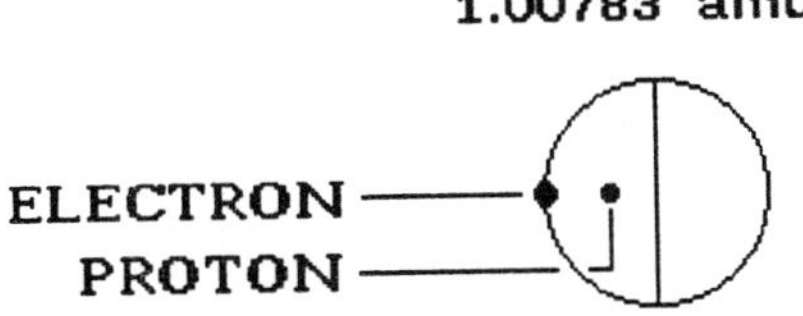

ELECTRON	0.00055 amu
NEGATRON	NONE
PROTON	1.00728 amu's
POSITON	NONE
NEUTRON	NONE

HYDROGEN 2

1H2

DEUTERIUM

2.01411 amu's

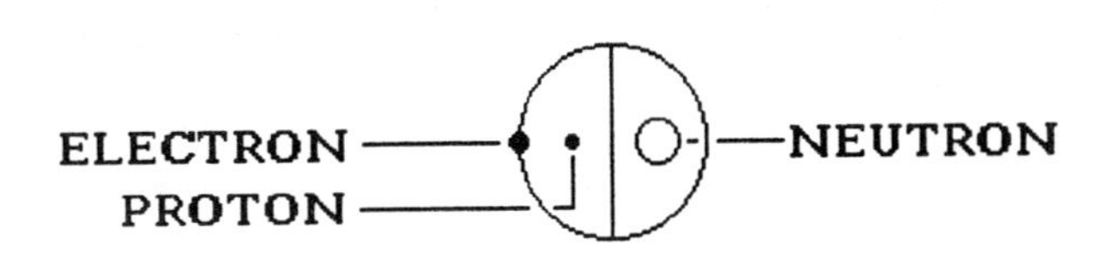

ELECTRON	0.00055 amu
NEGATRON	NONE
PROTON	1.00728 amu's
POSITON	NONE
NEUTRON	1.00628 amu's

DRAWINGS OF ALL THE STEPS

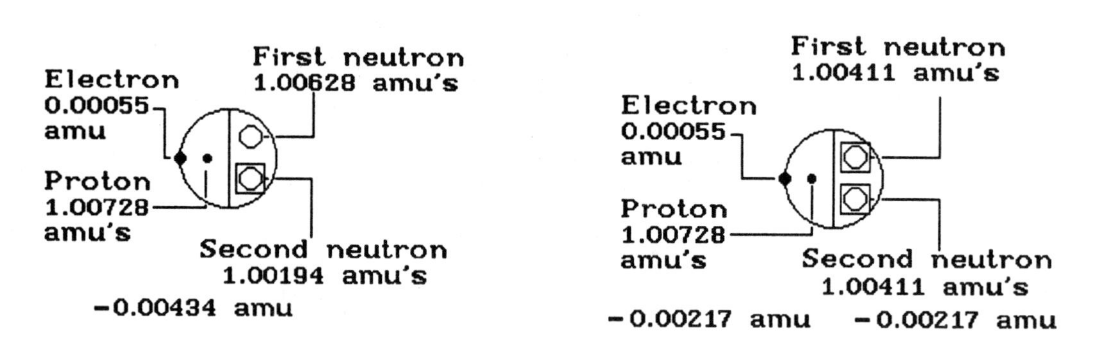

HYDROGEN 3

1H3

TRITIUM

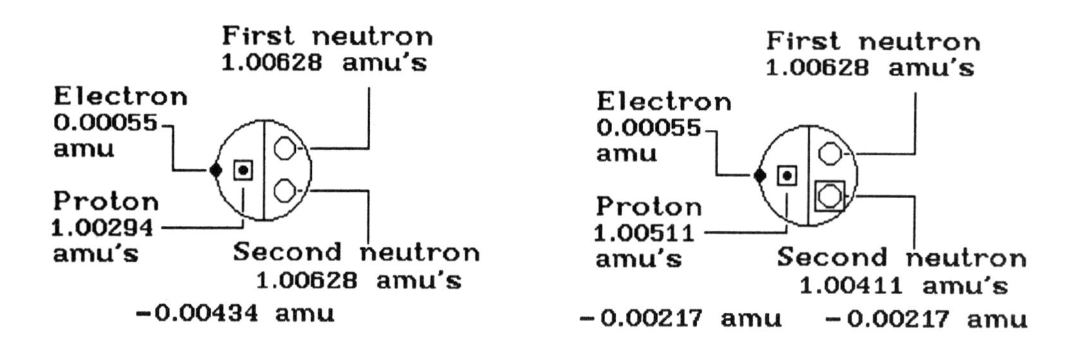

DRAWINGS OF ALL THE STEPS

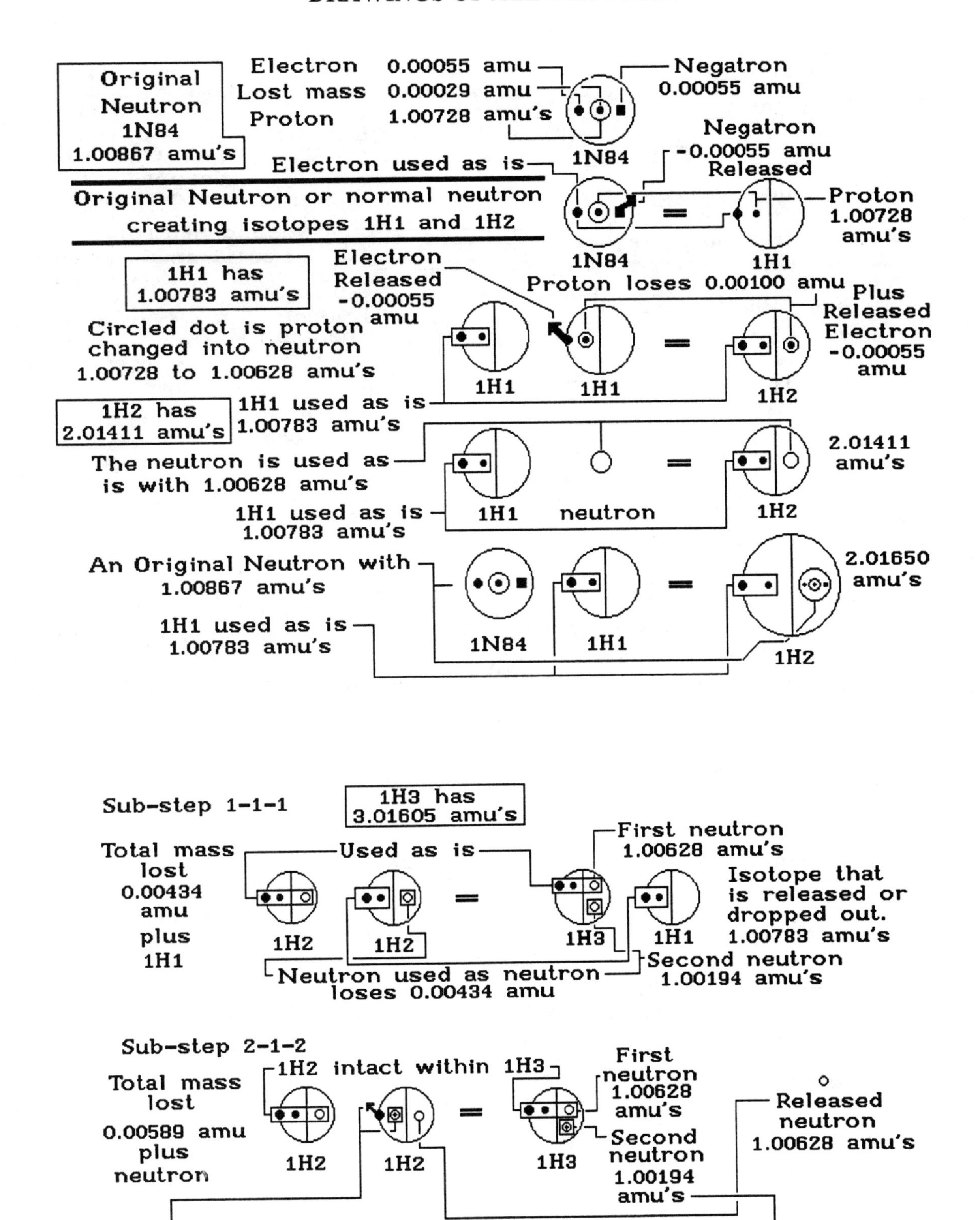

DRAWINGS OF ALL THE STEPS

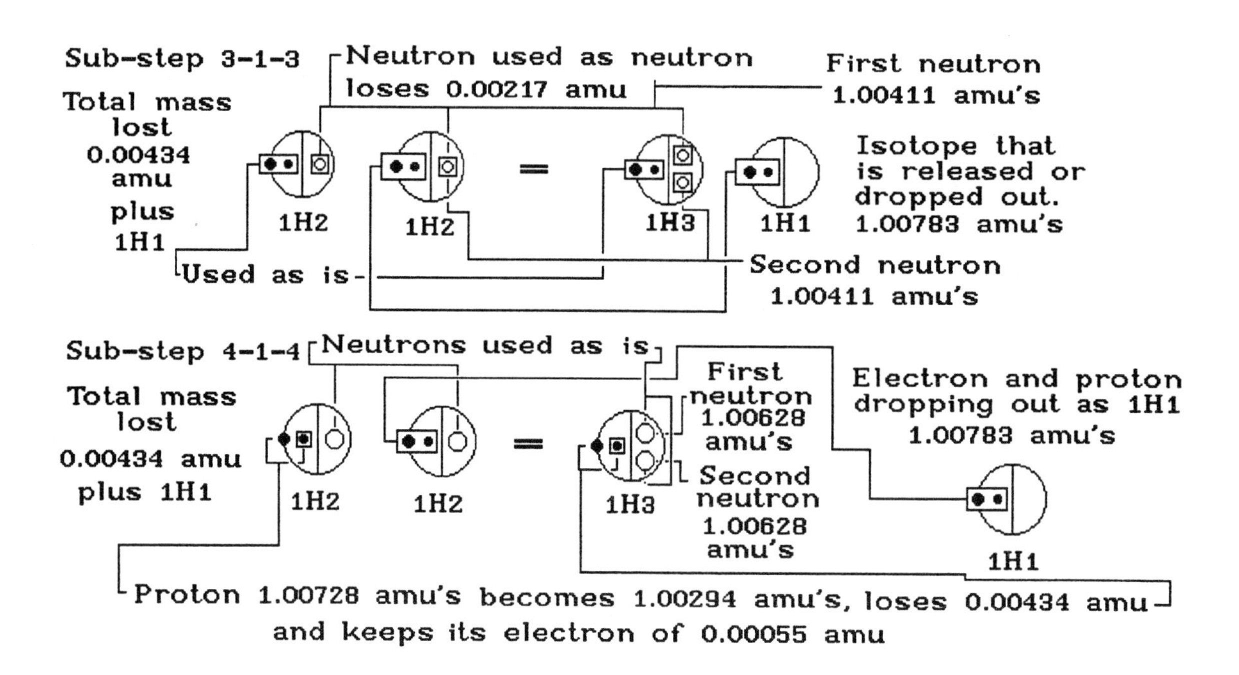

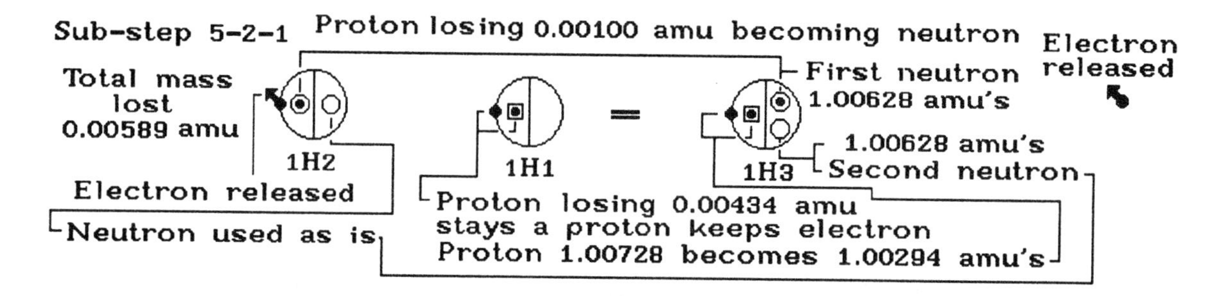

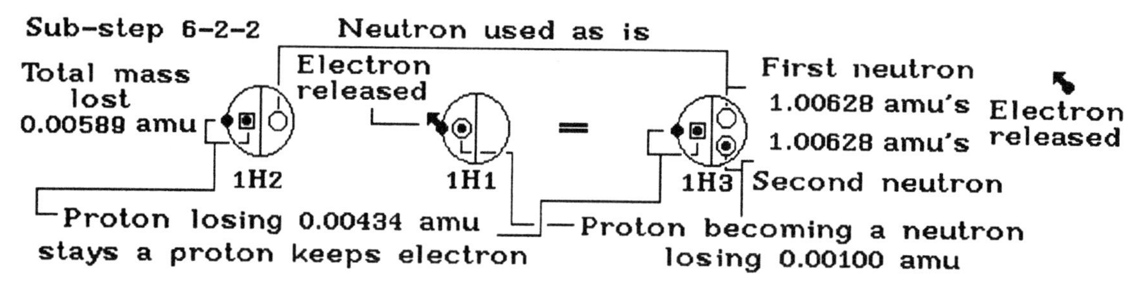

DRAWINGS OF ALL THE STEPS

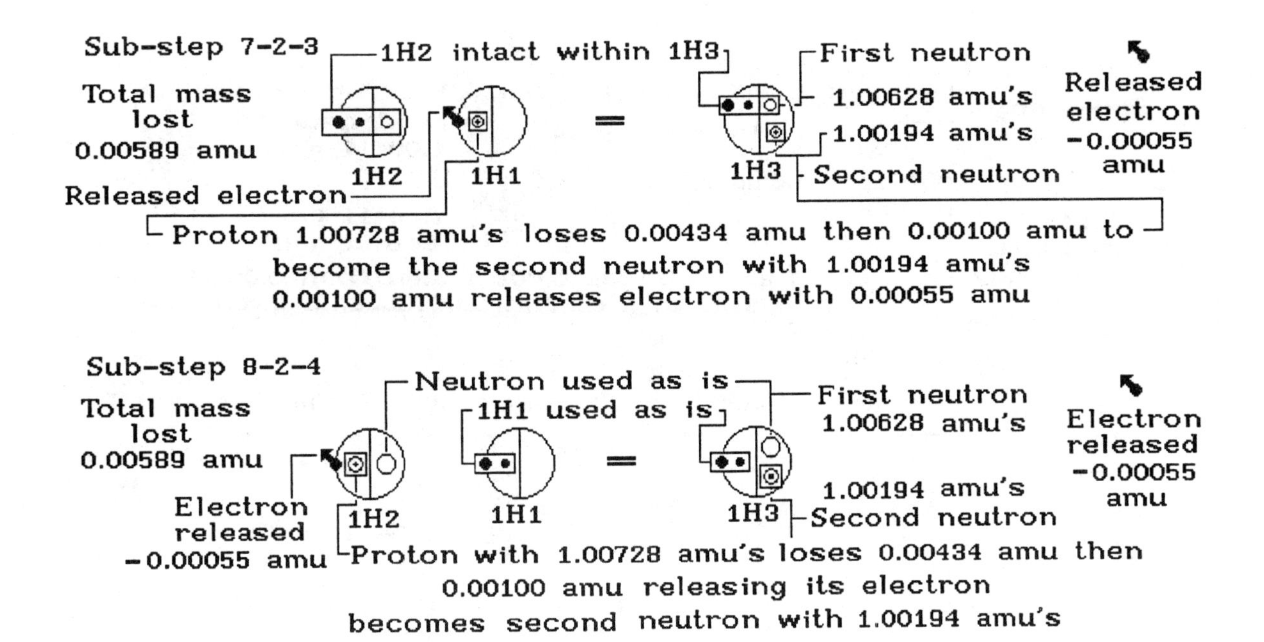

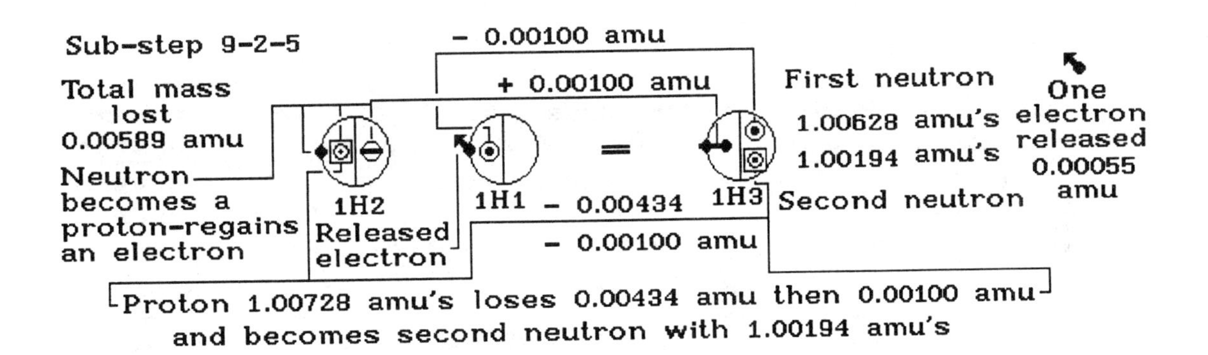

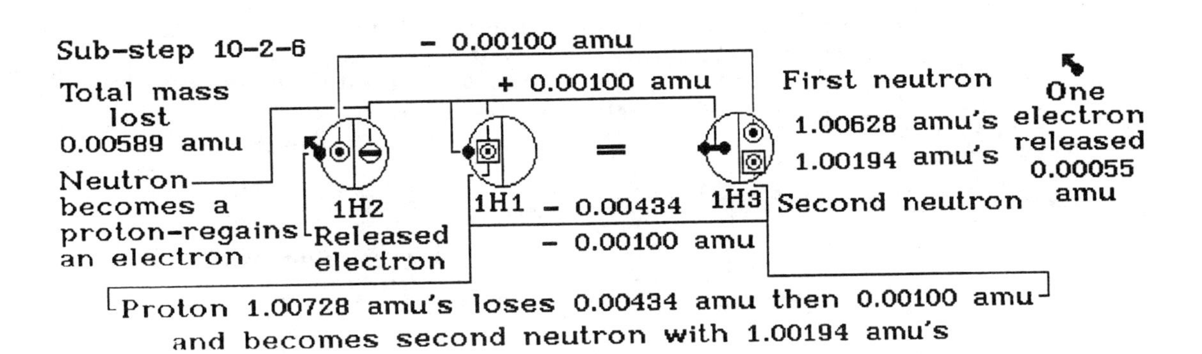

DRAWINGS OF ALL THE STEPS

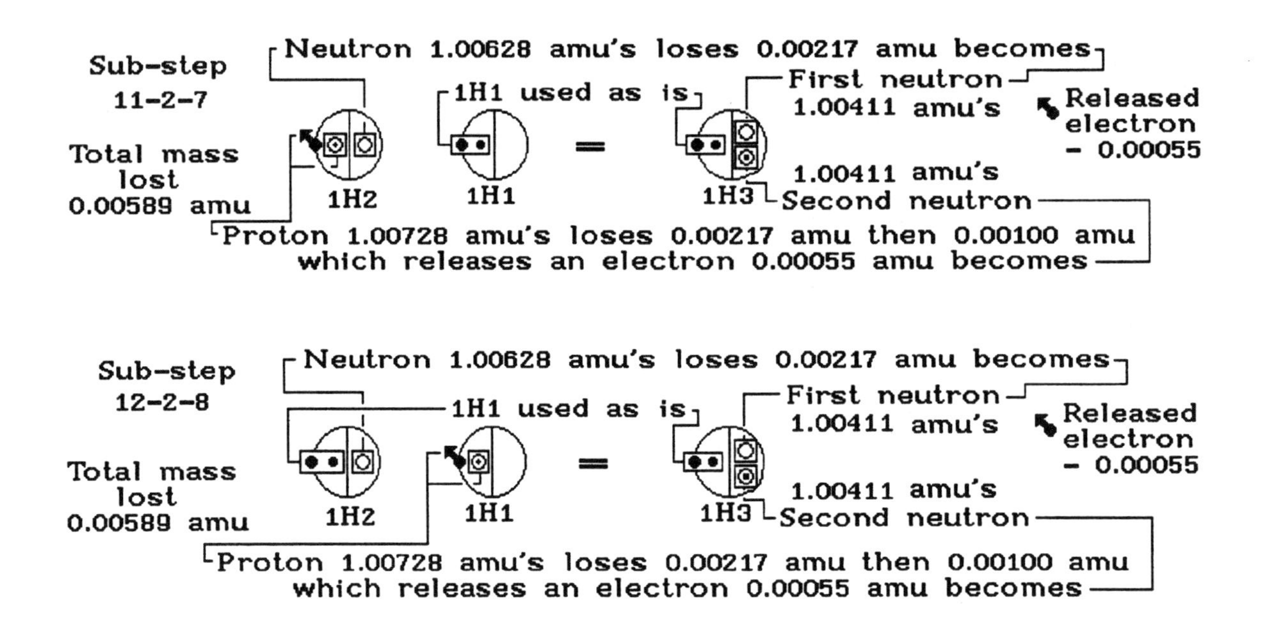

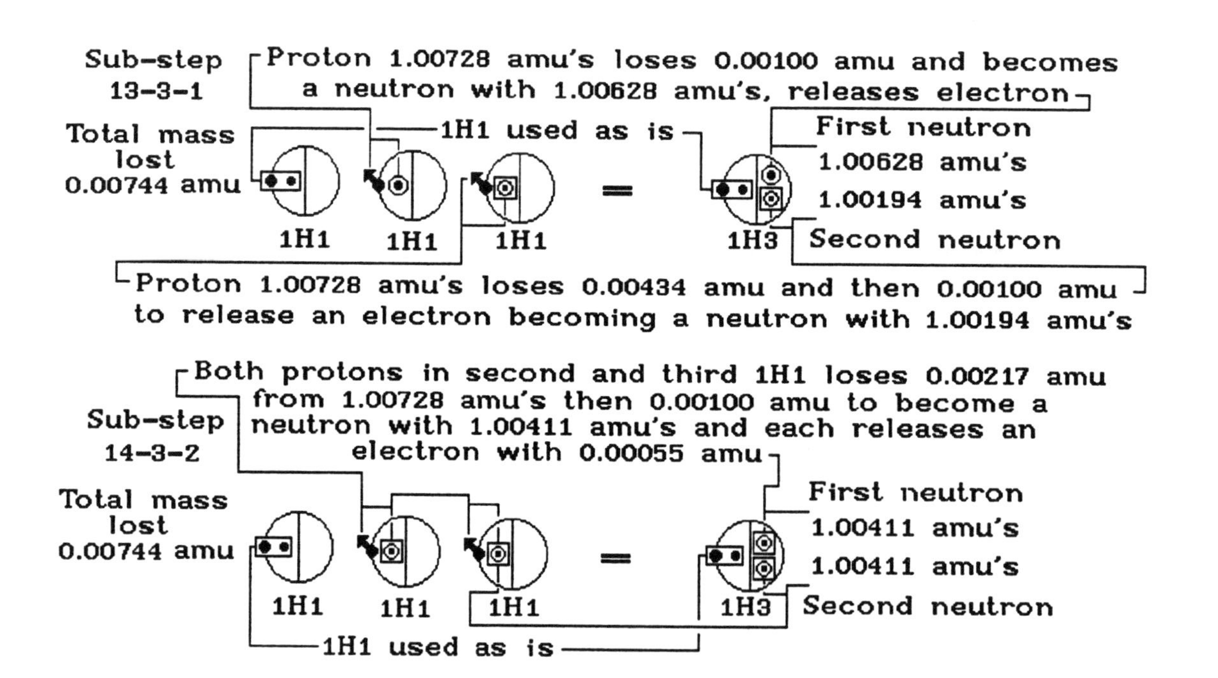

DRAWINGS OF ALL THE STEPS

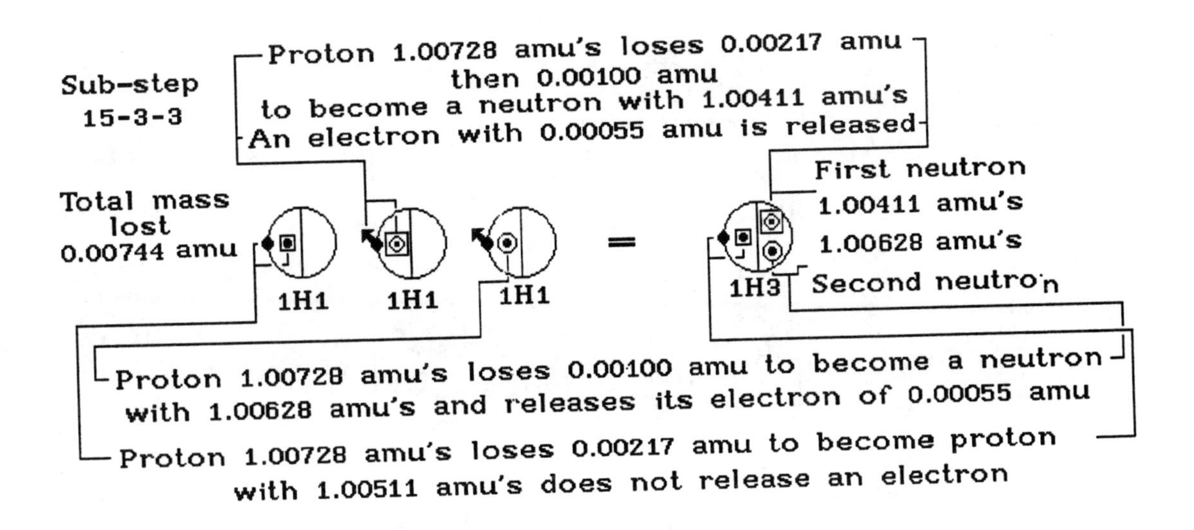

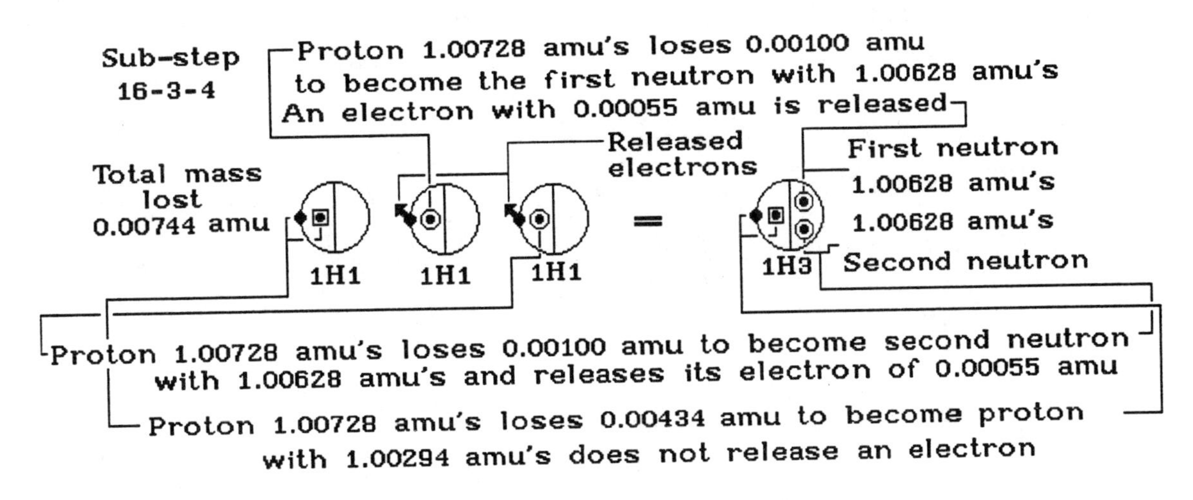

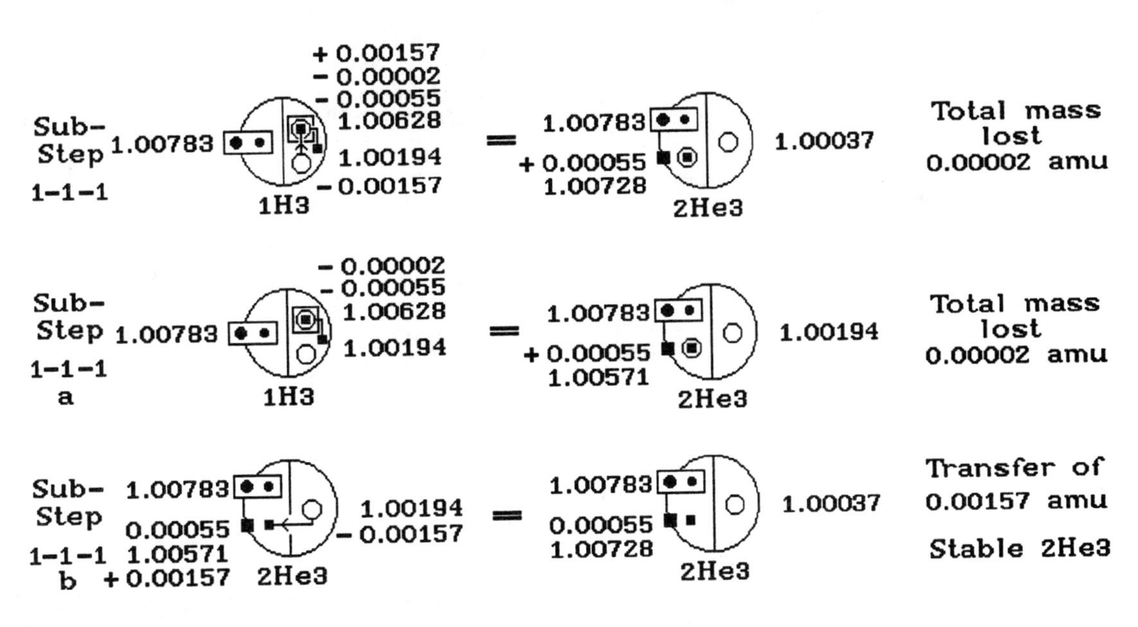

DRAWINGS OF ALL THE STEPS

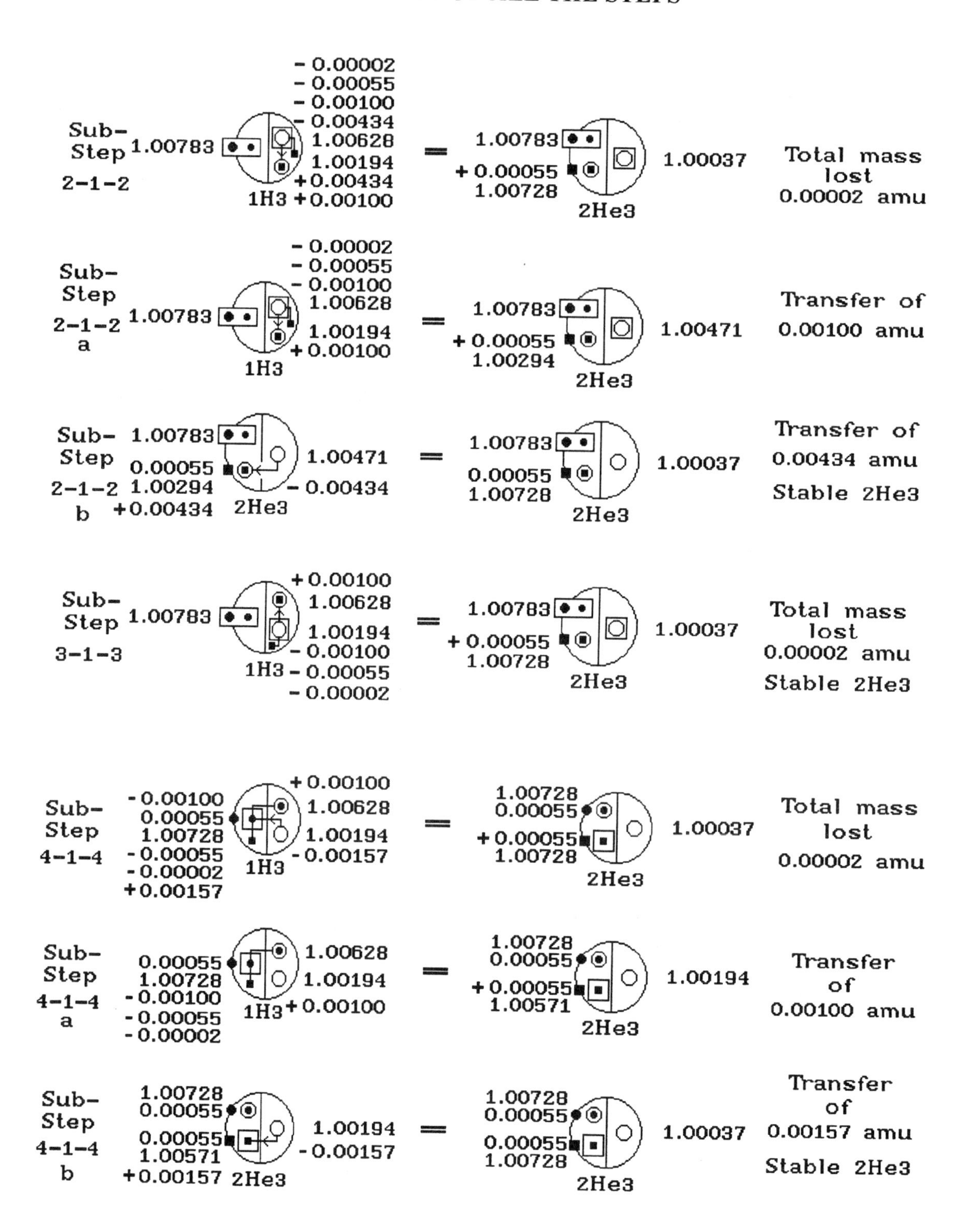

DRAWINGS OF ALL THE STEPS

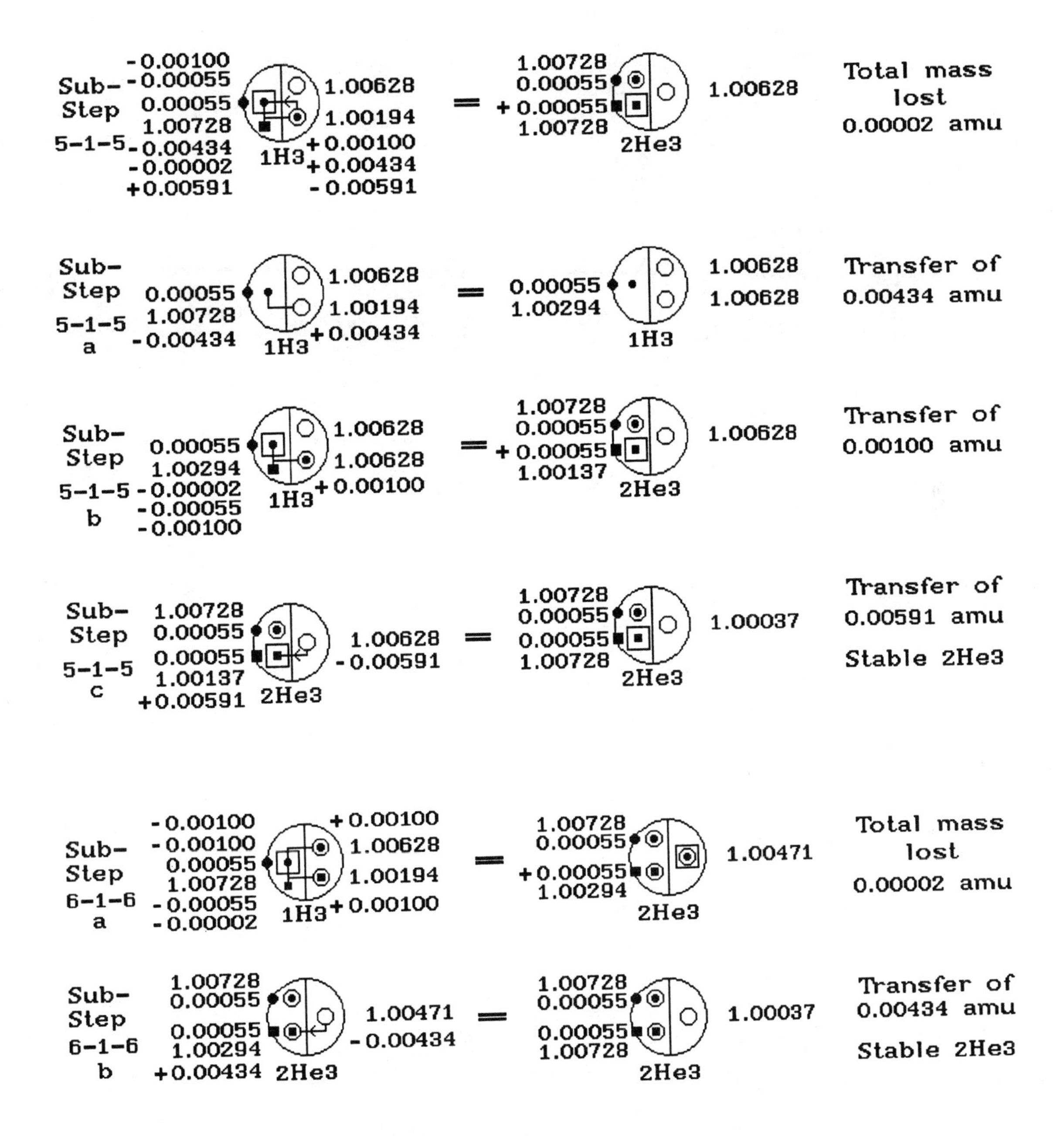

DRAWINGS OF ALL THE STEPS

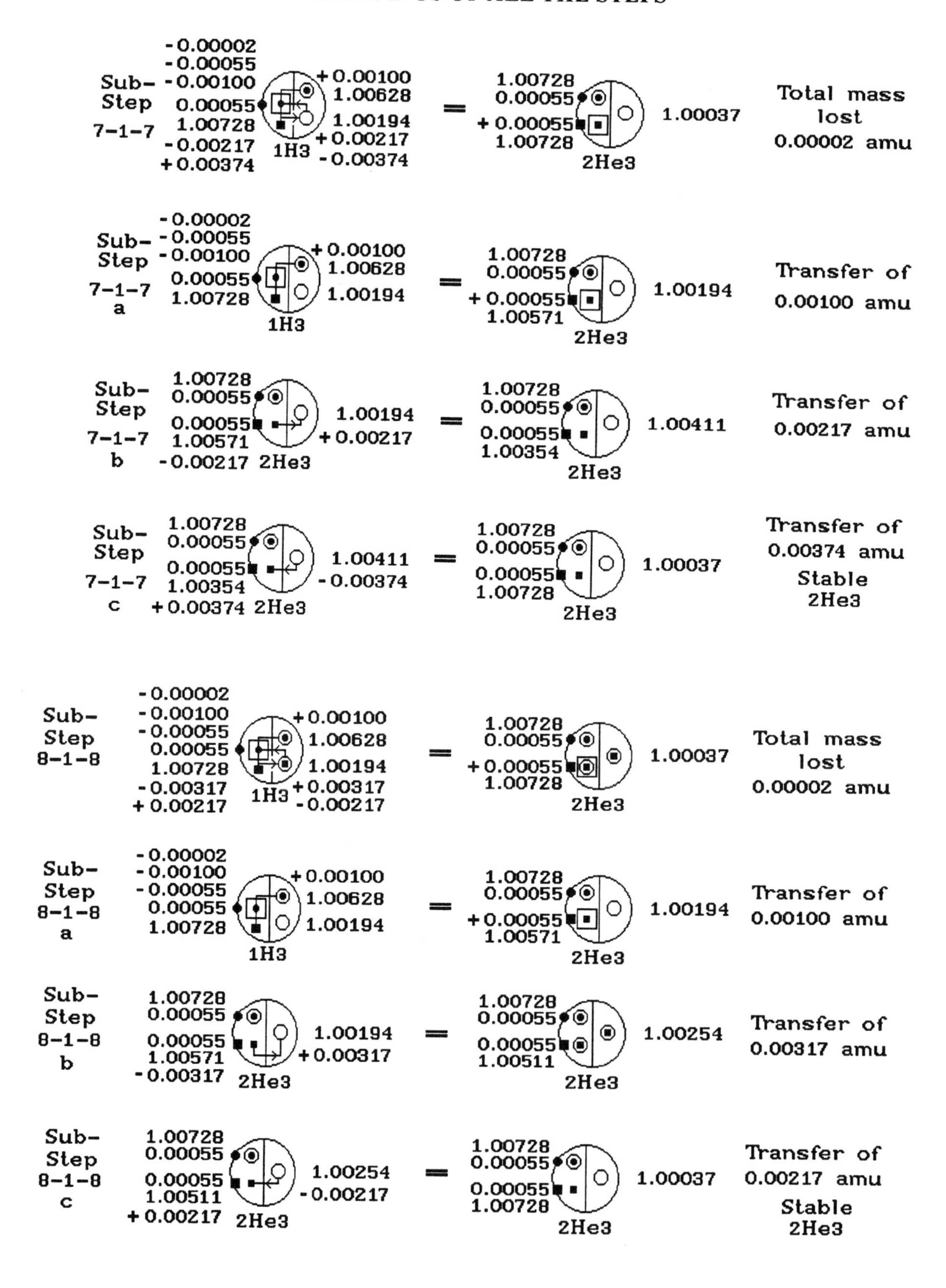

DRAWINGS OF ALL THE STEPS

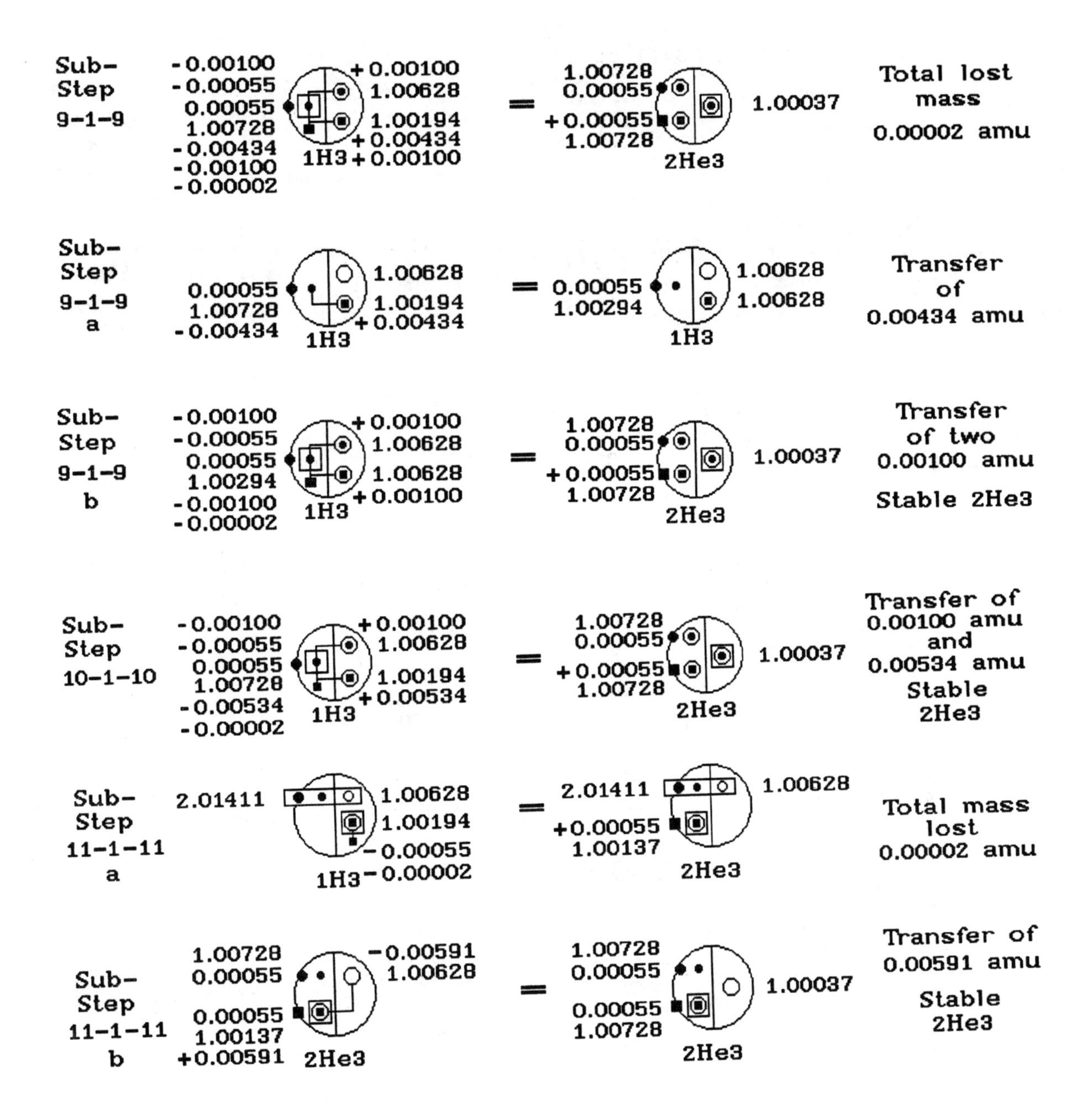

DRAWINGS OF ALL THE STEPS

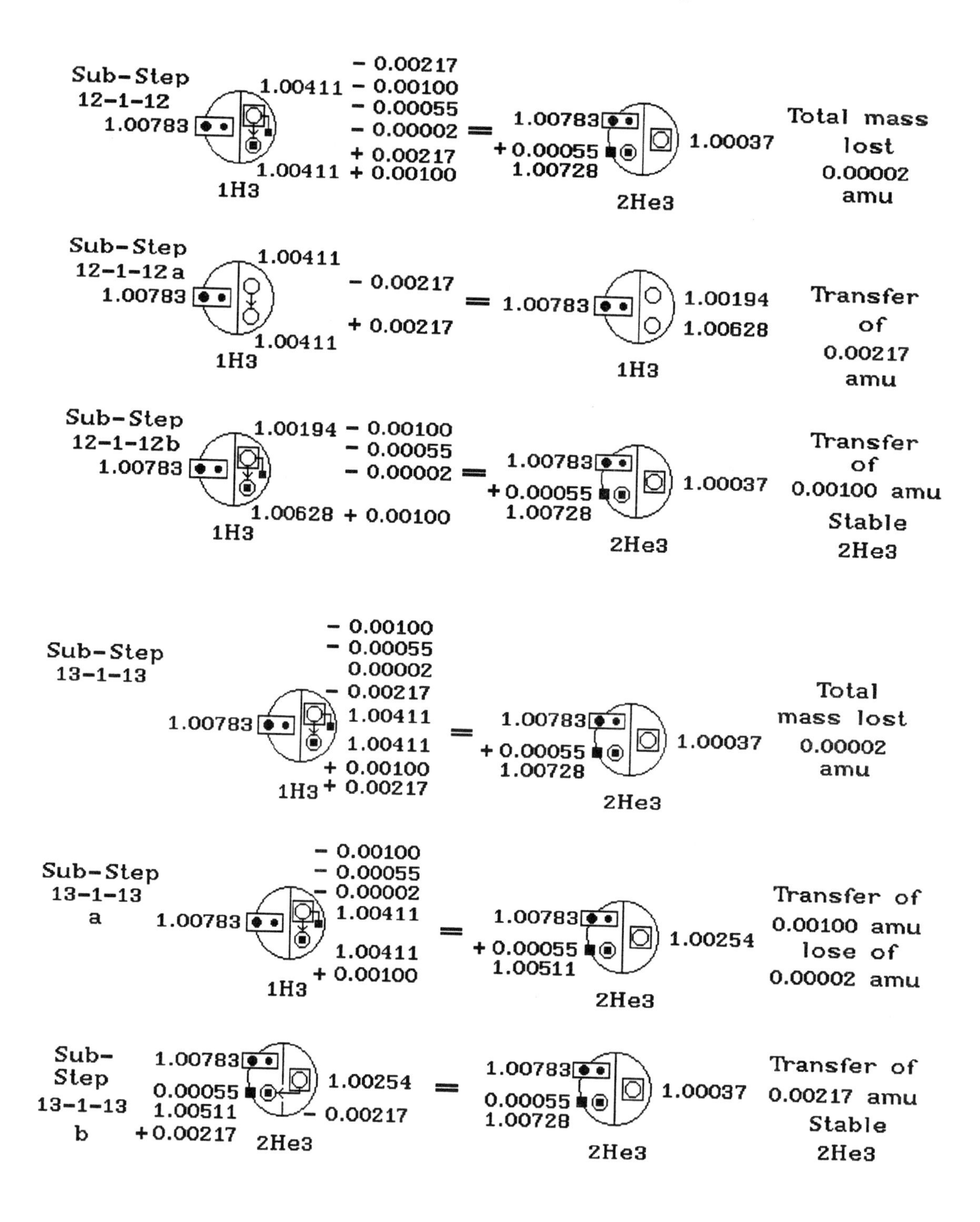

DRAWINGS OF ALL THE STEPS

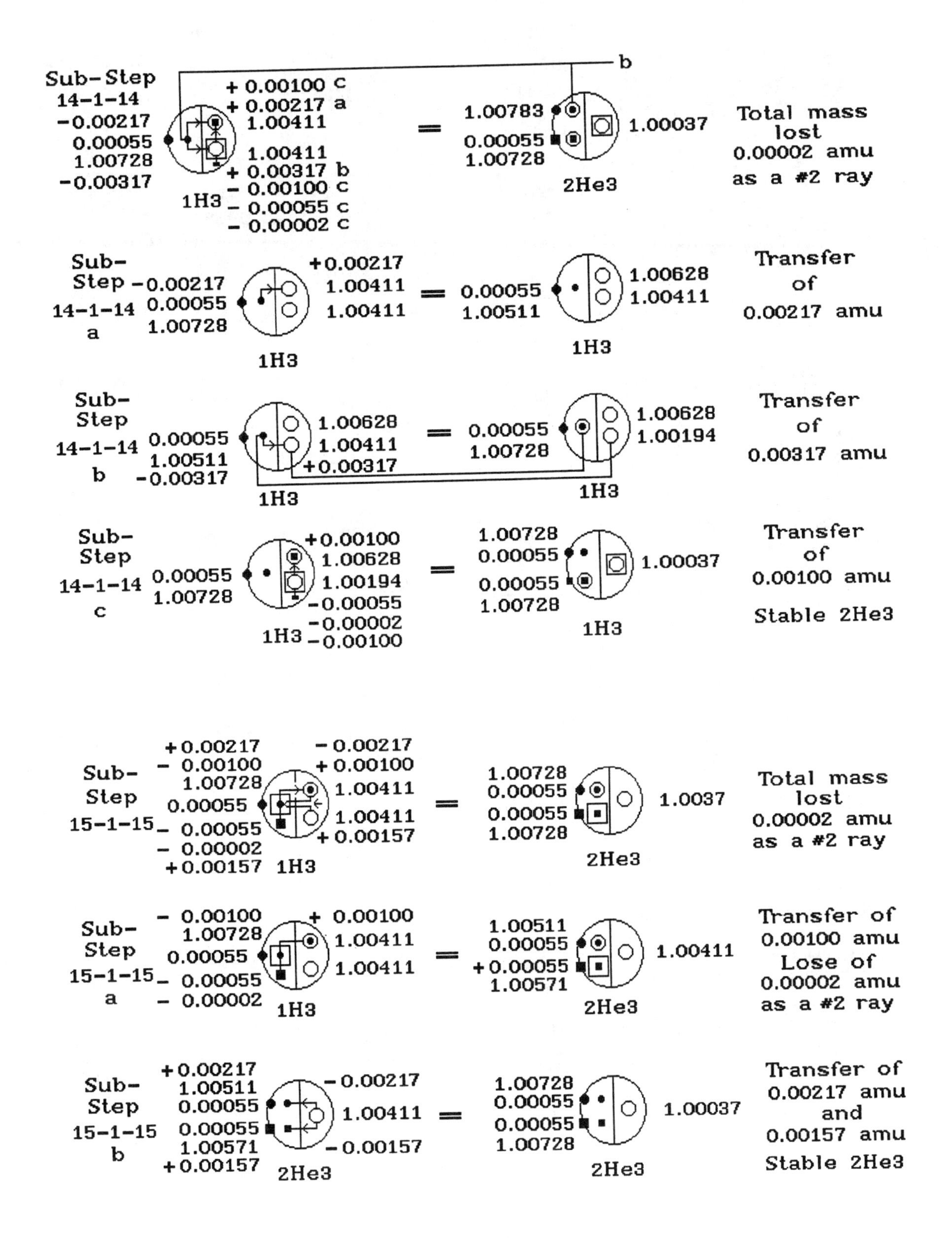

DRAWINGS OF ALL THE STEPS

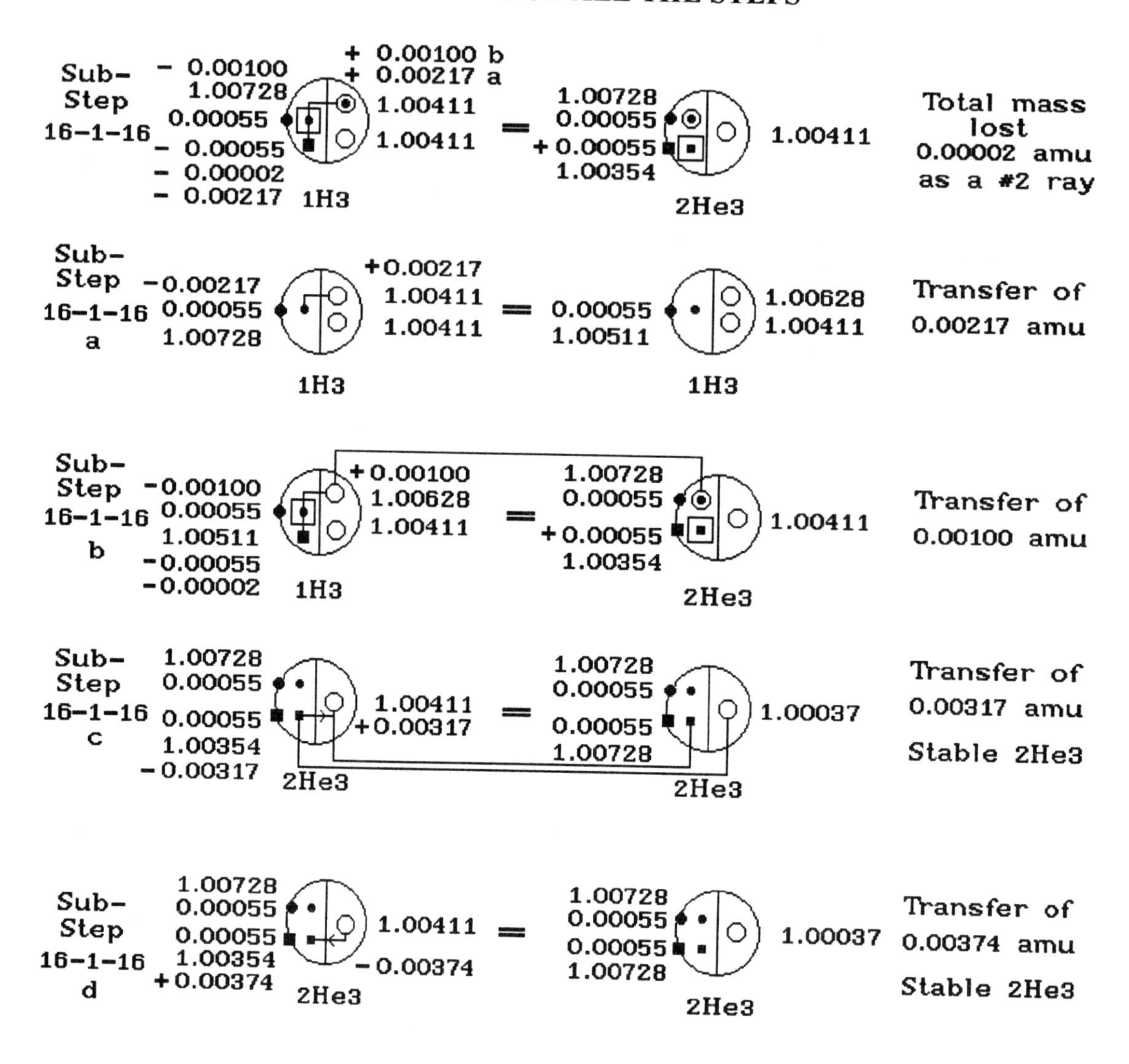

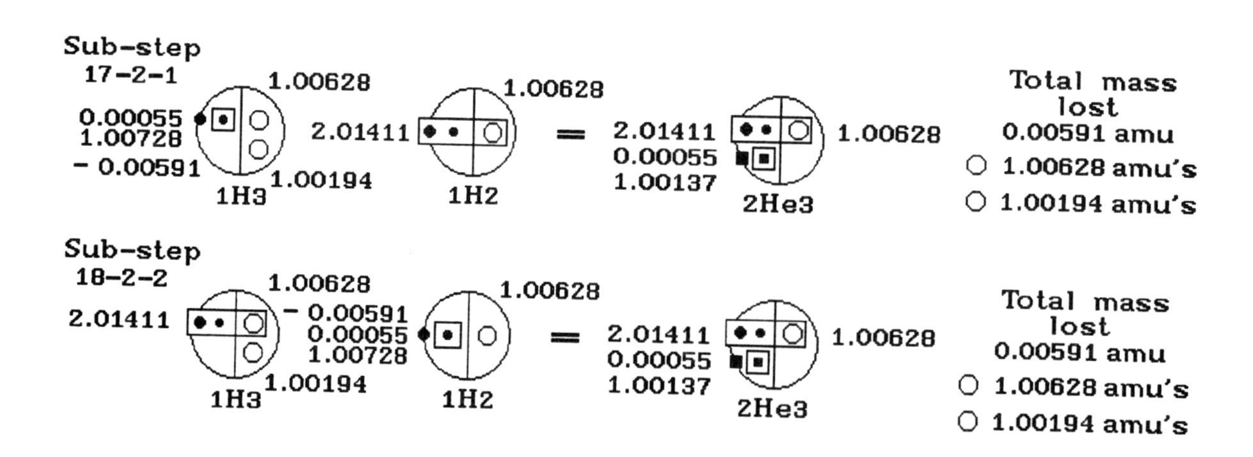

DRAWINGS OF ALL THE STEPS

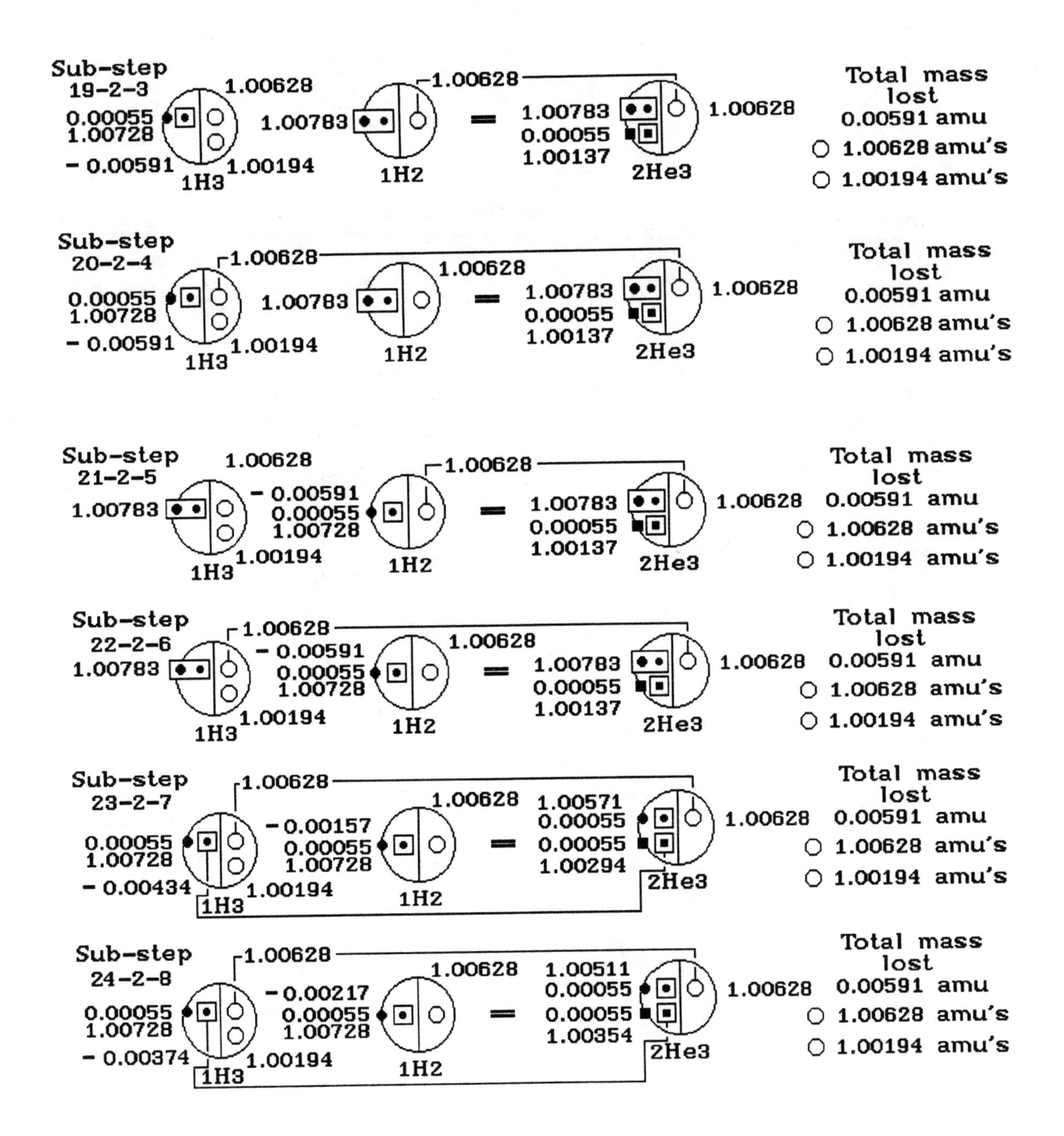

DRAWINGS OF ALL THE STEPS

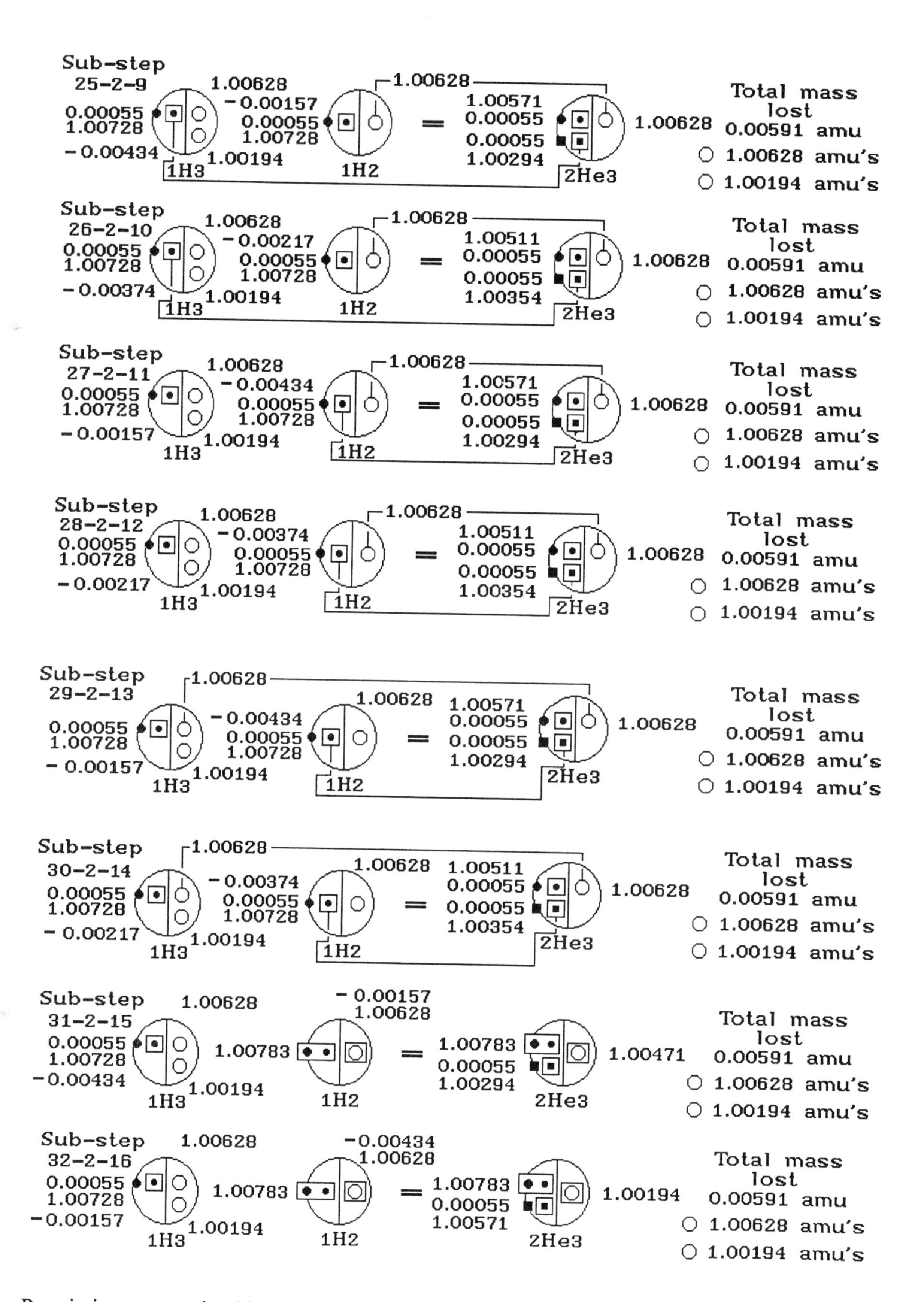

DRAWINGS OF ALL THE STEPS

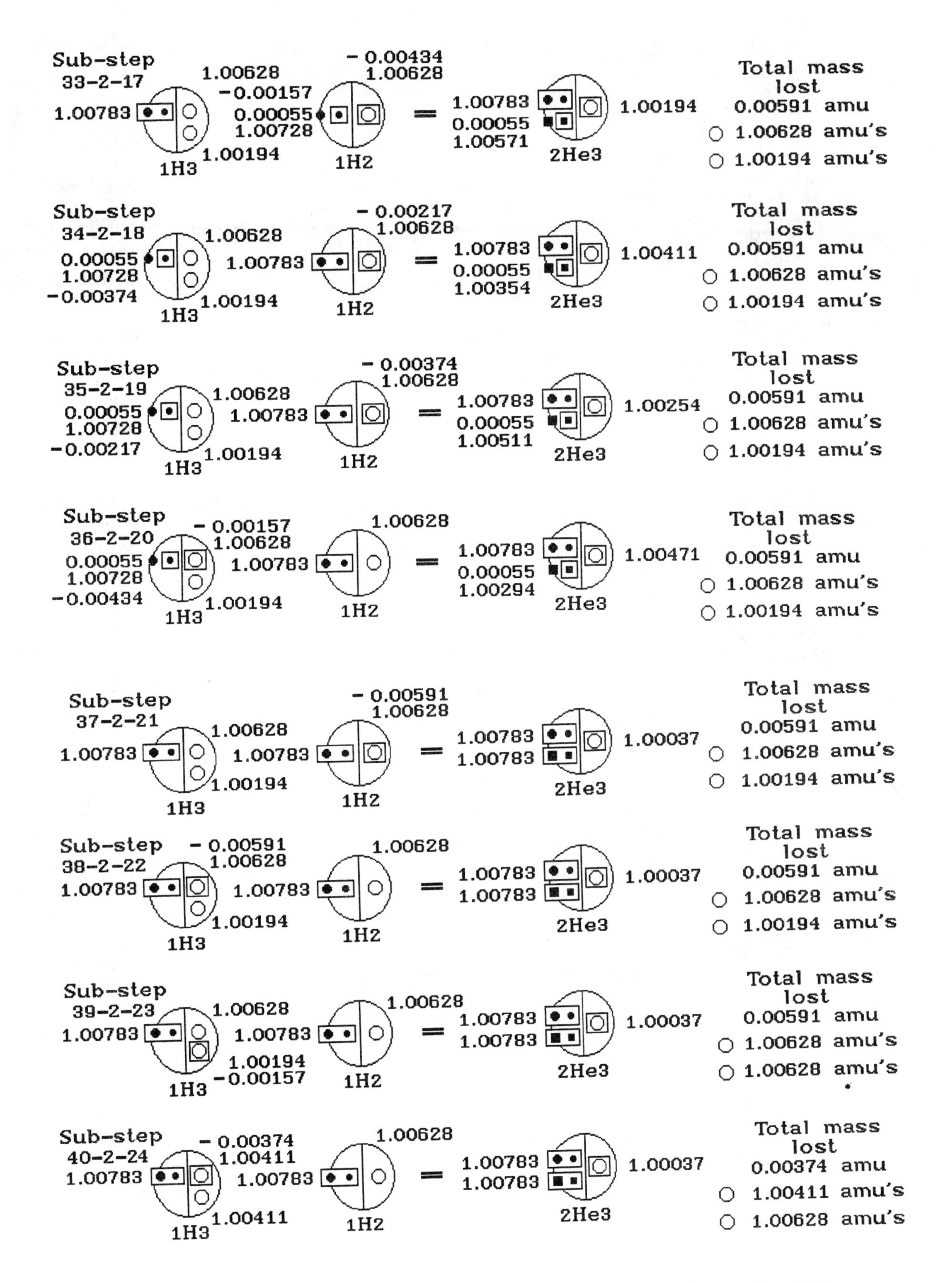

DRAWINGS OF ALL THE STEPS

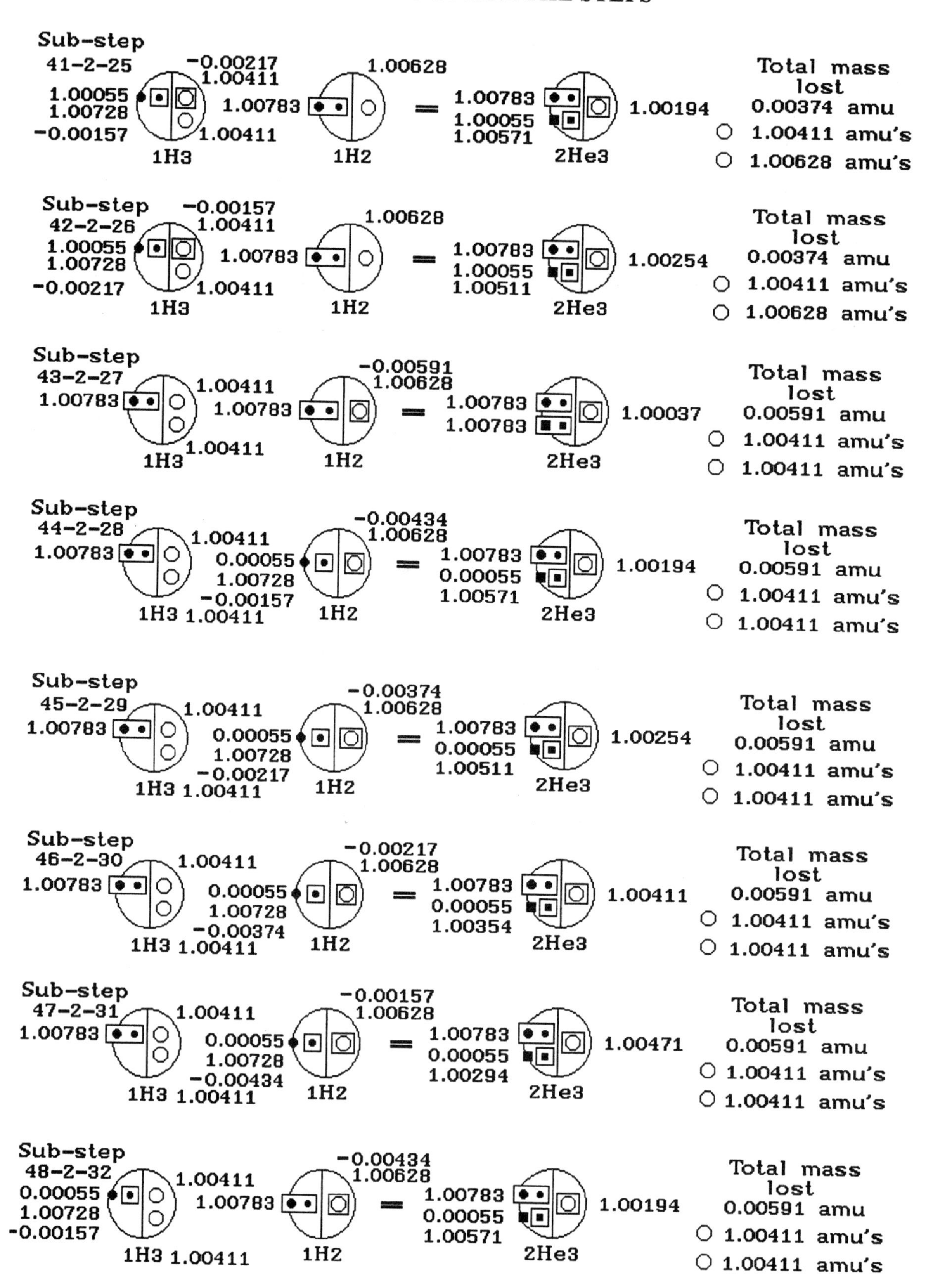

DRAWINGS OF ALL THE STEPS

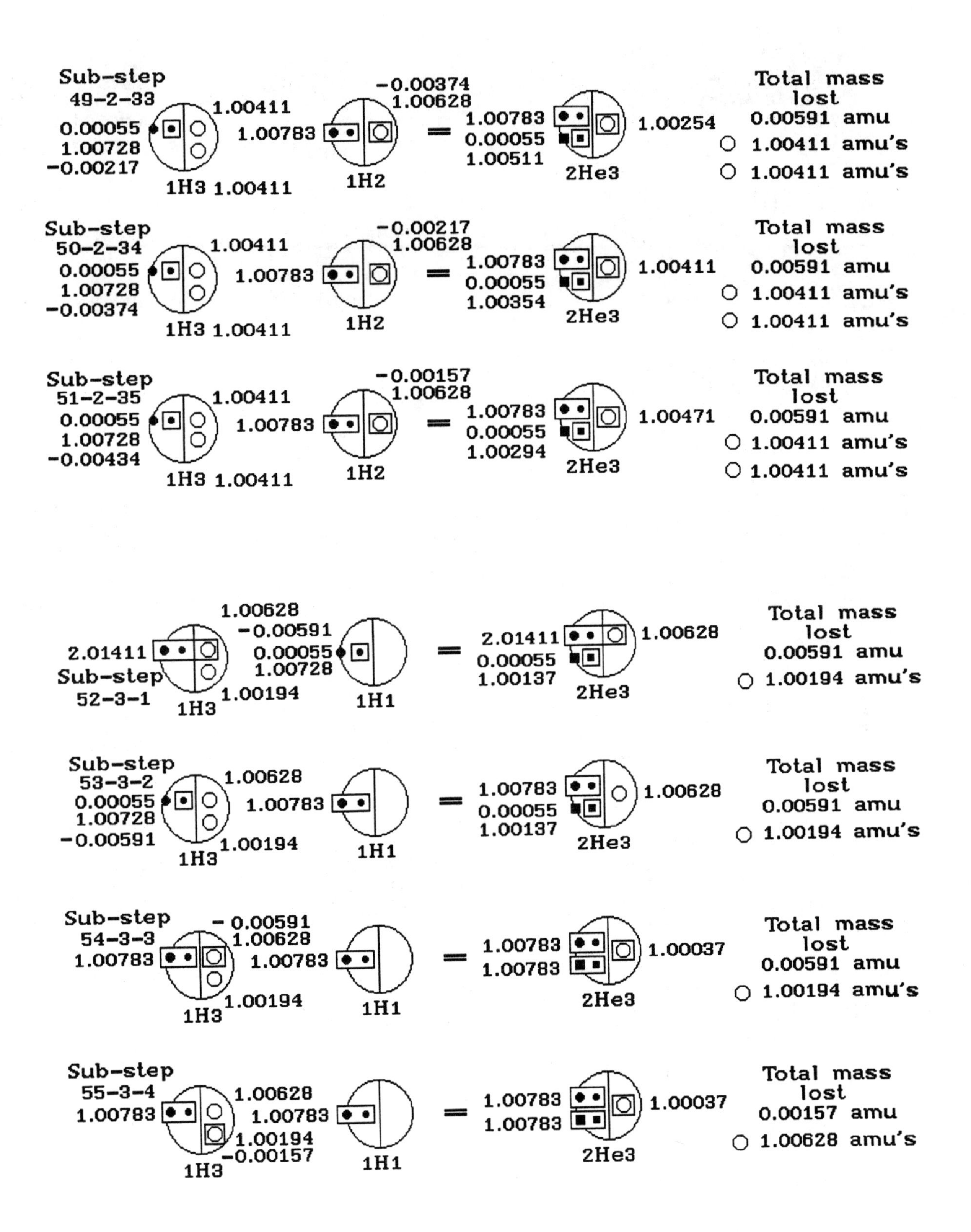

DRAWINGS OF ALL THE STEPS

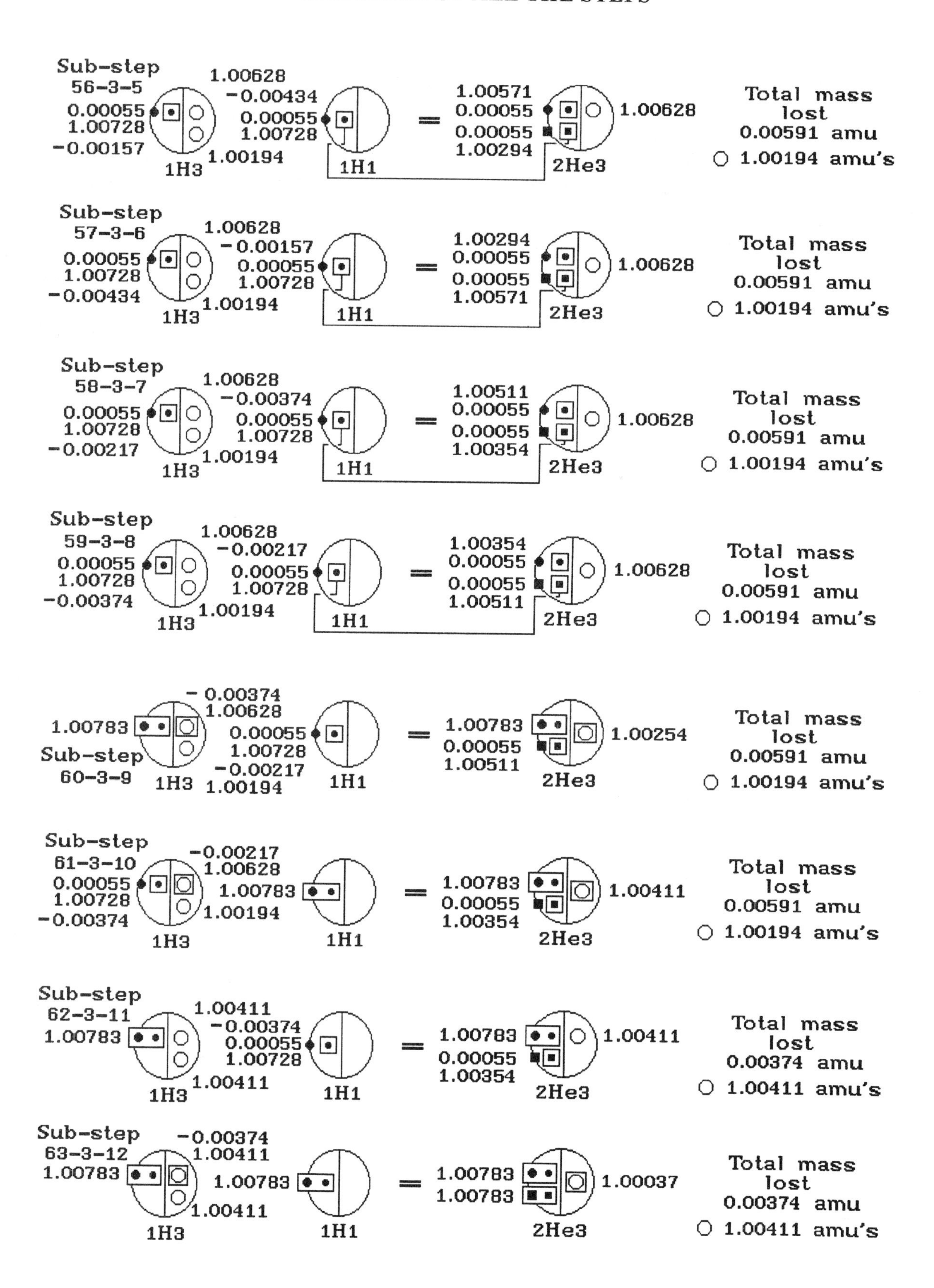

DRAWINGS OF ALL THE STEPS

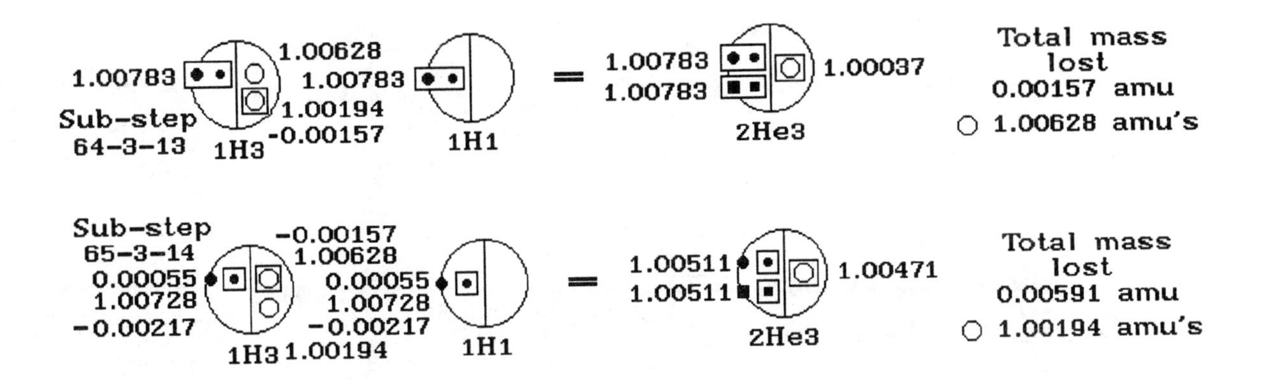

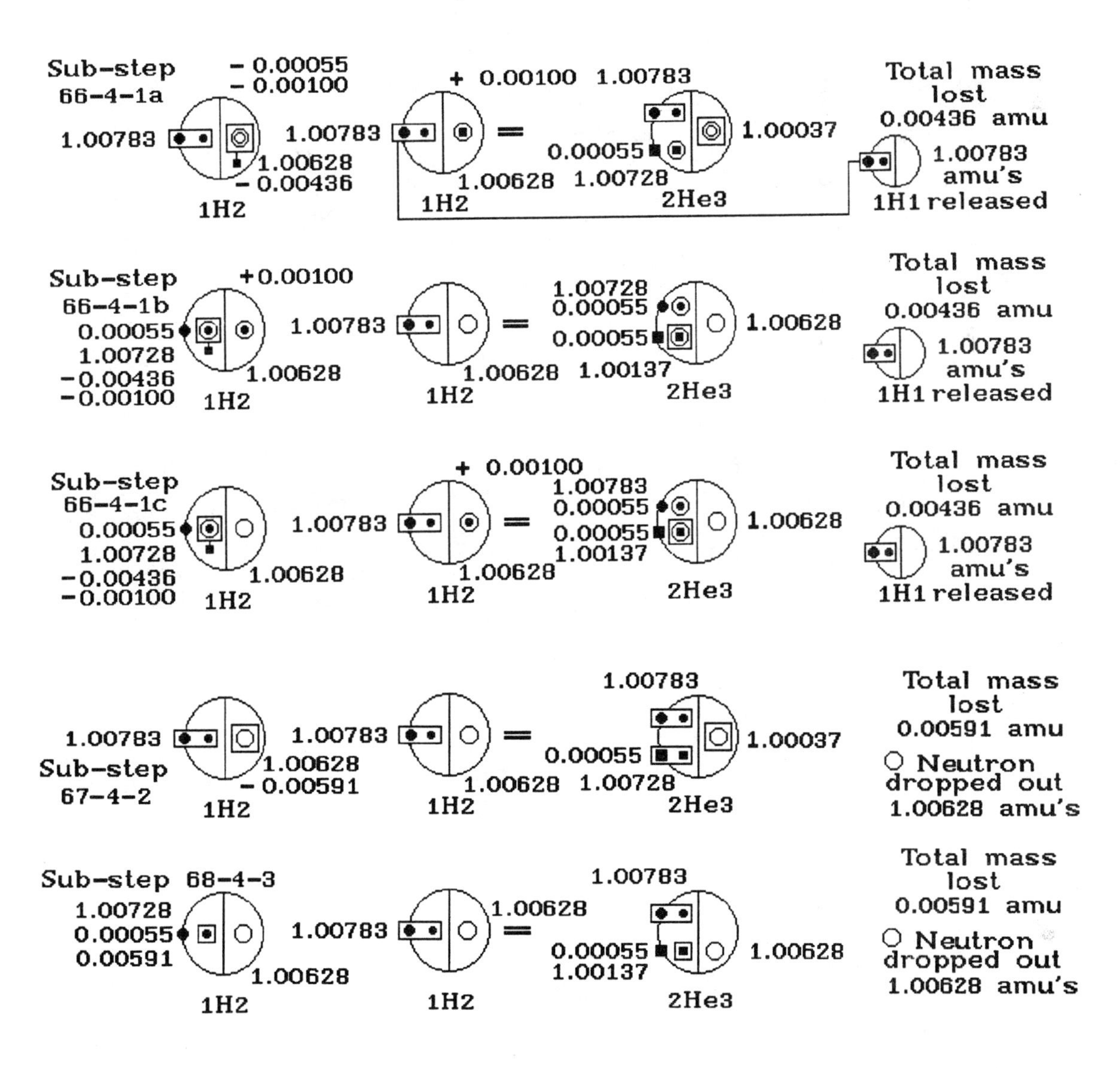

DRAWINGS OF ALL THE STEPS

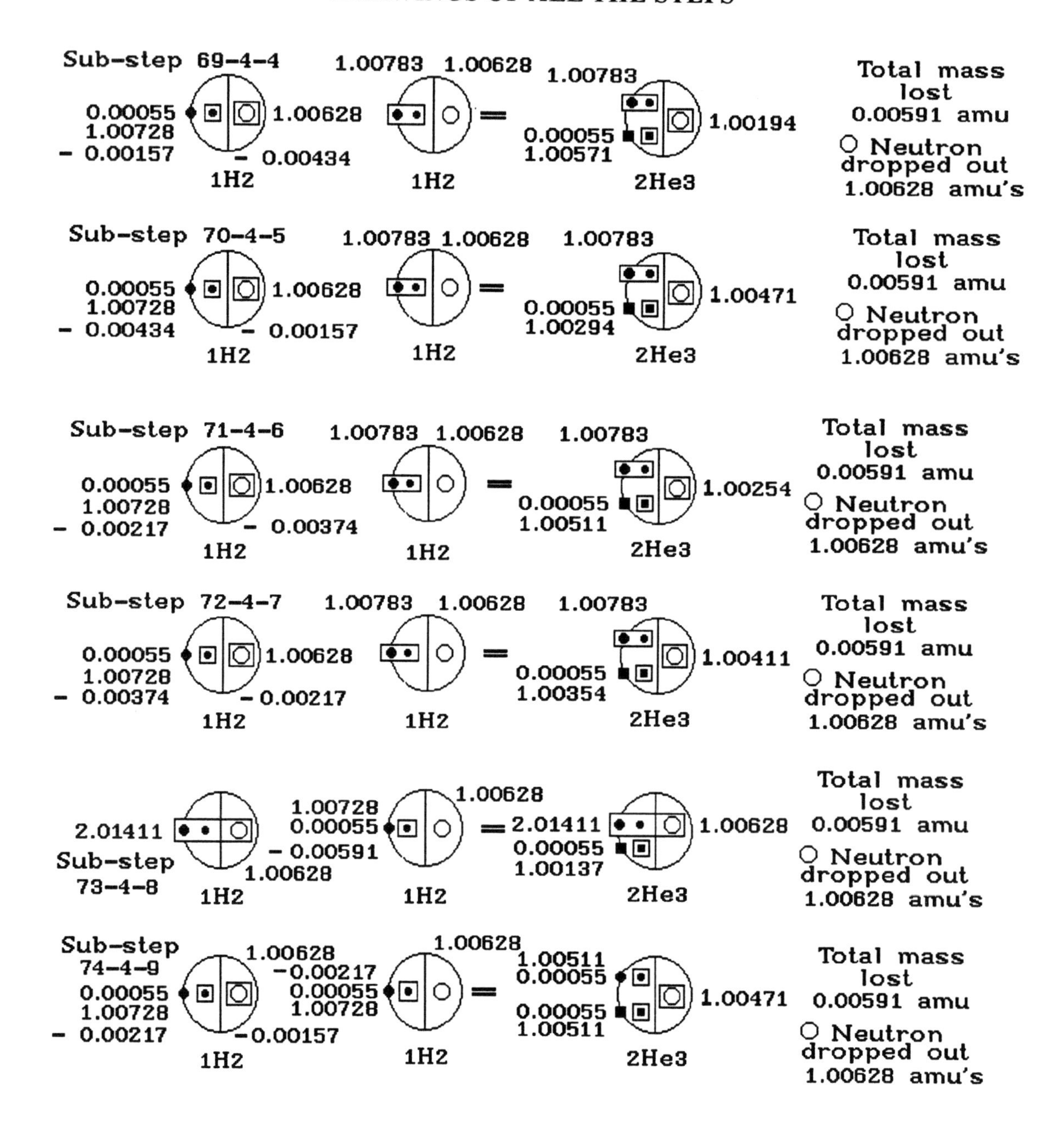

DRAWINGS OF ALL THE STEPS

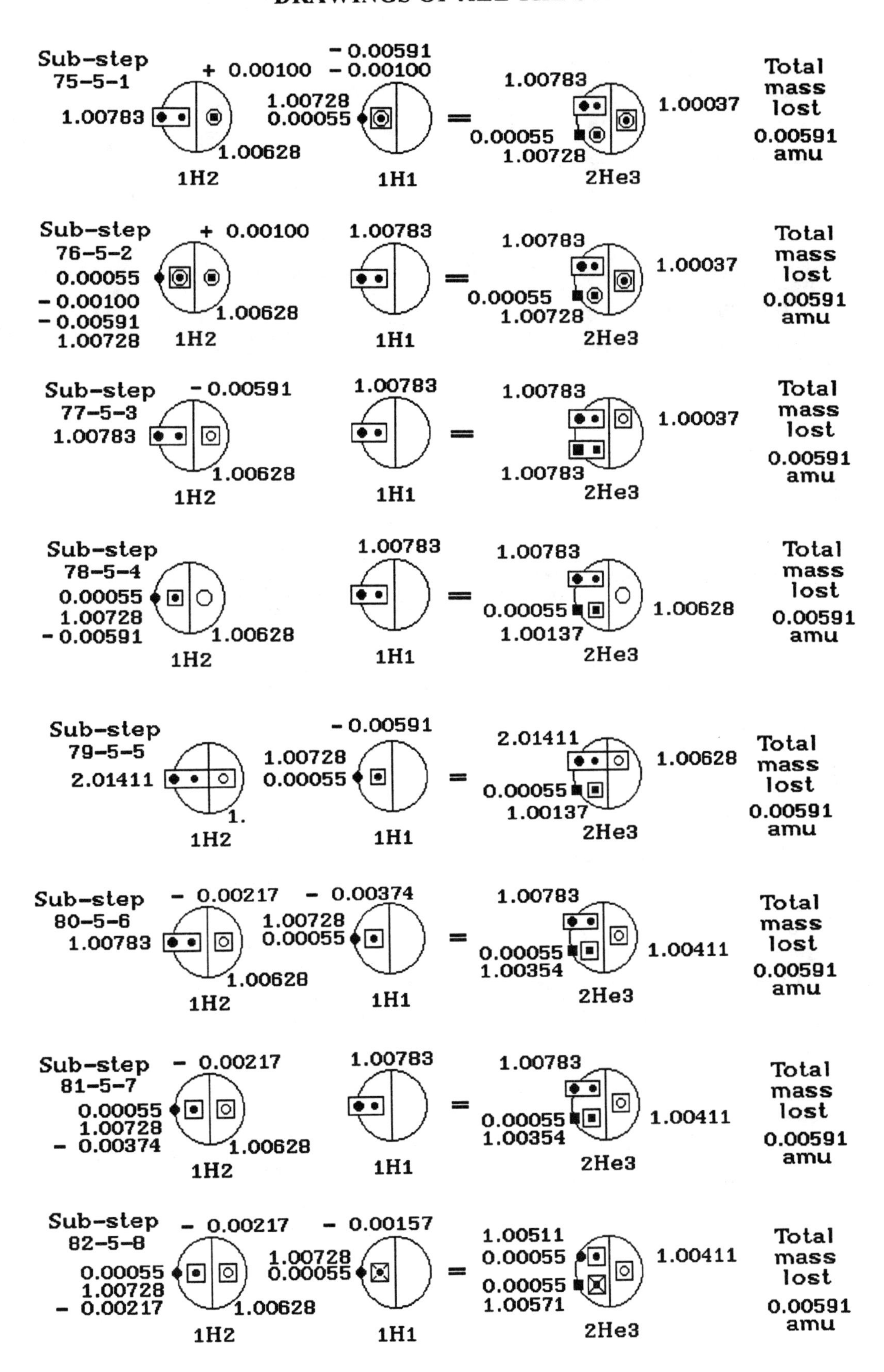

DRAWINGS OF ALL THE STEPS

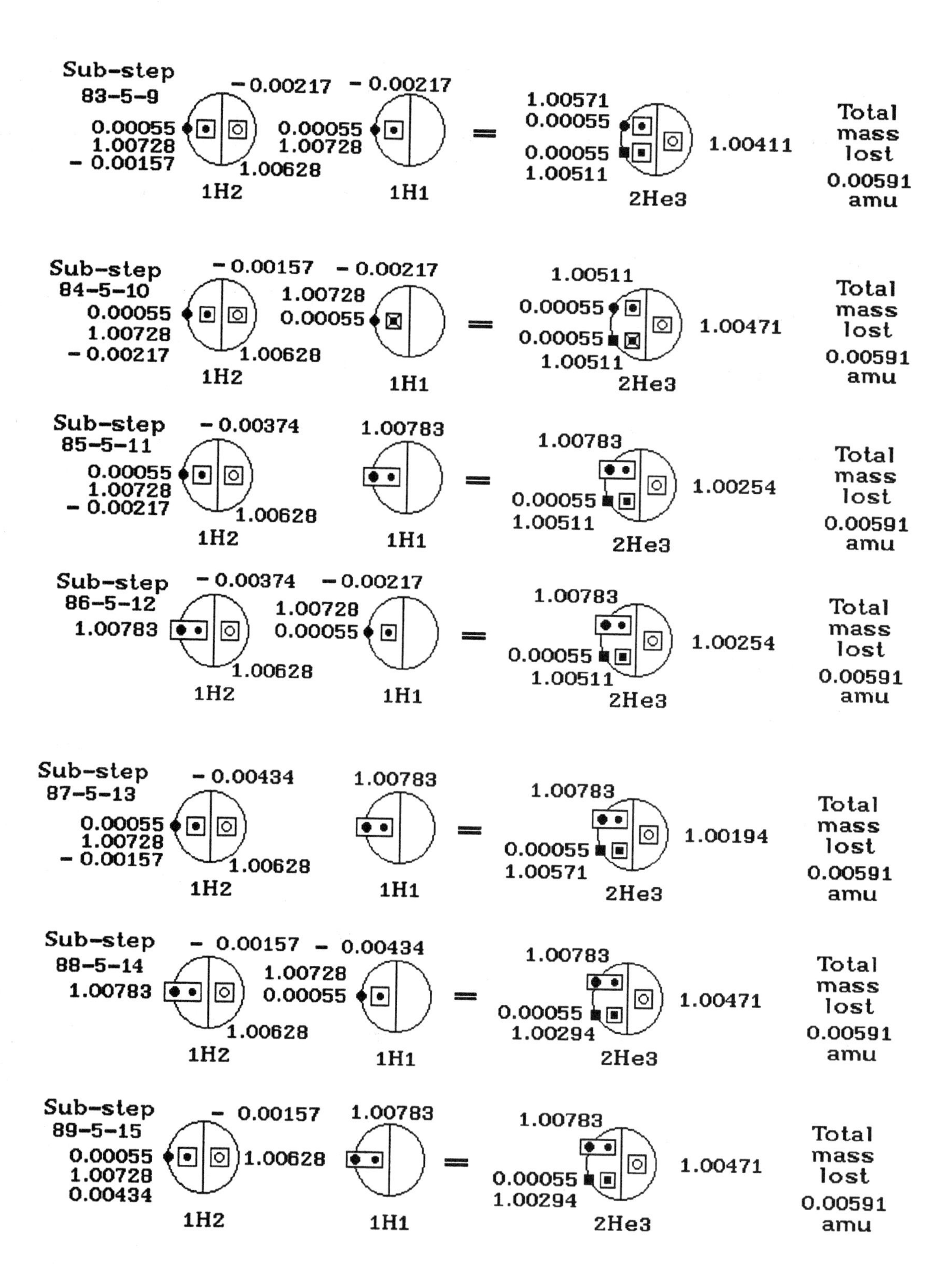

DRAWINGS OF ALL THE STEPS

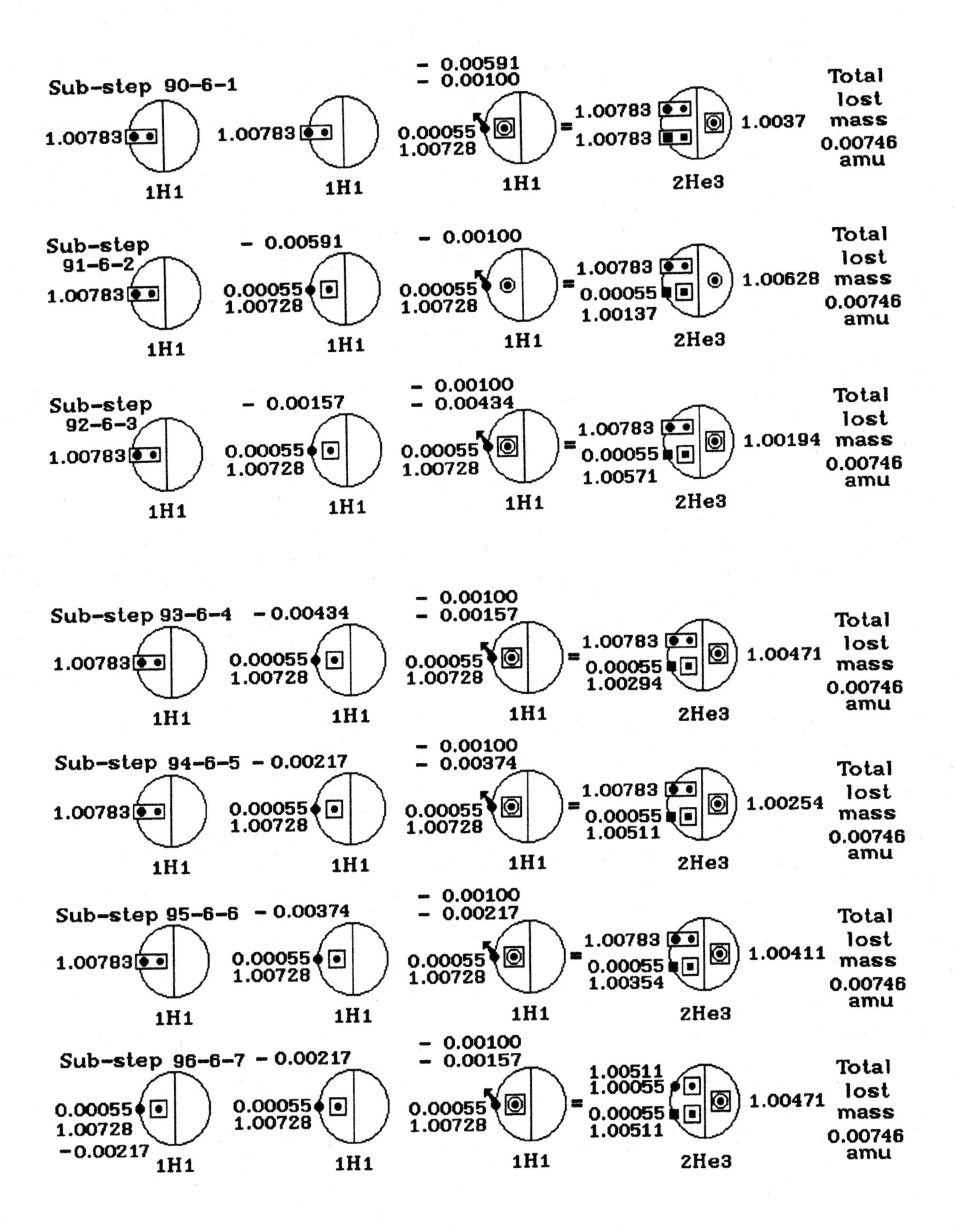

#2. HEILUM (He)

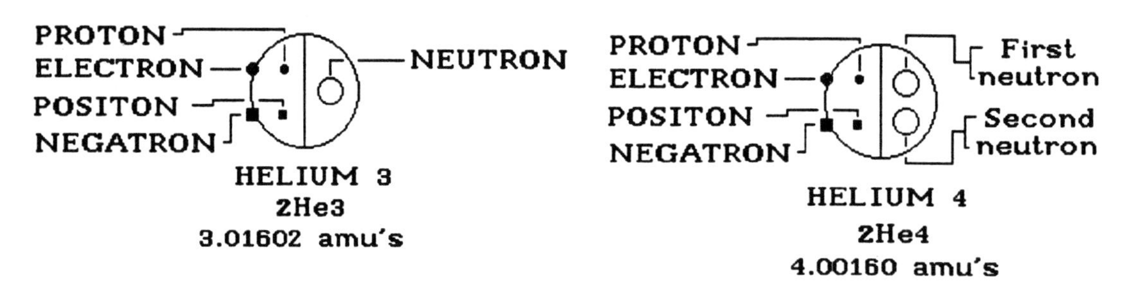

Electron	0.00055 amu
Negatron	0.00055 amu
Proton	1.00728 amu's
Positon	1.00728 amu's
Neutron	1.00037 amu's
2He3	3.01603 amu's

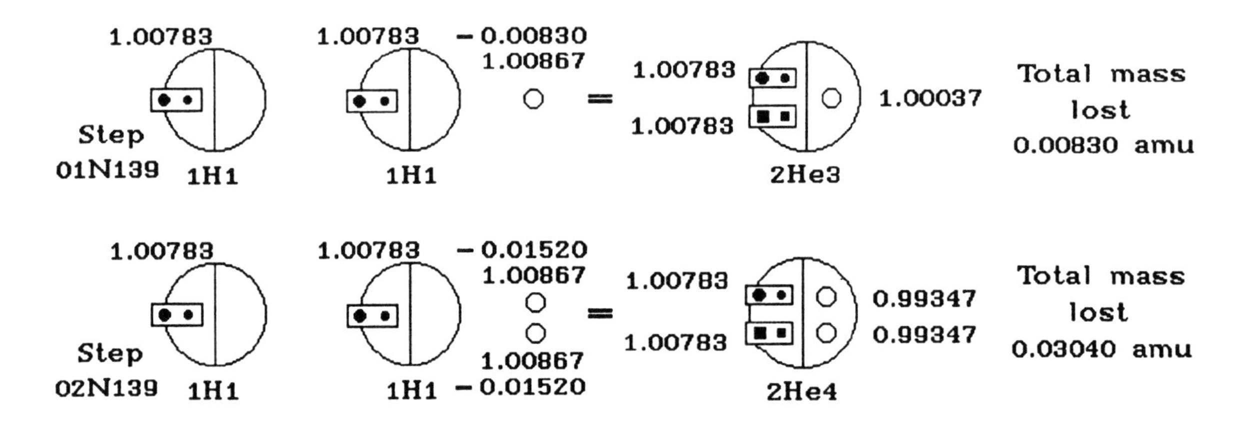

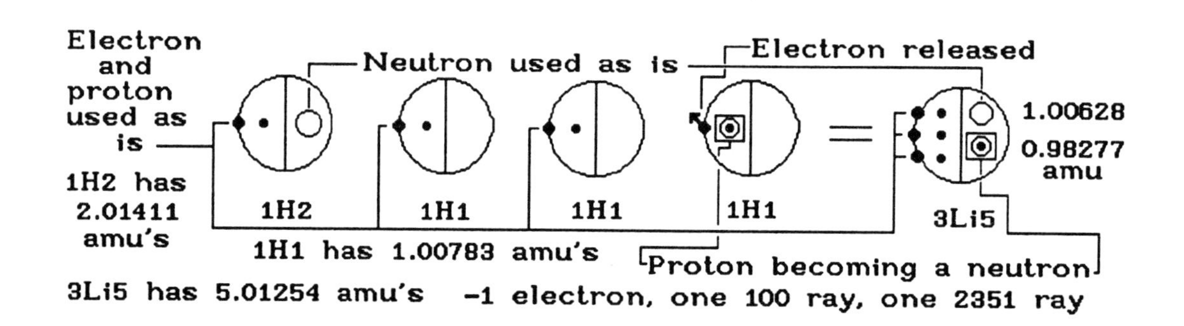

PART 7

Scientific American Library Books by copyright year

Powers of Ten by Philip and Phylis Morrison and The Office of Charles and Ray Emes, Copyright 1982

The Solar System by Roman Smoluchowki, Copyright 1983

The Discovery of Subatomic Particles by Steven Weinberg, Copyright 1983

The Science of Musical Sound by John R. Pierce, Copyright 1983

The Second Law by P.W. Atkins, Copyright 1984

Einstein's Legacy by Julian Schwinger, Copywright 1986

A Journey Into Gravity and Spacetime by John Archibald Wheeler, Copyright 1990

Beyond the Third Dimension by Thomas F. Banchoff, Copyright 1990

Stars by James B. Kaler, Copyright 1992

Short History of the Universes by Joseph Silk, Copyright 1994

Mathematics by Keith Devlin, Copyright 1994

BOOKS ON CHEMISTRY AND PHYSICS

A Text-book of Mineralogy by Edward Salisbury Dana and Professor James D. Dana, Copyright 1877
Organic and Medical Chemistry by Henry Leffmann, Copyright 1884
Dana's Manual of Mineralogy by William E. Ford, Fourteenth Edition, Copyright 1929
A Textbook of Mineralogy by Edward Salisbury Dana, Revised Fourth Edition by William E. Ford, 1949

Creative Chemistry by Professor E. H. S. Bailey and Julius Stieglitz, Copyright 1919

Theory of Electricity and Magnetism by J. J. Thomson, Copyright 1921
The Electron in Chemistry by J. J. Thomson, Copyright 1923
Electricity and Matter by J. J. Thomson, Copyright 1924

Practical Physiological Chemistry by Philip B. Hawk, Eight Edition, Copyright 1923

Microbe Hunters by Paul de Kruif, Copyright 1926

Introduction to Physiological Chemistry by Meyer Bodansky, Copyright 1927

Laboratory Manual of Biological Chemistry by Otto Folin, Copyright 1929

Introductory Theoretical Chemistry by G. H. Cartledge, Copywright 1929

The Story of Chemistry by Floyd L. Darrow, Copyright 1930

Second Year College Chemistry by William H. Chapin, Third Edition, Copyright 1933

Madame Curie Translated by Vincent Sheean, Copyright 1937

Chemistry At Work by William McPherson, William Edwards Henderson and George Winegar Flowler, Copyright 1938

General Chemistry by Thomas P. McCutcheon, Harry Seltz and J. C.. Warner, Third Edition, Second Printing, Copyright 1939

New Practical Chemistry by Newton Henry Black and James Bryant Conant, Copyright 1939

Outlines of Theoretical Chemistry by Frederick H. Getman and Farrington Daniels, Sixth Edition, Copyright 1940

Fundamental Chemistry by Horace G. Deming, Copyright 1940

General Chemistry by Harry N. Holmes, Fourth Edition, Copyright 1941

Fundamentals of Chemistry by L. Jean Bogert, Copyright 1946

Chemistry In Action by George M. Rawlins and Alden H. Struble, Copyright 1948

The Diffraction of Light, X-rays, and Material Particles by Charles F. Meyer, Copyright 1949

Chemistry Today by Harry C. Biddle and George L. Bush, Copyright 1949

Chemistry and You by Hopkins, Smith, McGill and Bradbury, Copyright 1949

A Practical Survey of Chemistry by Walter S. Dyer and Manfred E. Muelle, Copyright 1950

Electricity and Magnetism by Francis Weston Sears, Copyright 1951

Essentials of Chemistry by Alfred Benjamin Garrett, Joseph Fredric Haskins and Harry Hall Sisler, Copyright 1951

The Particles of Modern Physics by J. D. Stranathan, Copyright 1952

College Chemistry, A Systematic Approach by Harry H. Sisler, Calvin A. VanderWerf and Arthur W. Davidson, Copyright 1953

Chemistry by George W. Watt, Lewis F. Hatch and J. J. Lagowske, Copyright 1954

Physics, A Basic Science by Elmer E. Burns, Frank L. Verwiebe, Hergert C. Hazel and Gordon E Van Hooft, Copyright 1954

Elements of Physics by D. Lee Baker, Raymond B. Brownlee and Robert W. Fuller, Copyright 1954

Experimental Physical Chemistry by Farrington Daniels, Joseph Howard Mathews, John Warren Williams, Paul Bender and Robert A. Alberty, Copyright 1956

General Chemistry by Harry H. Sisler, Calvin A. Vanderwerf and Arthur W. Davidson, Copyright 1957

Concepts of Space by Max Jammer, Copyright 1954
Concepts of Force by Max Jammer, Copyright 1957

Principles of Physical Chemistry, Third Edition by Samuel H. Maron and Carl F. Prutton, Copyright 1958

Electron and Nuclear Physics by J. Barton Hoag, revised by S. A. Korff, 1958

The Pursuit of the Atom by Werner Braunbek, Copyright 1958

Principles of Geochemistry, Second Edition by Brian Mason, Copyright 1958

Elements of Radio by Abraham Marcus and William Marcus, Copyright 1959

The Proton in Chemistry by R. P. Bell, Copyright 1959

Modern Chemistry by Charles E. Dull, H. Clark Metcalfe and John E. Williams, Copyright 1958
Modern Physics by Charles E. Dull, H. Clark Metcalfe and John E. Williams, Copyright 1960

An Introduction to the Physics of Mass Length and Time by Norman Feather, Copyright 1959
An Introduction to the Physics of Vilbrations and Waves by Norman Feather, Copyright 1961

Atoms, Molecules and Chemical Change by Ernest Grunwald and Russell H. Johnsen, Copyright 1960

Twentieth Century Chemistry by Joseph I. Routh, Second Edition, Copyright 1960

The Nature of the Chemical Bond by Linus Pauling, Third Edition, Copyright 1960

The Abundance of the Elements by Lawrence H. Aller, Copyright 1961

The World of the Atoms by J. J. G. McCue and Kenneth W. Sherk, Copyright 1963

College Chemistry, Fourth Edition by G. Brooks King and William E. Caldwell, Copyright 1963

General Chemistry, Second Edition by Theodore L. Brown, Copyright 1963

Chemistry, Editor George C. Pimentel, Copyright 1963

The World of Elementary Particles by Kenneth W. Ford, Copyright 1963

Geology by William C. Putnam, Copyright 1964

The Architecture of Molecules by Linus Pauling and Roger Hayward, Copyright 1964

The Development of Modern Chemistry by Aaron J. Ihde, Copyright 1964

The Enjoyment of Chemistry by Louis Vaczek, Copyright 1964

The Nature of Solids by Alan Holden, Copyright 1965

Men and Molecules by John F. Henahan, Copyright 1966

The New Intelligent Man's Guide to Science by Isaac Asimov, Copyright 1965
Understanding Physics by Isaac Asimov, Copyright 1966

Niels Bohr by Ruth Moore, Copyright 1966

The World of the Atom, two volume set, Edited by Henry A. Boorse and Lloyd Motz, Copyright 1966

Physics by K. R. Atkins, Copyright 1966

College Chemistry by Bruce H. Mahan, Copyright 1966

General College Chemistry, Third Edition by Charles W. Keenan and Jesse H. Wood, Copyright 1966

Organic Insertion Reactions of Group IV Elements by Edmund Yanovich Lukevits and Mikhail Grigor'evich Voronkov, translated from Russian, Copyright 1966

The Encyclopedia of Chemistry Second Edition, Editor-in-Chief George L. Clark, Copyright 1966
The Encyclopedia of the Chemical Elements edited by Clifford A. Hampel, Copyright 1968

Chemistry by Theodore P. Perros, Copyright 1967

Optical Physics by S. G. Lipson and H. Lipson, Copyright 1969

Introductory Physical Chemistry by A. R. Knight, Copyright 1970

Methods of Science by E.L. Dellow, Copyright 1970

The Transuranium Elements by V.I. Gol'danskii and S. M. Polikanov, Copyright 1973

Rocks and Minerals by Cedric Rogers, Copyright 1973

Chemical Principles and Properties, Second Edition by Michell J. Sienko and Robert A. Plane, Copyright 1974

Chemistry, A Conceptual Approach, by Charles E. Mortimer, Copyright 1967
Chemistry, A Conceptual Approach, Third Edition by Charles E. Mortimer, Copyright 1975
Chemistry, A Conceptual Approach, Fourth Edition by Charles E. Mortimer, Copyright 1979

Basic Principles of Organic Chemistry, Second Edition by John D. Roberts and Marjorie C. Caserio, Copyright 1977

The Atlas of Scientific Discovery by Colin Ronan, Copyright 1983

Handbook of Chemistry and Physics, 66th Edition 1985 to 1986

Principles of Modern Chemistry by David W. Oxtoby and Norman H. Nachtrieb, Copyright 1987

The New Physics Edited by Paul Davies, Copyright 1989

Historical Atlas of Crystallography by J. Lima-De-Faria, Copyright 1990

Sargent-Welch Catalog on Biology-Chemistry- physics

BOOKS ON THE COSMOS, STARS AND UNIVERSE

A Dipper Full of Stars by Lou Williams Page, Copyright 1959

The Universe, Earth and Man, Complements of Disabled American Veterans

The View From A Distant Star by Harlow Shapley, Copyright 1963

Beyond the Observatory by Harlow Shapley, Copyright 1967

Man and the Cosmos by Gerald E. Tauber, Copyright 1979

Starseekers by Colin Wilson, Copyright 1980

Comets, Stars, Planets by Nigel Henbest, Copyright 1985

Astronomer by Chance by Bernard Lovell, Copyright 1990

Bound to the Sun by Rudolf Kippenhahn, Copyright 1990

The Arrow of Time by Peter Coveney and Roger Highfield, Copyright 1990

Frontiers II by Isaac and Janet Asimov, Copyright 1993

Hyperspace by Michio Kaku, Copyright 1994

Skywatching by David H. Levy, Copyright 1994
The Quest For Comets by David H. Levy, Copyright 1994

INDEX

This index is done in such a way that the words found will tell the reader how the word is used. All of the pages for a given word may not be found in this index because they are used so frequently.

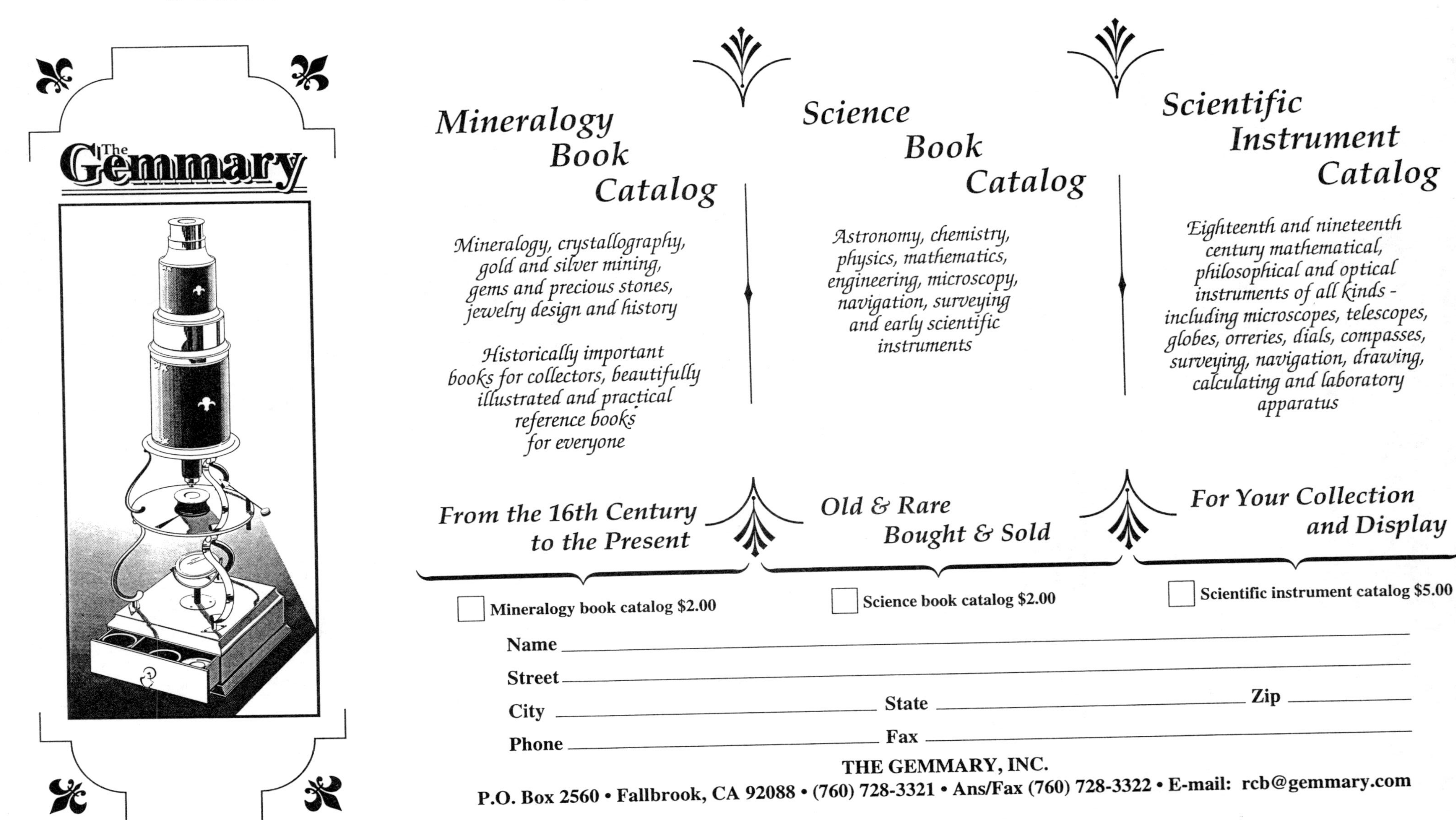

COPIES OF THIS BOOK MAY BE ORDERED FROM:

Glen C. DeLay
DeLay's Printing
1603 Aviation Blvd. #16
Redondo Beach, CA 90278-2855
Phone: (310) 374-4782 Fax: (310) 376-2787

Ordered by:

____
Name

____
Address

____
City State Zip

Send to: (IF DIFFERENT FROM ABOVE)

____
Name

____
Address

____
City State Zip

Quantity ______ ATOM ONE (Atomic Structure of the Atom)

(Price per book $38.00) ____ x 38.00 ____________

CA Res add 8¼ % Tax ____________

Plus Shipping & Handling ____________

TOTAL Amount Enclosed ____________

Shipping & Handling
U.S. residents: please add $6.50 for one book
$2.00 for each additional book
Shipping will be by U.S. Post Office 4th class book rate.

Foreign residents: please add $11.50 for one book
$5.00 for each additional book
Shipping will be by U.S. Post Office surface mail, with delivery times of six to eight weeks to be expected.

FORMS OF PAYMENT:
Check or Money Order in U.S. Currency.